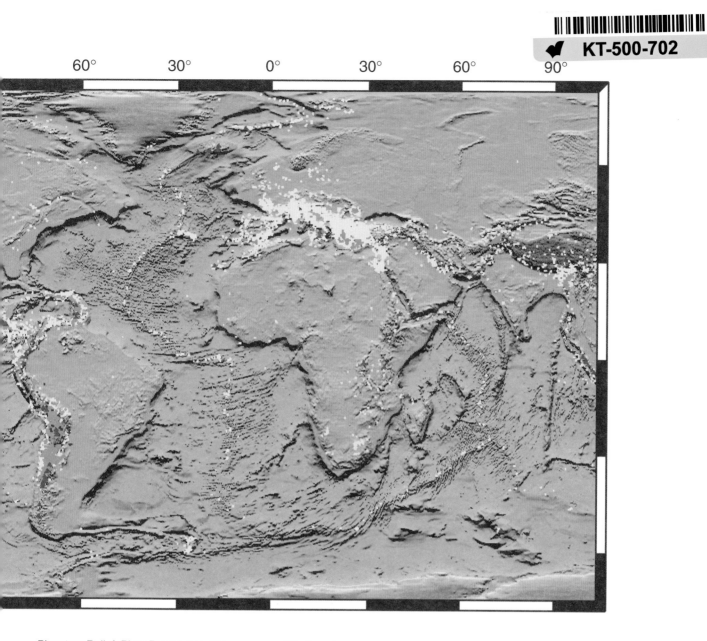

Planetary Relief, Plate Boundaries, Volcanoes, and Earthquakes

graphy: Mercator projection, ETOPO-5 data file, National Geographical Data
         Center, National Oceanographic and Atmospheric Administration,
         Washington, D.C. Darker blue indicates deeper water; darker yellow and
         orange indicate higher land.

triangles: Volcanic eruptions, post- 1964 (240 events) (Global Volcanism
           Program, Smithsonian Institution, Washington, D.C., courtesy T. Simkin).

e squares: Shallow earthquakes (<70 km) 1993; may include some large
           explosions (53,000 events).

n squares: Intermediate-depth earthquakes (70 to 300 km) 1993 (8300 events).

 squares: Deep earthquakes (>300 km) 1993 (730 events)

lified from Simkin, T., Unger, J.D., and 3 others, 1994, This dynamic earth:
         World map of volcanoes, earthquakes, impact craters, and plate tectonics
         [poster]: U.S. Geol. Survey, Washington D.C. Graphics courtesy D. Steer,
         Institute for the Study of the Continents, Cornell University, Ithaca, N.Y.

Geomorphology

Geomorphology

A Systematic Analysis
of Late Cenozoic Landforms

Third Edition

Arthur L. Bloom
Department of Geological Sciences
Cornell University

Prentice Hall
Upper Saddle River, New Jersey 07458

Bloom, Arthur L. (Arthur Leroy)
    Geomorphology : a systematic analysis of late Cenozoic landforms /
Arthur L. Bloom. — 3rd ed.
      p.   cm.
    Includes bibliographical references.
    ISBN 0-13-505496-6
    1. Geomorphology. 2. Geology, Stratigraphic—Cenozoic.
I. Title
GB401.5.B55 1998
551.41—dc21
                                 97-29109
                                     CIP

Cover: Postglacial gorge in Enfield Glen, Robert H. Treman State Park, Ithaca, N.Y. Stream was superposed from glacial sediments onto bedrock and has intrenched by a combination of joint-controlled hydraulic plucking and pothole drilling (see pp. 203-206). (Photo: A. L. Bloom)

Executive Editor: Robert A. McConnin
Executive Managing Editor: Kathleen Schiaparelli
Manufacturing Manager: Trudy Pisciotti
Production Project Coordinator: Thompson Steele Production Services
Full Service Liaison/Manufacturing Buyer: Benjamin D. Smith
Creative Director: Paula Maylahn
Art Director: Jayne Conte
Art Manager: Gus Vibal
Interior Design: Thompson Steele Production Services
Page Layout, Copy Editor, Art Studio, Text
       Composition: Thompson Steele Production Services
Cover Design: Bruce Kenselaar
Cover Photo: ©A. L. Bloom
Marketing Manger: Leslie Cavalier

© 1998, 1991 by Prentice-Hall, Inc.
Simon & Schuster/A Viacom Company
Upper Saddle River, New Jersey 07458

Printed in the United States of America
10 9 8 7 6 5 4 3 2

ISBN 0-13-505496-6

Prentice-Hall International (UK) Limited, *London*
Prentice-Hall of Australia Pty. Limited, *Sydney*
Prentice-Hall Canada, Inc., *Toronto*
Prentice-Hall Hispanoamericana, S.A., *Mexico*
Prentice-Hall of India Private Limited, *New Delhi*
Prentice-Hall of Japan, Inc., *Tokyo*
Simon & Schuster Asia Pte. Ltd., *Singapore*
Editora Prentice-Hall do Brasil, Ltda., *Rio de Janeiro*

Contents

Preface

In writing and revising this book, I have assumed that the student who uses it has had at least one introductory course in geology or physical geography. Elementary familiarity with the climatic zones, planetary winds and ocean currents, the geologic time scale, common rocks and minerals, and topographic maps is assumed. In my own course, the book has served mostly undergraduate majors and graduate students in geology with a stimulating addition of students from civil and environmental engineering, soils, crops, and atmospheric sciences, planetary science, archeology, city and regional planning, and natural resources. As a visiting professor at Australian and Japanese universities, I have also taught geography students. Each of these constituencies brings their own prior knowledge and experience to a geomorphology course, and their own deficiencies. Geography students who use this book might need to review plate tectonic theory and Cenozoic geologic history; geology students may find the principles of atmospheric and oceanic circulation new to them. It has been especially satisfying to introduce concepts of geomorphic time to students of soils science and soils engineering who are very skilled in subjects such as soil geochemistry and slope stability, but have not contemplated the longer-term evolution of landforms. Geomorphology is an excellent synthesizing and integrating subject for students training in other scientific fields, and I hope this edition reflects the eclectic nature of the subject.

Instructors who have used the earlier editions of this book will recognize familiar chapter and subchapter titles, though many are rearranged. A useful textbook should be as timeless as possible, but ideas, like landforms, change. Within these chapters are numerous changes of emphasis, new ideas from recent publications, and some major revisions. The well-received chapter on the energy of geomorphology has been moved to Part I, "Fundamentals of Geomorphology," as Chapter 2, which is where many instructors had been assigning it, anyway. The biggest change in this third edition is the inclusion in Part II of a new Chapter 4, "Cenozoic Climate Change," which, with Chapter 3, "Cenozoic Tectonism," reviews the history of the major tectonic and climatic changes during which our present landscape has taken shape. Especially compelling is the idea that many, perhaps most, erosional landscapes have been shaped by more than one assemblage of climate-controlled processes. To those of us who live in recently deglaciated areas, that idea is trite. However, relatively recent discoveries in paleontology, paleobotany, and oceanography make clear that within the Cenozoic Era our earth's climate has changed from pole-to-pole warmth (the "greenhouse") to our present latitudinally zoned temperature gradients with extensive regions of glaciers and permafrost (the "icehouse"). Superimposed on the general cooling trend of later Cenozoic time have been the cyclic oscillations of glacial–interglacial amplitude that have dominated the climatic system for at least the last two and a half million years. As a result, only the youngest landscapes have formed during a single set of climate-related processes, and all others show varying degrees of relict forms and evidence of processes no longer operative.

Chapters in previous editions that concerned climatic morphogenesis and alternating Quaternary morphogenetic systems have been reorganized as broader evaluations of the many changing environments that have shaped landscapes even within the last 1 or 2 million years of multiple glaciations, sea-level fluctuations, and low-latitude cooler–warmer and wetter–drier episodes. Although some of the examples of such changes have been retained from the earlier editions, newer examples illustrate that climate changes have been neither progressive nor cyclical, but complex. The resulting landscapes record these complex changes but are difficult to interpret. To appreciate the changing landscape of the late Cenozoic Era, students must go beyond the traditional geomorphic literature and study paleohydrology, paleoclimatology, paleobotany, and many other related fields. Similarly, to appreciate the vertical and horizontal movements of the continents and ocean floors that have contributed to the climatic changes, the literature of geophysics, tectonics, sedimentology, and geochemistry must be understood, or at least understandable. Thus, Chapter 3, "Cenozoic Tectonism," has been extensively revised as the background to Chapter 4, "Cenozoic Climate Change."

Part IV, "Subaerial Destructional (Erosional) Processes and Erosional Landforms," includes Chapter 7 to 15 and almost 50 percent of the text. Chapter 7, "Rock Weathering," has been greatly expanded to include more discussion of soils, soil formation, and soils chronosequences, which tell us so much about the evolution of the present landscape. Chapters 10 and 11 include new sections on bedrock river channels in addition to an updated description of the behavior of alluvial channels. Details of eolian processes and forms have been condensed and combined in Chapter 13 with other landforms of chronically or seasonally dry regions.

Regional landform assemblages typical of dry, periglacial, and glacial environments are analyzed in depth, not only because of their great present extent, but also to aid in their recognition in regions where such processes no longer operate. It is now amply clear that landscapes persist for intervals much longer than the time scale of major climatic changes. New techniques for dating landscapes by radioactive and cosmogenic isotopes are rapidly providing a numerical scale for landscape evolution. It is probably now time to turn from the past several decades of process-emphasized geomorphology to a new synthesis of regional landscape evolution during a time of dynamic changing processes.

Glaciers and glaciation are reviewed in Chapters 16 and 17 of Part V, followed by a new Chapter 18, "Late Quaternary Climatic Geomorphology," which stresses the geomorphic impact of the latest glacial–interglacial cycle and the brevity of Holocene time within which human civilization has risen. Readers should accept that change is a normal factor of life on earth. Rampant fear of any environmental changes, natural or human-induced, should be tempered by the knowledge that the 10,000 years of Holocene time are typical of only 5 to 10 percent of late Cenozoic time, during the remainder of which the earth has either been in or approaching an "ice age." Fear of change denies the evidence that both our species and our civilization are the product of environmental changes, the results of which are still apparent in our landscapes.

Coasts deserve separate treatment (Part VI) because the energy input that shapes them is so different from that which shapes subaerial landscapes. More important, the coastal zone is vital for human activities. Two-thirds of the world's people live in the 10 percent of the land defined as the coastal zone. The transience of sea level during late Quaternary time has been a major factor in the development of the estuaries, lagoons, deltas, and reefs upon which so much modern commerce and food production depend. Future trends of sea-level movement are critical to planning coastal-zone activities.

Regional descriptions are included to demonstrate how certain kinds of landscapes formed, but this is not a textbook of regional physiography. Examples and illustrations have been chosen from as wide a range as I have been able to find in my own travel and in published descriptions, in order to counter the natural tendency for provinciality that lessens the value of many geomorphology books.

At this writing, it seems likely that within a few years we will have orbiting space platforms monitoring our earth system on hourly or daily intervals and providing a massive data stream to computers and their operators. The geomorphologists who use this book as an introduction to the subject will have unprecedented opportunities to study the earth's scenery from space and to comprehend the land surface as the product of competing and varying internal and external processes.

I have spent many long and satisfying hours searching for the current, illustrative, authoritative, and reasonably available sources that are cited in the chapter references. Advanced students and especially teachers should not be content to accept the necessarily abbreviated descriptions in this or any textbook. A textbook should be the beginning, not the end, of understanding. Acknowledging the explosive growth of multiauthored scientific reports, I have limited full reference citations to papers by up to three authors; for larger groups, only the first two authors are cited and listed in the author index.

There is no glossary in this book. I prefer students to learn technical terms in context, so terms are set in boldface and defined at their first use. The subject index provides a quick reference to such words and also refers to additional usages that may be extensions of the basic definition. The subject index also contains many references to geographic localities and regions.

Students are always their professor's best critics, and I thank mine for all of their comments about the first two editions of this book. Numerous colleagues also have offered constructive comments concerning both previous editions. An unexpected pleasure of authorship has been experiencing the generous and enthusiastic responses to requests for illustrations. Many friends provided not only the photographs I requested, but also several others from which I could choose. Seeing one's name in a caption is a trivial repayment, so I further acknowledge here my debt of gratitude to all contributors. Photographs not otherwise credited are my own. Professional artists redrafted most of my line drawings. Various sections of this text have been critically reviewed by P. U. Clark, L. A. Derry, T. W. Gardner, J. M. Harbor, D. K. Hylbert, R. H. Keifer, J. P. Kennett, W. W. Locke, F. E. Nelson, A. N. Palmer, N. Pinter, and M. J. Woldenberg. I thank them for their constructive, and at times sharp, criticism. Numerous colleagues at Cornell University have offered advice and help on specific topics. Sally Carland typed most of the text and linked me to the modern world of electronic communication. Thank you, my friends.

Arthur L. Bloom

Again, to Donna

with Love

PART I

Fundamentals of Geomorphology

Geomorphology is the scientific study of landscapes and the processes that shape them. The science of geomorphology has two major goals. One is to organize and systematize the description of landscapes by intellectually acceptable schemes of classification. The other is to recognize in landscapes evidence for changes in the processes that are shaping and have shaped them.

There are many nonscientific ways to see a landscape. A painter sees three-dimensional shapes and surfaces defined by shades and hues of various colors that he or she then attempts to record on a flat surface. A photographer may wait hours to capture the right combination of light and shadow on film. The results of both techniques are a pleasure to view, and art galleries are full of contemplative people who return again and again to study a Constable English landscape, an O'Keeffe interpretation of an Arizona mountain, or an Ansel Adams photograph of Yosemite Canyon.

Geomorphologic interpretation can be no less satisfying, but relies on analytical rather than artistic skills, and portrays the results primarily in words rather than in pictures. Scientific analyses must never be allowed to destroy the esthetic pleasure of viewing a beautiful landscape. However, it can add to that pleasure a deeper understanding of how the landscape began and how it will change. A geomorphologist can visualize the other side of a mountain or its internal structure. Even better, he or she can visualize change, mentally running a motion picture of the scene backward and forward at will through millions of years.

To perform these wonders, one must study. The two chapters of Part I review some of the history of geomorphology, outline the methods of geomorphic analysis, and consider the energy that drives landscape evolution. Let the fun begin.

Chapter 1

The Scope
of Geomorphology

THE SYSTEMATIC STUDY OF LANDSCAPES

Many people, on viewing a mountain, a river valley, or any impressive landscape, instinctively wonder how that landscape came to be. More philosophical thinkers sense that no scenery is immutable but that everything changes with the passage of time. A few scientists of each generation choose to specialize in *the analysis, systematic description, and understanding of landscapes and the processes that change them.* These scientists are geomorphologists.

The description, analysis, and understanding of landforms (the component parts of landscapes) is a science only when it is done in a disciplined fashion. Many schemes of organized study can be conceived. **Regional description,** by district, nation, or natural provinces, is one logical approach (Thornbury, 1965; Hunt, 1974; Embleton, 1984; Short and Blair, 1986). The earliest geography books were largely regional descriptions of visible landmarks, passes, river valleys, and other landforms that helped guide travelers. Regional, descriptive geomorphology remains one of the accepted subdivisions of the science. A variant of growing importance, especially outside North America, is the preparation of geomorphic maps that show landform components by distinctive colors and symbols (Barsch and Liedtke, 1980).

The subject of landform analysis also can be organized on the basis of the processes that shape the land.

Many recent books deal with **process geomorphology** (for example, Leopold et al., 1964; Ollier, 1984; Embleton and Thornes, 1979; Birkeland, 1984; Drewry, 1986; Ritter et al., 1995). From the number of process-oriented books published in recent decades, one might conclude that geomorphology has been reduced to a study of processes, with only minimal consideration of the resulting landforms. Of course, the landforms are important, and as we learn more about processes, our appreciation of landform development will deepen. Process-oriented geomorphology is closely related to climatology because air temperature, precipitation, winds, and atmospheric humidity largely determine the response of rocks to subaerial exposure. **Climatic morphogenesis** is the approach to geomorphology that emphasizes the shaping of landforms in response to various climatic conditions (Birot, 1960, 1968; Tricart and Cailleux, 1965, 1972; Büdel, 1982). This is an important topic because many landscapes preserve the record of more than one set of climatic-controlled processes, and the correct interpretation of such a landscape provides evidence for reconstructing ancient climates.

Process geomorphology has evolved into subfields of specialization such as hillslope morphology, fluvial processes, karst, volcanic, arid, eolian, glacial, periglacial, tropical, and coastal geomorphology—to name only some of the better known. Even within the large field of fluvial geomorphology, specialized monographs are available that concern only meandering, braided (gravel-bed), or bedrock channels. As their names imply, many of these subfields are oriented toward a specific morphoclimatic region. A few, such as karst, volcanic, and coastal geomorphology, are concerned with a specific rock type, or a set of processes not directly related to a particular climate. **Tectonic geomorphology** (Chapter 5) is a relatively new and active subfield derived from the geologic roots of geomorphology. It has thrived in the current era of plate tectonics.

Much of the information about geomorphic processes is derived from other scientific fields. The elastic and plastic deformation of rocks under pressure and the thermal conductivity of rocks are studied by geophysicists, generally not with the primary intention of describing or analyzing landscapes. Seismologists are properly more concerned with the causes and prediction of earthquakes than with the landforms that earthquakes produce. Geochemists and mineralogists study the chemical reactions of rocks and minerals with hydrous solutions and thereby provide data about weathering processes. Pedologists study soil formation; hydrogeologists study the work of flowing water and groundwater; sedimentologists study depositional processes and sediments; meteorologists and climatologists study the circulating atmosphere and the resulting weather and climate patterns; glaciologists study the physical properties of ice; and oceanographers describe water chemistry and the submarine topography they see traced on their depth recorders. All these scientists and many others contribute to an understanding of the processes that shape the earth's surface. A geomorphologist must be at least conversant with these peripheral subjects, as well as with the techniques of remote sensing that have become essential for interpreting electronic images transmitted from spacecraft and aircraft.

Some earth scientists in the 1960s and 1970s specialized in **quantitative geomorphology** (Morisawa, 1971), by which numerical data concerning landforms and processes are subjected to mathematical analyses. At first, this was a significant departure from traditional descriptive techniques, but with the rapid increase in computer power, most geomorphologists have adapted "number crunching" to their analytical capabilities, and the entire profession has become more quantitative. Modern, computer-literate college students find it hard to comprehend that at about the time of their birth, geomorphologists were still using mechanical planimeters to measure drainage-basin areas, were counting contour lines on paper maps to calculate stream gradients, and were tediously calculating other parameters with mechanical desk calculators.

At the beginning of the nineteenth century, the science of geology became distinguishable from "natural philosophy." As facts and theories about the origins of rocks and mountain ranges accumulated, they were absorbed into regional or descriptive geomorphology to help understand the origins of landforms. Thus, **explanatory description** evolved as a component of geomorphology in which a landscape is not simply described by the heights of its hills and the steepness of its slopes but by the reconstructed geologic history of its evolution. Explanatory description is a higher form of analysis than regional description in that it requires inferences about past events. In a sense, a landscape cannot be understood until the entire geologic history of its rocks and slopes is known. Fortunately, for most landscapes it is possible to begin an explanatory description with a phrase like, "As the area emerged from the Cretaceous sea" or "Orogenic uplift in the Miocene Epoch produced a mountainous terrain." Whereas regional geomorphology, or physiography, is often regarded as part of the larger science of geography, explanatory or genetic geomorphology is mostly a geologic subject. Geomorphology is thus a part of two sciences, and geomorphologists may regard themselves as either geographers or geologists, or preferably as both (Dury, 1972). To be part of the growing field of "Earth System Science," future geographic geomorphologists will need to learn more about tectonics, and geologically trained geomorphologists will need to learn more climatology. Both groups will continue to need solid training in physics, chemistry, and computer science.

How can regional description, reconstruction of geologic history, and the numerical data on chemical and physical processes by which rock masses are altered by water and air be unified into a science of geomorphology? The best solution that has evolved is summarized by the trinity of *structure, process,* and *time.* Although generally attributed to the great American geomorphologist, William Morris Davis (1850–1934), the organization of geomorphic information under the three headings is implicit in many earlier writings. It is such a natural and intuitive organization, that it was and has been unconsciously used many times before and since Davis codified it with such great force and clarity.

Acknowledging that landscapes are not immutable leads to the immediate conclusion that if change is in progress, some rock mass (structure) is being altered by some process, and the alteration has proceeded to a definable extent (stage) or for a definite interval (time). Time has always been difficult to measure on the geologic scale, and rates of geomorphic change are extremely variable because of the wide range of processes and structures. For both of these reasons, the concept of a relative stage of development, rather than the absolute length of time a process has been acting, has been prevalent.

To illustrate the concept of geomorphic description in terms of structure, process, and time, consider the following two descriptive paragraphs. The first is empirical; it states clearly *what* and *where*. The second is explanatory; it connotes *how* as well as *what* and *where*.

1. "The city of Ithaca, New York, is built on low ground at the south end of Cayuga Lake. Steep slopes surround the town, above which is a broad, rolling upland that becomes gradually more rugged southward into the low mountains of Pennsylvania."
2. "The city of Ithaca, New York, is built on a delta plain at the south end of the Cayuga Lake trough. Glacially oversteepened valley walls around the city are in contrast to the maturely dissected, cuestaform upland of the glaciated northern Appalachian Plateaus."

In the second paragraph, the technical terms *delta plain, cuestaform,* and *plateaus* convey information about structure, process, and genesis. Geomorphic processes are explicit in the terms *trough, glacially oversteepened, dissected,* and *glaciated*. Time is implied in the term *maturely dissected* in that a degree or stage of erosion is specified. The first paragraph would be perfectly suitable for a tourist brochure or travel guide, but the second paragraph, in the same number of words, conveys much more information to a trained reader. True, technical terms are used that must be understood by the reader, but technical terms are a kind of shorthand notation to convey concepts and increase understanding in an efficient manner.

If empirical description were the only goal of geomorphology, any regional landscape description such as the two samples just given could be reduced (or expanded?) to sets of numerical coordinates. Only three numerical terms are needed to describe the altitude above sea level of any point on the landscape and locate the point by its latitude and longitude coordinates. Huge data banks of *digital topography* are now available, and even more precise ones are constantly being added. Realistic computer models of landscapes can now be generated and manipulated. A new subfield of **megageomorphology** has evolved that specializes in the regional-scale analysis of landscapes based on images from space. An integration of the changes in altitude with latitude and longitude would, in principle, give an equation that completely describes any specified area of land surface. But even

if such an equation could be written, no one, not even the most astute mathematician, could envision the land surface that it describes. A radar-scanning computer on a cruise missile can navigate over a landscape by comparing it to a digitized topographic model, but human minds require descriptive words to convey images of landscapes. Significantly, the most satisfying way to describe a landscape seems to be by describing its history or genesis. Geomorphic description explains *how* and *when,* as well as *what* and *where*. Only the ultimate question of *why* is left to the philosophers.

THE SCALES OF LANDSCAPE ANALYSIS

Spatial Dimensions

What makes a painting or a computer simulation of a landscape look "real"? A cloud is not a sphere, and most mountains are not cones or pyramids. No one is deceived by the simplified background scenery of animated cartoons, although computer programs are beginning to create extremely realistic, if totally imaginary, "landscapes" (Color Plate 1). The key to realistic landscape portrayal is the introduction of a statistical roughness parameter to the scene in a scale-invariant manner so that, whether viewed as an entirety or by microscopic examination of a small part, the scene always resolves into still smaller detail. For example, photographs of a rocky coastline or a drainage network taken from distances of 10 m, 1 km, or 100 km may be indistinguishable unless a scale bar is provided (Turcotte, 1997). Mandelbrot (1967, 1983) showed that many, and perhaps most, natural objects such as mountain sides, river drainage networks, and island shorelines do not resolve into simple geometric shapes on close examination but maintain systematic roughness at all scales. He called such objects *fractals* and developed an alternative to Euclidian geometry, which he called fractal (fractional) geometry. An essential measure of the fractal character of a surface is its *fractal dimension*. The fractal dimension of terrestrial relief is remarkably consistent over almost six orders of magnitude (Figure 1-1). Submarine abyssal hills, a 90 x 90 km terrain in Wales, and longer-wavelength submarine topography all show scale-invariant roughness. If raising each hill against gravity requires a certain amount of mechanical energy, it has been shown that the topography represents a statistical distribution in which the energy of formation is distributed uniformly over hill sizes. Smaller hills may require less energy of formation, but are more numerous, to ensure the uniform division of energy (Bell, 1975).

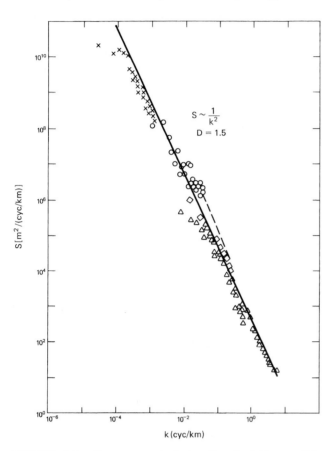

FIGURE 1-1. Power spectral density *S* of the earth's topography and bathymetry as a function of wave number *k* (the reciprocal of length). *S* is a measure of roughness, and its power law relation to *k* is defined by the fractal dimension *D* = 1.5. The relation holds over about six orders of magnitude from the spherical harmonics of the earth (x) through several scales of deep-sea topography (o, — —, Δ) to the roughness spectrum of a 90 km × 90 km square in northern Wales (◇). On the largest end (lowest *k* value) of each data set, the surface roughness deviates slightly from being a power function of the length of the transect (Turcotte, 1989, Figure 14, and references cited therein).

The shape and size of some landforms are not statistically distributed but are deterministic, controlled by the forces that built them and the inherent properties of their rock masses. The widths of fault blocks and grabens, for instance, can be related to the thickness of the brittle crustal layer that has been broken (Chapter 5). Scoria cones can be built only to a certain size, after which granular ejecta can no longer be expelled, and either they fail by lateral slumping or a basal lava flow emerges (Chapter 6). On the largest scale, planet earth is a spheroid of rotation determined by the density of its materials, its size, and its rotational velocity (p. 7).

Time Scales for Landform Development

In general, the larger the landform, the more durable it will be (Ahnert, 1988). The spheroidal shape of the earth was established early in its history and has not varied significantly since then. Continents have accreted over geologic time, but although the pieces have been repeatedly disrupted and reorganized, probably up to 90 percent of the continental crust had formed by 2.5 billion years ago, and the rate of accretion of new continental crust has been very slow since then (Taylor and McClennan, 1985, p. 234).

Smaller landforms can be both created and destroyed more rapidly than larger ones. Carey (1962, p. 98) suggested that geomorphic and other geologic processes have rates that are size-dependent over many orders of magnitude. For example, earthquake deformation and avalanches, with dimensions of a few meters to kilometers, are completed within seconds or minutes. Glaciers and ice sheets, ranging in size from hundreds of meters to a thousand kilometers across, require many years to many thousands of years to accumulate, depending on their size. Orogenic (mountain-building) deformation intervals typically range in time from 1 to 30 million years, with most events lasting between 1 and 5 million years (p. 38).

From the hypothesis that the rates of geomorphic processes are inversely proportional to the size of the affected area, one can deduce a corollary statement, that the rate of any process, as defined by the total change per unit of time, appears progressively more rapid when measured over progressively shorter intervals of time. This is in accord with many personal experiences during floods, avalanches, or earthquakes, when great changes are accomplished in brief intervals. By a similar rule, the total change per unit of area, or *intensity* of change, is greatest when viewed within a very small area, as for instance, at the head of an eroding gully (Figure 10-2). Rates established for any finite time interval, or intensities for a finite area, typically seem excessive when applied to a longer time scale (Gardner et al., 1987) or larger area. For example, sediment yields from small drainage basins always exceed the total sediment yield of the larger combined basin; local earth movements during single earthquakes, even in regions where the recurrence interval of comparable earthquakes is known, exceed the regional net change over longer periods. These observations that rates and intensities are time and area dependent create problems of geomorphic interpretation that require continued effort to comprehend.

SIZE OF TERRESTRIAL RELIEF FEATURES

A durable concept of both geomorphology and tectonism is that terrestrial relief features can be classified by size (Table 1-1). Over areas that range across many orders of magnitude, characteristic constructional and destructional processes shape landscape units of characteristic size. Table 1-1 and the back end paper also illustrate the point that larger landscape units persist for longer times. The large low-order features of the earth's relief are not usually analyzed in great detail, but a modernized size classification is introduced here as a prelude to the more complex details of landform analysis and description.

The Shape of the Earth and Other Planets

The best description for the shape of the whole earth is that it is an oblate, or slightly flattened, spheroid with minor bumps and hollows (King-Hele, 1976). The equivalent reference surface of gravity equipotential, which coincides with the earth's mean sea level and its extension beneath the land, is called the **geoid** (Gra-

farend, 1994). The oblate flattening due to earth rotation causes the polar radius (6357 km) to be 21 km less than the mean equatorial radius. As a result, Mt. Chimborazo, in Ecuador, (6310 m above sea level) is the point on the earth's surface farthest from the center of the earth. In addition to its fundamentally flattened spheroidal form, the geoid has a number of smaller undulations, related to unequal mass distribution in the crust and mantle (Figure 1-2). A boat on the ocean east of Papua New Guinea is 192 m farther from the center of the earth than one south of India, yet by definition, both are at sea level.

The geoid is an important reference surface for geomorphic processes. Rivers, for instance, cease to flow when they reach sea level, unless they drain into depressions below sea level in arid regions (Chapter 13). The geoidal projection of mean sea level under the subaerial landscape is the best definition of the term **ultimate base level**, which has been the subject of so much debate (Chapter 10).

Continents and Ocean Basins

The two largest units of the earth's rocky *relief* (altitude above or depth below the geoid) are the surfaces of the continental blocks and the abyssal basins of the

Table 1-1

A Classification of Terrestrial Geomorphological Features by Scale (Baker, 1986, Table 1-1)

Order	Approximate Spatial Scale (km^2)	Characteristic Units (with examples)	Approximate Time Scale of Persistence (years)
1	10^7	Continents, ocean basins	10^8–10^9
2	10^6	Physiographic provinces, shields, depositional plains	10^8
3	10^4	Medium-scale tectonic units (sedimentary basins, mountain massifs, domal uplifts)	10^7–10^8
4	10^2	Smaller tectonic units (fault blocks, volcanoes, troughs, sedimentary subbasins, individual mountain zones)	10^7
5	10–10^2	Large-scale erosional/depositional units (deltas, major valleys, piedmonts)	10^6
6	10^{-1}–10	Medium-scale erosional/depositional units or landforms (floodplains, alluvial fans, moraines, smaller valleys and canyons)	10^5–10^6
7	10^{-2}	Small-scale erosional/depositional units or landforms (ridges, terraces, sand dunes)	10^4–10^5
8	10^{-4}	Larger geomorphic process units (hillslopes, sections of stream channels)	10^3
9	10^{-6}	Medium-scale geomorphic process units (pools and riffles, river bars, solution pits)	10^2
10	10^{-8}	Microscale geomorphic process units (fluvial and eolian ripples, glacial striations)	

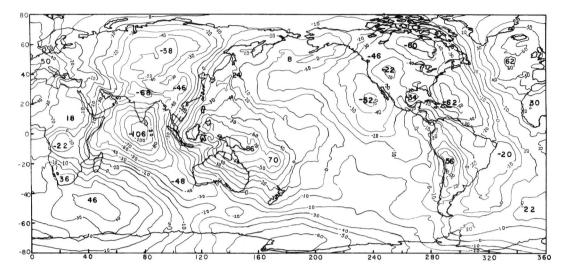

FIGURE 1-2. Geoid map of the earth. Contours show deviations from a spheroid of radius 6378.142 km with polar flattening of 1:298.255. Contour interval: 10 m. Improvements in laser satellite ranging now provide centimeter-level precision that is not practical to display on a full-earth map (Heirtzler and Frawley, 1994; LeProvost et al., 1995). (D. L. Turcotte, based on Goddard Earth Model GEM-8).

ocean floors. Continental crust underlies about 41 percent of the earth's surface area, but only 29 percent of the earth's area is now above sea level (Cogley, 1985). The average altitude of the land area is 823 m above sea level, but because of the extensive area of submerged continental crust under the continental shelves (front endpaper), the average altitude of continental crust is only 108 m (Cogley, 1985, Table 1). The average depth of the oceans is about 3700 m. Most of the average depth is formed by abyssal basins, which cover about 30 percent of the earth's surface. Smaller regions of the continents and ocean floors rise much higher and plunge much deeper than the averages, but it is a striking and important characteristic of our planet's primary relief that the **hypsographic curve** (the statistical cumulative curve of altitude and depth distribution) is strongly bimodal (Figure 1-3). It is this strong bimodality that justifies a primary geomorphic distinction between continents and abyssal ocean basins.

The earth's lithosphere is conventionally divided into ten major plates and several smaller ones as basic tectonic components (front endpaper). In the "new global tectonics," the continents are only incidental granitic masses that are being rafted along toward the subduction zones where lithospheric plates sink back down into the mantle. Because of their low-density rocks, the continental slabs are not drawn down but accumulate on top of the mantle.

The statement that on our earth "there is just about enough water to fill [the] ocean basins" (Wise, 1974, p. 45) either sounds silly or seems to express a

fortunate coincidence. In fact, it provides the explanatory link between the differentiation of continental and oceanic crust and the resulting primary planetary relief. Although closely related in composition to the other three inner, "terrestrial," planets (Mercury, Venus, and Mars) and our own moon, the earth is unique because it has a surface that is mostly water

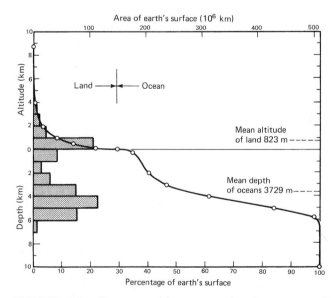

FIGURE 1-3. Hypsographic curve showing area of earth's solid surface above any given altitude or depth. Bar graph on left shows the distribution of area by 1-km intervals (from Menard and Smith, 1966; Kossinna in Fairbridge, 1968; see also Cogley, 1985).

covered. New oceanic crust forming along midocean ridges strongly interacts with water to form hydrous minerals. When these are subducted, they carry water deep into the mantle, which lowers the melting point there and has created large masses of nonsubducting granitic continental crust (Whitney, 1988; Meade and Jeanloz, 1991; de Wit et al., 1992). Lacking abundant water, the other inner planets and the moon developed thick primitive basaltic crusts, and plate motions ceased. Our planet continues to have a dynamic, evolving, tectonic landscape because of its abundant surface water (Campbell and Taylor, 1983). If subduction were not a submarine process, there would be no granitic continental crust, and our first-order relief would not exist. "No water, no granites — no oceans, no continents" (Campbell and Taylor, 1983; see also Fyfe, 1988). To a large extent, the amount of water on the primordial earth determined the eventual size of the continents and ocean basins.

The oceanic crust, experiencing geologically rapid creation and consumption, has a relatively simple exponential correlation of depth with age (Figure 1-4). At an accreting plate margin along a midocean ridge, the oceanic lithosphere is hot and buoyant. Progressive cooling and thermal contraction of a lithospheric plate as it moves away from an accreting margin accounts for both the observed geothermal heat flow and the oceanic depth, at least to a depth of about 6 km.

How the continents have maintained their altitudes near or above sea level throughout at least the one-half billion years of Phanerozoic time is one of the most important topics of geophysics. In the face of evidence for erosion to depths of 20 km or more into continental structures (p. 121), the mechanism of

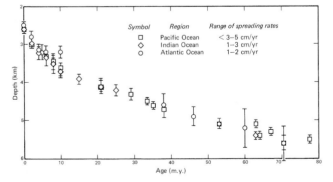

FIGURE 1-4. Average water depth to smooth oceanic crust as a function of the age of the seafloor. The crust at slowly accreting plate margins is shallower than predicted for the first 20 million years. Thereafter, depth is an accurate predictor of crustal age (data from Sclater et al., 1971).

long-term continental survival also is an important geomorphic topic.

Geomorphic (Physiographic) Provinces

Superimposed on the continental and oceanic structures are major relief units that are recognizable by their size, continuity, uniformity of structure and formational process, and geologic age. Continents are broadly subdivided into cratonic massifs or *shields, epicontinental seas, platforms* (thin, nearly flat-lying strata), and *orogenic belts* of several ages. Similarly, oceanic relief includes *midocean ridges, abyssal plains, island arcs,* and *back-arc basins*. The familiar geomorphic provinces of the United States (Figure 1-5), as organized by Fenneman (1917, revised 1928), are examples of such relief features. Each province was defined as a geomorphic (or physiographic) entity that has resulted from a set of processes or a succession of processes acting on a specific structure (in the broad sense of a regional geologic terrane, as noted on page 12) for a certain period of time or to a certain stage of completion. Fenneman's approach was thoroughly based on Davis's explanatory descriptive technique using structure, process, and time, as the preceding sentence emphasizes. Nevertheless, the boundaries that he drew formally defined "provinces" that had long been informally recognized in regional descriptions. Fenneman's provinces have been generally accepted for geomorphic descriptions of the United States. The provinces, subdivided into sections and grouped into larger divisions, are easily recognized as entities even though their exact boundaries may be matters for scholarly debate. Words such as "Colorado Plateau" and "Appalachian Valley and Ridge Province" communicate immediate understanding to the informed reader.

The relief features commonly identified as geomorphic provinces seem to be about the optimal size for regional or descriptive geomorphology. Books and maps of the regional geomorphology of the United States (Fenneman, 1931, 1938; Hammond, 1963; Thornbury, 1965; Hunt, 1974; Graf, 1987), Canada (Bostock, 1964), Alaska (Wahrhaftig, 1965), and the floor of the Atlantic Ocean (Heezen et al., 1959) are but a few examples of the use of geomorphic provinces to organize regional geomorphic description. Although the concept of geomorphic provinces was intended to facilitate regional descriptions of nations or large continental areas, it has been adopted, even by scientists not necessarily aware of the geomorphic tradition, as the framework for describing the ocean floor, the moon, and other planets. For example, Head (1990, p. 100) organized a description

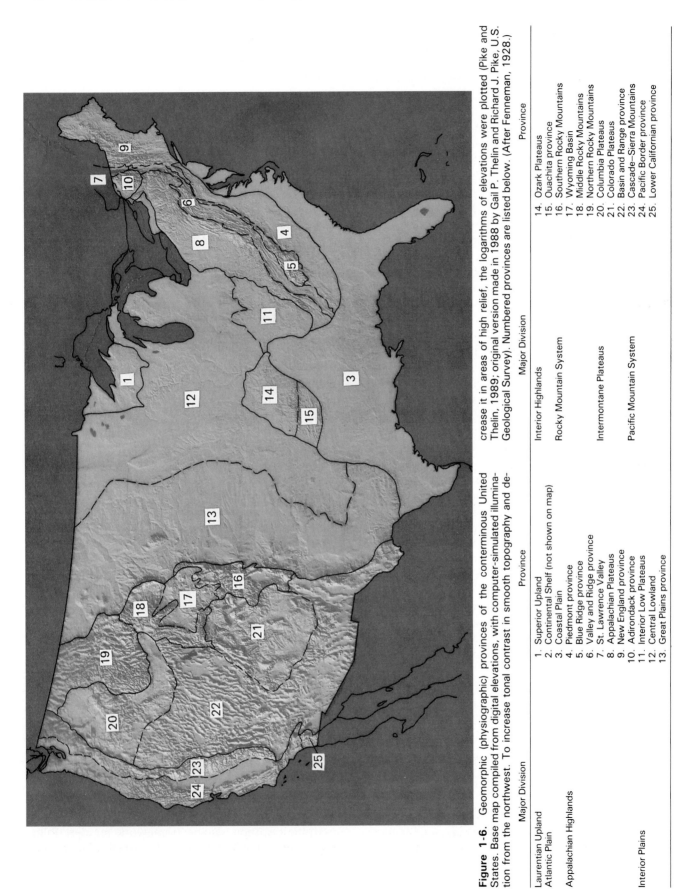

Figure 1-6. Geomorphic (physiographic) provinces of the conterminous United States. Base map compiled from digital elevations, with computer-simulated illumination from the northwest. To increase tonal contrast in smooth topography and decrease it in areas of high relief, the logarithms of elevations were plotted (Pike and Thelin, 1989; original version made in 1988 by Gail P. Thelin and Richard J. Pike, U.S. Geological Survey). Numbered provinces are listed below. (After Fenneman, 1928.)

Major Division	Province
Laurentian Upland	1. Superior Upland
Atlantic Plain	2. Continental Shelf (not shown on map)
	3. Coastal Plain
Appalachian Highlands	4. Piedmont province
	5. Blue Ridge province
	6. Valley and Ridge province
	7. St. Lawrence Valley
	8. Appalachian Plateaus
	9. New England province
	10. Adirondack province
Interior Plains	11. Interior Low Plateaus
	12. Central Lowland
	13. Great Plains province

Major Division	Province
Interior Highlands	14. Ozark Plateaus
	15. Ouachita province
Rocky Mountain System	16. Southern Rocky Mountains
	17. Wyoming Basin
	18. Middle Rocky Mountains
	19. Northern Rocky Mountains
Intermontane Plateaus	20. Columbia Plateaus
	21. Colorado Plateaus
	22. Basin and Range province
Pacific Mountain System	23. Cascade–Sierra Mountains
	24. Pacific Border province
	25. Lower Californian province

of a region on Venus into two large tectonic and geomorphic units, a *Plains Group* consisting of *Smooth Plains, Ridged Plains,* and *Mottled Plains,* and a *Linear Ridge Group* consisting of a *Banded Unit,* a *Ridged and Domed Unit,* and a *Linear Ridge and Groove Unit.* He based these divisions on structure, topography, and surface roughness as interpreted from radar images. Although neither the formative processes nor the relative age of the geomorphic units on Venus is known with certainty, relative surface roughness and stratigraphic relationships permitted reasonable inferences about both.

Morphotectonic Regions

Among structural geologists and tectonists, the terms *morphotectonic region* and *morphostructure* are sometimes used to define regions or landscapes characterized primarily by their tectonic or structural unity, ignoring their erosional modification through time. Morphostructures are large topographic features such as submarine trenches or midocean ridges, formed directly by tectonic activity. Morphotectonic regions usually are defined by the distribution, intensity, and type (style) of deformation of the earth's crust by tectonic processes. **Orogeny** (mountain building), **epeirogeny** (broad regional movements), and **extensional tectonism** are useful subdivisions. They are considered primarily in Chapter 3. The definition of geomorphic provinces includes tectonic origins, so a good geomorphic description is more comprehensive than a purely morphotectonic one.

Scenery

Von Engeln (1942) elegantly compared first-order relief features to theaters: "permanent" structures "in which geologic dramas are enacted." His second-order relief features, which would include Orders 2 to 4 of Table 1-1, were compared to theatrical stages, which are rebuilt or altered from time to time. His third-order relief features, which would include Orders 5 through 8 or 9 of Table 1-1, are the scenery, stage settings, and backdrops, which are mobile and temporary. Scenery is the assemblage of landforms that can be viewed from a single vantage point, the pleasure of the tourist and the raw material of the geomorphologist. Almost all the following chapters concern landscapes at this scale. Most scenery, excluding such notable landforms as recent volcanic cones (Figure 6-14), is erosional rather than constructional.

THE LANDFORM AS THE UNIT OF SYSTEMATIC ANALYSIS

Landscapes are surfaces composed of an assemblage of subjectively defined components. Each element of the landscape that can be observed in its entirety, and has consistence of form or regular change of form, is defined as a **landform**. The definition is necessarily subjective (Howard and Spock, 1940). To one geomorphologist, a hillside is a landform, continuous and homogeneous. Another observer sees gullies, ridges, and dozens of other small topographic features that either break the continuity of the hillside or establish its continuity, depending on the mental concept.

The mathematical concept of the fractal character of natural objects (p. 5) is based heavily on geomorphic examples. There is no limit to the scale of observation. For example, the simple question, "How long is the coast of Britain?" is impossible to answer without specifying the length of the measuring unit because at each closer scale of observation, new irregularities appear (Mandelbrot, 1967, 1983). In principle, to get the length of the British coast, the wetted perimeter of every pebble and sand grain on every beach at one instant would have to be summed! At another extreme, Landsat Thematic Mapper images of the earth (Color Plates 3, 8, 9, and 12) have a resolution of 30 m, and currently available radar images of Venus only allow landforms to be resolved at a scale of 1 to 3 km. Of necessity, a landform will be defined differently by an observer on the ground or one in an orbiting spacecraft or studying a computer-processed digital image.

Landforms may be *constructional,* as in the examples of fault scarps, volcanic cones, glacial moraines, coral reefs, and river deltas. However, because of the peculiar intensity of atmospheric weathering and erosion on our planet, most subaerial landforms are *destructional,* or *erosional.* The subaerial landscape should be viewed as an assemblage of residual landforms, each in the process of being consumed or transformed by erosion. One of the fundamental goals of geomorphologic training is to be able to see landscapes not only as an assemblage of slopes, ridges, and mountain peaks, but also of erosional valleys, gullies, and hollows, between which stand the residual hills and mountains. The eye must be trained to see that the missing rock is as significant as the rock mass that remains.

From these introductory comments, it can be appreciated how landforms came to be systematically evaluated and explained in terms of Davis's trinity of *structure, process,* and *time* (or *stage*). Each landform can be visualized as enclosing or covering a rock mass that has specific physical and chemical properties, imperfections, and geometrically disposed discontinuities (bedding planes, joints, faults, and so on). All these mineralogic, lithologic, and deformational factors are collected in the geomorphic concept of *structure.* Also included are active tectonic movements.

We observe that a complex series of reactions takes place when rocks are exposed to water, air, and organisms in our planet's gravitational field. Certain phenomena, such as downward movement of detached fragments by gravity, affect all matter. Other processes are unusually effective only on specific structures, such as solution of limestone, or plastic flow of glacier ice and rock salt. Thus, landforms are the result of constructive and destructive *processes* acting on *structures.*

An account of the interaction of surface processes with geologic structures does not completely explain landforms unless we include in the explanation the length of *time* during which the process has been operating or the degree of modification (*stage*) that has been reached. Relatively few landforms show purely constructional processes. Some geomorphic processes, such as landslides or earthquakes, are nearly instantaneous. The resulting piles of broken rock (and houses) represent a more stable configuration. Most geomorphic processes act much more slowly. A major topic of geomorphic study is whether processes continue as long as any rock mass remains, or whether stable equilibria are reached.

Many, perhaps most, present-day landforms were shaped by processes that are no longer acting on them. For example, many landforms in Canada, the northern United States, and northern Europe were shaped by former ice sheets. These are *relict* landforms, some of which are surprisingly pure records of former conditions. Other relict landforms have been buried for long periods and have only recently been *exhumed.* These landforms also resulted from the interaction of geologic structures and processes through an interval of time, but that time did not continue to the present.

New techniques that radiometrically date the exposure of land surfaces by measuring the accumulation in rock and soil of isotopes produced by cosmic radiation promise to alter fundamentally the geomorphic concept of time. Future trends of geomorphic research will be toward defining numerical rates of change and the true elapsed time during which specified processes have acted on specified structures (Jennings, 1973, p. 123).

FURTHER CONSIDERATIONS OF STRUCTURE, PROCESS, AND TIME

Structure

One of the first things a geomorphologist wants to know about a region is "What is the structure?" By this question, he or she inquires about the minerals that form the rocks, the structural and stratigraphic arrangement of rock layers and masses, the previous tectonic displacement of the rocks, and the present state of deformation.

A useful distinction can be made between the **terrane,** an area of a certain structure or rock type, as in the expression *granitic terrane* or *limestone terrane,* and the **terrain,** a landscape or "the lay of the land." Military tacticians must appreciate the terrain; they need not be concerned with the terrane. However, strategists need to know the terrane to evaluate its potential mineral resources, water supply, and similar factors.

Rocks at the earth's surface are normally regarded as rigid solids, not likely to flow under their own weight to produce landforms. Yet it is well known that rocks have finite yield strengths, which when exceeded will result in permanent deformation. Under sufficient shear stress, rocks "flow" or "creep." The property of a substance that determines whether it will behave as a solid or fluid has been defined as *rheidity.* Carey (1954) defined the rheidity of rock as the amount of time necessary for viscous flow to exceed by 1000 times the elastic (or recoverable) deformation. The concept of rheidity is more rigorously quantified by defining the conventional linear strain rate $\dot{e} = e/t$, where elongation e = change of length/original length, and t = duration of strain in seconds (Pfiffner and Ramsay, 1982). The dimensions of length cancel each other, so the dimension of strain rate is the reciprocal of time in seconds, s^{-1}. For example, measured strain rates for rock salt vary over eight orders of magnitude from 10^{-8} to 10^{-16} s^{-1}. At such rates, under specified conditions of temperature, confining pressure, and shear stress, a mass of salt would deformationally elongate by an amount equal to its original length in 10^8 to 10^{16} s (1 year = 3 x 10^7 s). On the time scale of tectonic deformation (1 million to 10 million years, or 10^{13} to 10^{14} s), the rheidity of deforming masses of salt is so low that their deformation is approximately viscous (Jackson and Talbot, 1986), meaning that their rate of strain is directly proportional to the shear stress. *Salt domes* or their upwardly arched cover strata form significant landforms on the Gulf coast of the United States, in northern Germany, and in Iran (p. 84). In the arid climate of Iran, the salt can break through to the surface and flow as a *salt glacier* down the flanks of the dome (Jackson and Talbot, 1986; Talbot and Alavi, 1996, pp. 95-98). When moistened with groundwater or precipitation, salt flows even more rapidly. Extruded salt at the heads of some Iranian salt glaciers flows at a rate of 2 m/yr.

Glacier ice deforms even more rapidly than salt; under low shear stress, it will deform by an amount equal to its original length within one year ($\dot{e} = 10^{-7}/s$) (Figure 16-9). It deforms readily to fill mountain valleys, and the profile of an ice sheet is elliptical or parabolic, steepest near the edges and flatter near the summit (Figure 16-5). It also fractures as a brittle substance to form open crevasses as deep as 50 m. Thus, the surface morphology of an ice sheet is largely the result of plastic, or rheid, flow, but it is also the result of fracturing and melting. Glacier ice is the second most common rock type at the surface of the earth (greatly exceeded by submarine basalt), forming a layer 1 to 2 km in thickness over one entire continent (Antarctica) and the world's largest island (Greenland).

Many volcanic landforms also have their shape determined by the rheidity of molten lava (Chapter 6). Even though the lava solidifies and becomes rigid on cooling, the landforms record the mobility of the former molten state.

Rocks in young orogenic zones, where intense deformation has occurred within the past 10 million years or less, must have conventional strain rates of between 10^{-13} and $10^{-15}\,s^{-1}$ (Pfiffner and Ramsay, 1982). Within the cores of orogenic belts, these rocks flow, as recorded by the intricate contortion of gneissic bands or the great folds exposed on eroding mountain sides (Color Plate 2). The strain rate of matter in the earth's interior must be very high in order for isostatic equilibrium between large areas of the surface to be so very nearly perfect. The shapes of other planetary masses are also fundamentally related to their masses, temperatures, and the strengths of their materials. Small icy satellites in our solar system, such as some of the moons of Jupiter, Saturn, and Uranus, are irregular in shape if their radii are less than 200 km. Larger icy satellites are more nearly spherical. Rocky satellites show less definite trends, but in general, if they are larger than 500 km in radius, they also are spherical (Thomas, 1989). All the larger objects in the solar system are spherical. Whereas a small icy satellite may have surface roughness equal to a significant percentage of its mean radius, the maximum surface relief on the earth's crust is only about 15 km, a trivial part of the planet's radius of 6370 km. Weisskopf (1975, p. 609) suggested that the silicon–oxygen bond in silicate minerals will begin to deform under terrestrial relief of about 10 km, and a mountain range cannot become

much higher than that before it spreads laterally or sinks to isostatic equilibrium in the mantle. The implication is that there have never been mountains on the earth much higher than the present ones.

On a smaller scale, the strength of any rock mass is dependent not only on the mineral bonds within a solid rock specimen, but also on the discontinuities that are omnipresent in rocks and minerals at all scales (Chapter 7). The strength of a rock mass thus decreases with increased size because of the increasing influence of these spatially distributed discontinuities. Therefore, mountainous relief is limited not only by the rates of uplift and erosional incision, but by the strength of the rock mass (Schmidt and Montgomery, 1995).

In summary, for most geomorphic considerations, rock masses are to be regarded as rigid solids weakened by numerous discontinuities. But in special cases, such as salt domes, glaciers, lava flows, and soil creep, rheidity of a rock mass is a factor in the description of landforms.

Process

A host of external processes contribute to produce a landscape. Air and water react with rocks to cause the familiar results of weathering. Organisms absorb mineral nutrients from the rocks on which they live. Gravity tends to pull down structures that rise above the general ground level. Moving water, air, and ice erode rocks and transport the eroded debris to depositional sites. The rates of geomorphic processes vary enormously, from a few centimeters per thousand years for the surface weathering of ancient monuments to 50 m/s or more for an avalanche roaring down a mountainside.

Geomorphic processes also vary in intensity from one region to another, depending on climate, vegetation, and altitude above or depth beneath the ocean. A large portion of any geomorphology book must be devoted to analyzing the processes that act on geologic structures. The vital feature of geomorphic processes is that like all chemical and physical processes, under the same set of environmental conditions, they have acted in the past and will act in the future exactly as they now can be observed to act. Hutton's great principle of uniformity of process, often called "uniformitarianism," is a keystone of the scientific method.

Rigid uniformitarianism is not useful when studying relict or exhumed landscapes, however. Throughout the decipherable history of the earth, life has been evolving, and living organisms are not the same as the organisms that lived in the past. Many geomorphic processes are controlled by organisms, either directly, as when the steepness of hillslopes is determined by the holding capacity of tree or grass roots, or indirectly, as when soil chemical reactions are determined by the microorganisms in the soil water. It is hard to envision a pre-Cenozoic landscape, for example, on which no grasses had yet evolved to anchor the sod. It is only since mid-Miocene time, about 15 million years ago, that landscapes have looked "modern" and have been evolving in response to modern climatic and biotic processes (Chapter 4). Paleogeomorphologic reconstructions of landscapes older than the late Cenozoic require a modified set of assumptions.

Mammals also had an explosive mid-Cenozoic evolution, based largely on grazing and running on the new open, grassy plains. Even within the Quaternary Period, when most of our modern mammals had already evolved, faunal migrations have influenced landscape development. Grazing and browsing by domesticated animals may be responsible for huge areas of semiarid landscapes (Chapter 13). *Homo sapiens*, a late Quaternary mammal, has reshaped many landscapes, especially in the past 200 years. It has been asserted that human activity now involves an annual flux of earth materials equal to that of plate tectonics (Chapter 2).

These examples illustrate the dangers of regional or temporal extrapolations of the processes that are assumed to shape landforms. Although physical and chemical laws are indeed uniform through geologic time, the oscillatory climatic changes of the late Cenozoic Era have caused repeated transpositions of climate-dominated processes on many regions (Chapters 4 and 18). Glaciated terrains in many midlatitude regions, and loess and sand sheets that record former eolian processes in areas that are now humid and vegetated, remind us that few contemporary landforms can be assumed to have developed under any single set of processes.

A related danger of using present landforms to interpret past processes is embodied in the concept of **equifinality,** the tendency for landforms created by very different processes to converge toward similar shapes. For example, broad parabolic or U-shaped valleys separated by knife-edge ridges are common tropical landforms and are also typical of valley glaciation. Tectonic grabens may resemble straight glacial troughs. Smaller landforms, such as scarps and benches on hillsides, might result from faulting, landslides, or animal tracks. Numerous examples of equifinality are noted in many of the following chapters.

Time

One of the fundamental axioms of geomorphology is that the processes that act on geologic structures at the earth's surface change the appearance of landscapes with the passage of time—in other words, landscapes evolve. Anyone who has seen or felt the weather-beaten surface of an ancient stone monument understands that rocks react with water, biosphere, and atmosphere, and only a slight extension of reasoning is necessary to deduce that entire landscapes are shaped by the collection of processes that we call weathering and erosion. The key word in the relationship between structure and process is "progressive." Although structures surely are affected by surface processes, are the changes progressive and sequential? Do landscapes pass through predictable evolutionary stages with the passing of time? These questions are at the heart of the science because, in order to be a valid science, geomorphology must generate hypotheses (predictions) that can be further tested and either falsified, in which case they must be modified, or corroborated, in which case they may rise to the level of theories, or even "laws of nature."

A corollary of the principle that processes of change alter structures through time is that *stages*, or degrees of completeness, of alteration can be defined. W. M. Davis conceived a geomorphic cycle[1] for river-eroded landscapes, in which specific stages could be recognized. His dangerously oversimplified terms, "youth," "maturity," and "old age," are familiar to every beginning student of geology and geography. They need careful reevaluation in light of our current understanding of the effectiveness of the processes and the time spans during which they have been operating (Ritter, 1988) (Chapter 15).

In process geomorphology, change through time is reduced as a factor. The goal is to study the actions and reactions between processes and rock masses ("structures"), here and now. Change is conceived as dynamic, fluctuating around some average condition, or as invoking a series of responses that tend to either restore a system to a preexisting equilibrium or shift it to a new state.

Schumm (1977, pp. 4–15) reviewed the various geomorphic perspectives on time as it applies to the fluvial system (Chapters 10 and 11) and proposed that the role of time changes when viewed on different scales (Figure 1-6). If one were measuring water or

---

[1]His term was "geographic cycle," but the growth of geography into humanistic studies soon rendered the term inappropriate.

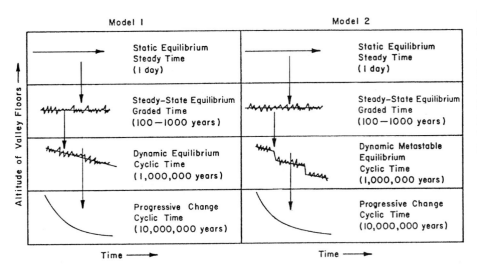

FIGURE 1-6. Models of landscape evolution. Model 1: Equilibrium components of Davis model of progressive erosion. Model 2: Equilibrium components of model based on episodic erosion (after Schumm, 1977, Figure 1-5, reprinted by permission; copyright © 1977, John Wiley & Sons, Inc.).

sediment discharge through a river cross section over a period of a few days, the altitude of the channel floor could be correctly assumed to be stable or in *static equilibrium* over the brief interval called *steady time.* However, if a record of 100 to 1000 years were available, the channel or valley floor might be seen to fluctuate about some average altitude in *steady-state equilibrium* over the interval called *graded time.* Over a much longer time interval of 1 to 10 million years, the steady-state equilibrium in turn would be seen as part of a changing *dynamic equilibrium* with a declining energy supply as the rock mass is worn away, and there would be *progressive change* (landscape evolution) over *cyclic time,* as conceived by Davis. Even within cyclic time, the dynamic equilibrium might not adjust uniformly but experience intervals of near steady state separated by abrupt, irreversible jumps to a new *metastable* equilibrium.

Practical or applied geomorphology emphatically needs prediction of changes through time. Will an excavation wall remain stable, or will it collapse as it adjusts to groundwater seepage or subaerial exposure? Will the soil on a hillside remain in place when cultivated with row crops, or should it be seeded with grass and used as pasture? These are practical geomorphic questions in which the time scale is measured in days or years. They introduce the concept of **thresholds** in geomorphic change (Coates and Vitek, 1980). If a certain process gradually increases in intensity, the rock mass may not be affected until a critical stress is exceeded, when the system suddenly and dramatically responds. For example, the velocity of flowing water might gradually increase without causing any erosion until the threshold of erosion is exceeded, when erosion suddenly begins (Figure 10-11). More-

over, as shown by the examples of predictive need, a rock or sediment mass may maintain a stable slope for a long time even though it gradually loses strength through weathering, then suddenly fail without any additional stress applied to it. In such a case, the threshold is *intrinsic* to the material, and not *extrinsic,* as when an imposed process such as water velocity exceeds some threshold of response.

Process-response models attempt to predict the behavior of geomorphic systems when extrinsic and intrinsic thresholds are crossed. The concept of **complex responses** includes time units such as *reaction time,* the interval after a perturbation before the system responds; *relaxation time,* during which the system grows into equilibrium with the new conditions; and *persistence time,* the duration of equilibrium state before the next perturbation (Figure 1-7). The *response time* is the sum of reaction time and relaxation time, and in the example illustrated by Figure 1-7, the perturbation is understood to be extrinsic, in the form of some climatic change (Bull, 1991, p. 12). The concepts of thresholds and complex responses have usually been applied to fluvial systems and are considered further in Chapters 10 and 11.

In the theoretical analysis of landscape evolution, time is typically measured in millions of years. How have landscapes evolved when exposed to the agents of weathering and erosion for a geologically significant interval of time such as the late Cenozoic Era? Traditionally, geomorphologists spoke of *relative* time and compared the development of landscapes only in terms of the relative stage of erosional evolution. The basis for this convention was that some landscapes are on weak or easily eroded terranes, have strong degradational processes at work, and evolve rapidly. Others, forming on resistant rocks or experiencing only

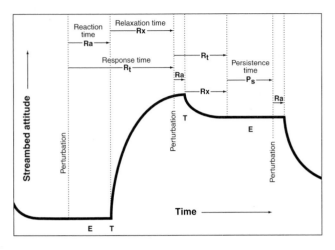

FIGURE 1-7. Threshold–equilibrium plot showing the components of response time R*t*, which is the sum of reaction time, R*a*, and relaxation time, R*x*. P*s* is the time of persistence of new equilibrium conditions, and T and E represent threshold and equilibrium, respectively (Bull, 1991, Figure 1.3, revised. © 1991 by Oxford University Press, Inc.).

weak processes, change very slowly. Little value was placed on specifying the rate of development of landscapes in terms of actual time.

With the development of methods for determining the ages of rocks and sediments and even the exposure ages of rock surfaces by radioactive disintegration of natural isotopes, the determination of real time, and therefore real rates of change, has now become possible. When applied to geomorphic problems, radiometric ages can answer such questions as when an ice sheet uncovered a region, when rivers began to dissect a volcanic ash bed, when sand dunes invaded a former river valley, or when a coral reef grew up to the ocean surface. Time is now specified to a precision that was not conceivable even a few decades ago, and the future of geomorphology lies in converting landscape analysis from describing relative ages to reporting real rates of change over a variety of temporal and spatial scales.

REFERENCES

AHNERT, F., 1988, Modelling landforms change, in Anderson, M. G., ed., Modelling geomorphological systems: John Wiley & Sons Ltd., Chichester, UK, pp. 375–400.

BAKER, V. R., 1986, Introduction: Regional landforms analysis, in Short, N. M., and Blair, R. W., Jr., eds., Geomorphology from space: A global overview of regional landforms: NASA Scientific and Technical Information Branch, Washington, D.C., Publication NASA SP-486, pp. 1–26.

BARSCH, D., and LIEDTKE, H., 1980, Principles, scientific value and practical applicability of the geomorphological map of the Federal Republic of Germany at the scale of 1:25,000 (GMK25) and 1:100,000 (GMK100): Zietschr. für Geomorph., Supp. no. 36, pp. 296–313.

BELL, T. H., JR., 1975, Statistical features of sea-floor topography: Deep-Sea Res., v. 22, pp. 883–892.

BIRKELAND, P. W., 1984, Soils and geomorphology: Oxford Univ. Press, Inc., New York, 372 pp.

BIROT, P., 1960, Le cycle d'érosion sous les différents climats: Univ. do Brasil Centro de Pesquisas de Geografia do Brasil, Rio de Janeiro, 137 pp.

———, 1968, Cycle of erosion in different climates (trans. by C. I. Jackson and K. M. Clayton): Univ. of California Press, Berkeley, 144 pp.

BOSTOCK, H. S., 1964, Provisional physiographic map of Canada: Canada Geol. Survey Paper 64-35, 24 pp.

BÜDEL, J., 1982, Climatic geomorphology (trans. from German by Fischer, L., and Busche, D.): Princeton Univ. Press, Princeton, N.J., 443 pp.

BULL, W. B., 1991, Geomorphic responses to climate change: Oxford Univ. Press, Inc., New York, 326 pp.

CAMPBELL, I. H., and TAYLOR, S. R., 1983, No water, no granites — no oceans, no continents: Geophys. Res. Ltrs., v. 10, pp. 1061–1064.

CAREY, S. W., 1954, The rheid concept in geotectonics: Geol. Soc. Australia Jour., v. 1, pp. 67–117.

———, 1962, Scale of geotectonic phenoma: Jour. Geol. Soc. India, v. 3, pp. 97–105.

COATES, D. R., and VITEK, J. D., eds., 1980, Thresholds in geomorphology: George Allen & Unwin Ltd., London, 498 pp.

COGLEY, J. G., 1985, Hypsometry of the continents: Zeitschr. für Geomorph., Supp. no. 53, 48 pp.

DREWRY, D., 1986, Glacial geologic processes: Edward Arnold (Publishers) Ltd., London, 276 pp.

DURY, G. H., 1972, Some recent views on the nature, location, needs, and potential of geomorphology: Professional Geographer, v. 24, pp. 199–202.

EMBLETON, C., ed., 1984, Geomorphology of Europe: The Macmillan Press Ltd. (published in the United States by Wiley-Interscience, a Division of John Wiley & Sons, Inc., New York), 465 pp.

———, and THORNES, J., ed., 1979, Process in geomorphology: Edward Arnold (Publishers) Ltd. (published in the United States by the Halsted Press, a Division of John Wiley & Sons, Inc., New York), 436 pp.

von ENGELN, O. D., 1942, Geomorphology: The Macmillan Company, New York, 655 pp.

FAIRBRIDGE, R. W., 1968, Continents and oceans—statistics of area, volume and relief, in Fairbridge, R. W., ed.,

Encyclopedia of geomorphology: Reinhold Book Corp., New York, pp. 177–186.

FENNEMAN, N. M., 1928, Physiographic divisions of the United States: Assoc. Am. Geographers Annals, v. 18, pp. 261–353.

———, 1931, Physiography of western United States: McGraw-Hill Book Company, New York, 534 pp.

———, 1938, Physiography of eastern United States: McGraw-Hill Book Company, New York, 714 pp.

FYFE, W. S., 1988, Granites and a wet convecting ultramafic planet: Trans., Royal Soc. Edinburgh: Earth Sciences, v. 79, pp. 339–346.

GARDNER, T. W., JORGENSEN D. W., and 2 others., 1987, Geomorphic and tectonic process rates: Effects of measured time intervals: Geology, v. 15, pp. 259–261.

GRAF, W. L., ed., 1987, Geomorphic systems of North America: Geol. Soc. America Centennial Spec. Vol. 2, 643 pp.

GRAFAREND, E. W., 1994, What is a geoid?, in Vaníček, P., and Christou, N. T., eds., Geoid and its geophysical interpretation: CRC Press, Inc., Boca Raton, Fla., pp. 3–32.

HAMMOND, E. H., 1963, Classes of land-surface form in the forty eight states, U.S.A: Map Supp. no. 4, Assoc. Am. Geographers Annals, v. 54 [1964].

HEAD, J. W., 1990, Formation of mountain belts on Venus: Evidence for large-scale convergence, underthrusting, and crustal imbrication in Freyja Montes, Ishtar Terra: Geology, v. 18, pp. 99–102.

HEEZEN, B. C., THARP, M., and EWING, M., 1959, Floors of the oceans, I. The North Atlantic: Geol. Soc. America Spec. Paper 65, 122 pp.

HEIRTZLER, J. R., and FRAWLEY, J. J., 1994, New gravity model for earth science studies: GSA Today, v. 4, pp. 269–270.

HOWARD, A. D., and SPOCK, L. E., 1940, Classification of landforms: Jour. Geomorphology, v. 3, pp. 332–345.

HUNT, C. B., 1974, Natural regions of the United States and Canada: W. H. Freeman and Company, San Francisco, 725 pp.

JACKSON, M. P. A., and TALBOT, C. J., 1986, External shapes, strain rates, and dynamics of salt structures: Geol. Soc. America Bull., v. 97, pp. 305–323.

JENNINGS, J. N., 1973, 'Any millenniums today, lady?' The geomorphic bandwaggon parade: Australian Geog. Studies, v. 11, pp. 115–133.

KING-HELE, D., 1976, The shape of the earth: Science, v. 192, pp. 1293–1300.

LEOPOLD, L. B., WOLMAN, M. G., and MILLER, J. P., 1964, Fluvial processes in geomorphology: W. H. Freeman and Company, San Francisco, 522 pp.

LePROVOST, C., BENNETT, A. F., and CARTWRIGHT, D. E., 1995, Ocean tides for and from TOPEX/POSEIDON: Science, v. 267, pp. 639–642.

MANDELBROT, B., 1967, How long is the coast of Britain? Statistical self-similarity and fractional dimension: Science, v. 156, pp. 636–638.

———, 1983, The fractal geometry of nature: W. H. Freeman and Company, New York, 468 pp.

MEADE, C., and JEANLOZ, R., 1991, Deep-focus earthquakes and recycling of water into the Earth's mantle: Science, v. 252, pp. 68–72.

MENARD, H. W., and SMITH, S. M., 1966, Hypsometry of ocean basin provinces: Jour. Geophys. Res., v. 71, pp. 4305–4325.

MORISAWA, M., ed., 1971, Quantitative geomorphology: Some aspects and applications: Second Annual Geomorphology Symposia Proc., Binghamton, N.Y., 315 pp.

OLLIER, C. D., 1984, Weathering, 2nd ed.: Longman Inc., New York, 270 pp.

PFIFFNER, O. A., and RAMSAY, J. G., 1982, Constraints on geological strain rates: Arguments from finite strain states of naturally deformed rocks: Jour. Geophys. Res., v. 87, no. B1, pp. 311–321.

PIKE, R. J., and THELIN, G. P., 1989, Shaded relief map of U.S. topography from digital elevations: Eos, v. 70, pp. 843, 853.

RITTER, D. F., 1988, Landscape analysis and the search for geomorphic unity: Geol. Soc. America, v. 100, pp. 160–171.

———, KOCHEL, R. C., and MILLER, J. R., 1995, Process geomorphology, 3rd ed.: Wm. C. Brown Communications, Inc., Dubuque, Iowa, 538 pp.

SCHMIDT, K. M., and MONTGOMERY, D. R., 1995, Limits to relief: Science, v. 270, pp. 617–620.

SCHUMM, S. A., 1977, The fluvial system: John Wiley & Sons, Inc., New York, 338 pp.

SCLATER, J. G., ANDERSON, R. N., and BELL, M. L., 1971, Elevation of ridges and evolution of the central eastern Pacific: Jour. Geophys. Res., v. 76, pp. 7888–7915.

SHORT, N. M., and BLAIR, R. W., JR., eds., 1986, Geomorphology from space: A global overview of regional landforms: NASA Scientific and Technical Information Branch, Washington, D.C., Publication NASA SP-486, 717 pp.

TALBOT, C. J., and ALAVI, M., 1996, The past of a future syntaxis across the Zagros, in Alsop, G.I., Blunell, D.J., and Davison, I., eds., Salt tectonics: Geol Soc. [London] Spec. Pub. No. 100, pp. 89-109.

TAYLOR, S. R., and McCLENNAN, S. M., 1985, The continental crust: Its composition and evolution: Blackwell Scientific Publications, Oxford, UK, 312 pp.

THOMAS, P. C., 1989, The shapes of small satellites: Icarus, v. 77, pp. 248–274.

THORNBURY, W. D., 1965, Regional geomorphology of the United States: John Wiley & Sons, Inc., New York, 609 pp.

TRICART, J., and CAILLEUX, A., 1965, Introduction a la géomorphologie climatique: Soc. d'édition d'enseignement supérieur, Paris, 306 pp.

———, 1972, Introduction to climatic geomorphology (trans. by C. J. K. de Jonge): Longman Group Ltd., London, 295 pp.

TURCOTTE, D. L., 1989, Fractals in geology and geophysics: PAGEOPH, v. 131, pp. 171–196.

TURCOTTE, D. L., 1989, Fractals in geology and geophysics: PAGEOPH, v. 131, pp. 171–196.

———, 1997, Fractals and chaos in geology and geophysics, 2nd ed.: Cambridge University Press, Cambridge, UK, 397 pp.

WAHRHAFTIG, C., 1965, Physiographic divisions of Alaska: U.S. Geol. Survey Prof. Paper 482, 52 pp.

WEISSKOPF, V. F., 1975, Of atoms, mountains, and stars: A study in qualitative physics: Science, v. 187, pp. 605–612.

WHITNEY, J. A., 1988, The origin of granite: The role and source of water in the evolution of granite magmas: Geol. Soc. America Bull., v. 100, pp. 1886–1897.

WISE, D. U., 1974, Continental margins, freeboard and the volumes of continents and oceans through time, in Burk, C. A., and Drake, C. L., eds., The geology of continental margins: Springer-Verlag New York Inc., New York, p. 45–58.

deWIT, M. J., ROERING, C., and 7 others, 1992, Formation of an Archaean continent: Nature, v. 357, pp. 553–562.

Chapter 2

Energy Flow in Geomorphic Systems

THE SYSTEMS CONCEPT IN GEOMORPHOLOGY

All physical phenomena may be conceived as being organized into systems, a system being defined as a collection of objects and the relationships between those objects. **Closed system**s are those that have boundaries across which no energy or matter moves; **open systems** have a flow of energy and matter through their boundaries. A closed system must change toward a static, time-independent equilibrium state, according to the second law of thermodynamics, in which the ratios of various physical phases remain constant, the amount of free energy (such as chemical potential, potential energy of position) is minimal, and the entropy (amount of disorder in the system, often visualized as the diffuse heat of random molecular motion) is maximal. An open system, on the other hand, may achieve a dynamic steady state, wherein the system and its phases remain in a constant condition even though matter and energy enter and leave it (von Bertalanffy, 1950; Strahler, 1950, p. 676).

Only the simplest geomorphic processes may be studied as closed systems. Certain weathering reactions, such as solution of limestone by acidified rainwater, can be duplicated by adding measured quantities of the reactants to a test tube and observing that, after a period of time, predicted amounts of new compounds have formed and are in stable equilibrium with each other. Hot, newly erupted lava radiates and conducts heat energy to its surroundings until the

temperature gradient is reduced to zero and the mass has solidified. Then it ceases to give off heat. Even in these simple examples of physical and thermal systems, it is obvious that the natural phenomena are much more complex than these brief statements imply.

The entire subaerial portion of the earth's surface can be profitably regarded as an open system (Chorley, 1962). Matter is added by tectonism or volcanism, with an additional minor net increase of mass from meteoritic infall. Energy is derived from solar radiation, gravitation, rotational inertia, and internal heat. Subaerial, subaqueous, and subice processes erode the rock structures, and the eroded debris is deposited elsewhere, usually below sea level. It is conceivable that a steady state could be reached in such a complex open system so that the form of the landscape would not change even though the mass that composes it was constantly changing and energy constantly being expended.

Compare a landscape (a "geomorphic system") with a rotary kiln used to manufacture portland cement (Figure 2-1). Simple raw materials, primarily limestone and shale, are fed into the upper end of a sloping, rotating, heated drum. The fragments tumble and are pulverized within the drum as they move toward the lower end, but in the presence of heat they react chemically to form the desired end product, a powder that reacts with water to form cementing calcium silicate minerals of great adhesive strength. The heat that is not stored in the newly created chemical

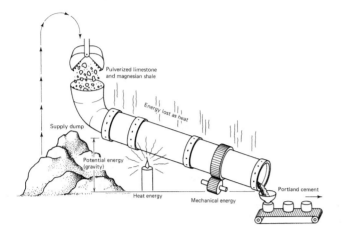

FIGURE 2-1. The rotary kiln analog to an open geomorphic system.

compounds is liberated as wasted energy, along with the mechanical frictional losses. Mechanical energy is provided by a gravity potential and also by the kinetics of rotation. If the supply of raw materials is adjusted to the supply of mechanical and thermal energy, a desired product is produced at a desired rate. If material is supplied at a more rapid rate, the energy supply also has to be increased to maintain the flow of the desired product. A steady state at many levels of net flow can be obtained by mutually adjusting the several variables of material and energy supply in this simple open system.

A landscape above sea level (or above the geoid) is similar to the rotary kiln model. Rainfall sweeps weathered rock fragments downhill to the sea in rivers. The rate of flow is a function of climatic variables such as temperature and precipitation, the gradient is established primarily by tectonism, and the intensity of chemical and physical change of the transported debris is a function of the time of travel through the system. As long as the energy and material flow are adjusted, the system stays in a steady state.

Suppose, however, that the intake end of a rotary kiln had been unwisely built on the supply dump. As the supply of raw material (the constructional landform) is consumed, the slope of the kiln is reduced. In order to maintain quality control during a longer residence time in the kiln, the heat must be reduced. The rate of cement production then declines. By proper control, even such a poorly designed system would appear to be in a steady state over any brief period of observation, although on an extended time scale the

potential energy of the system would be declining. This is analogous to erosional landscapes that are not being renewed by constructional processes (Figure 1-6). Although rivers may appear on a short time scale to be *graded*, that is, nicely adjusted open systems in steady states doing just the work they are required to do, on the longer scale of cyclic time, they are seen to be open systems of declining energy.

We assume that the mass of the earth is constant, ignoring the continuous small increment of meteoritic dust (Love and Brownlee, 1993) and the extremely rare impact of large extraterrestrial fragments. In principle, it should be possible to define the geoid (p. 7) as one boundary of an open system and to monitor the movement of rock material upward through it by tectonism and volcanism and back downward through it by tectonism and erosion (Figure 2-2). Because the geoid tends toward an equilibrium, the masses transported through this reference surface must balance, and lateral transport systems must also operate above and below the geoid to maintain isostatic equilibrium. The study of the geomorphic system has not, however, progressed to the point where such mass balances can be calculated.

In contrast to the geomorphic mass balance, the energy balance in the geomorphic system is becoming quite well known, owing to related research in biology, meteorology, astronomy, and geophysics. In the universe as a whole, gravity is the dominant form of energy (Dyson, 1971), greatly exceeding heat, light, and nuclear energy. Within our geomorphic system, solar radiation (in itself a result of the gravitational force that sustains thermonuclear fusion in the sun) dominates all other energy sources. The sun has been slowly increasing its output of luminous energy throughout the history of the solar system, but the geomorphic consequences of the slow change are not measurable. However, even though the total solar energy received by the earth is very nearly constant, varying by only about 0.2 percent over a cycle of about 95,000 years due to orbital variations (p. 56), the distribution of that energy over the earth's surface has fluctuated greatly, even during late Cenozoic time (Figure 4-6).

It is important for the next generation of geomorphologists to understand better the energy potentials available for any system they wish to study. Many poor generalizations have been made in the past that were not consistent with energy considerations. Trends toward quantification require that energy as well as material rates and balances be calculated. The remainder of this chapter is an evaluation of the energy flow in the geomorphic system, essentially a detailed review and discussion of Figure 2-2. A few rare or trivial forms of energy that are unlikely to produce land-

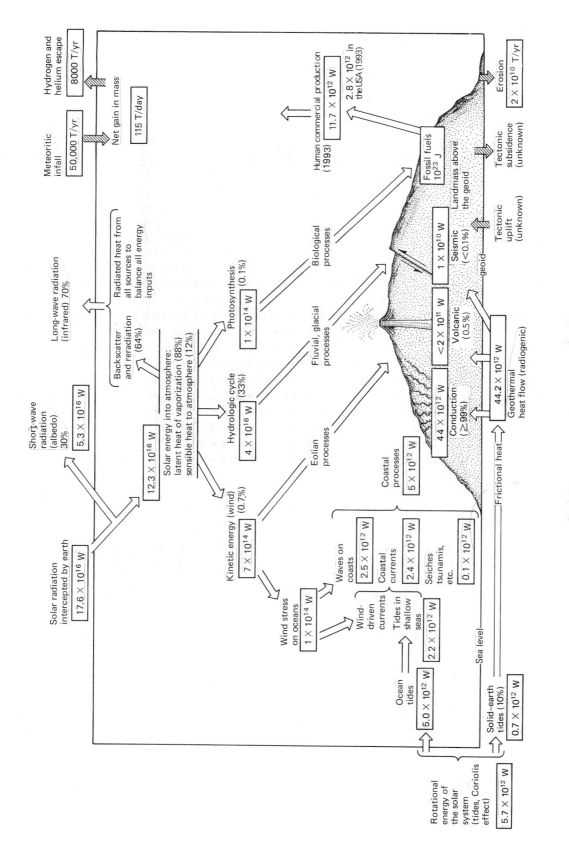

FIGURE 2-2. Energy and matter flux in the geomorphic system. System boundaries are the geoid and the air and water interfaces. Energy enters the system as solar radiation, rotational energy, and radiogenic heat. The rate of outgoing infrared radiation balances all incoming energy. Matter enters the system by volcanism and tectonic uplift above the geoid and leaves by tectonic subsidence and erosion. Minor amounts of matter enter as meteorites and escape as molecular hydrogen and helium (Hubbert, 1971; Inman and Brush, 1973; Love and Brownlee, 1993; United Nations, 1995, Table 4; other references cited in text).

forms, such as lightning discharge, cosmic radiation, and magnetic fields, are not considered further.

Human beings, who have become significant geomorphic agents during the last few thousand years (Nir, 1983), are now using energy on a scale comparable to that released by wind, tides, and geothermal heat. Ecologist Jared Diamond asserted that we now appropriate about 40 percent of the earth's net productivity (the net energy captured from sunlight) (Diamond, 1992, p. 3). Humanity uses 26 percent of the terrestrial evapotranspiration (Figure 2-5) to meet our water needs (Postel et al., 1996). We are processing as much rock annually as is being recycled by plate tectonic processes. We live on a land surface that is illuminated from above by radiation from a nuclear fusion reactor and is heated and deformed from beneath by energy from a thermal nuclear reactor driven by radioactive decay. When we learn how to harness safely the energy of nuclear fusion, which drives solar radiation, and nuclear fission, our role in the future modification of this and other planets will be almost limitless.

THE POWER AVAILABLE FOR GEOMORPHIC CHANGE

Power is defined as the rate of doing work, or of energy transferred per unit of time. Conventional units are watt (1 W = 1 joule/second = 1 newton-meter/second) or horsepower (1 HP = 746 W). Figure 2-2 outlines the rate of energy conversion (power) near the surface of the earth. Solar radiation, mostly in the visible and near-infrared wavelengths, supplies about 99.97 percent of the energy entering the geomorphic system. About 30 percent of the solar radiation is directly reflected, mostly by clouds (the planetary **albedo,** or reflective brightness, is the measure of this reflected energy). Some of the nonreflected energy is converted to thermal vibration of air, water, and mineral molecules as *sensible heat,* warming the planet, activating chemical, physical, and biological reactions, and possibly even heating rocks until they fracture. The largest fraction is absorbed as *latent heat of vaporization* by the abundant water on the surface of the earth and is transported by water vapor in the atmosphere until it is released by condensation in clouds. Much smaller fractions of the energy are converted directly to kinetic motion of air and water masses, driving convective and advective currents in the atmosphere and oceans, or are absorbed by photosynthesizing plants. A very small amount of the energy drives geomorphic processes. All this energy is ultimately returned to outer space by reradiation from the moderately warmed earth.

Solar radiation provides energy for geomorphic change at a rate 3300 times faster than all other energy sources combined, yet these other sources cannot be ignored. They include the gravitational and inertial forces associated with the mass and motion of the earth, moon, sun, and other objects of the solar system, expressed at the earth's surface by tides in both the lithosphere and the hydrosphere. Also included is the heat that flows outward through the surface from the interior of the earth.

Not shown in Figure 2-2 is the power of meteorite impacts (Melosh, 1989). More than 120 impact craters have now been identified on the earth, mostly in the continental shields where erosion is slow and the oldest parts of the continental crust are preserved. Half of the craters larger than 10 km in diameter have ages of less than 200 million years (Grieve, 1987; 1990). The energy of the iron meteorite 60 m in diameter that created the Barringer Crater in Arizona about 30,000 years ago was comparable to that of one of the most powerful modern nuclear weapons, and it created a "simple" crater 1.2 km in diameter and 170 m deep. Larger, complex craters with a central area of uplift and down-faulted margins range from 4 to 100 km or more in diameter. The largest were caused by impacts of kilometer-size meteorites. Lunar and terrestrial crater abundances indicate that from one to three craters wider than 20 km in diameter are created on earth every million years (Grieve, 1990, p. 68). More frequent or giant impacts in the early history of the earth may have been important in creating the moon, tilting the earth's rotational axis, and making the planet a uniquely habitable one (Stevenson, 1987; Laskar and Robutel, 1993). The possible role of a 10-km diameter meteorite impact in causing the mass extinctions that define the Cretaceous-Tertiary boundary (Alvarez et al., 1980) was debated at a major conference in 1981 (Silver and Schultz, 1982) and has continued to be a major controversial topic of earth history (Sharpton and Ward, 1990; Dressler et al., 1994; Ryder et al., 1996). Although the enormous power of such infrequent but catastrophic events can only be crudely estimated, their role in both creating individual landforms and affecting other global geomorphic processes cannot be ignored.

One difficulty with evaluating the power of the subaerial geomorphic system is that there is no simple way to measure the *power density,* or localized concentration of energy expenditure. For example, a meteorite impact could produce a huge crater in a few seconds. The small proportion of geothermal heat flow that is concentrated at volcanoes and faults creates spectacular landforms, whereas the other 90 to 95 percent drives all tectonic processes but makes no specific landforms.

Biological systems, especially human beings, have an exceptional ability to store and concentrate diffuse energy at the earth's surface (Häfele, 1974; Holdren and Ehrlich, 1974). The geomorphic impact of human beings is unique because we can concentrate power so intensely. Sunlight, when fixed by photosynthesis, accumulated as fossil fuel, burned in an engine, and focused at the tip of a drill bit, can alter the landscape along a highway right-of-way in a few months to an extent that only millions of years of normal incident radiant energy could accomplish. Solid-state lasers used for research in atomic and plasma physics can focus incredible power densities of 10^{18} W/cm^2 in bursts of energy of nanosecond (10^{-9} s) duration (Perry and Mourou, 1994). Holes can even be drilled through diamonds by such concentrated energy.

SOLAR RADIATION

A major goal of geomorphology is to understand how the immense but diffuse energy of solar radiation is converted into mechanical work that shapes landscapes. We can visualize the process in terms of a "geomorphology machine" (Figure 2-3) in which a steam engine fired by the sun powers a series of fans, saws, files, grinders, and hydraulic jets to reduce the landscape. The analogy with a steam engine is a good one because it is the phase changes in the abundant water on the earth's surface that convert solar energy

FIGURE 2-3.
The "geomorphology machine" (Bloom, 1969).

into geomorphically significant mechanical work. Of even greater significance is the fact that solar heat keeps the surface of our planet within the temperature range in which water naturally occurs in solid, liquid, and vapor phases. In a sense, we are inside the boiler and condenser of the machine rather than witnessing the final mechanical conversion of the solar energy.

The sun radiates energy over an electromagnetic spectrum of wavelengths that is closely approximated by a perfectly radiating black body at an absolute temperature of about 6000K (Peixoto and Oort, 1992, p. 98). The radiant energy intercepted by a disk perpendicular to the sun's rays at the earth's mean orbital radius from the sun is about 1360 W/m^2. This value is called the **solar constant**. The earth intercepts solar energy at a rate equal to the product of its cross-sectional area of about 1.28 x 10^{14} m^2 and the solar constant. This is the approximate value of incoming radiation shown in the upper-left part of Figure 2-2 (although derived from other sources). However, the earth is not a disk but approximately a rotating sphere, with a surface area of $4\pi r^2$ that is four times as large as the equivalent disk cross section. Thus, the average input of solar radiation on each square meter of the upper atmosphere is only about one-fourth of the solar constant, or 343 W/m^2 (Salby, 1992, p. 66).

In addition to reflecting 30 percent of the total incoming energy, the atmosphere acts as a selective filter to restrict further the wavelengths of solar radiation that reach the earth's surface. In the lower atmosphere, water vapor (H_2O), carbon dioxide (CO_2), and other less abundant gases strongly absorb incoming solar radiation in specific infrared wavelengths. It is one of the balances of nature that the earth is warmed by the sun to such a temperature that it reradiates energy out through the atmosphere into space at wavelengths that are much longer than the incoming radiation, and are in spectral "windows" between the wavelengths that are so strongly absorbed by water and by carbon dioxide. If earth had no atmosphere and radiated directly back into space all the energy it received from the sun, its *effective emission temperature* as a "black body" would be a chilly 255K (−18°C). The actual mean terrestrial surface temperature of 288K (15°C) is a result of atmospheric absorption and reradiation, the "greenhouse effect." Thus, thanks to our atmosphere, the earth's surface averages 33°C warmer than it otherwise would be (Peixoto and Oort, 1992, p. 118). As long as the earth has had its present atmospheric composition, the mean surface temperature cannot have fluctuated more than a few degrees. However, changes in the amount of atmospheric CO_2 and water vapor have a dramatic effect on surface temperatures, and there is good evidence that both have varied through geologic time (Chapters 4 and 18).

Clouds are a major regulator of the earth's heat budget, but their role is very poorly understood. They absorb reradiated long-wave energy from the earth's surface, but also reflect incoming solar radiation. The net effect of cloudiness is probably a slight planetary cooling, but the balance is small, and the amounts of energy absorbed and reflected are very large (Ramanathan et al., 1989). Tropical clouds especially seem to absorb far more short-wave solar radiation than has been predicted by theoretical models (Pilewskie and Valero, 1995). About one-half of the earth is cloud covered at any moment (Salby, 1992, p. 63), and the net effect of clouds on the earth's radiation budget has been estimated to be four times as large as the potential effect of a doubling of atmospheric CO_2 (Ramanathan et al., 1989). There is a great likelihood that changes in cloud cover as a result of other climate changes can have powerful feedback effects.

Terrestrial Thermal Gradients

Solar energy is absorbed and reradiated unequally by the lower atmosphere, hydrosphere, and lithosphere. Complex thermal gradients result, which cause atmospheric and oceanic circulation and produce climatic regions on the earth's surface. Because climate is such an important aspect of geomorphic processes, the major thermal energy gradients need specific discussion.

Latitudinal Gradients. For numerous reasons, most of the solar energy is received by the earth within the tropics (defined as the area between 23 1/2° N and S latitude). First, the sun is high in the tropical sky at all seasons, and the reflectance is least when the incoming radiation is at a high angle to the surface. Second, the tropics contain large oceanic areas, favoring absorption of heat by water. Third, the albedo of tropical regions is much lower than the planetary average. Fourth, the tropical atmosphere is cloudy and humid, absorbing both incoming solar radiation and reradiated earth heat. Fifth, the surface area between successive parallels of latitude is larger near the equator than near the poles. On a sphere, 50 percent of the area lies within 30° north and south of the equator. For all these reasons, in the lower latitudes more energy is received from the sun than is reradiated back into space.

The polar landscape receives very little of the total solar radiation. The low incident angle of the sunlight, even during the time of 24-hour daylight in high latitudes, promotes reflection. Both the Arctic Ocean and the Antarctic Continent are covered with ice and snow, which further aid in reflecting the incoming radiation.

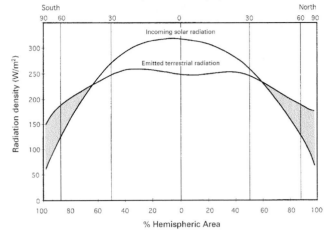

FIGURE 2-4. Average annual absorbed solar radiation and emitted terrestrial radiation profile (W/m²) from 80° N to 80° S. Area scale is linear. The total emitted terrestrial radiation is about 2.5 percent too small to balance the incoming solar radiation because of incomplete data (replotted from Peixoto and Oort, 1992, Figure 6.14 and Table 6.2).

It is true that for more than a month of midsummer when the sun shines continuously, each polar region in turn receives more radiation per day than is ever received per day in the tropics, but the season is very brief, and for the rest of the year the energy input at the poles is far lower than the output. Figure 2-4 summarizes the latitudinal distribution of incoming and outgoing radiation, and shows that at latitudes of about 38° north and 39° south, the annual heat budget of the earth is balanced. The hemispheres differ slightly because of differing land areas, albedo, and ocean currents. Approximately 60 percent of the area of each hemisphere absorbs more solar radiation than it emits. The smaller poleward areas of each hemisphere are net emitters. The excess heat received in the lower latitudes is carried poleward by winds and ocean currents, and it balances the net deficit between incoming and outgoing radiation at high latitudes.

Altitudinal Gradients. The atmosphere is reasonably transparent to incoming solar radiation of visible and near-infrared wavelengths but is strongly absorbent in certain wavelengths of the outgoing longer-wave infrared radiation from the solid earth. Thus, the lowest part of the atmosphere is heated from beneath and becomes convectively mixed, with warm air rising and cold air sinking. The lowest atmospheric layer is called the **troposphere** and ranges in thickness above sea level from about 16 km in the tropics to about 8 km near the poles. Within the troposphere, temperature decreases with height at a rate of about

6.5°C/1000 m. As water vapor condenses to form clouds, latent heat is released to the adjacent air, the air mass expands and rises, and powerful vertical air currents are generated. In extreme conditions, giant tropical storms develop, known variously as *hurricanes, cyclones,* or *typhoons* (p. 424).

With increasing altitude, the mean annual temperature on a mountainside steadily decreases to well below freezing. The altitude of the **regional snowline,** the altitude above which snow cover persists throughout the year, varies from nearly 6 km above sea level in the subtropical dry belts to sea level in the polar regions. Near the equator, the snowline dips to about 5 km because of the greater cloudiness and precipitation there than in the adjacent clear, dry, subtropical belts.

If a mountain is high and moist enough to intersect the regional snowline, it will carry a year-round snow cap and possibly glaciers (Chapter 16). The sediment loads in streams draining glaciated and nonglaciated mountain peaks imply that the presence of glaciers in its headwaters increases a stream's erosion rate by a factor of 100 or more (Table 15-2). Thus altitude alone can be a powerful factor in determining the effectiveness and even the kind of geomorphic processes acting on a landform.

Although mountain peaks are cold, they may nevertheless be subject to a very high energy flux. Unimpeded by most clouds, haze, and atmospheric dust, the incoming radiation intensity on a mountain greatly exceeds the average at sea level. Reradiation is also rapid. Sunlit rock surfaces at high altitudes may be much warmer than the air only a few centimeters away, but as soon as the sun sets or passes behind a cloud, the rocks reradiate into thin, cold, dry air and cool rapidly. In combination with freeze and thaw cycles, which also are frequent and intense, rocks are mechanically broken (p. 126).

Seasonal Gradients; Land-Sea Contrasts. Because the earth's equatorial plane of daily rotation is now inclined about 23 1/2° to the orbital plane, the sun is vertically overhead across a 47°-wide tropical belt during some part of each year. Seasonality, with its many geomorphic implications, results. Outside the tropics, the names *winter* and *summer* express the dominant seasonal role of temperature change. Within the tropics, where temperatures are generally high, seasonality is defined by precipitation maxima and minima so that rainy seasons, or wet monsoons, alternate annually with dry seasons of several months duration. In both high and low latitudes, the earth's seasonality slowly changes because of cyclic orbital variations (Figure 4-6).

Many geomorphic processes are associated directly with seasonality. The annual alternation of torrential rain and prolonged drought is associated with a distinctive savanna landscape. Subtropical desert landforms are shaped by flash floods and wind. Tundra regions have long, smooth slopes formed by soil creep and earth flow in a seasonally thawed, water-saturated active surface layer that overlies perennially frozen ground. As noted in Chapter 1, climate-controlled geomorphic processes give such distinctive shapes to landscapes that an entire subfield of descriptive geomorphology is organized by climatic regions.

Temperature contrasts and gradients between land and sea are also associated with seasonal climatic changes. Maritime climates are characterized by moderate daily and seasonal temperature contrasts because the abundance of water and water vapor provides a "buffer" of latent heat. The continental interiors, on the other hand, have extreme temperature ranges. The greatest seasonal temperature contrasts, defined by the difference between January and July monthly means, are about 20°C in the subtropical deserts, but exceed 40°C in northern Canada and 50°C in Siberia (Peixoto and Oort, 1992, Figure 7.4). In contrast, both the annual and daily temperature range on small tropical islands is commonly less than 3°C. There, temperature change is totally irrelevant to geomorphic processes. Despite their moderate temperatures, the general high humidity of maritime climates promotes chemical weathering. In cold maritime mountains, glaciers are likely to be significant geomorphic features.

Daily Temperature Fluctuations. In most tropical regions, the diurnal temperature range exceeds the annual range. Daily temperature changes are especially marked in tropical highlands and in subtropical deserts. In the Sahara, for instance, the daily temperature fluctuation in a shady location is over 60°C, but the daily temperature range of exposed rock surfaces exceeds 120°C. These fluctuations are reliably reported to shatter rocks (p. 123). In nondesert regions, night also is typically cooler than day, but reradiation back to earth from the warm humid atmosphere decreases the diurnal range. Geomorphic processes related to condensation of dew on rocks, or freezing and thawing, may be promoted by the daily temperature cycle.

The Hydrologic Cycle

A very large proportion of the solar radiation intercepted by the earth is isothermally absorbed as latent heat of vaporization by water (Figure 2-2). Ours is truly the hydrous planet, and water is as anomalous a compound in its thermal properties as in its chemical activity (p. 128). About 2.4 kJ are required to evaporate

1 g of water at 20°C. The large area of oceans in tropical regions ensures that much of the incoming solar radiation is absorbed not as an increase in sensible temperature but as latent, or insensible, heat by the change in state of water from liquid to vapor. When the water vapor condenses to form clouds of droplets, the latent heat is liberated to the surrounding atmosphere. Winds are generated by the turbulent mixing of air masses that have contrasting temperature and moisture content. Both vertical (convective) and horizontal (advective) circulation cells are active in the troposphere. The large, permanent patterns of tropospheric circulation create the familiar global regions of *intertropical convergence,* where persistent surface winds (the *tradewinds*) flow toward the equator, absorb water vapor from the oceans, and rise; the *subtropical arid belts,* where upper air sinks, compressionally warms, and is intensely evaporative; the midlatitude (40° to 60°) belts of prevailing *westerlies* that drag frontal storms eastward over the surface; and the high latitude *polar easterlies* that interact with the westerlies along a zone of strong advective shearing. Together, these tropospheric circulation systems distribute approximately 60 percent of the excess low-latitude solar energy to the polar regions of energy deficit (Figure 2-4). Furthermore, evaporation and surface winds drive the ocean currents that distribute the remaining 40 percent of the energy imbalance (Figures 4-2 and 4-3). It is a remarkable feature of our hydrous planet that almost exactly half of the tropospheric circulation (from the equator to about 30° N and S) distributes heat by large, stable convective cells, while over the remaining two quarters of the earth's surface area poleward of 30° latitude, shallow advective mixing and frontal storms are the dominant form of heat transfer. As will be noted in subsequent chapters, each of the earth's major climatic zones has obvious geomorphic character, ranging from features such as intense chemical weathering under tropical heat and rainfall to deserts in the subtropics, to intense snowfall and glaciers on west-coast mountains in the midlatitudes, to large ice sheets or perennially frozen ground above 60° latitude.

The **hydrologic cycle**, or interchange of water between oceans, atmosphere, land, glaciers, and organisms, is summarized in Figure 2-5. Two kinds of information are compiled: (1) an estimate of the amounts of water that are present in various reservoirs at or near the surface of the earth, and (2) the calculated annual exchange of water among the reservoirs. The figure portrays a steady-state open system. The energy inflow equals the outflow, and, with negligible exceptions, the amount of water in the system is constant.

First, examine the amounts of water in the system. In their order of importance, the reservoirs of water are (1) oceans, (2) glaciers, (3) groundwater, (4) lakes and rivers, (5) atmosphere, and (6) biomass (all living matter). Almost 97 percent of all the water near the surface of the earth is in the oceans, and most of the remainder is in glaciers. The interior of the earth must contain some water, chemically combined or dissolved in solid or molten rock, but for at least the last billion years, the amount of water at or near the surface of the earth has been approximately constant. Almost all the water that is incorporated into sediments by subduction at ocean trenches and is released by metamorphism and volcanism has been recycled through the earth's surface. Only a small amount of new water might be added annually by condensation of volcanic gases, and an equally small amount of water is probably lost from the earth annually by the photochemical dissociation of water vapor by solar radiation.

FIGURE 2-5. The hydrosphere, or the water on or near the surface of the lithosphere. For easy comparisons, all the amounts of water in the reservoirs are given in millions of cubic kilometers (10⁶ km³); all the annual exchanges among the reservoirs (the hydrologic cycle) are given in thousands of cubic kilometers per year (10³ km³/yr). The accuracy of the various estimates varies widely (based on Shiklomanov, 1993).

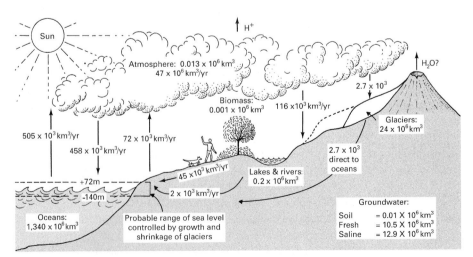

Next, consider the hydrologic cycle, usually shown as a rate of transfer or *flux,* commonly as cubic kilometers of water per year. It can be converted into energy or power units, too. Over the past several decades, various estimates of the water flux have increased by 20 to 25 percent, primarily because of increases in the estimated radiation budget of the earth. However, no high degree of accuracy is claimed even for the latest estimates (Gleick, 1993, Table A2).

Figure 2-5 shows that 505,000 km³ of water is evaporated from the sea annually, representing a layer 140 cm thick from the entire ocean. Most of the water precipitates back onto the ocean, but the excess falls on the land. Because more snow and rain fall on land every year than can be evaporated, 45,000 km³ of water annually drips and flows from the land back into the sea. In addition, each year approximately 2000 km³ of groundwater drains directly into the ocean, and about 3000 km³ of glacier ice either melts or breaks off into the ocean as icebergs.

Much of the water evaporated from the land is not simply evaporated from the exposed surfaces of lakes and streams but is used by plants and animals. A field of crops may annually use an amount of water equivalent to a layer 45 to 60 cm deep over the field. A forest of Douglas fir may annually pump into the atmosphere the equivalent of a layer of water 120 cm deep over its area. Over most of the land, rainwater falls on vegetation. Some evaporates, and some is absorbed by the plant roots and transpired through leaves. Because the variables cannot be clearly distinguished, the term *evapotranspiration* is used for the collective effect. It is likely that almost all the 72,000 km³ of water that annually evaporates from the land passes through a biologic cycle as a part of the larger hydrologic cycle. Deserts are hot not because they receive more solar radiation than vegetated regions at the same latitude, but because with only a trivial amount of energy absorbed by evapotranspiration, more of the incoming radiation is converted into sensible heat. Only a small part of the evapotranspired water is held at any time as living matter, and the turnover is continuous. The energy required for photosynthesis is only about 0.1 percent of the total received from the sun (Woodwell, 1970), but is almost 1 percent of the total absorbed by plants, the remainder of which warms the plant (Miller, 1981, p. 182). By combining water and carbon dioxide in photosynthesis, plants produce an estimated 2×10^{11} tons of dry organic matter annually. Some of this material builds landforms such as peat bogs. Photosynthesis is vital to geomorphology because it permits plants and soil to cover most land surfaces, but also because the use of fossil fuels permits human beings to achieve tremendous power densities (p. 22).

Glaciers store large amounts of water on land, temporarily removing it from the hydrologic cycle. If all the present glaciers were to melt, sea level would rise 60 to 70 m and submerge the most heavily populated areas of the earth. In the last 2 million years, continental ice sheets have repeatedly developed and melted again, each time temporarily upsetting the hydrologic cycle. During maximum glaciations, the amount of glacier ice on land may have been three times as great as the present ice volume, and sea level was lowered as much as 140 m. During the most recent ice age, about 20,000 years ago, sea level was about 120 m lower than present (Figure 18-4). The geomorphic effects are obvious in all coastal regions. Evidence for extremely dusty conditions during ice ages has led to the suggestion that the rate of water transfer by the hydrologic cycle might have been reduced by 50 percent during those times (p. 292).

All these aspects of the hydrologic cycle relate to geomorphology, but one aspect in particular must be emphasized. The average continental height is 823 m above sea level (Figure 1-3). If we assume that the 45,000 km³ of annual surface runoff flows downhill an average of 823 m, the mechanical power of the system can be calculated. Potentially, the runoff from all lands would continuously generate almost 12×10^9 kW. If all this power were used to erode the land, it would be comparable to having (in the English units of horsepower and area) one horse-drawn scraper or scoop at work on each 2.5-acre piece of land, day and night, year around. Of course, a large part of the potential energy of the runoff is wasted as frictional heat by the turbulent flow and splashing of water. Nevertheless, the "geomorphology machine" has a definable if low efficiency, eroding and transporting rock debris down to the oceans or into reservoirs and arid tectonic basins at a rate of about 2×10^{10} T/yr (Figure 2-2).

ROTATIONAL ENERGY OF THE EARTH-MOON SYSTEM

Terrestrial Gravity

Every mass has an inherent attraction for every other mass. This weak force, called *gravitation,* is an intrinsic property of mass. Although the nature of gravity remains undefined, we know something about its operation. It has been shown that the gravitational force between two large objects that are far apart operates as though the entire mass of each object were concentrated at the center of the mass, or the center of gravity. Because all objects at the earth's surface have small masses compared to that of the planet, we usually regard the force of gravitation as a tendency for a free-falling object at the surface to accelerate toward the center of the earth.

Objects moving in straight lines relative to the stars appear to curve to the right in the northern hemisphere and to the left in the southern hemisphere when viewed by an observer who is being carried in a circular path by the earth's rotation (Stommel and Moore, 1989, p. 12). This *Coriolis effect* is proportional to the sine of latitude and the velocity of the object. It has been claimed to cause northern-hemisphere rivers to erode their right banks preferentially, but the effect is actually much too weak. However, the Coriolis effect is of great importance to atmospheric and oceanic circulation, causing cyclonic circulation (counterclockwise in the northern hemisphere) around atmospheric low-pressure cells and the large, counter-rotating gyres of the surface ocean currents (Figure 4-2) that bring aridity to some coasts and mild, moist winters to others.

Gravity, as we commonly use the word, refers to the net force of gravitation as reduced by centrifugal and other lesser effects. The geoid is controlled purely by terrestrial gravitation and rotational forces (Figure 1-2). It is worth noting that the word *level*, as in "sea level," connotes a surface that is neither planar nor spherical, but rather a line or surface parallel to the geoid. A billiard table must be level, but if it were truly flat or planar, it would be tangent to the earth at only one point, and frictionless balls would, in principle, roll to that point and stop. We commonly diagram sea level as a straight line, as for example, in showing the concave long profile of rivers (Figure 10-9). In reality, of course, the bed of a major river has a convex-skyward longitudinal profile, just as has any line on the geoid. For instance, the Mississippi River originates in Lake Itasca, Minnesota, at an altitude of 450 m above sea level. In its 3800-km course southward to sea level, it follows a convex-skyward meridianal arc more than 15° in length. Because of the oblate spheroidal shape of the earth, the mouth of the Mississippi is actually several kilometers "higher" (with respect to the center of the earth) than its source. These amusing but real considerations are easily overlooked in most geomorphic thinking but should at least be mentioned once so that words like *up, down, horizontal,* and *level* can be seen in their true sense.

All water that falls as rain or snow on the 29 percent of the earth's surface that projects above the sea tends to move downhill by gravity, back to the ocean or into closed depressions. The few localities where dry land extends well below sea level (Death Valley,

California, at −86 m; the Dead Sea at −390 m; and others) are exceptions to the rule that sea level is the limit for the downhill flow of water. However, every basin that extends below sea level is in an arid region where few raindrops ever fall. Such basins never last very long in humid regions, because they are soon filled with water and sediment to overflowing, and sea level again becomes the ultimate downward limit of water flow from the basins. The tendency of water to flow downhill is the cause of most erosional landforms.

Rock particles as well as raindrops are attracted toward the center of the earth by gravity. When rock material moves downslope under the influence of gravity, but without a transporting agent such as flowing water or glacier ice, the process is called *mass wasting* (Chapter 9). Mass wasting includes not only such spectacular events as landslides and avalanches but also the slow, barely perceptible process of soil creep.

Extraterrestrial Gravity: Tides

The gravitational forces of the moon, sun, and nearby planets are expended as frictional energy in the earth's hydrosphere and lithosphere. The force of extraterrestrial gravity causes variations in the surface gravity field of only about 1 part in 5 million. Most of the energy generated by extraterrestrial gravitation is expended in the oceans, much of it in the turbulent friction of ebbing and flowing tides in shallow marginal seas (Hendershott, 1973; Wahr, 1988). It is hard to account for the 5.7 (estimates range from 3 to 6) terawatts (1 TW = 10^{12} W) of power consumed by tidal drag. Perhaps internal waves on boundary layers within the oceans contribute to the consumption. Because of the orbital motions of the planets and our moon, tides vary cyclically, but complexly. A related problem is that shallow oceanic basins have varied greatly in size and distribution, even within the late Cenozoic Era. Tidal friction in the oceans may be larger now than it has been during most of earth history (Wahr, 1988, p. 240).

Tidal energy is an important factor in coastal processes (Chapter 19), and for most purposes, tides are best studied in the context of the marine environment. However, *earth tides,* with a regular semidiurnal rhythm similar in period to ocean tides but smaller in amplitude, can be measured by sensitive strain meters and gravity meters. The lithosphere apparently has two high and two low tides, with an amplitude of about 30 cm, that follow the moon in its orbit. As the earth rotates daily through the tidal "bulges," rocks are constantly but minutely flexed. According to the estimates used in compiling Figure 2-2, the tidal energy expended on friction in the lithosphere contributes about 2 percent of the geothermal heat flow. The combined effect of oceanic and lithospheric tidal friction is to decrease the number of daily rotations of the earth per year by about 1 day every 10 million years (Wells, 1963). Our moon has long since "locked on" the earth with one rotation per month for the same reason, so that we always see the same side of the moon from the earth (Goldreich, 1972).

The role of lunar gravity as a geomorphic process (excluding the important tides in the hydrosphere) is not well known. Tidal friction causes the earth's axis to precess and cyclically changes the intensity of seasonal climates (Figure 4-6). Some seismologists suspect that earthquakes occur with greater frequency during the full and new phases of the moon, when the sun's gravitational force is additive to the moon's (Weems and Perry, 1989). The hypothesis was strengthened by the discovery that "moonquakes" were recorded by seismometers on the moon with greater frequency every 28.5 days when the moon was at perigee, its closest orbital approach to the earth (Latham et al., 1971).

One of the longer cyclic components of tides is the 18.6-year regression of the lunar nodes, a shift of the lunar orbit that causes a 10 percent variation in the average tide-producing force. Geyser activity in Yellowstone Park has been correlated with the 18.6-year cycle (Rhinehart, 1972). Attempts also have been made to correlate volcanic activity and tectonism with tide-producing forces (Shaw et al., 1971; Hamilton, 1973). On a very long time scale, it is possible that the slight flexing of crustal rocks by earth tides could lead to "fatigue" or brittle fracture. Joint patterns in young rocks may be propagated upward from older patterns beneath by such weak, but persistent, stresses.

INTERNAL HEAT

Geothermal Gradient and Heat Flow

The interior of the earth is hotter than the exterior. Most of the earth's internal heat is generated by radioactive decay of natural isotopes of uranium, thorium, and potassium, with a contribution from rotational or tidal friction, and an uncertain amount of primordial heat from the origin of the earth. Although small in comparison to the solar energy received by the earth, this internal heat must provide nearly the entire energy drive for tectonism and volcanism. There is no problem with potential sources of heat. Most estimates of radioactivity, rotational energy, and original heat provide an "embarrassment of riches," with radioactivity alone accounting for 60 to 80 per-

cent of the observed heat flow. This contribution would have been even greater during the earlier history of the earth, as would the original heat and the additional release of gravitational and rotational energy associated with core-mantle separation (Turcotte, 1995; Verhoogen, 1980, p. 28).

The increase in temperature with depth into the earth is the **geothermal gradient**, a typical near-surface gradient being about 20° to 30° C/km. Only one well penetrates more than 12 km into the earth, and beyond that depth, the gradient can only be inferred from indirect evidence (Kozlovsky, 1984). Clearly, the steep near-surface geothermal gradients must decrease sharply with depth; otherwise, the mantle would be molten and could not transmit seismic shear waves, as it does.

A steep geothermal gradient is obviously important for volcanic processes, including hot springs and geysers. However, any rock that solidifies or lithifies at depth and is later exposed by erosion has moved upward into a cooler near-surface environment that should cause thermal contraction. Fractures caused by elastic stresses in rocks are at least in part the result of cooling (p. 120).

The energetics of geomorphic processes are better understood in terms of the thermal energy that flows to the surface from the interior of the earth than from the temperature gradient within the earth. **Heat flow** is the product of the geothermal gradient and the thermal conductivity of the rocks through which the gradient is measured, plus the heat transported by convective fluid transport. The rate of energy flow to the earth's surface from internal heat is about 44×10^{12} watts, calculated from the measured average heat flow on continents and in ocean sediments, (Pollack et al., 1993). According to Figure 2-2, volcanoes and hot springs carry about 0.5 percent of the total internal heat flow to the surface by convection, a very small additional amount is released by earthquakes (Kanamori, 1988; Pye, 1995), and the remainder is conducted or convectively circulated to the surface of continents and ocean floors. Deep convection by fluids within both the oceanic and continental crust is an important means of heat transfer, especially within the young oceanic lithosphere (Fyfe, 1988; Pollack et al., 1993, pp. 270–272).

Perhaps as much as 70 percent of the heat flux from the interior of the earth is used to remelt the sinking slabs of lithosphere in subduction zones (Turcotte, 1995). Like ice cubes in a glass of water, these melting slabs cool the surrounding mantle and reduce the geothermal heat flow to the surface above them. Conversely, the geothermal heat flow is greater at midocean ridges, where new, hotter mantle rises.

Recent studies are just beginning to show the resulting heterogeneities in mantle temperature (Powell, 1991; Pollack et al., 1993). In ways not yet understood, heat energy and geothermal gradients in the interior of the earth, aided by kinetic rotational forces and gravity, drive all the tectonic activity in the lithosphere. Any comparison of internal and external energy sources for geomorphic processes must consider that although the power of the latter is 3300 times greater than the former, the internal energy is much more localized in both place and time, and is capable of constructing great mountain ranges or uplifting entire continental masses in spite of solar- and gravity-powered erosion.

Orogenic Energy

To uplift a mass of sedimentary rocks from a depositional site below sea level in order to build a mountain range, work must be done against gravity. However, the tectonic uplift necessary to replace the annual erosional mass requires only about 7×10^9 W (Verhoogen, 1980, p. 8). This is several orders of magnitude less than the power of volcanoes and earthquakes and a negligible part of the total geothermal heat flux (Figure 2-2). Along the 200 km length of a mountain belt in eastern Taiwan, only about 10 percent of the power input does useful work against gravity, a measure of the efficiency of orogeny (Barr and Dahlen, 1989). The remaining 90 percent of the tectonic energy is consumed by mechanical deformation of the rock mass and by friction on the subducting lithospheric slab. It is striking that the power (rate of energy consumption) required to build a mountain range totaling 10,000 km in length within 1 million years would be less than 1 percent of the power now being expended by humans (Masek, J. G., oral communication, 1994; Masek and Duncan, 1994).

Lithospheric plate motions (Chapter 3) are driven entirely by thermal and density gradients in the lithosphere and asthenosphere. Therefore, they require no energy not included in the geothermal heat flow.

Lattice Energy of Minerals; Metamorphic Heat

A difficult factor to evaluate in the energetics of geomorphic processes is the highly ordered atomic lattices of mineral phases that crystallize at the high temperatures within the earth. When exposed to weathering reactions, especially oxidation and hydrolysis, these minerals normally react exothermally and produce new phases that are more stable under surface conditions. The free energy of the original crystal lattice is degraded to lost heat at the earth's surface. When the debris of weathering and erosion is buried at great depths in subduction zones and recrystallized or melted, internal heat is again absorbed. The process has been repeated countless times in earth history. The absorption of geothermal heat by subducting relatively cold, wet, oceanic lithospheric plates is now thought to be a major method of internal heat transfer.

Degradation of Geothermal Energy

At a future distant time, the radiogenic energy source for tectonism will be expended, and when the land surface is worn down, no internal force will be able to renew it again. The half-lives of the radioactive isotopes that supply most of the heat are on the order of 10^9 or 10^{10} years, and the radioactive component of geothermal heat is now only about 50 percent of its value 3 billion years ago (Turcotte and Schubert, 1982, pp. 139–142). This is another grand example of an energy system that is slowly declining although it appears essentially in a steady state even when measured by the scale of geologic time.

THE RELATIVE ENERGY OF SURFACE PROCESSES

Earthquake magnitude is typically measured on a logarithmic scale in which each unit increase represents a factor of 10 increase in ground or instrument displacement, which is the equivalent of a factor of 32 in energy release. A recent revision of the earthquake scale, known as the *moment magnitude scale* (M) is algebraically related to the seismic energy radiated from an earthquake and can therefore be compared to other energy sources for geomorphic change and to human energy consumption (Figure 2-6). Note that Figure 2-6 does not measure the *rate* of energy release, or *power*. Rather, it records the total energy released, whether in seconds, as in a lightening bolt, nuclear bomb, or earthquake, or over a period of days or even a full year (Johnston, 1990). The inset of Figure 2-6, with the lower part of the energy scale plotted on a linear scale, is particularly helpful in reminding us of the enormous range of values that can be displayed on a logarithmic scale. It is instructive to see that the largest known earthquake (Chile, 1960, M = 9.5) released an amount of energy comparable to that generated by an average hurricane over ten days. Single major earthquakes (Alaska, 1964, M = 9.2) release more energy than the average annual global seismic energy release,

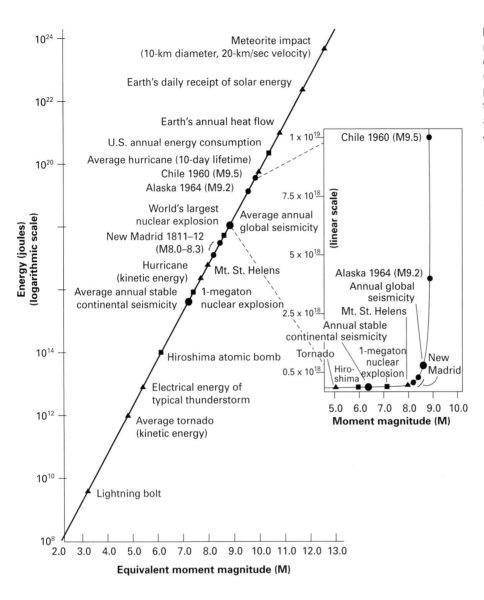

FIGURE 2-6. The equivalent moment magnitude of a variety of natural (triangles), man-made (squares), and seismic (dots) phenomena. The inset shows the true linear scaling between 10^{18} and 10^{19} joules (revised from Johnston, 1990).

and on a decade time scale, annual seismic energy release fluctuates tenfold (Kanamori, 1977, p. 2985). Such comparisons help explain how it is possible that impressive tectonic and volcanic landforms can be constructed quickly despite the overwhelming excess supply of solar energy to our geomorphic system, but also that all landforms are ultimately doomed to gravity- and solar-driven erosional degradation.

REFERENCES

ALVAREZ, L. W., ALVAREZ, W., and 2 others, 1980, Extraterrestrial cause for the Cretaceous-Tertiary extinction: Science, v. 208, pp. 1095–1108.

BARR, T. D., and DAHLEN, F. A., 1989, Brittle frictional mountain building 2. Thermal structure and heat budget: Jour. Geophys. Res., v. 94, pp. 3923–3947.

von BERTALANFFY, L., 1950, Theory of open systems in physics and biology: Science, v. 111, pp. 23–29.

BLOOM, A. L., 1969, The surface of the earth: Prentice-Hall, Inc., Englewood Cliffs, N.J., 152 pp.

CHORLEY, R. J., 1962, Geomorphology and general systems theory: U.S. Geol. Survey Prof. Paper 500-B, 10 pp.

DIAMOND, J. M., 1992, The third chimpanzee: The evolution and future of the human animal: HarperCollins Publishers, Inc., New York, 407 pp.

DRESSLER, B. O., GRIEVE, R. A. F., and SHARPTON, V. L., eds., 1994, Large meteorite impacts and planetary evolution: Geol. Soc. America Spec. Paper 293, 348 pp.

DYSON, F. J., 1971, Energy in the universe: Sci. American, v. 225, no. 3, pp. 50–59.

FYFE, W. S., 1988, Granites and a wet convecting ultramafic planet: Trans., Royal Soc. Edinburgh: Earth Sciences, v. 79, p. 339–346.

FYFE, W. S., 1988, Granites and a wet convecting ultramafic planet: Trans., Royal Soc. Edinburgh: Earth Sciences, v. 79, p. 339–346.

GLEICK, P. H., ed., 1993, Water in crisis: A guide to the world's fresh water resources: Oxford Univ. Press, Inc., New York, 473 pp.

GOLDREICH, P., 1972, Tides and the earth-moon system: Sci. American, v. 226, no. 4, pp. 42–52.

GRIEVE, R. A. F., 1987, Terrestrial impact structures: Ann. Rev. Earth, Planet. Sci., v. 15, pp. 245–270.

———, 1990, Impact cratering on the earth: Sci. American, v. 262, no. 4, pp. 66–73.

HÄFELE, W., 1974, A systems approach to energy: Am. Scientist, v. 62, pp. 438–447.

HAMILTON, W. L., 1973, Tidal cycles of volcanic eruptions: Fortnightly to 19 yearly periods: Jour. Geophys. Res., v. 78, pp. 3363–3375.

HENDERSHOTT, M. C., 1973, Ocean tides: EOS (Trans., Am. Geophys. Un.), v. 54, pp. 76–86.

HOLDREN, J. P., and EHRLICH, P. R., 1974, Human population and the global environment: Am. Scientist, v. 62, pp. 282–292.

HUBBERT, M. K., 1971, The energy resources of the earth: Sci. American, v. 225, no. 3, pp. 60–70.

INMAN, D. L., and BRUSH, B. M., 1973, The coastal challenge: Science, v. 181, pp. 20–32.

JOHNSTON, A. C. 1990, An earthquake strength scale for the media and the public: Earthquakes and Volcanoes, v. 22, pp. 214–216.

KANAMORI, H., 1977, Energy release in great earthquakes: Jour Geophys. Res., v. 82, p. 2981–2987.

———, 1988, Energy release in earthquakes and volcanic eruptions, in Proceedings, Kagoshima International Conference on Volcanoes: Kagoshima, Japan, pp. 21–23.

KOZLOVSKY, Y. A., 1984, The world's deepest well: Sci. American, v. 251, no. 6, pp. 98–104.

LASKAR, J., and ROBUTEL, P., 1993, The chaotic obliquity of planets: Nature, v. 361, pp. 608–612.

LATHAM, G., EWING, M., and 7 others, 1971, Moonquakes: Science, v. 174, pp. 687–692.

LOVE, S. G., and BROWNLEE, D. E., 1993, A direct measurement of the terrestrial mass accretion rate of cosmic dust: Science, v. 262, pp. 550–553.

MASEK, J. G., and DUNCAN, C. C., 1994, A least-work model for mountain building [abs.]: EOS (Trans., Am. Geophys. Un.), v. 75, p. 229.

MELOSH, H. J., 1989, Impact cratering: A geologic process: Oxford monographs on geology and geophysics no. 11, Oxford Univ. Press, Inc., New York, 245 pp.

MILLER, D. H., 1981, Energy at the surface of the earth: Academic Press, Inc., New York, 516 pp.

NIR, D., 1983, Man, a geomorphological agent: Keter Publishing House, Jerusalem, Israel, 165 pp.

PEIXOTO, J. P., and OORT, A. H., 1992, Physics of climate: American Institute of Physics, New York, 520 pp.

PERRY, M. D., and MOUROU, G., 1994, Terawatt to petawatt subpicosecond lasers: Science, v. 264, pp. 917–924.

PILEWSKIE, P., and VALERO, F. P., 1995, Direct observations of excess solar absorption by clouds: Science, v. 267, pp. 1626–1629.

POLLACK, H. N., HURTER, S. J., and JOHNSON, J. R., 1993, Heat flow from the earth's interior: Analysis of the global data set: Rev. Geophys. v. 31, pp. 267–280.

POSTEL, S. L., DALY, G. C., and EHRLICH, P. R., 1996, Human appropriation of renewable fresh water: Science, v. 271, pp. 785–788.

POWELL, C. S., 1991, Peering inward: trends in geophysics: Sci. American, v. 264, no. 6, p. 100–111.

PYE, D. M., 1995, Mass and energy budgets of explosive volcanic eruptions: Geophys. Res. Ltrs, v. 22, pp. 563–566.

RAMANATHAN, V., CESS, R. D., and 5 others, 1989, Cloud-radiative forcing and climate: Results from the earth radiation budget experiment: Science, v. 43, pp. 57–63.

RHINEHART, J. S., 1972, 18.6-Year earth tide regulates geyser activity: Science, v. 177, pp. 346–347.

RYDER, G., FASTOVSKY, D. E., and GARTNER, S., 1996, The Cretaceous-Tertiary event and other catastrophes in earth history: Geol. Soc. America Spec. Paper 307, 569 pp.

SALBY, M. L., 1992, The atmosphere, in Trenberth, K. E., ed., Climate system modeling: Cambridge Univ. Press, Cambridge, UK, pp. 53–115.

SHARPTON, V. L., and WARD, P. D., eds., 1990, Global catastrophes in earth history: An interdisciplinary conference on impacts, volcanism, and mass mortality: Geol. Soc. America Spec. Paper 247, 631 pp.

SHAW, H. R., KISTLER, R. W., and EVERNDEN, J. F., 1971, Sierra Nevada plutonic cycle: Part II, tidal energy and a hypothesis for orogenic-epeirogenic periodicities: Geol. Soc. America Bull., v. 82, pp. 869–896.

SHIKLOMANOV, I. A., 1993, World fresh water resources, in Gleick, P. H., ed., Water in crisis: A guide to the world's fresh water resources: Oxford Univ. Press, Inc., New York, pp. 13–24.

SILVER, L. T., and SCHULTZ, P. H., eds., 1982, Geologic implications of impacts of large asteriods and comets on the earth: Geol. Soc. America Spec. Paper 190, 528 pp.

STEVENSON, D. J., 1987, Origin of the moon—the collision hypothesis: Ann. Rev. Earth, Planet. Sci., v. 15, pp. 271–315.

STOMMEL, H. M., and MOORE, S. W., 1989, An introduction to the Coriolis force: Columbia Univ. Press, New York, 297 pp.

STRAHLER, A. N., 1950, Equilibrium theory of erosional slopes approached by frequency distribution analysis: Amer. Jour. Sci., v. 248, pp. 673–696, 800–814.

TURCOTTE, D. L., 1995, How does Venus lose heat?: Jour. Geophys. Res., v. 100, no. E8, pp. 16,931–16,940.

———, and SCHUBERT, G., 1982, Geodynamics: Applications of continuum physics to geological problems: John Wiley & Sons, New York, 450 pp.

UNITED NATIONS, 1995, Energy Statistics Yearbook, 1993: United Nations, New York, 494 pp.

VERHOOGEN, J., 1980, Energetics of the earth: National Academy of Sciences, Washington, D.C., 139 pp.

WAHR, J. M., 1988, The earth's rotation: Ann. Rev. Earth, Planet. Sci., v. 16, pp. 231–249.

WEEMS, R. E., and PERRY, W. H., Jr., 1989, Strong correlation of major earthquakes with solid-earth tides in part of the eastern United States: Geology, v. 17, p. 661–664.

WELLS, J. W., 1963, Coral growth and geochronometry: Nature, v. 197, pp. 948–950.

WOODWELL, G. M., 1970, Energy cycle of the biosphere: Sci. American, v. 223, no. 3, pp. 64–74.

Part II

Cenozoic Tectonism and Climates: The Modern Landscape Evolves

*A*t present, agents of geomorphic changes are retouching landscapes that are the result of a complex series of changing environments. Most subaerial landscapes are **palimpsests**, a word originally applied to parchment manuscripts that had been partially scraped clean and reused for later texts. As in a manuscript, one or more older texts can sometimes be read beneath the latest words; so in landscapes, evidence for previous processes and stages of development can be recognized above, beneath, or within the landforms that are now being shaped. The highest skill of the geomorphologist is in deciphering those fragments of ancient development almost obscured by present processes. Sometimes the task is easy, as where the striking landforms of glacial erosion and deposition are found in regions of currently temperate climate, or where branching systems of empty river valleys are found in deserts where no rain falls today. However, many landscapes do not record such obvious changes, and we must always apply the principle of uniformity of process with caution because the present dominant processes may not be the key to past landform development.

Although the premise is unproved, it is unlikely that the energy budget of the earth (Chapter 2) has significantly changed during the Cenozoic Era. However, the distribution of that energy over the earth's surface has varied greatly through even this relatively brief interval of geologic time. As a result, the climates of the earth have ranged from nearly global uniform warmth about 60 to 50 million years ago (the "greenhouse" earth) to extensive glaciation during Pleistocene cold times (the "icehouse" earth) (Chapter 4). These variations and their causes are complex, and many alternative hypotheses are being tested. Research on the topic of global climate change is currently motivated by concern over the role of our industrialized civilization in modifying climate by releasing particulate combustion products and atmospheric gases.

A map of continental paleogeography at the start of the Cenozoic Era, some 65 million years ago, is only broadly similar to a modern map. The Atlantic Ocean was then only about half of its present width, Australia was *still attached to Antarctica, and the Indian subcontinent was a large island moving northward across the middle of what is now the Indian Ocean.*

By about 30 million years ago, Australia and New Zealand were separating across the Tasman Sea, and both were moving equatorward from Antarctica. The Laramide Orogeny had raised the Rocky Mountains from Canada to Mexico. As India collided with and thrust under southern Asia, the Himalayan Mountains, Tibetan Plateau, and associated orogenic belts of southern and southeastern Asia were raised. Rifting of the Red Sea separated the Arabian Peninsula from Africa but began the folding and uplift of the Zagros and other Middle Eastern mountain ranges.

A series of major tectonic events began about 10 to 15 million years ago in the middle to late Miocene Epoch and brought the global landscape to its present form. Subduction was being replaced by right-lateral strike-slip faulting as the dominant tectonic style along the western margin of North America. Within the present Great Basin and Sonoran Desert regions, broadly east-west tectonic extension was beginning to widen the space between the Rocky Mountains and the Pacific coastal ranges. Subduction along the northern Mediterranean Sea created the east-west trending Pyrenees, Alps, and Balkan mountain chains. The tectonic and volcanic Andes Cordillera of western South America and the corresponding mountain ranges of Central and North America grew higher, creating important climatic boundaries. A series of young island arcs around the Pacific margin created archipelagos such as the Aleutians, Japan, the Philippine Islands, and Indonesia, separated from adjacent continents by shallow seas, as well as smaller island groups in Melanesia. Similar island chains rose in the Caribbean Sea. By three million years ago, the Panama isthmus had connected North and South America, promoting important terrestrial biotic interchanges and, with the contemporary opening of the Bering Strait, allowing the modern oceanic circulation to develop.

Of what geomorphic significance were the Cenozoic tectonic, climatic, and biologic changes? As illustrated at numerous places in subsequent chapters, many modern landscapes bear the imprint of processes no longer operative in the region. Deep chemical weathering in a warm, moist environment probably etched out the lithologic contrasts in the rocks of the Canadian Shield and prepared the surface for Pleistocene glacial erosion (p. 382). The smoothly sloping pediments (p. 283) that surround mountains of the Sonoran Desert are eroded across a truncated zone of deep weathering, also typical of humid, warm climates. The thick glacial-marine sediments of the Antarctic continental shelf indicate times of intense former glacial erosion and transport although today the Antarctic icebergs are remarkably free of sediment load. Almost wherever we study the landscape, we see a palimpsest that has been overwritten again and again. Only very young tectonic and volcanic landforms have been created within the modern climatic zonation, and even most of those have experienced the glacial-interglacial climatic fluctuations of the last few million years.

We must remember that tectonic and climatic changes have not stopped. We are amid ongoing changes such as those just described. The landscape on which you read these words is not the same as it was when they were written.

Chapter 3

Cenozoic Tectonism

Throughout the late Cenozoic Era, the modern continents have been generally emergent, with especially intense mountain building around the Pacific basin, across southern Europe and Asia, and around the Caribbean Sea (front endpaper). Much of our subaerial landscape is developed on relatively young tectonic structures although some of the rocks exposed by recent uplift and erosion are among the oldest known. Tectonism in the late Cenozoic Era has given us a landscape of exceptional diversity, complexity, and beauty. It may well be true that at any moment of geologic time orogenic belts have been actively rising on some part of the earth. However, other geologic intervals, when extensive continental emergence and high mountain ranges were as notable as those of the late Cenozoic, are not common in the stratigraphic record. The late Paleozoic Periods (Pennsylvanian, Permian) may be the best analog.

CENOZOIC PLATE MOTIONS, CONTINENTAL RELIEF, AND CLIMATE

The process of seafloor spreading, whereby new oceanic lithosphere is created at midocean ridges along diverging plate margins, generates the dominant relief pattern of the ocean floors, in which depth is primarily determined by age (front endpaper). On the scale of millions of years, the oceanic lithosphere cools and contracts as it moves away from the mid-ocean ridges, and subsides to a greater depth below sea level (Figure 1-4). Increases in spreading rates are believed to increase the volume of the midocean ridges, decrease the volume of the ocean basins, and cause marine transgressions over low-lying continental areas. Slower spreading rates should produce marine regressions from the continents. However, increased spreading rates may be offset by the corresponding increase in subduction rates, and the resulting changes in sea level are complex (Gurnis, 1990). Net increase in the volume of the ocean basins from reduced ridge volume and the late Cenozoic growth of the Antarctic ice sheet have combined with extensive continental tectonic emergence to lower sea level and expose extensive Miocene and younger coastal plains on many passive continental margins.

As described in Chapter 4, the strong latitudinal gradients of our modern climates, and the corresponding modern diversity of climate-determined erosional processes, are in part determined by tectonism. Cenozoic plate motions have moved large continental regions into middle to high latitudes of the northern hemisphere, preparing them for continental glaciation (Figure 3-1). In the process, oceanic circulation changed from domination by a globe-encircling warm low-latitude Tethys Ocean of the early Cenozoic Era to domination by a late Cenozoic cold circum-Antarctic ocean with north-trending Indian, Pacific, and Atlantic branches (Figure 4-3). Cenozoic mountain ranges and

Percent ocean area of 5° latitude bands

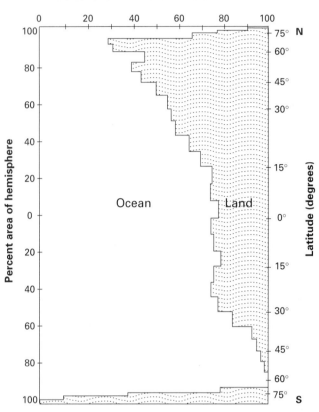

FIGURE 3-1. Percentage distribution of land and ocean in 5° latitude bands, plotted by equal areas (from Kossina in Fairbridge, 1968, Table 2).

plateaus have become powerful orographic barriers for global atmospheric circulation. Large areas of contemporary midlatitude continental shields and platforms with low relief have become arid or semiarid plains as a result of their size, flatness, and distance from moisture sources. Horizontal plate motions, vertical tectonism, and climates are intimately related.

CENOZOIC OROGENY

There are many regions of the earth where vertical crustal movements measured in millimeters per year have persisted throughout late Cenozoic time, and the cumulative uplift has outpaced erosional lowering. (Note that a rate of 1 mm/yr = 1 m/1000 yr = 1 km/million yr.) The subdiscipline of **tectonic geomor-**

phology is concerned with the constructional landscapes of such regions, and of erosional landscapes in which the tectonic origins are still obvious. This chapter adopts the "megageomorphic" approach (p. 5) of describing tectonic landscapes on the scale of entire mountain ranges. Individual tectonic landforms are analyzed in more detail in Chapter 5.

Orogeny clearly refers to mountain formation. Unfortunately, in geologic usage, the term has assumed much more complex and contradictory implications. Gilbert (1890, p. 340) formalized the term:

> "Displacements of the earth's crust which produce mountain ridges are called orogenic. . . . The process of mountain formation is **orogeny**, the process of continent formation is **epeirogeny**, and the two collectively are **diastrophism**."

This definition ignored Gilbert's earlier discoveries that the orogeny that deformed the rocks of the U.S. Basin and Range Province was much older than the block faulting that created the modern relief (p. 76). In fact, many mountain ranges are the result of "postorogenic" movements from which erosion has etched into mountainous relief the most resistant of the folded and metamorphosed rocks. Structural connotations have almost completely divorced the word from its obvious geomorphic ancestry so that orogeny is now used for "the process by which structures within fold-belt mountainous areas were formed" (Bates and Jackson, 1987; but see Cebull, 1973, 1976). As such, it has little geomorphic significance although for many general geomorphic descriptions the tectonic niceties of the term can safely be ignored. The dichotomous term *epeirogeny* has become archaic, and is redefined on p. 43. *Diastrophism* has largely been replaced by the synonym *tectonism*, ungrammatically transmuted into "tectonics."

Mountains and Orogenic Style

In plate-tectonic theory, all orogenic belts are on or near converging plate margins. A few plate boundaries are wide or diffuse, especially those that are embedded within continents, as in central Asia (Gordon and Stein, 1992, p. 334). Mountain ranges that are now embedded within continents, such as the Himalaya Mountains and the Ural Mountains of Asia, have been shown to have been near plate margins at their time of formation but have had new continental terranes accreted to them later. Orogenic vertical movements are not necessarily more rapid than broad, regional movements. However, they seem to be more continuous through time, and less likely to stop or reverse on a time scale of less than 1 to 10 million years; further, most are restricted to relatively narrow, elongate regions.

Cenozoic mountain ranges are not randomly distributed but form distinctive linear or arcuate belts across the earth's surface. Not all earthquake-prone regions are mountainous, but a map of recent earthquakes (front endpaper) outlines very well the most orogenic regions of the late Cenozoic Era, especially the circum-Pacific belt associated with active volcanoes along subduction zones (Simkin et al., 1994). Three distinct types of orogenic belts can be identified on converging plate margins: (1) island arcs and trenches, (2) cordilleran-type mountain ranges, and (3) collisional mountain belts. A fourth type of mountains are built by extensional and strike-slip plate motions in a variety of tectonic settings.

Island Arcs and Trenches.   These form distinctive morphotectonic assemblages that extend laterally for thousands of kilometers along the northern and western Pacific margin, through Indonesia, in the Caribbean Sea, and wherever subduction zones are between two portions of oceanic lithosphere (front endpaper). The downgoing lithospheric plate flexes upward in a broad *outer swell* seaward of the *trench,* into which it descends beneath the overriding plate (Figure 3-2). Sediments that accumulate in the trench and are scraped off the downgoing plate form a submerged *accretionary prism* of highly deformed mudstones and sandstones on the inner trench wall. Older and similarly deformed sedimentary rocks may rise above sea level as the *frontal arc* of mountainous

islands. In many tropical areas, the islands of the frontal arc are capped with uplifted coral limestone terraces.

Close behind the frontal arc may be a *volcanic arc* of andesitic composite cones (Chapter 6). Landward of the volcanic chain may be older, inactive *remnant arcs* and *back-arc* or *marginal basins* where the upper plate is spreading in response to the buoyancy of the subducted plate (Figure 3-2).

Cordilleran-Type Mountain Ranges.   Where oceanic lithosphere subducts under thicker continental lithosphere, more complicated mountain ranges evolve. On the *cordilleran type* of converging plate margin, exemplified by the Andes Mountains of South America (Figure 3-3) (Jordan et al., 1983), there may be narrow *coastal ranges* on the edge of the continent, related to the accretionary prism of trench sediments or to faulting and uplift of continental crustal blocks. Behind the coastal ranges, typically there is an elongate *longitudinal valley,* parallel to the coast. Next is a steep rise to the crest of the *magmatic* or *volcanic arc* surmounted by a line of andesitic composite cones. The volcanic peaks on a cordilleran type of mountain belt rise above a broad plateau or ridge of granitic batholiths and older sedimentary and metamorphic rocks that forms the axial part of the organic belt. This central region of the mountain range is sometimes called the *hinterland.* It is generally high because compressional shortening thickens the continental crust and causes it to rise in isostatic adjustment (p. 44). On the continental side of the central ranges is a *foreland* region, where sediments eroded from the central ranges accumulate on the continental platform and may later be folded and thrust

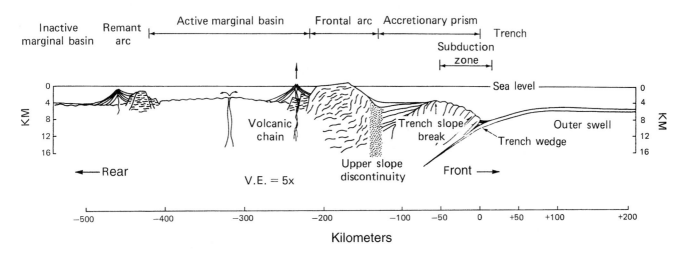

FIGURE 3-2.   Generalized geologic cross section of a typical island-arc system. Horizontal distances are measured from the trench axis; vertical exaggeration is 5:1 (Karig, 1974).

FIGURE 3-3. Generalized geologic model of the Andes Mountains of South America between 18° and 24° S latitude. This is a typical segment of a cordilleran-type mountain range. Width of the orogenic belt is about 600 km; vertical scale is imprecise (revised from Jordan et al., 1983, Figure 3).

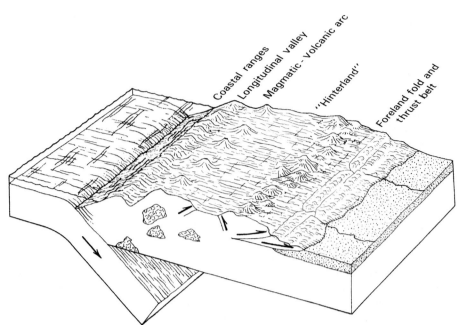

toward the continental interior. Crustal shortening by stacking of successive thrust sheets commonly amounts to 50 percent of the original foreland width (Allmendinger et al., 1990, p. 799). In the northwestern United States, the Olympic Mountains are part of the coastal ranges, the Puget Sound Lowland and Willamette Valley are part of the longitudinal valley, and the Cascade Range is the volcanic arc.

Collisional Mountain Belts. These are found where convergence and subduction have brought an island arc or another plate of continental lithosphere into contact with the overlying plate of the subduction zone. Because the low density of the converging island-arc and continental rocks prevents their being subducted back into the mantle, they crumple against, thrust over, or are thrust under the rocks of the continental organic belt, creating high mountains such as the Himalayas or Alps, with intensely deformed rocks of contrasting lithology in contorted or parallel ranges (Color Plate 2). A sedimentary foreland on a collisional plate boundary may form on the subducting plate, rather than on the upper plate of the hinterland. The Indo-Gangetic Plain of northern India is a good example.

Deformation of orogenic forelands can be subdivided into two main categories: *thin skinned*, in which sedimentary cover strata of the foreland basin and thin slices of basement rocks are folded or faulted above a low angle detachment zone that separates their deformation from the deeper continental basement, and *thick skinned*, in which the deeper basement rocks are massively included in the deformation by

riding upward on steeper thrust faults. Since the continental basement rocks are usually igneous and metamorphic rocks that lack the "layer cake" contrasts in strength and erodibility of sedimentary cover strata, thick-skinned deformation results in basement block faulting, often along preexisting zones of structural weakness, with or without cover strata draping over the displaced basement blocks.

The geomorphic expression of thin-skinned foreland orogeny is by linear anticlines and synclines or thrust slices, from which erosion carves a well-known set of structurally controlled landforms (Chapter 12). Thick-skinned deformation that includes resistant basement rocks is more likely to form bold fault-block mountain ranges arrayed in irregular chains, extending hundreds of kilometers into the continent from the actual orogenic belt (Rodgers, 1987). The Front Ranges of the central Rocky Mountains in the United States and the Sierras Pampeanas of Argentina (Color Plate 3) are good examples of thick-skinned deformation. Their geomorphic expression contrasts strongly to the landforms of foreland thin-skinned fold and thrust belts, such as the eastern Canadian Rocky Mountains, the Appalachian Valley and Ridge geomorphic province (Figure 12-11), and the Subandean Ranges of the Andes (Color Plate 12).

Extensional Mountains. The Basin and Range Province of the western United States (Figure 1-5) is a classic example of a late Cenozoic extensional, rather than compressional, tectonic landscape (Wernicke,

1992). After a long interval of Paleozoic and Mesozoic orogeny, crustal shortening, and accretion of exotic continental terranes, the entire Basin and Range Province between the Sierra Nevada and the Colorado Plateau has been undergoing lithospheric extension for the past 40 million years, most of which has occurred in the past 17 million years. In the central region, the present width of the Basin and Range Province is 350 km, and is the result of 250 km of extension (Wernicke, 1992, p. 575). The Sierra Nevada has moved 200 to 300 km to the west-northwest relative to the Colorado Plateau. It was moving at a rate of 20 km/million yr about 10 million years ago, slowing to 10 km/million yr in the last 5 million years (Wernicke, 1992, p. 504).

Rift-valley formation is sometimes given ranking as a special style of orogeny although the mountains that flank rift valleys are relatively localized (Chapter 5). The initial stages of continental breakup are characterized by an elongate swell or bulge, along which axial normal faulting creates a rift valley with inward facing fault scarps (p. 80).

Even more complex tectonic mountain ranges may grow along strike-slip faults. The historically famous Lebanon and Anti-Lebanon Mountains on the Syrian-Lebanon frontier have been uplifted by secondary compressional and extensional vectors oblique to the dominantly strike-slip faulting along the Dead Sea transform fault (Khair et al., 1993). The Transverse Ranges of southern California have a similar origin in their relation to the San Andreas Fault. The complex oblique forces that act along strike-slip faults are given compound names such as "transpressional" and "transtensional."

Orogenic Rates

Surface Uplift. As noted in Chapter 1, the rates of most geomorphic processes are inversely proportional to their spatial scale and duration, with small, brief events being generally the most rapid. The time intervals in the rate scales used in this book are chosen to represent the typical duration of the process, whether it be seconds or minutes for avalanches, hours or days for intense precipitation, or millions of years for tectonic uplift. For both constructional and destructional geomorphic processes, comparisons between rates that have been measured over unlike time intervals are not valid (p. 6).

Unless otherwise specified, in geomorphic studies *uplift* implies the displacement of a land surface in a direction opposite to the gravity vector and by an amount relative to the geoid (England and Molnar, 1990). This is *surface uplift,* and is not the same as *uplift*

of rocks, which is used to describe the displacement of rock masses relative to the geoid. Because of omnipresent erosion, surface uplift equals the uplift of a rock mass minus the amount of erosional lowering during the uplift episode.

Some uplift occurs abruptly, literally overnight. During the 1964 Alaskan earthquake, a sizable region of nearshore ocean floor was raised above tide level to become new land (Figure 3-4). The duration of the movement was probably only a few minutes or even seconds although no one witnessed the actual upheaval.

Rapid orogenic uplift is common on the converging margins of the Pacific Ocean. Much of Japan has been uplifted as much as 2 km during the Quaternary; the amount of tectonic uplift within the last 2 million years is commonly one-half to two-thirds of the highest altitude in any region (Research Group for Quaternary Tectonic Map, 1973). Uplift rates commonly range from 1 to 5 m/1000 yr.

All the major landforms of New Zealand are tectonic except for depositional forms. Numerous published reports confirm that the late Cenozoic Kaikoura Orogeny is still in progress there. On the South Island of New Zealand in the vicinity of Mount Cook, 7 to 8 km of rock uplift in the past 5 to 7 million years is inferred from fission-track dating (Kamp et al., 1989). Mount Cook is now 3754 m high (Figure 17-6), so at least 3 to 4 km of rock has been eroded from above the summit during uplift. On December 14, 1991, a large rock avalanche fell from the peak of Mount Cook and lowered its summit by 10 m (McSaveney et al., 1992). The wisdom of Powell's dictum (p. 118) is well illustrated, as is the distinction between uplift of rock and surface uplift. The regional surface uplift rate of the North Island of New Zealand during at least the last 100,000 years has been somewhat slower than on the South Island, based on the present altitude of coastal and river terraces. Uplift of as much as 4 m/1000 yr over the main axial mountain ranges is reported, with other areas of slower uplift or subsidence (Pillans, 1986).

Similar deformational rates are demonstrated for many other orogenic regions. Emerged late Quaternary reefs in Papua New Guinea (Figure 3-5) require uplift rates of 1 to 3 m/1000 yr during the past 200,000 years or more to carry them above the range of glacially controlled sea-level oscillations (Bloom et al., 1974).

Space Geodesy. Until the past decade, tectonic uplift could be measured only by displacement of a reference surface such as a shoreline, and that was complicated by glacially controlled fluctuations of sea level (Chapters 18 and 19). Calculating the rate of vertical uplift depends, in addition, on the availability of

FIGURE 3-4. Tectonic uplift and subsidence in south-central Alaska resulting from the earthquake of March 27, 1964. Changes in land level are shown by contours (in meters) that are dashed where approximate or inferred. Edge of continental shelf (−200 m) is shown by dotted line. Active and dormant volcanoes are identified by asterisks (Plafker, 1965; ©1965 by the American Association for the Advancement of Science).

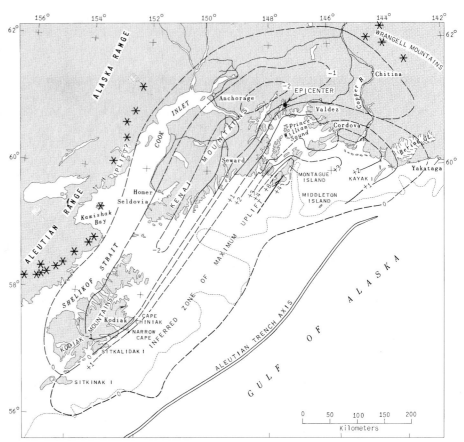

FIGURE 3-5. Emerged Pleistocene coral-reef terraces on the Huon Peninsula, Papua New Guinea. Terraces are on a fault block that has been rising at a rate of 1–3 m/1000 yr for at least 200,000 years. Figure 18-3 shows the terrace chronology for the last 140,000 years.

radiometric dates for the displaced reference surface. By contrast, horizontal tectonic motions could be measured by the offset of geomorphic, stratigraphic, or historical markers across faults, or with reference to the paleomagnetic patterns on the ocean floor. Now, however, several techniques are available using orbiting satellites to define both vertical and horizontal surface displacements to an accuracy of better than 2 mm. **Space geodesy** currently involves at least three techniques: (1) very long baseline radio interferometry (VLBI), (2) satellite laser ranging (SLR), and, of increasing importance, (3) the global positioning system (GPS) (Gordon and Stein, 1992; Smith and Turcotte, 1993). Each technique involves precise measurement of surface points with respect to each other or with respect to an artificial satellite or distant radio star. Repeated measurements at intervals of only a few years yield displacement vectors with extremely high reproducibility (Figure 3-6). Some of the important results of space geodesy are the verification (1) that lithospheric plate motions are steady so that rates and direction vectors of several centimeters per year measured over a time span of a few years are similar to the rates and directions inferred from magnetic anomaly patterns measured over time spans of millions of years; (2) that oceanic plate boundaries are very narrow, usually ranging in width from a few hundred meters to a few kilometers; and (3) that where plate boundaries pass through continental crust, deformation may be distributed through zones hundreds to thousands of kilometers in width (Gordon and Stein, 1992).

Additional specific geomorphic results of space geodesy include the doming of the Long Valley Caldera in California at a rate of 2.5 ± 1.1 cm/yr (Webb et al., 1995); seaward movement of the mountainside of about 10 cm/yr on the south flank of Kilauea volcano in Hawaii between 1990 and 1993 (Figure 5-5); and seismic creep on the San Andreas fault north of San Francisco at a rate of 33 ± 2 mm/yr (Williams et al., 1994). Until about 1994, most of the annual rates of plate motions measured by space geodesy seemed to be about 6 percent faster than the million-year average rates inferred from paleomagnetic anomaly patterns. However, revisions of the paleomagnetic timescale have reduced the average discrepancy to less than 2 percent, which impressively confirms both the accuracy of the space geodesy measurements and the constancy of plate motions (DeMets et al., 1994). The quality of the instrumentation and data processing are now such that differential ground motions of a few millimeters can be routinely and accurately measured within one or two years. The impact on geomorphic analysis is likely to be profound.

CENOZOIC EPEIROGENY

Gilbert's definition of epeirogeny as the process of continental formation (p. 38) in parallel to his definition of orogeny as the process of mountain formation is no longer tenable because the two sets of processes differ so enormously in the scales of space and time. Yet we need a word to describe nonorogenic tectonism, that is, tectonic movements not associated with mountain belts. Therefore, I redefine **epeirogeny** as *continental vertical tectonic movement of low amplitude relative to its wavelength, not within an orogenic belt, that does not deform rocks or the land surface to an extent that is*

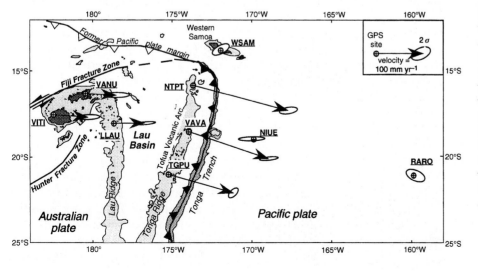

FIGURE 3-6. Crustal motion from GPS measurements made in July 1990 and July 1992 at nine stations in the Tonga-Lau-Samoa region. Velocity vectors are relative to a fixed reference frame on the Pacific plate, which minimizes motions of the three stations (Raro, WSAM, and Niue) on the Pacific plate. Velocity scale and 2σ (86 percent confidence) uncertainty ellipses are shown in the inset (Bevis et al., 1995, Figure 1; courtesy F. W. Taylor, Jr.).

measurable within a single exposure. Such broad regional tectonic movement without local deformation can be either positive (uplift) or negative (subsidence). One special type of epeirogeny is *glacial isostasy,* in which regional subsidence is caused by the weight of an ice sheet, and postglacial recovery has resulted from the removal of the load by deglaciation. It is described in Chapter 17.

Epeirogenic Processes and Landforms

Throughout Phanerozoic geologic history, prevailing isostatic equilibrium has maintained the average continental crustal "freeboard" within less than 100 m of sea level (Wise, 1974). By analogy with a boat at rest in water, the "freeboard" of a continent is the amount exposed above sea level, supported by isostatic buoyancy. With a crustal thickness of 30 km and density of 2.8 g/cm^3, a continental region will have no freeboard, but will be at isostatic equilibrium just at sea level (Kinsman, 1975, p. 94). Except in orogenic belts, continental crust is typically no thicker than 33 km. That is why relatively minor changes of sea level and epeirogenic movements have been able to flood extensive continental areas episodically and cover them with layers of sediments. Significantly, over very large areas of the continents, essentially undeformed marine sedimentary rocks as old as the early Paleozoic Era unconformably overlie igneous and high-grade metamorphic rocks of the continental crust. Although only 1000 to 2000 m in total thickness, these *platform* sedimentary rocks are separated by disconformities that have only minor relief, proving that they were never high enough above sea level to be subjected to deep fluvial dissection.

Much of the continental landscape is eroded from platform sedimentary rocks that are almost in their initial depositional attitude. Broad structural domes and basins with epeirogenic tectonic relief of a few hundred meters over distances of hundreds of kilometers are common, but the resulting dips of the strata are usually less than 1 percent and are barely detectable, as in panoramic views. The cuestaform erosional landscape that develops is analyzed in detail in Chapter 12. Commonly, epeirogenic domes are breached by erosion and their central regions become topographic basins. Uplifted former structural basins become low plateaus surrounded by outward-facing escarpments. The resulting *topographic inversion* by erosion on epeirogenic structures implies that the tectonic movements are either old or extremely slow relative to the rate of erosion (p. 263).

About one-third of the subaerial landscape is eroded from exposed ancient igneous and metamorphic rocks of the continental crust, without any sedimentary cover. These extensive areas are the *continental cratons,* or *shields.* Although the shields are regions of monotonous low relief not far above sea level, some of the exposed rocks must have solidified from magma or metamorphosed at depths of 10 to 20 km or more. Most shields are a complex of deeply eroded accreted orogenic terranes with a wide range of ages. Their present low altitude and low relief may be very old. For example, present relief on the Canadian Shield continues as an unconformity under the Paleozoic and Mesozoic sedimentary rocks on its periphery (p. 381). The Australian craton is a region of exceptionally low relief and tectonic stability, with landforms alleged to be 10^7 to 10^8 years old (p. 347).

If such large areas of continental terranes are either persistent low-lying crystalline shields or platforms of thin, old sedimentary rocks, what tectonic processes could cause the epeirogenic uplift of other large continental areas such as the Intermontane Plateaus of the western United States (Figure 1-5)? Vertical movements of nonorogenic regions are not explained by plate-tectonic theory even though horizontal plate motions measured in tens of centimeters per year are strongly confirmed. Currently, epeirogenic uplift is attributed to the emplacement of **mantle plumes** beneath broad areas of continental crust (Gurnis, 1992; Hill et al., 1992; Parsons et al., 1994; Parsons and McCarthy, 1995). For example, the northern Basin and Range Province lies at an average altitude of 1.5 km above sea level, but the crust is not thick enough to support that altitude by isostasy. Gravity anomalies are small throughout the province, therefore a low-density upper mantle must provide the isostatic buoyancy. If a large mushroom-shaped plume of relatively warm upward-convecting mantle with a diameter on the order of 1000 km rose beneath the western United States, it could cause regional uplift of 1 to 2 km (Parsons et al., 1994). The underplating and up-arching of the continental crust by the hypothetical plume head may also have caused the surface rifting and volcanism in the Columbia Plateau and Snake River Plain to the north. It is believed that the underplated mass subsequently moved west with the overlying continental crust even though the "tail" of the plume remained fixed in the mantle. Therefore, the plume tail appears to have migrated eastward under the continent from the Columbia Plateau through the Snake River Plain to its present position under the Yellowstone volcanic plateau (Pierce and Morgan, 1992). The role of mantle plumes in continental rifting is considered further in Chapter 5, and their volcanic significance in Chapter 6.

Epeirogenic Rates

The redefinition of epeirogeny (p. 43) intentionally excludes any reference to rates, which may range widely. Long-term epeirogenic subsidence rates derived from the thickness of sediments that have accumulated in epeirogenic basins or on subsiding continental margins are generally slow. For example, 6 km of Cenozoic sediment in the Gulf Coast geosyncline, deposited over an interval of some 60 million years, implies a subsidence rate of 100 m/million yr. Sedimentary basins such as the Paleozoic Michigan Basin also typically subsided at a rate of about 100 m/million yr. Rates such as these are comparable to the rates of ocean-floor subsidence (Figure 1-4), and are probably caused by the same process, thermal contraction (Sleep et al., 1980). Note that because of sediment deposition, these are rates of sediment, but not topographic, deformation. In both places, the ocean floor remained relatively shallow despite the subsidence of the crust beneath the sediment load.

Long-term epeirogenic uplift is harder to document than subsidence, for much of the evidence is removed by erosion. Pliocene and younger marine sediments now emerged on the Atlantic Coastal Plain of the United States record several million years of net uplift at rates averaging 10 to 30 m/million yr, comparable to the rate of subsidence farther offshore on the Atlantic continental shelf during Cenozoic sediment loading (Cronin, 1981). The Colorado Plateau is a stable, rigid block of continental crust that is in isostatic equilibrium. Nevertheless, it has risen from below sea level in Cretaceous time to its present average height of about 2 km during the Cenozoic Era, at an average rate of about 30 m/million yr.

With reference to the orogenic rates noted on p. 41, which are typically a few kilometers per million years, the epeirogenic uplift and subsidence rates complied in the previous paragraph are one or two orders of magnitude slower over similar time intervals. However, other measurements, some of which are highly controversial, suggest that epeirogenic movements can be as rapid as orogenic movements, at least when measured over time intervals of years or decades. These measurements are the basis for a theory called **neotectonism** (the "newest" tectonics—the study of both orogenic and epeirogenic earth movements of the Quaternary Period and the later part of the Tertiary Period, including those in progress at the present time) (Fairbridge, 1981).

Initial neotectonic hypotheses were based on repeated precise leveling surveys on long transects across continental shields and platforms, such as Russian surveys along the trans-Siberian railroad, that seemed to show wavelike undulations of land surface at rates of as much as 10 mm/yr. Geodesists in other countries soon reported similar discrepancies between repeated surveys. The subject has greatly expanded to include **seismotectonics**, the measurement of earthquake-related ground movements (Slemmons et al., 1991). Clearly, if large regions of supposedly "stable" continental crust are rising or subsiding at annual or decadal rates comparable to those in the most orogenic regions, the movements cannot be cumulative, or new tectonic landforms would be apparent within human lifetimes. Nevertheless, good quality resurveys continue to demonstrate vertical movement. Recent resurveys of the eastern United States show the Appalachians rising at 6 mm/yr relative to the Atlantic coast. Other leveling lines and tide-gauge studies show comparable rates of relative vertical motions (Figure 3-7). Some of the differences between such successive surveys are probably due to unrecognized instrumental errors. Many scientists are now attempting to verify these vertical movements by the new methods of space geodesy. Within the next decade, the reality of vertical epeirogenic movements of millimeters per year will very likely be confirmed or disproved. Meanwhile, the possibility of large-scale but episodic or oscillatory warping as a continuing feature of epeirogenic tectonics must be considered a possible factor in geomorphic analysis.

Vertical epeirogenic displacements might influence drainage networks, river terraces, lake shorelines, and many other geomorphic features. One example of the effects of a river system crossing an area of epeirogenic uplift is the Bogue Homo Creek across the Wiggins Anticline, on the southern border of Mississippi and Alabama (Figure 3-7). The river is being affected by uplift of nearly 5 mm/yr across its path. In response, its main channel upstream of the ridge has straightened in order to shorten its channel length and thereby maintain its downstream gradient against the rising structure. On the downstream flank of the anticline, it has incised to form a low terrace, increased its sinuosity to offset the steepened gradient, and divided into braided segments as a result of the increased sediment loads from the incision across the axis of uplift (Burnett and Schumm, 1983). The river is said to be *antecedent* to the uplift because its route was established prior to, and has persisted in spite of, the uplift. The tectonic significance of antecedent rivers is further discussed in Chapter 12.

FIGURE 3-7. Vertical movements (mm/yr) in the eastern United States, compiled from various sources. Tide gauges all show subsidence relative to sea level, but the cause could be either rising sea level, or sinking land, or both (Brown and Oliver, 1976, Figure 15).

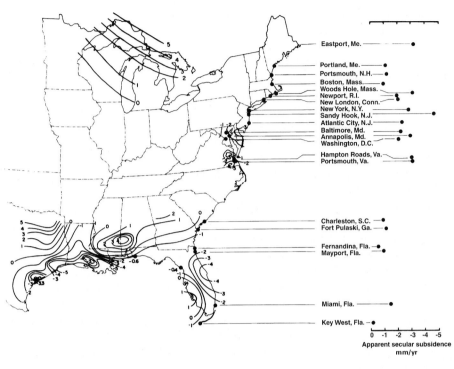

TECTONISM AND CLIMATE

No one doubts the influence of mountain ranges on local climates. Not so obvious are the possible effects of regional epeirogenic uplift on hemispheric-scale atmospheric circulation (Ruddiman et al., 1989; Kutzbach et al., 1989; Ruddiman and Kutzbach, 1989). It is well established that the south Asian monsoon depends on the high topography of the Himalayan Mountains and the Tibetan Plateau to provide the orographic lift that draws in Indian Ocean moisture during the wet season and to create the cooler, dry subsiding air mass that creates the offshore winds of the dry season. It is logical to propose that, because the present Tibet-Himalayan topography is the result of late Cenozoic orogeny, the present intensity of the monsoon circulation is also a late Cenozoic phenomenon (Prell and Kutzbach, 1992).

In the high- and midlatitude northern hemisphere, the distribution of oceans and landmasses as well as the regional height of the land are the major controls on the semistationary upper atmospheric *planetary waves* that steer surface air masses and their frontal storms. Ruddiman and Kutzbach (1989) hypothesized that late Cenozoic uplift of the Tibetan Plateau and the western third of North America together might have affected the paths of the northern hemispheric planetary circulation enough to be a contributing factor for Pleistocene continental glaciation. They tested a series of computerized atmospheric gen-

eral circulation models with "no mountains," "half mountains," and "full mountains," intended to approximate the northern hemisphere relief in the Eocene, Miocene-Pliocene, and present epochs, respectively. Because the Tibetan Plateau and western North America are almost 180° apart in longitude, they constructively interfere to reinforce the planetary wave system all around the northern hemisphere. With no mountains, the modeled upper atmospheric circulation is nearly zonal (along parallels of latitude). With half mountains and full mountains, the models predict progressively stronger planetary waves, which lead to predicted colder winters over North America, northwestern Europe, and northern Asia. The models also predict drier summers on the American Pacific coast, in the interior of Asia, and in the Mediterranean; winter drying on the northern plains of America and in the interior of Asia; and warm-wet conditions on the southeastern coasts of Asia and the United States (Ruddiman and Kutzbach, 1989, p. 18,423). All these model predictions are consistent with records of late Cenozoic climate change that led to Pliocene-Pleistocene ice ages in the northern hemisphere. Less satisfactory predictions for climate changes in Alaska and the U.S. southern plains and Rocky Mountains were attributed to the inability of the computer programs to predict the effects of narrow, high mountain ranges. But in general, 2 to 4 km of late Cenozoic surface uplift in Tibet and the Himalaya Mountains, and at least 1 km of uplift across a broad area of western North

America, including the Sierra Nevada, Basin and Range Province, Colorado Plateau, Rocky Mountains, and the western Great Plains, successfully produced changes in the model climates that are comparable to, although less extreme than, the inferred actual climatic changes of the late Cenozoic Era.

Others have argued that the evidence for late Cenozoic uplift, which includes geomorphic evidence of deep fluvial dissection, coarse sediment loads, and glaciation, has been misinterpreted. Rather than proving that uplift has occurred, the geomorphic changes could be the *result* of the onset of ice-age climates caused by other factors, that in turn modified preexisting mountain ranges and plateaus (Molnar and England, 1990). The increased erosion rates could also have caused or accelerated tectonic uplift because of isostatic response to the rock mass that was being removed (Small and Anderson, 1995).

This debate continues, but as is reviewed in more detail in Chapter 4, there is no doubt that the late Cenozoic Era has been a time of exceptional tectonism as well as strong climatic diversity, and the two conditions are obviously related. The details of cause and effect will become more clear as tectonic chronologies improve and climatic interpretations of erosion and sedimentation become more refined. Clearly, the geomorphic system we study today has an evolutionary history going back at least 15 or 20 million years, and its correct interpretation requires an understanding of that history.

REFERENCES

ALLMENDINGER, R. W., FIGUEROA, D., and 4 others, 1990, Foreland shortening and crustal balancing in the Andes at 30°S latitude: Tectonics, v. 9, pp. 789–809.

BATES, R. L., and JACKSON, J. A., 1987, Glossary of geology, 3rd ed.: American Geological Institute, Alexandria, Va., 788 pp.

BEVIS, M., TAYLOR, F. W., and 9 others, 1995, Geodetic observations of very rapid convergence and back-arc extension at the Tonga arc: Nature, v. 374, p. 249–251.

BLOOM, A. L., BROECKER, W. S., and 3 others, 1974, Quaternary sea level fluctuations on a tectonic coast: New ^{230}Th/^{234}U dates from the Huon Peninsula, New Guinea: Quaternary Res., v. 4, pp. 185–205.

BROWN, L. D., and OLIVER, J. E., 1976, Vertical crustal movements from leveling data and their relation to geologic structure in the eastern United States: Rev. Geophys. and Space Phys., v. 14, pp. 13–35.

BURNETT, A. W., and SCHUMM, S. A., 1983, Alluvial-river response to neotectonic deformation in Louisiana and Mississippi: Science, v. 222, pp. 49–50.

CEBULL, S. E., 1973, Concept of orogeny: Geology, v. 1, pp. 101–102.

———, 1976, Concept of orogeny: Reply: Geology, v. 4, pp. 388–389.

CRONIN, T. M., 1981, Rates and possible causes of neotectonic vertical crustal movements of the emerged southeastern United States Atlantic Coastal Plain: Geol. Soc. America Bull., v. 92, pp. 812–833.

DeMETS, C., GORDON, R. G., and 2 others, 1994, Effect of recent revision to the geomagnetic reversal time scale on estimates of current plate motions: Geophys. Res. Ltrs, v. 21, pp. 2191–2194.

ENGLAND, P., and MOLNAR, P., 1990, Surface uplift, uplift of rocks, and exhumation of rocks: Geology, v. 18, pp. 1173–1177.

FAIRBRIDGE, R. W., 1968, Continents and ocean—statistics of area, volume and relief, in Fairbridge, R. W., ed., Encyclopedia of Geomorphology: Reinhold Book Corp., New York, pp. 177–186.

———, 1981, The concept of neotectonics: An introduction, in Fairbridge, R. W., ed., Neotectonics: Zeitschr. für Geomorph., Supp. no. 40, pp. vii–xii.

GILBERT, G. K., 1890, Lake Bonneville: U.S. Geol. Survey Mon. 1, 438 pp.

GORDON, R. G., and STEIN, S., 1992, Global tectonics and space geodesy: Science, v. 256, pp. 333–342.

GURNIS, M., 1990, Ridge spreading, subduction, and sea level fluctuations: Science, v. 250, p. 970–972.

———, 1992, Long-term controls on eustatic and epeirogenic motions by mantle convection: GSA Today, v. 2, no. 7, p. 141–157.

HILL, R. I., CAMPBELL, I. H., and 2 others, 1992, Mantle plumes and continental tectonics: Science, v. 256, pp. 186–193.

JORDAN, T. E., ISACKS, B. L., and 2 others, 1983, Mountain building in the central Andes: Episodes, v. 1983, no. 3, pp. 20–26.

KAMP, P. J. J., GREEN, P. F., and WHITE, S. H., 1989, Fission track analysis reveals character of collisional tectonics in New Zealand: Tectonics, v. 8, pp. 169–195.

KARIG, D. E., 1974, Evolution of arc systems in the western Pacific: Ann. Rev. Earth, Planet. Sci., v. 2, pp. 51–75.

KHAIR, K., KHAWLIE, M., and 4 others, 1993, Bouger gravity and crustal structure of the Dead Sea transform fault and adjacent mountain belts in Lebanon: Geology, v. 21, pp. 739–742.

KINSMAN, D. J. J., 1975, Rift valley basins and sedimentary history of trailing continental margins, in Fischer, A. G., and Judson, S., eds., Petroleum and global tectonics: Princeton Univ. Press, Princeton, N.J., pp. 83–126.

KUTZBACH, J. E., GUETTER, P. J., and 2 others, 1989, Sensitivity of climate to late Cenozoic uplift in southern Asia and the American west: Numerical experiments: Jour. Geophys. Res., v. 94, pp. 18,393–18,407.

McSAVENEY, M. J., CHINN, T. J., and HANCOX, G. T., 1992, Mount Cook rock avalanche of 14 December 1991, New Zealand: Landslide News, no. 6, pp. 32–34.

MOLNAR, P., and ENGLAND, P., 1990, Late Cenozoic

uplift of mountain ranges and global climate change: Chicken or egg?: Nature, v. 346, p. 29–34.

PARSONS, T., and McCARTHY, J., 1995, The active southwest margin of the Colorado Plateau: Uplift of mantle origin: Geol. Soc. America Bull., v. 107, pp. 139–147.

PARSONS, T., THOMPSON, G. A., and SLEEP, N. H., 1994, Mantle plume influence on the Neogene uplift and extension of the U.S. western cordillera?: Geology, v. 22, pp. 83–86.

PIERCE, K. L., and MORGAN, L. A., 1992, The track of the Yellowstone hot spot: Volcanism, faulting, and uplift, in Link, P. K., Kuntz, M. A., and Platt, L. B., eds., Regional geology of eastern Idaho and western Wyoming: Geol. Soc. America Mem. 179, pp. 1–53.

PILLANS, B., 1986, A late Quaternary uplift map for North Island, New Zealand: Royal Soc. New Zealand Bull. 24, pp. 409–417.

PLAFKER, G., 1965, Tectonic deformation associated with the 1964 Alaskan earthquake: Science, v. 148, pp. 1675–1687.

PRELL, W. L., and KUTZBACH, J. E., 1992, Sensitivity of the Indian monsoon to forcing parameters and implications for its evolution: Nature, v. 360, pp. 647–652.

RESEARCH GROUP FOR QUATERNARY TECTONIC MAP, 1973, Explanatory text of the Quaternary tectonic map of Japan: Natl. Research Center for Disaster Prevention, Tokyo, 167 pp.

RODGERS, J., 1987, Chains of basement uplifts within cratons marginal to organic belts: Am. Jour. Sci., v. 287, pp. 661–692.

RUDDIMAN, W. F., and KUTZBACH, J. E., 1989, Forcing of late Cenozoic northern hemisphere climate by plateau uplift in southern Asia and the American west: Jour. Geophys. Res., v. 94, pp. 18,409–18,427.

RUDDIMAN, W. F., PRELL, W. L., and RAYMO, M. E., 1989, Late Cenozoic uplift in southern Asia and the American west: Rationale for general circulation modeling experiments: Jour. Geophys. Res., v. 94, pp. 18,379–18,391.

SIMKIN, T., UNGER, J. D., and 3 others, 1994, This dynamic planet: World map of volcanoes, earthquakes, impact craters, and plate tectonics: U.S. Geol. Survey Special Map.

SLEEP, N. H., NUNN, J. H., and CHOU, L., 1980, Platform basins: Ann. Rev. Earth, Planet. Sci., v. 8, pp. 17–34.

SLEMMONS, D. B., ENGDAHL, E. R., and 2 others, eds., 1991, Neotectonics of North America: Geol. Soc. America, Decade Map Volume 1, 498 pp.

SMALL, E. E., and ANDERSON, R. S., 1995, Geomorphically driven late Cenozoic rock uplift in the Sierra Nevada, California: Science, v. 270, pp. 277–280.

SMITH, D. E., and TURCOTTE, D. L., eds., 1993, Contributions of space geodesy to geodynamics: Crustal dynamics: Am. Geophys. Union Geodynamics Series, v. 23, 429 pp.

WEBB, F. H., BURSIK, M., and 4 others, 1995, Inflation of Long Valley Caldera from one year of continuous GPS observations: Geophys. Res. Ltrs., v. 22, pp. 195–198.

WERNICKE, B., 1992, Cenozoic extensional tectonics of the U.S. Cordillera, in Burchfiel, B. D., Lipman, P. W., and Zoback, M. L., eds., The Cordilleran Orogen: Conterminous U.S.: Geol. Soc. America, The Geology of North America, v. G–3, pp. 553–581.

WILLIAMS, S. D. P., SVARC, J. L., and 2 others, 1994, GPS measured rates of deformation in the northern San Francisco Bay region, California 1990–1993: Geophys. Res. Ltrs., v. 21, pp. 1511–1514.

WISE, D. U., 1974, Continental margins, freeboard and the volumes of continents and oceans through time, in Burk, C. A., and Drake, C. L., eds., The geology of continental margins: Springer-Verlag New York, Inc., New York, pp. 45–58.

Chapter 4
Cenozoic Climate Change

Global cooling, and with it a trend toward steeper latitudinal climate gradients (Figure 4-1) and increased seasonality and variability, seems to have been the climatic trend through the Cenozoic Era. The causes are so complex and interrelated that they defy simple explanation. Potential factors include lithospheric plate motions, vertical tectonism, weathering reactions, fluctuating atmospheric CO_2 content, volcanism, biologic evolution, changes in oceanic circulation, and cyclic variations in the earth's orbit around the sun. The geomorphic results have been spectacular, culminating during the last 2.5 million years in a series of ice ages that repeatedly covered as much as 30 percent of the land area with glaciers. Even today, in what some refer to as "postglacial" time (with excessive optimism), 10 percent of the land remains ice covered, including the entire continent of Antarctica and most of the largest island, Greenland. We are living in an ice age although all 6000 years of recorded human history have been within the Holocene Epoch, one of the numerous brief interglacial intervals when the extent of ice cover has been reduced. The geomorphic impact of the most recent glacial-interglacial cycle is reviewed in greater detail in Chapter 18.

CO₂ AND CLIMATE

The atmospheric CO_2 greenhouse effect (p. 23) has been a major control on the earth's climate throughout Phanerozoic time (Schwartzman and Volk, 1991; Berner, 1994, p. 88). Warming due to the slow increase of solar luminosity during the Paleozoic and Mesozoic Eras was largely offset by cooling due to decline in the CO_2 greenhouse gas by evolution of photosynthetic plants and the resulting deposition of carbon-rich sedimentary rocks. The cooling was by no means uniform, however. The other factors listed in the previous paragraph also had important climatic influence, and during various geological times there have not only been other ice ages but also warm, steamy coal-producing forests; shallow, warm epeirogenic seas; and large, arid continents. But throughout geologic history, the average surface temperature of the earth has remained within a relatively narrow range between 0°C and 100°C, in which water can change among its solid, liquid, and gaseous phases. Various feedback mechanisms involving atmospheric CO_2 have been important in maintaining that stability.

Carbon-isotope ratios in calcium carbonate concretions in palesols from a variety of Mesozoic and Cenozoic terrestrial strata have been analyzed to infer paleo-CO_2 atmospheric content (Cerling, 1991). The inferred amount of atmospheric CO_2 during the early Eocene and Miocene was 700 ppmV (parts per million by volume) or less, which at almost double this century's level seems very high, but was actually judged indistinguishable from current levels because of the large errors assigned to the analyses. Fossil European oak leaves ranging in ages from 10 million to 2.5 million

FIGURE 4-1. Estimated mean annual temperatures for the North Atlantic region through the Cenozoic Era, expressed as a function of latitude (based on Weidick, 1975, Figure 15, with revised scales). See back endpaper for a tabular Cenozoic time scale.

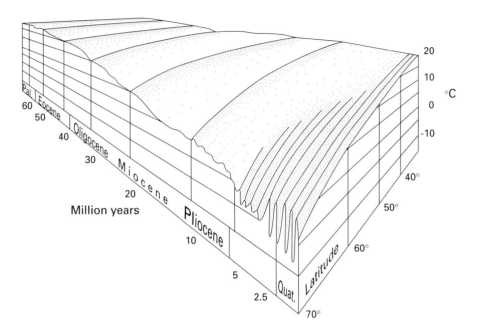

years show physiologic adaptation to an atmosphere with a CO_2 content that fluctuated between 280 and 370 ppmV, essentially the historical range (Van Der Burgh et al., 1993). Some causes for the fluctuations of atmospheric CO_2 and their geomorphic consequences are reviewed in the following paragraphs.

Tectonism, Weathering, Metamorphism, and Atmospheric CO_2

With orogenic uplift, erosion is accelerated. More rock debris is exposed to weathering, which generally consumes atmospheric CO_2 (Chapter 7). Decreased atmospheric CO_2 content should cool the planet, perhaps enough to induce glaciation. Although the cause-and-effect relationships between erosion, mountain building, and climatic cooling are vigorously debated (p. 46), their concurrent intensification in late Cenozoic time is well established.

In the great rock cycle through which all continental rocks have been repeatedly processed, the major contrasting processes of weathering and metamorphism can be represented by two simplified chemical reactions that do not presume any specific minerals, but simply show characteristic trends:

$$CaSiO_3 + 2CO_2 + H_2O \longrightarrow SiO_2 + Ca^{2+} + 2HCO_3^- \quad (1)$$

$$Ca^{2+} + 2HCO_3^- \longrightarrow CaCO_3 + H_2O + CO_2 \quad (2)$$

During weathering (reaction 1), silicate minerals react with atmospheric CO_2 in aqueous solution to form dissolved bicarbonate compounds that are carried to the sea by rivers and are eventually precipitated as carbonate rocks (reaction 2). Most of the silica reacts with other cations to form clay minerals (Chapter 7). The net loss is 1 molecule of CO_2 from the atmosphere to the rock mass. During burial and metamorphism the reactions are reversed. Silicate minerals reform and CO_2 is released to the atmosphere through volcanoes and hot springs. An even more simplified equation summarizes the reversible reactions:

$$CaSiO_3 + CO_2 \rightleftharpoons CaCO_3 + SiO_2 \quad (3)$$

a silicate + carbon $\rightleftharpoons$ calcite + silica in
mineral   dioxide   a mineral

To a degree, reaction (3) is self-limiting by negative feedback. As with most chemical reactions, weathering rates approximately double with a 10°C increase in temperature (Chapter 7). If atmospheric CO_2 decreases, the climate grows colder, and chemical weathering rates decrease. Contrarily, in a warm, CO_2-rich greenhouse climate, weathering rates increase, CO_2 is sequestered in sediments, and the greenhouse effect is decreased. By this reasoning, weathering and erosion have been powerful buffering factors in equilibrating the earth's climate. The landscape is the dynamic interface between the earth's crust and the hydrosphere, biosphere, and atmosphere. In reading the erosional history of a modern landscape, one may find a chronicle of changing environments that extends back in time for at least millions or tens of millions of years.

Evolution, Weathering, and Atmospheric CO_2

The importance of vegetation to weathering and erosion is discussed at length in later chapters. We do not know the geomorphic processes that operated on an early Paleozoic landscape before land plants evolved, for example. However, even during the later part of geologic history, when life forms were broadly similar to those of the present, evolutionary changes have produced some surprising adaptations that affected the carbon dioxide budget and therefore global climate and weathering rates. One such change was the late Mesozoic evolution of pelagic carbonate-secreting marine organisms. Prior to early Cretaceous time about 145 million years ago, up to 90 percent of marine carbonate sediments were deposited in shallow water and generally remained part of the continental platform stratigraphic succession. Through Cretaceous and Cenozoic time, deep-water carbonate deposition by pelagic organisms gradually has taken the place of shallow-water deposition so that at present, shallow-water carbonate accumulation is the lowest and pelagic sedimentation is the highest proportion of any Phanerozoic interval, with about 60 percent of all carbonate sediments resulting from pelagic organisms (Caldeira, 1992; Opdyke and Wilkinson, 1993, p. 218). Whereas shelf carbonates tend to be preserved as limestone, sea-floor spreading carries pelagic carbonate sediments into subduction zones where they are buried, metamorphosed, and "decarbonated" (reaction 3) to provide the increased CO_2 flux that increases weathering rates. This feedback mechanism is a relatively new factor in the global carbon cycle, and one that began to contribute to climate and, therefore, landscape evolution, primarily during the Cenozoic Era.

A second evolutionary development of direct significance to Cenozoic landscape evolution was the expansion of *deciduous angiosperms* (flowering plants that seasonally shed their leaves) during the Cenozoic Era. During the later Paleozoic Era and through most of the Mesozoic Era, terrestrial vegetation was dominated by evergreen gymnosperms such as conifers. By the early Cenozoic, however, deciduous trees were radiating from their original northern subpolar environment into large parts of the northern hemisphere, with a concurrent reduction of the conifer evergreen forests (Knoll and James, 1987). Neither the southern hemisphere land vegetation nor the tropical forests experienced a comparable radiation of deciduous trees. Deciduous angiosperms remove cations such as K^+, Mg^{2+}, and Ca^{2+} from their soil at a rate three to four times faster than do evergreen conifers, largely through the decay of the thick humus layer generated by fallen leaves (Knoll and James, 1987; Volk, 1989). Thus, midlatitude northern hemisphere soils have been weathering significantly faster during the Cenozoic than in earlier times. The increased weathering rates on the large northern hemisphere land area (Figure 3-1) may have consumed enough additional atmospheric CO_2 to cool the global climate by as much as 10°C (Volk, 1989). If seasonality increased with cooling, deciduous trees might have been a factor in producing the seasonal climatic cycle that permitted their own expansion and diversification.

A third example of the geomorphic impact of Cenozoic evolution concerns the youngest major evolutionary group of angiosperms, the grass family (Gramineae). "One of the most remarkable transformations of terrestrial ecosystems during the Tertiary was the spread of grasslands on all major continents of the world" (Retallack, 1982). Grass is first recorded by fossils in Eocene and Oligocene sediments in Argentina, and by late Oligocene time (ca. 25 million years ago), grasslands were widespread. Specialized grasses with metabolism adapted to tropical environments were present in Kenya by 15.3 My and in Pakistan by 9.4 My (Morgan et al., 1994). A savanna-type mosaic of grassy woodland and wooded grassland is recorded by Miocene fossils and paleosols in east Africa that are about 14 million years old (Retallack et al., 1990; Dugas and Retallack, 1993). Open plains, on which trees are susceptible to fire, are ideal for grasses, which can dry up, drop their tough seeds, and go dormant or be burned to stubble during dry or cold seasons. As soon as rain or warmth return, grasses sprout again. Mammals also had an explosive mid-Cenozoic evolution, as they adapted to grazing and running on the new open, grassy plains.

The geomorphic impact of grass is hard to underestimate. Grasses are unique among plants in their ability to form a tight, shallow mesh of roots called *sod* or *turf*. No one doubts the ability of sod to prevent gully erosion, so it is likely that major changes in mass-wasting (Chapter 9) and overland runoff (Chapter 10) attended upon the evolution of grass. Related geomorphic effects such as delta growth and submarine sedimentation can easily be inferred. Thus, it is only since mid-Miocene time, about 15 million years ago, that landscapes have looked "modern" and have been evolving in response to modern processes. The evolution of grass is the principal reason that the subtitle of this book specifies its applicability only to late Cenozoic landforms. Any systematic analysis of older landscapes requires many more assumptions and much greater uncertainty.

OCEAN CIRCULATION AND CLIMATE

In Chapter 2, it was noted that ocean currents combine with the planetary wind system to transport heat poleward from the topics. Both atmospheric and oceanic circulation have been affected by Cenozoic lithospheric plate motions and vertical tectonics. The tectonic impact on atmospheric circulation and climate was reviewed in Chapter 3. Here, the role of ocean currents, shallow and deep, in Cenozoic climate change is considered.

Surface Ocean Currents

The upper layer of ocean water, to a depth of a few hundred meters, flows in response to the shear stress applied to it from persistent winds. Near the equator, the resulting surface ocean currents flow more or less in the direction of the prevailing winds. Because of the increasing *Coriolis effect* with latitude, currents progressively deviate from the prevailing wind direction with increasing latitudes, to the right of the wind direction in the northern hemisphere and to the left in the southern hemisphere. The wind stress on the surface water and the Coriolis deflection is transferred to deeper layers at an exponentially decreasing rate, with the result that the net transport of water in the upper few hundred meters of the ocean is perpendicular to the surface wind vector, again to the right in the northern hemisphere and to the left in the southern hemisphere. The resulting ocean currents consist primarily of rotating surface gyres in the northern and southern hemisphere oceans (Figure 4-2). Their depth is limited to a few hundred meters by the **pycnocline,** a zone of steep density gradient caused by temperature and salinity contrasts between the surface mixed layer and the much larger, more stable, stratified deep ocean. The pycnocline is roughly coincident with the **thermocline,** or zone of steep temperature gradient.

The flux (flow rate) of surface ocean currents is enormous by comparison with the discharge of terrestrial rivers. For measuring ocean currents, a flux of 1×10^6 m³/s is defined as 1 Sverdrup (Sv) and is approximately equal to the combined discharge of all the world's rivers. The Gulf Stream transports between 75 and 115 Sv, and the largest surface current, the Antarctic Circumpolar Current, flows eastward with a flux of 233 to 240 Sv (Kennett, 1982, pp. 248–249). Where surface currents diverge from each other or from a coastline because of the Coriolis effect, deep water wells upward; where they converge, the surface layer thickens and bulges downward.

By upwelling and downwelling, the mixed surface layer interchanges with deep ocean water.

FIGURE 4-2. Major surface currents of the world ocean (Kennett, 1982, Figure 8-6).

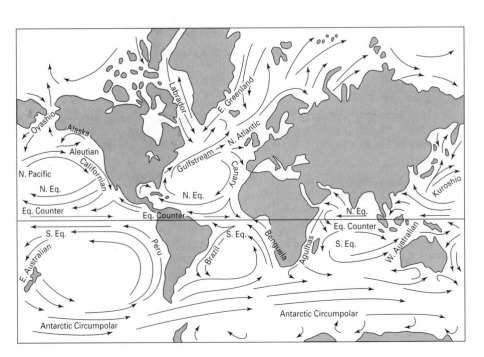

Onshore winds blowing over areas of upwelling cold water cause aridity along subtropical western continental margins (as in California and Mexico, Peru and northern Chile, northwestern and southwestern Africa, and western Australia) (Figure 13-1). Warm surface currents such as the Atlantic Gulf Stream and the Pacific Kuroshio Current create mild, moist, often snowy climates on midlatitude west coasts (Alaska, western Canada, Washington, and Oregon; Norway, Scotland, and Ireland; New Zealand; and southern Chile). The desert and glaciated landscapes of these regions are no older than the Cenozoic evolution of ocean circulation.

Deep Thermohaline Circulation

Below the pycnocline, water masses move slowly, governed by their relative densities. The density of seawater depends on its temperature and salinity, with warm, low-salinity water the least dense and cold, highly saline water the most dense. Density contrasts of only a few parts per thousand are sufficient to define deep water masses, which tend to spread laterally with ease, but mix with and displace each other vertically only slowly. The resulting pattern of currents is known as **thermohaline circulation**.

The controls of water density by temperature and salinity are nonlinear. In warm low-latitude water, temperature contrasts have a large effect on the density of water masses, and salinity contrasts are of lesser importance. High-latitude surface water, however, is almost as cold as deep water, and with the lack of temperature gradients, even a slight increase of surface salinity due to evaporation or freezing is sufficient to cause the water mass to sink. For this reason, all the deep water of the oceans today has a temperature of only 1°C to 2°C and originates in high latitudes. Most of the deep ocean water is derived from the circum-Antarctic region and is defined as Antarctic Bottom Water (AABW). The second major source is the north Atlantic region, which produces a water mass known as North Atlantic Deep Water (NADW). AABW flows along the ocean floor all the way north across the Pacific Ocean and to at least 40° N in the Atlantic Ocean. The slightly less dense NADW flows southward above the AABW all the way to the circum-Antarctic region where it flows eastward into the Indian Ocean and Pacific Ocean. As much as 59 percent of all ocean water is AABW (Kennett, 1982, p. 251), but perhaps as much as 25 percent of the water in the Pacific basin originates in the North Atlantic Ocean. Neither the Pacific nor the Indian Ocean contributes significant amounts of deep water.

Downwelling of dense high-latitude water masses such as AABW and NADW requires a shallower return flow of warmer or less saline water beneath the pycnocline (Figure 4-3). The thermohaline current velocities are an order of magnitude less than in surface currents; deep ocean currents flow at only 10 to 30 cm/sec or less. Nevertheless, the "great ocean conveyor belt" (Broecker, 1991) of thermohaline circulation has enormous significance to the global climate. In one simplified two-layer model (Figure 4-3), an eastward circumpolar deep water flux of 55 Sv in the southern ocean has left-veering loops of 20 to 24 Sv into the Indian and Pacific Oceans, from which warmer upper-level currents flow westward through the tropics (Schmitz, 1995). The circumpolar deep water is augmented by 14 Sv of NADW that flows south along the western boundary of the Atlantic basin. The upper-level Atlantic thermohaline return flow includes an estimated 10 Sv of circum-Antarctic water that enters the Atlantic through the Drake Passage between South America and Antarctica, and another 4 Sv that flows westward around southern Africa. A major feature of this conveyor belt is that in the South Atlantic, relatively warm upper-layer water flows toward the equator rather than away from it, then crosses into the Caribbean Basin to be further warmed before continuing its northward journey to provide the return flow for NADW in the northernmost Atlantic (Gordon, 1986).

At present, the Atlantic Ocean is saltier than the Pacific because of evaporation. In low latitudes, there is net evaporative moisture transport westward across Central America by the Trade Winds. Over the North Atlantic, evaporation by the westerlies blowing over the warm Gulf Stream carries precipitation into Eurasia, where it falls as precipitation beyond the Atlantic basin divide. Broecker (1991, p. 84) calculated that net evaporation of only 0.35 Sv from the surface water of the northern Atlantic basin is sufficient to drive the NADW thermohaline circulation (which he estimated at 20 Sv instead of the 14 Sv shown on Figure 4-3). This system has been operating in its present mode only for about the past 9000 years, and provides the Holocene high-latitude warmth and abundant moisture to western Europe. People are usually surprised to find that the latitude of sunny Rome, Italy (42° N) is comparable to that of Boston or Chicago in the United States. Paris, France (49° N), with mild and rainy winters, is at the latitude of International Falls, Minnesota, which regularly records the coldest winter daily temperatures in the United States. As described in Chapter 18, the great ocean conveyor belt may have shut down and restarted repeatedly during ice ages, either causing or being caused by major fluctuations in the northern hemisphere ice sheets, most of which were peripheral to the North Atlantic Ocean.

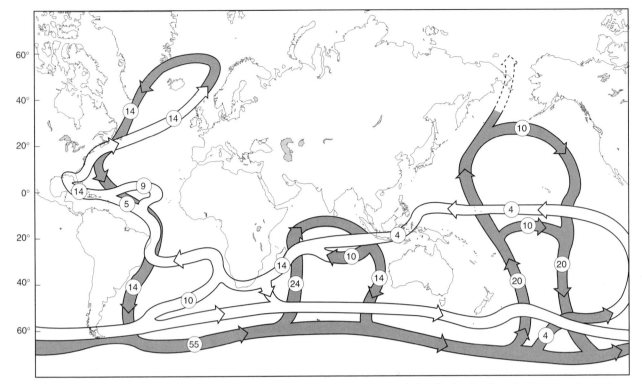

FIGURE 4-3. A simplified two-layer thermohaline circulation conveyor belt. Dark shading indicates abyssal water (deep and bottom layers), and light shading indicates uppermost layer flow (thermocline and intermediate water). Circles show fluxes in Sverdrups (10^6 m^3/s) (based on Schmitz, 1995, especially Figure 9).

The Thermal History of the Cenozoic Oceans

Only since mid-Cenozoic time, about 35 million years ago, has the modern ocean circulation been possible. By that time, plate motions had separated South America and Australia from Antarctica, making possible the modern powerful Antarctic Circumpolar Current and the equally important southern hemisphere circumpolar deep flow (Kennett, 1982, pp. 714 ff). The second most important component of the global thermohaline circulation, the North Atlantic Deep Water, is probably only about 3 million years old. That is the approximate date for the final tectonic closing of the Isthmus of Panama (Keigwein, 1982) and the opening of the Bering Strait, through which as much as 1 Sv of relatively low salinity northern Pacific water now "leaks" into the Arctic Ocean and from there to the North Atlantic NADW source area. Before the Panama isthmus closed, it is doubtful that the Atlantic Ocean had its excess surface salinity. The regional climate zonation would have been very different from today's (Burton et al., 1997).

The history of sea-surface and deep-ocean temperature is recorded by the oxygen-isotope ratios of *pelagic* (near-surface) and *benthic* (bottom-dwelling) foraminifera and other carbonate-secreting microorganisms. The oxygen in these calcium carbonate shells is deposited in equilibrium with the isotopic composition of the oxygen in the ocean H_2O molecules. About 99.8 percent of that oxygen is of the stable isotope ^{16}O; almost all the remainder is of the heavier, less common, stable isotope ^{18}O. The average ratio of $^{18}O/^{16}O$ in seawater is approximated by a carefully measured synthetic standard known as SMOW (standard mean ocean water). Deviations from SMOW of the $^{18}O/^{16}O$ ratio in water, ice, or organic samples are expressed in parts per thousand (‰) by the term $\delta^{18}O$. Values of $\delta^{18}O$ are routinely reported to a precision of ±0.1‰ or better by mass spectrometry. In principle, the $\delta^{18}O$ value for a marine shell is a measure of the water temperature in which it precipitated, with progressively higher or "heavier" values of $\delta^{18}O$ representing colder water (Figure 4-4). One of the more impressive results of oxygen-isotope analysis is the late Cretaceous and Cenozoic temperature history of the oceans recorded in long sediment cores (Figure 4-5). Since the early Cretaceous evolution of benthic foraminifera, their $\delta^{18}O$ values have recorded the irregular cooling of the deep oceans from a Cretaceous and early Cenozoic range of 15°C to 20°C to the present temperature of about 2°C (the freezing temperature of seawater is about –1.8°C). This probably reflects an early Cenozoic thermohaline circulation of warm but highly saline subtropical water sinking and

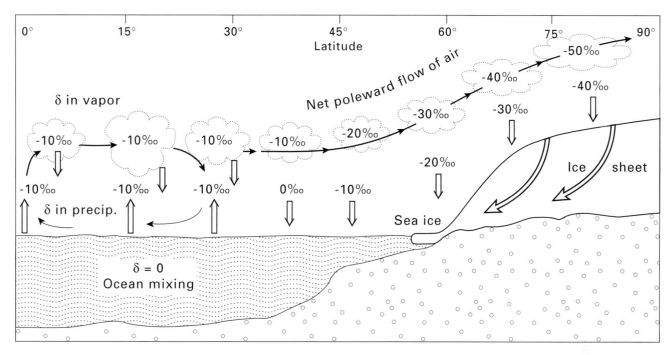

FIGURE 4-4. The process of oxygen isotope fractionation by evaporation, condensation, precipitation, and freezing, and typical resulting $\delta^{18}O$ values. Note that if ice sheets made of isotopically light water grow, the remaining ocean water will become isotopically heavier.

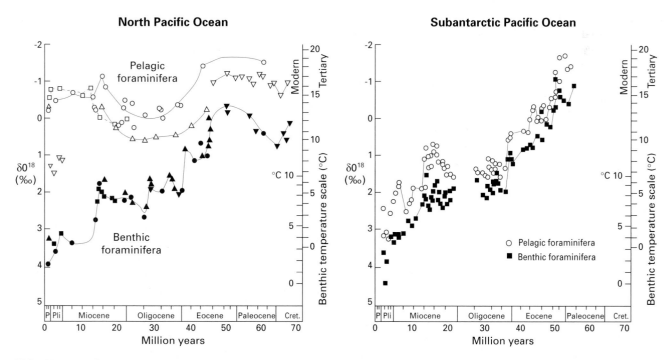

FIGURE 4-5. Cenozoic $\delta^{18}O$ trends for benthic and pelagic forminifera in low latitudes (left) and high latitudes (right). Modern and Tertiary benthic temperature scales are offset to compensate for the effect of modern ice sheets (Kennett, 1982, Figure 19-9).

filling the deep ocean basins at all latitudes, with the return upper flow from the poles toward the equator, almost a complete reversal of the present thermohaline circulation. With such a massive oceanic heat supply, even polar regions that had large solar energy deficits would have been as warm as the tropics of today (Kennett and Stott, 1990). Furthermore, because warm water holds less CO_2 in solution than cold water, the atmosphere of those times would have been CO_2-rich and warm. Isotopic evidence from marine sediment cores for this "greenhouse world," which persisted into the mid-Cenozoic, is supported by the terrestrial fossil record of palm trees and crocodilian reptiles in high polar latitudes, as described in following paragraphs.

A significant complication to oxygen-isotope thermometry is the growth of ice sheets. Water vapor $\delta^{18}O$ is usually about 10‰ lighter than the liquid from which it evaporated (Figure 4-4). Through multiple cycles of evaporation and precipitation, continental rainwater becomes isotopically lighter, with modern $\delta^{18}O$ for Great Lakes precipitation in the range of − 6 to −16‰, for example (Lawrence and White, 1991, Figures 1 and 2). The greatest fractionation occurs in the snow that falls over polar ice sheets, reaching extreme $\delta^{18}O$ values in the range of −55‰ over Antarctica and −35‰ over central Greenland. Although the total amount of the earth's water in ice sheets is small, the isotopic fractionation in those ice sheets is sufficiently extreme that the "ice volume" effect caused by the growth and shrinkage of Pleistocene ice sheets could be 30 to 70 percent of the total glacial-interglacial $\delta^{18}O$ fluctuation recorded in deep-sea cores (Bradley, 1985, pp. 179–180). For the late Cenozoic interval of time, global cooling and ice-sheet growth remain complementary but unresolvable factors in the marine $\delta^{18}O$ record. The sharp increase in $\delta^{18}O$ of marine fossils about 35 million years ago (Figure 4-5) could have been caused in part by the onset of continental glaciation in Antarctica.

THE ORBITAL PACEMAKER OF LATE CENOZOIC CLIMATE CHANGE

Assuming a certain output of solar energy, the amount and seasonal distribution of that energy entering the earth's atmosphere at any specified latitude depends on three variables: (1) the *eccentricity* of the earth's elliptical orbit, (2) the *obliquity* (tilt) of the earth's rotational axis relative to the plane of the orbit (currently 23° 27'), and (3) the *precession of the equinoxes,* an orbital phenomenon that is combined with eccentricity to produce a parameter known as **climatic precession**

(Figure 4-6). The orbital eccentricity varies cyclically from near zero (a perfectly circular orbit) to about 0.05, which results in a maximum variation of the incoming annual radiation of only 0.2 percent of the total over a cycle of about 95,000 years (Berger, 1988, p. 638). The obliquity of the earth's rotational axis varies between 22° and 24° 27', and controls the latitudinal distribution of incoming energy and the intensity of seasonality. It has a period of about 41,000 years. Climatic precession is a complex variable with principal periods of 23,000 and 19,000 years. The climate precessional parameter refers to the time of year when the earth is at *perihelion,* or closest to the sun, which is now in early January. Climatic precession controls the difference in length of the seasons and has an opposite effect on each hemisphere. At present, the northern hemisphere winter is relatively short and mild because the earth is near perihelion then and is moving rapidly along its orbit. By precession alone, in about one-half cycle or 11,000 years from now, the northern hemisphere winter season will be cooler, and more than 23 days longer than the summer season, because the northern hemisphere winter will occur while the earth moves more slowly along the part of its orbit most distant from the sun (Berger, 1978, p. 142).

Only eccentricity affects the total radiation received during a year, and the variation is small. Obliquity and precession act to redistribute that energy across latitudes and seasons. Obliquity has its largest effect in mid- to high latitudes, whereas precession, which is very important at all latitudes, has the dominant effect in low latitudes. The combined effect of all three variables is to change the flux of incident solar radiation by a maximum of 14 percent for any month.

Soon after the precession of the equinoxes was recognized in 1830, scientists began to speculate on the role of orbital fluctuations as a cause of ice ages (Berger, 1988). It remained for a Serbian engineer and astronomer, Milutin Milankovitch, to hand-calculate the variations in incoming solar radiation of various latitudes and seasons for the past 600,000 years. His first publication on the subject was in 1920, but the major compilation was published by the Royal Serbian Academy of Science in 1941 and translated into English only in 1969 (Berger, 1988, p. 630). Milankovitch, and others since, deduced that cool summers in the latitude range of 50° to 70°N were necessary to induce continental glaciation, arguing that winters over the large areas of northern hemisphere land are more than cold enough to accumulate snow, but cooler or shorter summers are necessary to prevent it from all melting each year. Further, if winters are relatively mild while summers are cool, it is likely that there will be more winter snowfall.

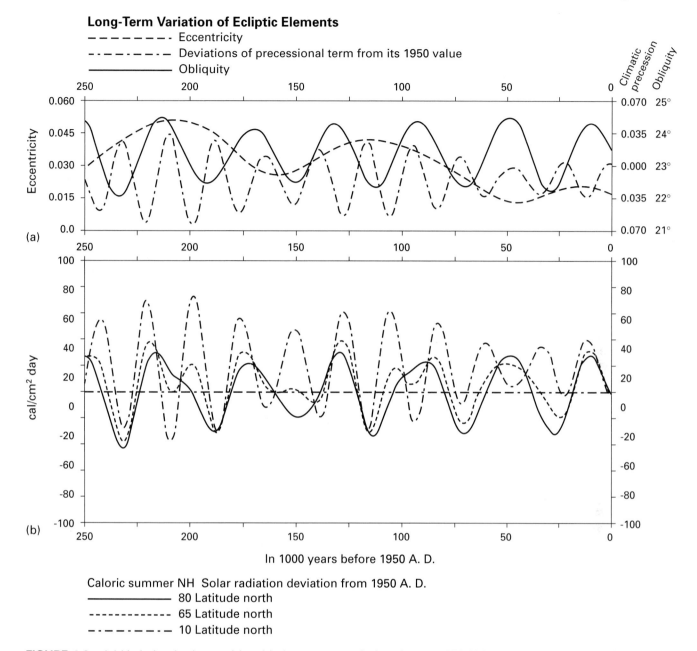

FIGURE 4-6. (a) Variation in the earth's orbital parameters during the past 250,000 years, and (b) the resulting deviations of incoming summer solar radiation flux (cal/cm²day) from the 1950 A.D. values at 80°N, 65°N, and 10°N (Berger, 1978, Figure 4a).

Variations of northern hemisphere solar radiation for 80°N, 65°N, and 10°N are shown in Figure 4-6b. With modern computer programs and better observations of the various parameters, solutions are now available for paleoclimate studies that are judged reliable for at least the past 3 million years (Berger, 1978, 1988; Berger and Loutre, 1992; Berger et al., 1992).

The triumph of the Milankovitch theory has been in its frequency rather than amplitude predictions

(Figure 4-6). As radiometric dating methods have improved, the timing of various climate-related landforms and sediments have been demonstrated to coincide with Milankovitch predictions. In particular, the late Cenozoic oxygen-isotope record in deep-sea carbonate sediments (Figure 4-7), clearly reproduces the complex spectra of Milankovitch curves.

Skeptics have argued that the amplitudes of the fluctuations of solar energy inferred by Milankovitch theory are inadequate to produce the documented climatic changes of ice ages. However, numerous positive feedback effects would enhance the relatively small changes of the radiation budget. For example, unusually persistent summer snow cover on land greatly increases the albedo and cools the land even more. Floating sea ice has the same effect on the high latitude sea surface. Less extreme cooling in lower latitude permits enough evaporation to feed the higher latitude snow precipitation. Even the earth's mantle may be involved by the slow isostatic recovery of deglaciated regions, draining shallow seas like the Baltic Sea and Hudson Bay to provide new land for snow accumulation. For these and other reasons, the Milankovitch cycles are regarded as "pacemakers" or *forcing functions* that, by their persistent regularity, force less systematic variables into harmonic regularity (Imbrie and Imbrie, 1979).

The Milankovitch orbital variables have presumably operated throughout geologic time. Only minor corrections are required to apply orbital frequencies to rhythmic layering in sedimentary rocks as old as 500 million years (Berger et al., 1992). But the Antarctic continental ice sheet is less than 35 million years old, and the multiple ice ages of the northern hemisphere go back in time only about 2.5 million years. Other conditions, such as the concentration of land areas in high northern latitudes by plate motions (Figure 3-1), had to be favorable for the orbital parameters to act as climatic pacemakers. Other tectonic, atmospheric, oceanographic, and biologic changes of the late Cenozoic Era, just reviewed, also created an environment in which the relatively weak orbital parameters could be expressed as major climate changes.

GEOMORPHOLOGY FROM THE GREENHOUSE TO THE ICEHOUSE

Tertiary Landscapes

Ancient climates are largely inferred from the fossil record of terrestrial plants and animals, which are primarily preserved in alluvium and lake beds. These are lowland and coastal deposits; they tell us little about the climate in adjacent highlands from which the sediments were being eroded. Therefore, the numerous reconstructions of early Cenozoic climates, which almost exclusively describe warm, moist conditions from pole to pole, describe lowland and coastal environments that are typically more equable than those found in highlands. Nevertheless, there is considerable agreement among the experts in a variety of fields that the earth's climate from late Mesozoic time until about

50 million years ago was uniform and mild, even in high latitudes with several months of 24-hour winter darkness (Figure 4-1). Inland and at higher altitudes there could have been colder, more continental climates (Sloan and Barron, 1990) with significant daily and annual temperature fluctuations as well as steeper precipitation gradients. However, vertebrate paleontologists maintain that the early Tertiary equability was not limited to coastal lowlands, but extended into continental interiors as well (Archibald, 1991; Wing, 1991; Wing and Greenwood, 1993). Eocene fossils from the northern Rocky Mountains at about latitude 45°N include nonburrowing tortoises, palms, tree ferns, and trees that lacked strong seasonal growth rings. The temperature probably stayed above freezing year-round, with only moderate seasonality (Sloan, 1994). Vegetation was probably broadleaf evergreen trees and shrubs.

With relatively warm temperatures from pole to pole, we must assume an early Tertiary landscape with little global diversity. Chemical weathering and soil development in Antarctica produced clay minerals that are typical of the humid tropics today (Robert and Kennett, 1994). Arctic rivers now frozen except for a brief but intense snowmelt runoff season would have had more sustained annual flow. Periglacial landforms (Chapter 14) were absent except possibly at high altitudes.

By early Oligocene time (35 Ma), the greenhouse warmth was weakening. Some authors refer to the time as a "crisis," whereas others insist the changes were transitional or fluctuating. The benthic $\delta^{18}O$ record shows sharp cooling of the deep oceans (Figure 4-5). The major climatic events were controlled by tectonism. By this time, Australia had moved equatorward from Antarctica far enough so that an unobstructed southern ocean current surrounded the polar continent, isolating it climatically from the other continents. Evidence of glaciers in West Antarctica can be traced back at least 27 million years. Lavas of that age are exposed above the ice plateau of Marie Byrd Land but were erupted under ice. Thus an ice cap probably covered the volcanic archipelago of West Antarctica as early as Late Oligocene time and perhaps earlier (p. 103). By about 15 million years ago, the East Antarctic Ice Sheet had begun to grow, and the modern thermohaline circulation was developing (Miller et al., 1991; Flower and Kennett, 1994). The spread of grasslands over the seasonally or perennially dry plains of the world at this time has been noted (p. 51).

A curiously provincial effect of the plate motions that carried Australia into lower latitudes was that much of the landscape of that island continent seems to have been fossilized. Australia has the lowest relief of all the continents, and it also now lies within the southern hemisphere zonal belt of aridity (Chapter 13). In many parts of the dry Australian interior, weathering profiles up to 100 m in thickness speak of former abundant precipitation and deep groundwater movement. One can almost visualize Australia as having been mummified or desiccated to preserve a mid-Tertiary landscape, overlain only lightly by younger dune fields or saline lake beds. This smallest, flattest, and driest continent is a veritable museum of ancient landscapes as well as of its unique Gondwana flora and marsupial fauna (p. 347).

In the latest part of the Miocene Epoch, the Mediterranean basin experienced a dramatic series of tectonically induced climate changes. Tectonic convergence of the African and Eurasian plates closed the Straits of Gibraltar and the eastward opening to the Red Sea. From the evidence of thick late-Miocene evaporite beds on the floor of the Mediterranean, it apparently dried up completely (Hsü, 1992), forming an arid basin more than 2 km below present sea level, into which surrounding rivers deeply incised their valleys. By the subsequent erosion of the Straits of Gibraltar, the Mediterranean basin catastrophically refilled with sea water. Former steep coastal river gorges became submarine canyons. Rivers with adequate sediment supply, such as the Rhône and the Po, built thick deltas into the restored sea. Beneath the Nile delta, for example, Pliocene marine sediments can be traced headward into the former entrenched valley all the way to southern Egypt at the site of the Aswan Dam, where a sill of resistant granite had limited the headward gorge erosion (Said, 1993, p. 39).

Sea-level changes caused by the late Miocene drying and subsequent refilling of the Mediterranean Sea are but one of many Tertiary marine events that helped shape the world's modern coastal landscapes. The early growth of the Antarctic ice sheet, perhaps in multiple episodes (Miller et al., 1991), plus changes in the thermal structure of the deep oceans, changes in spreading rates on midocean ridges, and other tectonic movements all contributed to a complex record of Tertiary sea-level oscillations, perhaps measured in hundreds of meters (Haq et al., 1988). Although the primary evidence used to infer Cenozoic sea-level fluctuations is stratigraphic, their geomorphic impact must have been significant (Chapter 19).

Quaternary Climatic Geomorphology

The latest part of the Cenozoic Era is the Quaternary Period, which as currently defined includes the Pleistocene and Holocene Epochs (back endpaper). The epoch boundaries are defined at 1.6 My and 10,000 years, respectively. Many Quaternary specialists would prefer an earlier time for the onset of "typical" Quaternary events such as the deposition of the Chinese loess (Chapter 13), northern hemisphere glaciation for about the past 2.7 million years, or the modernization of the western hemisphere mammalian fauna since the closure of the Isthmus of Panama about 3 million years ago (Marshall, 1988). The Greenland ice sheet may have formed 3 million years ago in part because of the emergence of the isthmus of Panama and the development of the Gulf Stream (Keigwein, 1982). Glaciation is documented in Iceland at 3.1 million years ago, but major horizons of ice-rafted sand in a deep-sea core south of Iceland and west of Ireland are abundant only during the last 2.4 million years.

The Pleistocene is conventionally subdivided into three parts: Early Pleistocene, extending from 1.6 My at the beginning of the epoch to the Brunhes/Matuyama paleomagnetic boundary at 788,000 years ago, Middle Pleistocene from 788,000 years ago to the boundary between deep-sea oxygen-isotope stages 5 and 6 at 132,000 years ago (Figure 4-7), and Late Pleistocene from 132,000 years ago to the Holocene boundary (Richmond and Fullerton, 1986). Major Early and Middle Pleistocene events of geomorphic significance are noted here. The better-known Late Pleistocene and Holocene are described in much more detail in Chapter 18. It is understandable that the world's landscapes will primarily record the most recently active processes, but the intensity as well as the frequency of Quaternary climatic changes has been exceptional in earth history.

A climatic chronology is provided by a long core of deep-Atlantic sediments rich in microfossils and well calibrated by paleomagnetic reversals (Figure 4-7). The most striking feature of this deep-sea record is that at the end of a long Tertiary cooling trend, for almost the entire last 2.7 million years global ice volume has been greater and deep-ocean temperature colder than at present. Only four or five brief interglacials have been warmer than the Holocene interglacial; during most of the older interglacials, glacier volume and deep-sea temperatures were more comparable to the relatively cold interstadial intervals that punctuated the latest ice age (Raymo, 1992, p. 212; Raymo, 1994). From about 2.7 to 1.2 million years ago, the glacial-interglacial fluctuations were less extreme than later, and the northern hemisphere continents supported only small ice sheets. Vegetation in northern midlatitudes probably fluctuated between hardwood forests and cold, dusty, grassy plains and tundra. During the past 1.2 million years, the amplitude of the ice volume-deep sea temperature curve

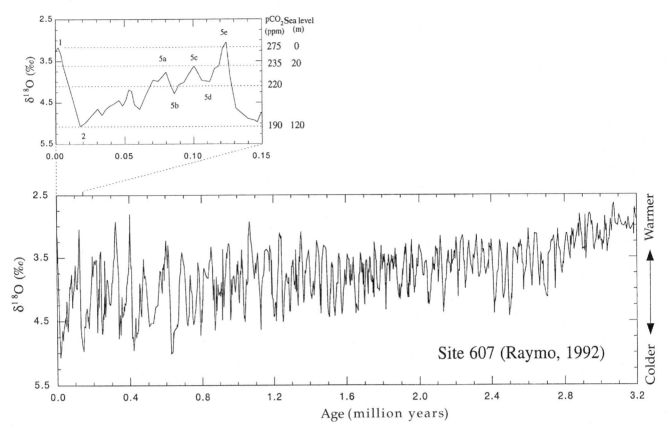

FIGURE 4-7. Benthic δ¹⁸O record from DSDP site 607 in the North Atlantic, calibrated by the paleomagnetic time scale. This long core shows almost 3.2 million years of δ¹⁸O changes that resulted from a combination of deep-water temperatures and global glacier ice volumes. The enlarged inset of the late Quaternary glacial-interglacial cycle of the past 150,000 years is described in Chapter 18. The four dashed reference lines represent the Holocene (top); oxygen-isotope state 5c, an early interstadial near the beginning of the latest ice age; the midpoint of the most recent transition from glacial to interglacial conditions (middle); and full-glacial conditions (bottom) (Raymo, 1992, Figure 2).

has increased, the 90,000 year Milankovitch frequency has become dominant, and glacial-interglacial fluctuations have been comparable to the latest cycle (Chapter 18).

Up to 30 percent of the earth's land area has been ice-covered repeatedly in at least the past 1.2 million years, three times the area currently affected. Most of the formerly glaciated areas are in eastern North America and northwestern Europe, on the borders of the northern Atlantic Ocean (Figure 16-1). Almost all landforms within the area of former glaciation are relict from glacial processes, with only relatively minor river incision or fluvial and eolian deposition (Chapters 17 and 18). Glaciated shield areas were stripped clean of their former warm-climate regolith, leaving the bare, polished rock ledges and deranged drainage patterns that we associate with the Canadian and Baltic Shields of today.

The chronology of the Early and Middle Pleistocene glaciations of North America and Europe is not

well established although progress has been made by fission-track and potassium-argon dating on volcanic tephras that are interstratified with drift (p. 112). In particular, several fine-grained tephras are widespread on the Great Plains of the United States from South Dakota to Texas. Formerly thought to be a single ash bed of Middle Pleistocene age known as the "Pearlette Ash" (Wilcox and Naeser, 1992), the tephras at numerous localities give ages that cluster in three groups. The oldest dates of about 2.1 million years are from tephra that is presumably younger than the oldest glaciation in central North America. Other ash beds with ages of 1.3 million years and 0.6 million years probably bracket at least one other glaciation. All three groups of ashes predate the very extensive Illinoian glaciation, the third of the traditional but now obsolete four "ice ages" of the North American Pleistocene chronology. The radiometric dates demonstrate that substantial regions in central North America were glaciated well before the Pleistocene Epoch began, which is about 1.6 million years ago as currently defined. Figure 4-7 demonstrates the probability for dozens of northern hemisphere glaciations of various intensities during the last several million years.

Successive Early and Middle Pleistocene ice-sheet advances from the Hudson Bay lowland moved southwestward up the gentle slope of the Great Plains, damming the ancestral rivers and forcing them to drain southeastward along the ice margin until they joined the Mississippi (Flint, 1971, pp. 232–235; Bluemle, 1972). Modern trenchlike valley segments of the upper Missouri River trend almost perpendicular to the regional slope and the ancestral, drift-filled valleys (Figure 4-8). Most of the diversions are correlated with the Illinoian glaciation, but some are younger. Almost all the tributaries of the upper Missouri River enter from its right side and are segments of a former Hudson Bay drainage network. The reservoirs of the large dams that have been built along the Missouri River (Oahe Dam, South Dakota; Garrison Dam, North Dakota; Fort Peck Dam, Montana) reproduce the ice-dammed lakes that occupied these valleys during late Quaternary glaciations. The Ohio River had a comparable but older origin. In preglacial time, the western Appalachian drainage probably flowed northwest, but was directed more nearly westward by one of the early ice sheets (Teller, 1973; Swadley, 1980). A buried valley comparable in size to the present Ohio River valley has been traced westward across Ohio, Indiana, and Illinois. The valley of the ancestral river, named the *Teays Valley*, has been nearly buried by sediments from several subsequent glaciations (Figure 4-8). At least two glaciolacustrine deposits in buried valleys of the Teays system have reversed magnetic polarity and are interpreted as having been deposited because of ice damming in Middle Pleistocene time, prior to 790,000 years ago. The western part of the Teays system, known as the Mahomet bedrock valley in Illinois, may be somewhat younger but contains the record of several pre-Illinoian glaciations (Melhorn and Kempton, 1991). The early ice sheets that buried the Teays drainage system may have spread southward across Michigan prior to the erosion of the modern Great Lakes basins (Johnson, 1985), which have been progressively deepened by succeeding glaciations.

Beyond the latitudinal and altitudinal limits of direct glacial action, the indirect effects of the Pleistocene icehouse also shaped the landscape. Periglacial landscapes, shaped primarily by processes acting above perennially frozen rock and soil (Chapter 14), smoothed uplands and filled gully systems with frost-shattered rubble in many midlatitude regions. Large areas of ice-marginal and other barren regions were mantled with thick accretions of wind-blown silt, or *loess*. Much of the modern cereal-grain-producing regions of the world owe their soil fertility to this relict eolian mantle, within which are multiple weathered layers and fossil soils. It has incrementally accumulated over the last several million years (Chapter 13).

The abstraction of water from the ocean to build the great ice sheets repeatedly lowered sea level to more than 100 m below its present level, only to rise again as the ice sheets melted. On a time scale of only 10^4 or 10^5 years, coastal river systems entrenched and aggraded, adjusting to the frequent and extreme changes in their base level as well as to their fluctuating discharges and sediment loads (Figure 4-9). Karstic solution in soluble terranes near coasts fluctuated as well, so that formerly air-filled caves with dripstone deposits are now found at depths of as much as 122 m below sea level (p. 154).

No-Analog Climates

One of the most difficult concepts of climate change is the "no-analog" climate. Imagine the scene on the shore of the Arctic Ocean, between 2 and 3 million years ago. On northern Greenland and on Meighen Island in the Canadian arctic, both barely 10 degrees from the North Pole, diverse conifer forests grew, richer in species (some now extinct) than the forest now found at the Arctic Circle, almost 2000 km farther south (Reppening and Browers, 1992; Funder et al., 1985). At 70°N on the Alaskan arctic coast, there was no shore ice at any season, no permafrost, and an open spruce-birch woodland, indicative of a climate warmer than at Anchorage, Alaska, today (Nelson and

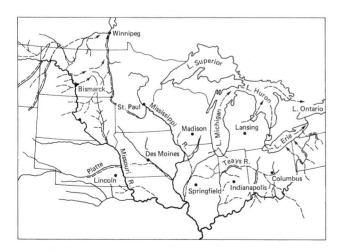

FIGURE 4-8. Preglacial drainage networks in the upper Mississippi drainage basin. Present rivers are shown in bold lines.

FIGURE 4-9. Interpretation of a seismic profile parallel to the shoreline of Pamlico Sound, northeastern North Carolina. Stratigraphic units are labeled E (early), M (middle), and L (late) Quaternary, based on fossils and various dating methods. Eighteen depositional sequences are interpreted; all boundaries contain fluvial paleochannels representing cut-and-fill events related to fluctuating Quaternary sea levels (Riggs et al., 1992, Figure 10).

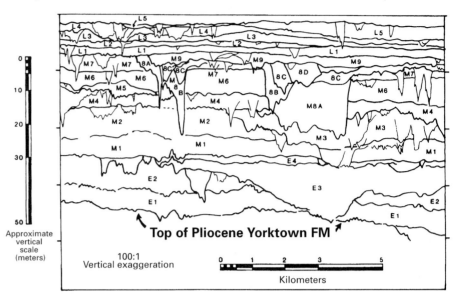

Carter, 1985). How could trees survive the total darkness of several months each year? What did the regional landscape look like? Did the coastal rivers flow year-round? Were there glaciers on the mountains farther inland?

Alternatively, imagine the climate at the edge of the ice sheet at 39°N latitude in central Illinois about 18,000–20,000 years ago. Abundant evidence (Chapter 14) demonstrates that the ground was perennially frozen, which requires an annual average temperature of –5°C or lower. The pollen record of the vegetation at the time confirms a cold, windy, dusty tundra with a few trees along river valleys and an upland ground cover of woody shrubs and herbs. But what kind of a tundra was it, with the sun high overhead for many hours of the day, even in winter? Today, most of the tundra zone is north of the Arctic Circle, and life there is adapted to winter months of near darkness followed by a summer of nearly continuous daylight. What adaptations did the tundra plants and animals make when they were forced more than halfway to the equator from their current habitats? How could a below-freezing average annual temperature be maintained with substantial summer solar radiation and without the long winter night? Clearly, the annual cycle of shallow freeze and thaw and the periglacial processes that accompany it, which gives the present tundra landscape its distinctive appearance, does not work in the same way that it did at a lower latitude.

These are but two examples of the difficulty we have in transposing climatic zones north and south, across a range of latitudes. Regardless of how ocean and air temperature and precipitation may change as climates shift, the length of day and night, seasonal energy flow, and sun angle are fixed within very narrow ranges by orbital parameters. We simply cannot find a modern analog for some former climates because none exists. To the extent that geomorphic processes are determined to a very significant degree by climatic conditions, it follows that we cannot use presently observed processes to explain landform evolution in a former climate at another latitude. We must learn how to extend our observations and hypotheses to some very different environments. This will be a repeated theme in several subsequent chapters.

REFERENCES

ARCHIBALD, J. D., 1991, Comment on: "Equable" climates during earth history?: Geology, v. 19, pp. 539.

BERGER, A. L., 1978, Long-term variations of caloric insolation resulting from the earth's orbital elements: Quaternary Res., v. 9, pp. 139–167.

———, 1988, Milankovitch theory and climate: Rev. Geophysics, v. 26, pp. 624–657.

———, and LOUTRE, M. F., 1992, Astronomical solutions for paleoclimate studies over the last 3 million years: Earth, Planet. Sci. Ltrs., v. 111, pp. 369–382.

———, LOUTRE, M. F., and LASKAR, J., 1992, Stability of the astronomical frequencies over the earth's history for paleoclimate studies: Science, v. 255, pp. 560–566.

BERNER, R. A., 1994, GEOCARB II: A revised model of atmospheric CO_2 over Phanerozoic time: Am. Jour. Sci., v. 294, pp. 56–91.

BLUEMLE, J. P., 1972, Pleistocene drainage development in North Dakota: Geol. Soc. America Bull., v. 83, pp. 2189–2194.

BRADLEY, R. S., 1985, Quaternary paleoclimatology: Methods of paleoclimatic reconstruction: Chapman and Hall, London, 472 pp.

BROECKER, W. S., 1991, The great ocean conveyor: Oceanography, v. 4, pp. 79–89.

BURTON, K. W., LING, H.-F., and O'NIONS, R. K., 1997, Closure of the Central American isthmus and its effect on deep-water formation in the North Atlantic: Nature, v. 386, pp. 382–385.

CALDEIRA, K., 1992, Enhanced Cenozoic chemical weathering and the subduction of pelagic carbonate: Nature, v. 357, pp. 578–581.

CERLING, T. E., 1991, Carbon dioxide in the atmosphere: Evidence from Cenozoic and Mesozoic paleosols: Am. Jour. Sci., v. 291, pp. 377–400.

DUGAS, D. P., and RETALLACK, G. J., 1993, Middle Miocene fossil grasses from Fort Ternan, Kenya: Jour. Paleontol., v. 67, pp. 113–128.

FLINT, R. F., 1971, Glacial and Quaternary geology: John Wiley & Sons, Inc., New York, 892 pp.

FLOWER, B. P., and KENNETT, J. P., 1994, The middle Miocene climate transition: East Antarctic ice sheet development, deep ocean circulation and global carbon cycling: Palaeogeog., Palaeoclimatol., Palaeoecol., v. 108, pp. 537–555.

FUNDER, S., ABRAHAMSEN, N., and 2 others, 1985, Forested arctic: Evidence from north Greenland: Geology, v. 13, pp. 542–546.

GORDON, A. L., 1986, Interocean exchange of thermocline water: Jour. Geophys. Res., v. 91, pp. 5037–5046.

HAQ, B. U., HARDENBOL, J., and VAIL, P. R., 1988, Mesozoic and Cenozoic chronostratigraphy and cycles of sea-level change, in Wilgus, C. K., Hastings, B. S., and 4 others, eds., Sea-level changes: An integrated approach: SEPM (Soc. Econ. Paleontol. Mineral.) Spec. Pub. No. 42, pp. 71–108.

HSÜ, K. J., 1992, Challenger at sea: A ship that revolutionized earth science: Princeton Univ. Press, Princeton, N.J., 417 pp.

IMBRIE, J., and IMBRIE, K. P., 1979, Ice ages: Solving the mystery: Harvard Univ. Press, Cambridge, Mass., 224 pp.

JOHNSON, W. H., 1985, Stratigraphy and correlation of the glacial deposits of the Lake Michigan lobe prior to 14 ka BP: Quaternary Sci. Rev., v. 5, pp. 17–22.

KEIGWEIN, L., 1982, Isotopic paleoceanography of the Caribbean and east Pacific: Role of Panama uplift in late Neogene time: Science, v. 217, pp. 350–353.

KENNETT, J. P., 1982, Marine geology: Prentice-Hall, Inc., Englewood Cliffs, N.J., 813 pp.

———, and STOTT, L. D., 1990, Proteus and Proto-Oceanus: Ancestral Paleogene oceans as revealed from Antarctic stable isotope results, ODP leg 113: Proc., Ocean Drilling Programs, Sci. Results, v. 113, pp. 865–880.

KNOLL, M. A., and JAMES W. C., 1987, Effect of the advent and diversification of vascular land plants on mineral weathering through geologic time: Geology, v. 15, pp. 1099–1102.

LAWRENCE, J. R., and WHITE, J. W. C., 1991, The elusive climate signal in the isotopic composition of precipitation, in Taylor H. P., Jr., O'Neil, J. R., and Kaplan, I. R., eds., Stable isotope geochemistry: A tribute to Samuel Epstein: Geochem. Soc. Spec. Pub. No. 3, pp. 169–185.

MARSHALL, L. G., 1988, Land mammals and the great American interchange: American Scientist: v. 76, pp. 380–388.

MELHORN, W. N., and KEMPTON, J. P., eds., 1991, Geology and hydrogeology of the Teays-Mahomet bedrock valley systems: Geol. Soc. America Spec. Paper 258, 128 pp.

MILLER, K. G., WRIGHT, J. D., and FAIRBANKS, R. G., 1991, Unlocking the ice house: Oligocene-Miocene oxygen isotopes, eustasy, and margin erosion: Jour. Geophys. Res., v. 96, no. B4, pp. 6829–6848.

MORGAN, M. E., KINGSTON, J. D., and MARINO, B. D., 1994, Carbon isotope evidence for the emergence of C4 plants in the Neogene from Pakistan and Kenya: Nature, v. 367, pp. 162–165.

NELSON, R. E., and CARTER, L. D., 1985, Pollen analysis of a late Pliocene and early Pleistocene section from the Gubik Formation of arctic Alaska; Quaternary Res., v. 24, pp. 295–306.

OPDYKE, B. N., and WILKINSON, B. H., 1993, Carbonate mineral saturation state and cratonic limestone accumulation: Am. Jour. Sci., v. 293, pp. 217–234.

RAYMO, M. E., 1992, Global climate change: A three million year perspective, in Kukla, G. J., and Went, E., eds., Start of a glacial: Springer-Verlag Berlin Heidelberg, 353 pp.

———, 1994, The initiation of northern hemisphere glaciation: Ann. Rev. Earth, Planet. Sci., v. 22, pp. 353–383.

REPPENING, C. A., and BROWERS, E. M., 1992, Late Pliocene-early Pleistocene ecologic changes in the Arctic Ocean borderland: U.S. Geol. Surv. Bull. 2036, 37 pp.

RETALLACK, G. J., 1982, Paleopedological perspectives on the development of grasslands during the Tertiary: Third North American Paleontological Convention, Proc., v. 2, pp. 417–421.

———, DUGAS, D. P., and BESTLAND, E. A., 1990, Fossil soils and grasses of a middle Miocene east African grassland: Science, v. 247, pp. 1325–1328.

RICHMOND, G. M., and FULLERTON, D. S., 1986, Summation of Quaternary glaciations in the United States of America: Quaternary Sci. Rev., v. 5, pp. 183–196.

RIGGS, S. R., YORK, L. L., and 2 others, 1992, Depositional patterns resulting from high-frequency Quaternary sea-level fluctuations in northeastern North Carolina, in Fletcher, C. H., III, and Wehmiller, J. F., eds., Quaternary coasts of the United States: Marine and lacustrine systems: SEPM (Soc. for Sedimentary Geol.) Spec. Pub. No. 48, pp. 141–153.

ROBERT, C., and KENNETT, J. P., 1994, Antarctic subtropical humid episode at the Paleocene-Eocene boundary: Clay-mineral evidence: Geology, v. 22, pp. 211–214.

SAID, R., 1993, The River Nile: Geology, hydrology and utilization: Pergamon Press, Inc., Tarrytown, N.Y., 320 pp.

SCHMITZ, W. J., JR., 1995, On the interbasin-scale thermohaline circulation: Rev. Geophysics, v. 33, pp. 151–173.

SCHWARTZMAN, D. W., and VOLK, T., 1991, Biotic enhancement of weathering and surface temperatures on earth since the origin of life: Palaeogeog., Palaeoclimatol., Palaeoecol., v. 90, pp. 357–371.

SLOAN, L. C., 1994, Equable climates during the early Eocene: Significance of regional paleogeography for North American climate: Geology, v. 22, pp. 881–884.

———, and BARRON, E. J., 1990, "Equable" climates during earth history?: Geology, v. 18, pp. 489–492 [and 1991, Reply to comments: Geology, v. 19, pp. 540–542].

SWADLEY, W C, 1980, New evidence supporting Nebraskan age for origin of Ohio River in north-central Kentucky: U.S. Geol. Survey Prof. Paper 1126-H, 7 pp.

TELLER, J. T., 1973, Preglacial (Teays) and early glacial drainage in the Cincinnati area, Ohio, Kentucky, and Indiana: Geol. Soc. America Bull., v. 84, pp. 3677–3688.

VAN DER BURGH, J., VISSCHER, H. and 2 others, 1993, Paleoatmospheric signatures in Neogene fossil leaves: Science, v. 260, pp. 1788–1790.

VOLK, T., 1989, Rise of angiosperms as a factor in long-term climatic cooling: Geology, v. 17, pp. 107–110.

WEIDICK, A., 1975, A review of Quaternary investigations in Greenland: Gronlands Geologiske Undersogelse (Geological Survey of Greenland) Miscel. Papers, no. 145, reprinted as Ohio State Univ., Instit. Polar Studies Rept. no 55, 161 pp.

WILCOX, R. E., and NAESER C. W., 1992, The Pearlette family ash beds in the Great Plains: Finding their identities and their roots in the Yellowstone country: Quaternary Internat., v. 13/14, pp. 9–13.

WING, S. L., 1991, Comment on: "Equable" climates during earth history?: Geology, v. 19, pp. 539–540.

———, and GREENWOOD, D. R., 1993, Fossils and fossil climate: The case for equable continental interiors in the Eocene: Philos. Trans., Royal Soc. London, Ser. B, v. 391, pp. 243–252.

PART III

Constructional Processes and Constructional Landforms

*C*ontinents include rocks with ages as great as almost 4 billion years because continental rocks tend to resist subduction and accumulate as thickened crustal slabs. But except for exhumed and other relict landscapes, the continental landscapes we now see are relatively young, mostly the products of processes that have been at work during late Cenozoic time.

Chapter 3 reviewed evidence that large, rigid lithospheric plates move over the surface of the earth at relative velocities of 10 cm/yr or more. Vertical movements measured in millimeters per year, especially at plate margins, were also demonstrated. In the face of such evidence for a mobile lithosphere, geomorphologists must constantly reevaluate the traditional concept that rock masses are passive structures from which landscapes are carved by erosion. The next two chapters establish the conceptual framework of landscape evolution during a time of great tectonic and volcanic activity.

Constructional landforms are defined here as all those that have been or are being built (that is, increasing in mass, height, or area), as compared to those that are primarily destructional, or erosional (that is, being reduced in size). Chapters 5 and 6 describe tectonic and volcanic constructional processes and landforms. Constructional landforms that grow by accretion of the waste products of erosion, such as alluvial fans and deltas, or by biologic processes, such as coral reefs, are deferred consideration until later chapters.

Tectonic and volcanic relief results from dynamic stresses and density changes, all of which are ultimately caused by thermal gradients within the earth. Nothing is ever really "constructed" on any planet because the total planetary mass remains very nearly constant. All constructional landforms are built by reprocessing or redistributing some other mass.

Chapter 5

Tectonic Landforms

Landforms that result from crustal movements are called **tectonic.** Some geologists have used "structural" in the same sense, but as noted in Chapter 1, "structure" in the geomorphic sense has a much broader connotation. *Structurally controlled* landforms that are shaped by differential erosion in response to rock structure are analyzed in Chapter 12.

It could be argued that no subaerial relief can occur until isostatic or tectonic uplift has raised land above sea level, and that therefore all subaerial landscapes are "tectonic" unless they are constructed by depositional (volcanic or sedimentational) processes. However, it is convenient to restrict the term to those landforms that are sufficiently undissected by erosion so that their shape defines the fractured or deformed surface. All degrees of transition are found between purely tectonic and totally erosional landscapes. The distinction is nevertheless one of the important topics of geomorphology because if a landform is classed as tectonic, we assume that the deformation has been either recent, or rapid, or both, compared to the erosion rate. The geomorphic recognition of tectonic landforms is of major value in working out the geologic history of a region (Keller and Pinter, 1996). Most tectonic landforms are the result of Quaternary, or at the oldest, late Cenozoic, land movements. In most regions of tectonic landforms, movement continues today.

As noted in Chapter 3, tectonic landforms abound in orogenic belts but are rarely discernible in epeirogenic regions (but see Figure 3-7). They occur in well-defined assemblages, appropriate to the style of tectonic deformation in their geomorphic province. Their scale ranges from global or continental, such as midocean ridges, transform faults, and rift-valley systems, to barely measurable offsets in stream channels, coastal terraces, or rock surfaces.

Whereas Chapter 3 reviews the regional and temporal contexts of tectonic landscapes, this chapter systematically explains individual tectonic landforms or groups of closely related forms. The origin and history of an orogenic belt can only be interpreted by careful analysis and integration of the assemblage of smaller landforms and structures.

TECTONIC SCARPS

A **scarp** (shortened form of "escarpment"; any steep, abrupt slope or cliff) may be formed by a variety of geologic causes, either tectonic, erosional, or depositional. It is well to retain a few purely descriptive topographic terms such as "scarp" and "bench" in the totally nongenetic sense and to add modifying adjectives to them as the need arises. Thus a **tectonic scarp** is any steep slope that results from differential movement of the earth's surface.

Fault Scarps

If a fault displaces the surface of the ground so that one side is higher or lower than the other, a **fault scarp** results. Very recent, high-angle faults may preserve the actual surface of faulting on the scarp complete with slickensides or fault breccia (Figure 5-1). Such fresh details of fault movement are rare, however. More commonly, fault scarps are weathered slopes, eroded or covered with sediment, rock, or soil. The original fault surface quickly slumps or collapses to a slope of gravitational stability and gives no clue to the attitude of the fault surface at depth (Figure 5-2). The name *fault scarp* is appropriate, even when subsequent erosion on the face of the raised block causes the scarp to retreat laterally some distance from the surface of fault movement, and when the scarp assumes a slope quite different from the dip angle, or becomes gullied and dissected. Some fault scarps include segments with varying degrees of degradation, if movement has been intermittent. The typical angle of repose (p. 169) for the surface sediments on scarps of all origins is between about 25° and 40°. Even historical fault scarps in humid regions quickly become grassy or brush-covered slopes (Figure 5-3).

The steeper the dip of the fault surface, the more linear the resulting fault scarp will be across the preexisting landscape. In the case of a low-angle or nearly horizontal thrust fault, even without erosional dissection, the scarp will follow the topography; if the landscape is irregular, the scarp will be as sinuous or crenulate as a contour line.

In seismic regions, many small scarps or "scarplets" are the direct result of displacement during earthquakes (Philip and Meghraoui, 1983; Baljinnyam et al., 1993). They may be actual fault scarps or secondary effects of differential compaction and slumping. Open fissures form during some earthquakes, probably due to consolidation of porous surface materials (Figure 5-2). Soil creep and erosion are likely to obliterate such features quickly. At the foot of recent fault scarps, closed basins or *sag ponds* may develop either by oblique movement or by ground settling. The rate at which soil creep and other mass-wasting processes modify fresh scarps can be modeled, and some success has been achieved at estimating the age of prehistoric fault scarps (Figure 9-20).

A variety of other small landforms are commonly observed along recent fault scarps. If the stress field on the fault has a compressional vector, *pressure ridges* (low elongate mounds) may parallel the surface scarp. Secondary scarps facing the main scarp across a shallow trench have been called *antithetic scarps, earthquake rents, reverse scarplets,* or *cicatrices* (Cotton, 1948, 1950b). Their significance is not clear, but they may record local or shallow buckling of one of the fault blocks adjacent to the major fault surface.

Many normal faults are *listric,* concave-up toward the downthrow block and flattening at depth. Their steep, near-planar topographic scarps are thus not a good indication of the relative movement of the opposite sides of the fault. For example, the *Pali,* great seaward-facing scarps on the south flank of Kilauea volcano in Hawaii (Figure 5-4), are the headscarps of great slumps (Figure 9-10) that pass into a low-angle fault at a depth of 7 to 9 km on which horizontal motion of as much as 10 cm/yr is in progress (Figure 5-5). The seaward movement of the volcanic flank may be the result of gravitational spreading under the weight of surface lava flows or dense magma injected within the volcano (p. 187).

A fault scarp may die out laterally, break into a series of *en echelon* splinters, or become a monoclinal scarp (Figure 5-6). A major fault or narrow fault zone at depth may break the land surface as a series of closely spaced scarps, none of which has any conspicuous continuity at the surface. If structural trends or preexisting fracture systems are prominent in surface rocks, individual segments of a fault scarp may follow these trends and only approximate the true fault orientation at depth. As a result, a variety of minor *en echelon* scarps and shallow depressions may mark a surface zone above the buried fault.

Fault scarps, like all high, steep landforms, are prone to rapid erosional dissection. Streams that formerly flowed across the faulted landscape either will be rejuvenated if they flow from the upthrown to the downthrown block, or they will be impounded against the scarp if they flow in the opposite direction. Since most dissected scarps were caused by late Cenozoic uplift, the first case is much more common. Initially, then, streams cascade down a fault scarp but quickly cut steep gorges that erode progressively headward from the scarp into the uplifted block (Figures 5-7 and 5-8).

The resulting valleys are narrow at their lower ends and widen upstream into the upthrown block (Figure 5-8). These distinctively shaped scarp valleys have been called *bottle-necked,* or *wineglass, valleys* (von Engeln, 1942, p. 377; Cotton, 1948, p. 87). They are not restricted to fault scarps but form on any previously undissected steep slope or scarp.

Dissection by many gullies or hillslope erosion causes a scarp to be segmented into a series of **triangular facets,** planar surfaces at an appropriate angle of repose for the weathered rock mass of the scarp, with their bases aligned on or parallel to the fault trace (Figure 5-9). Once they have assumed a stable angle of

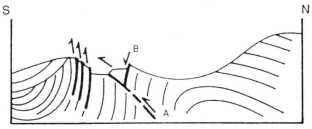

FIGURE 5-1. View west along the scarp of a near-vertical fault that resulted from the magnitude 7.3 El Asnam, Algeria, earthquake of October 10, 1980. The fault scarp is on a bedding plane of folded Pliocene strata (B in line rendering) and is offset in a direction opposite to the main earthquake-producing thrust fault (A in line rendering) (Philip and Meghraoui, 1983; photo courtesy M. Meghraoui).

FIGURE 5-2. Central segment of the Edgecumbe Fault, North Island, New Zealand (Beanland et al., 1989). Total scarp height is 3.3 m, of which 1.7 to 1.9 m of vertical motion occurred during an earthquake on March 2, 1987 (Photo: Quentin Christie, New Zealand Institute of Geological & Nuclear Sciences).

FIGURE 5-3.
View south along the Wairarapa Fault near Featherston, New Zealand. Parts of this scarp increased 2 m in height during the 1855 Wellington earthquake.

repose, these facets may persist for hundreds of thousands of years in dry climates (Wallace, 1978). If misinterpreted as remnants of the original fault surface, the facets imply that the fault was a low-angle normal fault because the angle of repose is rarely greater than about 25° to 35°. By contrast, low scarplets of recent movement at the base of the same facets may show by their straightness in map view to be the trace of a high-angle fault. Thus, two episodes of faulting, on two faults that are unrelated but coincide at the ground surface, are incorrectly postulated.

In the advanced stage of erosional dissection, a fault scarp may look no different from the face of any other dissected hill or mountain. If the dissecting streams discharge their alluvium into a desert basin, the aggradation may extend headward across the fault trace into embayments in the uplifted block. If a river flows along the foot of the scarp, the alluvium is more likely to be carried away and the tectonic relief preserved, but stream erosion may undercut the fault scarp and make its recognition difficult.

Thrust-Fault Scarps

Thrust faulting is now recognized as a major process of upper crustal shortening during orogeny. Thrusts may originate deep within the orogen, but ramp

FIGURE 5-4. View southwest along a coastal scarp of the Hilina fault system, south flank of Kilauea Volcano near Puu Kapukapu, Hawaii. The main Hilina Pali (scarp) forms the skyline (Photo: R. I. Tilling, U.S. Geological Survey).

upward through sedimentary cover strata and emerge at the land surface. Even while the thrust plate is advancing, an erosional topography develops on its upper surface, providing sediments that accumulate as breccia or conglomerate in front of the thrust and are then overridden by it (Vandervoort and Schmitt, 1990; Burbank and Vergés, 1994) (Figure 5-10). One of the best documented examples, the Muddy Mountain thrust in the southern Basin and Range Province, is part of the Sevier orogenic belt of late Cretaceous age that extends from southern Utah and Nevada northward to the Canadian Rocky Mountains. Palaeozoic carbonate and other rocks were thrust at least 40 km and possibly as much as 88 km horizontally eastward over a landscape eroded into the Jurassic Aztec Sandstone. The upper thrust plate was at least 4 to 5 km in thickness (Brock and Engelder, 1977). The prior subaerial landscape is preserved by deposits below the thrust, where hilltops of sandstone truncated by the thrust are separated by former valleys filled with fluvial sediments, some derived from the overriding plate. The overriding thrust was supported by the hilltops of resistant sandstone and minimally deformed the intervening fluvial sediments. Although the Muddy Mountain thrust is only a structural horizon today, it must have formed a spectacular tectonic landscape in late Cretaceous time, with a high mountain-

front tectonic scarp advancing rapidly eastward over a fluvial landscape.

Similar modern thrust-fault scarps can be seen in many regions of active tectonic compression. The south face of the San Gabriel Mountains is a thrust-fault scarp that dominates the north side of the Los Angeles basin in California. Recent earthquakes have resulted in reverse faults with a throw of 1 m or more, thrust to the south along the base of the range. A very linear scarp 1500 m high forms the west face of a thrust fault at the base of Sierra Ancasti, near Catamarca, Argentina (Color Plate 4). Here, the steep face is controlled by ancient joints in the Precambrian basement rocks of the thrust block. In broad outline, the entire south side of the Himalaya Range is the scarp of a south-facing thrust fault above the underthrusting Indian subcontinent, advancing at a rate of about 1 cm/yr (Meigs et al., 1995). These large, complex, highly dissected scarps are not sufficiently coherent to be seen as a landform except from an orbiting space craft.

Fault-Line (Erosional) Scarps

A scarp produced by structurally controlled erosion at an ancient fault is known as a **fault-line scarp** and is one of many kinds of structurally controlled ero-

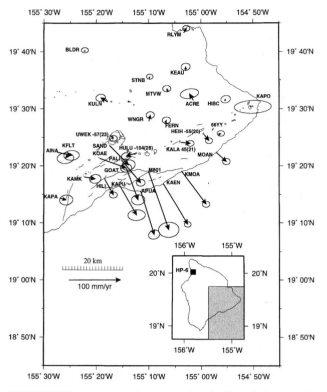

FIGURE 5-5. Average horizontal velocity of the southeastern part of the island of Hawaii relative to station HP-6 on northwestern Hawaii, determined by repeated GPS surveys (compare Figure 3-6) (Owen et al., 1995, Figure 2).

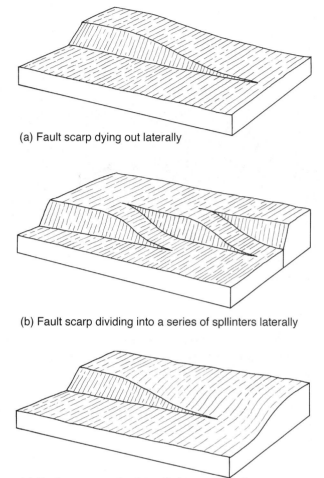

(a) Fault scarp dying out laterally

(b) Fault scarp dividing into a series of spllinters laterally

(c) Fault scarp passing laterally into a monocline

FIGURE 5-6. Varieties of lateral termination of fault scarps.

sional landforms (Chapter 12). The distinction between a fault scarp and fault-line scarp is not based on the observable morphology of a scarp but on full understanding of the regional geology. Two genetic classes of fault-line scarps are distinguished (Figure 5-11). **Obsequent** ("opposite to the consequent") scarps are unambiguous because the geologic relationships prove that movement on the fault would have produced a scarp facing in the opposite direction. **Resequent** ("renewed consequent") fault-line scarps are difficult to distinguish from tectonic fault scarps because, in both, the upthrown fault block is also the high side of the scarp. Geomorphic criteria alone cannot distinguish between the two.

The distinction between a fault scarp, of tectonic origin, and a fault-line scarp, formed by structurally controlled erosion, is not just a semantic matter. It is "the indispensable prerequisite to a correct interpretation of geological history" (Johnson, 1929, p. 356). The U.S. Nuclear Regulatory Commission stipulates that no nuclear powerplant site should by located in the vicinity of a "capable" fault. A "capable" fault is one that shows movement at or near the ground surface at

least once within the past 35,000 years or movement of a recurring nature within the past 500,000 years. Obviously, the correct interpretation of scarps in tectonic regions now has economic and political as well as theoretical significance. *Compound scarps* can result from various combinations of tectonic and erosional processes. If a fault scarp were increased in height by a river-cut bank at its base, the upper part of the scarp would be tectonic but the lower part erosional. Conversely, an exhumed fault-line scarp might be the locus of renewed faulting, and its lower part could thus become a true fault scarp (Cotton, 1950b).

Landforms Associated with Strike-Slip Faulting

A **strike-slip fault** is a fault of any scale along which movement is parallel to the strike of the fault. Very large strike-slip faults such as the San Andreas Fault in California do not move on a single surface, but are

FIGURE 5-7. Steep, short stream valleys dissecting a fault scarp, north flank of the Cromwell Range, Huon Peninsula, Papua New Guinea.

FIGURE 5-8. Continuation of the view in Figure 5-7. Face-on view of a bottle-necked scarp valley.

defined by a zone up to 10 km in width. Neither side of the zone need be higher than the other, and rather than forming a single continuous scarp, the surface expression of a strike-slip fault may be an assemblage of small landforms, arranged parallel to, or *en echelon* across the width of the fault zone. If a series of gullies is offset by a transcurrent fault, many low scarps will result. Typically, if the fault crosses a group of ridges, all the spur ends or ridge crests will be displaced in the same direction (Figure 5-12). Such half-displaced ridges are called **shutter ridges.** Continued or intermittent strike-slip faulting, especially if it developed a zone of crushed, erodible rock, could be expected to

FIGURE 5-9. Triangular facets aligned on the fault scarp of Spanish Peak, 15 km south of Provo, Utah. View east (Photo: H. J. Bissell).

encourage the erosion of a valley along the fault that would intersect and collect the drainage from a number of scarp gullies (Figure 5-13).

The San Andreas Fault in California was the first place where strike-slip fault movement was convincingly demonstrated (Hill, 1981; Sylvester, 1988, p. 1669; Wallace, 1990). It was studied extensively following the 1906 San Francisco earthquake, and is now being closely monitored for precursory evidence of the next damaging earthquake. Some segments moved almost 5 m during the 1906 San Francisco earthquake.

Stratified, organic-rich Holocene sediments were deposited in the marshy alluvial fan of Pallett Creek where it crosses the San Andreas Fault 55 km north of Los Angeles. By measuring the deformation of the sediments and dating the interbedded peat layers by radiocarbon, Sieh et al. (1989) documented ten earthquakes between A.D. 671 and 1857 that caused lateral displacement on this segment of the San Andreas Fault. At least five of the earthquakes caused movement comparable to the 1857 displacement of 3 to 4 m. They estimated the average recurrence interval of such large earthquakes to be about 132 years. Farther north, in the Carrizo Plain of central California, similar techniques demonstrated 128 m of right-lateral offset in the modern channel of Wallace Creek, which flows eastward across the San Andreas Fault between a series of shutter ridges (Figure 5-14) (Sieh and Wallace, 1987). The rate of right-lateral strike-slip movement (the west side of the fault moving north relative to the east side) has averaged 34 mm/yr for the past 13,000 years. During each of the last three great earthquakes, slip ranged from 9.5 to 12.3 m, giving a recurrence

FIGURE 5-10. Four phases of the sedimentary–tectonic model for sedimentation adjacent of Laramide basement thrusts. These phases include early thrusting (a), rapid thrusting (b), slow thrusting (c), and postthrusting (d). Irregular oval outline pattern indicates conglomeratic facies; fine stippling indicates sandstone/mudstone/coal facies; coarse stippling; dashed line pattern indicates mudstone/coal/carbonate/ evaporite facies; dark patterned areas in cross section indicate pre-late Cretaceous strata; crosses indicate Precambrian basement. Structure is schematic only (Beck et al., 1988, Figure 3).

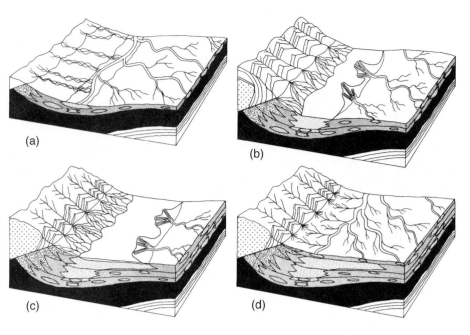

interval of 240 to 450 years for such events. The long-term slip rate at Wallace Creek, although consisting of a discrete major event every few centuries, is indistinguishable from the slip rate on a segment of the fault farther north where continuous creep of a few centimeters per year is demonstrated by deforming buildings and offset rows of orange trees.

Intermontane basins in New Zealand and elsewhere have been attributed to gaps that opened in the zone of tension where a major strike-slip fault makes an abrupt bend or breaks into *en echelon* segments (Cowan, 1990; Wood et at., 1994). The Hanmer basin in the northern part of the South Island, New Zealand, is the best studied locality. There, northeast-trending, right-lateral strike-slip faulting across a zone 6 to 7 km in width has created an intermontane basin 15 km long. Recent earthquakes have produced numerous scarps, especially on the north side of the basin. The western part of the basin is actively subsiding under continuing tension although previously deposited basin sediments are now being compressionally deformed in the eastern end of the basin (Wood et al., 1994, pp. 1472–1473). A similar explanation has been offered for the origin of Death Valley in California and the Dead Sea basin in the Middle East (Color Plate 5). Such tectonic depressions are called **pull-apart basins** (Mann et al., 1983; Sylvester, 1988, pp. 1683–1684). They are similar to grabens, but their regional control is strike-slip rather than dip-slip faulting (Figure 5-13). Anticlines or horsts are forced up in comparable cir-

cumstances where the local stress is compressional rather than tensional. The Transverse Ranges that bound the northern margin of the Los Angeles Basin are an impressive example.

In 1965, a new type of strike-slip fault was devised to explain the offset of submarine topography and paleomagnetic anomalies (Wilson, 1965). The topographic offset of a spreading ridge across a **transform fault** is in the opposite sense from the lateral movement (Figure 5-15). The transform fault was an entirely new concept and has been one of the major contributions to the theory of plate tectonics. Initially, the transform fault was a structure thought to be unique to the thin, basaltic oceanic crust. Subsequently, several major continental strike-slip faults such as the San Andreas Fault have been mapped as plate-boundary transform faults (Sylvester, 1988; Powell et al., 1993). The oceanic crust has been displaced hundreds of kilometers laterally along great transform faults (front endpaper). Some of them form submarine scarps or ridges several thousand kilometers long and 1 to 2 km in height (Mammerickx, 1989, pp. 7–9). Their height and continuity is not due to active seismicity, but to the fact that the oceanic lithosphere on opposite sides of the transform fault is of different ages, and therefore has cooled and contracted to different depths (Figure 1-4). Transform faults are typically seismic only on the segments between two spreading ridges (Figure 5-15). Where an aseismic ocean-floor transform fault is subducted, the irregular

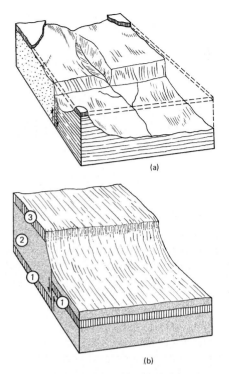

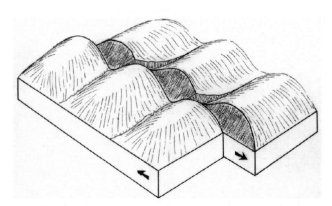

FIGURE 5-11. Two examples of fault-line scarp. In both examples, the stipulation is made that no tectonic movement has occurred in the present erosion cycle. (a) Resequent fault-line scarp. The former extent of a postfaulting, but dissected, lava flow has been added to suggest that evidence can sometimes be found to prove that such a landform is a fault-line scarp rather than a fault scarp. (b) Obsequent fault-line scarp. Erosion has removed resistant bed 3 from the front half of the block and the scarp faces toward the uplifted side of the fault. At a still later stage of erosion, the scarp might become resequent if bed 1 resists erosion on the front half of the block and beds 2 and 3 are eroded from the rear half.

FIGURE 5-12. Shutter ridges caused by right-lateral movement along a strike-slip fault across ridges and gullies. Streams are deflected along the fault trace or may be captured by the downstream segment of an adjacent gully.

relief on the subducting plate can produce differential uplift on the overlying plate margin (Hsu et al., 1989; Taylor et al., 1990).

FAULT VALLEYS AND FAULT-BLOCK MOUNTAINS

By a peculiar accident of geography and history, the earliest and most studied examples of block-faulted landscapes are in an arid region, where vegetation is sparse and geologic structures are boldly exposed. The regional assemblage of fault-block mountains in the Great Basin section of the Basin and Range Province of the western United States (Figure 1-5) led to the original name "basin ranges" for all the mountains of the region. Subsequently, "basin-range structure" became the general term for block faulting. Many deductions about the erosional evolution of landscapes on faulted structures are really based on an assumption of aridity. Although it is true that few places in the world have such an array of well-exposed fault-block landforms as the U.S. Basin and Range Province, the systematic description of these tectonic landforms must not be overly conditioned by the climatically controlled erosional processes that act on them.

Horsts and Grabens

A **horst** is a fault block that has been uplifted relative to the blocks on either side; a **graben** is a depression produced by subsidence along faults. Horsts do not seem to be particularly common landforms. Grabens, on the other hand, are well-known spectacular landforms in many tectonic regions (Figure 5-16).

Tilted-Block Mountains and Half Grabens

Rather than being horsts, uplifted fault blocks are commonly tilted, and bounded on only one of their sides by fault scarps. The asymmetric valley or basin between adjacent elongate blocks that are tilted in the same direction is a **half graben.** An assembly of tilted blocks, half grabens, grabens, and horsts is called simply a **fault-block landscape.**

Fault-block landscapes composed of originally flat-lying rocks are called the *Oregon type* (Davis, 1903; Cotton, 1950a, 1950b), named for the block-faulted basalt terrane around Klamath Lake in southern Oregon. According to Davis (1903, p. 130): "There was good evidence for regarding the Oregon lava-block ranges as types of the youngest, most elementary mountain forms known to geographers." The faulting is so recent, the erosional dissection so slight, and the displaced flow

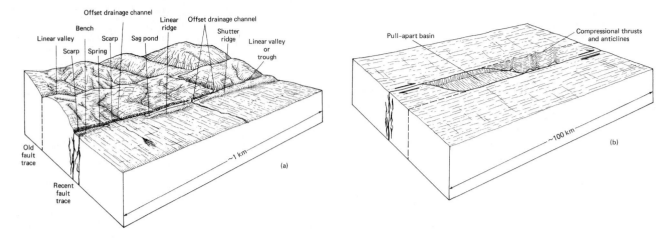

FIGURE 5-13. (a) Small (<1 km) landforms developed along recently active strike-slip faults (Wesson et al., 1975). (b) Large (10 to 100 km) extensional (pull-apart) basins and compressional thrusts and anticlines between *en echelon* segments of strike-slip faults.

sheets so strongly in control of surface slopes that photos of the region approach block diagrams in their simplicity (Donath, 1962). The relief is entirely tectonic. Small alluvial fans and lakes bury or drown some of the grabens and half grabens, but horsts and tilted-block mountains stand in bold profile. Oregon-type landscapes are also found in Iceland (Figure 5-16) and in the Afar region of Africa, near the Red Sea (Tazieff, 1970; Bannert, 1972).

The *basin-range type* of fault-block landscape is characterized by tilted, lifted, or dropped blocks with an internal structure of complexly deformed rocks in

an arid climate, where drainage is perpendicular to the ranges and their bounding faults. Rivers rise near the summits of the ranges and terminate in closed basins between them. Very early in the exploration of the Great Basin, Gilbert (1874, 1928) recognized that the faulting that produced the range-and-basin topography was younger than the orogeny that had intensely deformed and metamorphosed the rocks within the ranges. As noted in Chapter 3 (pp. 40–41), the Basin and Range Province experienced multiple and intense orogenies in the Paleozoic, Mesozoic, and Cenozoic Eras, which strongly deformed the rocks of

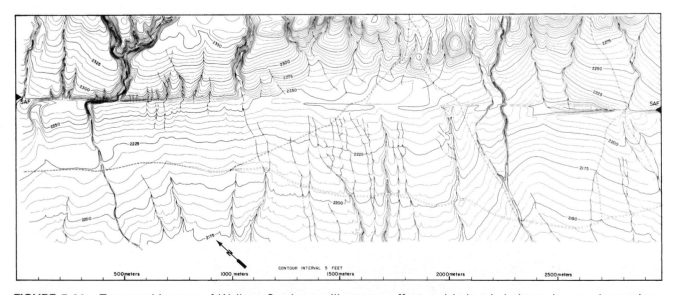

FIGURE 5-14. Topographic map of Wallace Creek area illustrates offset and beheaded channels, sag depressions, shutter ridges, and other tectonic landforms produced by recurring strike-slip movement along San Andreas Fault. Triangles (SAF) at left and right margins mark location of San Andreas Fault (Sieh and Wallace, 1987, Figure 3).

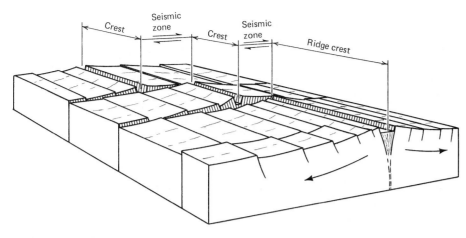

FIGURE 5-15. Transform faults that offset a spreading ridge crest in a left-lateral sense although the seismic portions of the faults are zones of right-lateral slip.

the region. In the late Cenozoic Era, especially in the last 17 million years, the region has undergone east-west extension that has created the present fault-block landscape. This is the reason that the word orogeny is no longer used for the creation of topography but is restricted to the deformation of rocks (p. 38). Louderback (1904, 1923) demonstrated that some tilted blocks are capped by lava flows that are neither folded nor thrusted but simply rest in sharp unconformity on erosional surfaces of low relief and are tilted because of the latest episode of extensional block faulting. In western Nevada, an interval of tectonic quiescence between 7.5 and 3.5 million years ago produced widespread erosion surfaces across older structures, which were then covered with latest Pliocene and Pleistocene

volcanic rocks and broken by Quaternary faulting into the modern topography (Gilbert and Reynolds, 1973).

Since about 1980, it has been shown that the normal faults that bound the basin ranges are listric (p. 68) and merge with or are truncated by **detachment faults,** very large subhorizontal faults at midcrustal depths of about 15 km along which the brittle upper crust has been extended by 300 to 400 percent in certain "highly extended" domains (Wernicke, 1992, p. 560) (Figure 5-17). These regions form a complex mosaic with contrasting "stable" domains in which extension has been only a few tens of percent or less, generally along steeply dipping normal faults. The tectonic deformation has been compared to a jelly sandwich, with the upper brittle crust breaking and

FIGURE 5-16. The central graben of Iceland (Photo: S. Thorarinsson).

FIGURE 5-17. Stepwise-restored cross sections through a highly extended domain in the Snake Range, eastern Nevada. Preextension depths of 7 and 15 km are not intended to correspond to any particular stratigraphic horizon (Wernicke, 1992, Figure 4b).

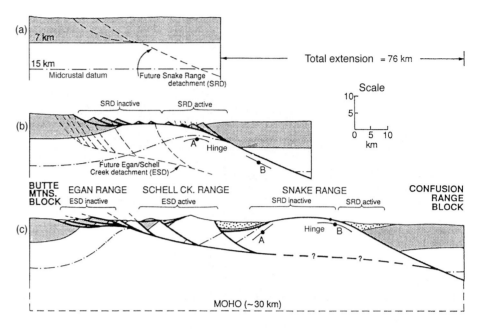

sliding along a highly plastic deep crustal layer (Wernicke, 1992, pp. 567–568). Thermal doming of the upper mantle (p. 44) may have provided the gravitational gradient for the extreme stretching and rotation of the upper crustal slices that form the present basin ranges (Figure 5-17). Such extreme crustal extension can result in tectonic unroofing, or **tectonic denudation,** of many kilometers of upper crustal rocks, exposing much deeper metamorphic terranes. Other high-grade metamorphic terranes now at the earth's surface also may be the result of such tectonic, rather than fluvial, erosion (p. 121).

The Sierras Pampeanas, or Pampean Ranges, of northwestern Argentina are an interesting variant of the basin-range type of fault-block mountains (Jordan and Allmendinger, 1986; Jordan et al., 1989). They are composed of Paleozoic and older igneous and metamorphic rocks, which are truncated by an erosion surface of low local relief (Color Plates 3 and 4). Within the last 10 million years, this ancient continental terrane has been caught up in the east-west compressional stresses of the Andean orogeny. The block mountains have been pushed up along reverse faults. The cover strata are mostly continental arkosic sandstone and conglomerates that are highly erodible. The great contrast in erodibility between the basement blocks and the cover strata has resulted in extensive exhumation of the ancient erosion surface, with inherited structures such as joints (Chapter 12) being etched out in the semiarid climate (Color Plate 3). Structural

relief between the crests of the blocks and the floors of the adjacent sediment-filled basins is as much as 8 km. Many of the range-bounding faults have been active in the last 3 million years, and recent earthquake scarps have been identified across alluvial fans at the bases of some ranges. Although they are bounded by reverse faults, the Pampean Ranges superficially look like the basin ranges of the western United States because the scarps in both provinces have eroded to the appropriate stable angle for similar rocks in similar climates. Tectonically, the Pampean blocks are a modern analog of the early Cenozoic Laramide fault blocks in the Rocky Mountains of the United States (Figure 5-10).

The basin-range type of fault-block landscape has been described in some detail because of its significance to the historical development of American geology, and because drawings and photographs of basin ranges are featured in almost every American text. Lacking personal familiarity with fault-block landscapes, many geologists and geographers mentally associate them with deserts, forgetting that the arid or semiarid climate and internal drainage of most of the U.S. Basin and Range Province impose special geomorphic processes on the tectonic landforms. In particular, almost all deductions about the sequential erosional development of fault-block mountains have started from the assumption that alluvium eroded from the ranges would accumulate in the adjacent basins and progressively bury the base of the range.

FIGURE 5-18. View southwest along the Awatere fault scarp, South Island, New Zealand. Mountain block on the right has moved up and toward the camera relative to the valley along a high-angle reverse fault (Photo: D. L. Homer, N. Z. Geological Survey).

To correct the climatic bias toward arid and semi-arid regions in geomorphic descriptions of fault-block mountains, Cotton (1948, 1950a, 1950b, 1968) designated a third type of fault-block landscape now known as the *New Zealand type*. The Awatere fault scarp (Figure 5-18) is representative of fault scarps in New Zealand, Venezuela, Japan, and other humid regions, where through-flowing rivers dominate the erosional processes. The Awatere fault scarp "differs in one important respect from those of the basin-range type. A vigorous consequent river has been present parallel to the scarp, at or near its base, during much of its long erosional development—a river that has been capable of carrying away the debris furnished by smaller streams dissecting the scarp" (Cotton, 1950a, p. 196). "Distinction of the Awatere from the basin-range type [of fault scarp] depends on the absence of a bahada fringe, all the debris derived from rapid early wasting has long ago been removed, and newer debris, though in abundant supply, continues to be carried off progressively by active river transportation" (Cotton, 1950b, p. 727).

New Zealand type fault-block landscapes are characterized, then, by half grabens and grabens that are river valleys or stream-dissected lowlands rather than alluvium-filled desert basins (Stevens, 1974). Where near the sea, New Zealand type half grabens may be drowned to form excellent harbors, such as Port Nicholson, at Wellington, New Zealand (Figure 5-19), Spencer and St. Vincent Gulfs near Adelaide, Australia, and the "Port Phillip Sunkland," the outer harbor of Melbourne, Australia (Twidale, 1968).

Early and extensive stream dissection of fault scarps and back slopes is the other diagnostic feature of the New Zealand type of fault-block landscape. Instead of planar triangular facets, rounded or blunt-ended ridges descend to the fault trace, where they are sharply cut off (Figures 5-18 and 5-19). The dissecting valleys between the scarp segments are large and are occupied by vigorous perennial streams that have substantial catchment areas in the interior of the upthrown block.

The unique features of New Zealand type fault-block landscape are not tectonic, but erosional. Many major New Zealand faults are high-angle reverse faults, but as in the case of all fault scarps, the steepness of the scarp face is controlled by rock-mass strength, not the dip of the fault. The fault scarp

FIGURE 5-19. View northeast over Port Nicholson (Wellington Harbor), the drowned portion of a half graben. Wellington, New Zealand, in foreground (Photo: Royal New Zealand Air Force. Crown Copyright reserved).

shown in Figures 5-7 and 5-8 is presumably of the New Zealand type because it stands in a humid tropical climate, is heavily forested, and lacks alluvial fans at the mouths of the dissecting creeks.

Other fault-block landscapes in the humid tropics, such as the great graben that forms the Markham Valley westward from Lae, Papua New Guinea, also are of the New Zealand type. Wherever precipitation exceeds evaporation, the excess water must drain along the tectonic depressions to the sea, creating river valleys along the lowest axes of half grabens or on the floors of grabens. Continued uplift on a scarp of the New Zealand type might create flights of uplifted river terraces on the scarp face, in contrast to the multiple earthquake scarplets across alluvial fans at the base of a basin-range scarp. Considering that most of the world's landscapes are drained to the sea by rivers (p. 198), it is tempting to conclude that the New Zealand type of fault-block landscape is more common than the basin-range type. Perhaps we have been educated to expect the basin-range type of fault-block landforms and have therefore overlooked or ignored fault-block landforms in humid regions. Highly tectonic regions in southeast Asia, Japan, Alaska, and the Caribbean provide many additional examples of the New Zealand type of tectonic landscape.

Rift Valleys

The great *rift valleys* of the Rio Grande, Rhine, Lake Baikal, West Antarctica, the Afro-Arabian rift system, and elsewhere (Figure 5-16) so completely dominate their tectonic and geomorphic settings that they deserve special treatment as tectonic landforms, although descriptively they are little more than large, branching, complex grabens and half grabens. Most authorities have emphasized the association of rift valleys with plateaus produced by broad epirogenic updoming, crustal and lithospheric thinning, profuse volcanism, and high heat flow (Illies, 1981). The amount of crustal extension is insufficient to account for the lithospheric thinning, and mantle plumes (p. 44) that thin the lithosphere from beneath are now regarded as the basic causal mechanism (Keller et al., 1994, p. 47). Experiments (Figure 5-20) show that excellent models of rift-valley morphology and tectonics, including the high plateaus that rise toward the rims of the depression and other details, result if moist clay layers are arched or updomed from beneath. The models develop inward-facing normal faults, their width approximates the thickness of the deformed layers, and no isostatic or mechanical inconsistencies develop. Instead of resulting from simple lateral ten-

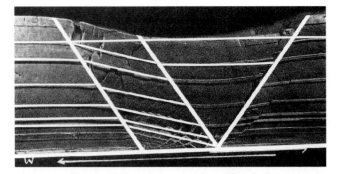

FIGURE 5-20. Clay model of a half graben produced by crustal extension with straight-line geometry superimposed. The clay model is from Cloos (Groshong, 1989, Figure 10, and references therein).

sion, most rift valleys are now interpreted as "keystones" subsiding along the crests of thermally uparched plateaus that are underlain by unusually thin continental crust and lithosphere.

The Rio Grande rift is a series of grabens and half grabens with bordering uplifts through which the Rio Grande flows from central Colorado southward across New Mexico to the Texas-Mexico border (Keller and Cathes, 1994). The Rio Grande rift has opened since mid-Miocene time (13 to 10 million years ago). The boundary faults are listric normal, some following reactivated older faults. Integration of the Rio Grande drainage along the rift system dates only from Pliocene time (Christiansen and Yeats, 1992, pp. 357–359). Unlike most of the great continental rifts, broad doming did not accompany or precede rifting; rather, postorogenic east-west extension similar to that in the Basin and Range Province, accompanied by lithospheric thinning and high heat flow, seems to be the cause.

The Rhine graben is a small rift valley, compared to those of Africa and the Middle East. It is 30 to 40 km in width (Illies, 1981) and, including several branching segments, about 300 km long. It cuts obliquely across crystalline rocks or older folded structures, and Mesozoic and Cenozoic formations within the graben are strongly deformed and broken by secondary faults of Miocene and younger age. It is strongly asymmetric, with the graben-bounding master fault on the eastern flank in the north and on the western flank in the south (Brun et al., 1992, p. 145). Rifting and thermal doming ceased by the end of the Miocene Epoch (Glahn and Granet, 1992), but minor seismicity continues. The fault scarps of the Rhine graben are of the New Zealand type: Dissected, rounded spurs end abruptly at the alluvial plain of the Rhine River on the graben floor. Hot springs and health spas mark the boundary faults today.

The Afro-Arabian rift system extends along a 60° arc of the earth's crust, for a length of 6500 km from southeastern Turkey to Mozambique. It includes the Dead Sea, Red Sea, Gulf of Aden, and eastern African rifts. Along various segments, it exhibits combinations of epeirogenic doming, strike-slip faulting, and normal faulting that have been in progress for at least the Cenozoic Era.

The rift valleys of eastern Africa are a branching, *en echelon* series of grabens and half grabens that divide into two major groups, the eastern or Kenya rift and the longer western rift system (Figure 5-21). Seismic profiles show that only about 20 percent of the Lake Tanganyika sector of the central western rift is a true or full graben. The remainder of the rift is an alternating series of half grabens 50 to 70 km in length and 20 to 40 km in width. The half grabens are arcuate or crescent shaped in plan view so that the rift has a sinuous shape (Rosendahl et al., 1986). Most of the major basins in the eastern or Kenya rift are also half grabens (Baker 1986, p. 52), as are the 32 or more basins that form the southern and northern sectors of the western rift system (Ebinger, 1989). Opposed or overlapping half grabens probably are the dominant structural form of all rift systems. They are separated by complex structural zones generally called *accommodation zones* or *transfer faults* (Chapin and Cathes, 1994, pp. 14–15). Accommodation zones may segment the floor of a rift so that in semiarid regions such as eastern Africa, closed lake basins form in each segment.

In the midst of the western rift valley stands the Ruwenzori horst, the highest nonvolcanic mountain in Africa. At 5100 m, its summit is nearly 4 km higher than the plateau rims on each side of the rift valleys, which in turn are 400 to 500 m above the floors of the rifts. Clearly, the Ruwenzori massif is an uplifted horst, not simply a fault block that did not subside. It is said to be topped by the same mid-Cenozoic erosion surface that is arched upward toward the rift valleys across the adjacent plateaus (McConnell, 1972).

The plateau surface around Lake Victoria, between the eastern and western rift valleys, has been strongly upwarped in the Quaternary (Bishop and Trendall, 1967). By Miocene time, a widespread erosion surface of low relief had beveled the metamorphic terrane of the plateau region. Miocene volcanic rocks buried this erosion surface although much of it has subsequently been exhumed. Pre-Miocene rivers may have drained westward across Africa, perhaps via the Congo River, but by Miocene time the western rift had developed and the plateau drainage entered it (Ollier, 1991, p. 40). An uplifted "shoulder" just east of the rim of the western rift valley defeated and reversed streams that formerly crossed it. The head-

FIGURE 5-21. Tectonic setting of Lake Victoria and Lake Kyoga between the eastern and western rift valleys in eastern Africa. Former west-flowing streams were defeated by the uplifted shoulder of the western rift and reversed. Their drowned headwaters now form the lakes and drain north to the Nile (after Holmes, 1965; Bishop and Trendall, 1967).

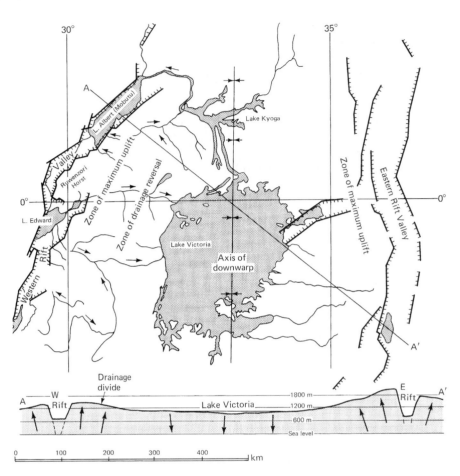

waters of these streams ponded to form Lakes Victoria and Kyoga (Figure 5-21), which deepened against the uplifted plateau rim until they spilled around it to the north. The present Victoria Nile River resulted, draining into Lake Albert (Mobutu) over an exhumed fault-line scarp at Murchison Falls. Lake Kyoga is a drowned dendritic drainage system on the warped early Pleistocene surface. The warping continues: 90 m of relative uplift across the defeated drainage systems may have occurred between 50,000 and 20,000 years ago (Bishop, 1965, p. 317). The differential uplift rate of 3 m/1000 yr thus calculated approaches that of strongly compressional orogenic regions (p. 41) although the plateau movements are certainly extensional and of epeirogenic scale.

The Dead Sea occupies a pull-apart basin on a rift valley 600 km long and only 10 to 20 km wide that extends northward from the Red Sea to Lebanon (Girdler, 1990) (Color Plate 5). The eastern scarp is straight and continuous along a single border fault. The western side of the rift is more complex, with several step faults or fault splinters involved. Fault scarps along the Dead Sea rift valley are predominantly of

the basin-range type. The region is one of extreme aridity.

The Dead Sea rift has a more complex structure than the Rhine graben and the eastern African rift systems (Quennell, 1958; Hatcher et al., 1981; Ten Brink and Ben Avraham, 1989). It is part of the left-lateral, 1100-km Dead Sea transform fault, the plate boundary between the Levantine (eastern Mediterranean) and Arabian plates (front endpaper). Left-lateral movement totaling 100 to 120 km along a crooked strike-slip fault is believed to have opened the pull-apart basin of the Dead Sea. Late Cenozoic vertical movements along the boundary faults and several oblique transverse faults have created the present basins (Marco and Agnon, 1995). The shore of the Dead Sea is the lowest subaerial place on the earth's surface at −392 m.

The Red Sea rift basin has become oceanic only in the past 5 to 6 million years although rifting began at least 18 million years ago (Davison et al., 1994). On the northeast shoulder of the rift in Yemen, 3.6 km of surface relief has been generated by doming. The "Great Escarpment" of Yemen has retreated by erosion 1 to 2 km from the series of normal faults that form the rift margin. Thus, the Great Escarpment is an impressive example of the basin-range type of fault scarp geomorphology. However, the narrow coastal plain between the Great Escarpment and the Red Sea may have resulted from seaward gravity sliding on salt layers within the older rift-valley sediments. Today, spreading is occurring primarily along a narrow axial spreading ridge on the floor of the Red Sea, which has thus become a true ocean basin in structure.

Lake Baikal, the world's deepest freshwater lake (1600 m), lies in a rift valley north of Mongolia along the crest of the Baikal arch. The tectonic setting is similar to that of the eastern African rift lakes. Domal uplift of 2 to 3 km is inferred to be caused by an upper mantle intrusion beginning in the Oligocene and continuing into Quaternary time (Logatchev and Zorin, 1992; Hutchinson et al., 1992). It is possible that the Baikal rifting is also influenced by the collision of India with Asia, far to the south. The fault scarps are conspicuously straight, primarily controlled by older structural trends.

A median rift valley is a characteristic feature of many submarine midocean ridges or rises (Figures 5-15 and 5-22). The rift valleys of midocean ridges are bounded by a series of parallel normal faults, each with a throw of several hundred meters. The fault blocks that parallel the central rift valley are commonly lifted and rotated so that their upper surfaces slope away from the rift axis. The morphology of median rift valleys is consistent with their tectonic setting on the crests of spreading ridges at accreting plate margins.

LANDFORMS MADE BY FOLDING

Fault scarps are bold, obvious landforms that attract the eyes of geomorphologists and sightseers. Much less has been written about tectonic landforms that result from warping or flexing of the earth's surface without fracture and displacement on faults. A few examples can be cited from regions where earth movements are obvious. Perhaps by calling attention to known examples, this review will stimulate readers to consider other possibilities, especially in regions of humid climate, heavy vegetation cover, active soil-forming processes, and mild tectonism.

A major difficulty in recognizing tectonic landforms produced by folding is that a visible reference surface is not generally available. Whereas every structure, as well as the landscape itself, is sharply displaced across a fault scarp, warping or flexing of an irregular landscape or homogeneous structure is extremely difficult to recognize. Only where approximately planar erosional and depositional surfaces such as shore platforms or river terraces are exposed is warping likely to be visible. Even then, surveyed profiles may need to be taken along the surfaces to prove that they are warped.

Monoclinal Scarps

Tectonic scarps with height, steepness, and lateral continuity comparable to fault scarps are formed by steep monoclines (Figure 5-23). As is the case with fault scarps, monoclinal scarps are regarded as tectonic

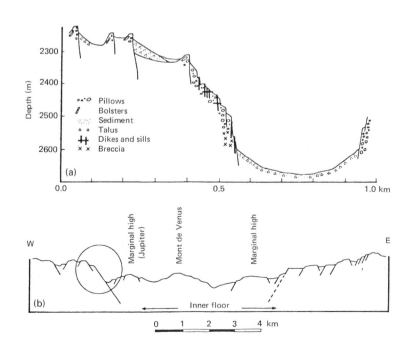

(a)

Pillows
Bolsters
Sediment
Talus
Dikes and sills
Breccia

(b)

W   E

Marginal high (Jupiter)   Mont de Venus   Marginal high

Inner floor

0 1 2 3 4 km

FIGURE 5-22. Interpretive cross sections of the median rift valley in the Mid-Atlantic Ridge (ARCYANA, 1975, Fig. 3). (a) Surface geology observed from a submersible vessel; (b) regional cross section showing the location of (a) (Copyright 1975 by the American Association for the Advancement of Science).

landforms only if their relief can be shown to be due to tectonic movement rather than differential erosion or exhumation, and if substantial parts of the original surface of deformation can be recognized. Monoclines, as other folds, are described by the attitude of the deformed strata, not by the shape of the displaced landscape.

Monoclines are the most distinctive structure of the Colorado Plateau. The longest is more than 400 km in length; strata are displaced vertically as much as 4300 m. In the Colorado Plateau, many land surfaces coincide with particularly resistant formations, so structural benches are flexed sharply along the monoclines to produce imposing scarps. The major period of deformation in the region was the late Cretaceous-Eocene Laramide orogeny. The monoclines are in sedimentary rocks that overlie older high-angle faults in basement rocks that were reactivated by the Laramide orogeny (Miller et al., 1992, p. 216).

No systematic distinction of monoclinal landforms analogous to the distinction between fault scarps and fault-line (erosional) scarps has ever been made. It would seem that the great monoclines of the Colorado Plateau were folded at some depth. Tertiary strata that postdate the folding have been partly stripped away. If so, such scarps are better considered as examples of structurally controlled erosion (Chapter 12) than as tectonic landforms.

Topographic Domes

A variety of processes can produce domal tectonic landforms. Shallow intrusions of ice, salt, mud, or molten rock can up-arch the overlying rock or sediment. Active anticlines or blind thrust faults in rock structures normally produce topographic domes in the landscape above. All these landforms are tectonic

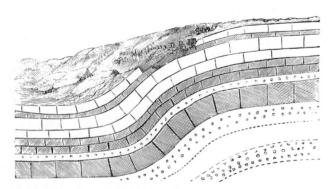

FIGURE 5-23. First illustration of a monoclinal fold (Powell, 1875, Figure 67).

in the sense that they are built by earth movements, even if the movements are at a very shallow depth.

Salt Tectonics.   More than 400 domal structures with salt in their cores are known from the northern part of the Gulf of Mexico (Jackson and Talbot, 1986). The salt rose hydrostatically or under tectonic pressure (p. 12) from source beds of Jurassic or older age that are now buried under many kilometers of sediment. Some of the U.S. Gulf Coast salt domes have risen through 7500 m or more of covering strata. They are circular or ridgelike in plan view and may be up to 10 km in diameter. Similar structures are known or suspected in the Colorado Plateau, the Canadian Arctic, many parts of the continental shelves, Saudi Arabia, France, Germany, Iran, and the Soviet Union. Because of the potential for petroleum production from structural and stratigraphic traps on their flanks, exploration geologists actively seek salt domes (Alsop et al., 1996).

Only five salt domes on the U.S. Gulf Coast have positive topographic relief. Although the salt penetrates Pleistocene and even Holocene sediments, deltaic progradation and fluviatile deposition have obliterated most of the salt-dome relief. Avery Island, Louisiana, has a highest point of 46 m and forms a dome about 2.3 km in diameter. Salt is found at a depth of only about 4 m below its surface. Several salt springs and ponds drain radially off the dissected dome.

The Paradox basin of east-central Utah and adjacent western Colorado includes an area of impressive grabens and breached, elongate, doubly plunging anticlines formed by salt tectonics and subsequent solution. The best known of the eight large salt-anticline valleys in the region is the Moab Valley, which trends northwest-southeast across the Colorado River at the town of Moab, Utah (Figure 5-24). The river does not follow the valley, but emerges from a deep gorge on its northeast side and exits to the southwest through an equally spectacular gorge. The floor of the valley is along the axis of an anticline cored by salt and other evaporites of the Paradox Formation of late Paleozoic age. As Triassic and Jurassic sediments accumulated, the salt flowed as much as 5500 m upward along older structural lineaments to dome the overlying strata (Figure 5-25). The anticlinal doming was complete in the early Cenozoic (Baars and Doelling, 1987, p. 276). Subsequent groundwater solution of the salt caused the collapse of the overlying sedimentary rocks, creating the grabenlike valley along the crest of the anticline (Figure 5-24). Blocks of the Mesozoic sandstones that form the in-facing escarpments are found in the detrital sediments on the valley floor, 1000 m below their projected original position at the crest of the anticline.

FIGURE 5-24. Aerial view toward the southeast with the Moab Fault in the foreground, Moab Valley salt-intruded anticline in the middle distance, and the Tertiary igneous-intrusive La Sal Mountains in the distance. See Figure 5-25 for explanation (Photo: D. L. Baars) (Baars and Doelling, 1987, Figure 2).

Geomorphologists debate whether the breached salt-anticline valleys should be classified as collapse karst (Chapter 8), structurally controlled valleys, or tectonic landforms. They are included here as tectonic forms by the argument that their relief is the direct result of differential displacement of rocks, although the steep escarpments are the result of differential solution rather than faulting. The valleys are impressive examples of inverted relief, because they are lowlands that form above 5500 m of deformed, upwardly displaced salt.

Early explorers of the region noted that the main rivers of the region cross, rather than follow, the salt-anticline valleys. This paradox led to the name of the region, and subsequently to the naming of the Pennsylvanian evaporite sequence as the Paradox Formation. The issue, discussed further in Chapter 12, is whether the Colorado and other rivers of the region were antecedent to (that is, predated the final uplift of) the salt anticlines, or whether they are superposed across the structures from former gradients across undistributed overlying strata. Hunt (1969) concluded that the modern river valleys are a combination of both types, in that they formerly flowed across overlying sediments prior to the uplift of the Colorado Plateau, but that they are primarily antecedent with respect to the final stages of salt tectonics. Continuing lateral flow in the salt below nearly 500 m of overlying sandstone beds causes the Needles District in Canyon-

lands National Park (Figure 5-26), where the Colorado River has cut through the brittle sandstones into the ductile underlying evaporite rocks and removed their lateral support (McGill and Stromquist, 1979; Huntoon, 1982).

In the Canadian Arctic archipelago, Tertiary or older folding forced Paleozoic anhydrite and gypsum into anticlinal cores and domes. In contrast to the rapid solution of evaporites in humid regions, in the polar deserts these massive, unjointed, monomineralic rocks are resistant ridge- and dome-formers (St. Onge, 1959). Whether the highlands of gypsum in the Sverdrup Islands of northernmost Canada are tectonic or structural has not been determined (Stott, 1969, p. 37), nor has the role of Cenozoic climate change been evaluated, but aerial reconnaissance shows that many of the higher hills are underlain by massive white gypsum, intrusive into darker sedimentary rocks that form surrounding lowlands.

Igneous Domes. Shallow igneous sills may swell to a plano-convex lens shape and arch overlying sedimentary strata to produce topographic domes. The name **laccolith** (lens-shaped intrusion) was given for such forms in the Henry Mountains of Utah in one of the classic works of early American geology (Gilbert, 1877). Some of the larger types of examples subsequently have been claimed to be igneous stocks (Hunt et al., 1953) and not laccoliths, but the debate continues (Jackson and Pollard, 1988).

The laccoliths of the Colorado Plateau were emplaced in late Eocene or Miocene time, probably at a depth in excess of 1600 m (Hunt et al., 1953, p. 147). Because some of the laccoliths are thicker than that, surface arching must have resulted. The regional drainage pattern suggests that major rivers, flowing on an erosion surface 1000 m or more above the present landscape, were diverted by the rising laccoliths and migrated laterally off the domes (Hunt et al., 1953, p. 212). The subsequent geomorphic history of the Henry Mountains laccoliths, as is true of the monoclines of the region, is one of structurally controlled erosion (Chapter 12).

Numerous instrumental records demonstrate that volcanoes bulge in a domal shape prior to an eruption. The movements are undetectable except by precise tilt-meters or geodetic surveys, but are a useful indication of an impending eruption. The doming is attributed to magma rising into shallow reservoirs within or beneath the volcano. On a larger scale, the Yellowstone Plateau, within Yellowstone National Park, updomed almost 1 m in 50 years prior to 1975 (Figure 5-27). As is the case with neotectonic movements (Chapter 3), such rapid uplift rates cannot continue for very long without

FIGURE 5-25. Highly generalized cross section of the Moab salt-intruded anticline. Basement fault blocks originated in late Precambrian time and were episodically rejuvenated throughout the Phanerozoic. Middle Pennsylvanian evaporites, including salt, buried the structure as salt flowage was initiated. Upward growth of the salt bulge continued through late Paleozoic and Mesozoic time, causing excess thickness of clastic sediments in flanking synclines and thinning along the rising salt core. Tertiary to Holocene near-surface groundwater dissolution of salt created a residual "leached gypsum cap" and subsequent collapse of overlying strata. The valley surface is now largely covered by fluvial and eolian Holocene deposits (Baars and Doelling, 1987, Figure 4).

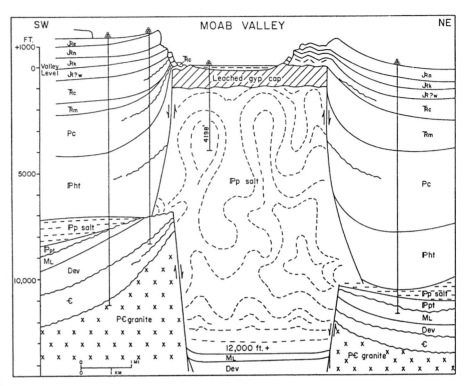

major geomorphic consequences. There was no uplift in 1984 and 1985, and from 1985 to 1993, the area subsided at an average rate of 19 ± 1 mm/yr. The domal

FIGURE 5-26. The graben system in the Needles District of Canyonlands National Park, Utah, looking south. The upland surface is underlain by heavily jointed sandstone that has stretched by horizontally gliding to the right (west) toward the Colorado River, which has intrenched into salt layers underlying the sandstone. The two prominent grabens in the foreground, Devils Lane and Cyclone Canyon, are 1 km apart (Photo: G. E. McGill).

uplift of the Yellowstone Plateau has been attributed to either hydrothermal pressure or a recent influx of magma into the upper crust within an ancient volcanic caldera (Pelton and Smith, 1982; Dzurisin et al., 1994; Locke and Meyer, 1994).

Rising or Live Anticlines. Orogenic regions such as the margin of the Pacific Ocean have domes or arches in the landscape over the crests of structural anticlines, or at compressional segments of curved strike-slip faults (Figure 5-13b). Warped coastal and river terraces (Figure 5-28) are a common form of evidence. The Ventura River in southern California flows south across the active Ventura Avenue anticline (Rockwell et al., 1988). For the past 200,000 years, based on radiocarbon dating, soil stratigraphy, and correlation with dated marine terraces, the anticline has been rising across the path of the river. As terraces were cut, they were progressively arched upward (Figure 5-28). The rate of uplift along the anticlinal axis may have decreased from about 20 m/1000 yr for the first 100,000 years to about 5 m/1000 yr at the present time.

In New Zealand, several active anticlines warp a coastal-plain landscape (Te Punga, 1957). One of them, the Pohangina anticline, can be traced for 30 km. Recent alluvium and Pleistocene coastal-plain strata dip 2° to 4° off the west flank of the anticline, but on the steeper eastern flank, beds dip as steeply as 60°. On this rising anticline, in a humid climate, vigorous

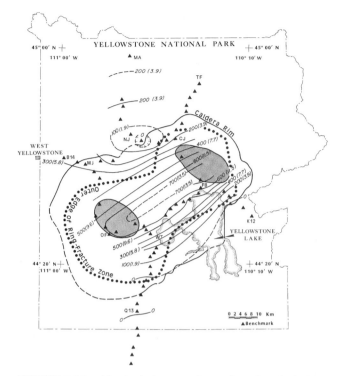

FIGURE 5-27. Vertical changes in surface levels in Yellowstone National Park from surveys done in 1923 and 1975. Contour interval is 100 mm with corresponding velocity value given in parentheses in mm/yr. Resurgent domes shaded. Caldera rim and outer edge of ring-fracture zone are also shown (Pelton and Smith, 1982, Figure 7).

gullies migrate rapidly headward on the steeper flank, and stream divides have shifted laterally beyond the crest of the structure onto the less-steep dip slope. Many drainage diversions have been noticed in the century since the land was cleared and fenced for agriculture. Seismic surveys and the regional geologic setting suggest that active anticlines in both New Zealand and Japan (Ota, 1975) are surface manifestations of rising fault blocks of indurated basement rock. Each anticline is thought to be in sedimentary cover strata that are draped over the highest edge of an upfaulted basement block.

Numerous examples of rising anticlines can be cited from other orogenic regions. Lees (1955) described a Persian *Qanat,* or horizontal tube well, that had been arched by a rise of 18 m in the middle part of its 4-km course across an anticline in the Khuzistan Plain at Shaur, Iran. The tunnel was dug 1700 years ago, giving a differential uplift rate across the structure of 10 m/1000 yr. Prehistoric irrigation canals in Peru, built between about A.D. 900 and 1400, show successive repairs and rerouting that

were at least in part made necessary by tectonic tilting and faulting (Ortloff et al., 1985). In both Japan and California, Holocene deformation of river terraces corresponds well with continuing deformation documented by geodetic releveling (Sugimura, 1967; Castle et al., 1974).

REFERENCES

ALSOP, G. I., BLUNDELL, D. J., and DAVISON, I., eds., 1996, Salt Tectonics: Geol. Soc. [London] Spec. Pub. No. 100, 310 pp.

ARCYANA, 1975, Transform fault and rift valley from bathyscaph and diving saucer: Science, v. 190, pp. 108–116.

BAARS, D. L., and DOELLING, H. H., 1987, Moab salt-intruded anticlines, east-central Utah: Geol. Soc. America Centennial Field Guide, v. 2, pp. 275–280.

BAKER, B. H., 1986, Tectonics and volcanism of the southern Kenya rift valley and its influence on rift sedimentation, in Frostick, L. E., Renault, R. W., and 2 others, eds., Sedimentation in the African rifts: Geol. Soc. (London) Spec. Pub. no. 25, pp. 45–57.

BALJINNYAM, I., BAYASGALAN, A., and 9 others, 1993, Ruptures of major earthquakes and active deformation in Mongolia and its surroundings: Geol. Soc. America Mem. 181, 62 pp.

BANNERT, D., 1972, Afar tectonics from space photographs: Am. Assoc. Petroleum Geologists Bull., v. 56, pp. 903–915.

BEANLAND, S., BERRYMAN, K. R., and BLICK, G. H., 1989, Geological investigations of the 1987 Edgecumbe earthquake, New Zealand: New Zealand Jour. Geol. Geophys., v. 32, pp. 73–91.

BECK, R. A., VONDRA, C.F., and 2 others, 1988, Syntectonic sedimentation and Laramide basement thrusting, Cordilleran foreland; timing of deformation, in Schmidt, C. J. and Perry, W. J., Jr., eds., Interaction of the Rocky Mountain foreland and the Cordilleran thrust belt: Geol. Soc. America Mem. 171, pp. 465–487.

BISHOP, W. W., 1965, Quaternary geology and geomorphology in the Albertine Rift Valley, Uganda: Geol. Soc. America Spec. Paper 84, pp. 293–321.

———, and TRENDALL, A. F., 1967, Erosion-surfaces, tectonics and volcanic activity in Uganda: Geol. Soc. London Quart. Jour., v. 122, pp. 385–420.

BROCK, W. G., and ENGELDER, T., 1977, Deformation associated with the movement of the Muddy Mountain overthrust in the Buffington Window, southeastern Nevada: Geol. Soc. America Bull., v. 88, pp. 1667–1677.

BRUN, J. P., GUTSCHER, M.-A., and DEKORP-ECORS teams, 1992, Deep crustal structure of the Rhine graben from DEKORP-ECORS seismic reflection data: a summary, in Ziegler, P. A., ed., Geodynamics of rifting, v. 1. Case history studies on rifts: Europe and Asia: Tectonophysics, v. 208, pp. 139–147.

BURBANK, D. W., and VERGÉS, J. 1994, Reconstruction of topography and related depositional systems during active thrusting: Jour. Geophys. Res., v. 99, pp. 20,281–20,297.

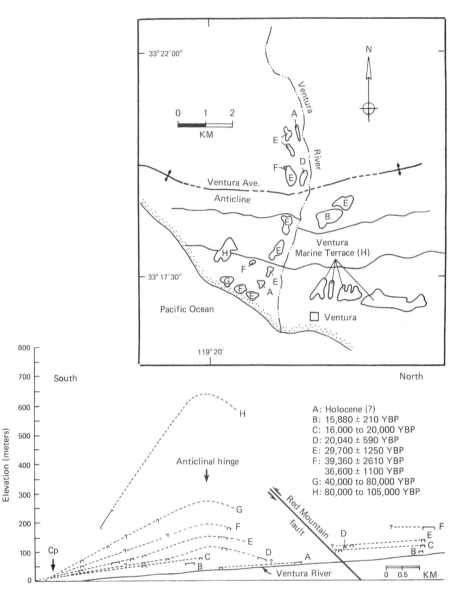

FIGURE 5-28. Ventura River terrace profiles over the Ventura Avenue Anticline and Red Mountain Fault. For location, see map. Age control is from radiocarbon dates (terrace B, D, E, F) and correlation to a marine terrace east of the Ventura River (terrace H) dated by amino-acid racemization. YBP = years before present (Rockwell et al., 1988, Figures 3 and 4).

A: Holocene (?)
B: 15,880 ± 210 YBP
C: 16,000 to 20,000 YBP
D: 20,040 ± 590 YBP
E: 29,700 ± 1250 YBP
F: 39,360 ± 2610 YBP
    36,600 ± 1100 YBP
G: 40,000 to 80,000 YBP
H: 80,000 to 105,000 YBP

CASTLE, R. O., ALT, J. N., and 2 others, 1974, Elevation changes preceding the San Fernando earthquake of February 9, 1971: Geology, v. 2, pp. 61–66.

CHAPIN, C. E., and CATHES, S. M., 1994, Tectonic setting of the axial basins of the northern and central Rio Grande rift, in Keller, G. R. and Cathes, S. M., eds., Basins of the Rio Grande rift: Structure, stratigraphy, and tectonic setting: Geol. Soc. America Spec. Paper 291, pp. 5–25.

CHRISTIANSEN, R. L. and YEATS, R. S., 1992, Post-Laramide geology of the U.S. Cordilleran region, in Burchfiel, B. C., Lipman, P. W., and Zoback, M. L., eds., The Cordilleran orogen: Conterminous U.S.: Geol. Soc. America, The Geology of North America, v. G–3, pp. 261–406.

COTTON, C. A., 1948, Landscape as developed by the processes of normal erosion, 2nd ed: Whitcombe and Tombs Ltd., Wellington, New Zealand, 509 pp.

———, 1950a, Tectonic scarps and fault valleys: Internat. Geog. Cong., 16th, Lisbon 1949, Comptes rendus, pp. 191–200.

———, 1950b, Tectonic scarps and fault valleys: Geol. Soc. America Bull., v. 61, pp. 717–758.

———, 1968, Tectonic landscapes, in Fairbridge, R. W., ed., Encyclopedia of geomorphology: Reinhold Book Corp., New York, pp. 1109–1116.

COWAN, H. A., 1990, Late Quaternary displacements on the Hope Fault at Glynn Wye, North Canterbury: New Zealand Jour. Geol. Geophys., v. 33, pp. 285–293.

DAVIS, W. M., 1903, Mountain ranges of the Great Basin: Mus. Comp. Zool. Harvard Bull. 42, pp. 129–177 (reprinted 1954 in Geographical essays: Dover Publications, Inc., New York, pp. 725–772).

DAVISON, I., Al-KADASI, M., and 12 others, 1994, Geological evolution of the southeastern Red Sea rift margin, Republic of Yemen: Geol. Soc. America Bull., v. 106, pp. 1474–1493.

DONATH, F. A., 1962, Analysis of basin-range structure, south-central Oregon: Geol. Soc. America Bull., v. 73, pp. 1–15.

DZURISIN, D., YAMASHITA, K. M., and KLEINMAN, J. W., 1994, Mechanisms of crustal uplift and subsidence at the Yellowstone caldera, Wyoming: Bull. Volcanol., v. 56, pp. 261–270.

EBINGER, C. J., 1989, Tectonic development of the western branch of the East African rift system: Geol. Soc. America Bull., v. 101, pp. 885–903.

von ENGELN, O. D., 1942, Geomorphology: The Macmillan Company, New York, 655 pp.

GILBERT, C. M., and REYNOLDS, M. W., 1973, Character and chronology of basin development, western margin of the Basin and Range Province: Geol. Soc. America Bull., v. 84, pp. 2489–2509.

GILBERT, G. K., 1874, Preliminary geological report, expedition of 1872: U.S. Geog. Geol. Survey W. 100th Mer. (Wheeler), Progress Rept., pp. 48–52.

———, 1877, Report on the geology of the Henry Mountains [Utah]: U.S. Geog. Geol. Survey Rocky Mtn. Region (Powell), pp. 18–98.

———, 1928, Studies of basin-range structure: U.S. Geol. Survey Prof. Paper 153, 92 pp.

GIRDLER, R. W., 1990, The Dead Sea transform fault system, in Kovach, R. L., and Ben-Avraham, S., eds., Geologic and tectonic processes of the Dead Sea rift zone: Tectonophysics, v. 180, pp. 1–13.

GLAHN, A., and GRANET, M., 1992, 3-D structure of the lithosphere beneath the southern Rhine graben area, in Ziegler, P. A., ed., Geodynamics of rifting, v. 1. Case history studies on rifts: Europe and Asia: Tectonophysics, v. 208, pp. 149–158.

GROSHONG, R. H., JR., 1989, Half-graben structures: Balanced models of extensional fault-bend folds: Geol. Soc. America Bull., v. 101, pp. 96–105.

HATCHER, R. D., JR., ZIETZ, I., and 2 others, 1981, Sinistral strike-slip motion on the Dead Sea Rift: Confirmation from new magnetic data: Geology, v. 9, pp. 458–462.

HILL, M. L., 1981, San Andreas Fault: History of concepts: Geol. Soc. America Bull., v. 92, pp. 112–131.

HOLMES, A., 1965, Principles of physical geology, 2nd ed.: Ronald Press, New York, 1288 pp.

HSU, J. T., LEONARD, E. M., and WEHMILLER, J. F., 1989, Aminostratigraphy of Peruvian and Chilean Quaternary marine terraces: Quaternary Sci. Rev., v. 8, pp. 255–262.

HUNT, C. B., 1969, Geologic history of the Colorado River: U.S. Geol. Survey Prof. Paper 669C, pp. 59–130.

———, AVERITT, P., and MILLER, R. L., 1953, Geology and geography of the Henry Mountains region, Utah: U.S. Geol. Survey Prof. Paper 228, 234 pp.

HUNTOON, P. W., 1982, The Meander anticline, Canyonlands, Utah: An unloading structure resulting from horizontal gliding on salt: Geol. Soc. America Bull. v. 93, pp. 941–950.

HUTCHINSON, D. R., GOLMSHTOK, A. M., and 4 others, 1992, Depositional and tectonic framework of the rift basins of Lake Baikal from multichannel seismic data: Geology, v. 20, pp. 589–592.

ILLIES, J. H., 1981, Epilogue: Mechanism of graben formation, in Illies, J. H., ed., Special Issue: Mechanism of graben formation: Tectonophysics, v. 73, pp. 249–266.

JACKSON, M. D., and POLLARD, D. D., 1988, The laccolith-stock controversy: New results from the southern Henry Mountains, Utah: Geol. Soc. America Bull., v. 100, pp. 117–139.

JACKSON, M. P. A., and TALBOT, C. J., 1986, External shapes, strain rates, and dynamics of salt structures: Geol. Soc. America Bull., v. 97, pp. 305–323.

JOHNSON, D. W., 1929, Geomorphic aspects of rift valleys: Internat. Geol. Cong., 15th, South Africa, 1929, Comptes rendus, v. 2, pp. 354–373.

JORDAN, T. E., and ALLMENDINGER, R. W., 1986, The Sierras Pampeanas of Argentina: A modern analogue of Rocky Mountain foreland deformation: Am. Jour. Sci., v. 286, pp. 737–764.

JORDAN, T.E., ZEITLER, P., and 2 others, 1989, Thermochronometric data on the development of the basement peneplain in the Sierras Pampeanas, Argentina: Jour. South American Earth Sci., v. 2, pp. 207–222.

KELLER, E. A., and PINTER, N., 1996, Active tectonics: Earthquakes, uplift, and landscape: Prentice-Hall, Inc., Upper Saddle River, N.J., 338 pp.

KELLER, G. R. and CATHES, S. M., eds., 1994, Basins of the Rio Grande rift: Structure, stratigraphy, and tectonic setting: Geol. Soc. America Spec. Paper 291, 304 pp.

KELLER, G.R., PRODEHL, C., and 11 others, 1994, The east African rift system in the light of KRISP 90: Tectonophysics, v. 236, pp. 465–483.

LEES, G. M., 1955, Recent earth movements in the Middle East: Geol. Rundschau, v. 43, pp. 221–226.

LOCKE, W. W., and MEYER, G. A., 1994, A 12,000 year record of vertical deformation across the Yellowstone caldera margin: The shorelines of Yellowstone Lake: Jour. Geophys. Res., v. 99, pp. 20,079–20,094.

LOGATCHEV, N. A., and ZORIN, YU. A., 1992, Baikal rift zone: Structure and geodynamics, in Ziegler, P. A., ed., Geodynamics of rifting, v. 1. Case history studies on rifts: Europe and Asia: Tectonophysics, v. 208, pp. 273–286.

LOUDERBACK, G. D., 1904, Basin Range structure of the Humboldt region: Geol. Soc. America Bull., v. 15, pp. 289–346.

———, 1923, Basin range structure in the Great Basin: Calif. Univ. Pub. Geol. Sci., v. 14, pp. 329–376.

MAMMERICKX, J., 1989, Large-scale undersea features of the northeast Pacific, in Winterer, E. L., Hussong, D. M., and Decker, R. W., eds., The eastern Pacific Ocean and

Hawaii: Geol. Soc. America, The Geology of North America, v. N, pp. 5–132.

MANN, P., HEMPTON, M. R., and 2 others, 1983, Development of pull-apart basins: Jour. Geology, v. 91, pp. 529–554.

MARCO, S., and AGNON, A., 1995, Prehistoric earthquake deformations near Masada, Dead Sea graben: Geology, v. 23, pp. 695–698.

MCCONNELL, R. B., 1972, Geological development of the rift system of east Africa: Geol. Soc. America Bull., v. 83, pp. 2549–2572.

MCGILL, G. E., and STROMQUIST, A. W., 1979, The grabens of Canyonlands National Park, Utah: Geometry, mechanics, and kinematics: Jour. Geophys. Res., v. 84, pp. 4547–4563.

MEIGS, A. J., BURBANK, D. W., and BECK, R. A., 1995, Middle-late Miocene (>10 Ma) formation of the Main Boundary Thrust in the western Himalaya: Geology, v. 23, pp. 423–426.

MILLER, D. M., NILSEN, T. H., and BILODEAU, W. L., 1992, Late Cretaceous to early Eocene geologic evolution of the U.S. Cordillera, in Burchfiel, B .C., Lipman, P. W., and Zoback, M. L., eds., The Cordilleran orogen: Conterminous U.S.: Geol. Soc. America, The geology of North America, v. G–3, pp. 205–260.

OLLIER, C., 1991, Ancient landforms: Belhaven Press (a division of Pinter Publishers), London, 233 pp.

ORTLOFF, C. R., FELDMAN, R. A., and MOSELEY, M. E., 1985, Hydraulic engineering and historical aspects of the pre-Columbian intravalley canal system of the Moche Valley, Peru: Jour. Field Archeol., v. 12, pp. 77–98.

OTA, Y., 1975, Late Quaternary vertical movement in Japan estimated from deformed shorelines: Royal Soc. New Zealand Bull. 13, pp. 231–239.

OWEN, S., SEGALL, P., and 6 others, 1995, Rapid deformation of the south flank of Kilauea Volcano, Hawaii: Science, v. 267, pp. 1328–1332.

PELTON, J. R., and SMITH, R. B., 1982, Contemporary vertical surface displacements in Yellowstone National Park: Jour. Geophys. Res., v. 87, pp. 2745–2761.

PHILIP, H., and MEGHRAOUI, M., 1983, Structural analysis and interpretation of the surface deformations of the El Asnam earthquake of October 10, 1980: Tectonics, v. 2, pp. 17–49.

POWELL, J. W., 1875, Exploration of the Colorado River of the west and its tributaries: U.S. Govt. Printing Office, Washington, D.C., 291 pp.

POWELL, R. E., WELDON, R. J. II, and MATTI, J. C., 1993, The San Andreas Fault system: Displacement, palinspastic reconstruction, and geologic evolution: Geol. Soc. America Mem. 178, 332 pp.

QUENNELL, A. M., 1958, Structural and geomorphic evolution of the Dead Sea Rift: Geol. Soc. London Quart. Jour., v. 114, pp. 1–24.

ROCKWELL, T. K., KELLER, E. A., and DEMBROFF, G. R., 1988, Quaternary rate of folding of the Ventura Avenue

anticline, western Transverse Ranges, southern California: Geol. Soc. America Bull., v. 100, pp. 850–858.

ROSENDAHL, B. R., REYNOLDS, D. J., and 6 others, 1986, Structural expressions of rifting: Lessons from Lake Tanganyika, Africa, in Frostick, L. E., Renaut, R. W., and 2 others, eds., Sedimentation in the African rifts: Geol. Soc. [London] Spec. Pub. no. 25, pp. 29–43.

ST. ONGE, D. A., 1959, Note sur l'érosion du gypse en climat périglaciaire: Rev. Canadienne de Geog., v. 13, pp. 155–162.

SIEH, K., STUIVER, M., and BRILLINGER, D., 1989, A more precise chronology of earthquakes produced by the San Andreas Fault in southern California: Jour. Geophys. Res., v. 94, pp. 603–623.

SIEH, K., and WALLACE, R. E., 1987, The San Andreas Fault at Wallace Creek, San Luis Obispo County, California, in Geol. Soc. America Centennial Field Guide v.1, pp. 233–238.

STEVENS, G. R., 1974, Rugged landscapes: The geology of central New Zealand: A. H. & A. W. Reed Ltd., Wellington, New Zealand, 286 pp.

STOTT, D. F., 1969, Ellef Ringnes Island, Canadian Arctic Archipelago: Canada Geol. Survey Paper 68–16, 44 pp.

SUGIMURA, A., 1967, Uniform rates and duration period of Quaternary earth movements in Japan: Osaka City Univ. Jour. Geosci., v. 10, pp. 25–35.

SYLVESTER, D. G., 1988, Strike-slip faults: Geol. Soc. America Bull., v. 100, pp. 1666–1703.

TAYLOR, F. W., JR., EDWARDS, R. L., and 2 others, 1990, Seismic recurrence intervals and timing of aseismic subduction inferred from emerged corals and reefs of the central Vanuatu (New Hebrides) frontal arc: Jour. Geophys. Res., v. 95, pp. 393–408.

TAZIEFF, H., 1970, The Afar triangle: Sci. American, v. 222, no. 2, pp. 32–40.

TEN BRINK, U. S., and BEN AVRAHAM, Z., 1989, The anatomy of a pull-apart basin: Seismic reflection observations of the Dead Sea basin: Tectonics, v. 8, pp. 333–350.

TE PUNGA, M. T., 1957, Live anticlines in western Wellington: New Zealand Jour. Sci. Technol., sec. B., v. 38, pp. 433–446.

TWIDALE, C. R., 1968, Geomorphology with special reference to Australia: Thomas Nelson (Australia) Ltd., Melbourne, Australia, 406 pp.

VANDERVOORT, D. S., and SCHMITT, J. G., 1990, Cretaceous to early Tertiary paleogeography in the hinterland of the Sevier thrust belt, east-central Nevada: Geology, v. 18, pp. 567–570.

WALLACE, R. E., 1978, Geometry and rates of change of fault-generated range fronts, north-central Nevada: U.S. Geol. Survey Jour. Res., v. 6, pp. 637–650.

———, ed., 1990, The San Andreas Fault system, California: U.S. Geol. Survey. Prof. Paper 1515, 283 pp.

WERNICKE, B., 1992, Cenozoic extensional tectonics of the U.S. Cordillera, in Burchfield, B. C., Lipman, P. W., and

Zoback, M. L., eds., The Cordilleran Orogen: Conterminous U.S.: Geol. Soc. America, The Geology of North America, v. G–3, pp. 553–581.

WESSON, R. L., HELLEY, E. J., and 2 others, 1975, Faults and future earthquakes, in Borcherdt, R. D., ed., Studies for seismic zonation of the San Francisco Bay region: U.S. Geol. Survey Prof. Paper 941, pp. 5–30.

WILSON, J. T., 1965, A new class of faults and their bearing on continental drift: Nature, v. 207, pp. 343–347.

WOOD, R. A., PETINGA, J. R., and 3 others, 1994, Structure of the Hanmer strike-slip basin, Hope Fault, New Zealand: Geol. Soc. America, v. 106, pp. 1459–1473.

Chapter 6
Volcanoes

Just as tectonic processes and landscapes can be studied systematically either in the temporal and spatial context of plate tectonics (Chapter 3) or as discrete landforms created by tectonic deformation (Chapter 5), so volcanoes can be studied either on a "megageomorphic" scale or as a special group of landforms built by a set of localized, dramatic, and often dangerous processes. This chapter describes and explains the plate-tectonic setting of volcanoes. However, the emphasis is on volcanoes as constructional landforms—parts of the modern landscape that either are being built or were built recently enough for their constructional form to be recognized in spite of erosion. Volcanic activity ("eruptions," "volcanicity," and "volcanism" are synonyms) must be described, as must the products of eruptions, in order to understand the structure and genesis of volcanic landforms. Posteruptive erosional development on volcanic structures is deferred and considered with other forms of structurally controlled erosion (Chapter 12). By this arrangement, full attention can be given here to volcanoes as regional assemblages of dramatic, often strikingly beautiful, constructional landforms, built by processes that have had legendary impact on human senses.

Volcanoes have always had an uncertain place in systematic geomorphology. Establishing the pattern, W. M. Davis (1905) treated volcanism as an "accident" that occurs so arbitrarily in time and place and is so disruptive to the erosional development of landscapes that the landforms cannot be treated in a systematic

manner. A common arrangement of recent geomorphology textbooks is, if they are mentioned at all, to review both constructional and erosional volcanic landforms in a single chapter late in the book.

Some books define volcanoes as the vents or fissures from which hot subsurface materials are transferred to above the former ground surface. Others define volcanoes as the mountains or hills of erupted rock around the vents. Literally, then, there is no general agreement on whether volcanoes should be defined as holes or hills! Cotton (1944, preface) summarized the issue as follows: "An approach to volcanism may be made through either petrology or geomorphology, a volcano and its mechanism being regarded either as a geological crucible or as a builder of landforms."

Information on volcanoes, their activity, and their products can be found in petrology books as well as in geomorphology books (Cotton, 1944; Rittmann, 1962; Blong, 1984; Fisher and Schmincke, 1984; Short, 1986; Ollier, 1988; Decker and Decker, 1989, 1991; Francis, 1993; Scarth, 1994). In addition, several recent monographs catalog and document the volcanoes of the world (Simkin and Siebert, 1994), North America (Wood and Kienle, 1990), the central Andes (de Silva and Francis, 1991), Hawaii (Decker et al., 1987), Antarctica (LeMasurier and Thomson, 1990), and the Asian mainland (Whitford-Stark, 1987).

Although volcanic landforms and geomorphic processes are intrinsically included in all the aforementioned sources, none of them are devoted to volcanic geomorphology. Francis (1993, p. 340) included in his excellent book a chapter entitled "Volcanoes as landscape forms" after the title of Sir Charles Cotton's 1944 book, which though 50 years old, remains the only authoritative book on volcanic geomorphology.

For at least 200 years, scientists have speculated about the impact of volcanic eruptions on the earth's climate (Francis, 1993, pp. 368ff). In recent decades, major explosive eruptions of Agung in Indonesia (1963), el Chichón in Mexico (1982), and Pinatubo in the Philippines (1991) all ejected dust and acidic aerosol condensates into the stratosphere where they spread over the entire earth. Global cooling of about 0.2°C was documented for one or two years after each event (Jones and Kelly, 1996). These well-studied eruptions, and others that have occurred since weather-monitoring satellites have been in orbit, have forced climatologists to include volcanoes as potent factors in at least short-term climate change. Much of the current scientific interest in volcanoes concerns the influence of particulate and gaseous ejecta on climate (Fiocco et al., 1996).

It is generally understood that volcanoes are characterized by eruptive activity. That is, hot gases, liquids, molten rock, and shattered rock fragments are forcibly or rapidly ejected from openings in the earth's surface. Classification of volcanoes beyond that level of understanding has been based on a wide variety of criteria, such as (1) chemical composition and temperature of the ejecta; (2) physical state of the ejecta— whether a gas, liquid, or solid phase is dominant; (3) historic record of volcanicity—active, dormant, or extinct; (4) shape of the aperture—whether a central vent or linear fissure, for instance; (5) nature of the volcanic activity—exhalative gas discharge, effusive lava flows, explosive ejection of fragments, frothing of gas-charged liquid, fire fountains, and so on; or (6) shape of the resulting landforms—mounds, rings, craters, domes, cones, shields, or plateaus. For geomorphic analysis, both volcanic processes and landforms must be understood. For evaluating volcanic hazards and the impact of volcanoes on climate, understanding the nature and variability of volcanic activity is absolutely essential.

GEOLOGIC AND GEOGRAPHIC SIGNIFICANCE OF VOLCANOES

Planetary Volcanology

More than 25 years of exploration of Mercury, Venus, Mars, and the Earth's moon has stimulated much contemporaneous study of terrestrial volcanoes (Frankel,

1996; McGuire et al., 1996). The smaller planets (Mercury, Mars) and our moon are probably "single plate" objects whose surfaces were shaped by meteorite impact and voluminous lava flows in their early history. Their surface landforms may have frozen in place more than 3 billion years ago, however. Their dark areas are probably plains (mare) formed by basalt magma that welled upward to fill impact craters. Mars has some great basaltic shield volcanoes with summit calderas and a variety of other volcanic landforms (Hodges and Moore, 1994).

Venus may have had mantle convection during its early history, but without an ocean to chemically react with and differentiate mantle rock into continental crust (pp. 8–9), only a continuous basaltic crust evolved, probably similar in composition to earth's oceanic basalt. However, Venus has a spectacular abundance of volcanic landforms similar to those built by terrestrial basalts (Head et al., 1991), and may have had active volcanism until 1 billion years ago.

Spacecraft exploration of the outer planets and their satellites has been especially challenging and exciting for volcanologists (Mursky, 1996). The numerous moons that orbit Jupiter, Saturn, Uranus, and Neptune have diverse compositions ranging from sulfur to water ice. Volcanic landforms dominate the landscape of the Jovian satellite Io, and eruptions have been observed, probably of molten sulfur. Other icy satellites show surface forms that may be related to breakup of large surface slabs and upwelling of internal material. Since these satellites are probably made largely of water ice, a special term, *cryovolcanism*, has been proposed to describe their active surface processes (Francis, 1993, p. 432).

Plate Tectonics and Volcanism

Large-scale motions in the earth's interior, involving both the crust and mantle, determine the location of volcanoes. Cylindrical rising segments of convection cells or upwelling plumes, perhaps from deep within the mantle adjacent to the core-mantle boundary, probably determine the location of semipermanent "hot spots" beneath the lithosphere (front endpaper) (Richards et al., 1989; White and McKenzie, 1989a, 1989b; Hill et al., 1992). A chain of basalt volcanic islands and seamounts such as the Hawaiian Islands may be built as an oceanic lithospheric plate moves over the hot spot (Ihinger, 1995), or a continent may be stretched and rifted with an outpouring of flood basalt. Where slabs of relatively cool, dense, oceanic lithosphere sink to a depth of about 125 km in a subduction zone, andesitic magmas are generated to form volcanic arcs, either as part of an island arc on oceanic lithosphere (Figure 3-2) or as part of a cordilleran or collisional orogenic belt (Figure 3-3). The most abundant effusive volcanism, which accretes new oceanic basalt lithosphere at diverging plate margins along midocean spreading ridges, is probably generated by shallow upper mantle differentiation, the result of passive separation of oceanic lithospheric plates under the pull of their margins sinking in subduction zones.

The annual production rate of volcanic ejecta is estimated to be as much as 20 km^3, most of which is basalt accreted at midocean spreading ridges. Only minor amounts of this rock appear above sea level in islands such as Iceland and the Azores. Another 2 km^3 per year may be added to the subaerial landscape although single major eruptions may greatly exceed the annual average (Francis, 1993, p. 3).

Abundance and Geographic Distribution of Subaerial Volcanoes

Even though only 10 percent of the typical annual volcanic output appears above sea level, and the volcanoes that we call mountains are small compared to their submarine counterparts (Figure 6-1), volcanic contribution to the terrestrial landscape is significant. Entire geomorphic provinces are volcanic in origin. The great basalt plateaus (p. 107) are among the largest constructional relief features of the subaerial

landscape (although exceeded greatly by the ice caps of Greenland and Antarctica). An estimated 2 million km^2 of land surface have been covered to depths of 500 to 3000 m by lava flows in geologically recent time (Rittmann, 1962, p. xiii). Layers of fragmental volcanic ejecta (ash and dust), as recorded by their frequent occurrence in deep-sea sediment cores and ice cores, have probably covered the entire earth. More than 500 volcanic eruptions have been recorded historically, and more than 1300 have probably erupted in Holocene time (Simkin and Siebert, 1994). Tens of thousands of extinct volcanoes can be recognized by their form or structure. About 62 percent of the active subaerial volcanoes are on the subduction zones around the Pacific basin, in the circum-Pacific "ring of fire" (front endpaper). Another 22 percent are in Indonesia, 10 percent are in the Atlantic Ocean (including the Caribbean region), and a few percent each are in Africa, the Mediterranean, and the Middle East. The remaining few form the Hawaiian Islands and other midocean islands, and a few others are on the continents (for example, Yellowstone).

Volcanoes have attracted renewed geomorphic interest in recent years because new techniques of geochronometry permit precise dating of eruptive rocks. It is now possible, by measuring the volume of rock eroded from a dated volcanic landscape, to calculate the absolute rates of landscape modification (p. 336). Because many volcanic landforms have sym-

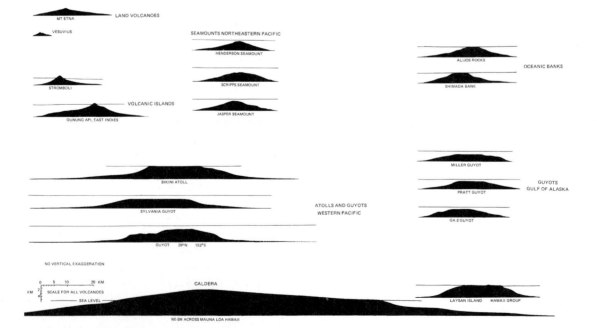

FIGURE 6-1. Size comparisons for some subaerial volcanoes, volcanic islands, and guyots (flat-topped, volcanic seamounts). No vertical exaggeration (Menard, 1964, Figure 4-8).

metric geometry, their original constructional shape can be restored with much greater confidence than is possible with tectonic landforms. Furthermore, key marker beds of volcanic ash, dated by radioactive isotopes and chemically identified as to place of origin, can blanket millions of square kilometers of landscapes formed by other processes. It is hard to overstate the scientific value of identifiable ash beds or aerosol-rich ice layers that were inserted at an instant of geologic time into a stratigraphic sequence or onto an evolving landscape (pp. 60, 112). Sulfate-rich ice layers, derived from aerosol precipitation following major Holocene explosive eruptions, form key reference horizons in the Greenland ice sheet (Chapter 18). The nearly perfect preservation of annual layers in the Greenland ice has permitted a precise chronology of explosive, sulfur-rich volcanism to be established for at least the past 9000 years (Zielinski et al., 1994). The full significance of these layers to climatic history has yet to be established.

A final point about the geomorphic importance of volcanoes concerns their wide distribution. Volcanic landforms are constructed independently of any climatically controlled processes. Volcanic edifices are built in or on the Antarctic ice cap, in the tropical forests of Melanesia and Indonesia, in deserts, and in every other geomorphically significant climate. In each instance, the initial structure and shape of the constructed landforms are similar. A great opportunity is thus available for geomorphologists to study contrasting erosional processes and rates in many climatic regions on constructed volcanic landforms that are not only similar but datable. This aspect of geomorphic volcanology alone justifies an extended systematic treatment of volcanic landforms.

VOLCANIC ACTIVITY AND PRODUCTS

The products of eruption can be described in terms of three types of activity: *exhalative* (gas), *effusive* (lava), or *explosive* (tephra). Alternatively, characteristic modes of activity and products can be defined by describing active volcanoes that behave in certain ways. Both devices are used in the following pages.

Exhalative Activity and Mud Volcanoes

Hot springs, geysers, fumaroles, solfateras, and many other terms are used to describe *hydrovolcanic* vents that continuously, intermittently, or cyclically discharge hot water, steam, and other gases but rarely any solid or molten rock. Such hydrovolcanic vents are generally not considered volcanoes even though their activity is similar to volcanic eruption. They build minor landforms such as *sinter mounds* and *cones* of precipitated minerals.

Mud volcanoes make impressive small landforms over exhalative vents. Several in the Copper River Basin of Alaska were described by Nichols and Yehle (1961). They are between 45 and 95 m in height and discharge mineralized warm water and gas, including light hydrocarbons probably derived from the decay of buried peat beds or coal. One of the largest cones contains abundant fragments of lava and volcanic glass. Travertine crusts intermingle with the hardened mud and angular rock fragments of the crater rims. Cones such as those in the Copper River Basin are probably the minimum accepted "volcanic" landform to most geomorphologists.

In 1951, a spectacular mud volcano in northeastern California suddenly erupted steam, gases, and mud to an estimated height of 1000 m (White, 1955). Fine debris, including pellets of mud, fell on a town and farms 7 km downwind from the eruption. The entire episode ended within four days. After a few months of quiescence, probably with some ground subsidence, the area was flooded with boiling water and mud springs. This kind of mud volcano draws its energy from volcanic heat, but most of the water seems to be of near-surface origin.

Lake Nyos, in the west African nation of Cameroon, occupies a gas maar (p. 110) only a few centuries old. It has a surface area of about 1.5 km^2 and a depth of about 200 m. During the evening of August 21, 1986, a large volume of a gas heavier than air (probably CO_2) was released from Lake Nyos. It flowed downhill through valleys as far as 25 km from the lake as a layer initially 120 m thick above lake level. Countless animals and more than 1700 people were killed, presumably by a lethal mixture of 10 to 20 percent CO_2 in the air. An international conference called to evaluate the disaster could not conclude whether a *phreatic* (p. 103) eruption in the maar had occurred or whether the lake had become slowly supersaturated with exhalative CO_2 gas until it convectively overturned and spontaneously degassed (Sigvaldason, 1989, and related papers in the same volume; Zhang, 1996). Certainly, volcanic activity of some kind was the cause of the disaster. The tragedy has alerted scientists to the possibility of similar hazards in other recently active volcanic regions. It also offered an intriguing explanation for widespread "bone beds" of mammalian remains in strata that are rich in volcaniclastic sediments, especially some that are common in eastern Africa. Even in more predictable Hawaiian basalt volcanoes, lethal exhalative activity is a continuing hazard.

Effusive Activity: Lava

Lava is the name given to erupted molten rock and to its cooled, solid equivalent, provided that it has an obvious flowlike surface form. More often, solidified lava is simply called "volcanic rock," along with solidified ejecta, because the fine grain sizes or glassy textures of eruptive rocks may not permit a clear conclusion about their origin. **Effusive** (as contrasted to explosive) activity refers to copious outpourings of lava from a vent or fissure.

Most lavas are molten silicates although silica-free carbonatite (sodium-calcium carbonate) melts are known, especially in the rift systems of eastern Africa (Dawson et al., 1994). Commonly, volcanic vents and fumaroles are surrounded or lined by thick exhalative precipitates of elemental sulfur, which may be rich enough for commercial mining. At times, the sulfur masses may be remelted and flow like lava (Naranjo, 1985). The Japanese name for the volcanic island of Io Jima (Iwo Jima) means "sulfur island." It is a curious linguistic coincidence that the Jovian moon Io, named for a Greek goddess, also has huge sulfur volcanoes and flows (p. 93). The ordinary range of SiO_2 in volcanic rocks is from about 35 to 75 percent, and the viscosity of natural silicate melts increases with silica content through 13 orders of magnitude. At any given temperature, acidic (silica-rich) lavas are more viscous than basic (silica-poor) lavas (Fisher and Schmincke, 1984, p. 53; Ryan and Blevins, 1987). The composition of the molten phase is a major factor in landform development because of the control it exerts on viscosity. Low-silica Hawaiian basalt freely flows down slopes of only a few degrees. The resulting landforms are likely to have low relief compared to their lateral extent, with smooth, domal form. On the other extreme, degassed silica-rich dacite and rhyolite may not flow from a vent but form a solid, protruding plug or spine.

Temperature and dissolved volatiles are the two other factors that determine the viscosity of lava. Confining pressure alone normally elevates the melting temperature of solids, but when water and other volatile components such as CO_2 and SO_2 are dissolved in a silicate melt, the melting temperature is lowered. Although acidic magmas can dissolve more volatile compounds than basic magmas, and thereby be highly mobile at depth, they become highly viscous at surface pressure, when the volatiles have boiled off. In the process of degassing, they become violently explosive.

Surface pools and flows of Hawaiian basalt have a surprisingly narrow range of temperatures between 1100°C and 1200°C. The viscosity increases by a factor of 10 with cooling of only 100°C (Ryan and Blevins, 1987, pp. 458–459). On cooling below 1150°C, some of the minerals in a silica-poor melt begin to crystallize, and the viscosity increases rapidly with only slightly lower temperature. The heat of fusion of mineral components is probably the reason that most lava has such a relatively narrow temperature range. These relations between temperature, volatile pressure, and composition clarify why effusive activity is almost totally confined to silica-poor, or basic, melts. By far the most characteristic effusive lava is basalt although intermediate-composition lavas such as andesite and dacite also form stiff, rough, slow-moving flows.

Lava emerges from vents or fissures as white-hot lobes or tongues that flow downhill (Figure 6-2). Each lobe or tongue is called a **lava flow.** Hawaiian lava flows generally advance at slower than human walking speed (3 to 5 km/h), but may double or triple that rate on steep slopes. Lava streams in open channels on steep slopes may reach speeds of 55 km/h. They range in thickness from less than 1 to 20 m (Mullineaux et al., 1987, p. 601).

Lava flows follow preexisting topography downhill. On open slopes, they can be diverted by walls or dikes (Barberi et al., 1993). A flow front may congeal and stop when a lateral lobe breaks out higher upslope and follows an easier path.

The upper surface of a basalt lava flow is usually highly *vesicular* or porous due to gas bubbles that escape during the final stages of cooling. If sufficiently rough in texture, the rock is called **scoria.** Two contrasting surface textures develop on flows, both commonly known by the traditional Hawaiian terms: **aa** for angular, blocky, scoriaceous surfaces, and **pahoehoe** (or ropy) for smoothly twisted, convoluted surfaces. Rough aa surfaces on flows result from the cooled and broken surface layer being carried forward by laminar flow and then being "plowed under" at the front of the advancing lava. The motion is analogous to that of the endless tread on a tracked vehicle. *Blocky lava* is sometimes distinguished from aa surfaces on the basis of a slablike or blocky character. Pahoehoe surfaces develop on hotter, more fluid, flows (Figure 6-3). An extreme variant of pahoehoe lava occurs when hot, fluid lava flows into or erupts under water. Blobs or lobes of lava up to a meter in diameter form steam-jacketed, tough, flexible skins under a film of steam and pile up like pillows or sandbags while still molten in their interiors. The result is **pillow lava** (Figure 6-4), a diagnostic feature of subaqueous eruption.

Pahoehoe flows remain hot and fluid inside even though their top and bottom surfaces and lateral margins have solidified (Hon et al., 1994). One result of

FIGURE 6-2. Active lava flows, Mauna Loa, Hawaii (photo: Hawaiian Volcano Observatory, U.S. Geol. Survey).

FIGURE 6-3. Pahoehoe lava draped over a former sea cliff south of Kilauea Volcano, Hawaii (photo: D. A. Swanson, U.S. Geol. Survey).

fluid lava becoming encased in cooler margins is the formation of *lava tunnels* (Figure 6-5). These caves, many of which are several kilometers in length and large enough for humans to enter, mark the draining out of the molten lava from within a solidified surface. They provide highly efficient routes for lava to flow, maintaining heat and mobility where a surface flow would rapidly cool and halt. Tunnel flows are capable of thermal erosion, remelting and transporting preexisting rock around them.

Lava flows, in common with intrusive dikes and sills and thick sheets of fused tephra, develop distinct sets of intersecting contraction joints upon cooling that result in the familiar *hexagonal* or *columnar* jointing (Figure 12-1). These and other primary structures exert powerful controls on the landforms eroded from volcanic terranes (Chapter 12).

Explosive Activity: Tephra

Explosive volcanism along converging plate boundaries (front endpaper) supplies only 10 to 13 percent of the magma that reaches the earth's surface annually, but because of its impact on human beings, includes 63 percent of all known Holocene volcanoes and 84 percent of all historical eruptions (Simkin and Siebert, 1984, p. 110). To classify the "bigness" of historic volcanic eruptions, a *volcanic explosivity index* (VEI) was devised (Newhall and Self, 1982; Simkin and Siebert, 1994). The VEI combines the volume of ejecta, height of the ejecta and gas cloud above the volcano, descriptive terms, and other criteria on a simple scale of 0 to 8

FIGURE 6-4. Submarine lava pillows in the median valley of the Mid-Atlantic Ridge. Pillows show effects of repeated surface quenching during expansion (photo: R. D. Ballard).

of increasing explosivity (Table 6-1). More than 4800 historical eruptions were used to compile the index. A voluminous lava flow on Hawaii might be locally important, yet have an explosivity index of 0. Nevertheless, the VEI scale is the best available measure of the relative size of 80 percent of the world's known eruptions.

Obviously, explosive volcanism has a human impact far out of proportion to its role in shaping the earth's surface. Explosive eruptions are capable of killing millions of people and devastating thousands of square kilometers. They may give little warning. Of 205 well-documented historical eruptions with a VEI ≥ 3 (Table 6-1), 45 percent experienced the major eruption within one day of the first indication of activity, and half the first-day eruptions occurred within the first hour (Simkin and Siebert, 1984, p. 112). Active magma chambers under large calderas such as Yellow-

FIGURE 6-5. Unnamed lava tube, Lava Beds National Monument, California (Waters et al., 1990) (photo: 11097ct, U.S. Geol. Survey).

Table 6-1
Criteria for Estimation of the Volcanic Explosivity Index (VEI)

Criteria	VEI:	0	1	2	3	4	5	6	7	8
Description		nonexplosive	small	moderate	mod–large	large	very large	→		
Volume of ejecta (m³)		$<10^4$	10^4–10^6	10^6–10^7	10^7–10^8	10^8–10^9	10^9–10^{10}	10^{10}–10^{11}	10^{11}–10^{12}	$>10^{12}$
Column height (km)*		<0.1	0.1–1	1–5	3–15	10–25	>25 →			
Qualitative description		← "gentle, effusive" → ← "explosive" → ← "cataclysmic, paroxysmal, colossal" →								
					← "severe, violent, terrific" →					
Classification (see Table 6-2)			← "Strombolian" → ← "Plinian" →							
		← "Hawaiian" → ← "Vulcanian" → ← "Ultraplinian" →								
Duration of continuous blast (hours)		← <1 → ← >12 →								
				← 1–6 →						
					← 6–12 →					
Tropospheric injection		negligible	minor	moderate	substantial →					
Stratospheric injection		none	none	none	possible	definite	significant →			
Eruptions (total in file)		699	845	3477	869	278	84	39	4	0

Simplified from Newhall and Self, 1982, Table 1, with additions from Simkin and Siebert, 1994, Table 4.
*For VEI's 0–2, uses km above crater; for VEI's 3–8, uses km above sea level.

stone, Wyoming, and Long Valley, California, are now monitored for warnings of possible eruptions. Tectonic movements (Figure 5-27) are important clues to possible eruptions, as are certain kinds of rhythmic seismicity, increased heat flow, and fumarolic activity. Unfortunately, the largest explosive volcanoes have the longest intervals between eruptions, typically hundreds to thousands of years, and people may forget or be unaware that they are living in a hazardous region.

In an explosive eruption, rock may be fragmented by a sudden, foaming loss of volatiles, by abrupt contact with groundwater or surface water, by grinding abrasion in the volcanic neck, or by dispersion of molten droplets. The resulting solid volcanic ejecta that settle out of air or water are called *pyroclastic* or *volcaniclastic* sediments or rocks. **Tephra** is a less cumbersome collective term for all fragmental ejecta from volcanoes. It is derived from the Greek word for "ash" and is intended to be used in parallel with the terms *lava* and *magma* (Thorarinsson, 1951, p. 5; Fisher and Schmincke, 1984, p. 89). Tephra consists of fragments with a wide range of grain sizes and shapes. *Ash* is the most common term for sand-size and finer tephra. Larger fragments or aggregates are called *lapilli* (gravel size; ejected either molten or solid), *blocks* (cobble or boulder-size solid ejecta), or *bombs* (twisted, air-cooled masses ejected in a molten state) (Fisher and Schmincke, 1984, p. 89–96). The terminology is elaborate and extensive because explosive volcanic eruptions are among the most feared and discussed phenomena on earth.

Tephra is strongly size-sorted during air transport. Large bombs and blocks (up to 5 m diameter) are propelled to distances proportional to their original blast velocities. A theoretical horizontal limit of about 12 km for a 5-m block has been calculated although no blocks are known to have traveled more than half that distance (Fagents and Wilson, 1993, p. 370). Intermediate-size particles travel shorter distances because they are subjected to relatively large frictional and inertial losses. Lapilli and ash are airborne for many kilometers, the grain size and thickness of the resulting tephra layer decreasing with increasing distance from the source (Porter, 1972; Fisher and Schmincke, 1984, pp. 125–162). Their transport is controlled more by the effective winds than the explosive force of the volcano.

A variety of explosive volcanic activity has been recorded by direct observations of relatively small eruptions. In the eruption of Mount St. Helens, Washington, on May 18, 1980, the north flank of the volcano swelled, oversteepened, and collapsed as an avalanche. The sudden pressure drop in the central vent led to a *directed blast,* a violent lateral explosion that flattened forests as far as 25 km from the vent (Figure 6-6) (Lipman and Mullineaux, 1981). A much larger flank collapse about 7500 years ago on Socompa Volcano in northern Chile covered about 600 km² with avalanche debris (deSilva and Francis, 1991).

Since the Mount St. Helens eruption, the terms *blast* or *blast explosion* have been used with increasing frequency to describe laterally directed eruptions

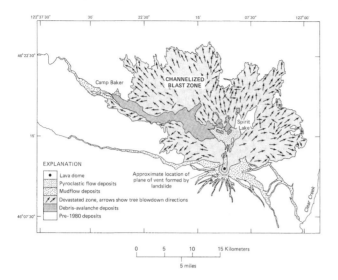

FIGURE 6-6. Major kinds of volcaniclastic deposits, excluding fallout ash, from the May 18, 1980 eruption of Mount St. Helens, Washington (after Lipman and Mullineaux, 1981, Plate 1 and Figure 219).

(Fisher, 1990; Francis, 1993, p. 181; Macias and Sheridan, 1995). In addition to the Mount St. Helens and Socompa eruptions, other great blast explosions include the 1902 eruption of Mont Pelée in Martinique and a previously little-known 1907 eruption in Kamchatka, Russia (Macias and Sheridan, 1995). Initial blast debris may travel at supersonic speed, destroying everything in its path, with the ground-hugging debris spreading out fanlike until it slows to a halt. The debris may overtop ridges and go airborne like a giant avalanche (p. 185), or it may turn and flow down radial valleys. A blast is one variety of a *pyroclastic surge*, in which turbulent, dense, hot, debris-rich masses flow outward and generally downward on the flanks of explosive volcanoes (Francis, 1993, pp. 235ff). Surge deposits are usually found at the bottom of an eruptive depositional sequence, proving that they arrive at distant sites well ahead of other eruptive debris. A pyroclastic surge was responsible for the widespread death and destruction at Herculaneum by the A.D. 79 eruption of Vesuvius (Francis 1993, p. 241).

A **nuée ardente,** or glowing cloud, is the most terrifying form of pyroclastic activity. In May 1902, a nuée ardente from Mt. Pelée, Martinique, killed 30,000 people in the city of St. Pierre (Smith and Roobol, 1990). The eruption, the first of its kind to be scientifically studied, gave the name Peléean to such peculiarly explosive activity (Table 6-2). The phenomenon can be produced either by a lateral, or directed, eruption of superheated steam and glowing hot tephra or by "boiling over" of foaming lava. Beneath the tower-

ing, glowing cloud of hot, fine tephra and gases, the leading edge of a nuée ardente may speed downhill at 50 m/s. The base of the fluidized mass is heavier than air, so it remains in contact with the ground and flows in a turbulent fashion as a *pyroclastic flow.*

Volcanologists emphasize that pyroclastic flows beneath nuées ardentes are soundless, evidence that despite their forward velocity comparable to the winds of a hurricane, the glowing fragments in them are cushioned by hot gas and are not abraded by the rapid, turbulent motion. The temperature in the nuée ardente that devastated St. Pierre is estimated to have been between 650°C and 1060°C. It melted glass.

Most of the volume of a nuée ardente is hot gas, primarily superheated steam. One model for the emplacement of the Taupo ignimbrite in the central plateau of North Island, New Zealand, has it deposited by a pyroclastic flow that was only 0.3 percent solids by volume (Dade and Huppert, 1996). The tephra on the ruins of St. Pierre after the 1902 disaster was less than a meter in thickness, generally only about 30 cm. A nuée ardente in the Philippines charred trees to a height of 4.5 m, clearly marking the thickness of the flowing layer, yet left only 13 mm of ash on the ground (Macdonald, 1972, p. 5). On cooling, the steam condenses to make scalding hot mud of the tephra. Although the cloud of airborne tephra that towers above and obscures the fast-moving ground layer is technically the nuée ardente, the basal pyroclastic flow is responsible for the mass of the resulting deposit (Fisher and Schmincke, 1984, p. 187).

The eruption of Nevada del Ruiz, Colombia, on November 13, 1985, was the culmination of almost a year of continuing explosive activity. In the final explosion, relatively small hot pyroclastic flows on the glaciated summit melted part of the summit glacier, which generated rapidly moving **lahars,** or volcanic mudflows (p. 179), that eroded additional masses of rock from mountain valleys and destroyed numerous villages along lower river valleys with the loss of 25,000 lives (Naranjo et al., 1986; Mileti et al., 1991).

The terrible lesson that was learned from the Nevada del Ruiz eruption is that a relatively minor eruption on a snow- and ice-covered volcano can release massive lahars that can travel more than 100 km from their source, devastating valley floors; destroying habitations, dams, and water supplies; and endangering large human populations. In the United States, attention has been drawn to the proximity of ice-covered Mount Rainier to the heavily populated, highly industrialized Puget Sound lowland near the Seattle-Tacoma, Washington, metropolitan area. Mount Rainier last erupted approximately 150 years ago and is regarded as active. More than 100,000

Table 6-2

Types of Volcanic Eruptions (modified from Short, 1986, Table 3-1)

Type	Characteristics
1. Icelandic	Fissure eruptions, releasing low-viscosity basaltic magma; non-explosive, gas-poor; great volumes of lava issued, flowing as sheets over large areas to build basalt plateaus.
2. Hawaiian	Fissure, caldera, and pit crater eruptions; mobil basalt lavas, with some gas; quiet to moderately explosive eruptions (VEI: 0 or 1); occasional rapid emission of gas-charged lava produces fire fountains; only minor amounts of tephra; builds lava domes.
3. Strombolian	Summit crater with lava lake; moderate rhythmic to nearly continuous explosions resulting from spasmodic gas escape (VEI:1–3); clots of lava ejected, producing bombs and scoria; periodic more intense activity with outpourings of lava; light-colored clouds (mostly steam) reach upward only to moderate heights.
4. Vulcanian	Stratocones (central vents); viscous lavas; lavas crust over in vent between eruptions, allowing gas buildup below surface; eruptions increase in violence until lava crust is broken up, clearing vent, ejecting bombs, pumice, and lava flows from top of flank after main explosive eruption; dark ash-laden clouds, convoluted, cauliflower-shaped, rise to moderate heights depositing tephra along flanks of volcano. (Note: Ultraplinean eruption has similar characteristics but results when other types (e.g., Hawaiian) become phreatic and produce large steam clouds, carrying fragmental matter.)
5. Vesuvian	More paroxysmal than Strombolian or Vulcanian types; extremely violent expulsion of gas-charged intermediate-composition magma from stratocone vent; eruption occurs after long interval of quiescence or mild activity; vent tends to be emptied to considerable depth; lava ejects in explosive spray (glow above vent), with repeated clouds (cauliflower) that reach great heights and deposit tephra.
6. Plinian*	More violent form of Vesuvian eruption (VEI: 3 or more); last major phase is uprush of gas that carries cloud rapidly upward in vertical column for kilometers; narrow at base but expands outward at upper elevations; cloud generally low in tephra.
7. Peléean	Results from high-viscosity silicic lavas; delayed explosiveness; conduit of stratovolcano usually blocked by dome or plug; gas (some lava) escapes by lateral blasts or by destruction or uplift of plug; gas, tephra, and blocks move downslope in one or more blasts as nuées ardentes producing directed deposits.
8. Katmaian	Variant of a Peléean eruption characterized by massive outpourings of fluidized ashflows; accompanied by widespread explosive tephra; ignimbrites are common end products, also hot springs and fumaroles.

*named for the Roman scholar, Pliny the Younger, who wrote a detailed narrative of the Vesuvian eruption of A.D. 79.

people live on extensive lahar deposits on the valley floors around Mount Rainier. Even a moderate eruption of Mount Rainier would cause enormous personal and economic losses to the region (U.S. Geodynamics Committee, National Research Council, 1994; Scott et al., 1995).

The largest known historical eruption was the explosion of Tambora volcano in Indonesia in A.D. 1815 (Self et al., 1984; Stothers, 1984) although it was only about one-tenth as large as the prehistoric Toba eruption or the latest eruption of Yellowstone (p. 112). The volcanic cone more than 1400 m in height was totally destroyed. The earth-circling dust cloud caused spectacular sunsets for months after the eruption and probably was responsible for the colder-than-normal temperatures recorded in the northern hemisphere during the following two years. Another large explosive eruption occurred when the Greek island of Thera (Santorini) exploded, probably in 1628 B.C. (Kuniholm et al., 1996), and covered the eastern Mediterranean and Black Sea basins with pumice and tephra (Stanley and Sheng, 1986; Newhall and Dzurisin, 1988, pp. 129–134; Gulchard et al., 1993). The event has been blamed for the decline of the Minoan culture, the darkness and plagues in Egypt during the exodus of the Jewish people, and the legends of lost Atlantis.

Less thick than pyroclastic flows and lahars but even more widespread, *ash showers* result from some explosive eruptions. Explosive force cannot propel tephra more than a few kilometers, so wind must be the transporting force for widespread ash showers. They are cold when they reach the ground. The entrainment of tephra by wind is governed by the same principles that control sediment transport in water (Figure 10-11) and other kinds of fine sediment in the atmosphere (Figure 13-12). However, glassy shards of tephra have extremely irregular shapes that allow them to be carried much farther than is predicted for more compact, spherical particles. When a tephra sheet is mapped in detail (Figure 6-7), the direction of prevalent wind during the eruption is obvious.

Widespread ash showers are the great tools of **tephrochronology.** Individual ash beds can be identified by a variety of analytical methods, including the refractive index of the glass shards, chemical analyses, especially of the rare-earth elements, and mineralogy. Ash from a huge eruption of Toba volcano in Sumatra about 75,000 years ago has been identified in numerous submarine cores in the eastern Indian Ocean at distances of 500 to 2100 km from the volcano, and in a sample from India, 3100 km from the volcano (Rose and Chesner, 1987; Chesner et al., 1991).

Tephrochronology has become an important tool for measuring rates of constructional and erosional landscape development, especially in the volcanic terranes of Japan, New Zealand, Iceland, and western North America. One example from Japan suggests the potential of the method. West of Tokyo, beds of lapilli and ash totaling 100 m in maximum thickness compose the Older Fuji Tephra, which was deposited by hundreds of ash falls over a long period, probably between 80,000 and 10,000 years ago (Machida, 1967; Arculus et al., 1991). It contains preceramic artifacts in its upper part. The Younger Fuji Tephra, which also accumulated during multiple ash falls, is 6 m thick in the vicinity of Mt. Fuji but less than 1 m thick near Tokyo, 80 km to the east (Figure 6-8). It overlies archeologic sites with pottery, indicating a young age. The uppermost tephra layer is known to date from historical eruptions since A.D. 781. The weathered zone, fine ash, and soil between the Older and Younger Fuji Tephras shows that the latest historic activity of Fuji began after a long period of quiescence. These widespread tephra layers are used to measure tectonic displacement on faults, warping of coastal terraces, subsidence in the metropolitan Tokyo area, and many other tectonic movements of geomorphic interest. They have also been used to study rates of soil formation and for dating archeologic sites. In Arctic Canada, pumice that floated ashore 5000 years old marks a distinctive emerged beach from which postglacial isostatic uplift can be accurately measured (p. 388).

Subaqueous Volcanism: Hyaloclastite

Pillow lavas have been noted as one diagnostic feature of subaqueous eruptions of fluid lava. They are especially common on the deep ocean floor, where basalt is the only eruptive rock, and where the high hydrostatic

FIGURE 6-7. Thickness map of air-fall ejecta from Mount St. Helens on May 18, 1980. Lines represent uncompacted thickness in mm; +, light dusting of ash; x, no ash observed; circles, observation sites (modified from Sarna-Wojcicki et al., 1981, Figure 336).

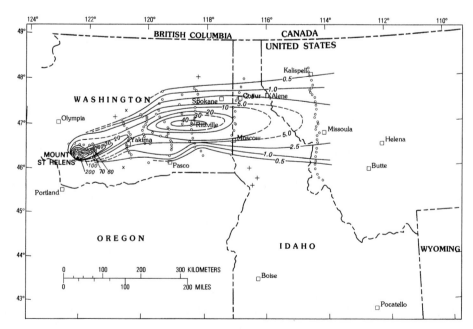

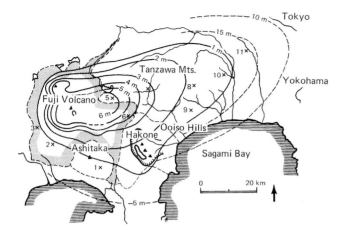

FIGURE 6-8. Isopach map of the Younger Fuji Tephra Formation (solid contours) and the Older Fuji Tephra Formation (dashed contours). Locations of measured sections are shown by numbered crosses (Machida, 1967).

pressure prevents degassing of the lava and minimizes fracturing (Figure 6-4).

More viscous lavas, and those erupted at lesser depths, develop shattered glassy margins on pillows and flow surfaces. The related volcanic product is **hyaloclastite** (literally: glassy-fragment rock). Hyaloclastites are commonly found in association with pillow lavas. They are also common in Iceland and Antarctica, where basalt eruptions under glaciers or through glaciers have been frequent. In Marie Byrd Land of Antarctica, hyaloclastites make up significant proportions of several volcanic peaks that protrude through the ice sheet. The abundance of hyaloclastites in the volcanic rocks demonstrates that a thick ice sheet has been present in West Antarctica during most of the late Cenozoic Era (LeMasurier, 1990, p. 15).

Characteristic Eruptive Types

Volcanologists have traditionally used the names of certain well-known volcanoes to classify characteristic modes of eruptive activity. Since no two volcanoes are alike, such a classification could become nothing less than a description of all known eruptions. A summary of eruptive types, based on eight named volcanoes or other widely recognized modes of activity, is widely used for descriptions (Tables 6-1 and 6-2). None of the named volcanoes erupt with only the activity described, however.

Fire fountains are the hallmark of the **Hawaiian** eruptive type. They are curtains or fountains of glowing lava driven by gas pressure within the rift or vent. The gas may be derived directly from the magma or from limited contact with groundwater. In **Strombolian** activity, the lava is depleted of most gas before it

reaches the surface, and a glowing lava lake forms in the crater. The lake is episodically disrupted by large escaping gas bubbles so that larger blobs of lava are thrown from it (Parfitt and Wilson, 1995). A "wet" variant of Strombolian activity called **Surtseyan** (Francis, 1993, pp. 129–130) is named for the new island of Surtsey that grew off the coast of Iceland in 1963–1964. It was highly explosive because of the continued contact of lava and water until it built above sea level, after which effusive lava poured out of the vent and armored the hyaloclastite mound. Since the eruption ended in 1965, the island has become a laboratory for the study of plant and animal colonization of new land.

Not included in Table 6-2 are violent gas or steam explosions that erupt shattered old rock, with little inclusion of new lava or tephra. These explosions are called either **Ultraplinian** (Table 6-1), because their violence exceeds the Plinian type, or **phreatic,** because they are caused by groundwater or seawater entering the magma chamber below a vent. The enormous 1883 eruption of Krakatau, in Indonesia, was a major phreatic eruption (Simkin and Fiske, 1983), as was part of the 1815 Tambora eruption (p. 101).

Katmaian eruptions are characterized by massive foaming eruptions of glowing hot pumice and other gas-rich tephra. The 1912 eruption of Alaskan volcano Katmai, which created the Valley of Ten Thousand Smokes due to steaming groundwater venting through the ash flows, gave this eruptive type its name.

VOLCANIC LANDFORMS

Having extensively described and classified the types of volcanic activity and the products of volcanic eruptions, we now turn to the description of constructional volcanic landforms. The viscosity of the magma is the major factor that determines the type of eruptive activity and the products that are erupted. (Viscosity, in turn, is determined by fundamental properties of composition, temperature, and dissolved gas and water content of the magma rising within the feeder conduits.) Viscosity and the size of the various edifices can be combined as the basis for a classification of volcanic landforms (Table 6-3; Figure 6-9).

The vertical components of Table 6-3 are the usual type of volcanic rock, the viscosity of the magma, and as an equivalent, the type of activity (effusive, mixed, or explosive). For a horizontal component, the table uses the volume of erupted matter. Increasing quantities of lava and tephra do not simply produce larger constructional landforms of the same shape. The

Table 6.3

Classification of Volcanic Landforms

Volcanic Rock	Quality of Magma	Type of Activity	Quantity of Eruptive Material — Small ← → Great				
Basalt	Fluid, very hot, basic	Effusive	Lava flows	Exogenous domes	Basalt plateaus and shield volcanoes		
						Icelandic	Hawaiian
Andesite	*(Increasing viscosity, water and gas content, and silica percentage)*	Mixed	Scoria cones with large craters and flows	Composite cones (Strato-volcanoes)			Volcanic fields with multiple cones
			Tuff rings, tephra cones, and thick flows				
Dacite			Endogenous domes (plug domes, flows; tholoids, spines)	Ruptured endogenous domes with thick lava flows (coulees)			
Rhyolite	Viscous, relatively cool, acidic	Explosive	Maars with tephra	Maars with ramparts	Collapse and explosion calderas		Ignimbrite sheets
	Extremely viscous, abundant crystals	Explosive, mostly gas	Gas maars	Explosion craters			Resurgent calderas

Modified from Rittman (1962), Tables 4 and 5.

intriguing concept of *morphological capacity* was introduced by Rittmann (1962, p. 113) to account for changes in landform with increasing size. *Scoria cones* (Figure 6-10), for instance, can only grow to a certain size before they can no longer support the column of lava rising in the throat of the volcano. Then flows break out on the flank or from beneath the base of the cone (Figure 6-11, Color Plate 7), and subsequent growth of the cone follows a different trend. The change in the relative importance of specific volcanic landforms as the total edifice grows larger is similar to the biologic concept of *allometry,* the study of changing body proportions as an organism grows (Mosley and Parker, 1972).

In the following paragraphs, the forms that are more common and more significant to a general explanatory-descriptive scheme are described in a sequence beginning with the forms built by the least viscous, effusive lava eruptions, followed by consideration of the forms built by extremely viscous lava and explosive tephra eruptions. Within each eruptive type, a sequence of forms from small to large is reviewed. Most of the small landforms result from a single erup-

tion; multiple eruptions are necessary to build the large edifices.

Basalt Flows, Domes, and Shield Volcanoes

The basic geomorphic unit of a lava eruption is the *flow,* described previously. Flows may reduce the preexisting relief by filling valleys or may increase it by building elongate, flat-topped lobate sheets and tongues of lava on top of the previous landscape (Figures 6-2, 6-11). Their edges and upper surfaces may have pahoehoe or aa texture or may be a jumble of lava blocks. On a flow, a variety of small landforms may be found, such as *tumuli,* formed by small-scale laccolithic swellings on a pahoehoe surface; *pressure ridges,* trenches from collapsed lava tubes; and *spatter cones* or *hornitos,* built where secondary gas eruptions from within or beneath the flow throws up small cones of lava clots. *Lava blisters* are created by steam pockets beneath thin flows that have spread over marshy ground. Thorarinsson (1951, pp. 63–73) reviewed the origins of these minor

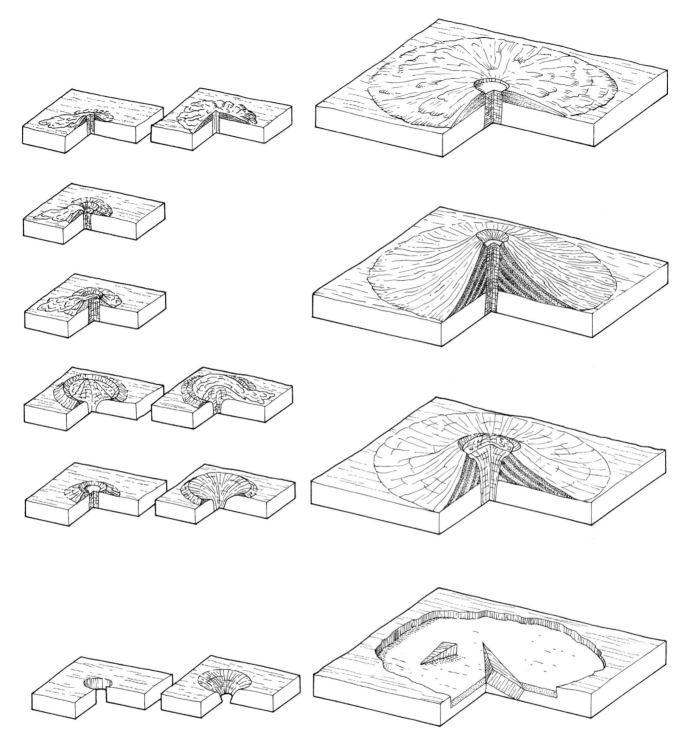

FIGURE 6-9. Schematic block diagrams of simple central volcanoes. The blocks are arranged according to the classification scheme of Table 6-3 except that the largest volcanic structures are not represented. Scale is only approximate (from Rittmann, 1962, Figure 57).

FIGURE 6-10. Scoria cones partly burying a lava flow, Sheeptrail Butte, Craters of the Moon National Monument, Idaho (photo: U.S. National Park Service).

landforms and introduced a new term, *pseudocraters,* for lines or clusters of explosion craters along valley-filling lava flows. He demonstrated that the crater fields and lines of Iceland are not vents along the fissures that fed the flows but result from steam erup-

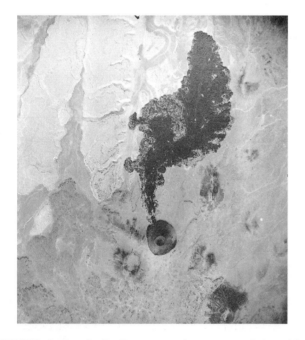

FIGURE 6-11. S. P. Crater scoria cone and basaltic andesite flow, San Francisco volcanic field, Arizona. Flow preceded final stages of cone building. Cone is 1 km in diameter (photo: U.S. Geol. Survey).

tions when flows spread over water-saturated alluvium.

Effusive flows from a central vent in flat terrain may build a low dome over the vent. A dome will develop only if slopes are nearly flat because of the low viscosity of basalt flows. These domes built of successive flows piled around eruptive vents are called **exogenous domes,** to distinguish them from **endogenous domes,** or cumulo-domes of viscous lava that swell within a skin of obsidian and pumice by slow, continued extrusion of highly viscous lava (Cotton, 1944, pp. 158–160). The latter forms are described on page 109.

Exogenous domes grade upward in size to become **shield volcanoes.** No definitive size criteria have been suggested. Shield volcanoes of the Icelandic type are between 100 and 1000 m in elevation and have base diameters about 20 times their height. Surface slopes on shield volcanoes average only a few degrees though slopes of 20° may surround the summit crater. Late-stage caldera subsidence may flatten the summit profile. A genetic sequence for the Icelandic-type shield volcanoes in the Galápagos Islands has been deduced by Nordlie (1973).

Shield volcanoes of the Hawaiian type rank with basalt plateaus and ignimbrite sheets as the largest volcanic landforms. They are not simply scaled-up Icelandic shield volcanoes or exogenous domes; again, the morphological capacity of the domal form is involved. When a lava dome has grown to a critical size by central eruption, it is no longer strong enough to support a central lava lake or carry a central conduit up to the summit crater. Flank eruptions break out, sometimes along the huge slump headscarps that scar the growing mass (Figure 5-4), and a new dome grows peripherally with only slow increase in height. The island of Hawaii includes at least seven coalesced domes. Subaerial slopes range from only 3° to 10°; somewhat steeper submarine slopes range up to 20° (Peterson and Moore, 1987, p. 161). The crest of each dome becomes a broad, convex-upward plateau around a rimless *pit crater,* or a *summit caldera* (Figure 6-12).

Because of hydrostatic balance in the internal magma chambers, a pit crater may temporarily hold a lava lake. When the lake drains because of a flank eruption, the crater rim subsides as giant slump blocks along concentric normal faults.

Hawaiian-type shield volcanoes are enormous in size (Decker et al., 1987). Mauna Kea and Mauna Loa, the two major shields of the seven that merge to form the island of Hawaii, have summit elevations of 4206 and 4175 m above sea level. More significant, they rise smoothly from a submarine base about 400 km in diameter, and the total height of the multiple shield volcano is about 10 km, higher than Mt. Everest (Fig-

FIGURE 6-12. Snow-covered summit of Mauna Loa shield volcano, Hawaii, viewed from the southwest. Included are the large Mokuaweoweo Caldera, South Pit, and smaller pit craters to the south. Summit of Mauna Kea in distance (photo: HV0341ct, U.S. Geol. Survey).

ure 6-1). Recent submarine surveys have demonstrated that very large submarine landslides have occurred throughout the submarine and subaerial growth phases of Hawaiian shield volcanoes. Slumps and debris avalanches (pp. 186–187) covering hundreds of square kilometers surrounded the growing shields and were incorporated into their structure.

Basalt Plateaus

During various intervals of geologic time, continental regions measuring from 0.2×10^6 to 2×10^6 km² have been engulfed by massive multiple eruptions of fluid basalt, totaling 1 km in thickness. These *flood basalts* are probably caused by mantle plumes, or hot spots, breaking through from beneath continents (pp. 44, 80). Many of them are now near the margins of continents and were emplaced at the time of a continental breakup (Macdougall, 1988, p. 388; Richards et al., 1989; White and McKenzie, 1989a). There are no central vents; rather, *fissure eruptions* of regional extent provided the lava. Dikes can be seen that cut individual flows but rarely are continuous through the entire thickness of multiple flows. The former land surface subsided beneath the accumulating sheets of basalt, either because of isostatic compensation or because of the enormous volume of magma that was withdrawn from beneath.

The results of such massive lava effusions are plateaus of considerable size and monotonous structure. The largest such area in the United States is the 164,000 km² of the Columbia Plateau in Washington and Oregon. The Columbia Plateau flood basalts define an entire geomorphic province of the United States (Figure 1-5). A total volume of 170,000 km³ of basalt erupted during the Miocene Epoch, between 17 million and 11 million years ago. The most voluminous eruptions were within a brief interval of about 2 million years and accounted for 85 percent of the total volume (Mangan et al., 1986; Hooper, 1988; Reidel et al., 1989). Adjacent to the south is the smaller Snake River Plain, with an area of 52,000 km² extending across southern Idaho and merging into the Columbia Plateau. The basalt flows of the Snake River Plain are of Quaternary age though no historic eruptions are known.

The largest known flood basalt region is the Deccan district of northwestern India, which now covers 500,000 km² and originally may have been three times as large (Mahoney, 1988, p. 152). The Deccan flood basalts were entirely emplaced within an interval of 2 million years sometime between 65 and 69 million years ago. Most of the eruptions occurred within an even shorter interval of 0.5 million years (Richards et al., 1989, p. 103). Similar basalt plateaus of recent enough age to form constructional landscapes are present in southern Brazil, Manchuria, and central Siberia, and in fragments around the North Atlantic basin on Greenland, Iceland, Ireland, the Faeroe Islands, and Jan Mayen.

Except for exogenous domes or minor scoria mounds, the constructional morphology of a basalt plateau is extraordinarily flat. Flows have initial slopes of 1° or less, barely detectable to the eye. The impression is created that the entire regional landscape is the surface of a single lava flow, which is, of course, not the case. However, individual flows within the Deccan Plateau have been mapped over several thousand square kilometers and traced laterally for more than 160 km without showing signs of dying out (Choubey, 1973, 1974; Hooper, 1988, p. 155). Many flows within the Columbia Plateau basalt covered distances exceeding 500 km (Reidel et al., 1989, p. 29). The extensive flows were made possible by extremely low viscosity, profuse eruption from long fissures, and their very large volume, which helped them retain their heat.

Erosional dissection of basalt plateaus is strongly controlled by the sheetlike continuity and columnar jointing of the flows. Flat benches and vertical cliffs dominate the landscape (Figure 12-1). Hills are commonly flat topped. The Germanic name for the steplike landscape (*treppen*) has survived in English not as a geomorphic term but as the petrographic term "trap" or "traprock" for all dense, dark igneous rocks that look like basalt.

Tephra Cones and Composite Cones (or Strato-Volcanoes)

Historic subaerial volcanoes have been characterized by explosive rather than effusive activity (Table 6-1). The percentage of tephra in the total ejecta (tephra plus lava) for historic circum-Pacific, Indonesian, and Caribbean volcanoes generally has exceeded 90 percent (Rittman, 1962, p. 153). However, these volumetric estimates exclude the enormous amount of quiet extrusive basalt volcanism along the midocean ridges.

Small *tephra cones* have straight, steep sides at the angle of repose of tephra and scoria (Porter, 1972) (Figures 6-10, 6-11, and Color Plate 7). They are permeable enough to resist erosional dissection by streams, but straight gullies may descend radially from their rims.

Their bases are wide relative to their height, and their craters are large. Larger cones cannot be built entirely of scoria or tephra because the material is not strong enough to support continued summit eruptions (Figure 6-13).

All proportions of tephra and lava are found in the **composite cones** or **strato-volcanoes** that form the earth's most perfectly symmetric large landforms (Figure 6-14). Lava flows that mantle the slopes help support the cone. More important, several sets of systematic intrusive dikes and sills (Figure 6-15) are integral parts of large volcanic edifices, and these, more than the eruptive lava flows, provide a strong framework to the cone. The intermingling of tephra, lava flows, intrusive pipes, dikes, and sills makes the structure of large volcanic cones too complex for any simple designation of lava or tephra cone, which is

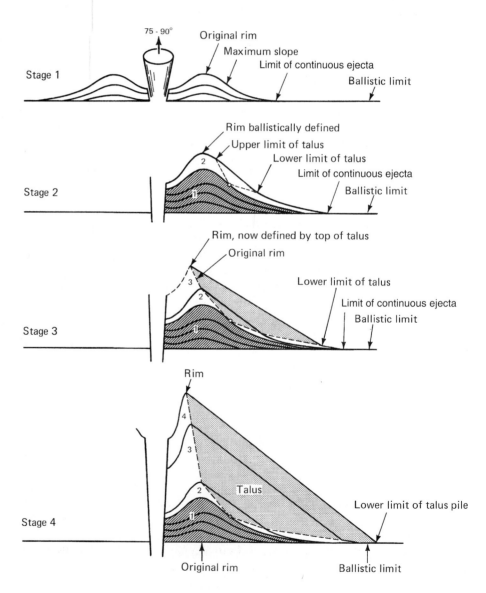

FIGURE 6-13. Four major stages in the development of tephra cones. Stage 1: Tephra ring with shape controlled by ballistic trajectories of particles. Stage 2: Angle-of-repose talus forms a straight midslope segment. Stage 3: As cone grows, talus extends upward to crater rim. Stage 4: Talus extends from rim to beyond ballistic limit. From this stage until the morphological capacity of the cone is exceeded, the slope angle remains constant (redrawn from McGetchin et al., 1974, Figure 14; © 1974 by the American Geophysical Union).

FIGURE 6-14. Mount Mayon (height: 2400 m) in southern Luzon, Philippines; reputedly the most perfectly symmetric volcanic cone (photo: Philippine Department of Tourism).

why they are called composite cones. Their complex internal structure plays an important controlling role in their subsequent erosional dissection (Chapter 12). Composite cones have complex eruptive histories. They can be largely destroyed by major eruptions and rebuilt in a relatively brief time. Monte Somma, a crescent-shaped ridge on the flank of Vesuvius, is the "stump" of an older edifice slightly larger than the modern cone. The modern cone of Fuji has buried the stump of the older cone within the last 10,000 years (p. 102).

The magnificent symmetry of composite cones has been the subject of much scientific and esthetic inquiry. Their smooth, concave-up profiles (Figure 6-14) are best fitted by logarithmic or exponential curves and are attributed to summit lava flows, slumping, and angle-of-repose conditions (Cotton, 1944, pp. 235–243). Complexities of tephra accumulation by size

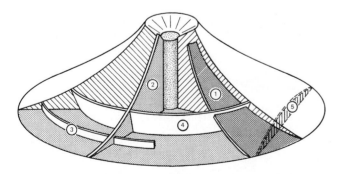

FIGURE 6-15. Systems of dikes and sills combine with lava flows to strengthen large composite cones: (1) mantle sill or buried flow, (2) radial dike, (3) ring dike, (4) cone sheet, (5) peripheral dike (redrawn from Rittmann, 1962, Figure 73).

sorting, fallout rates, and shifting wind patterns complicate the pattern (Fisher and Schmincke, 1984, pp. 125–162).

Plug Domes, Cumulo-Domes, Tholoids, and Spines

Very viscous lava of highly silicic composition may surge upward into a crater but be too stiff to flow and too degassed to explode. The result is an **endogenous dome** (Table 6-3), a "skinned-over," bulbous protrusion of lava from the vent. These domes vary greatly in size and shape from the awesome "shark-fin" spine that gradually extruded from Mount Pelée nine months after the St. Pierre disaster and eventually reached a height of more than 300 m above the crater, to the mass of Mount Lassen, a dome about 1.6 km in diameter and more than 600 m high. After the Mount St. Helens eruption of May 1980, a succession of plug domes formed in the breached crater, the largest of which had grown to more than 90 m in height by the end of that year (Moore et al., 1981). **Tholoid** (dome shaped) is the term generally used for a flattened hemispheroidal dome within a crater rim. **Cumulo-domes,** also known as **plug domes,** are more irregular or cylindrical masses of rigid lava that form hills or low domal mountains (Smith, 1973; Blake, 1989; Francis, 1993). Sometimes their surfaces are cratered, or covered with small flows and spinelike viscous extrusions, proving that they are not entirely endogenous. If they sag or flow, they are called *coulées* (Color Plate 8).

The surface of an endogenous dome is constantly stretched by continued inflation, so a talus of blocks or breccia accumulates around its base or on its flanks and may nearly bury the dome. Rarely, the striated, polished surface of an extruded plug will be visible above its talus, and may solidify, preserving the marks of extrusion on its flanks.

Plug domes should not be confused with **volcanic necks,** which are the result of long-continued erosion of a composite cone to expose the solidified lava-filled conduit or neck. A plug dome is a constructional landform; a volcanic neck (Chapter 12) is a structurally controlled erosional remnant. Both may be present in the same volcanic region, as in the Auvergne region of France. There, the same term, *Puy,* is given both to domes and to eroded necks.

Maars; Tuff Rings and Cones

Extremely explosive eruptions blast out broad, shallow craters below general ground level with tuff or pumice rings around them. The Eifel district of Germany, at the northern end of the Rhine graben (p. 81), is the type locality for these landforms. The name **maar** (not to be confused with the lunar *mare*) refers to the shallow lake that fills the crater in a humid climate such as that of western Europe. Maars (Germanic plural: *maare*) and their enclosing ring walls are surrounded by widespread thin sheets of tephra, especially in the prevailing downwind direction. If there is no central lake, such an explosive eruption results in no geomorphic expression other than a low rampart or **tuff ring,** grading radially outward into thin tephra, and inward into a tephra-covered shallow crater underlain by shattered country rock (Figure 6-16). The sequence of small volcanic landforms ranging from scoria cones to tuff rings, tuff cones, and maars, as shown in Table 6-3, gives a false impression that they are exclusively compositionally controlled. In fact, many of these forms are of basaltic tephra. The increased explosiveness is determined not so much by magma composition as by the amount of water in contact with the magma. Hot, dry tephra builds scoria cones; if the magma encounters groundwater on its way up the vent, it erupts in the Strombolian mode (Table 6-2) with violent steam explosions that shower hot tephra in a tuff ring around the crater. With increased water contact, cooler, wetter tephra surges out of the crater or falls from the air in the Surtseyan mode, building a **tuff cone** (Wohletz and Sheridan, 1983). Tuff rings and cones are broad and relatively flat as viewed from the ground, but aerial views graphically illustrate their origin (Figures 6-16 and 6-17).

Purely gas explosions produce craters surrounded by little or no ejecta. A lake may form in the resulting crater (a *gas maar*), or the geologic evidence for the eruption may be nothing more than an area of shat-tered rock. If volcanic breccia can be identified, the dike or stocklike intrusive mass is called a *diatreme.* Some large circular areas of shattered sedimentary rocks in the central United States were once thought to have resulted from gas eruptions, but because of the total lack of volcanic rocks, they were called *cryptovolcanoes.* Most of them are now interpreted as ancient meteorite impact scars.

Ignimbrite Sheets: Large Explosion Craters and Calderas

Pyroclastic flows are believed to be responsible for some of the largest constructional volcanic landforms. If the ejecta is hot enough to fuse into a layer of porous volcanic rock, the resulting **ignimbrite,** or **welded tuff,** sheets can cover thousands of square kilometers, as in New Zealand, the Yellowstone Plateau, Katmai National Monument in Alaska, the Lake Toba region of northern Sumatra, the Altiplano Plateau of the central Andes (Figure 13-15), and elsewhere. The ignimbrite may be hundreds of meters in thickness (Francis, 1993, pp. 208ff).

Plinian or Katmaian types of explosive activity, or the even more explosive Ultraplinian and phreatic types, may create a "constructional" landscape of catastrophically destructive appearance. A number of large composite cones either have had their entire summits blown off or subside inward following a major explosion. The subsidence hypothesis is favored because the size of many calderas is larger than the accountable volume of ejecta. Furthermore, the structure of ancient calderas now exposed by deep erosion demonstrates the subsidence of the central part of the

FIGURE 6-16. Crater Elegante, a maar in the Pinacate volcanic field, Sonora, Mexico (Arvidson and Mutch, 1974). Preexplosion tephra and lava beds are exposed on the crater walls, overlain by a tephra ring (photo: R. E. Arvidson).

FIGURE 6-17. Diamond Head, Honolulu, Hawaii, viewed toward the southwest. An asymmetric tuff ring with its highest, distant point built by tephra blown by the northeast Trade Winds (photo: P. Francis).

caldera floor, surrounded by ring dikes and inward-facing arcuate normal faults (Francis, 1993, pp. 292ff).

The distinction between craters and calderas was arbitrarily set at a diameter of 1 mile (1.6 km) by Macdonald (1972, p. 290). Although arbitrary, this dimensional distinction is related to the morphological capacity concept mentioned earlier, in that summit craters up to a mile in diameter are normal components of the constructional morphology of composite cones, whereas calderas larger than that are probably all due to subsidence during or following a Katmaian-type terminal eruption.

Large explosion craters and calderas should be distinguished from the pit craters and summit calderas of basalt shields (Figure 6-12) that are the site of nonexplosive lava lakes or lava fountains. Explosion craters and calderas are surrounded by tephra or ignimbrite sheets that thicken and become coarser toward the crater. The inward-facing walls of explosion craters are typically slump scars or fault scarps of blocks that subsided into the funnel-shaped opening prepared by an explosion.

Calderas can be huge basins, ranging in diameter from the defined minimum of 1.6 km to one of the largest yet recognized, 70 km long and 45 km wide, largely buried beneath the thick rhyolite flows and ignimbrite sheets of the Yellowstone Plateau (Figure 5-27). At least 138 modern calderas are defined as "large" ($\geq$ 5 km in diameter) (Newhall and Dzurisin, 1988). Collectively, they have experienced 1299 known episodes of unrest in historic time though not all the episodes of activity have been eruptions. Calderas may be on the summits of composite cones or in volcanic plateaus (Figure 6-18). Many contain, or formerly contained, lakes. The site of Crater Lake,

Oregon, is a caldera 10 km in diameter and 1200 m deep that was formed by the Mount Mazama tephra eruption 6900 years ago. The Mazama tephra is a useful mid-Holocene stratigraphic marker for geologic and archeologic studies across 1 million km^2 of western North America and the adjacent ocean floor (Nelson et al., 1988).

The largest calderas are those associated with massive ignimbrite sheets (Francis, 1993, pp. 208ff). The best hypothesis for this relationship is that a large subsurface reservoir of magma persists long enough to become compositionally stratified, with the most silicic, viscous, and gas-rich fraction near the top. When the magma begins to subside within the chamber (by a cause, and to a place, unknown), the pressure drop causes a massive effusion of pumice and glowing hot tephra along fissures caused by subsidence over the magma chamber. In a geyserlike self-perpetuating sequence, the explosive eruption lowers the pressure still more, and further eruptions occur. In some of the great ignimbrite sheets, the later eruptives are more basic than the earlier, showing that deeper levels of more basic magmas were progressively expelled.

The quantities of tephra ejected from calderas are staggering. Whereas the 1980 eruption of Mount St. Helens ejected an estimated 0.6 km^3 of lava and tephra and left a crater 2 km in diameter, the most recent eruption of the Yellowstone caldera, 600,000 years ago, erupted 10^3 km^3 of tephra. Toba caldera, in western Sumatra, Indonesia, erupted an estimated 2800 km^3 of tephra in its most recent eruption about 75,000 years ago. The Toba caldera is about 100 km long and 30 km wide, and is the youngest and largest caldera known (Rose and Chester, 1987; Chesner et al., 1991). It now contains a large lake. The harbors of Naples, Italy, and

FIGURE 6-18. Aniakchak caldera on the Alaskan Peninsula, viewed from the northwest. Caldera rim is about 10 km in diameter. A former lake in the crater drained catastrophically through the eroded gap on the southeast rim less than 3400 years ago (Waythomas et al., 1996, Figure 2).

Rabaul, Papua New Guinea, are both in calderas. Both show recurring volcanic and seismic activity, and are potentially sites of major disasters.

Following eruption and collapse, the floors of many large calderas then slowly rise, probably caused by the intrusion of fresh magma into the chamber beneath. The result is a central broad dome, or an island in the lake of a flooded caldera, known as a *resurgent dome*. A large example is Cerro Galán, a volcanic complex in the northwestern Argentine Andes (Francis, 1993, pp. 300ff). The broad central resurgent dome rises 1 km above the surrounding caldera floor and occupies almost half of the elliptical caldera, which is 34 km long from north to south (Color Plate 9). Because of the extreme aridity of the high Andean plateau, only a small lake remains in the caldera. Ignimbrite sheets from the Cerro Galán eruption 2.5 million years ago are up to 500 m in thickness and can be traced 70 km from the caldera. At least 100 km³ of ignimbrite erupted in this youngest of several major events at Cerro Galán.

The Yellowstone Plateau was built during three great cycles of Katmaian volcanism, each of which culminated in the eruption of massive ignimbrite sheets. They occurred approximately 2.1, 1.3, and 0.6 million years ago (p. 60). The huge caldera was buried by several thousand meters of tephra and rhyolite flows.

Individual ignimbrite sheets totaling hundreds of meters in thickness cooled as single units, with columnar joints as well developed as in lava flows. Tephra from the multiple Yellowstone eruptions is well known in the older Pleistocene deposits of the Great Plains from Texas to Kansas. Other great calderas in the United States are the Valles caldera near the Rio Grande rift in New Mexico, which erupted about 1 million years ago, and the Long Valley caldera in eastern California, which last erupted 760,000 years ago. Both show ominous current activity. Eruption of any one of the giant calderas in the modern world would create a catastrophe comparable to that of a large meteorite impact.

REFERENCES

ARCULUS, R. J., GUST, D. A., and KUSHIRO, I., 1991, Fuji and Hakone: National Geographic Res. and Explor., v. 7, no. 3, pp. 276–309.

ARVIDSON, R. E., and MUTCH, T. A., 1974, Sedimentary patterns in and around craters from the Pinacate volcanic field, Sonora, Mexico: Some comparisons with Mars: Geol. Soc. America Bull., v. 85, pp. 99–104.

BARBERI, F., CARAPEZZA, M. L., and 2 others, 1993, The control of lava flow during the 1991–1992 eruption of Mt. Etna: Jour. Volcanol. and Geothermal Res., v. 56, pp. 1–34.

BLAKE, S., 1989, Viscoplastic models of lava domes, in Fink, J. H., ed., Lava flows and domes: Emplacement mechanisms and hazard implications: IAVCEI Proceedings in Volcanology 2, Springer-Verlag, Berlin, p. 88–126.

BLONG, R. J., 1984, Volcanic hazards: Academic Press Inc., Orlando, Florida, 424 pp.

CHESNER, C. A., ROSE, W. I., and 3 others, 1991, Eruptive history of earth's largest Quaternary caldera (Toba, Indonesia) clarified: Geology, v. 19, pp. 200–203.

CHOUBEY, V. D., 1973, Long-distance correlation of Deccan basalt flows, central India: Geol. Soc. America Bull., v. 84, pp. 2785–2790.

———, 1974, Long-distance correlation of Deccan basalt flows, central India: Reply: Geol. Soc. America Bull., v. 85, pp. 1008–1010.

COTTON, C. A., 1944, Volcanoes as landscape forms: Whitcombe and Tombs Ltd., Christchurch, New Zealand, 416 pp. (republished with minor corrections, 1952).

DADE, W. B., and HUPPERT, H. E., 1996, Emplacement of the Taupo ignimbrite by a dilute turbulent flow: Nature, v. 381, pp. 509–512.

DAVIS, W. M., 1905, Complications of the geographical cycle: Internat. Geog. Cong., 8th, Washington, 1904, Repts., pp. 150–163. (reprinted 1954 in Geographical essays: Dover Publications, Inc., New York, pp. 279–295).

DAWSON, J. B., PINKERTON, H., and 2 others, 1994, June 1993 eruption of Oldoinyo Lengai, Tanzania: Exceptionally viscous and large carbonatite lava flows and evidence for coexisting silicate and carbonate magmas: Geology, v. 22, pp. 799–802.

DECKER, R. W., and DECKER, B. B., 1989, Volcanoes: Revised and updated ed.: W.H. Freeman and Co., New York, 285 pp.

———, 1991, Mountains of fire: The nature of volcanoes: Cambridge University Press, Cambridge, UK, 198 pp.

DECKER, R. W., WRIGHT, T. L., and STAUFFER, P. H., 1987, eds., Volcanism in Hawaii: U.S. Geol. Survey Prof. Paper 1350, 1667 pp.

FAGENTS, S. A., and WILSON, L., 1993, Explosive volcanic eruptions—VII. The range of pyroclasts ejected in transient volcanic explosions: Geophys. Jour. Internat., v. 113, pp. 359–370.

FEDOTOV, S. A., AND MARKHININ, YE. K., eds., 1993, The great Tolbachik fissure eruption: Geological and geophysical data, 1975–1976: Cambridge University Press, Cambridge, UK, 341 pp.

FIOCCO, G., FUÁ, D., and VISCONTI, G., 1996, The Mount Pinatubo eruption: Effects on the atmosphere and climate: Springer-Verlag Berlin Heidelberg, 310 pp.

FISHER, R. V., 1990, Transport and depositions of a pyroclastic surge across an area of high relief: The 18 May 1980 eruption of Mount St. Helens, Washington: Geol. Soc. America Bull., v. 102, pp. 1038–1054.

———, and SCHMINCKE, H.-U., 1984, Pyroclastic rocks: Springer-Verlag Berlin, 472 pp.

FRANCIS, P., 1993, Volcanoes: A planetary perspective: Oxford University Press, Oxford, UK, 443 pp.

FRANKEL, C., 1996, Volcanoes of the solar system: Cambridge University Press, Cambridge, UK, 232 pp.

GULCHARD, F., CAREY, S. and 3 others, 1993, Tephra from the Minoan eruption of Santorini in sediments of the Black Sea: Nature, v. 363, pp 610–612.

HEAD, J. W., CAMPBELL, D. B., and 6 others, 1991, Venus volcanism: Initial analysis from Magellan data: Science, v. 252, pp. 276–288.

HILL, R. I., CAMPBELL, I. H., and 2 others, 1992, Mantle plumes and continental tectonics: Science, v. 256, pp. 186–193.

HODGES, C. A., and MOORE, H. J., 1994, Atlas of volcanic landforms on Mars: U.S. Geol. Survey Prof. Paper 1534, 194 pp.

HON, K., KAUAHIKAUA, J., and 2 others, 1994, Emplacement and inflation of pahoehoe sheet flows: Observations and measurements of active lava flows on Kilauea Volcano, Hawaii: Geol. Soc. America Bull., v. 106, pp. 351–370.

HOOPER, P. R., 1988, The Columbia River basalt, in Macdougall, J. D., ed., Continental flood basalts: Kluwer Academic Publishers, Dordrecht, The Netherlands, pp. 331–341.

IHINGER, P. D., 1995, Mantle flow beneath the Pacific plate: Evidence from seamount segments in the Hawaiian-Emperor chain: Am. Jour. Sci., v. 295, pp. 1035–1057.

JONES, P. D., and KELLY, P. M., 1996, Effect of tropical explosive volcanic eruptions on surface air temperatures, in Fiocco, G., Fuá, D., and Visconti, G., eds., The Mount Pinatubo eruption: Effects on the atmosphere and climate: Springer-Verlag Berlin Heidelberg, pp. 95–111.

KUNIHOLM, P. I., KROMER, B., and 4 others, 1996, Anatolian tree rings and the absolute chronology of the eastern Mediterranean, 2220–718 BC: Nature, v. 381, pp. 780–783.

LEMASURIER, W. E., 1990, Late Cenozoic volcanism on the Antarctic plate: An overview, in LeMasurier, W. E., and Thomson, J. W., eds., Volcanoes of the Antarctic plate and the southern oceans: Am. Geophys. Union, Antarctic Research series, v. 48, pp. 1–17.

———, and THOMSON, J. W., eds., 1990, Volcanoes of the Antarctic plate and the southern oceans: Am. Geophys. Union, Antarctic Research series. v. 48, 487 pp.

LIPMAN, P. W., and MULLINEAUX, D. R., eds., 1981, The 1980 eruptions of Mount St. Helens, Washington: U.S. Geol. Survey Prof. Paper 1250, 844 pp.

MACDONALD, G. A., 1972, Volcanoes: Prentice-Hall, Inc., Englewood Cliffs, N.J., 510 pp.

MACDOUGALL, J. D., 1988, Continental flood basalts and MORB: A brief discussion of similarities and differences in their petrogenesis, in Macdougall, J. D., ed., Continental flood basalts: Kluwer Academic Publishers, Dordrecht, The Netherlands, pp. 331–341.

MACHIDA, H., 1967, The recent development of Fuji Volcano, Japan: Tokyo Metropolitan Univ. Geog. Repts., no. 2, pp. 11–20.

MACIAS, J. L., and SHERIDAN, J. F., 1995, Products of the 1907 eruption of Shtyubel' volcano, Ksudach Caldera, Kamchatka, Russia: Geol. Soc. America Bull., v. 107, pp. 969–986.

MAHONEY, J. J., 1988, The Deccan traps, in Macdougall, J. D., ed., Continental flood basalts: Kluwer Academic Publishers, Dordrecht, The Netherlands, pp. 151–194.

MANGAN, M. T., WRIGHT, T. L., and 2 others, 1986, Regional correlation of Grande Ronde Basalt flows, Columbia River Basalt Group, Washington, Oregon, and Idaho: Geol. Soc. America Bull., v. 97, pp. 1300–1318.

MCGETCHIN, T. R., SETTLE, M., and CHOUET, B. A., 1974, Cinder cone growth modeled after northeast crater, Mount Etna, Sicily: Jour. Geophys. Res., v. 79, pp. 3257–3272.

McGUIRE, W. J., JONES, A. P., and NEUBERG, J., 1996, Volcano instability on the earth and other planets: Geol. Soc. [London] Spec. Pub. No. 110, 388 pp.

MENARD, H. W., 1964, Marine geology of the Pacific: McGraw-Hill Book Company, Inc., New York, 271 pp.

MILETI, D. S., BOLTON, P. A., and 2 others, eds., 1991, The eruption of Nevada del Ruiz volcano, Colombia, South America, November 13, 1985: Natural Disaster Studies, v. 4, National Academy Press, Washington D.C., 109 pp.

MOORE, J. G., LIPMAN, P.W., and 2 others, 1981, Growth of lava domes in the crater, June 1980-January 1981, in Lippman, P. W., and Mullineaux, D. R., eds., The 1980 eruptions of Mount St. Helens, Washington: U.S. Geol. Survey Prof. Paper 1250, pp. 541–547.

MOSLEY, M. P., and PARKER, R. S., 1972, Allometric growth: A useful concept in geomorphology?: Geol. Soc. America Bull., v. 83, pp. 3669–3674.

MULLINEAUX, D. R., PETERSON, D. W., and CRANDELL, D. R., 1987, Volcanic hazards in the Hawaiian Islands, in Decker, R. W., Wright, T. L., and Stauffer, P. H., eds., Volcanism in Hawaii: U.S. Geol. Survey Prof. Paper 1350, pp. 599–621.

MURSKY, G., 1996, Introduction to planetary volcanism: Prentice-Hall, Inc., Upper Saddle River, N.J., 293 pp.

NARANJO, J. A., 1985, Sulphur flows at Lastarria volcano in the northern Chilean Andes: Nature, v. 313, pp. 778–780.

NARANJO, J. L., SIGURDSSON, H., and 2 others, 1986, Eruption of the Nevado del Ruiz volcano, Colombia, on November 13, 1985: Tephra fall and lahars: Science, v. 233, pp. 961–963.

NELSON, C. H., CARLSON, P. R., and BACON, C. R., 1988, The Mount Mazama climactic eruption (~6900 yr B. P.) and resulting convulsive sedimentation on the Crater Lake caldera floor, continent, and ocean basin, in Clifton, H. E., ed., Sedimentologic consequences of convulsive geologic events: Geol. Soc. America Spec. Paper 229, pp. 37–57.

NEWHALL, C. G., and DZURISIN, D., 1988, Historical unrest at large calderas of the world: U.S. Geol. Survey Bull. 1855, 1108 pp.

NEWHALL, C. G., and SELF, S., 1982, The volcanic explosivity index (VEI): An estimate of explosive magnitude for historical volcanism: Jour. Geophys. Res., v. 87, pp. 1231–1238.

NICHOLS, D. R., and YEHLE, L. A., 1961, Mud volcanoes in the Copper River Basin, Alaska, in Raasch, G. O., ed., Geology of the Arctic: Univ. of Toronto Press, Toronto, pp. 1063–1087.

NORDLIE, B. E., 1973, Morphology and structure of the western Galápagos volcanoes and a model for their origin: Geol. Soc. America Bull., v. 84, pp. 2931–2955.

OLLIER, C. D., 1988, Volcanoes: Basil Blackwell Ltd., Oxford, UK, 228 pp.

PARFITT, E. A., and WILSON, L., 1995, Explosive volcanic eruptions—IX. The transition between Hawaiian-style fountaining and Strombolian explosive activity: Geophys. Jour. Internat., v. 121, pp. 226–232.

PIERI, D. C., KHRENOV, A. P., and 11 others, 1997, Joint effort results in first TIMS survey of Kamchatka volcanoes: EOS, v. 78, pp. 125, 128.

PETERSON, D. W., and MOORE, R. B., 1987, Geologic history and evolution concepts, island of Hawaii, in Decker, R. W., Wright, R. L., and Staufer, P. H., eds., Volcanism in Hawaii: U.S. Geol. Survey Prof. Paper 1350, pp. 149–189.

PORTER, S. C., 1972, Distribution, morphology, and size frequency of cinder cones on Mauna Kea volcano, Hawaii: Geol. Soc. America Bull., v. 83, pp. 3607–3612.

REIDEL, S.P., TOLAN, T. L., and 4 others, 1989, The Grande Ronde Basalt, Columbia River Basalt Group; stratigraphic descriptions and correlations in Washington, Oregon, and Idaho, in Reidel, S. P. and Hooper, P. R., eds., Volcanism and tectonism in the Columbia River basalt province: Geol. Soc. America Spec. Paper 239, pp. 21–53.

RICHARDS, M. A., DUNCAN, R. A., and COURTILLOT, V. E., 1989, Flood basalts and hot-spot tracks: Plume heads and tails: Science. v. 246, pp. 103–107.

RITTMANN, A., 1962, Volcanoes and their activity (transl. by E. A. Vincent): Wiley-Interscience, New York, 305 pp.

ROSE, W. I., and CHESNER, C. A., 1987, Dispersal of ash in the great Toba eruption, 75 ka: Geology, v. 15, pp. 913–917.

RYAN, M. P., and BLEVINS, J. Y. K., 1987, Viscosity of synthetic and natural silicate melts and glasses at high temperatures and 1 bar (10^5 Pascals) pressure and at higher pressures: U.S. Geol. Survey Bull. 1764, 563 pp.

SARNA-WOJCICKI, A. M., SHIPLEY, S., and 3 others, 1981, Areal distribution, thickness, mass volume, and grain size of air-fall ash from the six major eruptions of 1980, in Lippman, P. W., and Mulineaux, D. R., eds., The 1980 eruptions of Mount St. Helens, Washington: U.S. Geol. Survey Prof. Paper 1250, pp. 577–600.

SCARTH, A., 1994, Volcanoes: An introduction: Texas A&M University Press, College Station, Tex., 273 pp.

SCOTT, K. M., VALLANCE, J. W., and PRINGLE, P. T., 1995, Sedimentology, behavior, and hazards of debris flows at Mount Rainier, Washington: U.S. Geol. Survey Prof. Paper 1547, 56 pp.

SELF, S., RAMPINO, M. R., and 2 others, 1984, Volcanological study of the great Tambora eruption of 1815: Geology, v. 12, pp. 659–663.

SHORT, N. M., 1986, Volcanic landforms, in Short, N. M., and Blair, R. W., Jr., eds., Geomorphology from space: NASA Spec. Pub. 486, pp. 185–253.

SIGVALDASON, G. E., 1989, International conference on Lake Nyos disaster, Yaoundé, Cameroon, 16–20 March, 1987: Conclusions and recommendations, in Guern, F. Le., and Sigvaldason, G. E., eds., The Lake Nyos event and natural CO_2 degassing, I: Jour. Volcanol. and Geothermal Res., v. 39, pp. 97–107.

deSILVA, S. L., and FRANCIS, P. W., 1991, Volcanoes of the central Andes: Springer-Verlag, Berlin, 216 pp.

SIMKIN, T., and FISKE, R. S., 1983, Krakatau 1883: The volcanic eruption and its effects: Smithsonian Institution Press, Washington, D.C., 464 pp.

SIMKIN, T., and SIEBERT, L., 1984, Explosive eruptions in space and time: Durations, intervals, and a comparison of the world's active volcanic belts, in National Research Council Geophysics Study Committee, Explosive volcanism: Inception, evolution, and hazards: National Academy Press, Washington, D.C., pp. 110–121.

———, 1994, Volcanoes of the world, second edition: Geoscience Press Inc., Tucson, Ariz., 348 pp.

SMITH, A. L., and ROOBOL, M. J., 1990, Mt. Pelée, Martinique; A study of an active island-arc volcano: Geol. Soc. America Mem. 175, 105 pp.

SMITH, E. I., 1973, Mono Craters, California: A new interpretation of the eruptive sequence: Geol. Soc. America Bull., v. 84, pp. 2685–2690.

STANLEY, D. J., and SHENG, H., 1986, Volcanic shards from Santorini (Upper Minoan ash) in the Nile Delta, Egypt: Nature, v. 320, pp. 733–735.

STOTHERS, R. B., 1984, The great Tambora eruption in 1815 and its aftermath: Science, v. 224, pp. 1191–1198.

THORARINSSON, S., 1951, Laxárgljúfur and Laxárhraun: A tephrochronological study: Geografiska Annaler, v. 33, pp. 1–88.

U.S. GEODYNAMICS COMMITTEE, National Research Council, 1994, Mount Rainier active Cascade volcano: Research strategies for mitigating risk from a high, snow-clad volcano in a populous region: National Academy Press, Washington D.C., 114 pp.

WATERS, A. C., DONNELLY-NOLAN, J. M., and ROGERS, B. W., 1990, Selected caves and lava-tube systems in and near Lava Beds National Monument, California: U.S. Geol. Survey Bull. 1673, 102 pp.

WAYTHOMAS, C. F., WALDER, J. S., and 2 others, 1996, Catastrophic flood caused by drainage of a caldera lake at Aniakchak Volcano, Alaska, and implications for volcanic hazards assessment: Geol. Soc. America Bull., v. 108, pp. 861–871.

WHITE, D. E., 1955, Violent mud-volcano eruption of Lake City Hot Springs, northeastern California: Geol. Soc. America Bull., v. 66, pp. 1109–1130.

WHITE, R. S., and MCKENZIE, D.P., 1989a, Magmatism at rift zones; The generation of volcanic continental margins and flood basalts: Jour. Geophys. Res., v. 94, pp. 7685–7729.

———, 1989b, Volcanism at rifts: Sci. American, v. 261, no. 1, pp. 62–71.

WHITFORD-STARK, J. L., 1987, Survey of Cenozoic volcanism on mainland Asia: Geol. Soc. America Spec. Paper 213, 74 pp.

WOHLETZ, K. H., and SHERIDAN, M. F., 1983, Hydrovolcanic explosions II. Evolution of basaltic tuff rings and tuff cones: Am. Jour. Sci. v. 283, pp. 385–413.

WOOD, C. A., and KIENLE, J., 1990, Volcanoes of North America: United States and Canada: Cambridge University Press, Cambridge, UK, 354 pp.

ZHANG, Y., 1996, Dynamics of CO_2-driven lake eruptions: Nature, v. 379, pp. 57–59.

ZIELINSKI, G. A., MAYEWSKI, P. A., and 7 others, 1994, Record of volcanism since 7000 B.C. from the GISP2 Greenland ice core and implications for the volcano-climate system: Science, v. 264, pp. 948–952. (See also technical comments, 1995, Science, v. 267, pp. 256–258.)

PART IV

Subaerial Destructional (Erosional) Processes and Erosional Landforms

*H*owever they may be constructed, subaerial landforms cannot stand immutable. The rocks that make them were formed in physical and chemical environments that are grossly out of equilibrium with the oxygen-rich, low-pressure, hydrous atmosphere into which they are thrust or erupted by constructional processes, and they are sure to crumble or dissolve. Landscapes are carved from structures by agents of physical and chemical weathering, and erosion. Water, the only compound that occurs naturally at earthly temperatures and pressures as a gas and liquid, both chemically reactive, and as an erosive solid, is the primary agent. The energy drive is provided by solar radiation, converted into heat at the earth's surface and eventually reradiated into space. Gravity and internal heat are secondary sources of energy for geomorphic change (Chapter 2). We call these processes **destructional, degradational,** or **erosional** because they tend to destroy the edifices that are constructed by tectonism and volcanism. The eroded debris accumulates in tectonic basins on the continents or is washed to the sea by rivers. Distinctive landforms are built by the accumulated waste (Chapter 11), but the far greater proportion of the subaerial landscape consists of erosional remnants, still in the process of reduction.

Present-day erosional processes conform to universal thermodynamic laws but are also influenced by the steady evolution of the earth's lithosphere, hydrosphere, atmosphere, and biosphere. We can see a geologic record of at least 1 billion years of weathering and erosional processes similar to those now operating although different to the degree that different forms of life were involved in the processes.

The certainty of geomorphic change by subaerial processes acting on geologic structures has been the subject of philosophical comment since the beginning of written history, at least. A Chinese scholar, in 240 B.C., observed:

"Nothing in the world is softer or weaker than water. But when it attacks what is hard and strong none of them can win out" (LaFargue, 1992, p. 162).[1] *One of the oldest geomorphic deductions is "Every valley shall be exalted and every mountain and hill shall be made low; and the crooked shall be made straight, and the rough places plain"* (Isaiah, 40:4).

J. W. Powell, one of the great geomorphologists who learned his trade while exploring the mountains of the western United States, summarized this philosophical tradition of skepticism about the dubious durability of the "everlasting" hills in a single statement. **Powell's dictum** *is "We may now conclude that the higher the mountain, the more rapid its degradation; that high mountains cannot live much longer than low mountains, and that mountains cannot remain long as mountains: they are ephemeral topographic forms. Geologically all existing mountains are recent; the ancient mountains are gone"* (Powell, 1876, p. 193).[1]

New knowledge of the absolute rates of erosion makes it clear how true Powell's dictum really is. All mountains except volcanoes are carved from either late Cenozoic orogenic belts or areas of more ancient terranes that were uplifted by late Cenozoic epeirogeny. The sediment loads carried in modern rivers demonstrate that the watersheds ought to be reduced to near sea level in 10 to 12 million years. Yet we find in many regions landscapes that are 25 million years old or older that have survived almost unchanged although they have not been buried and exhumed (p. 347). Especially on the fragments of the old Gondwana supercontinent such as Australia, many landforms seem to be fossilized relics. The great issue before geomorphologists today is not whether erosional processes can accomplish the work they are alleged to have done but how continents and even specific geomorphic provinces have managed to survive despite such enormous denudation.

Orogenic belts create special climates such as cold highlands, regions of heavy precipitation on windward slopes, and rainshadow deserts. Conversely, climate-controlled erosion can exert powerful controls on orogeny. A mountain belt of erodible sedimentary rocks in a humid tropical zone, as for example in eastern Taiwan, may never be more than 1 to 2 km in height despite rapid uplift rates because erosion rates are also high (Dahlen and Suppe, 1988, p. 169).[1] In contrast, the late Cenozoic central Andes mountains rose by crustal shortening and thickening at a converging plate margin in a hyperarid climate (Isacks, 1988).[1] Because of rapid uplift and aridity, the high Altiplano Plateau was created without significant erosion by externally draining rivers. The lack of erosion can, through isostatic equilibrium, prevent the Altiplano from rising further and either stop the orogeny or widen the orogenic belt. The Tibetan Plateau may be a similar example of climate imposing limits on orogenic processes.

The following nine chapters consider the processes and landforms of subaerial weathering and erosion, and offer deductions about sequential changes in these forms through time. Stressed repeatedly is the role of climate in determining the relative intensity of the various destructional processes, with special consideration of the distinctive landforms that evolve under climatic extremes such as aridity and continuous freezing. The unique processes of glaciation and those at work on coasts are deferred until later chapters.

1. See **References**, *Chapter 7.*

Chapter 7

Rock Weathering

The processes that alter the physical or chemical state of rocks at or near the surface of the earth, without necessarily eroding or transporting the products of alteration, are collectively called **rock weathering.** Except for minor surface roughening, weathering alone does not make landforms. Instead, it provides altered or broken rock from which landforms are shaped. Rock weathering must always be studied in detail as a set of mineralogic or geochemical processes because rocks are assemblages of minerals, and every mineral has specific physical and chemical responses to the near-surface environment. However, for many geomorphic generalizations, it is sufficient to speak of weathering at the scale in which an entire rock mass is the unit of alteration.

Weathering is an assemblage of rock-altering processes that are powered by exogenic, essentially solar, energy. The depth of weathering is thereby restricted by the depth to which exogenic-powered processes can operate. For some purposes, it is important to distinguish weathering from other types of rock alteration, such as by hydrothermal liquids and gases. Although the chemical reactions may be similar, weathering can be expected to decrease with depth, whereas hydrothermal alteration might increase. The engineering implications of such a distinction are obvious.

Silicate minerals are strong and brittle solids in the environment of the shallow continental crust as long as aqueous solutions are not present. However, numerous observations and experiments verify that in the presence of water, rock-forming minerals lose strength and are much more easily broken and deformed (Kirby, 1984, and references therein). These mineralogic studies become geomorphically significant with the demonstrations that *meteoric water* (derived from the hydrologic cycle) circulates to depths of 10 to 20 km. Fracture permeability permits groundwater to circulate to depths of 10 km or even 20 km along fault zones (Costain et al., 1987; Nesbitt and Muehlenbachs, 1989). Circulation to depths greater than 3 km is a widespread phenomenon in the Appalachian provinces of the eastern United States (Tillman, 1980), and water with the isotopic composition of rainfall circulated around Eocene-age plutons in Idaho to depths of 5 to 7 km (Criss and Taylor, 1983). Thus rocks can be altered, at least in part by exogenic processes and therefore weathering, at depths of many kilometers and for millions of years before they are exposed by uplift and erosion to create a landscape.

Weathering is distinguished from other destructive processes by the inclusion in its definition of the concept of *in situ*, or nontransported, alteration. Mass wasting and erosion, which are considered in later chapters, always involve translocation or transportation of material. Of course, when a mineral grain reacts with water, the soluble products are carried away in solution. Yet the basic structures of the rock mass, including bedding and other primary layered structures, granularity, and intrusive dikes or veins, are not obliterated until extremely advanced stages of weathering (Figure 7-1).

Weathering is the precursor of mass wasting and erosion although weathering continues even after a rock fragment has become dislodged from a hillside ledge or pried loose from the bed of a stream. Indeed, weathering processes do not end until a rock fragment has been finally dissolved, transformed to a stable compound, or buried or submerged beyond contact with the atmosphere and circulating groundwater.

A common effect of weathering is the detachment of slabs, sheets, spalls, or chips from rock surfaces. **Exfoliation** is the general term for the loosening or separation of concentric shells or layers of rock. Exfoliation is caused by chemical, thermal, and physical processes, but regardless of the details, the tendency is to reduce the weathering mass to a subspherical shape, which may weather entirely away or break out of the weathered matrix as a *corestone* (p. 139).

Any classification of weathering processes will be arbitrary. Traditionally, weathering processes were subdivided into a mechanical, or physical, group and a chemical group. The recently recognized importance of organic processes in weathering now requires the additional consideration of biologic weathering, which includes both macrobiota and microbes. Mechanical weathering is also called **disintegration,** implying that pieces of rock are taken apart or disaggregated without alteration. In contrast, chemical weathering is also called **decomposition,** emphasizing the breakdown of the chemical composition of the mineral grains that make a rock. The more that is learned about weathering processes, the less sharp becomes the distinction among the mechanical, chemical, and biologic categories because the same thermodynamic principles govern all three. The most rational geomorphic approach is to consider mechanical weathering processes first because mechanical fracturing of rocks usually is necessary before air, water, and organisms can begin their largely chemical attack.

MECHANICAL WEATHERING PROCESSES

Fractures in Rock

The very useful engineering concept of **rock-mass strength,** as opposed to rock strength, has not been sufficiently used by geomorphologists (Selby, 1980). "The complete specification of a rock mass requires descriptive information on the nature and distribution in space of both the materials that constitute the mass (rock, soil, water, and air-filled voids) and the discontinuities which divide it" (Bell, 1992a, p. 54). It is the discontinuities rather than the intact rock pieces that largely determine the weathering as well as the engineering behavior of a rock mass. Selby (1980) defined eight parameters for assessing rock-mass strength for geomorphic purposes: (1) strength of intact rock, (2) state of weathering of the rock, (3) spacing of joints, (4) orientation of joints with respect to the hillslope, (5) width of the joints, (6) lateral or vertical continuity of the joints, (7) infilling of the joints, and (8) movement of water out of the rock mass. It is notable that of the eight parameters, all but the first refer to either weathering or the fractures that divide the rock. Only the first refers to the strength of the rock material itself.

All rocks have fractures and pores, some on the submicrometer scale (Figure 7-2). Crystalline rocks are dominated by microcracks, whereas a variety of intergranular pores and tubes on many scales are inherent to sedimentary processes (Fredrich et al., 1995). Microcracks in granite begin to form during magma cooling, when earlier-formed grains such as plagioclase are forced into contact and crack, and later-formed minerals, especially quartz, intrude the cracks (Bouchez et al., 1992). In deformed crystalline rocks, there is a strong preferred orientation of microfractures, which are primarily tensional features, parallel to the maximum principal compressive stress (Moore and Lockner, 1995). Tensile microcracks grow much more rapidly and easily in ceramics, glasses, and minerals if moisture is present, again emphasizing the merging of physical and chemical processes in weathering (Kirby, 1984, p. 3992, and references therein). Many physical rock properties such as permeability, seismic velocity, electrical resistivity, elasticity, and strength are determined by the number, orientation, and interconnectedness of pores and fractures.

Whenever a rock mass is fractured, whether by internal or external stresses, the resulting fragments have sizes with a statistical distribution that follows *Rosin's law* (Bennett, 1936). Rosin distribution produces an excess of fine fragments, with particle abundance in inverse proportion to their size, following a fractal distribution (Figure 1-1). The theory of crushing has important applications to mining and quarrying operations, but it is also instructive as a rigorous mathematical explanation of why rocks broken by any mechanical process yield many small fragments with a large total surface area.

The most important processes by which rock masses are mechanically broken or disintegrated are (1) differential expansion with pressure release; (2) thermal expansion and contraction, including fire

FIGURE 7-1. Deep weathering of granite, San Diego, California. Exfoliation weathering has reduced most of the originally cubic blocks defined by quartz veins to spheroids, but one vein has not been affected. Spherical corestone on the left is about 75 cm in diameter (photo: William W. Locke).

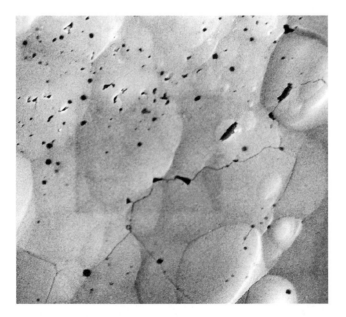

FIGURE 7-2. Photomicrograph of the boundary between grains of feldspar (upper left) and quartz (lower right) in Rutland quartzite. Feldspar grain is full of pores or tubes about 1 µm in diameter. On the grain boundary is a nearly continuous crack with larger triangular cavities aligned on it (photo: W. F. Brace).

damage; and (3) growth of foreign crystals in cracks and pores. The mechanical pressures generated by growing and moving organisms might be important enough to rate a place on this list as well. Each of these processes affects different rock types in different ways, and the second and third processes depend strongly on climatic conditions.

Pressure Release on Unloading: Sheeting

Several examples will demonstrate that some rocks now at the earth's surface were formerly buried at depths of 20 km or more. A combination of both tectonic denudation (p. 78) and subaerial erosion may be responsible for the exposure. The reconstructed geometry of structures in the Valley and Ridge Province of the Appalachian Mountains requires that at least 8 km of folded sedimentary rocks have been removed since late Palaeozoic time. Apatite fission-track ages corrected for an assumed geothermal gradient, and the metamorphic rank of anthracite coal in eastern Pennsylvania, both require the removal of more than 9 km of former strata from the eastern Appalachians (Roden and Miller, 1989; Slingerland and Furlong, 1989, p. 35). From the metamorphic grades of certain mineral assemblages, 24 km of rock is estimated to have been removed from the southeastern Adirondacks (Whitney and McLelland, 1973). Post-Cretaceous erosion to a depth of 10 km has been required to expose the igneous rocks of the Sierra Nevada (Cowan and Bruhn, 1992, p. 180); 16 to 24 km of post-Triassic erosion has exposed the tightly folded sedimentary rocks of the Wellington District, New Zealand (Stevens, 1974, p. 26). The Alps in Switzerland have been denuded by 30 km in the last 30 million years (Clark and Jager, 1969).

The confining pressure that is typically released by erosion of the magnitudes just listed is on the order of 1.5 to 8 x 10^2 MPa (1.5 to 8 kilobar). The coefficients of elastic expansion of typical rocks are such that expansion of 0.1 to 0.8 percent should result (Birch, 1966). However, contraction due to cooling and other factors more than compensate for the expansion due to unloading, and the predicted state of stress in uplifted, cooled, and unloaded rock masses would be strongly tensile except for the growth of microcracks within and between mineral grains due to their anisotropic contraction (Haxby and Turcotte, 1976; Bruner, 1984). The net result is that massive rocks are under horizontal compressive stresses measured as high as 41 MPa (0.4 kilobar) in surface exposures. Newly quarried blocks of Stone Mountain granite (Figure 7-3) expand 0.1 percent in the direction of greatest confinement (Hopson, 1958, p. 70). If a massive

FIGURE 7-3. Sheeting joints on the southeast flank of Stone Mountain, a granite dome near Atlanta, Georgia. An estimated 12–15 km of overlying rock have been removed from Stone Mountain in 300 million years (Whitney et al., 1976) (photo: C. A. Hopson).

rock is confined on all sides but one, corresponding to the land surface, the expansion must be toward the free face.

Quarry workers and miners are well aware of the dangers of *rock bursts* in a freshly excavated quarry or mine (Yatsu, 1988, pp. 144–145). The abrupt pressure drop at the surface can cause slabs of rock to bulge or burst into the mined-out space, sometimes with explosive force. It is common practice in mining and quarrying operations to detonate explosive charges at the end of a work shift, allowing time for the immediate and most dangerous expansive stresses to decay before the next crew enters the area.

Sheeting joints or simply **sheeting,** the separation of rock into layers parallel to the land surface, is a near-surface phenomenon, most common in massive rocks such as igneous intrusives (Figure 7-3), quartzites, and thickly bedded sandstones. Rocks with other types of fractures, such as columnar joints or bedding planes, are able to expand upon unloading without developing sheeting even though they may develop minor sets of relief joints parallel to gorge walls and excavations. Granite in quarries near Boston, Massachusetts, has an average thickness of joint-bounded sheets of 10 cm to 1 m near the surface, but the average sheet thickness increases rapidly to about 5 m at a depth of 20 m, and to more than 10 m at depths of 30 to 40 m. At greater depths, the sheeting joints become so widely spaced as to be insignificant

(Jahns, 1943, p. 97). Sheeting joints on granitic monoliths in Yosemite National Park, California, are spaced from a few centimeters apart at the surface to a few meters apart at depth. The aggregate thickness of the sheets rarely exceeds 30 m (Matthes, 1930, p. 116). In tunnels through sandstone at the Glen Canyon dam site on the Colorado Plateau, all joints parallel to the canyon wall were encountered within 10 m of the surface (Bradley, 1963, p. 522).

Of the various proposed causes of sheeting, only pressure release by unloading generally satisfies the observed field relations (Jahns, 1943). A minority opinion favors tectonic stress, however (Twidale, 1982, pp. 150–158). Sheeting cuts across primary structures and intrusive contacts within igneous rocks; the curvature and orientation are too local to be related to regional tectonic stresses; it is similar in climatic regions of great insolation variation, such as polar and tropical, and in wet and dry regions; the minerals within the sheets show negligible chemical alteration; and sheeting invariably conforms to the present or a previous land surface (Hopson, 1958; Twidale, 1982, p. 152; Yatsu, 1988, pp. 145ff).

Probably, sheeting joints develop only near the land surface because the geothermal gradient decreases to much less than normal in the outer 200 m of the crust and the temperature of the rock becomes nearly constant. The overburden pressure, however, continues to decrease at a rate proportional to the rate at which overlying rock is peeled off by erosion. No compensating contraction due to decreased temperature is possible at shallow depths, so the expanding outer layers of rock are bowed upward or outward toward the land surface, and sheeting joints develop. Preexisting microcracks determine the final fracturing, and because they are so abundant in rocks, slight directional stresses can cause large, continuous joints to propagate. *Continuity* is one of the eight parameters used to define rock-mass strength (p. 120). Short, discontinuous joints have numerous "bridges" of intact rock between them; continuous, parallel sets of joints such as sheeting reduce cohesive strength, provide zones of shear, permit water circulation, and may become filled with detrital clay that may allow sliding on steep-dipping joints, especially if they dip outward from a topographic slope.

The size of monolithic domes is determined largely by the spacing of faults and joints other than sheeting, but their shape is determined by sheeting. If a rock mass is irregularly fractured at depth, the most fractured portions adjust readily as the superincumbent load is removed, but portions with few or widely spaced fractures store the stress of expansion until the overburden pressure is so low that failure occurs. Such failure is parallel to the land surface at a depth of 100 m or less, so the sheeting joints conform closely to local topography. In glaciated mountains, sheeting on cirque walls and floors is concave skyward, parallel to the bowl shape of the cirques, whereas the sheeting joints are convex on the adjacent convex summits. At the rims of cirques, the upland and cirque-wall joint sets intersect (Figure 7-4).

Thermal Expansion and Contraction

Mechanical weathering of rocks by temperature changes can result from *thermal stress fatigue* either by repeated or cyclic heating and cooling or by *thermal shock,* extreme, sudden temperature change (Yatsu, 1988, p. 120). The thermal volumetric expansion of quartz is three times that of feldspar. In addition, quartz, feldspar, and many other common minerals expand under heat in a highly anisotropic fashion, with expansion along certain crystallographic axes being as much as 20 times greater than along other axes. Therefore, temperature changes create stresses both within and between mineral grains, sufficient to cause disaggregation. Furthermore, rocks in general are poor thermal conductors, so extreme temperature

FIGURE 7-4. Cirque headwall eroded in granite, Angel Pass, Wind River Range, Wyoming. Intersecting sets of sheeting joints from the cirque (left) and summit (right) both increase in spacing inward (photo: William W. Locke).

gradients develop in the outer centimeters of a rock exposed to intense surface heat. Ground temperature in deserts can exceed 85°C with diurnal fluctuation greater than 50°C (Goudie, 1989, pp. 12–13). Because the heating decreases rapidly with depth, during daytime the outer few centimeters of a rock expand relative to the deeper parts, but as the outer shell cools, it contracts faster than the still-warm interior. This lag effect (Warke and Smith, 1994, p. 63) creates diurnal alternating compressive and tensional stresses, especially in the outermost few millimeters of rock where the changes are most extreme. Microcracks begin to open, as measured by detectable acoustic emissions (snap, crackle, pop), when the temperature exceeds 60°C to 70°C , and the effect rapidly increases with further heating (Yatsu, 1988, pp. 134-135). However, there is a kind of annealing that decreases the rate of acoustic emissions during later cycles that do not exceed the maximum temperature of earlier cycles, so the effect is not entirely cumulative.

Thermal effects are not limited to low-latitude deserts. If anything, they are even more pronounced in regions of extremely low air temperatures, such as Antarctica. Simulations of intense solar radiation on rocks in a refrigerated chamber demonstrate that the thermal shock related to lag effects of surface and sub-surface heating and cooling at rates greater than 2°C/minute will break rocks (Hall and Hall, 1991). The rapid temperature changes are realistic simulations of the effect of a passing cloud. In addition to the likelihood of thermal shock, thermal fatigue is a probable longer-term result. A problem with cold-region thermal shock experiments is that they cover the same range of temperature over which freezing and thawing are possible if water is present. The cited experiments were conducted on thoroughly dried specimens to simulate conditions in the Antarctic interior, but water content may be significant in field conditions.

Despite considerable experimental and observational evidence that rocks spall or break under solar insolation, doubt remains about the effectiveness of dry thermal stress (Yatsu, 1988, pp. 120–137; Goudie, 1989, pp. 11–24; Cooke et al., 1993, pp. 28–31; Smith, 1994, pp. 40–44; Warke and Smith, 1994, pp. 57–70). Older experiments on small specimens seemed to show no detectable weathering of rock by the equivalent of centuries of diurnal dry heating and cooling, but perhaps the samples were too small to build up the necessary thermal stresses. All authorities agree that in the presence of moisture and the various soluble salts that commonly precipitate on rock surfaces, terrestrial ranges of temperatures can fracture rocks.

Similarly, there is no doubt about the effect of fire on rocks (Yatsu, 1988, pp. 137–138; Cooke et al., 1993,

p. 31). Thermal shock generated by brush fires and forest fires has been observed to spall 5 to 50 percent (Bierman and Gillespie, 1991) or 70 to 90 percent (Zimmerman et al., 1994) of the surface areas of boulders. Spalls, ranging in thickness from a few millimeters to a few centimeters, remove areas of several hundred cm² from boulders, mostly on their sides (Figure 7-5). The depth of spalling is consistent with observations that heating generates the steepest thermal gradients in the outer few centimeters of exposed rock faces. As it does with other thermal effects, the presence of water in rock pores and fractures has an important but variable influence on fire spalling. On the one hand, by boiling away, it can delay and reduce the heating of the rock. On the other hand, the steam can be a potent mechanical and chemical reactant (Allison and Goudie, 1994).

Growth of Foreign Crystals in Cracks and Pores; Salt Weathering

Certain minerals, such as pyrite (FeS_2) in shale or slate, oxidize readily on exposure to oxygen-bearing groundwater or a moist atmosphere. The newly formed iron oxides are of lower density and larger volume than the original minerals, and the volume increase can be sufficient to split weak rocks or cause spalls or "pop-outs" in strong rocks.

Of greater significance to weathering processes are the numerous salts that are water soluble (Evans, 1970). When pore water percolates to or near the surface of a rock and evaporates, any dissolved salts precipitate. Rainwater may also accumulate salts, especially when it flows over soot-stained city buildings and through masonry joints. The force of crystallization around nuclei of precipitating salts is substantial. Halite (NaCl), for example, when precipitating over a range of temperature between 0°C and 50°C from supersaturated solutions, generates *crystallization pressure* ranging from 54 to 366 MPa (554 to 3737 kb/cm²) (Winkler and Singer, 1972, Table 1). Such pressures exceed the strength of almost all rocks. Supersaturation is a necessary condition for the growth of salt crystals but is easily achieved either by cooling of a saturated solution or by evaporation.

Chemically distinct from crystallization pressure, but similar in its weathering effect, is *hydration pressure* (Yatsu, 1988, pp. 88–90; Goudie, 1989, pp. 18–21). Dendritic salt crystals, or even crusts, may form in cracks, along mortar joints in buildings, or on grain boundaries within a permeable but otherwise coherent rock. Many water-soluble salts precipitate as hydrates or invert to hydrated compounds within the normal ranges of atmospheric temperature and humidity. Hydration pressure is more effective than simple crystal growth because, as temperature and humidity change seasonally or daily, the salts may hydrate and dehydrate repeatedly (Winkler and Wilhelm, 1970). The hydration pressures of common salts range from 10 MPa to more than 200 MPa (100 to more than 2000 bar). In theory, the pressure should be highest at low temperature and high relative humidity.

Mirabilite (Glauber's salt, $Na_2SO_4 \cdot 10\ H_2O$) dehydrates to thenardite (Na_2SO_4) readily, quickly, and reversibly. At 39°C, the dehydration is complete in about 20 minutes (Winkler and Wilhelm, 1970, p. 570). Desert temperatures commonly range across the transition temperature (Goudie, 1989, p. 19). The standard test for building-stone durability makes use of this hydrated sodium sulfate salt. Blocks of construction-grade granite that can be subjected to 5000 cycles of alternate freezing (6 hours at −12°C) and thawing (1 hour in water at 20°C) with barely visible disintegration crumble after an average of only 42 cycles when soaked in a saturated solution of sodium sulfate (17 hours at room temperature) and dried (7 hours at 105°C) (Kessler et al., 1940; Evans, 1970).

Until the last century, salt precipitation and hydration were weathering phenomena of concern only in desert regions with a high water table, as, for example, along the Nile River or the Rio Grande. However, many of the same salts are leached out of bricks and mortar and attack stone and concrete masonry in ancient and modern cities. The growing use of deicing salts on highways is causing serious maintenance problems (Cody et al., 1996). The physical force of

FIGURE 7-5. Fire-induced exfoliation of granitic boulder on a moraine at Fremount Lake, Wyoming. Photograph was taken 6 years after a major range fire in July 1988. Note burned trees, and piles of spalls of the boulder (photo: E. B. Evenson).

Table 7-1

Composition of the Earth's Crust in Order of Chemical and Mineralogic Abundance (From Richardson and McSween, 1989, pp. 328–330).

Chemical Composition	Weight % as Oxides	Mineralogic Composition	Volume %
SiO_2	59.3	Plagioclase	39
Al_2O_3	15.9	Orthoclase	12
CaO	7.2	Quartz	12
FeO	4.2	Pyroxenes	11
MgO	4.0	Amphiboles	5
Na_2O	3.0	Micas	5
Fe_2O_3	2.5	Clay minerals and chlorite	5
K_2O	2.4	Olivine	3
TiO_2	0.9	Calcite, aragonite, and dolomite	2
		Others	6

hydrating crystals is a major cause for the growing concern over ways to maintain and preserve historic monuments in heavily populated industrialized regions (Figure 7-6) (Committee on Conservation of Historic Stone Buildings and Monuments, 1982; Winkler, 1994; Viles, 1994). A particularly troublesome salt is hydrous calcium sulfate $CaSO_4 \cdot 2H_2O$ (the mineral gypsum). In industrial regions, where the air is polluted by sulfurous smoke from fossil fuels, rainwater becomes a dilute sulfuric acid (Likens and Bormann, 1974; Schwartz, 1989) that corrodes limestone and marble buildings. The reaction product of the corrosion is gypsum, a relatively insoluble hydrated salt

FIGURE 7-6. Detail of the entrance vault at Cathedral St. Coretin, Quimper, France. Original carvings of the late fifteenth century are in Caen Stone, a soft Jurassic limestone widely used in western European architecture. Three inner archivaults have been restored recently, with similar stone. Outer carvings have been nearly obliterated by 500 years of weathering.

that crystallizes in rock cracks, flakes off thin pieces of rock, and accelerates chemical attack. Salt weathering, whether by crystallization or hydration, is regarded as a mechanical weathering process, but the close relationship with chemical processes is obvious.

Tafoni, or honeycomb weathering pits, are a common feature of deserts and semiarid regions, especially near coasts or wherever salts can precipitate from dew or fog (Mustoe, 1982; Young, 1987). They are also common in the spray zone just above high tide level on many coasts, both on bedrock and on breakwaters and other structures (Mottershead, 1994). Evaporation draws solute-laden pore water toward the surface, where the salts precipitate and cause exfoliation. Surface *flaking* of granular rock under projecting ledges in arid parts of Australia (Dragovich, 1967; Bradley et al., 1978; Bradley, 1980) occurs because dew and other atmospheric moisture is uniformly absorbed to a depth of 15 mm or less, and small chips or flakes of slightly hydrated and salinized but otherwise fresh rock are spalled off. Hemispheric pits or small caves form on the undersides of overhanging blocks or in exposed cliffs of a variety of rocks, including granite, sandstone, and volcaniclastic rocks (Figure 7-7). Some of the pits have overhanging visors or rims that partly enclose the opening. In permeable sandstones, volcaniclastic rocks, and limestone, secondary cementation by precipitates in the outer layers of rock may contribute to the resistant visors or "monks hoods" that partly enclose tafoni (Martini, 1978; Winkler, 1980).

Frost Cracking and Hydrofracturing

A foreign crystal of peculiar abundance in surface rocks is ice. Water volumetrically expands 9 percent on freezing at 0°C, but the expansion ratio increases to

FIGURE 7-7. Tafoni and surface flaking in eolian sandstone, Canyonlands National Park, Utah (photo: William W. Locke).

13.5 percent for confined water at $-22°C$. The maximum pressure generated is 207 MPa (2070 bar). At higher pressures, more dense phases of ice crystallize (Tharp, 1987, p. 94). Intuitively, the impressive pressures of freezing that break water pipes in houses and soft-drink containers in freezers are also assumed to break rocks, and indeed, the details of weathering by ice are all related to the fundamental fact of the anomalously large expansion of water upon freezing (Yatsu, 1988, p. 74). However, two popular concepts—(1) that expansion of confined water by freezing breaks rocks, and (2) that the process is enhanced by frequent cycles through the freezing temperature—are both probably wrong. Especially, freeze-thaw caused by changes in air temperature alone are inadequate to cause fracturing (Hall and Hall, 1991).

Few rocks could withstand the high pressure generated by freezing confined water. However, total confinement is required, and there is no natural way for water to be totally confined in rock. A more likely cause of *frost cracking*, supported both by observation and thermodynamic considerations, is that a capillary film of water along the ice-rock interface resists freezing because of the strong molecular adhesion of water to rock. As the temperature is lowered and the freezing front advances into the rock, some of this water freezes, and adjacent pore water migrates to the area of freezing to maintain the thickness of the capillary film and preserve thermodynamic equilibrium. The crack walls are forced apart by the increasing volume of water and ice, but the expansion is limited by the availability of unfrozen water and the permeability of the adjacent rock. The maximum pressure generated is an order of magnitude less than the pressure at the

same temperature in a confined system (Tharp, 1987, p. 95). Frost cracking is thus promoted by a rather slow but steady cooling rate of $-0.1°C/hr$ to $-0.5°C/hr$, maintained in a temperature range of $-4°C$ to $-15°C$. At warmer temperatures, thermodynamic limitations prevent significant ice growth; at colder temperatures or with more rapid cooling, the necessary migration of pore water to the growing ice wedge is inhibited (Walder and Hallet, 1985). In such an open system, water is drawn toward the freezing zone, not driven away from it, as the pressure increase in a closed system would cause. Only in the special closed system of a block of permeable rock freezing inward from all surfaces would water be driven into the unfrozen interior and perhaps cause *hydrofracturing* along pressurized water-filled joints or microcracks. Normally, capillary water will not generate sufficient stresses to propagate cracks (Tharp, 1987).

Thin films of water that form a semicrystalline "ordered" molecular structure on mineral surfaces (especially on clays) can pry apart microfractures and propagate cracks (Dunn and Hudec, 1966; Winkler, 1968). The powerful molecular forces in films of pure water and in saline solutions are special cases of hydration (Yatsu, 1988, p. 77) but are analogous to those involved in freezing, drawing more water to the region where the ordered structure is developing (Evans, 1970, p. 167). Studies of "ordered films" of water once more emphasize the disappearing distinction between mechanical and chemical weathering processes.

Many mountain slopes above the treeline are wastelands of shattered rock rubble that is gradually being broken down to particle sizes that can be attacked by other weathering and erosional agents (Figures 14-3 and 14-9). In humid subpolar regions of perennially frozen ground, freezing and hydrofracturing become the dominant weathering processes. The landforms that result from intense frost action are described in detail in Chapter 14.

Plants and Animals as Agents of Mechanical Weathering

The prying or wedging action of growing plant roots, especially of trees, is often described as mechanical weathering. Two-dimensional networks or sheets of interlaced roots can be followed for many meters along bedding planes or joints, deep into fresh rock. It has been supposed that the growing roots exert a pressure on the rock and force cracks to open. However, the efficacy of roots as agents of mechanical weathering probably has been overestimated, just as their importance as agents of chemical weathering has been underestimated (p. 132). Roots follow the paths of least resistance and conform to each little irregularity of a crack, but they do not exert much force on the rock. Maximum axial pressure by growing roots is reported to be in the range of 3 MPa, and radial pressure is only about 1 MPa (Yatsu, 1988, pp. 365–366). Although such low pressures may exceed the tensile strength of weak rocks, the cross-sectional area of the applied pressure is so small that the total disruptive force is trivial. Cracks opened by other processes can be maintained by roots, however, and decaying vegetable matter and washed-in dirt can keep rock surfaces wet and chemically active. In addition, when trees sway in a strong wind, their roots pry apart rocks.

The "strangler fig" (*Ficus* sp.) of tropical regions has the ability to envelop other trees, masonry, and bedrock with snakelike, anastomosing stems (Figure 7-8). As the stems grow and cross each other, they fuse, encasing a host tree with a wooden latticework sufficiently strong to prevent its further growth and cause its eventual death (Argo, 1964). They also pry apart masonry and natural joint blocks, but it is doubtful that they could crush or pry apart sound rock. The mature *Ficus* may grow to huge size, however, and when such a tree is toppled during a storm, large masses of rock may be displaced. Under temperate and boreal forests, soils usually include many mounds and pits formed by tree roots that were torn up when trees fell (Fanning and Fanning, 1989, pp. 54–61). Because translocation is so obviously involved, *root throw* is better treated as a category of mass wasting or erosion than as weathering.

Soils teem with animal as well as plant life. It is claimed that the species diversity of soil animals, even in a small area, equals that of a coral reef. Almost all the animal phyla have representatives that spend at least part of their life within the soil layer (Hole, 1981). They digest, abrade, and mix mineral grains, admit water and air to the soil through their burrows, and supply chemically active humus and excrement to the soil. Most of their activity merges mechanical and chemical weathering with mass wasting and erosion (Yatsu, 1988, pp. 375–379). Larger animals, including human beings, are obvious excavators of soil and rock, and are primarily erosional agents.

CHEMICAL WEATHERING

Physical or mechanical weathering is generally credited with being about six times more effective than chemical weathering in preparing surface rocks for removal by erosion (Lasaga et al., 1994, p. 2376). However, few rock fragments or mineral grains that

FIGURE 7-8. Stems of *Ficus* encasing prehistoric stone blocks, Kosrae Island, Micronesia. Blocks are 30–50 cm high.

are disaggregated by mechanical processes escape simultaneous or subsequent chemical reactions with water, air, and the biota. Through the history of the earth, chemical weathering has been a major buffer in the ocean-atmosphere-biosphere-lithosphere system, maintaining atmospheric oxygen and CO_2 content and global temperature within narrow limits (pp. 49, 51).

The general trends of chemical weathering can be predicted from the conditions under which rock-forming minerals crystallize or precipitate. If they formed in an environment of high thermal energy and high pressure, the minerals in rocks at the surface of the earth tend to weather by chemical reactions that produce new compounds of greater volume and lower density. **Oxidation** is one of the typical volume-increasing reactions between minerals and the wet atmosphere; especially common is the reaction of iron-bearing minerals with oxygen dissolved in water. Other typical weathering reactions are **carbonation,** the reaction of minerals with dissolved CO_2 in water; **hydrolysis,** the decomposition and reaction with water; **hydration,** the addition of water to the molecular structure of a mineral; **base exchange,** the exchange of one cation for another between a solution and a mineral solid; and **chelation,** the incorporation of cations from the mineral into organic compounds. Hydration was discussed in an earlier section as a primarily mechanical process, as demonstrated by the volumetric expansion of hydrated salts that enter cracks and pores, yet it is certainly a chemical process when the water is absorbed by or reacts with the minerals that make the rock.

An important principle of chemical weathering is that mineral dissolution is concentrated along sites of excess surface energy such as crystal dislocations, cleavage planes, microcracks, and other imperfections. In photomicrographs, grains of weathered feldspars, pyroxenes, and amphiboles all show micron-size etch pits aligned on crystal defects (Berner et al., 1980). Chemical weathering is governed by the kinetics of chemical reactions at these activated sites on mineral surfaces, and not by diffusion rates. However, rapid chemical reactions require steep gradients and continued disequilibrium between the reactants (Brantley and Stillings, 1996). If reactants are not renewed and equilibrium is approached, the reactions slow down.

Water in Chemical Weathering

Dissolution of mineral species in water is by far the most effective process of chemical weathering (Yatsu, 1988, p. 204). Dissolution can be either *congruent,* in which the components of the solid are equally soluble and occur in the solution in the same proportion as in the solid (NaCl dissolving in water is a simple example), or *incongruent,* if certain soluble components go into solution and other less soluble compounds form residual solids. Many of the weathering reactions illustrated on the following pages are examples of incongruent dissolution. Dissolution in water is not considered here a separate process because it is part of all chemical weathering.

Water is by far the most abundant material near the surface of the earth. In the outer 5 km of the earth, water is about three times as abundant as the next most common substance, the feldspar group of minerals (Kuenen, 1955). Its general solvent power and its surface tension are greater than those of any other fluid. Its heat of vaporization is the highest of all sub-

stances; recall the significance of this property in the conversion of solar energy to geomorphic work (Chapter 2). Its maximum density at 4°C in the liquid phase, with expansion toward the freezing temperature as well as away from it, is a strange and almost unique property. We could review the properties of water at length, but this summary is enough to suggest that the outermost few kilometers of the earth are generally saturated with an abundant and chemically active compound, made available in unending supply by the hydrologic cycle, that readily attacks and reacts with rock-forming minerals. The chemical potential of rain is largely expended by various reactions with minerals soon after the drops penetrate the ground. River water has just about the dissolved chemical load that could be predicted from analyzing the rocks through and over which it has flowed.

The availability of water for weathering reactions is not well indicated by atmospheric climatic conditions (p. 140). In hot, dry deserts, diurnal heating and cooling can cause condensation of dew on rocks, where it can be absorbed. On desert coasts, fog is a source of moisture. As noted in connection with thermal stresses in rock (p. 123), even rocks exposed to direct sunlight in very cold climates are heated to above the melting point of water, and it is difficult to separate the purely thermal stresses from those produced by freezing and thawing. Water in rocks is in at least four physical states: (1) water chemically combined in hydrated minerals; (2) hygroscopic, retained, or bound water adsorbed on mineral surfaces as ordered films on the order of 0.1 μm in thickness, with density and viscosity much greater than normal and not mobile under gravity; (3) capillary water, filling or wetting cracks and voids, intermediate in mobility between hygroscopic and gravitational water; and (4) gravitational water that moves through larger pores and cracks under gravity and other pressure gradients (Bell, 1992b, p. 78; Pope et al., 1995, pp. 48–49). In each of these states, water facilitates weathering in a variety of ways: as a transport medium for reactants and solutes, as a solvent, by exerting mechanical pressure, as a chemical reactant, and as a chemical buffer (Pope et al., 1995, p. 41).

Oxidation: Weathering of Iron-Bearing Minerals

Iron-bearing minerals commonly contain iron in the ferrous state. This is true for the major iron sulfide (pyrite, FeS_2), iron carbonate (siderite, $FeCO_3$), and various iron silicates. On contact with water that has oxygen in solution, the ferrous iron oxidizes to the ferric state and forms nearly totally insoluble ferric oxides or hydroxides. Only oxygen-free, chemically reducing, alkaline water permits iron to remain in the more soluble ferrous state. Thus, when aerated water reacts with iron-bearing minerals, the reaction is more complex than simply solution. *Oxidation-reduction reactions* are inevitable, with the result that instead of the mineral going into hydrous solution as ionic species, new compounds are formed, some of which are nearly insoluble.

Weathering by oxidation takes place with water as the intermediary. Unprotected iron surfaces stay clean and bright in extremely dry air, as is proved by the condition of tools and equipment found in polar regions or in the Libyan desert decades after they were abandoned. But iron rusts, or oxidizes, quickly beneath even a dew-thick film of moisture. There is always enough dissolved oxygen in rainwater and circulating groundwater to oxidize metallic iron and to change the ferrous iron in mineral compounds to the more oxidized ferric state. As long as water is in contact with atmospheric molecular oxygen on the one hand and incompletely oxidized iron on the other hand, oxygen dissolves from the air, diffuses through the water, and combines with the iron. When the iron or other elements from a mineral grain combine with oxygen, the original mineral structure is destroyed, and the remaining mineral components are free to participate in other chemical reactions. The oxidation-reduction reactions of water with compounds of aluminum, magnesium, manganese, chromium, and some less common metals are similar to the reactions with iron compounds. The several oxides often occur together.

Iron oxides have intense yellow or red colors. Rocks and soil are stained various shades of red or brown by mere traces of iron oxide. The reactions are so rapid that a brown surface layer is the first indication of weathering on most rocks. Black iron-sulfide muds from the anaerobic conditions of a swamp, for instance, oxidize and develop a brown color within a few minutes after exposure to air. Iron-rich rock-forming minerals such as pyroxenes, amphiboles, and biotite are usually rust colored on all but the most recent natural exposures.

The oxides of iron and related metals are exceptionally stable chemical compounds, even on the geologic time scale. Precambrian rocks rich in iron oxides are widely distributed on the continents, most of them dating from the time about 2.5 billion years ago or less when our evolving atmosphere began to be oxidizing (Pinto and Holland, 1988). They are important sources of iron ore today.

Hydrolysis and Base Exchange: Reactions of Carbonated Water with Silicates

Hydrolysis, like oxidation, is more than just a mineral dissolving in water; both the mineral and water molecules decompose and react to form new compounds. It is the most important chemical weathering reaction of silicate minerals.

Pure water ionizes only slightly, but it does react congruently with some easily weathered silicate minerals in the following fashion:

$$Mg_2SiO_4 + 4H^+ + 4OH^- \longrightarrow 2Mg^{2+} + 4OH^- + H_4SiO_4$$

olivine (forsterite) + four ionized water molecules = ions in solution + silicic acid in solution

The result of such complete hydrolysis is that the mineral is entirely dissolved, assuming that a great excess of water is available to carry the ions in solution. Silicic acid, one of the reaction products, is such a weak acid that we can disregard its name and simply think of it as silica (SiO_2) dissolved in water.

The foregoing reaction is deceptively simple because it implies that water is an ionized compound. Such is not the case. Pure water is known to be a very poor H^+ donor and electrolyte. Its electrical conductivity is extremely low, proving that it has few free ions. Of great importance to weathering processes, therefore, is the fact that carbon dioxide gas dissolves readily in water to form a weak acid, or H^+ donor (p. 148). Even though the terrestrial atmosphere contains only 0.03 percent CO_2 (by weight), raindrops absorb CO_2 as they fall, so that rainwater is usually slightly acidic, with a pH of 5.7 or less.

Cold water dissolves more carbon dioxide gas than does warm water; water under pressure also dissolves more gas. For every condition of temperature and pressure, an equilibrium is established. Addition of a strong mineral acid such as H_2SO_4 from atmospheric pollution prevents absorption of atmospheric CO_2 and decreases the amount of bicarbonate ion in solution (Johnson et al., 1972).

Air that fills pore spaces in soil is greatly enriched in carbon dioxide by the decay of humus. Soil air drawn from the biologically active upper layer of a soil profile may have from 10 to 1000 times more CO_2 than that in the free atmosphere; that is, CO_2 may constitute as much as 30 percent of soil air compared to the 0.03 percent of dry normal atmosphere. Carbon dioxide dissolving in soil water is the most common and important way that water is provided with H^+ ions, or acidified, for hydrolysis.

We can elaborate the chemical equation previously given for the reaction between water and olivine by the following equation, also a congruent reaction:

$$Mg_2SiO_4 + 4CO_2 + 4H_2O \longrightarrow$$

olivine (forsterite) + carbon dioxide + water =

$$2Mg^{2+} + 4HCO_3^- + H_4SiO_4$$

magnesium and bicarbonate ions in solution + silicic acid in solution

This hydrolysis reaction is much more common than the one with pure water shown previously. Note that the H^+ ions that react to form silicic acid and bicarbonate are derived from water molecules through the carbonation process and the accompanying formation of bicarbonate ions in solution. Carbonic acid is consumed by silicate weathering, and the resulting solutions become more alkaline because of the dissolved bicarbonate. The common cations freed by silicate weathering are quite soluble in solution with bicarbonate ions, and most of the dissolved load of rivers is in the form of bicarbonates. The important role of weathering in buffering the amount of CO_2 in the earth's atmosphere, and thereby affecting climate, is described in Chapter 4 (p. 50).

Quartz (SiO_2) is a common silicate mineral (Table 7-1), but it is chemically inert at the earth's surface and SiO_2 can be classed with the oxides of iron, aluminum, and titanium as extremely stable compounds. Silica glass and polymorphs of quartz are significantly more soluble in water than is quartz, but they are not common except in volcanic terranes. The world average for the dissolved silica concentration in river water is 13 parts per million, about twice the calculated solubility of quartz (Livingstone, 1963). Therefore, reactions such as those given previously for olivine and those that follow for feldspars must be the principal sources for silica in solution.

Feldspars, including orthoclase and the plagioclase family of minerals, compose more than half the minerals in the crust (Table 7-1). Therefore, probably the most common weathering reaction on earth is the hydrolysis of feldspars by carbonic acid in water. A typical, though simplified, weathering reaction between potassium feldspar (orthoclase) and carbonated water is as follows:

$$2KAlSi_3O_8 + 2H_2CO_3 + 9H_2O \longrightarrow$$

orthoclase + carbonic acid + water =

$$Al_2Si_2O_5(OH)_4 + 4H_4SiO_4 + 2K^+ + 2HCO_3^-$$

kaolinite— a clay mineral + silicic acid in solution + potassium and bicarbonate ions in solution

In carbonated water, calcium and sodium feldspars (plagioclase) hydrolyze even more readily than does orthoclase. In the foregoing example of incongruent dissolution, as in many others, the initial mineral and acid react, and the end products are (1) a detrital clay mineral, (2) silica in solution, and (3) bicarbonate of potassium, sodium, or calcium in solution.

Clay minerals are an abundant group of aluminosilicate minerals at the earth's surface. They are also known as *phyllosilicates* for their sheetlike structures. Clay minerals are secondary products of weathering, especially of feldspars. They are hydrolyzed and commonly hydrated, and because they form by weathering, they are among the more stable minerals at the earth's surface, ranking with the oxides of silica, iron, and aluminum. Their sheetlike structure is created by layers of silica tetrahedra $[Si_2O_5]^{2-}$ and aluminum (or magnesium) hydroxide octahedra $[Al_2(OH)_6]$, linked by hydroxyl $[OH]^-$ ions, other cations, or even organic molecules (Evans, 1992, p. 108). *Kaolinite,* the clay mineral created by feldspar weathering in the reaction shown above, is the most common clay mineral with 1:1 layering, in which the silica and alumina layers alternate and Si and Al are equally abundant. In 2:1 layer clay minerals, which include *illite, smectite* (commonly *montmorillonite*), *vermiculite,* and *chlorite,* two silica sheets sandwich each alumina sheet. They have a higher Si:Al ratio, and the layers are weakly bonded, especially with K^+ cations, so they expand when wet. Illite is similar to mica in its molecular structure.

Even within the clay mineral group, there are degrees of stability. It is hypothesized that the weathering of feldspar may proceed in steps, first producing a muscovite structure, then illite, then kaolinite, as cations and silica are progressively lost (Figure 7-9).

Kaolinite when intensely leached by water may finally become the stable hydrated aluminum hydroxide called gibbsite:

$$Al_2Si_2O_5(OH)_4 + 5H_2O \longrightarrow 2Al(OH)_3 + 2H_4SiO_4$$

kaolinite + water = gibbsite + silicic acid in solution

This reaction is common only in wet tropical climates (Tole, 1987). The kinds of clay minerals that form during weathering may be used, with caution, for interpreting climatic conditions: illite and mixed-layer clays are common in cool temperate climates; montmorillonite and other swelling clays are associated with arid climates where Na and Ca accumulate, and kaolinite with the humid tropics. As discussed in subsequent paragraphs, the survival of clay minerals in paleosols may give important clues about the climate at the time of the soil formation. Clay mineralogy has evolved into an important subfield of geology

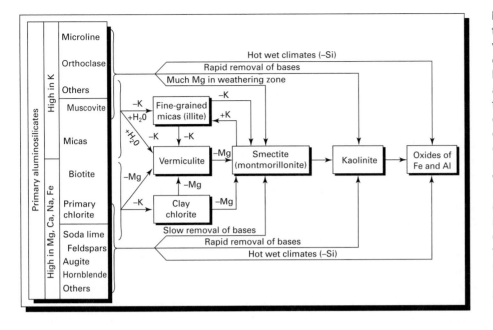

FIGURE 7-9. General conditions for the formation of the various layer-silicate clays and oxides of iron and aluminum. Fine-grained micas, chlorite, and vermiculite are formed through rather mild weathering of primary aluminosilicate minerals, whereas kaolinite and oxides of iron and aluminum are products of much more intense weathering. Conditions of intermediate weathering intensity encourage the formation of smectite. In each case, silicate clay genesis is accompanied by the removal in solution of such elements as K, Na, Ca, and Mg (Brady and Weil, 1996, Figure 8-10).

and soil science, with its own literature (Evans, 1992, is a good recent review).

Soil scientists in recent years have recognized the importance of another chemical weathering reaction, *base exchange*. Base exchange involves a mutual transfer of cations such as Ca^{2+}, Mg^{2+}, Na^+, or K^+ between an aqueous solution rich in one cation and a mineral rich in another. The exchange of cations between minerals and groundwater may expand or collapse the mineral structure and free other chemical components. As in other chemical reactions, if one mineral grain in a rock is so destroyed, adjacent grains are detached and exposed to the same and other weathering processes.

Weatherability of Silicate Minerals

By comparing the rates of chemical reactions between water and various powdered minerals in controlled laboratory experiments, it is possible to predict which minerals will weather most rapidly under natural conditions. Extensive studies of this sort, especially on the silicate minerals that form most of the earth's crust, have led to the recognition of a mineral stability series or **weathering series,** a list of common or representative silicate minerals arranged in order of relative susceptibility to chemical weathering. A dramatic way to present such a list is to calculate the average lifetime of a single 1-mm crystal of each mineral (Table 7-2).

The silicate minerals that crystallize at the highest temperatures and that have the lowest silicon-oxygen ratios in general weather most rapidly.

Quartz (SiO_2), by far the most resistant to chemical dissolution of the silicate minerals, is also hard enough to resist mechanical abrasion, and it commonly crystallizes from silicate magma in millimeter-size grains. These three facts explain the abundance of quartz in alluvium, beach, and dune sand, and eventually in sandstones, which in turn can be mechanically weathered and the quartz grains recycled. It is notable that the second-most resistant mineral in Table 7-2, kaolinite, is a common clay mineral produced by incongruent weathering reactions.

Table 7-2

Mean Lifetime of a 1 mm Crystal at 25°C and pH 5 (From Lasaga et al., 1994, Table 1 and References Therein).

Mineral	Lifetime (years)
Quartz	34,000,000
Kaolinite	6,000,000
Muscovite	2,600,000
Epidote	923,000
Microcline	921,000
Prehnite	579,000
Albite	575,000
Sanidine	291,000
Gibbsite	276,000
Enstatite	10,100
Diopside	6,800
Forsterite	2,300
Nepheline	211
Anorthite	112
Wollastonite	79

The weathering of micas illustrates some of the complexities of any simple weathering series. Although superficially similar, biotite and muscovite weather at very different rates. Both weather by losing interlayer K^+ ions to solution by hydrolysis, but the different orientation of the hydroxyl ions in the two micas binds the K^+ ions more firmly in muscovite than in biotite (Nahon, 1991, pp. 27–30). As a result, muscovite is a common detrital mineral in soil and sediment (Table 7-2), whereas biotite weathers rapidly. It has also been noted that biotite weathers faster than might be expected because the iron oxidizes and the layered mica structure expands greatly on hydrolysis. Biotite weathering is a large contributor to the granular disaggregation of granite to *grus* (Isherwood and Street, 1976).

Another way to study relative chemical weathering rates is to make a ratio of the concentration of each cation in solution in groundwater or rivers with the abundance of that cation in the regional terrane as illustrated in Table 7-1. Feth et al. (1964) established a **cation mobility series** in this way for a granitic terrane in California; Na, Ca, and Mg are the most mobile elements, K and Si are less mobile, and Al and Fe are by far the least mobile. Colman (1982, p. 36) identified a similar mobility sequence by analyzing the weathering rinds on basalt and andesite cobbles from soils on glacial moraines of various Quaternary ages in the western United States. Based on the abundance of each element in the weathered rinds compared to the abundance in the unaltered rock, the sequence of mobility is:

$$Ca \geq Na > Mg > Si > Al \geq K > Fe > Ti$$

A similar mobility sequence has been developed for Icelandic basalt (Gíslason et al., 1996, pp. 867ff). In general, the cations that are lost to solution in greatest proportion are those that are derived from easily weathered plagioclase feldspars and mafic minerals. An exception is iron, which is abundant in the dark mafic silicate minerals that weather most easily, but which oxidizes to an insoluble oxide that does not move away in solution. The iron and titanium in magnetite and titanomagnetite in basic igneous rocks are extremely stable, and remain in weathered rock as part of the original grains (Colman, 1982, pp. 13–14).

Another anomalous result of weathering involves potassium, which is of comparable crustal abundance and chemical behavior to sodium (Table 7-1), yet whose chemical mobility is of a much lower order, comparable to that of silica. Potassium is "recycled" into the crystal lattice of micalike clay minerals such as illite (Figure 7-9), whereas the other alkalic ions are carried away in solution (Colman, 1982, p. 36).

BIOLOGIC WEATHERING

Chelation: Chemical Weathering by Plants

Plants and their associated microbiota accelerate weathering in a variety of ways, but not always in the obvious ways. For example, deciduous angiosperms remove K^+, Mg^{2+}, and Ca^{2+} from soil at rates three to four times faster than the rates of uptake by gymnosperms (p. 51). Plants promote weathering in a variety of ways: by complexing essential mineral nutrients into organic compounds, by releasing organic acids, by generating CO_2, which increases acidity in the soil, by retaining and recycling water, and by physically creating open pores and channels in the soil. The microbial population on the roots of higher plants also is very important in weathering.

One caution about plant-induced chemical weathering is that to be effective, the reactants must be removed from the system in order to maintain the disequilibrium that causes the weathering. If, under heavy vegetation and thick soils, the removal of reactants is inhibited, chemical weathering rates may decrease (Bluth and Kump, 1994, p. 2354). It has been claimed that, as a result, the weathering rate on very similar basalt is lower under thick soils in Hawaii than under poorly vegetated thin soils in Iceland (Drever, 1994, p. 2330).

Chelation is a complex organic process by which metallic cations are incorporated into hydrocarbon molecules (Yatsu, 1988, pp. 319–321). The word *chelate*, which means "clawlike," refers to the tight chemical bonds that hydrocarbons may impose on metallic cations. To function, many organic processes require metallic-organic chelation. The role of iron in hemoglobin to carry oxygen from the lungs is a good example. Synthetic chelating agents are widely used in analytical chemistry laboratories.

Soils in which plants are growing and decaying contain complex organic compounds that can chelate metallic cations. By analogy with laboratory procedures, if natural organic molecules in soil can remove metallic cations directly from mineral grains by chelation, the mineral lattices are disrupted, and possibly fragments of crystal lattices, including otherwise insoluble elements such as aluminum, can be mobilized.

Vascular plants promote chemical weathering by mutually reinforcing interactions among chelation, hydrolysis, and base-exchange reactions. Each plant rootlet maintains a negative electrical charge on its surface (Robert and Tessier, 1992, p. 79) and a surrounding field of H^+ ions. Hydrolysis and base exchange are strongly promoted by the availability of

H+ ions in the vicinity of the root hair, and organic compounds in the roots absorb and chelate the essential nutrient cations as they become available by hydrolysis. Vascular plants use the energy derived from sunlight falling on green leaves to weather minerals many meters underground.

An impressive demonstration of the ability of plants to weather rock was provided by Lovering and Engel (1967). In controlled-environment greenhouses, they grew *Equisetum* (scouring rush) and three grasses in pots of freshly crushed, sterilized rock samples. The silica uptake rate by the plants, all of which are known to be silica accumulators, was equivalent to removing all the silica from a basalt substrate 30 cm thick in only 5350 years. This remarkable rate of withdrawal of silica from igneous rocks was exclusive of any chemical weathering or leaching by percolating water although some of the silica taken up by the plants might have been first dissolved by other weathering reactions. Even though the rocks in the experiment had been mechanically crushed in advance, and the plants were growing under optimal conditions, the amounts of translocated mineral matter are impressive.

On mainland Antarctica, where no higher plants live, weathering is also very slow. The lack of organic activity is compounded by the fact that liquid water, although available as thin films at ice-rock interfaces (p. 126), does not permeate through rocks and transport dissolved compounds (Ugolini, 1986). Meteorites that were collected from the surface of the Antarctic ice sheet showed slightly less oxidation and hydration than the weathering on chemically similar meteorites that fell in Arizona in 1912 and were collected in 1931 and 1968. However, the Antarctic meteorites fell in middle Pleistocene time, so a few hundred thousand years of weathering on the barren Antarctic ice sheet had less effect than 19 to 56 years of weathering in central Arizona (Gooding, 1981, p. 1121). The weathering of the Antarctic specimens was probably restricted to intervals when the black meteorites were exposed on the ice surface and absorbed enough solar radiation to melt some nearby ice.

Microbial Weathering

Microorganisms such as bacteria, fungi, and algae have the ability to dissolve elements from minerals (Viles, 1995). Fungi (*Penicillium* sp.) grown in a nutrient solution mixed with finely crushed rock greatly increased the quantity of Si, Al, Fe, and Mg entering solution, compared to the amount of elements entering solution in similar, but uninoculated control solutions (Silverman and Munoz, 1970). During the incubation period of the experiment, the pH of the nutrient solution dropped from 6.8 to below 3.5 in seven days. As the acidity increased, the rate of mineral solution increased correspondingly. Other experiments were conducted with a variety of organic and mineral acid solutions of equivalent concentrations to the fungal growth medium, with essentially similar effects on rocks. Thus fungal weathering was proved to be due not to some special biogenic process but simply to chemical weathering in acidified water.

Even in regions devoid of macroscopic plants, algae and fungi contribute to weathering processes. Cyanobacteria (blue-green algae) have been discovered living in near-surface pores and cracks in sandstone and quartzite in both the Sahara and Antarctica, with no known source of liquid-phase water (Friedmann and Ocampo, 1976). These cyanobacteria are representatives of the oldest known forms of cellular life on our planet. Their well-preserved remains have been found in early Archean rocks that are 3.3 to 3.5 billion years old (Schopf and Packer, 1987). The fossil record of oxygen-producing photosynthetic microorganisms (microbes) demonstrates that biologic weathering processes have been active for essentially the entire geologic history of the earth (deRonde and Ebbesen, 1996). All the purely chemical or thermodynamic weathering processes described in this chapter are enormously enhanced or facilitated by microbial activity. For example, the oxidation of ferrous iron, one of the most obvious chemical reactions, is accelerated by a factor of 10^5 to 10^6 in the weathering of pyrite (FeS_2) by *thiobacilli* (colorless sulfur bacteria) (Jorgensen, 1983, p. 117). Even quartz dissolution accelerates in anoxic, organic-rich groundwater (Hiebert and Bennett, 1992).

Microbes are now being cultivated and genetically engineered to leach commercial heavy metals from mine dumps, to immobilize radioactive and toxic wastes from leachates, and to increase food production by higher plants (Lindow et al., 1989). The importance of symbiotic mycorrhizal fungi and associated bacteria that live on the roots of higher plants and facilitate their mineral uptake has led to the introduction of the term **rhizosphere** for that part of the soil weathering profile around roots in which microbes are active (Berthelin, 1983, p. 249; Yatsu, 1988, pp. 366–375; Robert and Tessier, 1992, pp. 78–79). It is even appropriate to redefine all weathering as "the chemical and mineralogic modifications of the different types of rocks and minerals within the biosphere . . ." (Berthelin, 1983, p. 223), with the demonstration of microbial activity as deep as 4.2 km within the earth's crust in salt domes and petroleum reservoirs, at the vents of submarine volcanoes, and in the most hostile earth surface environments known (Berthelin, 1988, p. 34; Fyfe, 1996; Ghiorse, 1997; Madigan and Marrs, 1997). Life on Venus and Mars, if found, is most likely to be microbial.

ROCK WEATHERING AND SOIL FORMATION

Definitions

Soil formation is an obvious result of weathering, and soils form the active surface layer that mantles most rock masses. Pedology (soil science) and geomorphology have always been closely linked (Knuepfer and McFadden, 1990). The term **soil** is used to describe the rock detritus at the surface of the earth that has been sufficiently weathered by physical, chemical, and biologic processes so that it supports the growth of rooted plants. This is an agricultural definition, emphasizing that soil is a biologic as well as a geologic material. Engineers are less specific about their definition of a soil. To them, all the loose, unconsolidated, or broken rock material at the surface of the earth, whether residual from weathering at that place or transported by rivers, glaciers, or wind, is soil. However, for most of the engineers' "soils," genetic terms such as "alluvium" or "glacial drift" are available. As a general term for the surficial layer of broken rock, **regolith** is also useful. In most contexts, the engineering or agricultural implication of the term is clear enough so that tedious definitions are unnecessary.

Soils are characterized by **horizons:** distinctive weathered zones, approximately parallel to the surface of the ground, that are produced by soil-forming processes. The five principal horizons, from the surface down into unaltered rock, are conventionally given the capital-letter symbols O, A, E, B, and C (Table 7-3). A **soil profile** is a vertical cross section through these horizons (Figure 7-10). The horizons in a soil profile illustrate the principle that many mechanical, chemical, and biologic weathering processes operate simultaneously near the surface of the earth but at different rates and to various depths. By color, chemical analyses, grain size, and other diagnostic criteria, soil profiles are divided into horizons and subhorizons that clearly record the intensity and duration of the various soil-forming, or weathering, processes. Despite the strong genetic implications of horizons, they are intended to be defined by observable facts, not genetic inferences.

Below the five master horizons is the underlying parent material, such as bedrock, alluvium, or other material from which the soil has formed. It is sometimes referred to as the **R horizon.** Although not specified in the definition of pedologists, most geologists would observe the effects of mechanical weathering on the rock mass in the upper part of the R horizon. Its lower limit is not specified. To pedologists, the R horizon or regolith usually implies the rock mass beneath the soil profile, in the geologists' realm of interest rather than their own. However, growing concern about pollutant transport from the land surface through the soil and regolith to the zone of groundwater saturation has caused pedologists to extend their view considerably deeper than their traditional 1- to 1.5-meter-deep soil pit. A 1992 joint symposium of the Soil Science Society of America and the Clay Minerals Society (Cremeens et al., 1994) demonstrated the broader view of pedology and the effect of surficial weathering on rocks even hundreds of meters below the surface. Some of this new outlook is reviewed in the remainder of this chapter and elsewhere in the book.

Soil Genesis

Five key factors of soil formation have been long recognized by pedologists: *parent material, climate, organisms, topography,* and *time.* The kind of soil that forms is the result of chemical and mechanical weathering of some parent rock type ("structure," in

TABLE 7-3

Principle Soil Horizons (From Brady and Weil, 1996, pp. 52–53)

Horizon	Properties
O	Uppermost horizon; fresh to partly decomposed organic matter.
A	Mineral soil mixed with humus; dark colored.
E	Zone of maximum *eluviation* (leaching) of clay, iron, and aluminum, leaving residual minerals such as quartz in sand and silt sizes. Lighter color than A horizon.
B	Zone of *illuviation* (accumulation) of clay, iron, and aluminum compounds from above, with development of distinctive structures such as granular, blocky, or prismatic aggregates. Colors more intense than overlying and underlying horizons.
C	Relatively unweathered unconsolidated mineral matter below the zone of major biologic activity. May be mechanically fractured, stained by oxides, or loosely recemented by calcium carbonate, gypsum, iron oxide, or more soluble salts.

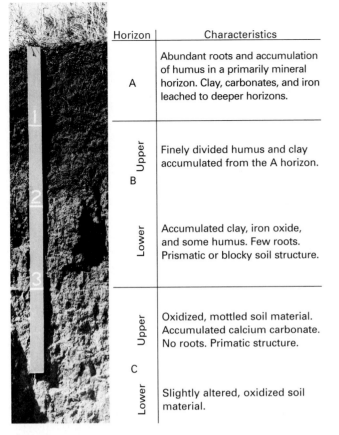

Horizon		Characteristics
A		Abundant roots and accumulation of humus in a primarily mineral horizon. Clay, carbonates, and iron leached to deeper horizons.
B	Upper	Finely divided humus and clay accumulated from the A horizon.
	Lower	Accumulated clay, iron oxide, and some humus. Few roots. Prismatic or blocky soil structure.
C	Upper	Oxidized, mottled soil material. Accumulated calcium carbonate. No roots. Primatic structure.
	Lower	Slightly altered, oxidized soil material.

FIGURE 7-10. A Mollisol soil profile in Nebraska showing horizons A, B, and C and the notable characteristics of each. Depth scale is in feet. The R horizon is not exposed in this profile (photo: W. M. Johnson, U.S. Department of Agriculture—Soil Conservation Service).

the geomorphic sense defined in Chapter 1) under a certain climatic regimen of temperature and precipitation in the presence of plants, animals, and microorganisms (that are at least in part controlled by geologic and climatic factors) on some kind of landform through an interval of time. Under certain conditions, each of the five variables can exert the controlling influence. Three of the factors—climate, organisms, and topography—are likely to change during the progress of soil genesis.

Geologists are inclined to assume that parent material is the dominant soil-forming factor. A residual soil cannot contain minerals that are not present in, or cannot be made by weathering of, the parent material. We assume that a soil on weathered limestone will be calcareous and that one on quartz sand will be siliceous. However, the weathering processes that have been reviewed earlier in this chapter should demonstrate that the residual or new compounds may be in very different proportions from

the primary minerals. For instance, the Paleozoic Beekmantown limestone in Tennessee is composed of 90 percent calcium and magnesium carbonate, 7 percent quartz sand, and 3 percent detrital clay. A simple calculation shows that weathering by solution (Chapter 8) of 5 m of this limestone produces a residual parent material 1 m thick if the residuum has one-half the bulk density of the original rock. The residuum is 70 percent sand and 30 percent clay and hydrated iron oxides. Unless we comprehend the relative solubilities of limestone, quartz, and clay minerals, it is hard to believe that a typical red, sandy, clay loam soil on the Beekmantown limestone can result from *in situ* weathering. (It is possible that some of the detrital minerals were blown in by wind, however.)

An even more extreme example of the contrast between bedrock and residual soil is the great surficial deposit of lateritic bauxite (hydrated aluminum and iron oxides) at Weipa, Queensland, Australia (Loughnan and Bayliss, 1961). The bedrock is sandstone consisting of 90 percent quartz and 10 percent kaolinite (Figure 7-11). Under a hot monsoonal climate, with strongly seasonal rainfall averaging more than 1500 mm/yr, most of the quartz has been leached to a depth of more than 8 m, and the upper 5 m is mostly hydrated aluminum oxides stained by iron oxides. Only 5 percent of the thick surface layer is quartz, yet the bedrock is dominantly quartz, which is a relatively insoluble mineral even under the intense heat and rainfall of northern Queensland. A long interval of weathering must be assumed in order to explain the Weipa bauxite ore because the continuity of the weathering profile shown in Figure 7-11 is clear evidence that the surface bauxite ore is the residuum of the quartz sandstone at depth.

The two examples just cited demonstrate that residual soils can be very different from their parent material, given the right combination of the other four soil-forming factors. Nevertheless, structural or parent-material control of weathering is as important in soil genesis as it is in landscape development.

Climate must compete with parent material for the status of "first among equals" in the list of soil-forming factors. Like geomorphologists, pedologists can probably be divided into two groups: (1) those who believe that weathering processes (primarily a function of climate) give the distinctive character to a soil or a landscape despite lithologic or structural differences, and (2) those who believe that structure always shows through despite climate-controlled differences in the intensity of soil-forming and weathering processes. It must be noted that the atmospheric conditions usually used to define climate may not

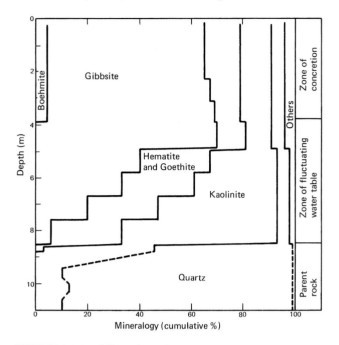

FIGURE 7-11. Mineralogy in relation to depth at Weipa, Queensland, Australia (from Loughnan and Bayliss, 1961, Table 1).

accurately describe the local environment within the soil profile (p. 128).

Organisms as a factor in soil formation have been reviewed in their role as chemical and biologic weathering agents. Some pedologists view plants as a secondary response to climate and parent material. Others emphasize the role of soil microorganisms and give reduced emphasis to the rooted macroflora. Still others see organisms as little more than the creators of a certain chemical soil environment in which essentially chemical weathering reactions form the various horizons of the profile.

The topographic influence on soil formation is emphasized more by pedologists than by geologists. The geologist sees slopes as the result of processes acting on structures through time (Chapter 9). The pedologist sees topographic relief as a preexisting factor in the formation of soil. If a slope is steep, runoff is rapid, erosion removes the soil as fast as it forms, little water enters the soil, and the profile is thin and poorly developed. On more level terrain, runoff is inhibited and more water enters the ground to weather minerals and translocate clay and other mobile components of the profile. Soils that are similar except for slope or drainage-induced properties are grouped as a *drainage catena* (p. 192). Yet while a soil is forming, so is the slope around it. Pedologists need the geomorphic appreciation of the origin of slopes, lest they assume

that the primary condition of a region is of soil-less but otherwise fully shaped landforms. This cannot be true because weathering, and therefore soil formation, is as much a factor in slope development as slope is a factor in soil development. No simple cause-and-effect relationship is possible.

The role of the fifth soil-forming factor, time, is at the heart of both geomorphology and pedology. Simply and directly asked: Do soils and landscapes continue to evolve through time at rates appropriate to the energy expended on them? If so, *soil chronosequences* (p. 142) can provide valuable clues to the relative or even numerical ages of various landforms. Alternatively, do soils and landscapes reach some end form in a closed system and cease to evolve, or do they reach some dynamic equilibrium between processes and form and maintain thereafter a steady-state morphology even though energy and matter continue to move through their open physical systems? These underlying questions of both process-oriented and explanatory geomorphology are evaluated again and again in this book.

Soil Classification

The very definition of soil as the abode of rooted plants emphasizes the practical reason for studying and classifying soils. Each developed nation has a scheme of soil mapping that enables its agronomists and economists to plan the effective use of available land. Although genetic assumptions are useful and unavoidable, modern soil classifications attempt to classify according to definable, observable chemical and physical properties of the horizons in soil profiles.

In 1965, the U.S. Department of Agriculture adopted a totally new classification of soils to replace the classification that had been in use since 1938 (Soil Survey Staff, 1975). The new system had been under development since 1951 as a series of "approximations" until it was formally adopted for use by the U.S. Soil Conservation Service and state cooperative programs. Although created to meet the needs of describing and mapping soils in the United States, the new U.S. soil taxonomy was designed for worldwide applicability. Special attention was given to tropical soils, which received woefully inadequate treatment in previous soils classifications. The classification is outlined here in the confident belief that it is the best and most comprehensive classification of soils and soil-forming processes yet devised. Students of geomorphology must become familiar with the sometimes curiously compounded terminology. It is regrettable that some recent textbooks of geology and geography have persisted in using the once highly useful but now obsolete terminology of the 1938 U.S. classification.

The nomenclature of the U.S. soil taxonomy (Soil Survey Staff, 1975) is systematic. The hierarchy of terms, in descending order of rank, is *order, suborder, great group, subgroup, family,* and *series*. Eleven soil orders are defined, based on the presence or degree of development of the various horizons in soil profiles (Figure 7-12). The name of each soil order (Table 7-4) ends in *sol* (L. *solum,* soil) and contains a formative element that is used as the final syllable in the names of taxa in suborders, great groups, and subgroups. Thus the name of each of the 55 suborders consists of 2 syllables, a formative element from a list of terms (Table 7-5) prefixed to the formative element of an order. The name of each great group, now about 230 in number, consists of 3 or 4 syllables, created by prefixing a suitable element from a list (Table 7-6) to the name of a suborder. The names of the 1240 subgroups are created by an adjective modifying the name of a great group. At first glance, the resulting names are totally exotic, but all can be decoded quickly and precisely.

The formative elements are derived from Greek or Latin roots, are as short as possible, and are designed for use in any modern language that uses the Latin alphabet. By compounding the several ranks of formative elements, any soil type on earth can be classified, and in the definition of the formative elements, the soil is also described.

For example, the Dunkirk Series of central New York State is a well-drained soil formed in silty, calcareous glaciolacustrine deposits of late Pleistocene age. The Dunkirk Series is classified at the subgroup level as a *Glossoboric Hapludalf.* Decoded, using the formative elements from Tables 7-4, 7-5, and 7-6, the Dunkirk Series belongs to a subgroup of the great group *Hapludalf,* the suborder *Udalf,* and the order *Alfisol.* Translating the formative elements of the name, the soil has a gray to brown surface horizon, moderate to high base (cation) saturation, and clay accumulation in a subsurface horizon (formative element *alf;* Table 7-4). It is of the suborder of Alfisol that forms in humid regions (formative element *ud;* Table 7-5). More specifically, as described in the sources cited in the next paragraph, it is one of the

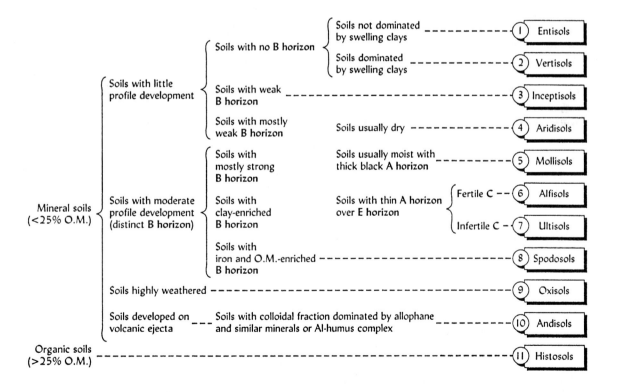

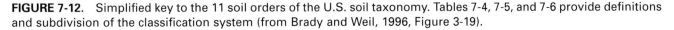

FIGURE 7-12. Simplified key to the 11 soil orders of the U.S. soil taxonomy. Tables 7-4, 7-5, and 7-6 provide definitions and subdivision of the classification system (from Brady and Weil, 1996, Figure 3-19).

Table 7-4

Names of Soil Orders, with Simplified Definitions

ALFISOL	Soil with gray to brown surface horizon, medium to high base supply, and a subsurface horizon of clay accumulation. Formative element: *alf.*
ANDISOL	Soil derived from volcanic ejecta, dominated by allophane (an amorphous clay mineral derived from weathered volcanic rock) or Al-humic complexes. Formative element: *and.*
ARIDSOL	Soil with pedogenic horizons, low in organic matter, usually dry. Formative element: *id.*
ENTISOL	Soil without pedogenic horizons. Formative element: *ent.*
HISTOSOL	Organic (peat and muck) soil. Fomative element: *ist.*
INCEPTISOL	Soil with weakly differentiated horizons showing alteration of parent materials. Formative element: *ept.*
MOLLISOL	Soil with a nearly black, organic-rich surface horizon and high base supply. Formative element: *oll.*
OXISOL	Soil that is a mixture principally of kaolin, hydrated oxides, and quartz. Formative element: *ox.*
SPODOSOL	Soil that has an accumulation of amorphous materials in subsurface horizons. Fomative element: *od.*
ULTISOL	Soil with a horizon of clay accumulation and low base supply. Formative element: *ult.*
VERTISOL	Cracking clay soil. Formative element: *ert.*

alfisols of temperate and tropic regions that are usually moist but may be intermittently dry in some horizons during summer. It is of the great group that has the least complex horizons of the suborder (formative element *hapl;* Table 7-6); specifically, it is described in greater detail as a brownish or reddish Udalf with only a slight loss of clay from the A horizon and a moderate accumulation of clay in the B horizon. It is without a dense, clay-rich crust (fragipan) in the profile. It is in an area characterized by mean summer and mean winter soil temperatures that differ by 5°C or more and a mean annual tem-

TABLE 7-5

Principal Formative Elements to be Used in Naming Suborders

Names of suborders consist of two syllables, for example, Aqualf. Formative elements are as follows:

alb	L. *albus,* white; soils from which clay and iron oxides have been removed.
aqu	L. *aqua,* water; soils that are wet for long periods.
arg	Modified from L. *argilla,* clay; soils with a horizon of clay accumulation.
bor	Gr. *boreas,* northern; cool.
fluv	L. *fluvius,* river; soils formed in alluvium.
ochr	Gr. base of *ochros,* pale; soils with little organic matter.
orth	Gr. *orthos,* true; the common or typical.
psamm	Gr. *psammos,* sand; sandy soils.
ud	L. *udus,* humid; of humid climates.
umbr	L. *umbra,* shade; dark colors reflecting much organic matter.
ust	L. *ustus,* burnt; of dry climates with summer rains.
xer	Gr. *xeros,* dry; of dry climates with winter rains.

TABLE 7-6

Principal Firmative Elements to be Used in Naming Great Groups

Names of great groups consist of more than two syllables and are formed by adding a prefix to the suborder name, for example, Cryoboralf. Some of the formative elements are as follows:

cry	Gr. *kryos,* icy cold; cold soils.
dystr, dys	Modified from Gr. *dys,* ill; infertile.
eutr, eu	Modified from Gr. *eu,* good; fertile.
frag	Modified from L. *fragilis,* brittle; a brittle pan.
gloss	Gr. *glossa,* tongue; deep, wide tongues of albic materials into the argillic horizon.
hapl	Gr. *haplous,* simple; the least advanced horizons.
quartz	Ger. *quarz,* quartz; soils with very high content of quartz.
torr	L. *torridus,* hot and dry; soils of very dry climates.
trop	Modified from Gr. *tropikos,* of the solstice; humid and continually warm.

perature of 8°C or more (temperate climate). The great group Hapludalf is equivalent to the Gray-Brown Podzolic soils of the 1938 classification. To classify the Dunkirk Series to the subgroup level, the compound adjective *glossoboric* precedes the great group name. *Glossic* refers to minor but characteristic tongues of the clay-depleted A horizon that extend down into the clay-enriched B horizon; if this tonguing were more pronounced, the soil would be a *Glossudalf* instead of a *Hapludalf*. *Boric* is from the root *boreal*; as defined for this subgroup, the mean annual soil temperature is less than 10°C.

The decoding exercise of the preceding paragraph is tedious and complex for a novice, but it illustrates that approximately a half page of written description about a soil and its climatic environment is contained in the name of the soil at the subgroup level. If additional modifiers concerning texture, mineralogy, grain size, permeability, and soil temperature are prefixed to describe the soil at the family level, a comprehensive impression of the soil, its genesis, and its agricultural potential can be gained just from the name alone. A map of the principal orders, suborders, and great groups of soils found in the United States is published in the U.S. National Atlas (U.S. Geol. Survey, 1970, Plate 38) with a useful accompanying descriptive text. If you are unable to consult the definitive document (Soil Survey Staff, 1975) or appropriate textbooks of soil classification and morphology (Buol et al., 1989; Fanning and Fanning, 1989; Brady and Weil, 1996), the U.S. National Atlas map and text is a good introduction to the U.S. soil taxonomy.

STRUCTURE, PROCESS, AND TIME IN ROCK WEATHERING

To what extent does structure influence or control rock weathering? To what degree are landforms shaped by climate-controlled weathering processes? Does weathering continue indefinitely, or gradually equilibrate with the process of alteration? In the remainder of this chapter, we consider these three questions, in review and in anticipation of expanded treatment in later chapters where structural and climatic control of mass wasting and erosion are also considered, and where landscape evolution through time is the dominant theme.

Structural Influences in Rock Weathering

Lithology, the physical character of intact rock, exerts an obvious structural control on weathering. It is the first factor needed to define rock-mass strength

(p. 120). Some rocks are massive, homogeneous, and nearly isotropic in their lithologic character, at least superficially. Most have lithologic *fabrics* of various kinds: stratification or other forms of layering; metamorphic foliation; or a pattern of differential stresses imposed by cooling history, tectonic deformation, or unloading (p. 121).

Close in importance to the lithology of the intact rock are the discontinuities that separate the rock mass into blocks of various sizes and shapes. Ranging from microscopic pores and microcracks to meter-spaced sheeting joints and open bedding planes, these discontinuities provide access for water, air, and organisms, and allow weathering reaction products to be carried away.

During weathering, joint-bounded blocks become progressively more rounded. **Corestones** of fresh rock, initially surrounded by weathered spheroidal exfoliation shells (Figures 7-1 and 7-13), may weather free to become loose boulders on the surface or in regolith. It has been proposed that the large boulders of crystalline rocks that are found in glacial drift are not necessarily quarried and rounded by glacial erosion but might be corestones from a preglacial or interglacial weathering profile (p. 382).

Mineralogic factors are also part of the structure that controls weathering. Chemical composition is obviously important, as noted by the fact that each example of chemical weathering cited earlier in the chapter used a particular chemical compound (mineral) in the equation. Two other less obvious mineralogic factors also influence rock weathering. One is grain size, and the other can be called "crystallinity," or the amount of crystalline, as opposed to amorphous or glassy, minerals in a rock. For example, different tex-

FIGURE 7-13. Granite corestones weathering free from saprolite, Xiamen, southeastern China. Corestone in right background is being split for construction stone.

tural types of compositionally identical granite have varying resistance to weathering and erosion. In the humid tropical climate of Hong Kong, coarse porphyritic granite is least resistant, coarse-grained granite is stronger, and medium-grained granite, fine-grained granite, and finally granite porphyry dikes are the most resistant (Ruxton and Berry, 1957, p. 1273). Obsidian, basalt glass, and amorphous silica (chert and chalcedony) are notably more soluble in water than their crystalline equivalents (Colman, 1982, p. 10). The older basalt rocks of Iceland erupted under extensive glacial ice cover, and the resulting *hyaloclastites* (p. 102) weather about ten times faster than do younger, post-glacial basalts of similar composition that are more crystalline (Gíslason et al., 1996). Hydrated rinds form on the surface of obsidian and basalt-glass artifacts only a few thousand years old. Their thickness can be calibrated as an archeologic dating method.

The poor crystallinity of volcanic rocks makes them much more subject to chemical weathering than are their better crystallized intrusive equivalents. In addition to crystallinity, the complex flow structures, voids, gas tubes, cooling joints, and other physical features of eruptive rocks greatly increase their surface area and reactivity.

Rock types, like minerals and cations, can be ranked in order of their chemical weatherability by measuring the flux of bicarbonate and silica from small, homogeneous watersheds. From the water chemistry of 101 rivers across the United States, Puerto Rico, and Iceland, weathering rates were shown to increase in the following sequence: sandstones, granites, basalts, shales, and carbonate rocks (Bluth and Kump, 1994). The sequence generally follows the relative proportions of silicon to other cations: silicate rocks (especially quartz-rich sandstones) have the lowest weathering rates, and carbonate rocks the highest. The rapid weathering of shale was attributed to the fact that most shales have a significant carbonate component. Basalt was relatively close to granite in weatherability.

Almost every rock type has a distinctive weathering character. Limestones are often smooth, gray, and lichen covered, or deeply pitted by solution (Chapter 8). Shales split and crumble. Granites form quartz-rich sandy grus. The list could easily be lengthened. The weathered appearance of rock is so useful that it is included among the diagnostic criteria for defining formations, the basic units of geologic mapping (Dunbar and Rodgers, 1957, pp. 267, 273).

Climatic Influence on Rock Weathering

The relation of climate to weathering is taken for granted, perhaps too much so. Three excellent recent books on weathering (Nahon, 1991: Martini and Chesworth, 1992; Robinson and Williams, 1994) use the major terrestrial climatic zones as an organizational framework for analyzing weathering. Most geomorphologists will readily and rightly equate salt weathering with arid regions, freeze and thaw with subpolar or high mountain environments, and deep chemical weathering with the wet tropics.

The traditional viewpoint of the role of climate is well represented by an idealized meridianal cross section of the earth's weathered layer (Figure 7-14). From polar to equatorial latitudes, the figure not only shows the varying depths of weathering but illustrates that certain sets of weathering and soil-forming processes follow each other in prescribed order. Mechanically fractured rocks, but little altered chemically, are shown at the surface in polar and subtropical deserts. Either because of extreme cold or lack of precipitation, or both, these latitudinal belts are represented as regions of minimal chemical and biologic weathering.

These and other generalizations are being tempered by new observations on the microenvironments in which chemical weathering progresses (p. 128). The assumed minor role of chemical weathering in dry regions has been challenged by Pope et al. (1995). They argue that the importance of climate to weathering is at the micrometer to millimeter scale, at the interface between water films and the rock mass. At such scales, water-mediated chemical weathering may proceed at a significant rate. The relative humidity of soil air approaches 100 percent at depths below 25 cm, even in extremely hot, dry deserts (Friedman et al., 1994, p. 188). By this logic, the reason for so much exposed and slightly weathered rock in hot and cold deserts is not so much that weathering is slow, but that without vegetative ground cover, erosion is rapid. The supply of sediment becomes "weathering limited," whereas in humid regions it is "transport limited."

In a humid midlatitude climate with seasonal freezing, cold weather brings a dormant period for both vegetation and soil microorganisms. Trees shed their leaves or needles, and the humus layer at the surface of the ground accumulates faster than microorganisms, worms, and insects can consume it. Rainwater or snowmelt that soaks through the rotting humus collects organic compounds and carbon dioxide that chelate and hydrolyze metallic cations from the underlying minerals and generally leave a residue of quartz and mixed-layer clay minerals. Iron oxides and clay minerals that are washed from the A and E horizons accumulate in the B horizon of soil. Clay minerals that accumulate in the soil prevent water from freely penetrating the ground and encourage overland runoff. The resulting landscape develops broad, gentle, soil-covered slopes shaped by soil creep

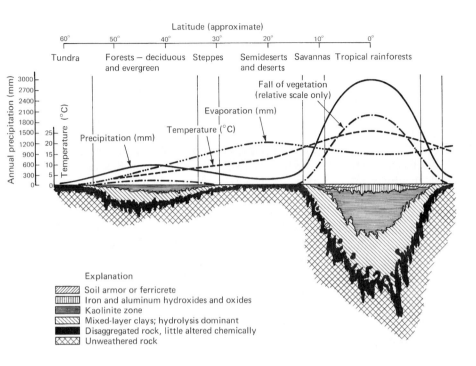

Explanation

- Soil armor or ferricrete
- Iron and aluminum hydroxides and oxides
- Kaolinite zone
- Mixed-layer clays; hydrolysis dominant
- Disaggregated rock, little altered chemically
- Unweathered rock

and stream erosion. Ridge crests commonly have thin soils or exposed bedrock ledges.

In the humid subtropics and tropics, hydrolysis and residual clay-mineral formation reach to depths of 100 m or more (Figure 7-14). The great depth of tropical weathering is due in part to temperature, but abundant and intense precipitation is an even greater climatic factor (Thomas, 1994, pp. 4, 126). Despite the enormous biologic productivity of a tropical rainforest, little humus accumulates because of continuous microbial and fungal attack and rapid recycling of nutrients. The heavy rainfall literally flushes the soil profile of all soluble compounds.

Under these conditions, a thick layer of weathered *saprolite* forms in the regolith on crystalline rocks (Nahon, 1991, pp. 208-227; Tardy, 1992; Robinson and Williams, 1994, Section 4). **Saprolite** is defined as isovolumetrically weathered bedrock that retains the structure and fabric of the parent rock mass (Figure 7-13). It is the parent material for the overlying soil, but is not classified as part of the soil profile (Stolt and Baker, 1994, p. 1). It is typically 10 m or more in thickness. The importance of saprolite to groundwater pollution, engineering design, and soil-nutrient studies has led to more intensive study of it, especially by soil scientists (Cremeens et al., 1994). As much as 95 percent of the eastern Piedmont Province of the United States (Figure 1-5) is underlain by saprolite.

In latitudinal belts peripheral to the tropical rainforests are the regions of tropical savanna climate

(p. 285). During the wet season, normally when the sun is overhead, the savannas are flooded, but during the dry half of the year, almost no rain may fall. The seasonal wetting promotes deep weathering, but the dry season accelerates generally irreversible oxidation of aluminum and iron hydroxides. Perhaps 25 percent of the African landscape has soils in the Oxisol or Ultisol orders (Lal, 1987, p. 285). Massive *lateritic* crusts may form cap rock that exerts a strong structural control on the landscape (Chapter 13). In some regions, the layer is called *soil armor,* or *ferricrete* (Nahon, 1986, 1991, pp. 228–240).

Tropical soils are complex and as yet relatively unstudied. At least some of the soils stay soft and clay-rich as long as they are under forest cover and remain wet throughout the year. If they are allowed to dry, either seasonally or by exposure through agricultural use or road construction, they irreversibly harden, apparently within a few years (Tardy, 1992; Tardy and Roquin, 1992). These are the true **laterites,** defined by their tendency to harden. In sub-Saharan western Africa, hardened laterites may cover 2.5 million km[2], an area equal to 30 percent of the conterminous United States (Lal, 1987, p. 1070). Large areas of laterite in southeast Asia are believed to have been created by traditional slash-and-burn agriculture. After being gardened for a few years, a patch of burned forest floor becomes hard and infertile, and a new patch must be cleared. Geographers and geologists have tended to exaggerate the extent of laterite and lateritic soils in the tropics, incorrectly equating the common reddish color of iron oxide-rich soils with the tendency to harden. Perhaps only 7 percent of the tropical soils are true laterites, and many of the other soils can be productively farmed with improved technology (Sanchez and Buol, 1975; Lal, 1987).

In general, the intensity of tropical weathering minimizes the effects of structural control. The depth of the saprolite is equal to the local relief on hills, escarpments, and other landforms so that entire landscapes may be within the weathered zone, rather than the weathered zone being a rind on the surface of the landforms. Geomorphologists trained in the cool, humid climates of Europe and North America have always been profoundly impressed with the scale of climate influence in tropical landscapes. As the developing nations of the tropics have become more influential, and as tropical mineral resources such as bauxite and laterite have become primary raw materials for industrial nations, tropical weathering has been given much greater attention.

Time and Rock Weathering; Soil Chronosequences

A basic issue of pedology is whether soils continue to evolve as some function of time, which might be linear, logarithmic, or exponential, or whether they reach equilibrium with land-surface lowering so that a "steady-state" soil profile and weathered zone move downward through the rock masses as the surface is lowered. Recent studies show that even soils several million years old continue to evolve (Machette, 1985; Harden, 1986–1988; Harden et al., 1991; Birkeland, 1992, p. 276; Markewich et al., 1994). Such long-term weathering must have been in progress during very significant late Cenozoic climate changes (Chapter 4). Demonstration of continuing evolution in old soil profiles has led to the concept of *soil chronosequences*, in which a succession of coastal or river terraces, glacial moraines, alluvial fans, or other landforms of progressively greater age but with similar parent materials, vegetation, slopes, and present climate, are shown to have soil profiles with progressively greater development. The typical long-term pattern seems to be of exponentially decreasing rates of soil development (Harden et al., 1991). With adequate calibration by studies of soil profiles on land surfaces of known age, soil chronosequences provide a valuable technique for establishing the age of other landscapes with similar rock types and climates.

Three sets of processes are seen in soil chronosequences. In arid regions, especially, the *addition* of silt and clay to the soil by wind, with significant additions of Ca^{2+} from weathered silicate and carbonate dust,

and Na^+ from seawater, are important. The *transformation* of minerals, especially the progressive alteration of clay minerals and silicates to stable hydrous oxides, is a long-continued process, as is the *transfer* of various soil components to progressively greater depths (Birkeland, 1992).

A useful technique is to study the mass balance of elemental gains and losses with depth in soil profiles. On a succession of marine terraces in northern California, which range in age from Holocene to an estimated 240,000 years, net progressive losses of Na, Ca, Mg, K, and Si are substantial, in some profiles approaching 100 percent. Fe and Al show only minor losses, by comparison (Merritts et al., 1992). On correlative terraces in southern Oregon, soil chronosequences show a progression from Spodosols to Ultisols (Bockheim et al., 1996). In a series of regional studies, soil chronosequences were examined that ranged in age from modern to mid-Pleistocene on glacial drift in Washington State, from 500 to 80,000 years on marine terraces in California, from 700,000 to 2 million years on granite alluvium in Montana, and from 200 to 3 million years on granite alluvium in central California (Harden, 1986–1988). Even older soil chronosequences are described from the Piedmont and Coastal Plain Provinces in southeastern United States, where soil chronosequences above deep saprolite cannot be younger than early Pleistocene and have probably been forming since late Miocene to late Pliocene time (between 6 and 2 million years ago) (Markewich et al., 1994).

Chronosequences in desert soils can be based on the extent of secondary mineral accumulation, especially of calcium carbonate. **Calcrete** is a massive shallow accumulation of soil carbonate that may be a meter or more in thickness and may effectively control scarp formation. A series of six stages of calcium carbonate accumulation in arid and semiarid soils has been established, ranging from a few percent $CaCO_3$ that forms coating on pebbles (Stage I), through layers of coalesced nodules (Stage III), to massive cemented surface layers 0.5 to 2 m in thickness in which 25 to 75 percent of the soil mass is pedogenic calcrete (Stages IV to VI) (Machette, 1985, p. 5). Stage I soils range from Holocene to late Pleistocene in age. Stage IV and older calcrete layers are found on land surfaces that range in age from 100,000 years to several million years. If the relative degree of calcification of soils on a series of river terraces or alluvial fans can be calibrated by radiometric dating of overlying lava flows or other means, it becomes very useful in reconstructing the late Cenozoic geomorphic history of the region. Most of the carbonate that has accumulated in dry soils is probably derived from windblown dust-size fragments of calcium-bearing minerals that were previously weathered by hydrolysis to calcium bicarbonate in solution (p. 130), then precipitated as calcium carbonate and blown by the wind (Reheis et al., 1995). Lesser amounts are derived from weathering in place and from upward or laterally moving groundwater (Machette, 1985, p. 8).

Relict Soils and Paleosols

The general acceptance of evidence that some soils and thick saprolites are not the result of very rapid processes, but rather are very old, raises profound questions about landscape evolution. Earlier in this chapter (pp. 121), it was noted that many modern landscapes are on rock masses that have been exhumed from depths of 20 km or more. The rates of denudation required to accomplish such deep erosion are evaluated at length in Chapter 15, where it is concluded that orogenic regions with high relief are likely to be reduced to near sea level with greatly reduced relief on time scales of 10 million years or less. Yet weathering studies and new cosmogenic techniques for dating land surfaces make it clear that in regions of low relief, many present landscapes with their weathered regolith have been largely unchanged for as long as 5 to 6 million years. Within such a time span, the earth has experienced major tectonic events and climatic changes (Chapters 3 and 4). How can some landscapes have been so passive? Are they, in fact, relict from former conditions, or are they still evolving, but at extremely slow rates? These are exciting geomorphic, pedologic, and climatologic questions that will challenge the best future earth scientists.

REFERENCES

ALLISON, R. J., and GOUDIE, A. S., 1994, The effect of fire on rock weathering: An experimental study, in Robinson, D. A., and Williams, R. B. G., eds., Rock weathering and landform evolution: John Wiley & Sons Ltd., Chichester, UK, pp. 41–56.

ARGO, V. N., 1964, Strangler fig: Native epiphyte: Nat. Hist., v. 73, no. 9, pp. 26–31.

BELL, F. G., 1992a, Description and classification of rock masses, in Bell, F. G., ed., Engineering in rock masses: Butterworth-Heinemann Ltd, Oxford, UK, pp. 54–77.

———, 1992b, Groundwater in rock masses, in Bell, F. G., ed., Engineering in rock masses: Butterworth-Heinemann Ltd, Oxford, UK, pp. 78–100.

BENNETT. J. G., 1936, Broken coal: Jour. Inst. Fuel, v. 10, no. 49, pp. 22–39.

BERNER, R. A., SJÖBERG, E. L., and 2 others, 1980, Dissolution of pyroxenes and amphiboles during weathering: Science, v. 207, pp. 1205–1206.

BERTHELIN, J., 1983, Microbial weathering processes, in Krumbein, W. E., ed., Microbial geochemistry: Blackwell Scientific Publications, Oxford, UK, pp. 223–262.

———, 1988, Microbial weathering processes in natural environments, in Lerman, A., and Meybeck, M., eds., Physical and Chemical Weathering in Geochemical Cycles: NATO Advanced Science Institutes Series C, v. 251, Kluwer Academic Publishers, Dordrech, The Netherlands, pp. 33–59.

BIERMAN, P., and GILLESPIE, A., 1991, Range fires: A significant factor in exposure-age determination and geomorphic surface evolution: Geology, v. 19, pp. 641–644.

BIRCH, F., 1966, Compressibility; elastic constants, in Clark, S. P., Jr., ed., Handbook of physical constants: Geol. Soc. America Mem. 97, pp. 97–173.

BIRKELAND, P. W., 1992, Quaternary soil chronosequences in various environments-extremely arid to humid tropical, in Martini and Chesworth, eds., Weathering, soils and palesols: Elsevier Science Publishers B.V., Amsterdam, pp. 261–281.

BLUTH, G. J. S., and KUMP, L. R., 1994, Lithologic and climatic controls of river chemistry: Geochimica et Cosmochimica Acta, v. 58, pp. 2341–2359.

BOCKHEIM, J. G., MARSHALL, J. G., and KELSEY, H. M., 1996, Soil-forming processes and rates on uplifted marine terraces in southwestern Oregon, USA: Geoderma, v. 73, pp. 36–62.

BOUCHEZ, J. L., DELAS, C., and 3 others, 1992, Submagmatic microfractures in granites: Geology, v. 20, pp. 35–38.

BRADLEY, W. C., 1963, Large-scale exfoliation in massive sandstones of the Colorado Plateau: Geol. Soc. America Bull., v. 74, pp. 519–528.

———, 1980, Role of salts in development of granitic tafoni, South Australia: A reply: Jour. Geology, v. 88, pp. 121–122.

———, HUTTON, J. T., and TWIDALE, C. R., 1978, Role of salts in development of granitic tafoni, South Australia: Jour. Geology, v. 86, pp. 647–654.

BRADY, N. C., and WEIL, R. R., 1996, Nature and properties of soils, eleventh edition: Prentice-Hall, Inc., Upper Saddle River, N.J., 740 pp.

BRANTLEY, S. L., and STILLINGS, L., 1996, Feldspar dissolution at 25°C and low pH: Am. Jour. Sci., v. 296, pp. 101–127.

BRUNER, W. M., 1984, Crack growth during unroofing of crustal rocks: Effects on thermoelastic behavior and near-surface stresses: Jour. Geophys. Res., v. 89, pp. 4167–4184.

BUOL, S. W., HOLE, F. D., and McCRACKEN, R. J., 1989, Soil genesis and classification, 3rd ed.: Iowa State Univ. Press, Ames, 446 pp.

CLARK, S. P., JR., and JAGER E., 1969, Denudation rate in the Alps from geochronology and heat flow: Am. Jour. Sci., v. 267, pp. 1143–1160.

CODY, R. D., CODY, A. N., and 2 others, 1996, Experimental deterioration of highway concrete by chloride deicing salts: Environmental and Engineering Geoscience, v. II, pp. 575–588.

COLMAN, S. M., 1982, Chemical weathering of basalts and andesites: U.S. Geol. Survey Prof. Paper 1246, 51 pp.

COMMITTEE ON CONSERVATION OF HISTORIC STONE BUILDINGS AND MONUMENTS, 1982, Conservation of historic stone buildings and monuments: National Materials Advisory Board, National Research Council Publication NMAB–397, National Academy Press, Washington, D.C., 365 pp.

COOKE, R., WARREN, A., and GOUDIE, A., 1993, Desert geomorphology: UCL Press Limited, London, 526 pp.

COSTAIN, J. K., BOLLINGER, G. A., and SPEER, J. A., 1987, Hydroseismicity—a hypothesis for the role of water in the generation of intraplate seismicity: Geology, v. 15, pp. 618–621.

COWAN, D. S., and BRUHN, R. L., 1992, Late Cretaceous geology of the U.S. cordillera, in Burchfiel, B. C., Lipman, P. W., and Zoback, M. L., eds., The Cordilleran orogen: Conterminous U.S.: Geol. Soc. America, The Geology of North America, v. G–3, pp. 169–203.

CREMEENS, D. L., BROWN, R. B., and HUDDLESTON, J. H., eds., 1994, Whole regolith pedology: Soil Science Soc. America Special Pub. No. 34, 136 pp.

CRISS, R. E., and TAYLOR, H. P., JR., 1983, An $^{18}O/^{16}O$ and D/H study of Tertiary hydrothermal systems in the southern half of the Idaho batholith: Geol. Soc. America Bull., v. 94, pp. 640–643.

DAHLEN, F. A., and SUPPE, J., 1988, Mechanics, growth, and erosion of mountain belts, in Clark, S. P., Jr., Burchfiel, B. C., and Suppe, J., eds., Processes in continental lithospheric deformation: Geol. Soc. America Spec. Paper 218, pp. 161–178.

DRAGOVICH, D., 1967, Flaking, a weathering process operating on cavernous rock surfaces: Geol. Soc. America Bull., v. 78, pp. 801–804.

DREVER, J. I., 1994, The effect of land plants on weathering rates of silicate minerals: Geochimica et Cosmochimica Acta, v. 58, pp. 2325–2352.

DUNBAR, C. O., and RODGERS, J., 1957, Principles of stratigraphy: John Wiley & Sons, Inc., New York, 356 pp.

DUNN, J. R., and HUDEC, P. P., 1966, Water, clay and rock soundness: Ohio Jour. Sci., v. 66, no. 2, pp. 153–167.

EVANS, I. S., 1970, Salt crystallization and rock weathering: A review: Rev. Geomorph. Dynamique, v. 19, no. 4, pp. 153–177.

EVANS, L. J., 1992, Alteration products at the earth's surface-the clay minerals, in Martini, I. P., and Chesworth, W., eds., Weathering, soils, and paleosols: Developments in earth surface processes 2: Elsevier Science Publishers B.V., Amsterdam, pp. 107–125.

FANNING, D. S., and FANNING, M. C. B., 1989, Soil: Morphology, genesis, and classification: John Wiley & Sons, Inc., New York, 395 pp.

FETH, J. H., ROBERSON, C. E., and POLZER, W. L., 1964, Sources of mineral constituents in water from granitic rocks, Sierra Nevada, California and Nevada: U.S. Geol. Survey Water-Supply Paper 1525-I, 70 pp.

FREDRICH, J. T., MENÉNDEZ, B., and WONG, T.-F., 1995, Imaging the pore structure of geomaterials: Science, v. 268, pp. 276–279.

FRIEDMAN, I., TREMBOUR, F. W., and 2 others, 1994, Is obsidian hydration dating affected by relative humidity?: Quaternary Res., v. 41, pp. 185–190.

FRIEDMANN, E. I., and OCAMPO, R., 1976, Endolithic blue-green algae in the dry valleys: Primary producers in the Antarctic desert ecosystem: Science, v. 193, pp. 1247–1249.

FYFE, W. S., 1996, The biosphere is going deep: Science, v. 273, p. 448.

GHIORSE, W. C., 1997, Subterranean life: Science, v. 275, pp. 789–790.

GÍSLASON, S. R., ARNÓRSSON, S., and ARMANNSSON, H., 1996, Chemical weathering of basalt in southwest Iceland: Effects of runoff, age of rocks and vegetative/glacial cover: Am. Jour. Sci., v. 296, pp. 837–907.

GOODING, J. L., 1981, Mineralogical aspects of terrestrial weathering effects in chondrites from Allan Hills, Antarctica: Proc., Lunar Planet. Sci., v. 12B, pp. 1105–1122.

GOUDIE, A. S., 1989, Weathering processes, in Thomas, D. S. G., ed., Arid zone geomorphology: Halsted Press: A division of John Wiley & Sons, Inc., New York, pp. 11–24.

HALL, K., and HALL, A., 1991, Thermal gradients and rock weathering at low temperatures: Some simulation data: Permafrost and periglacial processes, v. 2, pp. 103–112.

HARDEN, J. W., ed., 1986–1988, Soil chronosequences in the western United States: U.S. Geol. Survey Bull. 1590, sections A, B, C, D, F.

————, TAYLOR, E. M., and 6 others, 1991, Rates of soil development from four soil chronosequences in the southern Great Basin: Quaternary Res., v. 35, pp. 383–399.

HAXBY, W. F., and TURCOTTE, D. L., 1976, Stresses induced by the addition or removal of overburden and associated thermal effects: Geology, v. 4, pp. 181–184.

HIEBERT, F. K., and BENNETT, P. C., 1992, Microbial control of silicate weathering in organic-rich ground water: Science, v. 258, pp. 278–281.

HOLE, F. D., 1981, Effects of animals on soil: Geoderma, v. 25, pp. 75–112.

HOPSON, C. A., 1958, Exfoliation and weathering at Stone Mountain, Georgia, and their bearing on disfigurement of the Confederate Memorial: Georgia Mineral Newsletter, v. 11, pp. 65–79.

ISHERWOOD, D., and STREET, F. A., 1976, Biotite-induced grussification of the Boulder Creek Granodiorite, Boulder County, Colorado: Geol. Soc. America Bull., v. 87, pp. 366–370.

ISACKS, B. L., 1988, Uplift of the central Andean plateau and bending of the Bolivian orocline: Jour. Geophys. Res., v. 93, no. B4, pp. 3211–3231.

JAHNS, R. H., 1943, Sheet structure in granite: Its origin and use as a measure of glacial erosion in New England: Jour. Geology, v. 51, pp. 71–98.

JOHNSON, N. M., REYNOLDS, R. C., and LIKENS, G. E., 1972, Atmospheric sulfur: Its effect on the chemical weathering of New England: Science, v. 177, pp. 514–516.

JORGENSEN, B. B., 1983, The microbial sulphur cycle, in Krumbein, W. E., ed., Microbial geochemistry: Blackwell Scientific Publications, Oxford, UK, pp. 91–124.

KESSLER, D. W., INSLEY, H., and SLIGH, W. H., 1940, Physical, mineralogical, and durability studies on the building and monumental granites of the United States: U.S. Nat. Bureau of Standards Jour. Res., v. 25, pp. 161–206.

KIRBY, S. H., 1984, Introduction and digest to the special issue on chemical effects of water on the deformation and strengths of rocks: Jour. Geophys. Res., v. 89, pp. 3991–3995.

KNUEPFER, P. L. K., and MCFADDEN, L. D., eds., 1990, Soils and landscape evolution: Proceedings of the 21st Binghamton Symposium in geomorphology: Geomorphology, v. 3, pp. 197–575.

KUENEN, P. H. H., 1955, Realms of water: John Wiley & Sons, Inc., New York, 327 pp.

LAFARGUE, M., 1992, The Tao of the Tao Te Ching: A translation and commentary: State University of New York Press, Albany, N.Y., 270 pp.

LAL, R., 1987, Managing the soils of sub-Saharan Africa: Science, v. 236, pp. 1069–1076.

LASAGA, A. C., SOLER, J. M., and 3 others, 1994, Chemical weathering rate laws and global geochemical cycles: Geochimica et Cosmochimica Acta, v. 58, pp. 2361–2386.

LIKENS, G. E., and BORMANN, F. H., 1974, Acid rain: A serious regional environmental problem: Science, v. 184, pp. 1176–1179.

LINDOW, S. E., PANOPOULOS, N. J., and McFARLAND, B. L., 1989, Genetic engineering of bacteria from managed and natural habitats: Science, v. 244, pp. 1300–1307.

LIVINGSTONE, D. A., 1963, Data of geochemistry, Chapter G, Chemical composition of rivers and lakes: U.S. Geol. Survey Prof. Paper 440-G, 64 pp.

LOUGHNAN, F. C., and BAYLISS, P., 1961, Mineralogy of the bauxite deposits near Weipa, Queensland: Am. Mineralogist, v. 46, pp. 209–217.

LOVERING, T. S., and ENGEL, C., 1967, Translocation of silica and other elements from rock into Eqisetum and three gases: U.S. Geol. Survey Prof. Paper 594-B, 16 pp.

MACHETTE, M. N., 1985, Calcic soils of the southwestern United States, in Weide, D. L., and Faber, M. L., eds., Soils and Quaternary geology of the southwestern United States: Geol. Soc. America Spec. Paper 203, pp. 1–21.

MADIGAN, M. T., and MARRS, B. L., 1997, Extremophiles: Sci. American, v. 276, no. 4, pp. 82–87.

MARKEWICH, H. W., PAVICH, M. J., and 5 others, 1994, Genesis and residence times of soil and weathering profiles on residual and transported parent material in the Pine Mountain area of west-central Georgia: U.S. Geol. Survey Bull. 1589-E, 69 pp.

MARTINI, I. P., 1978, Tafoni weathering, with examples from Tuscany, Italy: Zeitschr. für Geomorph., v. 22, pp. 44–67.

———, and CHESWORTH, W., eds., 1992, Weathering, soils, and paleosols: Developments in earth surface processes 2: Elsevier Science Publishers B.V., Amsterdam, 618 pp.

MATTHES, F. E., 1930, Geologic history of the Yosemite Valley: U.S. Geol. Survey Prof. Paper 160, 137 pp.

MERRITTS, D. J., CHADWICK, A. O., and 3 others, 1992, The mass balance of soil evolution on late Quaternary marine terraces, northern California: Geol. Soc. America Bull., v. 104, pp. 1456–1470.

MOORE, D. E., and LOCKNER, D. A., 1995, The role of microcracking in shear-fracture propagation in granite: Jour. Structural Geol., v. 17, pp. 95–114.

MOTTERSHEAD, D. N., 1994, Spatial variations in intensity of alveolar weathering of a dated sandstone structure in a coastal environment, Weston-super-Mare, UK, in Robinson, D. A., and Williams, R. B. G., eds., Rock weathering and landform evolution: John Wiley & Sons, Ltd., Chichester, UK, pp. 151–174.

MUSTOE, G. E., 1982, The origin of honeycomb weathering: Geol. Soc. America Bull., v. 93, pp. 108–115.

NAHON, D. B., 1986, Evolution of iron crusts in tropical landscapes, in Colman, S. M., and Dethier, D. P., eds., Rates of chemical weathering of rocks and minerals: Academic Press, Inc., Orlando, Fla., pp. 169–191.

———, 1991, Introduction to the petrology of soils and chemical weathering: John Wiley & Sons, Inc., New York, 313 pp.

NESBITT, B. E., and MUEHLENBACHS, K., 1989, Origins and movement of fluids during deformation and metamorphism in the Canadian cordillera: Science, v. 245, pp. 733–736.

PINTO, J. P., and HOLLAND, H. D., 1988, Paleosols and the evolution of the atmosphere; Part II, in Reinhardt, J., and Sigleo, W. R., eds., Paleosols and weathering through geologic time: Geol. Soc. America Spec. Paper 216, pp. 21–34.

POPE, B. A., DORN, R. I., and DIXON, J. C., 1995, A new conceptual model for understanding geographical variations in weathering: Annals, Assoc. American Geographers, v. 85, pp. 38–64.

POWELL, J. W., 1876, Report on the geology of the eastern portion of the Uinta Mountains: U.S. Geol. and Geog. Survey Terr. (Powell), Washington, D.C., 218 pp.

REHEIS, M. C., GOODMACHER, J. C., and 6 others, 1995, Quaternary soils and dust deposition in southern Nevada and California: Geol. Soc. America Bull. v. 107, pp. 1003–1022.

RICHARDSON, S. M., and McSWEEN, H. Y., JR., 1989, Geochemistry: Pathways and processes: Prentice-Hall, Inc., Englewood Cliffs, N.J., 488 pp.

ROBERT, M., and TESSIER, D., 1992, Incipient weathering: Some new concepts on weathering, clay formation and organization, in Martini, I. P., and Chesworth, W., eds., Weathering, soils and paleosols: Elsevier Science Publishers B.V., Amsterdam, pp. 71–105.

ROBINSON, D. A., and WILLIAMS, R. B. G., eds., 1994, Rock weathering and landform evolution: John Wiley & Sons Ltd, Chichester, UK, 519 pp.

RODEN, M. K., and MILLER, D. S., 1989, Apatite fission-track thermo-chronology of the Pennsylvania Appalachian basin: Geomorphology, v. 2, pp. 39–51.

deRONDE, C. E. J., and EBBESEN, T. W., 1996, 3.2 b.y. of organic compound formation near sea-floor hotsprings: Geology, v. 24, pp. 791–794.

RUXTON, B. P., and BERRY, L., 1957, Weathering of granite and associated erosional features in Hong Kong: Geol. Soc. America Bull., v. 68, pp. 1263–1292.

SANCHEZ, R. A., and BUOL, S. W., 1975, Soils of the tropics and the world food crisis: Science, v. 188, pp. 598–603.

SCHOPF, J. W., and PACKER, B. M., 1987, Early Archean (3.3-billion to 3.5-billion-year-old) microfossils from Warrawoona, Australia: Science, v. 237, pp. 70–73.

SCHWARTZ, S. E., 1989, Acid deposition: Unraveling a regional phenomenon: Science, v. 243. pp. 753–763.

SELBY, M. J., 1980, A rock mass strength classification for geomorphic purposes: with tests from Antarctica and New Zealand: Zeitschr. für Geomorph., v. 24, p. 31–51.

SILVERMAN, M. P., and MUNOZ, E. F., 1970, Fungal attack on rock: Solubilization and altered infrared spectra: Science, v. 169, pp. 985–987.

SLINGERLAND, R., and FURLONG, K. P., 1989, Geodynamic and geomorphic evolution of the Permo-Triassic Appalachian Mountains: Geomorphology, v. 2, pp. 23–37.

SMITH, B. J., 1994, Weathering processes and forms, in Abrahams, A. D., and Parsons, A. J., eds., Geomorphology of desert environments: Chapman and Hall, London, pp. 39–63.

SOIL SURVEY STAFF, 1975, Soil taxonomy: A basic system of soil classification for making and interpreting soil surveys: U.S. Dept. Agr. Handbook 436, 754 pp.

STEVENS, G. R., 1974, Rugged landscape: the geology of central New Zealand: A.H. & A.W. Reed, Wellington, New Zealand, 286 pp.

STOLT, M. H., and BAKER, J. C., 1994, Strategies for studying saprolite and saprolite genesis, in Cremeens, D. L., Brown, R. B., and Huddleston, J. H., eds., Whole regolith pedology: Soil Science Soc. America Special Pub. No. 34, pp. 1–19.

STRAKHOV, N. M., 1967, Principles of lithogenesis, v. 1 (transl. by J. P. Fitzsimmons): Oliver & Boyd Ltd., Edinburgh, 245 pp.

TARDY, Y., 1992, Diversity and terminology of lateritic profiles, in Martini, I. P., and Chesworth, W., eds., Weathering, soils and paleosols: Elsevier Science Publishers B.V., Amsterdam, pp. 379–405.

———, and ROQUIN, C., 1992, Geochemistry and evolution of lateritic landscapes, in Martini, I. P., and Chesworth, W., eds., Weathering, soils and paleosols: Elsevier Science Publishers B.V., Amsterdam, pp. 407–443.

THARP, T. M., 1987, Conditions for crack propagation by frost wedging: Geol. Soc. America Bull., v. 99, pp. 94–102.

THOMAS, M. F., 1994, Geomorphology in the tropics: A study of weathering and denudation in low latitudes: John Wiley & Sons Ltd., Chichester, UK, 460 pp.

TILLMAN, J. E., 1980, Eastern geothermal resources: Should we pursue them?: Science, v. 210, pp. 595–600.

TOLE, M. P., 1987, Thermodynamic and kinetic aspects of formation of bauxites: Chem. Geol., v. 60, pp. 95–100.

TWIDALE, C. R., 1982, Granite landforms: Elsevier Scientific Publishing Company, Amsterdam, 372 pp.

UGOLINI, F. C., 1986, Processes and rates of weathering in cold and polar desert environments, in Colman, S. M., and Dethier, D. P., eds., Rates of chemical weathering of rocks and minerals: Academic Press, Inc., Orlando, Fla., pp. 193–235.

U.S. GEOLOGICAL SURVEY, 1970, National atlas of the United States of America: U.S. Dept. of Interior, Geol. Surv., Washington, D.C., 417 pp.

VILES, H. A., 1994, Time and grime: Studies in the history of building stone decay in Oxford: Research Paper 50, School of Geography, University of Oxford, Oxford, UK, 27 pp.

———, 1995, Ecological perspectives on rock surface weathering: Toward a conceptual model: Geomorphology, v. 13, pp. 21–35.

WALDER, J., and HALLET, B., 1985, A theoretical model of the fracture of rock during freezing: Geol. Soc. America Bull., v. 96, pp. 336–346.

WARKE, P. A., and SMITH, B. J., 1994, Short-term rock temperature fluctuations under simulated hot desert conditions: Some preliminary results, in Robinson, D. A., and Williams, R. B. G., eds., Rock weathering and landform evolution: John Wiley & Sons Ltd., Chichester, UK, pp. 57–70.

WHITNEY, J. A., JONES, L. M., and WALKER, R. L., 1976, Age and origin of the Stone Mountain Granite, Lithonia district, Georgia: Geol. Soc. America Bull., v. 87, pp. 1067–1077.

WHITNEY, P. R., and McLELLAND, J. M., 1973, Origin of coronas in metagabbros of the Adirondack Mts., N.Y: Contr. Mineral. and Petrol., v. 39, pp. 81–98.

WINKLER, E. M., 1968, Frost damage to stone and concrete: Geological considerations: Eng. Geology, v. 2, pp. 315–323.

———, 1980, Role of salts in development of granitic tafoni, South Australia: Discussion: Jour. Geology, v. 88, pp. 119–120.

———, 1994, Stone in architecture: Properties, durability, 3rd ed.: Springer-Verlag New York, Inc., New York, 313 pp.

———, and SINGER, P. C., 1972, Crystallization pressure of salts in stone and concrete: Geol. Soc. America Bull. v. 83, pp. 3509–3014.

———, and WILHELM, E. J., 1970, Salt burst by hydration pressures in architectural stone in urban atmosphere: Geol. Soc. America Bull., v. 81, pp. 567–572.

YATSU, E., 1988, The nature of weathering: An introduction: Sozosha, Tokyo, 624 pp.

YOUNG, A. R. M., 1987, Salt as an agent in the development of cavernous weathering: Geology, v. 15, pp. 962–966.

ZIMMERMAN, S. G., EVENSON, E. B., and 2 others, 1994, Extensive boulder erosion resulting from a range fire on the type-Pinedale moraines, Freemont Lake, Wyoming: Quaternary Res., v. 42, pp. 255–265.

Chapter 8

Karst and Speleology

In the preceding chapter, solution (dissolution) was excluded as a specific process of chemical weathering, on the argument that all chemical processes of rock weathering involve aqueous solutions. In one special case, however, solution of massive quantities of abnormally soluble rock gives rise to a terrain so distinctive that a special set of descriptive terms is needed for the landforms. The Germanized form of an ancient Slavic term for such a region of solution topography in Slovenia is *karst.* The term connotes both an assemblage of landforms and a set of processes. It has become a truly international term, and although an enormous and provincial vocabulary has accumulated to describe karst processes and forms, including caves and their deposits, the one word offers a central and unifying concept for the special landforms on exceptionally soluble rocks.

DEFINITIONS AND HISTORICAL CONCEPTS

Karst is "terrain with distinctive hydrology and landforms arising from a combination of high rock solubility and well-developed secondary porosity" (Ford and Williams, 1989, p. 1). This sensible and practical definition stresses geomorphic form, but emphasizes *process* in stressing the role of groundwater hydrology, and emphasizes *structural control* in stressing that abnormally soluble rocks with well-developed sec-

ondary porosity are involved. Solution becomes a process of both weathering and erosion inasmuch as the weathered products are removed from the surface and carried away in solution.

Attention to the definition is important. For some students, karst is synonymous with only limestone landscapes. However, other soluble-rock terranes may be karst, and limestone weathering in both hot and polar deserts is not characterized by intense solution. Dolomite, gypsum, and salt all show karst landforms in some regions. The term has been deplorably extended (with modifying prefixes, as in "glacier karst," "thermokarst," and "pseudokarst") to a variety of landforms, such as those formed by compaction of fine-grained volcanic tephra, melting of glacier ice and permafrost, and subsidence of desert soils by a lowering of the water table. Under very long intervals of humid tropical weathering, even the dissolution of silica from quartz-rich rocks (Figure 7-11) has produced karst analogs (Bögli, 1980, p. 1; Twidale, 1987), but the usual concept of karst is restricted to limestone, dolomite, and much less commonly, gypsum and salt.

No branch of geomorphology is so plagued by complex and confusing terminology as "karstology." Part of the problem is psychological in that every language and dialect has special, often highly imaginative, terms to describe caverns and other unusual karst landforms. The other part of the problem is historical. The first systematic studies of karst were carried out in the Dinaric Alps of Slovenia and Croatia in a region traditionally known as The Karst (Demek et al., 1984, Figure 16-4). Albrecht Penck (the author of the first textbook of geomorphology) and his students popularized the study of karst landforms (Cvijić, 1893; Penck, 1894). They translated local Serbo-Croatian terms for the forms first into German, then into French, then Italian, and subsequently into other languages. In each country, the words were given slightly different equivalents. Only since the end of World War II has there been international integration of terms and theories of hydrogeologists, speleologists, and geomorphologists to create a body of karst literature.

Even before Penck and his students began their systematic study of the Slovenian karst, descriptions of tropical karst had been published that emphasized solution landforms unlike those in the Mediterranean basin and other temperate karst regions. Sawkins (1869) described the "cockpits" of Jamaican karst, a maze of solution depressions separated by steep ridges. Climatic control of karst processes received greater attention in the decades of 1920 and 1930 with studies of tropical karst in south China and in Java (Roglić, 1972). The *tower karst* landscape with isolated, steep- or vertical-sided hills rising from flat plains in the humid tropics was proposed to be a fundamentally different, climate-controlled karst landscape created by processes unique to warm, humid regions (p. 160).

KARST PROCESSES AND HYDROGEOLOGY

Limestone Solution

Except for a few unusual caves that were probably dissolved by sulfuric acid derived from deep crustal sources such as sulfur-rich coal, petroleum, or hydrothermal sulfur (known as **hypogenic** caves: Palmer, 1991, pp. 16–19), most solution, as the term is used in the context of limestone karst, is essentially the hydrolysis of $CaCO_3$ by downward moving water (p. 129).

This is an **epigenic** (near-surface) process, driven by the hydrologic cycle. Water readily absorbs CO_2 from the atmosphere and especially from soil air (p. 129) to form a weak acid, *carbonic acid,* which dissociates to hydrogen and bicarbonate ions in solution. The reversible reactions can be simplified as:

$$H_2O + CO_2 \rightleftharpoons H_2CO_2 \rightleftharpoons H^+ + HCO_3^-$$

In the groundwater of limestone terranes, HCO_3^- is the dominant anion species in solution. In contact with $CaCO_3$, usually the mineral calcite, a series of reversible, congruent reactions can take place, summarized as:

$$CaCO_3 + H_2O + CO_2 \rightleftharpoons Ca^{2+} + 2HCO_3^-$$

The Ca^{2+} ion from dissolved calcite does not form a hydroxide but remains in solution, in ionic equilibrium with the bicarbonate (Drever, 1988, pp. 48ff; Ford and Williams, 1989, pp. 42ff). The dissolved ions are the same as those that enter solution when a calcium-rich silicate rock weathers by hydrolysis (p. 50). The reversible reaction is forced to the right, and more calcite is dissolved, if carbon dioxide in solution is increased. If carbon dioxide is removed from solution, the reaction moves to the left, and a saturated bicarbonate solution will become cloudy with finely divided $CaCO_3$ precipitate, which is much less soluble than calcium bicarbonate. Thus, limestone solubility is controlled primarily by addition or loss of CO_2 gas to a great excess of water. Changes of temperature or pressure, mixing of unlike water masses, and biologic process can all promote solution or deposition.

The solubility of $CaCO_3$ in water over a wide range of partial pressures of CO_2 is graphed in Figure 8-1. More detailed graphs, including the relations of temperature, pH, and concentration of CO_2 in solution are given in Palmer (1991, Figure 7). The partial pressure of CO_2 in an aqueous solution exposed to sea-level atmosphere is only 30 Pa, and the related solubility of $CaCO_3$ is very low, only about 63 mg/L. However, during decay of humus in a soil, the partial pressure of CO_2 may increase to 30 kPa and result in an increase of $CaCO_3$ in solution to 700 mg/L, more than a tenfold increase over the amount in aqueous solution at equilibrium with normal atmosphere. Most groundwater contains dissolved $CaCO_3$ in a concentration of between 200 and 400 mg/L, indicating CO_2 partial pressures well within the range of observed composition of soil air (Figure 8-1). Limestone solution, and therefore karst, is clearly accentuated by biologically generated CO_2 in decaying humus. We consider the geomorphic implications of this statement in subsequent sections.

Biologic Effects.   In addition to providing the decaying humus that enriches soil air in CO_2, plants and animals may corrode limestone directly. Folk et al. (1973) designated a distinctive minor karst landform on Grand Cayman Island as **phytokarst.** The surface is intricately pitted, with pits intersecting to form a sharp-edged, spongy network of pinnacles more than 2 m in height (Figure 8-2). *Cyanobacteria* (blue-green algae) are the principal microorganisms that live on

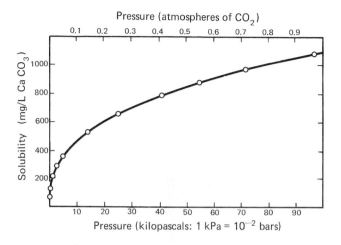

FIGURE 8-1. Solubility of $CaCO_3$ in water at various partial pressures of CO_2 (t = 16.1°C) (after Adams and Swinnerton, 1937).

the surface (epilithic), within preexisting cavities (chasmolithic), and within the outer millimeter (endolithic) of limestone in tropical regions (Viles, 1987). In all three microenvironments, the algae weather limestone, but they may also create biochemical conditions that inhibit solution or promote calcite deposition. The endolithic cyanobacteria penetrate the outer millimeter of rock with micron-size borings that are the pathway for water and acids from respiration and decay to enter the rock, and the chemically active surface area is greatly increased. However, relict boreholes are usually filled with fine-grained high-magne-

sium calcite (*micrite*) that is resistant to weathering and sometimes forms a crust that protects more soluble coralline limestone. Like the related cyanobacteria that live within the outer rind of quartz boulders in deserts and in Antarctica (p. 133), the algae in and on limestone produce a complex microenvironment for weathering within 1 mm beneath the surface. The weathering environment for limestone is further complicated by snails and other "grazing herbivores" that eat the endolithic algae and simultaneously erode impressive amounts of limestone (p. 428; Shachak et al., 1987).

On certain tropical limestone coasts, flocks of sea birds have nested for many generations and accumulated thick deposits of droppings. This *guano* is rich in phosphate from the scales and skeletons of the small fish that are the birds' primary food. Decay and leaching of the guano makes phosphoric acid, which strongly corrodes the limestone and reacts with it to form commercial deposits of **rock phosphate** (earthy calcium phosphate of variable composition). Rock phosphate has been mined extensively, especially on Nauru and Ocean Island in the Pacific (Figure 8-3). It fills karst cavities in limestone to a depth of 20 m (Power, 1925). On many atolls, the location of former groves of trees in which sea birds roosted can be inferred from areas of phosphatized reef limestone. The acid humus of trees also corrodes peat-filled depressions into limestone at their bases.

Temperature Effect. Water at 10°C dissolves about twice as much atmospheric CO_2 as water at 30°C;

FIGURE. 8-2. Spongelike phytokarst on Pleistocene coral limestone, Huon Peninsula, Papua New Guinea. One-meter axe handle is nearly concealed.

FIGURE 8-3. Karst pinnacles on Nauru island exposed by removing the overlying phosphate rock (photo: British Phosphate Commissioners).

water at 0°C dissolves almost three times as much as at 30°C. Therefore, cold water saturated with CO_2 should be more "aggressive" than warm water in dissolving soluble rocks. However, although CO_2 is less soluble in warmer water, the CO_2 production in warm soil may be greater, so the two effects can cancel each other to varying degrees. Furthermore, even though carbonate minerals are less soluble at high temperatures, even at the same dissolved CO_2 content in water, the rate of solution is greater at higher temperatures. The reduced viscosity of water at higher temperatures is an additional factor in the complex effects of temperature on karst development. The net effect is that warm temperature favors the karst process, providing that adequate groundwater is available. Karst is not a conspicuous feature of cold, dry regions.

Pressure Effect. CO_2 derived from soil diffuses downward into air-filled voids in regolith and the rock mass above the water table. Groundwater moving downward into zones of greater hydrostatic pressure will continue to absorb available CO_2 in proportion to the partial pressure graph (Figure 8-1) and can dissolve more $CaCO_3$. Therefore, solution can be an active process throughout the aerated zone above the water table. However, the release of pressure when percolating groundwater under hydrostatic pressure

enters the free atmosphere of a cave is an important factor in dripstone deposition.

Related to the pressure effect is turbulence. The movement of groundwater through narrow fractures and granular materials is not believed to be turbulent, but dissolved CO_2 gas may be exsolved by splashing or turbulent flow in larger openings, and $CaCO_3$ thereby may be deposited.

Mixing Effect. The nonlinear solubility of calcium carbonate in carbonated water gives rise to an interesting phenomenon when two unlike water masses mix. For clarity, the extreme lower-left portion of Figure 8-1 is enlarged (Figure 8-4) to show the *mixing effect.* Suppose that deep groundwater, saturated with $CaCO_3$ at a low CO_2 partial pressure, and water from the aerated zone, also saturated but at a higher CO_2 partial pressure, meet and mix at the water table. Because of the nonlinear solubility graph, mixing of any two saturated but unlike water masses invariably results in an undersaturated solution (Dreybrodt, 1988, Figure 2-16; Ford and Williams, 1989, Figure 3-10). Mixing is also very important in the coastal zone, where groundwater and seawater meet and mix (p. 427). The presence of other common ions such as Mg^{2+}, Na^+, K^+, SO_4^{2-}, and Cl^- in seawater changes the solubility of calcite and can either increase or decrease dissolution (Dreybrodt, 1988, pp. 30–34).

An interesting concept has been proposed that relates cave formation in the eastern midcontinental United States to changes in groundwater flow paths at the margin of Pleistocene ice sheets, which enhanced

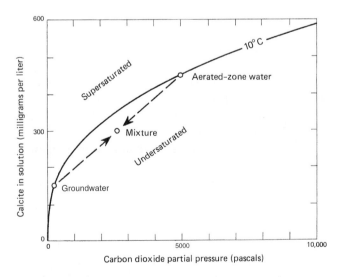

FIGURE 8-4. The result of mixing unlike water masses that are each saturated with respect to calcite is a water mass that is undersaturated (diagram courtesy G. W. Moore).

mixing of glacier-derived meltwater with older, saline-rich groundwater (Panno and Bourcier, 1990). The concept has been criticized on several grounds, including calculations that the additional solution by mixed water would be only minor, and that caves in the glaciated region do not display forms attributable to mixing (Mylroie, 1991). Caves in glaciated New York State do show abundant effects of flooding by glacial meltwater, however (Rubin, 1991).

Flow-Rate Effect. Somewhat related to the mixing effect are the solubility changes related to variations in rate of flow of surface and underground water. If water is moving slowly over or through soluble rock, it should approach saturation equilibrium. However, during rapid runoff periods, river water is notably undersaturated because the water is not in contact with mineral soil long enough. If such undersaturated water, normally with a low dissolved CO_2 partial pressure, enters a karst terrane and mixes with saturated groundwater, the extreme difference in their composition greatly enhances the mixing effect. This process may be especially important in enlarging caves beneath plateaus that normally drain outward to a surface river but become backflooded when the river is in flood. The deep levels of Mammoth Cave, Kentucky, are notable for this process (Thrailkill, 1968, p. 36; Hess et al., 1989, p. 45).

Structural Control of Karst Processes

The most obvious structural factors in karst are lithology and permeability. Most karst is on limestone, an extremely diverse rock type. To be defined as a limestone for industrial purposes, a rock should contain more than 80 percent calcium and magnesium carbonate, but many limestones contain sand, silt, and clay. Karst limestones, in general, are quite pure. For example, the Dinaric karst is in limestones that range from 80 to 98 percent $CaCO_3$, with many of the analyses in the upper part of the range (Herak, 1972, p. 28). The Annville Limestone of Ordovician age that underlies a karst area near Hershey, Pennsylvania, is more than 95 percent $CaCO_3$ (Foose, 1953, p. 624). By contrast, the Luray Caverns of Virginia are in coarsely crystalline Ordovician dolomite that contains as much as 10 percent quartz, chert, feldspar, and clay, but no calcite (Hack and Durloo, 1962, p. 8).

More than 60 variables have been identified that affect the porosity and permeability of carbonate rocks (Brahana et al., 1988, Table 1). Highly permeable limestone such as chalk, or reef limestones that have massive primary voids through which water can freely drain, do not develop extensive surface karst. The landscape of southeast England and northwest France, for instance, is underlain by thick beds of Cretaceous chalk, a soluble, fine-grained, soft, and massively permeable limestone. Almost all the chalk terrain lacks surface streams except during peak runoff intervals, yet the chalk does not form the unique landscape of karst. The uniform permeability precludes it. As noted on p. 147 in the definition of karst, well-developed secondary porosity, or the tendency of solution to create and enlarge voids in the rock mass, is essential for karst. The intersections of vertical joint sets are conducive to concentrating the downward movement of groundwater and are dissolved to make open *shafts*. Joints and bedding planes in flat-lying rocks localize lateral movement and dissolve lateral *galleries*. Pressure-release joints parallel to valley walls strongly influence karst solution in the Appalachian Plateau (p. 175).

Karst Hydrogeology

In the classical theories of hydrogeology, precipitation percolates into soil and rock and infiltrates downward through an unsaturated zone of variable and discontinuous intergranular water and air. This zone is known as the **zone of aeration,** or *vadose zone*. At some depth, ranging from zero to hundreds of meters, pore spaces are full of water. This is the **zone of saturation,** or *phreatic zone*. The surface of the zone of saturation is the **water table** (Figure 8-5) (Freeze and Cherry, 1979, p. 39). Groundwater moves slowly enough that it establishes local hydraulic gradients, and in humid regions the water table is a subdued replica of the topography, rising beneath hills and intersecting lower parts of the landscape at the surface of rivers, lakes, or springs.

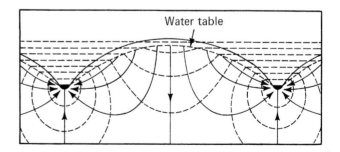

FIGURE 8-5. Flow lines (solid lines) in uniformly permeable material between an intake area at the upland water table and discharge areas on valley floors. Flow lines are orthogonal to surfaces of pressure equipotential (dashed lines) (Hubbert, 1940, Figure 45: © Univ. Chicago Press).

Water beneath the water table in isotropic granular sediments moves in accordance with laws of hydraulics along flow lines, perpendicular to equipotential surfaces (Figure 8-5). From each point on the water table, there is but one unique flow path to an outlet. The general hydrogeologic equation, known as *Darcy's law,* is fully analogous to Ohm's law for electricity because the amount of flow per unit area of cross section (amperes) is proportional to the pressure-potential difference (volts) and inversely proportional to the resistance (ohms) and the length of the conduit.

The fundamental problem of karst hydrogeology is scale. Models of classical groundwater flow (Figure 8-5) are large compared to the grain size or pore size of the granular rock mass through which water permeates. However, dense, well-crystallized limestone and marble have extremely low primary permeability (Brahana et al., 1988, Table 2). *Fracture permeability* is the only way water can move through such rock masses. Initial openings such as bedding planes and joints may be no wider than 10 to 50 μm. Water moves slowly through the fractures as laminar films so that only a very large volume of the rock is statistically similar to a uniform porous material. The slowly moving water is generally saturated, or at most only slightly unsaturated, with respect to $CaCO_3$.

A definitive criterion of karst-prone rocks is that their permeability increases with time (Ford and Williams, 1989, p. 133; Palmer, 1991, p. 8–10). As solution slowly widens fractures, three thresholds are passed, which change the nature of flow (Watson and White, 1985, p. 121). First, in a widened fracture, flow changes from laminar to turbulent. Second, the rate of solution increases rapidly because a larger volume of undersaturated water moves through the fracture. Third, mechanical transport of abrasive clastic particles begins when the flow velocity exceeds a certain level (Figure 10-11). All three thresholds occur under typical groundwater conditions when fractures are opened to a width of 5 to 10 mm. A time span of 3000 to 10,000 years is suggested for opening an initial tight joint to the critical threshold size. The subsequent opening of selected conduits to meter-size cave passages may require another 10,000 to 1 million years (Watson and White, 1985, p. 122). The slowness of the initial solution is supported by the opinion of Bögli (1980, p. 58) that shallow surface fractures may require more time to widen to 2 mm than they require subsequently to widen from 2 to 200 mm.

Carbonate rocks have the most variable hydrogeologic conditions of all major rock types (Brahana et al., 1988), and groundwater movement in developed karst is highly irregular and complex. Open solution shafts and galleries are so much larger than the intergranular

pore spaces through which water moves in granular sediments and rocks that some authors have even questioned the applicability of the fundamental concept of a water table beneath karst terrains (Jennings, 1985, p. 66; Ford and Williams, 1989, pp. 129, 142–148).

Instead of diffusing downward through the zone of aeration as in granular materials, karst water enters the ground through shafts beneath sinkholes, which act as *point sources.* From these points, the flow can be modeled as a network of intersecting pipes of various sizes, from which the water eventually emerges at springs (Figure 8-6).

Applied Karst Hydrogeology

Applied karst geology and hydrogeology are relatively new fields for geotechnical engineers, karst hydrologists, and geologists. As population grows in regions such as Florida, the Yucatan Peninsula of Mexico, and extensive regions of the Mediterranean basin, the demand is strong for special engineering techniques to avoid or remediate collapse or subsidence and to safeguard water resources from pollution.

An estimated 25 percent of the earth's human population is supplied with most or all of their water from karst aquifers (Ford and Williams, 1989, p. 6). Most of that water flows rapidly through pipelike networks, derived from point sources in the floors of sinkholes, and discharges at springs or into rivers or the sea. In rural farming communities, as well as in urban and mining areas, the temptation to use sinkholes as dumps is very strong. They are regarded as a nuisance and a danger that most landowners are only too happy to have filled in, even with junked cars and construction debris. Many groundwater pollution problems derive directly from the non-Darcy flow sys-

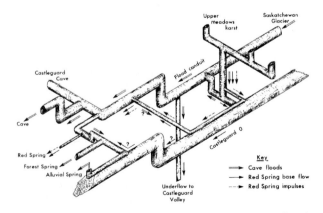

FIGURE 8-6. Pipe model of the meadows karst, Castleguard Meadows, Banff National Park, Alberta, Canada (Smart and Ford, 1986, Figure 9).

tems in karst. If road salt, septic-tank wastes, or industrial pollutants enter the system, there is almost no filtration (Hughes et al., 1994).

The standard way to trace pollutant flow in a karst system is to add a water-soluble dye to a possible point source of groundwater, and monitor local wells, springs, and creeks for signs of color. Usually, the tracer dye travels several km in a few hours (numerous case studies are described in Beck and Wilson, 1987; Günay et al., 1993; Beck, 1993). It is not unusual for nearby wells and springs to be unaffected by dye tracer injection while outlets farther away show prompt response. Obviously, the complex pipe systems of open karst aquifers (Figure 8-6) are not all interconnected.

Dams constructed in karst terrane are notoriously dangerous. Deep groundwater circulation from a reservoir may reemerge downstream of a dam or even in adjacent river valleys. Thornbury (1969, Figure 22-12) showed a cross section of rubble-filled collapse karst that extended at least 190 ft (58 m) below the foundation of a dam on the Tennessee River. Extensive and unanticipated sealing of the dam foundation by cement grout pumped into the deep openings was required. Similar problems were encountered at many dam sites in Ohio, Kentucky, Tennessee, and adjacent regions during the massive Tennessee Valley Authority (TVA) dam building projects of the 1930s. An added difficulty of building dams in limestone terranes is that some of the karst features may be on ancient solution unconformities within the stratigraphic section, totally unrelated to modern conditions (pp. 162–163). The sediment that filled the collapse karst below the Tennessee River dam described by Thornbury was interpreted as the Ripley Formation, a Gulf of Mexico coastal plain unit of Pliocene age. If true, that implies at least 100 m of epeirogenic uplift of the Interior Low Plateaus (Figure 1-5) of the United States since the Pliocene, a tectonic movement that has not been otherwise considered in the geomorphic history of the region.

Quarrying operations, common in limestone and marble terranes, are also complicated by karst conditions. A quarry near Hershey, Pennsylvania, intersected a large water-filled karst pipe that threatened to flood the quarry. When large pumps were installed to dewater the quarry, private and municipal wells went dry, sinkholes began collapsing, and the entire area was in a state of chaos until the conduit was successfully sealed (Foose, 1953).

The Mediterranean basin has a long history of karst-related hydrogeologic and geologic problems. Limestone and dolomite of the ancient Tethys Ocean, and their metamorphosed marble correlatives, are widespread. Much of the architecture and sculpture of classical Greece and Rome and of the European Renaissance was based on the ready availability of marble in the region. The latest Miocene "Messinian crisis" (Chapter 4, p. 159), when the Mediterranean Sea dried up, caused deep entrenchment of river valleys around the basin and a corresponding lowering of groundwater level, with accelerated karst development, some of which is presently submerged. Numerous descriptions of karst-related problems in Turkey, Greece, Italy, France, and Spain are included in Günay et al. (1993). Middle East and northern African countries undoubtedly have similar problems.

The most notorious karst region in the United States for construction and groundwater problems is in Florida and adjacent parts of Georgia and Alabama. The Florida aquifer system that supplies much of the freshwater to the region is a thick sequence of Cenozoic carbonate rocks (Johnston, 1993). In much of the region, the carbonate rocks are covered by younger sandy strata. Limestone dissolution has been intense because of the warm climate, heavy precipitation, low relief that encourages infiltration, and multiple Pleistocene sea-level oscillations of 100 m or more. During droughts, when groundwater pumping is at high rates, many sinkholes suddenly appear in the overlying cover strata. Lakes may go dry when a plug of clay or organic-rich mud suddenly fails. The karst lake district of north-central Florida is a real estate lawyer's paradise, full of "shore-front" property that suddenly becomes property at the edge of a sawgrass "prairie" or a mud flat (for case studies, see Beck and Wilson, 1987; Beck, 1993).

Climatic Control of Karst Processes

Temperature, precipitation, biologic activity, and amount and seasonality of runoff are all climatic factors that affect the intensity of karst processes. Therefore, it might be supposed that a climatic model of karst processes would have been developed, similar to the latitudinal belts of weathering and soil-forming processes (Figure 7-14). However, the chemical solution of carbonate and other soluble rocks is such a complex process that only a few extreme climatic "end members" are easily deduced (Jennings, 1985, pp. 194–214).

One obvious climatic extreme is the hot dry climate, where low precipitation, rapid evaporation, and lack of vegetation produce a minimal flow of groundwater in which biogenic carbon dioxide is negligible. The high temperature further reduces the amount of atmospheric CO_2 in solution. Deserts are not karst prone. Massive limestone and dolomite formations assume the role of ridge- and cliff-formers, perhaps rivaling granite in structural control of landforms.

Caves in deserts are not uncommon, but most of them are hypogenic (p. 148), and the others are probably relict from former more humid climatic conditions.

The lack of conspicuous karst in polar regions is due to the slow rate of chemical reactions in cold water, the negligible bacterial decay of humus, the short season of runoff, and permafrost that limits infiltration. Corbel (1959a, 1959b) emphasized the importance of karst solution in cold humid regions, but few other authors have supported him (Jennings, 1985, p. 200; White, 1988, p. 215). Ford (1971) made extensive measurements of dissolved components in limestone regions of the Canadian Rockies. He found that glacial meltwater above the treeline was saturated with respect to calcite at only 50 to 90 mg/L $CaCO_3$, indicating that little CO_2 was in solution. Below the treeline, groundwater was not saturated with respect to calcite until it contained 100 to 265 mg/L $CaCO_3$, and even creeks and lake water had enough dissolved CO_2 to carry 100 to 140 mg/L $CaCO_3$ in solution at saturation. Ford concluded that groundwater on forested mountains in a subarctic climate zone is as capable as groundwater in temperate or tropical regions of dissolving limestone but that there is no particular "aggressiveness" associated with glacier or snowbank meltwater.

The cool-humid midlatitude zone of increased weathering (Figure 7-14) has well-developed karst on suitable rocks. Precipitation is greater than in polar regions, more of the precipitation falls as rain, infiltration is not inhibited by frozen ground, and the seasonal leaf fall and dormant period promote humus accumulation. Groundwater is generally abundant. Valley systems eroded by rivers dominate the landscape, but karst landforms mingle with landforms of surface erosion. This common landscape is called *fluviokarst,* distinguished from *holokarst,* in which the entire regional drainage is underground (Ford et al., 1988, p. 401; White, 1988, p. 112).

The "botanic hothouse" climatic extreme of the humid tropics has the most diverse and dramatic examples of karst landscape although some have argued that the controls are lithologic and tectonic rather than climatic (Jennings, 1985, p. 201; Ford et al., 1988, p. 403). The partial pressure of CO_2 in soil air measured during the growing season at a number of worldwide sites was correlated with temperature, precipitation, and evapotranspiration (Brook et al., 1983). The extrapolated or predicted high values of CO_2 in soil air broadly coincide with subtropical and tropical areas of active karst development such as the southeastern United States, Caribbean Islands, Mexico and Central America, southeast Asia, and the limestone islands of the southwestern Pacific.

KARST LANDFORMS

Regional and Temporal Significance of Karst

Karst landforms are worthy of study not only because of their dramatic form and unusual origin but because of their abundance. Limestones and dolomites constitute between 5 and 15 percent by weight of the total sedimentary rock mass (Garrells and Mackenzie, 1971, p. 40); evaporites constitute an additional 5 percent. About 15 percent of the area of the conterminous United States has karst-prone rocks at or near the surface (Davies and LeGrande, 1972, p. 469), and about 12 percent of the earth's land area has exposed carbonate rocks although not all of it shows karst landforms (Ford et al., 1988, p. 401; Ford and Williams, 1989, p. 6).

The late Cenozoic Era has been a time of unusually extensive karst landscapes. Both the Indonesian–Philippines–New Guinea orogenic belt and the West Indies–Caribbean belt are in climatic zones of submarine carbonate deposition and coral-reef growth where active tectonism has uplifted coral limestone (Figures 3-5 and 20-1 through 20-3). Furthermore, the Alpine orogenic belt, extending from the Pyrenees Mountains through southern Europe, Turkey, and Iran, includes thick sections of recrystallized, fractured, and deformed carbonate rocks in which karst processes have been active in late Cenozoic time.

Quaternary glacier-controlled fluctuations of sea level (Figure 18-3) have alternately exposed and drowned karst landscapes on all carbonate-rock coasts. When sea level was lowered by expanding glaciers, coastal regions developed deeper groundwater circulation, and karst landforms and caves actively developed. When, as at the present time, the level of the sea is at a high interglacial level, coastal karst terranes are drowned. Speleothem (dripstone; p. 164), which formed in aerated caves, has been collected by divers from caves 55 m below sea level in Bahama (Harmon et al., 1983; Lundberg and Ford, 1994; Richards et al., 1994), and observed at a depth of 122 m in a "blue hole" in a British Honduras reef (Dill, 1977). Radiometric dates on speleothem from submerged caves in Bahama and elsewhere, which record times of glacially lowered sea level, are combined with dates on corals, which document times of relatively high sea levels, to construct sea level curves similar to those in Figures 4-7 and 18-3. They show that sea level passed through a depth zone of –10 to –15 m numerous times in the last million years. Only rarely in that time interval has sea level been higher than it is now (Chapter 18). Hundreds or thousands of blue holes below sea level today are almost certainly sinkholes that formed during glacial-age low sea levels (Backshall et al.,

1979). The shape of atolls also depends to a large extent on karst solution during Pleistocene low sea levels (Chapter 19).

Karst Denudation Rates

The rate of limestone solution is most commonly measured by mass balance equations involving the water discharge per unit area and the concentration of dissolved ions in the water. Corrections are required for the percentage of the discharge area underlain by noncarbonate rocks (which also weather to bicarbonate species in solution; p. 130), the bulk density of the rocks, the amount of the dissolved load reprecipitated within the drainage area, seasonal variations in discharge and dissolved load, and the underground discharge area, which may not coincide with the surface drainage area. Results are usually reported in volume loss per unit area per year (m³/km² yr), which is equivalent to a regional *denudation rate,* or average landscape lowering, measured in mm/1000 yr (Figure 8-7).

Denudation rates for limestone terranes are deceptive in that much of the surface volume loss may be redeposited in the subsurface, but on the scale of geologic time, the net result must be the lowering of the landscape toward sea level. The denudation rates compiled in Figure 8-7 fall well within the range of regional fluvial denudation rates (Chapter 15), which is not surprising since a significant percentage of all landscapes consists of soluble rocks. Clearly, the most important variable in limestone denudation is runoff, or the excess of precipitation over evapotranspiration (Ford and Williams, 1989, p. 97). This is the water that is available to carry away the dissolved compounds. Denudation rates are also significantly correlated with total precipitation. Much less important is the climatic variable of temperature. More important than temperature is the proportion of bare and soil-covered ground. Although arctic and alpine environments with little or no soil cover show significantly less rapid denudation rates for a given amount of runoff than do vegetated and soil-covered terranes, temperate and tropical humid climates have similar denudation rates for any specified runoff (Jennings, 1985, pp. 192–194; White, 1988, pp. 215–219) though runoff is generally greater in tropical than in temperate humid climates.

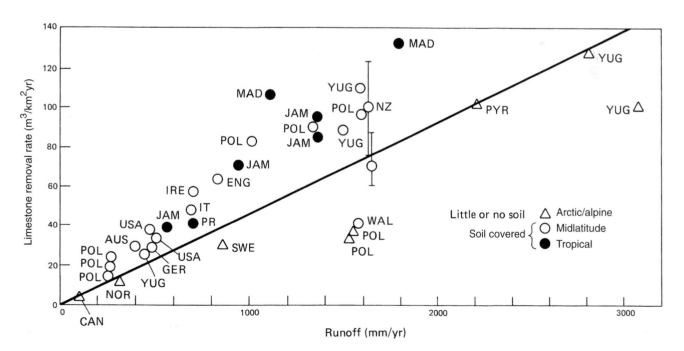

FIGURE 8-7. Limestone removal rate as a function of runoff: AUS, Australia; CAN, Canada; ENG, England; GER, West Germany; IRE, Ireland; IT, Italy; JAM, Jamaica; MAD, Madagascar; NZ, New Zealand; NOR, Norway; POL, Poland; PR, Puerto Rico; PYR, Pyrenees; SWE, Sweden; USA, United States; WAL, Wales; YUG, Yugoslavia. Units of limestone removal rate are equivalent to denudation rate in mm/1000 yr (Jennings, 1985, Figure 6-7).

FIGURE 8-8. Well-developed, but small, clint-and-grike topography on a dolomite pavement. This example is on Silurian reefal dolomite in the Bruce Peninsula, Ontario, Canada (Ford et al., 1988, Figure 1).

Minor Karst Landforms[1]

Limestone and dolomite, and to a lesser degree other soluble rocks, commonly develop distinctive surface and near-surface weathering features sometimes termed *epikarst.* Surfaces may be smoothly rounded forms (Figures 8-8 and 8-9), or hemispheric pits that intersect as razor-sharp points and ridges. In the extreme, a spongelike framework of rock is all that remains (Figure 8-2). Phytokarst (p. 148) is independent of gravity, with surface pits corroding equidimensionally beneath an algal film, whereas forms created by flowing water tend to elongate vertically or downslope.

Solution Pits, Facets, Runnels, Flutes, and Karren.
On exposed rock (bare karst, pavement karst), horizontal surfaces develop shallow **solution pits** (rainpits, makatea). On surfaces inclined a few degrees from the horizontal, sheet runoff produces flat **solution facets** (solution bevels). **Solution runnels** (Rinnenkarren) are branching channels in networks on gently inclined bare surfaces. Many runnels probably first form under soil cover but, having formed, are perpetuated by runoff on bare rock surfaces. On steeply inclined karst surfaces, elongate **solution**

---

[1]Because of the multilingual terminology of karst forms, a single boldface term is defined. At the first use, synonyms follow in parentheses but are not defined. The reader is urged to consult glossaries or expanded texts for the detailed shades of meaning (see Monroe, 1970; Sweeting, 1972; Jennings, 1985; White, 1988; Ford and Williams, 1989).

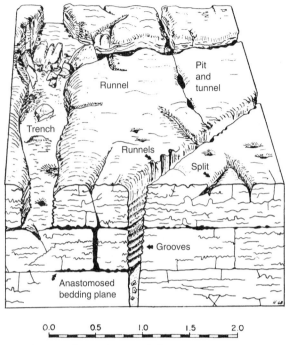

0.0   0.5   1.0   1.5   2.0

Approximate scale in meters

FIGURE 8-9. Labeled sketch of minor karst forms on Silurian dolomite near Hamilton, Ontario (after Pluhar and Ford, 1970, Figure 2).

flutes (Rillenkarren, lapies) are common landforms. These are straight, parallel-sided grooves with sharp crests separating them. They follow the steepest slope down a rock face.

Covered Karst: Grikes, Clints, Solution Pipes, and Solution Notches.
If soil or an organic mat covers a limestone surface, the effectiveness of solution increases manyfold. Solution pits are even more common under soil cover than on bare ledges, but the divides between pits are likely to be more rounded and smooth than the knife-edged ridges between intersecting pits on bare karst.

For vertical openings such as joints that have been widened by solution, the British term is **grike** (Kluftkarren, solution slot). Grikes intersect as circular, vertical **solution pipes** (karst wells, potholes). The remnant pavement surfaces between open grikes are called **clints,** on which minor solution features develop (Figures 8-8 and 8-9) (Pluhar and Ford, 1970).

At the level where bare rock projects above the soil level, a **solution notch** may develop. A distinct undercut measures the greater effectiveness of karst processes below ground level. An extreme example of a solution notch has been named a **swamp slot** (Wilford and Wall, 1965) or *foot cave* (Jennings, 1985, p.

208). It is probably restricted to the humid tropics. Acid swamp water corrodes a narrow slit a meter or more horizontally into the base of limestone cliffs. Residual clay or an organic detrital layer may prevent downward solution, especially if the water table is at the swamp surface and if percolation is slow. A swamp slot or solution notch may be a basic component of tropical tower karst (p. 160), characterized by marginal corrosion at the base of steep-sided residual karst hills surrounded by swamps or alluvial plains.

Major Karst Landforms

Karst is essentially a near-surface process in vegetated and soil-covered landscapes, where most of the solution takes place in or just beneath the soil profile. On bare rock, solution is concentrated at the water table, in the zone of mixing. Thus, there is a continuum of landforms from the minor surface features just described to distinct landforms comparable to the hills and valleys of fluvial landscapes. The forms described in this section are of a scale that can be studied on aerial photographs and topographic maps. Portraying karst topography on contour maps poses special problems for the cartographer. Hachured depression contours, contour lines that cross to show overhanging cliffs, and dotted contours in natural bridges are some of the unusual mapping conventions that are required to contour karst landforms.

Dolines (Sinkholes, Sinks, Swallow Holes, Cockpits, Blue Holes, Cenotes). The fundamental geomorphic component of karst topography is the **doline,** or limestone sink (Figures 8-10 and 8-11) (Ford and Williams, 1989, p. 396). Whether due to joint control (the most common cause), differential solubility, or random events such as cave collapse, localized areas of karst terrain are lowered more rapidly than the surrounding area and form closed depressions. Their sides slope inward at the angle of repose of the adjacent material, typically 20° to 30°. Some have much more gentle side slopes, and others, especially those forms known as *collapse sinks,* have vertical or overhanging cliffs. They are usually circular in plan view and less commonly elongate or oval. They range in size from shallow soil depressions a few meters in diameter and a meter deep to major landforms several kilometers in diameter and hundreds of meters in depth.

Doline and its synonyms are genetic terms. They describe a landform produced by the karst process. Identically shaped closed depressions can form by subsidence, vulcanism, wind deflation, or glaciation—in fact, by any process that selectively translocates a mass of rock and permits the surrounding material to slump into the excavation. In areas such as southern Indiana,

FIGURE 8-10. Doline about 120 m in diameter and 45 m deep, which abruptly collapsed on the night of December 2, 1972, near Montevallo, Alabama (photo: U.S. Geological Survey).

FIGURE 8-11. Five major classes of dolines (from Jennings, 1985, Figures 37 and 43).

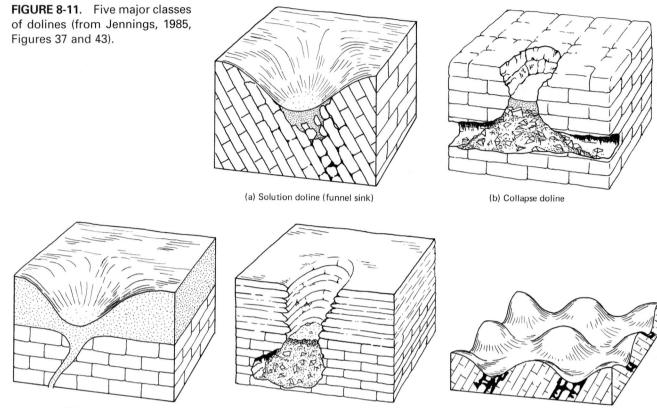

(a) Solution doline (funnel sink)

(b) Collapse doline

(c) Subsidence doline

(d) Subjacent karst collapse doline

(e) Cockpits (intersecting star-shaped dolines)

where glacial moraines cross a region of karst topography, real judgment and careful investigation are required to distinguish dolines from glacial kettles (p. 387).

Dolines and cave systems are the karst analog of river valleys. They are the fundamental unit of erosion as well as weathering. As localized sources of surface water, dolines concentrate and maximize subsurface dissolution and later abrasion (p. 152). The bottom of a doline may open into a subterranean passage, down which water pours during a rainstorm. More commonly, the subsurface passage is partly blocked by rock rubble, soil, or vegetation, and temporary lakes form until the water percolates away.

Five major classes of dolines are recognized (Figure 8-11). The two most contrasting types are the funnel-shaped *solution doline* and the steep or cliffed *collapse doline*. In the field, distinction between the two varieties of doline is rarely as obvious as the sketches of Figure 8-11 suggest because both solution and collapse are common processes in nearly all dolines. *Subsidence dolines* and *subjacent karst collapse dolines* are surface forms in nonsoluble rock, caused by solution of buried karst. A thick residual soil, alluvium, or loess may mantle an active karst landscape, and the surface

material will gradually settle or be carried into a buried doline. Frequently, small dolines (*swallow holes, ponors*) form in a stream bed, and although alluvium completely conceals the buried karst relief, the stream abruptly disappears and abandons the downstream segment of its former channel. These have been called *alluvial streamsink dolines*.

Dolines in tropical regions develop rapidly and grow very large. Adjacent dolines intersect, fuse, or engulf smaller ones to form compound, reticulate, honeycomb, or even star-shaped depressions. If they are of similar size and closely spaced, the intersection ridges will form a polygonal pattern (Williams, 1972; Monroe, 1976; Williams, 1993). Their floors are permanent or seasonal swamps at the water table. The *cockpit country* of Jamaica is the classic locality for tropical dolines (Versey, 1972). One doline or cockpit in Puerto Rico is so nearly a perfect hemisphere in cross profile that it has been surfaced with metal mesh as a giant radio telescope (Figure 8-12).

Uvalas (Compound Sinkholes, Karst Windows, Blind Valleys).   A genetic progression can be visualized in which dolines progressively abstract surface runoff into groundwater circulation and leave net-

FIGURE 8-12. The radio telescope at the Arecibo Observatory, Puerto Rico. The reflector surface in a doline is 305 m in diameter and 51 m deep. The floor of the doline probably drains through a cave into the Tanama River, seen entering a natural bridge in the right foreground. The Arecibo Observatory is part of the National Astronomy and Ionosphere Center, which is operated by Cornell University, under contract with the National Science Foundation (photo: R. C. Hamilton, Cornell University Office of Public Information).

works of dry valleys as relict surface forms. The Serbo-Croatian term **uvala** was given to such a compound doline, or a chain of intersecting dolines (Cvijić, 1960). A **karst window** is similar, in that an unroofed segment of an underground stream channel becomes a surface valley in the window and passes underground again downstream.

Allogenic rivers originate in a terrane or climate unlike the one through which they flow, as for example, the Colorado or Nile. In the karst context, an allogenic river may have a network of valleys on insoluble rocks, then flow onto karst-prone terrane. If the exposed segment has normal stream valley form and an alluvial floor but disappears downstream into a swallow hole, the term **blind valley** is often used. A natural bridge (Figure 8-12) may mark a segment of unroofed cave or may form by progressive underground abstraction of surface flow, either across the neck of an intrenched stream meander or from a high-level stream to a lower adjacent stream.

A **karst valley** is a special form of blind valley that occurs where a surface drainage network first developed on insoluble rock and later eroded into soluble rock, whereafter subsurface drainage totally defeated the surface flow. It is a relict form that has all the properties of a stream valley except that it lacks a stream. Karst valleys are common in areas of nearly flat-lying, interbedded limestone, shale, and sandstone. Many good examples are shown on topographic maps of the Mammoth Cave area in Kentucky.

Poljes (Plans, Wangs, Hojos). The Dinaric karst region of Slovenia and Croatia is in a belt of Alpine folding (Figure 20-6). The karst topography is on carbonate rocks that are folded and faulted, often in proximity to insoluble rocks. Structural control becomes a major factor of landscape evolution. Large blind valleys or karst valleys, enclosed by either soluble or insoluble rocks, usually elongated along tectonic axes, are called **poljes** there. Many are grabens or half grabens, but others are simply large-scale examples of structurally controlled karst processes (Figure 8-13). Their floors truncate folded limestone formations or deep sedimentary infills at a local base level determined by a narrow gorge through insoluble rock or a series of swallow holes (ponors) at the downstream end of the polje (White, 1988, p. 41; Ford and Williams, 1989, pp. 428ff). Poljes are large closed valleys; the largest in Croatia are up to 60 km in length and 6 to 8 km wide. In general, uvalas are measured in hundreds of meters of lateral extent; poljes in kilometers (Herak, 1972, p. 37).

Poljes have broad, flat floors that abut sharply against steep enclosing walls. On occasion, during floods, the swallow holes by which some valley floors are drained become choked with debris, and the entire polje floor becomes a shallow lake. Some Croatian poljes flood annually when the water table is high and are fertile agricultural land during drier seasons. Their hydrology is the subject of sophisticated Croatian research (Herak, 1972; White, 1988; Ford and Williams, 1989, p. 428–432).

Many geomorphologists tend to consider karst landforms as a genetic sequence: doline → uvala → polje. Dolines grow and merge into uvalas through time in a genetic sequence. However, poljes are more than large, complex uvalas. They have strong structural control that does not lend itself to a genetic sequence following the much smaller, process-controlled forms.

Residual Karst

The concept of a karst "cycle of erosion," beginning with the genetic sequence of doline and uvala and ending with a plain of solution dotted with residual limestone hills, can be traced to the historical impact of the visit with Albrecht Penck to the Dinaric karst by

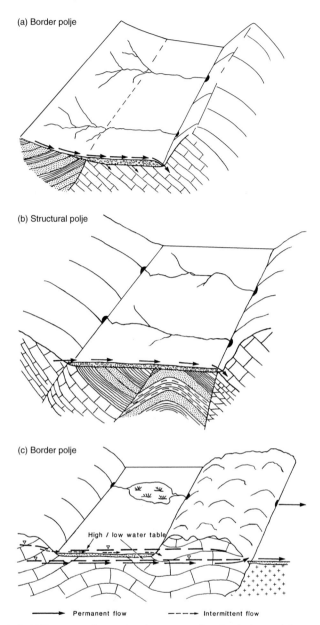

(a) Border polje

(b) Structural polje

(c) Border polje

High / low water table

→ Permanent flow   - - - → Intermittent flow

FIGURE 8-13. Some basic types of polje (adapted from Ford and Williams, 1989, Figure 9.30).

W. M. Davis in 1899 (Davis, 1901). At the time, the Davisian concept of a cycle of erosion was achieving great popularity. J. Cvijić (1893) had previously published a distinguished monograph on karst under the supervision of Penck. Penck and Davis inspired him to rework his observations into a cyclic concept, which he published in France while he taught there during World War I. The paper was not well known because of the wartime communication problems in Europe,

but a review was published in English (Sanders, 1921) reproducing Cvijić's original block diagrams. A major manuscript by Cvijić (1960) was lost by his publisher and not rediscovered and published until 33 years after his death.

The deduced residual landforms of the karst cycle were not well illustrated by the Dinaric karst. Small limestone hills (hums) rise above the floors of Croatian poljes, but the structural complexities of the region made it unsuitable as an area for a model of a karst erosion cycle. Later authors used the tropical karst of Jamaica and southeast Asia as models for the more advanced stages of a deduced karst cycle, neglecting the obvious differences in lithologic and climatic controls. As a result, the deduced karst "cycle of erosion" starts in a Mediterranean climate but ends in the tropics. Rather than persist in following the cyclic concept, it is more productive to review residual karst landforms simply as the products of extreme karst processes, especially common in tropical humid climates (Sweeting, 1993).

Lehmann (1936) introduced the term **kegelkarst** (cone karst) for a landscape in Java where the terrain is approximately equally shared by dolines and residual hills. In fact, both in Jamaica and in Java, the typical kegelkarst residual hills are more nearly hemispheric than conic, and the dolines are star-shaped or valley-like among the hills.

Tower karst (turmkarst, mogotes, pepino hills) has been distinguished from kegelkarst in that the residual hills are steep-sided or vertical and are separated from each other by swamps or alluvial plains (Figure 8-14). Marginal solution at the edge of the swampy plains may be largely responsible for the steep walls of tower karst. The residuals are riddled with caves; a swamp slot may deeply notch the base of the towers. In China at least, a level limestone floor underlies the swamps at a shallow depth (Zhu, 1988, p. 25).

The term **tower karst** has become the general term for all residual karst hills, including those with conical or hemispheric shapes, that rise above adjacent plains (Williams, 1987; Ford and Williams, 1989, pp. 440–447). The plains are usually erosional surfaces that truncate the same soluble rocks that form the towers, but they can equally well be erosional plains across other rock types, or aggradational plains. In coastal regions and along the valleys of large rivers, late Quaternary sea-level fluctuations have caused alternate cycles of aggradation and degradation, and the tower karst of these regions is clearly part of the *palimpsest* landscape (p. 35) (Williams, 1987).

FIGURE 8-14. Tower karst in Guilin, Guangxi Province, China. Compare with Figure 8-15 (photo: J. H. Ferger, M.D.).

The tower karst of northern Vietnam extends northwestward from near Haiphong harbor, where it is drowned by the postglacial rise of sea level, along the Red River to the Chinese frontier. It continues into southern China, especially in the Guangxi Zhuang Autonomous Region in the vicinity of the city of Guilin (Zhu, 1988). Many art historians believe that the strange, needlelike mountains of classical Chinese art originated in the subtropical karst landscape of the southern provinces of China (Figure 8-15). The scenes seem exotic and dreamlike to Western eyes, but they are actually reasonable geomorphic sketches. Later classical Chinese art idealized the forms of mountains so that all look equally exotic. The incident portrayed by Figure 8-15, "Emperor Ming-Huang's Journey to Shu," is known to have occurred in the karst region of southern China.

Tower karst requires thick formations of pure limestone. Some of the towers in China are more than 500 m in height, but the Devonian to Carboniferous limestone in which it forms has an aggregate thickness of as much as 4600 m (Zhu, 1988, p. ii). Most tower karst is in the tropics, but a small example has been described in the cold, dry Mackenzie Mountains of northwestern Canada (Brook and Ford, 1978; Ford et al., 1988, p. 405). Sweeting (1993) cited evidence that some of the Chinese karst is older than 1.5 million years; the role of climate change in its formation has yet to be explored.

The *mogotes* (haystack hills) of Sierra de los Organos, Cuba, are residual karst towers of classic beauty (Lehmann, 1960, Plate 1; Gèze and Mangin, 1980). The pepino hills of Puerto Rico are comparable (Meyerhoff, 1938; Monroe, 1976; Guisti, 1978), but many show distinct asymmetry, with a gentle slope toward the northeast from which most of the wind-driven rain strikes them (Figure 8-16). Thorp (1934) proposed that solution on the windward slopes lowered that side of the pepino hills more rapidly. Subsequently, Monroe (1966; 1976, p. 45) proposed that solution and redeposition of surficial secondary carbonate has "case hardened" the windward slopes, whereas the lee slopes are subject to more extensive basal solution notching and develop steeper cliffs on that side.

Residual karst landforms have a curious relevance to the political and social histories of the regions in which they are found. Bandits, partisans, guerrilla troops, and fugitives who are native to a karst region are able to live safely in the many caves that penetrate karst towers. The Viet Minh operated against French colonial troops from the karst of northern Vietnam for decades; Fidel Castro and his revolutionaries based their eventual control of Cuba on the karst of the Sierra Maestra; Yugoslav partisans prevented motorized German troops from controlling large parts of the Dinaric karst; the examples could be multiplied at length. On a less significant level, illegal distilleries are traditionally hidden in the Appalachian karst of eastern United States, and Jesse James, a notorious bandit, hid his gang in the caves of Missouri. Whatever the merit of a cause, its supporters are likely to find a karst

FIGURE 8-15. A south China landscape as painted by an unknown artist of the T'ang dynasty (A.D. 618–907). Compare with Figure 8-14 (collection of the National Palace Museum, Taipei, Taiwan, Republic of China).

terrain that they know intimately an overwhelming advantage against numerically superior and better equipped, but foreign, opponents.

Relict Karst and Paleokarst

Relict karst is a landscape that is no longer forming by the processes that created it although it is still being modified by contemporary processes (Ford and Williams, 1989, p. 507). A common example in relatively undisturbed, flat-lying strata is structurally controlled, where karst develops in a limestone formation, then erodes downward onto underlying insoluble sedimentary rocks and evolves into a fluvial network of surface streams. The relict karst, left "high and dry," will be subject to mass wasting and erosion by the fluvial system, but karstic processes will no longer be effective.

Increased aridity in a formerly moist climate will also make karst terrains relict, but climate change is not easy to infer from a karst landscape. There will always be the danger of the false assumption that a well-developed karst terrain must have formed in a humid or warm climate, and if that is not the present condition, the climate must have changed. Many karst terrains are old enough to have been forming throughout the changes from the "greenhouse" to the "icehouse" Cenozoic world (Chapter 4). The word *relict* must be used only with great caution for such landscapes.

Paleokarst is clearly abandoned and usually buried by younger sediment or rocks although it may have been *exhumed* and reexposed (Ford and

FIGURE 8-16 Mogotes of Puerto Rico (photo: W. H. Monroe 380, U.S. Geological Survey).

Williams, 1989, p. 509–512; James and Choquett, 1988; Bosak et al., 1989; Palmer and Palmer, 1995). Karstic unconformities are common in stratified carbonate rocks. They create major petroleum reservoirs and may control the subsequent deposition of ore deposits, especially lead, zinc, silver, bauxite, and phosphate. Karst can be buried by volcanic eruptions, progradation of alluvial fans, river aggradation, glaciation, or any other subaerial depositional process. A marine transgression can bury karst beneath a variety of marine sediments, including more limestone. Paleokarst is found in carbonate rocks of all geologic ages, and later episodes of solution may follow or cut across older solution unconformities (Palmer and Palmer, 1995). The multiple sea-level changes of the Quaternary Period have been especially important in alternately exposing carbonate coastal sediments to subaerial weathering and erosion, and submerging them. Blue holes and submarine caves (p. 154) should be classed as relict forms, still a landscape but submerged. Other, older karst surfaces such a those deeply submerged or buried within subsiding sedimentary basins, are truly paleokarst.

LIMESTONE CAVERNS AND SPELEOLOGY

A limestone cavern (cave)[2] is arbitrarily defined as a solution cavity large enough for people to enter. Access implies topologic continuity with the subaerial terrain, and therefore caverns are part of the karst landscape, even when they are hundreds of meters below the generally accepted "surface." Even the largest cavern is volumetrically insignificant as a landform, but the beauty of the deposits and the mysterious thrill of cave exploration give caverns a geomorphic significance out of proportion to their size. Especially in Europe, North America, and China, the scientific study of caves, **speleology,** has an active following and excellent publications.

Theories of Cave Development

This chapter has repeatedly stressed that karst is essentially a near-surface process. The maximum solution takes place immediately beneath or in the soil

---

[2]Caverns form in a variety of ways other than by solution. Lava tubes, tubes of headward "piping" by streams, and unusually wide overhanging ledges of any strong rock may be classed as caverns, or caves. Here the terms are meant to imply limestone solution features unless specifically stated otherwise.

profile. How, then, can caves form hundreds or even thousands of meters below the other landforms of a karst landscape?

Early theories of cave formation were vague about the chemistry of the process but implied that caves developed in the zone of aeration by downward-moving vadose water. Beginning with solution enlargement along bedding planes and joints, open conduits developed that were large enough to guide underground rivers and waterfalls. Great emphasis was placed on erosion by underground rivers. A surface river was necessary to accept the drainage from underground streams, and as the *vadose theory* developed, it became closely tied to theories of multiple erosion surfaces that had been rejuvenated by tectonic uplift. As the surface streams intrenched their valleys in each new cycle of erosion, the water table was lowered and caves were excavated in the newly drained vadose zone (Watson and White, 1985, p. 110).

The vadose theory of cave formation was accepted without serious question until 1930, when W. M. Davis wrote an extended evaluation of old theories and new observations and pointed out serious theoretical deficiencies. A fundamental problem was that many caverns contain extensive dripstone deposits, which are observed to grow. If vadose water is actively filling caverns with dripstone, how could it be the agent that formerly excavated them? Davis (1930, p. 549) revived interest in long-standing hydrogeologic theories of deep flow paths in the zone of saturation (Figure 8-5) by proposing that because the phreatic water circulated to great depths, caverns could be excavated below the water table, perhaps far below it. Later, when the regional surface drainage network had eroded downward into the limestone terrane, the caves would be in the zone of aeration, rivers could flow through them, and dripstone could form. This became known as the *two-cycle theory* of cavern formation.

Two years after Davis published his study, A. C. Swinnerton (1932) proposed that the maximum solution takes place at or just beneath the water table, especially in the narrow zone within which the water table fluctuates. He called his variant the *water-table hypothesis.* Swinnerton postulated that water at the water table has a variety of paths along which it can flow toward a spring, but most of the volume moves along the shortest path, which is along or just below the water table.

Swinnerton's water-table hypothesis explained observations that many caverns are nearly horizontal, even in deformed strata (Figure 8-17). The water-table hypothesis permitted the observed development of three-dimensional networks of cavern passages, with

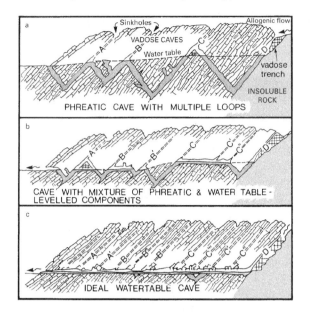

FIGURE 8-17. A variety of sites for cave development. Vadose caves (A, B, etc.) form in the zone of aeration by water moving downward by gravity, following the easiest structurally controlled paths to the water table. Phreatic caves form in the zone of saturation by water moving along hydraulic pressure gradients. (a) If permeability is uniform and restricted, a complex maze may form; (b) when one passage opens to a critical size, it becomes dominant and others may be abandoned. (c) If flow at the water table is strong, a water-table cave may develop (based on Ford and Williams, 1989, Figure 7-14).

blind passages that cannot be explained by through-flowing rivers or vadose percolation. It permitted, but did not require, second-cycle episodes of dripstone deposition and dissection by underground rivers as the regional water table and landscape were lowered. Few further theoretical studies were done until after World War II, except for an excellent extension of Davis's theory by Bretz (1942).

The present opinion is that caverns form at, above, and below the water table. Commonly, cavern formation is in progress at all three levels simultaneously (Figure 8-17). The favored cause for renewed aggressiveness of groundwater is the mechanism of mixing (Figure 8-4). Vadose water begins the task, by slowly enlarging discontinuities in the rock mass. As long as multiple narrow fractures provide equivalent permeability, Darcy-like flow with a defined water table is a reasonable regional model. Above the water table in the vadose zone, water moves essentially by gravity along any available discontinuities. Once in the phreatic zone, it moves toward an outlet along the hydraulic gradient. When the critical stage is reached

in the vadose zone where certain fractures become open enough for turbulent flow and abrasive sediment transport, or even sooner along fractures that provide the most direct path to the outlet (Groves and Howard, 1994), those channels dominate vadose circulation and a pipe model of flow is more appropriate. At the same time, solution at or beneath the water table may also be enlarging fractures along trends of most efficient hydraulic flow.

Cave passages show by their morphology whether they formed in the vadose or phreatic zone (White, 1988, Chapter 3; Ford and Williams, 1989, Chapter 7; Palmer, 1991, pp. 1–5). Those enlarged by flowing water typically have continuous downward gradients with floors entrenched below the level of the original solution-enlarged fractures. Phreatic passages are more likely to be tubular, demonstrating three-dimensional solution. Their gradients are discontinuous and irregular. From their morphology, it can be shown that vadose segments form in the upstream part of the hydrogeologic system, at the same time as phreatic segments are forming in deeper parts of the system. According to Palmer (1991, p. 2): "The question posed by early workers (for example, Davis, 1930; Swinnerton, 1932; Bretz, 1942) as to whether caves originate above, at, or below the water table is not pertinent."

Speleothem and Other Cave Deposits

Dripstone and **flowstone** are descriptive terms for cave deposits from water; the equivalent term **speleothem** is gaining popularity. In special conditions, speleothem may coat the surfaces of phreatic openings, but most is deposited in the vadose zone (Babić et al., 1996). A variety of delicate and beautiful shapes are assumed by speleothem (Figure 8-18). As limestone is dissolved by vadose water, the pH of the solution rises but in equilibrium with a relatively high CO_2 partial pressure. When the water enters an aerated cave, the excess dissolved CO_2 diffuses into the cave atmosphere, and calcite is deposited. The popular idea that cave deposits are the result of evaporation should be dispelled because wet caves have essentially 100 percent relative humidity, and evaporation is negligible. There is a possibility that biologic processes may cause some speleothem. Fungal threads have been observed at the tips of growing stalactites in Lehman Caves, Nevada (Went, 1969).

The principal speleothem shapes are *stalactites* and *helictites,* which grow downward or outward from cave roofs and walls, and *stalagmites,* which build upward from the floors. *Helictites* grow in a variety of directions because of slow accumulation, capillarity, and surface tension in addition to gravity. Ribbons of

FIGURE 8-18. Speleothem in the Postojna Caves, Slovenia (formerly northern Yugoslavia). This is one of the most popular tourist caves in the famous Dinaric karst region (photo: Yugoslav State Tourist Office).

flowstone may coat cave walls or form a curtain beneath a prominent joint-controlled seepage zone. In cave pools, a variety of delicate skeletal mineral "flowers" may form. Tufa rims and terraces line underground streams.

In addition to speleothem, a variety of clastic deposits may accumulate in caves. Caves that were occupied by animals and prehistoric people may have layers of fossiliferous or artifact-bearing litter on their floors. Rock falls or weathering spalls from cave roofs and walls sporadically add to the thickness of cave-floor deposits. Caves that have flowing streams contain a variety of alluvial deposits. Many limestone caves have thick layers of red clay that was formerly thought to be the insoluble residue of subsurface solution. However, the volume of dissolved limestone has been judged insufficient to account for the amount of red clay, so it must also include washed-in surface sediment (Jennings, 1985, p. 166). Windblown silt (loess), tephra, and organic detritus may accumulate in caves to produce stratigraphic records of great significance. Slack-water flood deposits that have backed up into caves from rivers record exceptional flood events.

Speleothem and clastic deposits in caves offer excellent opportunities to date cave evolution. Radiocarbon dating of organic-rich biogenic and archeologic strata can date occupation history and intervals of climate changes back to at least 50,000 years (Solecki, 1963, 1972). Oxygen-isotope variations in layered speleothem record climatic changes within the cave watershed, and uranium-series dating of the same layers of calcite gives a chronology back to 250,000 years or more (Winograd et al., 1992; Edwards and Gallup, 1993; Ludwig et al., 1993; Coplen et al., 1994). Classic sediments in several levels of Mammoth Cave show magnetic-polarity reversals that document the downcutting of the cave for at least 1 million years and possibly 2 million years (Schmidt, 1982; Palmer, 1989) related to Pleistocene glacial diversions of the Ohio River (p. 61) and possibly to epeirogenic uplift (p. 153). Multilevel caves in unglaciated southern Indiana record comparable development with the added complication that local valleys were deepened by glacial meltwater from the nearby ice margin (Johnson and Gomez, 1994). Similar studies in caves on the Appalachian Plateau in Tennessee and West Virginia provide a measure of valley incision during the last 900,000 years, and by extrapolation, dissection of the plateau for the last 4.6 million years (Sasowsky et al., 1995; Springer et al., 1997). Cave deposits are becoming of increased importance in documenting Pleistocene climatic changes on the continents, to correlate with the deep-sea, ice-core, and coral-reef records (Chapter 18).

REFERENCES

ADAMS, C. S., and SWINNERTON, A. C., 1937, Solubility of limestone: Am. Geophys. Un. Trans., pt. 2, pp. 504–508.

BABIĆ, L., LACKOVIĆ, D., and HORVATINČIĆ, N., 1996, Meteoric phreatic speleothems and the development of cave stratigraphy: An example from Tounj cave, Dinarides, Croatia: Quaternary Sci. Rev., v. 15, pp. 1013–1022.

BACKSHALL, D. G., BARNETT, J., and 7 others, 1979, Drowned dolines-the blue holes of the Pompey Reefs, Great Barrier Reef: BMR Jour. Australian Geol. & Geophys., v. 4, pp. 99–109.

BECK, B. F., ed. 1993, Applied karst geology: A.A. Balkema, Rotterdam, 295 pp.

————, and WILSON, W. L., eds., 1987, Karst hydrology: Engineering and environmental applications: A.A. Balkema, Rotterdam, 467 pp.

BÖGLI, A., 1980, Karst hydrology and physical speleology (transl. by J. C. Schmid): Springer-Verlag, Berlin, 284 pp.

BOSAK, P., FORD, D. C., and 2 others, eds., 1989, Paleokarst: A systematic and regional review: Elsevier Science Publishers, Amsterdam, 725 pp.

BRAHANA, J. V., THRAILKILL, J., and 2 others, 1988, Carbonate rocks, in Back, W., Rosenshein, J. S., and Seabor, P. R., eds., Hydrogeology: The geology of North America, v. O-2, Geol. Soc. America, Boulder, Colorado, pp. 333–352.

BRETZ, J. H., 1942, Vadose and phreatic features of limestone caverns: Jour. Geology, v. 50, pp. 675–811.

BROOK, G. A., FOLKOFF, M. E., and BOX, E. O., 1983, A world model of soil carbon dioxide: Earth Surface Processes and Landforms, v. 8, pp. 79–88.

BROOK, G. A., and FORD, D. C., 1978, Origin of labyrinth and tower karst and the climatic conditions necessary for their development: Nature, v. 275, pp. 493–496.

COPLEN, T. B., WINOGRAD, J. J., and 2 others, 1994, 500,000-year stable carbon isotope record from Devils Hole, Nevada: Science, v. 263, pp. 361–365.

CORBEL, J., 1959a, Vitesse de l'érosion: Zeitschr. für Geomorph., v. 3, pp. 1–28.

————, 1959b, Erosion en terrain calcaire: Annales de Geog., v. 68, pp. 97–120.

CVIJIĆ, J., 1893, Das Karstphänomen: Geographische Abhandlungen Wein, herausgegeben von A. Penck, v. 5, no. 3, pp. 218–329.

————, 1960, La géographie des terrains calcaires (French transl. by E. de Martonne): Serbe Acad. Sci. Arts Mon., v. 341, 212 pp.

DAVIES, W. E., and LEGRAND, H. E., 1972, Karst of the United States, in Herak, M., and Stringfield, V. T., eds., Karst. Important karst regions of the northern hemisphere. Elsevier Publishing Co., Amsterdam, pp. 467–505.

DAVIS, W. M., 1901, An excursion in Bosnia, Hercegovina, and Dalmatia: Geog. Soc. Philadelphia Bull., v. 3, pp. 21–50.

————, 1930, Origin of limestone caverns: Geol. Soc. America Bull., v. 41, pp. 475–428.

DEMEK, J., GAMS, I., and VAPTSAROV, I., 1984, Balkan Peninsula, in Embleton, C., ed., Geomorphology of Europe: John Wiley & Sons, Inc., New York, pp. 374–386.

DILL, R. F., 1977, The Blue Holes-geologically significant submerged sink holes and caves off British Honduras and Andros, Bahama Islands: Third Internat. Coral Reef Symp., Proc., v. 2, pp. 237–242.

DREVER, J. I., 1988, Geochemistry of natural waters, 2nd ed.: Prentice-Hall, Inc., Upper Saddle River, N.J., 437 pp.

DREYBRODT, W., 1988, Processes in karst systems: Physics, chemistry, and geology: Springer-Verlag, Berlin, 288 pp.

EDWARDS, R. L., and GALLUP, C. D., 1993, Dating of the Devils Hole calcite vein: Science, v. 259, p. 1626.

FOLK, R. L., ROBERTS, H. H., and MOORE, C. H., 1973, Black phytokarst from Hell, Cayman Islands (B.W.I.): Geol. Soc. America Bull., v. 84, pp. 2351–2360.

FOOSE, R. M., 1953, Groundwater behavior in the Hershey Valley, Pennsylvania: Geol. Soc. America Bull., v. 64, pp. 623–646.

FORD, D. C., 1971, Characteristics of limestone solution in the southern Rocky Mountains and Selkirk Mountains, Alberta and British Columbia: Canadian Jour. Earth Sci., v. 8, pp. 585–609.

————, PALMER, A. N., and WHITE, W. B., 1988, Landform development; karst, in Back, W., Rosenshein, J. S., and Seaber, P. R., eds., Hydrogeology: The geology of North America, v. O-2, Geol. Soc. America, Boulder, Colorado, pp. 401–412.

————, and WILLIAMS, P. W., 1989, Karst geomorphology and hydrology: Unwin Hyman Ltd., London, 601 pp.

FREEZE, R. A., and CHERRY, J. A., 1979, Groundwater: Prentice-Hall, Inc., Englewood Cliffs, N.J., 604 pp.

GARRELLS, R. M., and MACKENZIE, F. T., 1971, Evolution of sedimentary rocks. W. W. Norton & Company, Inc., New York, 397 pp.

GÈZE, B., and MANGIN, A., 1980, Le karst de Cuba: Revue de Géologie Dynamique et de Géographie Physique, v. 22, pp. 157–166.

GROVES, C. G., and HOWARD, A. D., 1994, Early development of karst systems 1. Preferential flow path enlargement under laminar flow: Water Resources Res., v. 30, pp. 2837–2846.

GUISTI, E. V., 1978, Hydrogeology of the karst of Puerto Rico: U.S. Geol. Survey Prof. Paper 1012, 68 pp.

GÜNAY, G., JOHNSON, A. I., and BACK, W., eds., 1993, Hydrogeological processes in karst terranes: IAHS Publication no. 207, International Association of Hydrological Sciences, Wallingford, UK, 412 pp.

HACK, J. T., and DURLOO, L. H., 1962, Geology of Luray Caverns, Virginia: Va. Div. Min. Resources Rept. Invest. 3, 43 pp.

HARMON, R. S., MITTERER, R. M., and 7 others, 1983, U-series and amino-acid racemization geochronology of Bermuda: Implications for eustatic sea-level fluctuation over the past 250,000 years: Palaeogeog., Palaeoclimatol., Palaeoecol., v. 44, pp. 41–70.

HERAK, M., 1972, Karst of Yugoslavia, in Herak, M., and Stringfield, V. T., eds., Karst: Important karst regions of the northern hemisphere: Elsevier Publishing Co., Amsterdam, pp. 25–83.

HESS, J. W., WELLS, S. G., and 2 others, 1989, Hydrogeology of the south-central Kentucky karst, in, White, W. B., and White, E. L., eds., Karst hydrology: Concepts from the Mammoth Cave area: Van Nostrand Reinhold, New York, pp. 15–63.

HUBBERT, M. K., 1940, Theory of ground-water motion: Jour. Geology, v. 48, pp. 785–944.

HUGHES, T. H., MEMON, B. A., and LaMOREAUX, P. E., 1994, Landfills in karst terrains: Bull., Assoc. Engineering Geologists, v. 31, pp. 203–208.

JAMES, N. P., and CHOQUETTE, P. W., eds., 1988, Paleokarst: Springer-Verlag, New York Inc., New York, 416 pp.

JENNINGS, J. N., 1985, Karst Geomorphology, 2nd ed.: Basil Blackwell, Inc., New York, 293 pp.

JOHNSON, P. A., and GOMEZ, B., 1994, Cave levels and cave development in the Mitchell Plain following base-level lowering: Earth Surface Processes and Landforms, v. 19, pp. 517–524.

JOHNSTON, R. H., 1993, Historical development of concepts of regional groundwater flow in the Floridan aquifer system, southeastern United States, in Günay, G., Johnson, A. I., and Back, W., eds., Hydrogeological processes in karst terranes: IAHS Publication no. 207, International Association of Hydrological Sciences, Wallingford, UK, pp. 351–357.

LEHMANN, H., 1936, Morphologische studien auf Java: J. Engelhorn, Stuttgart, 114 pp.

———, 1960, International atlas of karst phenomena, Sheets 1, 1a, and 1b, "Sierra de los Organos, Cuba": Zeitschr. für Geomorph., Supp. no. 2, plates in pocket.

LUDWIG, K. R., SIMMONS, K. R., and 3 others, 1993, Dating of the Devils Hole calcite vein: Response: Science, v. 259, pp. 1626–1627.

LUNDBERG, J., and FORD, D. C., 1994, Late Pleistocene sea level change in the Bahamas from mass spectrometric U-series dating of submerged speleothem: Quaternary Sci. Rev., v. 13, pp. 1–14.

MEYERHOFF, H. A., 1938, Texture of karst topography in Cuba and Puerto Rico: Jour. Geomorphol., v. 1, pp. 279–295.

MONROE, W. H., 1966, Formation of tropical karst topography by limestone solution and reprecipitation: Caribbean Jour. Sci., v. 6, pp. 1–7.

———, 1970, Glossary of karst terminology: U.S. Geol. Survey Water-Supply Paper 1899-K, 26 pp.

———, 1976, Karst landforms of Puerto Rico: U.S. Geol. Survey Prof Paper 899, 69 pp.

MYLROIE, J. E., 1991, Cave development in the glaciated Appalachian karst of New York: Surface-coupled or saline-freshwater mixing hydrology?, in Kastning, E. H., and Kastning, K. M. eds., Appalachian karst: Appalachian karst Sympos. Proc., Radford, Va., Mar. 23–26, 1994, pp. 85–90.

PALMER, A. N., 1989, Geomorphic history of the Mammoth Cave system, in White, W. B., and White, E. L., eds., Karst hydrology: Concepts from the Mammoth Cave area: Van Nostrand Reinhold, New York, pp. 317–337.

———, 1991, Origin and morphology of limestone caves: Geol. Soc. America Bull., v. 103, pp, 1–21.

———, and PALMER, M. V., 1995, The Kaskaskia paleokarst of the northern Rocky Mountains and Black Hills, northwestern U.S.A.: Carbonates and Evaporites, v. 10, pp. 148–160.

PANNO, S. V., and BOURCIER, W. L., 1990, Glaciation and saline-freshwater mixing as a possible cause of cave formation in the eastern midcontinent region of the United States: A conceptual model: Geology, v. 18, pp. 769–772.

PENCK, A., 1894, Morphologie der Erdoberfläche: J. Engelhorn, Stuttgart, v. 1, 471 pp.; v. 2, 696 pp.

PLUHAR, A., and FORD, D. C., 1970, Dolomite karren of the Niagara escarpment, Ontario, Canada: Zeitschr. für Geomorph., v. 14, pp. 392–410.

POWER, F. D., 1925, Phosphate deposits of the Pacific: Econ. Geology, v. 20, pp. 266–281.

RICHARDS, D. A., SMART, P. L., and EDWARDS, R. L., 1994, Maximum sea levels for the last glacial period from U-series ages of submerged speleothems: Nature, v. 367, pp. 357–360.

ROGLIĆ, J., 1972, Historical review of morphologic concepts, in Herak, M., and Stringfield, V. T., eds., Karst: Important karst regions of the northern hemisphere: Elsevier Publishing Co., Amsterdam, pp. 1–18.

RUBIN, P. A., 1991, Modification of preglacial caves by glacial meltwater invasion in east-central New York, in Kastning, E. H., and Kastning, K. M. eds., Appalachian karst: Appalachian karst Sympos. Proc., Radford, Va., Mar. 23–26, 1994, pp. 91–99.

SANDERS, E. M., 1921, Cycle of erosion in a karst region (after Cvijić): Geog. Rev., v. 11, pp. 593–604.

SAWKINS, J. G., 1869, Reports on the geology of Jamaica: Great Britain Geol. Survey Mem., Longmans, Green, and Co., London, 339 pp.

SASOWSKY, I. D., WHITE, W. B., and SCHMIDT, V. A., 1995, Determination of stream-incision rate in the Appalachian plateaus by using cave-sediment magnetostratigraphy: Geology, v. 23, pp. 415–418.

SCHMIDT, V. A., 1982, Magnetostratigraphy of sediments in Mammoth Cave, Kentucky: Science, v. 217, pp. 827–829.

SHACHAK, M., JONES, C. G., and GRANOT, Y., 1987, Herbivory in rocks and the weathering of a desert: Science, v. 236, pp. 1098–1099.

SMART, C. C., and FORD, D. C., 1986, Structure and function of a conduit aquifer: Canadian Jour. Earth Sci., v. 23, pp. 919–929.

SOLECKI, R. S., 1963, Prehistory in Shanidar Valley, northern Iraq: Science, v. 139, pp. 179–193.

———, 1972, Shanidar: The humanity of Neanderthal man: Allen Lane, London, 322 pp.

SPRINGER, G. S., KITE, J. S., and SCHMIDT, V. A., 1997, Cave sedimentation, genesis, and erosional history in the Cheat River Canyon, West Virginia: Geol. Soc. America Bull., v. 109, pp. 524–532.

SWEETING, M. M., 1972, Karst landforms. Columbia Univ. Press, New York, 362 pp.

———, 1993, Reflections on the development of karst geomorphology in Europe and a comparison with its development in China: Zeitschr. für Geomorph., Supp. no. 93, pp. 127–136.

SWINNERTON, A. C., 1932, Origin of limestone caverns: Geol. Soc. America Bull., v. 43, pp. 663–693.

THORNBURY, W. D., 1969, Principles of geomorphology, 2nd ed.: John Wiley & Sons, Inc., New York, 594 pp.

THORP, J., 1934, Asymmetry of the "Pepino Hills" of Puerto Rico in relation to the trade winds: Jour. Geology, v. 42, pp. 537–545.

THRAILKILL, J., 1968, Chemical and hydrologic factors in the excavation of limestone caves: Geol. Soc. America Bull., v. 79, pp. 19–45.

TWIDALE, C. R., 1987, Sinkholes (dolines) in lateritised sediments, western Sturt Plateau, Northern Territory, Australia: Geomorphology, v. 1, pp. 33–52.

VERSEY, H. R., 1972, Karst of Jamaica, in Herak, M., and Stringfield, V. T., eds., Karst: Important karst regions of the northern hemisphere: Elsevier Publishing Co., Amsterdam, pp. 445–466.

VILES, H. A., 1987, Blue-green algae and terrestrial limestone weathering on Aldabra Atoll: An S.E.M. and light microscope study: Earth Surface Processes and Landforms, v. 12, pp. 319–330.

WATSON, R. A., and WHITE, W. B., 1985, History of American theories of cave origin, in Drake, E. T. and Jordan, W. M., eds., Geologists and ideas: A history of North American geology: Geol. Soc. America Centennial Special Volume 1, pp. 109–123.

WENT, F. W., 1969, Fungi associated with stalactite growth: Science, v. 166, pp. 385–386.

WHITE, W. B., 1988, Geomorphology and hydrology of karst terrains: Oxford Univ. Press, New York, 464 pp.

WILFORD, G. E., and WALL, J. R. D., 1965, Karst topography in Sarawak: Jour. Tropical Geog., v. 21, pp. 44–70.

WILLIAMS, P. W., 1972, Morphometric analysis of polygonal karst in New Guinea: Geol. Soc. America Bull., v. 83, pp. 761–796.

———, 1987, Geomorphic inheritance and the development of tower karst: Earth Surface Processes and Landforms, v. 12, pp. 453–465.

———, 1993, Climatological and geological factors controlling the development of polygonal karst: Zeitschr. für Geomorph., Supp. no. 93, pp. 159–173.

WINOGRAD, I. J., COPLEN, T. B., and 6 others, 1992, Continuous 500,000-year climate record from vein calcite in Devils Hole, Nevada: Science, v. 258, pp. 255–260.

ZHU, XUEWEN, 1988, Guilin karst: Shanghai Scientific and Technical Publishers, Shanghai, P.R. China, 188 pp.

Chapter 9

Mass Wasting and Hillslopes

On the scale of individual landforms, where rheid flow is not a significant factor of deformation, an unweathered rock mass resists the forces that act on it. Only when the inherent discontinuities in the rock mass have been opened by mechanical stresses or have reacted with water and the atmosphere can the fragments be mobilized. Weathering, then, is a necessary precondition for the movement of rock fragments downslope and for the development of the slopes. A variety of forces at the earth's surface can move loose rock particles. Whenever particles move, ubiquitous gravity adds a downward component to the motions produced by other forces so that they move preferentially downhill.

MASS WASTING, GRAVITY, AND FRICTION

The collective term for all gravitational or downslope movements of weathered rock debris is **mass wasting.** The term implies that gravity is the sole important force and that no transporting medium such as wind, flowing water, ice, or molten lava is involved.

Although flowing water is excluded from the process by definition, water nevertheless plays an important role in mass wasting by oversteepening slopes through surface erosion at their bases, by adding weight to the rock mass that might exceed a threshold of failure, and by generating seepage pressures through groundwater flow. Sometimes, this third role of water is incorrectly referred to as "lubrication" of rock or weathered rock debris on a slope. In fact, water has a complex effect on the sliding friction between surfaces of rock-forming minerals, increasing the friction for quartz, feldspar, and calcite, but decreasing it for the sheetlike mica minerals (Mitchell, 1976, pp. 310–313). It is the very important seepage pressures generated by groundwater flow that have such great influence on slope instability and mass wasting.

Angle of Friction

A series of simple experiments illustrates the relative importance of gravity and pore-fluid pressure in determining downslope movement. A cone of noncohesive dry sand can be built by pouring sand onto a horizontal surface through a funnel. The cone will have a surface *angle of repose* of approximately 35°, depending on the size, shape, surface roughness, and other properties of the sand grains. The tangent of the angle of repose of dry granular materials is slightly greater than, but approximately equal to, the coefficient of sliding friction of the material, or its *mass friction* (ϕ). The balance of forces between gravity and friction determines both. The coefficient of sliding friction of a particle is equal to the ratio between the downslope component of weight and the component of weight acting perpendicular to the slope when the particle is moving (Figure 9-1a).

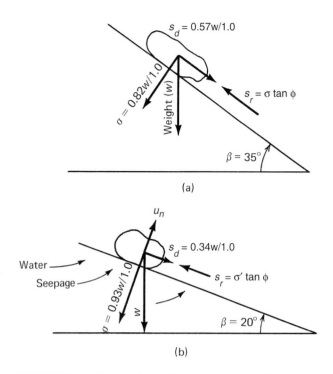

FIGURE 9-1. Resolution of forces on a cohesionless block of unit basal area resting on a slope. (a) σ (stress component perpendicular to surface) = $w \cos \beta$; s_d (stress component parallel to surface) = $w \sin \beta$. Shear strength (s_r) resists motion. When $\beta = \phi$ (angle of repose), $s_d = s_r$, and the block is ready to move. (b) As above, but with u_n (component of seepage pressure normal to the surface) counteracting σ ($\sigma' = \sigma - u$) and reducing s_r. Particle will move when $\beta << \phi$.

If in a second experiment, the same sand is deposited in a saturated condition under water, the angle of repose of the cone will be virtually identical to the angle of repose of the dry sand cone. In neither example will there be any fluid flow through the sand; consequently, neither cone will be subjected to seepage pressures tending to decrease its stability.

A third experiment demonstrates the influence of water flow. If water is poured gently onto the top of a cone of dry sand and allowed to seep downward and outward (the water must infiltrate and not erode the surface), failure occurs as soon as the water reaches the slope face. It can be demonstrated by experiment, or by adding a vector representing seepage pressure to a simple diagram (Figure 9-1b), that even the flow of rainwater runoff in the surface soil on a slope reduces the stable angle of repose of the soil to less than half the dry value.

Cohesion

Fine-grained sediments, especially clay-size[1] particles, have strong *cohesive* surface forces between them, determined by electrostatic charges on grain surfaces that immobilize the water films between them. These cohesive forces act in addition to the mass friction to determine the angle at which slope failure occurs. Referring to the first experiment described previously, if the noncohesive dry sand is moistened but not saturated, it can be molded and shaped to a vertical slope, as any child on a beach will verify. Strength in that sense depends on the tension in the capillary water between sand grains.

Pore water can either add or subtract from the mass strength. If voids in the mass are only partly filled with water, the capillary surface tension of the water films tries to draw the mineral grains closer, and the pore pressure is described as negative. When the void space is fully saturated, however, cohesion is lost because there is no surface tension, and part of the overburden weight is transferred to the water from the individual grains. Pore pressure is then positive, and has a buoyancy effect, reducing the strength of the mass.

In 1776, C. A. Coulomb first summarized the previous discussion as an equation, in which

$$s = c + \sigma \tan \varphi$$

where s is the shear strength (in stress, or force per unit area, units), c is cohesion (also in stress units), σ is the stress component perpendicular to the surface, and ϕ is the mass angle of friction. This important equation, which relates the shear strength of a soil mass just at the threshold of failure to the stresses acting on it, is now called the *Mohr-Coulomb failure criterion* (Holtz and Kovacs, 1981, p. 453; Nash, 1987).

An equivalent form of the Mohr-Coulomb equation takes into account the excess pressures in the pore fluids, including the seepage pressures. This concept is called the *principle of effective stress*. It is common practice to add a superscript to the symbols in the Mohr-

---

[1]"Clay" is the name used for a group of residual minerals (Chapter 7) as well as for the fine-grained fraction of sediment regardless of its mineralogy. The distinction between "clay minerals" and "clay-size sediment" should always be made clear. The geologists' classification defines clay-size particles as less than 4 μm mean diameter; the engineers' classification defines clay particles as all those less than 2 μm in diameter.

Coulomb equation when working with effective stresses:

$$s = c' + \sigma' \tan \varphi'$$

where c' is the effective stress cohesion, $\sigma' = (\sigma - u)$ is the effective stress component normal to the surface, as reduced by positive fluid pressure u, and ϕ' is the effective stress angle of friction. In noncohesive sand, gravel, or large talus blocks, $c' = 0$, and shear stress is proportional to the tangent of the angle of friction although the relationship is complicated by shape factors such as the tendency of large angular blocks to stack and support steeper-than-predicted angles. Fine-grained silt and clay make cohesive soils for which c' can be significant. As cohesion increases, so does shear strength. Many moist clays hold vertical faces of modest height without failure, which noncohesive materials can never do.

In silt- and clay-rich sediments, the water content becomes a major parameter of deformational behavior. *Water content* is usually expressed as a percentage, the ratio of the weight of water in the sample divided by the weight of the dried sediment. Some clay minerals can absorb water equal to several hundred percent of their dry weight. Because various clay minerals and various mixtures of silt- and clay-size sediment behave differently at comparable water contents, a set of arbitrary but definable indices has been established to describe the behavior of wet fine-grained sediments. The *liquid limit* is based on the rate at which a standard conical weight sinks into a paste of water and sediment. It defines the water content at which the sediment passes from a plastic to a liquid state. The liquid limit is a good measure of cohesion because, at the liquid limit, cohesion is essentially zero. Clay-rich soils typically have liquid limits in the range of 40 to 60 percent.

The *plastic limit* is another measure of water content, in this case the water content at which the mass changes from a solid to a plastic state. It is defined as the water content at which a sample can be rolled into a 3-mm diameter "soda straw" rod without crumbling. It is a good measure of the clay content of the sample because silty or sandy soils with little clay content cannot be rolled into such thin rods. The *plasticity index* is the numerical difference between the liquid limit and the plastic limit. If it is low (~5 percent) the sediment will change readily from a plastic semisolid to a liquid. A plasticity index of 20 percent or more implies that the material can accept much water before it loses shear strength and liquefies.

Engineers use an alternative form of the Mohr-Coulomb equation when studying short-term stability problems such as the initial cutting of a clay slope (Holtz and Kovacs, 1981, p. 556). In this case, complex fluid-pressure distributions can be developed within the soil pores during loading or unloading. These fluid pressures can be difficult to calculate or predict, but they can be simulated fairly easily in the laboratory with *undrained strength* tests. The strength measured in these laboratory test simulations is the *undrained shear strength* (s_u). For this alternative form of the Mohr-Coulomb equation, s_u replaces s and includes c' σ', ϕ', as well as the additional developed fluid pressures in the simulation. Although these matters are of importance to geomorphology, their technical details are the subject of the related field of geotechnical engineering and are best studied in appropriate engineering texts (for example, Mitchell, 1976; Holtz and Kovacs, 1981; Haneberg and Anderson, 1995).

Rates of Downslope Movement

The geometry of Figure 9-1 demonstrates that the shear stress acting parallel to a slope, which causes a particle to move downslope, is proportional to the sine of the slope angle ($S_d = w \sin \beta$). Intuitively, we would expect that the rate of downslope movement of small, loose rock fragments responding to that stress would also be proportional to the sine of the slope angle (Strahler, 1956, p. 577). Most data support this inference (Figure 9-2). Schumm (1967) measured the rate of migration of small plates of sandstone down a shale hillslope in semiarid western Colorado by marking 110 pieces with paint and measuring their distance from reference stakes driven into bedrock. Over a period of seven years, the rate of downslope movement ranged from 2 to 3 mm/yr to almost 70 mm/yr in proportion to the sine of the slope angle, which ranged from 3° to 40°. Rates of periglacial gelifluction (p. 176; see also Chapter 14) over perennially frozen ground in eastern Greenland are much higher than the sliding rate of dry rock chips on a semiarid talus (Figure 9-2), and the modes of movement are also very different; nevertheless, the rates of surface movement on both rapidly flowing solifluction lobes and "undifferentiated wet slopes" in Greenland are proportional to the sine of the slope angles.

Note that for angles of less than 10°, the sine and tangent functions differ by only a few percent, and Figure 9-2 would not look much different if the movement rates were plotted against the tangent of slope angle as an approximation of the mass friction ($\tan \phi$).

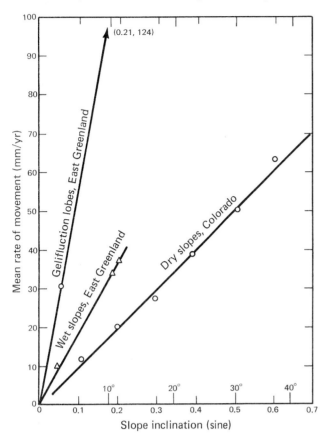

FIGURE 9-2. Rate of downslope movement is proportional to the sine of the slope inclination under a wide range of conditions (data from Schumm, 1967; Washburn, 1967, Figure 43).

Whichever function is chosen, the conclusion is obvious that downslope movement is slowest on gentle slopes and increases rapidly with increase in slope.

Slope angles on unvegetated, noncohesive weathered debris such as grus or loose rock rubble are usually between 25° and 40°, somewhat less than the measured or calculated angles of repose for the debris (Selby, 1993, p. 354). Natural slopes steeper than 40° are usually nearby barren of rock debris, and in British terminology are classed as *cliffs*. On vertical slopes, a loosened rock particle drops under the free acceleration of gravity, its rate of fall limited only by air resistance.

DESCRIPTIVE CLASSIFICATION OF MASS WASTING

The explanation of mass wasting is so simple and obvious that a genetic, or explanatory-descriptive, classification of the processes and forms is not neces-

sary. The structure is a weathered or weakened rock mass; the primary process is gravitational instability; time ranges from seconds to millenniums. Instead, a simplified descriptive classification is generally adopted, which emphasizes certain features as diagnostic and recognizes other features as parts of continuous series. No combination of categories has been devised to classify totally the many landforms and processes of mass wasting, nor is such a classification likely or desirable. The essence of mass-wasting is that it bridges the gap between *weathering,* which is defined as occurring in place, and *erosion,* which requires as one element of its definition transport by some agent or medium. Mass wasting combines elements of weathering and erosion, yet separates those two large categories of processes. In every sense of cause and effect, mass wasting is a category of transitional phenomena (Turner and Schuster, 1996).

There is broad international agreement that the definitive criteria of mass wasting should include (1) the type of material in motion, including its coherence and dimensions, and (2) the type and rate of movement, whether falling, toppling, sliding, spreading, or flowing (Cruden and Varnes, 1996, p. 37). An abbreviated classification (Table 9-1, Figure 9-3) of **landslides** (simply defined as perceptible movement of rock, debris, or earth down a slope) is the latest version of multiple published revisions by teams of engineering specialists (Cruden and Varnes, 1996). By simple definitions useful to geologists and engineers, the material in a landslide is either *rock,* a hard firm mass that was intact and in place before movement began, or *soil,* an aggregate of solid rock and mineral particles with interstitial liquids and gases that either had been previously transported or had weathered in place. Soil is further subdivided into *earth,* with 80 percent or more of the particles smaller than 2 mm (sand, silt, and clay), and *debris,* coarser material in which 20 to 80 percent of the particles are larger than 2 mm (Cruden and Varnes, 1996, pp. 52–53). The name of any landslide is then a combination of a material name and a name for the type of movement (Table 9-1). Other modifying terms can be added to describe the state of activity of the landslide (active, dormant, relict, and so on), its water content, the rate of movement, and other attributes. No uniform spelling or definitions of the various terms have been established. Some authors prefer single words such as "mudflow" or "earthflow," whereas others use two words, as in "debris flow." Still others use various combinations of terms not listed in Table 9-1 such as "creep," "debris avalanche," and "slump-earthflow." In the following paragraphs, we review the better-known terms, generally progressing from slow to rapid movements.

Table 9-1

Abbreviated classification of slope movements (landslides)

	TYPE OF MATERIAL		
TYPE OF MOVEMENT	*BEDROCK*	*ENGINEERING SOILS PREDOMINANTLY COARSE*	*PREDOMINANTLY FINE*
Fall	Rock fall	Debris fall	Earth fall
Topple	Rock topple	Debris topple	Earth topple
Slide	Rock slide	Debris slide	Earth slide
Rotational (Slump)			
Translational			
Spread	Rock spread	Debris spread	Earth spread
Flow	Rock flow	Debris flow	Earth flow

Based on Cruden and Varnes, 1996, Table 3-1.

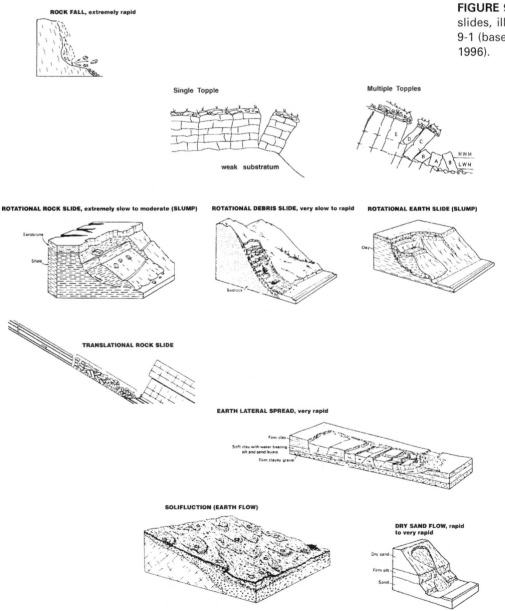

FIGURE 9-3. Main types of landslides, illustrated. Compare Table 9-1 (based on Cruden and Varnes, 1996).

Creep

Creep is barely perceptible and nonaccelerating downslope movement. It is most common, in fact, ever-present, in regolith on slopes, but the term is also applied to the slow movement of otherwise unweathered joint blocks. The material moving downslope is called **colluvium,** whether it be derived from *in situ* weathering or from transported sediment such as glacial drift.

Particle Creep. Individual surface pebbles or clods of soil are constantly shifting because of wetting and drying, heating and cooling, and freezing and thawing. In frost climates, shale fragments and pebbles are lifted several centimeters above the surrounding soil surface by ice crystals (p. 302). Volumetric expansion by any cause displaces particles toward the free face of the expanding mass, or perpendicular to the ground surface. On contraction, however, the particle is not pulled back into its former position but settles downslope with the gravitational component. Only rarely are the cohesive forces of soil and water strong enough to draw particles back into the ground during contraction without some net downslope motion. The result is *particle creep* (Figures 9-2 and 9-4). Wind, raindrop impact, and disturbance by plants and animals complicate any simple isolation of particle creep as a discrete process.

Soil Creep. Defined as imperceptible except over a period of years, soil creep is more difficult to measure than particle creep. In some climates, it is strongly seasonal; in others, it is intermittent (Selby, 1993, pp. 255–258). In theory, soil creep should be essentially laminar, with each layer of soil carried downhill by the motion of the layer beneath it. The effect should be cumulative, with the maximum rate at the surface exponentially decreasing to zero with depth (Kirkby, 1967). As a result, soil creep should not shear across an immobile substrate at depth and should not abrade a buried surface. Measured profiles of creeping soil show many variations, however. *Depth creep* is pluglike, with the entire surface layer, perhaps bound by plant roots, moving over a narrow shear zone at depth (Selby, 1993, Figure 13-2).

A variety of surface phenomena are usually attributed to creep, such as downslope tilted trees, walls, and utility poles (Figure 9-5). Tree trunks on steep slopes are commonly concave uphill although their curvature is more likely because of initial light-seeking growth perpendicular to the slope and later more nearly vertical growth as saplings outgrow shading ground cover (Phipps, 1974; Ishii and Higashi, 1997). Strata in unweathered bedrock thin abruptly as they enter the surface layer that is creeping, and instead of breaking the surface as outcrops, they are drawn out in a downhill direction (Figure 9-6). Creep is recognized as both a problem and an aid in prospecting. If fragments of the desired coal or ore mineral are found in the "float," or colluvium, the buried layer must be in the subsoil at some point farther upslope.

Grass may be able to retain a continuous sod cover over an area of soil creep because movement is typically only a few millimeters per year. The highest rates of creep are measured on steep slopes under

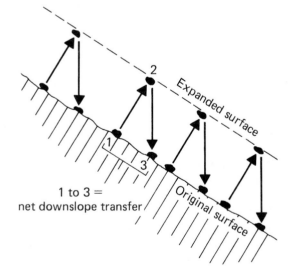

FIGURE 9-4. Path of a surface particle during expansion and contraction of soil on a slope. See Figure 9-7 for more complex motions.

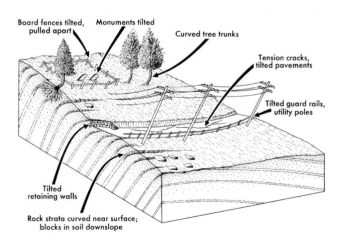

FIGURE 9-5. Common effects of creep. Not all will be present in one place.

tropical rain forests, where the entire surface layer of interlocking roots moves downhill at rates of up to 4 mm/yr. Anomalously rapid depth creep was measured in a Puerto Rico rainforest, just below the root zone (Lewis, 1974). Large eucalyptus trees in Rwanda were carefully measured and demonstrated to be "floating" in a thick, creeping regolith (Selby, 1993, p. 257). Yet observations in Poland failed to detect any creep at all over a 12-year interval under old forests, even on 30° slopes (Jahn, 1989).

A complicating factor in measuring soil creep under forest cover is the potentially greater effect of tree uprooting (Norman et al., 1995). When trees fall on slopes, they usually fall downhill. The pit and mound topography that results may persist for as long as 2000 years, disrupting soil profile development, bringing relatively unweathered rock to the surface, and accelerating downslope movement. The process has been claimed to work at rates comparable to, or even three orders of magnitude faster than, soil creep (Schaetzl et al., 1990, p. 285). An important secondary effect of tree uprooting is the resulting microrelief, which ponds water on hillslopes and promotes infiltration, perhaps leading to other more rapid forms of mass wasting. Even the impact of falling trees can trigger spontaneous debris flow on a rain-saturated hillside.

Soil creep in deserts is not the same as in humid climates, primarily because the surface layers are not laced together with a turf of interconnecting roots. Further, desert soils are likely to be broken into prisms of weakly cemented calcrete (p. 142) by vertical desiccation cracks so that creeping masses are blocks rather than granular layers. Because of the coarse, angular colluvium, hillsides are likely to be steeper than those of humid regions.

Soil creep tends to maintain smooth, ungullied hillsides by filling the small channels eroded by seeping groundwater or surface rill wash (199). It strongly depends on soil moisture conditions, and creep rates fluctuate seasonally, so creep is as much climate controlled as it is gravity controlled. Creep is responsible for some of the net downslope movement even on terrains where slide and flow are more obvious, by recharging the sediment supply for areas of repeated, but intermittent, slides and flows (Sidle et al., 1985, p. 14).

Rock Creep and Topple.   To geophysicists and engineers, rock creep connotes slow, permanent internal rock deformation (strain) under low stress. As a process of mass wasting, rock creep is less precisely used to describe slow movement of large coherent rock masses over a discrete basal surface. Most often, rock creep is observed on a cliff face, where a massive rock such as sandstone overlies shale. Large joint-bounded sheets or blocks, slowly creep or lean outward until they become unstable and topple over or slide downhill (de Frietas and Watters, 1973).

The best-documented history of toppling failure that resulted from rock creep (Schumm and Chorley, 1964) is "the fall of Threatening Rock" in 1941 in Chaco Canyon National Monument, New Mexico. For five years prior to the fall, the gap between the cliff face and a free-standing monolith of sandstone 46 m long, 30 m high, and 9 m thick had been measured at monthly intervals. For three years, the gap widened at approximately 25 mm/yr. During the next two years, the rate doubled, culminating in a violent rock fall that partly destroyed an ancient pueblo at the base of the cliff. The annual movement was concentrated during the wet winter season and was greater during years of more precipitation. Movement had been in progress or at least recognized as possible for at least nine centuries; this was known because the prehistoric occupants of Pueblo Bonito had built an earth and masonry terrace at the base of Threatening Rock and wedged logs beneath its base about A.D. 1000. By extrapolation, Schumm and Chorley (1964, p. 1052) estimated the time required for each foot (30.5 cm) of movement away from the cliff as follows:

FIGURE 9-6.   Rock creep in roadcut along Yarra Boulevard, Melbourne, Australia (photo: Renate R. Hodgson).

Movement	Years Required
First foot	1600
Second foot	650
Third foot	200
Fourth foot	60
Fifth foot	8
Sixth foot	0.2

Rock creep illustrates the importance of pressure-release joints, or sheeting (p. 121), parallel to cliff faces or gorge walls in massive rocks. Rapid stream erosion can unbalance the stress distribution in rocks so that elastic expansion is outward toward the valley axis (Ferguson, 1967; Sasowsky and White, 1994). When open joints have formed, it is only a matter of time until weathering destroys the support and a large mass of rock collapses from the cliff. Robinson (1970, p. 2802) demonstrated that the crushing strength of massive sandstone is sufficient to support a free-standing joint column several hundred meters high on a cliff face and that failure occurs primarily by loss of basal support. Even though these events seem random and catastrophic, and may occur only a few times in a thousand years, they are part of the "normal" process of valley-side evolution.

Flow

Incoherent rock debris may be mobilized sufficiently so that it flows like a viscous fluid. The criteria for defining flow are evidence of internal turbulence and either discrete boundaries or narrow marginal zones of shear. In both criteria, flow is distinct from creep. Subdivision of flow can be based either on the kind of material, the degree of saturation, or the speed of advance. The factors of sediment concentration in water and the flow rate can be combined in a rheologic classification, ranging from fluid flow with low sediment concentrations, through plastic flow with a finite yield stress but essentially fluid behavior above that stress level, to granular flows that are dry or nearly dry. Flow rates for all rheologic behaviors can range through many orders of magnitude (Pierson and Costa, 1987). Flows typically move as lobes or tongues that follow the preexisting topography. Some are known to erode channels. The transitional nature of mass-wasting terminology is nowhere better illustrated than in attempts to distinguish between a water-saturated mudflow and a mud-laden stream. Thus far, attempts at rheologic quantification use only approximate ranges of sediment content (Pierson and Costa, 1987, p. 6).

Solifluction. If soil is saturated with water, the soggy mass may flow downhill a few millimeters or a few centimeters per day or per year. This type of movement is called **solifluction** (literally, "soil flow"). The original definition of solifluction excluded any connotation of climate but was included in a discussion of cold-climate landforms (Andersson, 1906;

Washburn, 1967, pp. 10–14) and has frequently been misdefined as a cold-climate process. To clarify the confused terminology, Washburn (1967, p. 14) urged adoption of **gelifluction** for solifluction associated with frozen ground. That usage is adopted here. Solifluction is not restricted to frozen ground. It is a form of mass wasting common wherever water cannot escape from a saturated surface layer of soil by percolation into deeper levels. A clay hardpan in a soil or an impermeable bedrock layer can promote solifluction as effectively as a frozen substratum. Under tropical rain forests, steep slopes in deeply weathered rock are scarred by large debris slides, and lesser slopes are masses of solifluction lobes.

Gelifluction and Frost Creep. Long before the term *periglacial* (Chapter 14) had been proposed, European geologists were aware that thick sheets of poorly sorted angular debris mantled their landscape beyond the limits of glaciation. The material was given quaint colloquial terms such as "head," or "coombe rock," and had been hypothesized to be the result of cold-climate, but not glacial, processes. During an expedition to the Falkland Islands, J. G. Andersson observed fluid sheets of debris moving down gentle slopes over perennially frozen ground and producing deposits similar to the "head" of European authors. He named the process solifluction (later modified to **gelifluction**). The process is generally accepted as being responsible for the deposits of "head," and in fact, massive deposits of relict gelifluction debris are accepted as evidence of a former periglacial climate (Cotton and Te Punga, 1955).

During the brief high-latitude summer thaw, a surface active layer a meter or so in thickness, composed of tundra peat, rock rubble, and other weathered debris, may flow down slopes of almost negligible gradient because meltwater saturates the active layer but cannot penetrate the frozen ground beneath. The mass of saturated gelifluction debris may flow with a kind of rolling motion like the endless tread of a tracked vehicle. Arcuate ridges and troughs mark the toe, or lower part of the mass. The rate of downhill motion in gelifluction lobes in East Greenland is several times more rapid than the nonlobate creep on wet slopes of similar gradient (Figure 9-2) (Washburn, 1967, pp. 94–95). The entire movement takes place during the brief summer season when net movement may be several millimeters per day on slopes of less than $10°$ (Figure 14-6).

Washburn accurately surveyed lines of conical targets driven into the soil across Greenland hillsides, and by repeated resurveys that extended over a period of nine years, he was able to separate the components of gelifluction and **frost creep** (soil creep due to the ratchetlike motion of particles during alternating freeze-thaw cycles) (Figure 9-7). His periodic resurveys proved that frost creep was quantitatively slightly more important on most slopes than gelifluction. An unexpectedly large component of *retrograde motion* due to cohesion caused his targets to move uphill during the summer thaw season, with respect to the vertical plane through the targets at the time of maximum frost heave. The retrograde motion partly or entirely cancels downslope movement by gelifluction. Points P_1 to P_4 in Figure 9-7 trace a typical path of a soil particle on a Greenland hillside during an annual cycle. The diagram should be examined step by step, from the "jump" (P_1-P_2) during frost heave, through the downhill movement (P_2-P_3) during thaw and gelifluction, to the interaction between retrograde motion and gelifluction (P_3-P_4) that determines the net annual downslope motion. In no instance was the net motion uphill. Although retrograde motion in some years largely negated the role of gelifluction and frost creep, it cannot exceed them. By freezing insulated trays of saturated silt tilted at various low angles, Higashi and Corte (1971) were able to reproduce in a laboratory the various combinations of frost creep, gelifluction, and retrograde motion reported by Washburn.

Debris Flow, Earthflow, Mudflow. These three terms are applied to types of mass wasting very similar to solifluction. They are somewhat more rapid, and they commonly flow along valleys, whereas solifluction sheets or lobes cover an entire hillside with moving debris. Some authors (Johnson and Rodine, 1984, p. 257) lump all three terms under the general term *debris flow* to stress the similarity of their mechanisms, despite differences in their flow rate or percentage of fine particles. The transition between a debris flow and a muddy river is illustrated by measurements of densities of 2.0 to 2.4 g/cm³ in debris flows (dense enough to buoyantly support large blocks), densities of 1.3 to 1.8 g/cm³ in more fluid "hyperconcentrated" flows, and densities of only slightly greater than 1.0 g/cm³ in a stream with a heavy suspended load (Hooke, 1987, p. 508). The distinction is that in debris flows, earthflows, and mudflows, gravity acts directly on the particles that drag along interstitial fluid, rather than on a fluid that then entrains the particles (Hooke, 1987, p. 505).

Flows almost always result from excessive rainfall, and oversaturation seems to be the major factor in their formation. They may form on planar hillsides or at the toe of large slumps, but most often they develop at the heads of small gullies, where groundwater seepage or surface runoff concentrates to initiate streams. Unusually deep weathering along "seepage lines" at the head

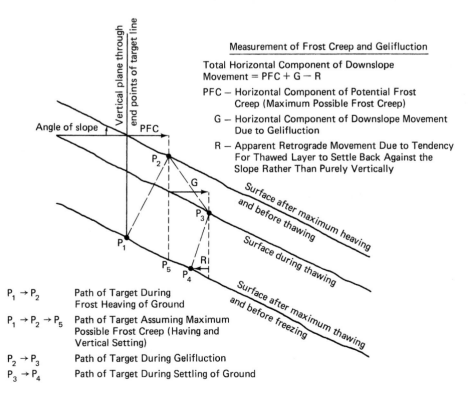

FIGURE 9-7. Diagram of frost creep, gelifluction, and retrograde components of movement by a survey target on a perennially frozen hillside in northeast Greenland (Washburn, 1967, Figure 5).

of incipient drainage networks may be the reason for the high frequency of flow initiation there (Bunting, 1961).

Case histories of flows from around the world (Brabb and Harrod, 1989; Allison, 1995) emphasize the combination of steep slopes, weak or easily weathered rock or surficial cover, and abnormal precipitation in triggering them. Artificial alteration, especially excavation at the foot of a slope, and vegetation clearing are very often additional factors. In seismic regions, earthquakes may trigger flows and avalanches (Sidle et al., 1985, pp. 66–72). Quite commonly, aerial photographs of flows show many scars of earlier flows of the same general form. The Slumgullion earthflow in the San Juan Mountains of Colorado has been established as a natural laboratory for monitoring flow. Parts of it are at least 700 years old, and the younger part has been continuously flowing for the past 300 years (Varnes and Savage, 1996).

A measure of the mobility of a flow is the *sensitivity* of the mass, defined as the ratio of the undisturbed or undrained shear strength (p. 171) to the remolded shear strength, the cohesiveness that remains after the mass has been thoroughly disturbed or remolded while retaining the same water content. Sensitivity measures the loss of shear strength with severe deformation. Low sensitivity is defined as a ratio of 2 or less; medium sensitivity is in the range of 2 to 4; sensitive and extrasensitive in the range of 4 to 8 and 8 to 16; and *quick clay* (Figure 9-8) has sensitivity greater than 16 (Selby, 1993, pp. 111–114). A Norwegian classification defines quick clay as having a sensitivity greater than 30 and a remolded shear strength of less than 0.5 kPa, which guarantees complete fluidity of the landslide mass (Lefebvre, 1996, p. 607).

Earthflows in quick clay are common along the valley of the St. Lawrence River and in British Columbia, Canada, and in Scandinavia, where thick deposits of glaciomarine silt and clay (p. 373) have been uplifted by postglacial isostatic movements and now form most of the lowlands (Lefebvre, 1996). The clay-size particles were deposited rapidly in saline water and were flocculated into large agglomerates that have chaotic internal structure. Instead of being dense and laminated, the glaciomarine clay has unusually high porosity and is initially held together by electrostatic forces between clay-size particles and the saline pore water. After subaerial leaching for 10,000 years or more, the soluble salts have been removed, and the clay may become sensitive or quick (Figure 9-8). It retains shear strength if it is not excessively or abruptly stressed, but with excessive vibration or shock, the pore water mobilizes, the weak framework of poorly arranged mineral grains collapses, and in extreme cases an apparently stiff plastic silty clay suddenly becomes a thick, milky fluid with no residual shear strength. Along coastal river valleys in Canada

FIGURE 9-8. Demonstration of quick clay. Undisturbed sample (left) supports 11 kg. Another sample of the same clay is poured from the beaker after being stirred. No water was added (photo: Div. Building Research, National Research Council of Canada).

and Scandinavia, such sediment in its cohesive unleached state forms vertical river banks tens of meters high. The strength degenerates, however, and heavy rain or sudden shocks such as earthquakes, thunder, highway traffic, blasting in excavations or log jams, well drilling, or the spring breakup of ice on a river may cause abrupt slope failure (Bentley and Smalley, 1984; Torrance, 1987).

The form of an earthflow in quick clay is a characteristic pear or bottleneck shape, with a relatively narrow neck at the point of initial failure. As each slump mass collapses, spreads, liquefies, and flows toward the adjacent lowland, new concentric slices collapse. The flow margin retrogresses to a bowl shape from the initial narrow flow, which becomes the gorge through which a slurry of liquefied mud flows. The pattern on aerial photographs (Figure 9-9) has a distinctive ribbed pattern of blocks that retain their cohesive strength briefly and then collapse. The earthflow bowls may continue to grow for hours or weeks to a

FIGURE 9-9. Landslide of June 1993 at the evacuated village site of Lemieux on the South Nation River, Ontario, Canada. View up to the north. Head of the flow is 680 m from the river channel. The entire event took place in about 1 hour (Brooks et al., 1994; Evans and Brooks, 1994) (photo: Geological Survey of Canada, photo no. GSC 1993-254f).

diameter of several kilometers and are as much as 100 m in depth. Their floors are hummocky, including tilted slices and pinnacles of the former surface layer that retained some strength (Evans and Brooks, 1994). In 1971, a disastrous flow at Saint-Jean-Vianney, Quebec, claimed 30 lives in a series of expanding flows that progressively engulfed houses in a small village (Tavenas et al., 1971). Personal narratives recount incidents of people outrunning the retrogressive expansion of earthflow scarps in the St. Lawrence River lowland. It appears that the sod and soil profile stay coherent even after liquefaction has begun at depth, and people who have felt the road collapse beneath their automobile have yet had time to abandon the car and run uphill to safe ground.

A remarkable feature of Norwegian quick clay is the abruptness with which it regains strength if salt is restored and mixed into the flowing mass. Throwing bags of sodium or magnesium chloride into the fluid mass and churning through it with a tracked vehicle can stop some earthflows across Norwegian roads. Almost instantly, the water-clay bonds are restored, the mass stabilizes, and it can be cleared with standard earth-moving equipment.

Torrential rains or *cloudbursts* may send flash floods roaring through mountain canyons. Dams of debris, including mud, rocks, logs, and stumps, temporarily impede the flood water, then fail and surge downstream to strike the next obstacle with even greater force. Such processes, aided by landslides and bank collapse into the floodwater, quickly convert the mass into a high-density debris flow (Johnson and Rodine, 1984; Hooke, 1987). The deposits are poorly sorted and unstratified; they may choke valley floors or spread onto piedmont slopes as lobes. The role of flash floods and debris flows in shaping alluvial fans in arid regions is reviewed at greater length in Chapter 13.

Lahar is an Indonesian word that describes a debris flow on the flank of a volcano (p. 100). Torrential rains, sometimes triggered by pyroclastic eruptions, saturate the unstable tephra on the slopes of composite cones. Other sources of mobilizing water are breached crater lakes, melted snow, and slide-dammed rivers. Lahars may be hot or cold. Their water content is comparable to that of wet concrete. One of the most serious lahars from Mount St. Helens in 1981 began more than four hours after the initial eruption, and was probably triggered when the large amount of ice in the initial debris avalanche had melted and a magnitude 5 earthquake within the volcano liquefied the mass (Fairchild, 1987, pp. 57–60). Numerous debris flows from destabilized watersheds continued for months following the 1981 eruption of Mount St. Helens (Figure 6-6). Most followed former river channels, but rapidly spilled over adjacent floodplains. Their mode of flow typically involved a frontal lobe of boulders that rode the crest of each surge, were carried forward on the surface of the flow like a conveyor belt, then were overridden. The surface boulders moved forward faster than the internal mass, ensuring their concentration at the front of each surge (Pierson, 1986, p. 279). Lahars triggered by intense tropical storms continue to be a major hazard for thousands of people who live near Mt. Pinatubo in the Philippines since its June 1991 eruption. In Japan, more than 25 percent of the nearly 5000 people who died in natural disasters between 1967 and 1987 died as a result of debris flows, including lahars (Takahashi, 1991).

Spread

A **spread** is defined as a lateral extension of a cohesive rock or soil mass combined with general subsidence of the fractured cohesive material into a softer substrate (Cruden and Varnes, 1996, p. 62). The cause is usually liquefaction of the substrate while the surface retains cohesive strength, as illustrated by the previous descriptions of quick clay flows. *Rock spreads* are typically extremely slow, as when brittle layers are rafted along by slow deformation of underlying ductile materials such as shale or salt (Figure 5-26). Spreads are a common form of landslide on low-gradient terrains during earthquakes, when rapid shaking liquefies sediments below the water table and displaces overlying blocks of drier material.

FIGURE 9-10. A slump-earth-flow with principal parts labeled. Compare Figure 9-11.

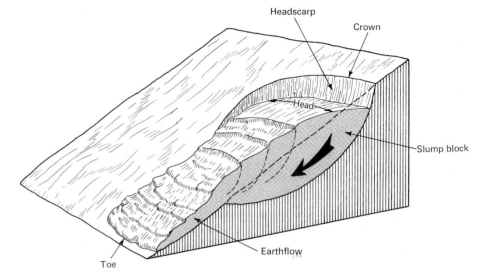

Slide

Mass wasting wherein a mass of rock or weathered debris moves downhill along discrete shear surfaces is defined as a **slide**. Slides are subdivided into **rotational** and **translational** categories based on the shape of the shear surface. The distinction is important for engineering analytical and control procedures (Cruden and Varnes, 1996, p. 56). Subcategories of slide in various classifications include *slump, rock slide, block glide, mud slide, debris slide,* and *debris avalanche* (Brunsden, 1984).

Slump. Figure 9-10 is a composite diagram of the nomenclature of the parts of a typical rotational slide, or **slump** passing downhill into an earthflow (compare Figure 9-11). This is an extremely common form of mass wasting although some of the named components might be absent in any single example.

Slump is the form of slide most common in thick, homogeneous, cohesive sediments. The surface of failure beneath a slump block is spoon shaped, concave upward or outward (Figure 9-10). The upper surface of a slump block commonly is tilted backward because the entire mass rotates as the lower part moves outward and downhill. Vegetation or even houses may be carried intact on the surface of a large slump block. Slumps are tens or hundreds of meters wide and may be single blocks or consist of multiple slices. Ponds often form in the angle between the base of a headscarp and top of the rotated slump block. This water, percolating down along the surface of rupture, may cause renewed or persistent instability on old slumps.

Slumps may be caused by water erosion undercutting the foot of a slope. They are also a common result of faulty engineering design of cut embankments. They are sometimes mitigated by loading the base of an unstable slope with a heavy layer of coarse rock rubble, which permits water to drain off the hill but offsets the weight of unstable earth higher on the slope. An elaborate engineering technology has been developed to predict the surface of rupture beneath a slump in order to drill into it and drain the water from the vicinity (Krohn, 1992; Proffer, 1992; Slosson et al., 1992).

The most diagnostic feature of an ancient slump area is the hummocky or chaotic landforms on it (Figure 9-12). Unless the material moved as a single slump block or is thoroughly fluidized as a mudflow, the surface is a mass of broken rock rubble in mounds and pits, sometimes in systematic transverse ridges or lobes, but often totally unsystematic. The debris is poorly sorted and can easily be mistaken for glacial till in a mountain valley (Porter, 1970, p. 1421; Porter and Orombelli, 1981). The surface topography of slump blocks can be easily mistaken for lateral moraines.

Mud Slide. Depending on the slope gradient, degree of saturation, and grain size of sediment emerging from the toe of a slump, it can be called a mudflow or a *mud slide* (Brunsden, 1984). A mud slide moves over or between discrete shear surfaces in a lobate shape, somewhat slower than a more turbulent mudflow. At its terminus, it may form a bulbous lobe (Figure 9-11).

Rock Slide, Block Glide. The simplest form of translational slide is a rock slide or block glide. The movement is relatively rapid and most commonly occurs where steeply dipping strata or sheeting nearly parallels the surface slope (Figure 9-13). Rock slides are generally shallow. A heavy rain or freezing and

FIGURE 9-11. Slump-earthflow on highway 24 near Orinda, California, December 9, 1950. Failure occurred in highly deformed mudstone after six days of rain totaling 250 mm (photo: E.I.B. 278, U. S. Geological Survey).

FIGURE 9-12. Ancient landslide in Bailey Basin, Columbia River valley, northeastern Washington (Jones, et al., 1961) (photo: F. O. Jones, U.S. Geological Survey).

FIGURE 9-13. Rock slide on dipping sandstone strata near Glenwood Springs, Colorado. Bedding has the same effect as sheeting in massive rocks (Figure 7-3) (photo: D. J. Varnes, U.S. Geological Survey).

thawing provides fluid pressure, or vibration breaks off obstructions and reduces the coefficient of friction on the glide plane, and a detached slab or block slides down. It may shatter at the base of the slope, or it may remain intact. Rock slides or block glides have no specified size, but their thickness is normally only about 10 percent of their downslope length.

The dip of strata or sheeting is an obvious factor in rock slides (Figure 9-13). Other than the inherent strength of massive rock, the controlling factor in such rock slides is the presence, spacing, and orientation of joints and other discontinuities in the rock mass (p. 120). Sheeting joints in massive rocks roughly conform to the shape of the hill or mountain and create rock slabs whose movement is controlled by the surface friction on noncohesive joint faces (Moon, 1986; Selby, 1993, pp. 338–344). If they dip less steeply than the hill slope, individual slabs are unsupported at their downhill edges and are more subject to sliding (Figure 7-3). If they dip more steeply than the hillside,

they may fail by buckling outward or toppling (Lee, 1989, Figure 13). Rock slides can be deadly if large masses slide downhill along a sloping joint or bedding surface. Such a plane of weakness was involved in the Vaiont Reservoir disaster of October 1963 in northern Italy (Kiersch, 1964; Hendron and Patton, 1985). On the night of October 9, a rock slide 2 km long, 1.6 km wide, and 250 m thick moved suddenly down the south wall of the Vaiont Canyon and completely filled the 270-m-deep reservoir for 2 km upstream from the dam to heights of 180 m above the former water level (Figure 9-14). The movement took less than a minute, so rapid that the water in the reservoir was ejected 235 m up the north canyon wall and propelled in great waves both upstream in the reservoir and downstream over the dam, where it fell vertically 435 m onto the valley floor below. The resulting floods killed 2043 people, mostly around the town of Longarone, more than 2.5 km downstream from the dam and across a broad valley from the mouth of the Vaiont Canyon. The dam itself survived, though damaged (Figure 9-14).

The before and after geologic cross sections of the reservoir are sketched in Figure 9-15. The most obvious feature of the geology is the bowl-shaped structure of the rocks, which dip inward toward the valley axis from the south wall. The rocks are mainly impure limestone, with gypsum and thin clay layers at intervals. The limestone and gypsum on the upper slopes are full of caves and smaller solution channels so that large amounts of rainwater can penetrate the rock and spread along the clay layers. A whole series of natural and man-made factors contributed to the disaster. The principal factors were that (1) the steeply dipping limestone beds and clay layers offered little frictional resistance to sliding; (2) all the rock types are inherently weak; (3) the river had eroded the steep inner canyon across the rock structure and removed lateral support long before the dam was built; (4) older landslides along a similar slide surface had deformed and weakened the rock mass; (5) two weeks of heavy rainfall had raised the water level in the cavernous rocks and increased both fluid pressure and weight in the potential slide mass; and (6) the high water level in the reservoir had saturated the lower part of the slide, weakening the rock mass and increasing buoyancy.

The Vaiont Reservoir disaster was a surprise only in its severity. The scars of ancient slides mark the canyon. In 1960, a smaller slide had occurred, and a pattern of cracks and slumps developed on the south valley wall that ultimately outlined the great slide of 1963. For six months prior to the slide, precise records of survey stations on the slide area showed that rock creep of 1 cm/week was in progress. By three weeks prior to the disaster, the rate of creep had increased to

FIGURE 9-14. Vaiont dam, landslide, and former reservoir, view eastward. Dam is 310 m high and 185 m long at its crest (Hendron and Patton, 1985, Figure G14A).

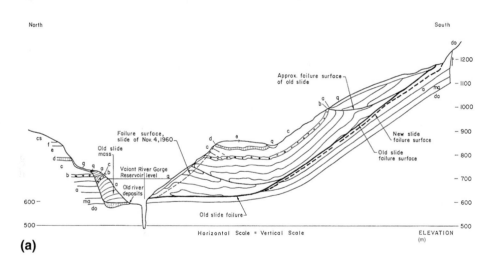

(a)

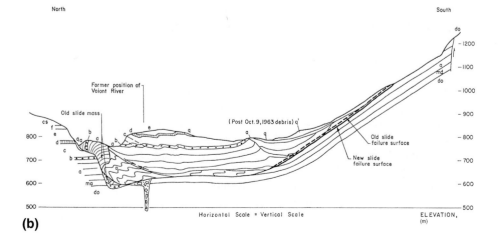

(b)

FIGURE 9-15. (a) Before, and (b) after, geologic cross sections of the Vaiont reservoir area. Note evidence of previous bedding-surface failures (Hendron and Patton, 1985, Figures 15 and 16).

1 cm/day; during the last week of heavy rains prior to the slide, the rate of creep had increased to 20 cm/day. Wild animals that had grazed on the south wall of the valley sensed the danger and moved away about October 1. On the night of the disaster, desperate measures were underway to lower the reservoir level, but the rock creep was apparently pinching the reservoir and actually raising the water level despite open outlet gates. No earthquake or other "trigger" for the Vaiont slide has been identified. The rock along the principal slide plane simply failed under excessive weight and excessive pore-water pressure. Evidence collected after the disaster suggested that by draining water from the mass that eventually failed, it could have been stabilized (Hendron and Patton, 1985, p. 98).

Avalanches and Debris Slides. Any large, catastrophic landslide may be called an **avalanche.** They can include broken rock, ice, and snow, usually so mixed that the material term *debris* is most appropriate. Debris avalanches in the temperate forested region of the eastern United States have claimed many lives and produced major geomorphic changes during torrential rainfalls (Williams and Guy, 1973; Clark, 1987; Kochel, 1987; Jacobson et al., 1989; Gryta and Bartholomew, 1989). Intense storms, usually in the summer and of local extent, are characteristic of the Appalachian region, and their results must be regarded as part of the normal land-forming processes of the region (Clark, 1987; Neary and Swift, 1987).

Debris avalanches of a 1969 torrential summer storm originated at the heads of first-order tributary streams on slopes of 16° to 39° (Figure 9-16). The bedrock in the area is massive crystalline gneiss and granite, with regolith as much as 6 m in thickness (Gryta and Bartholomew, 1989). The debris avalanches stripped off the regolith but moved relatively little fresh rock. Bedrock, commonly composed of a single joint sheet, is exposed the entire length of many avalanche scars, which are as much as 250 m long downslope and 10 to 25 m wide. The avalanche debris was carried down the initial slope and onto the adjacent valley floor, where it became a debris fan or flow or was carried away later as alluvium (Figure 9-16).

The obvious general cause for humid-climate debris avalanches is saturation by heavy rain, but the precise trigger may never be known (Wieczorek, 1996). Summer storms are known to fluctuate manyfold in intensity, even over a time span of hours or less, and during the brief intervals of heaviest precipitation, water alone may be the ultimate cause of failure (Allison, 1995, pp. 265–266). Lightning striking trees, the vibration of thunder, and tree throw are also suggested trigger mechanisms. They are tragic examples of some threshold (Figure 1-7) being exceeded.

Very large terrestrial debris slides and avalanches have received renewed interest in recent years because the shape and size of their deposits and their probable mode of origin are similar to those of debris lobes below crater walls and cliffs on Mars and the moon. For example, the deposit of the largest subaerial Quaternary debris avalanche known on earth is a lahar that extends 43 km northwest from the base of Mount Shasta in northern California (Crandell et al., 1984). Radiometric

FIGURE 9-16. Debris-avalanche scars in the headwaters of Davis Creek, Nelson County, Virginia. Most of the slides occurred during a few hours near the end of a torrential rainstorm on August 19, 1969 (Williams and Guy, 1973, Figure 11) (photo: Edwin Roseberry).

FIGURE 9-17. Sherman landslide, Alaska. Debris fell from the high peak on the right, overtopped a spur in the right foreground, and spread more than 2.5 km across the Sherman Glacier during the earthquake of March 27, 1964 (photo: Austin Post, U.S. Geological Survey).

ages from volcanic rocks within the debris and from a basalt flow that overlies it bracket the age of the event between about 300,000 and 360,000 years ago. The surface morphology of hills, mounds, and ridges that covers 450 km² of valley floor had been interpreted previously as an area of glacial moraine. All the rock debris is volcanic and must have been derived from the ancestral Mount Shasta. The flat surface of lahar deposits that surround the mounds and hills has a slope of only about 5 m/km (0.3°).

The well-documented Sherman slide (Figure 9-17) was triggered by the March 27, 1964, magnitude 9.2 Alaskan earthquake (Shreve, 1966; Marangunic and Bull, 1968). A sheet of debris averaging 1.3 m in thickness traveled 5 km from its origin across the gently sloping surface of the Sherman Glacier yet did not disturb the 2 m of fresh snow on the surface of the glacier. After the initial rock slide down a 40° cliff for a vertical height of 600 m, the slide crossed a bedrock ridge 150 m high and then spread out over 8.5 km² of glacier surface. The debris was thoroughly pulverized, probably by crossing the rock ridge if not in the initial slide. The ridge may have served as the "launching platform" to send the debris slide airborne over the glacier at a top speed of at least 185 km/h (Shreve, 1966, p. 1640).

In these and other large debris slides, the ratios of vertical drop to horizontal "runout" distance were small, equal to the tangents of angles ranging from 0.3° to 7°. These angles are typically one-tenth or less of the angle of mass friction (p. 169) of the rock debris.

So some kind of lubrication or additional transporting energy was required for their motion. Furthermore, their debris slid like a flexible sheet over the landscape, rather than flowing turbulently. Shreve (1966, 1968) argued that these large debris slides moved on a layer of trapped and compressed air, but later experiments demonstrated that the debris sheets would be too permeable. Slides of similar dimensions on Mars and the moon were emplaced in a vacuum or near vacuum, so support on an air layer cannot explain them, either. An alternative hypothesis of elastic impact between particles, or *dispersive grain flow*, proposes that particles in the rapidly moving debris are separated by open space but support themselves by elastic collision in a manner analogous to the molecules in a gas. A third hypothesis of *acoustic fluidization* is now favored (Melosh, 1983, 1987). In this hypothesis, the particles remain in contact with each other and transmit pressure (sound) waves by elastic compression and rarefaction. The energy of the initial fall is converted into ultrasonic vibrational energy that supports the mass, but does not turbulently mix it. It is called acoustic fluidization because, like the molecules in a fluid, the particles are in contact and can efficiently transmit sound energy, yet the mass can freely flow. Acoustic fluidization is effective only in very large debris flows, in which the ratio of volume to surface area is very large. Smaller slides, with large surface areas relative to their masses, dissipate energy too fast (Melosh, 1983, p. 162). The implication is that the larger the flow, the more gentle can be its gradient. For a few dozen measured terrestrial, Martian, and lunar landslides, this relationship seems to be confirmed (Melosh, 1987, Figure 2).

Two of the worst debris avalanches in history destroyed the region around Ranrahirca and Yungay, Peru, on January 10, 1962, and again on May 31, 1970 (Cluff, 1971; Plafker and Concha, 1971; Plafker and Ericksen, 1978). Observers witnessed both catastrophes from the time huge ice cornices fell from the north peak of Huascaran (6700 m) until the debris came to rest against the opposite valley wall, 14.5 km away and 4 km lower in altitude. The initial rock and ice masses, of an estimated 3 million m^3 in 1962 and 50 to 100 million m^3 in 1970, tore loose other millions of tons of rock as they roared down the valley. The shock waves produced a noise like continuous growing thunder and stripped hillsides bare of vegetation. Rocks and ice were pulverized by the turbulent flow.

The 1962 avalanche required only seven minutes to travel 20 km. It bounced from one side of a narrow gorge to the other at least five times before it emerged onto the fertile, heavily populated valley floor at the base of the mountain. As it spread to 1 km wide over the villages and fields of the valley, the mass slowed to an estimated 100 km/h and thinned to about 20 m. When the avalanche stopped, air and water spouted from the settling debris. Later, melting blocks of ice created pockets of soft mud in the flow that were added hazards to the nearly hopeless search operations. An estimated 3500 people were killed.

Eight years later, a major earthquake triggered another debris avalanche from the north face of Huascaran. The 1970 avalanche followed the same path as the 1962 avalanche for most of the 14 km down the gorge, but at the edge of the main valley a lobe of it jumped a bedrock ridge 200 to 300 m high. Within three minutes after the original rock fall, the debris avalanche had obliterated the town of Yungay, which had been protected from the earlier avalanche by the ridge. The average velocity must have been between 280 and 335 km/h. Possibly 40,000 people were killed in Yungay and neighboring towns by the 1970 avalanche, but because the total casualties of the earthquake in central Peru were estimated at 70,000 killed and 50,000 injured, the exact numbers are unknown.

Fall

The examples just cited prove that many avalanches in high mountains are initiated by a mass of rock or ice breaking off and free-falling from a glacier or cliff. Three years before the 1970 Yungay disaster, a French mountain-climbing team reported that Huascaran peak had well-developed vertical joints behind the nearly vertical north face. The earthquake may have done nothing more than topple or shake down some huge joint blocks, which in turn shattered the lower cliffs and started the avalanche. Fall is a distinct category of mass wasting, but it is rarely independent of subsequent events.

Long, steep slopes are especially characteristic of mountains that have been glaciated, with their diagnostic sharp peaks and glaciated valley forms collectively termed *alpine topography* (Chapter 17). Their present slopes are oversteepened, relief joints have opened parallel to recently deepened valleys, and ice fields or glaciers cover summit areas. Whether in the Canadian Rocky Mountains, the Alps, the Andes, the Himalayas, or any other high mountains, such relief promotes frequent and severe avalanches (Porter and Orombelli, 1981; Hewitt, 1988; Shroder, 1989). Fully 137 destructive debris avalanches have been documented during the recorded history of settlement in the European Alps alone (Eisbacher and Clague, 1984).

Submarine Mass Wasting

The term *landslide* is restricted neither to an event on land nor to the specific process of sliding. The development of high-quality side-scan sonar ocean floor imaging instruments in the past decade has lead to spectacular discoveries of enormous submarine landslides on the flanks of the Hawaiian volcanic chain as well as on the continental shelf and slope of the U.S. continental margins. These submarine landslides, mostly slumps and debris avalanches, are by far the largest such features on our planet and are exceeded only by those inferred on Mars.

Landslides on Basalt Domes. On the submerged flanks of the Hawaiian Islands about 70 major landslides more than 20 km in length have been identified. Some are longer than 200 km and have volumes exceeding 5000 km^3 (Moore et al., 1994). Collectively, they cover about 50 percent of the ridge that is surmounted by the volcanic islands, and they have occurred repeatedly, but at intervals of perhaps 100,000 years, during the growth of the volcanoes, both before and after the volcanoes grew above sea level. Single landslides may have removed 10 to 20 percent of their source volcano (Figure 10-6; Color Plate 6).

The Hawaiian submarine landslides are typically either slumps with well-defined headscarps, or debris avalanches that have fragmented into smaller masses. Those interpreted as slumps are short, wide, and occur on submarine slopes of greater than 3° (Normark et al., 1993). The Hilini fault scarp (Figure 5-4) may be the head of one such slump. The slump surfaces are very deep, in the range of 10 km, roughly equal to the full thickness of the volcanic edifices on the former ocean floor. Movement on these deep slump surfaces may be responsible for occasional Hawaiian earthquakes.

The debris avalanches around Hawaii are even more spectacular than the slump scars. They cross ocean floor topography with average slopes of less than 3°, and although they are disaggregated into a jumble of debris, individual blocks of 1 to 10 km in length are common. The largest yet mapped, the Nuuanu debris avalanche, is on the northeastern side of Oahu. It extends for 23 km across the floor of the Hawaiian trough to a depth of 4600 m, then rises 300 m up the far side of the trough at its toe. Giant blocks tens of kilometers in length have been transported 50 km or more. The largest block had been mapped as the Tuscaloosa Seamount, 90 km northeast of Oahu. It is 30 km long, 17 km wide, and has a flat summit 1800 m above the ocean floor (Normark et al., 1993, pp. 192–193). The upslope flow of the Nuuanu debris avalanche required the momentum of rapid emplacement; it is likely that such avalanches produced giant waves comparable to tsunamis (p. 424) (Moore and Moore, 1988). The Nuuanu debris avalanche involved a mass of about 5000 km^3. It predates another debris avalanche that has been dated indirectly at about 1.4 million years. Not only do the Hawaiian submarine debris avalanches raise fascinating questions about mass wasting processes, such as the mechanism that can carry kilometer-size blocks across a very gentle ocean floor and even uphill, but also fundamental issues of the role of landslides in building oceanic shield volcanoes (Chapter 6). Gravity sliding may be one of the important ways that the rifts form along the axes of the shields. Alternatively, the intrusion of magma into chambers near the base of the edifice may force the deep gravitational spreading and slumping. At the probable temperature of the magma, olivine masses that settle to the floor of the magma chamber have rheologic properties similar to glacier ice (Chapter 16) and may provide a weak surface for sliding (Clague and Denlinger, 1994). Forms comparable to those reviewed here have been reported from many other regions, including Reunion Island in the Indian Ocean, Mt. Etna and Stromboli in the Mediterranean Sea, along other hot-spot volcanic islands in the Pacific Ocean, and on spreading-ridge volcanoes in the Atlantic Ocean (Moore et al., 1994; Masson, 1996).

Continental Margin Landslides. All along the U.S. continental margins, large slumps and debris-flow scars mark the continental slope. They might be triggered by rapid sedimentation, storm waves, earthquakes, methane gas generated in organic-rich sediments, and probably other causes. A selection of fascinating specific studies, based mostly on side-scan sonar images, has been published (Schwab et al., 1993). The features seem to be an integral part of the growth of passive continental margins, whereby sediment carried to the continental shelf by rivers eventually progrades across the shelf to the upper continental slope, then slumps, slides, or flows down the slope at an average gradient of only 5° and spreads across the adjacent continental rise and abyssal basins. The headscarps of some slumps are gullied to a degree that suggests they are 10,000 years old or more. Perhaps some of the landslides were triggered by the massive sediment loads carried by late-glacial rivers from the retreating margins of Pleistocene ice sheets, at a time when sea level was lower, the continental shelves were largely emerged, and glacially derived sediment was dumped directly on the upper slope.

At least one Atlantic continental slope landslide is known to have been earthquake induced. A 1929 earthquake of magnitude 7.2 had its epicenter on the continental slope of Newfoundland, Canada (Piper et al., 1988). Numerous telegraph cables crossed the area, and their time of breakage has provided a means to reconstruct the mass wasting. All cables within a radius of 100 km from the epicenter broke at the same time, defining the zone of seismic-induced slumping. Others downslope broke at times up to 13 hours after the earthquake, by a submarine debris flow and *turbidity current* from the epicentral region. The flow followed preexisting submarine valleys and traveled at a rate of approximately 20 m/s (72 km/hr) out onto the Sohm Abyssal Plain. As much as 200 km^3 of sediment was carried to the submarine plain. Gravel beds were deposited 300 km from their source in glacial sediments on the upper slope.

Off the Mississippi delta in the Gulf of Mexico, massive loads of clay- and organic-rich sediments accumulate on the shallow continental shelf. The rapid accumulation of fine-grained sediment traps pore fluids in the sediment and with compaction from added overburden, pore pressure approaches the weight of the overlying mud so that surface layers are

FIGURE 9-18. Types of sub-aqueous mass wasting in the offshore part of the Mississippi River delta (Coleman et al., 1993, Figure 4).

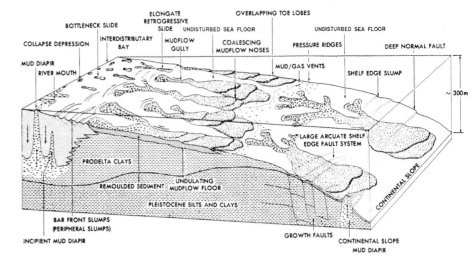

almost "floating" on very gentle slopes. Water depth is shallow enough so that storm waves cause cyclic pressure fluctuations on the mud that trigger a variety of mass wasting features (Coleman et al., 1993) (Figures 9-18 and 9-19). Several offshore drilling platforms have been destroyed or damaged by the slides.

Clearly, we are entering a period of exciting ocean-floor exploration that will change our understanding of the role of mass wasting in shaping undersea landscapes (Hampton et al., 1996). Many more new discoveries are certain to be forthcoming.

ing ridge will have a significance very different from that of a profile measured along the bed of an adjacent gully. Slopes are irregular surfaces that cannot be described by simple mathematical equations. The best topographic maps are only approximations of the infinite irregularities of hillside, which perhaps will be best described by fractal geometry (Figure 1-1) (Culling and Datko, 1987, pp. 371–372). We do not yet know what degree of irregularity is significant in the stability of slopes, so we are never sure that we are measuring the correct angles and distances.

HILLSLOPE DEVELOPMENT AND EVOLUTION

Most of a landscape consists of curved, sloping surfaces, largely shaped by mass wasting. How these slopes form, how they are maintained, and how they change with time are major topics of geomorphic research. It is surprisingly difficult even to describe the geometry of a natural slope. One is commonly provided with only a profile surveyed down the steepest part of a hillside as a description of the slope, but obviously a profile measured along the crest of a descend-

FIGURE 9-19. Side-scan sonar image of about 1.5 km² of Mississippi delta-front ocean floor, showing retrogressive slides and mudflows moving down a very gentle gradient (photo: D. B. Prior).

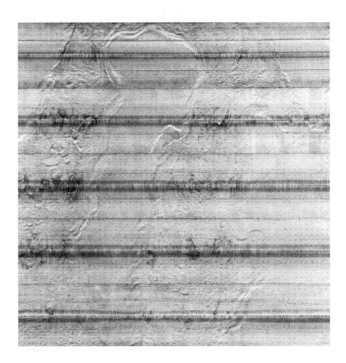

Hillslope Analysis: Techniques and Theories

There have been two philosophies about studying slopes. An older school deduced the systematic changes of slope form that would accompany long-continued subaerial weathering and erosion. Because landscape evolution is too slow to be witnessed, deductions concerning the changes of slope form with time were based on assumptions that could not be tested until we had radioactive and cosmogenic isotopic techniques for dating old land surfaces. It is no wonder that the deductive approach to slope analysis has enriched geologic literature with some remarkably opinionated and authoritarian writings. A majority of deductive geomorphologists have held that slopes, especially in humid regions, become lower and more broadly rounded with time. A vocal minority have insisted that slopes are stable forms with angles controlled by rock type and weathering processes, and when a stable slope has evolved, it persists through time, migrating backward parallel to itself unless it is eliminated by the intersection of other slopes. For an excellent review of the classic debate, see von Engeln (1942, pp. 256–267). The issues are evaluated further in Chapter 15.

Another group of geomorphologists have concerned themselves with the empirical description of slopes. With less regard for theoretical projection into the future, they have studied the processes of slope formation and the geometry of slopes. Innumerable slope profiles and descriptive texts have been published, but empirical study has suffered from the lack of a guiding theory.

Progress has been made in modeling the evolution of simple scarps a few meters high, such as those made by faulting (Figures 5-1 and 5-2) or river erosion in unconsolidated sediment. Typically, a fresh scarp will be steeper than the angle of repose, but when it is abandoned, the scarp *ravels,* or collapses back, as masses of sediment from the upper part of the slope break loose and fall or slide to its base (Figure 9-20). Within a few months or a few years, a straight slope is established at or a little less than at the angle of repose, which is approximately equal to the mass angle of friction and is usually about 25° to 35° (Colman and Watson, 1983; Nash, 1984; Pierce and Colman, 1986). Thereafter, the evolution of the slope through time by creep, raindrop impact, and slope wash can be described by the *diffusion equation:*

$$\partial y / \partial t = c \ (\partial^2 y / \partial x^2)$$

At any point on the initial slope, the rate at which the elevation y changes with time t is the product of a diffusion constant c and the local slope curvature. Erosion occurs on the portion of the scarp that is convex

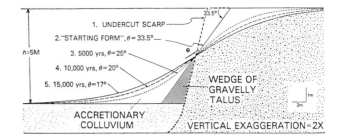

FIGURE 9-20. Stages in the evolution of a terrace scarp. 1. Lateral undercutting of the terrace gravels by stream to form a scarp steeper than the angle of repose. 2. This oversteepened slope rapidly ravels to form a rectilinear scarp at the angle of repose, here inferred to be 33.5°, which is the starting form for diffusion-equation modeling. 3, 4, 5. Scarp profiles predicted by diffusion-equation model at stated time intervals (degradation rate coefficient, c, of 12×10^{-4} m²/yr) (Pierce and Colman, 1986, Figure 1).

upward and deposition on the portion that is concave upward. The rate of creep is proportional to the sine or tangent of the slope at any point (Figure 9-2), so as the local curvature decreases, so does the further rate of change (Figure 9-20). The equation is identical in form to the diffusion equation used to describe chemical diffusion, conductive heat flow, and fluid flow through a permeable medium. With time, the upper and lower part of the slope become less sharply curved and the midslope angle decreases. By the simplest assumptions of the diffusion equation, the midpoint of the slope does not change position and the initial scarp height does not decrease. If the diffusion constant for a particular sediment type in a particular climate can be established by measuring the profiles of scarps of various known ages and correcting for scarp height and exposure direction, the diffusion equation can be used to predict the age of an unknown scarp by its shape (Figure 9-20). The method has had some success in predicting the ages of recent fault scarps (p. 68).

Hillside slope profiles generally have an upper segment convex to the sky and a lower concave segment, some slope profiles show a straight segment between the upper and lower curves, and if a cliff interrupts the slope, an additional segment marked by free fall of weathered debris is introduced. A total of nine slope segments are recognized in a later classification (Figure 9-21). The straight segment of a slope below a free face (Figure 9-21, unit 5) is usually a **talus.** A talus is a slope landform, not a type of material. It is redundant to speak of a "talus slope," and the coarse rock debris on a talus should be called **sliderock** or **scree** (the common British usage). The angle of the talus is a function of fragment size and angularity, climate, vegetation, rate of cliff recession, and rate of

FIGURE 9-21. Diagrammatic representation of a hypothetical nine-unit land-surface model (Dalrymple et al., 1968, Figure 1).

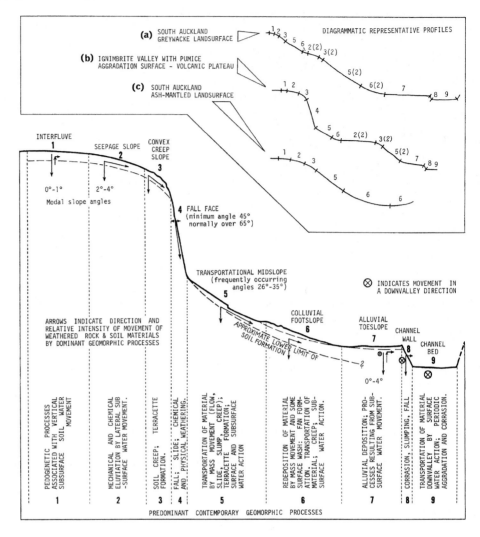

sliderock supply and removal (Statham and Francis, 1986; Selby, 1993, pp. 366–369).

Gentler slopes with straight intermediate segments (not taluses at the bases of cliffs as in Figure 9-21) seem to form where erosion is unusually rapid. In the extreme examples of gullies cut on artificial embankments by single intense storms, most of the slopes are straight. Natural landscapes scored by closely spaced V-shaped gullies with straight sides that intersect as knife-edged ridges are called **badlands** (Figure 9-22). Strahler (1950, 1956) demonstrated that straight valley-side slopes are common in areas of recent or continuing tectonic uplift where downcutting by rivers is active. Mass wasting, especially soil creep, largely controls the upper convex and straight segments of slope profiles (Figure 9-21, units 1 to 3). Gilbert (1909) offered an explanation of summit convexity that is largely deductive but is still a useful explanation. He observed a reasonably uniform thickness of soil or regolith over the convex surface and assumed that in a given inter-

val of time a uniform thickness of the weathered material is removed from the entire summit area. Under these conditions, larger quantities must move through cross sections progressively farther downhill. In other words, with the stated assumption, the amount of material that creeps past any point is proportional to the distance of the point from the summit. Since creep is primarily a gravitational phenomenon, the slope angle must increase radially from the summit in order to move the progressively greater amount of debris. The summit curvature becomes convex to the sky. Carson and Kirkby (1972, pp. 108–109, 306–307) strongly supported Gilbert's conclusion that the hilltop convexity of humid temperate regions is primarily the result of soil creep, but they presented additional theoretical arguments that Gilbert's deduction is unnecessarily restrictive.

The mathematical treatment of creep as a simple diffusion phenomenon is consistent with Gilbert's qualitative explanation of hillslope profile convexity.

FIGURE 9-22. Natural badland slopes in the Mancos Shale badlands near Caineville, Utah. Note the nearly linear slope profiles and very narrow divides (Howard, 1994, Figure 9) (photo: Alan D. Howard).

As drawn in Figure 9-20, the diffusion curve is symmetrical around the center point on the profile, and the lower half simply records the accumulation of an equivalent cross section of debris removed from the upper half. On slopes higher than a few meters, other processes of mass wasting and erosion remove some of the accumulated debris from the lower half of the profile, the midpoint of the curve migrates with the retreating slope, and the simple diffusion curve becomes asymmetric. Nevertheless, the fundamental relationship between the rate of creep and convexity of the local upper slope is retained.

Selby (1993, pp. 252–258) expressed a number of doubts about the role of soil creep in the convexity of slopes. He noted that few actual measurements have been repeated over long enough time intervals to distinguish soil creep from more specific or seasonal processes such as heave associated with wetting and drying or freezing and thawing although many geomorphologists would include such processes within the general concept of soil creep. Some rounded, convex slopes could be relict landforms, for example, the result of gelifluction (p. 176). Decade-length measurements that show no creep on steep slopes under old forests (p. 174) should at least create healthy skepticism about attributing all convex or ungullied slopes to soil creep without a careful examination of the evidence.

An unreasonably large volume of sediment would have to move through successive downhill cross-sectional areas of slope to lower an uphill surface appreciably. Apparently, the major function of creep is to maintain smooth upland slopes by removing soil from local mounds and filling the heads of incipient gullies and hollows (Young and Saunders, 1986, p. 8). The process is slow enough to span significant intervals of Quaternary climate change; colluvium-filled hollows in California have been accumulating creep debris for 9000 to 14,000 years, since the end of a late Pleistocene episode of gully erosion under different climatic conditions (Reneau et al., 1990).

Downhill (Figure 9-21, units 6 and 7 and part of unit 5), transportation by flowing water assumes dominance over creep. Seepage pressure of groundwater emerging from lower slopes subtracts from the shear strength of the rock and sediment and encourages weathering to finer grain sizes, reducing the slope angle. Small rills also collect seepage water and rainwater and assume characteristic concave-up profiles. Two little rills flowing down a bare hillside during a rainstorm require a certain slope to keep flowing with their suspended-sediment load. When the two rills join, the resulting rivulet has a greater proportional increase of mass than its increase in wetted surface area. Friction is reduced in proportion to the discharge, and the larger trickle of water can transport the joint loads of the two lesser streams with no loss of velocity but on a more gentle slope. Thus slopes controlled by seepage, rain wash, sheet wash, or rill wash are generally concave skyward. At some position on a slope, rain wash becomes dominant over soil creep, and the slope profile inflects from convex near the top to concave near the base. (For an extended discussion of slope concavity, see Carson and Kirkby, 1972, pp. 310–315.)

Thus far we have considered only slope profiles. Even from profiles alone, one can see how soil creep near the top of a slope increases the gradient downhill until water begins to flow over the surface instead of penetrating and saturating the creeping soil. At that level on the hillside, sheet wash and slope concavity begin. To date, most of the research on hill slope development has been confined to analyses of profiles (Schumm, 1967; Carson and Kirkby, 1972).

Slopes also curve in directions other than downhill, and these other curvatures also affect water movement. Where contour lines bulge convexly outward on a hillside around sloping spurs or **noses,** water is spread laterally as it flows downhill. Noses and ridge crests tend to be drier than adjacent **hollows** where the contour lines swing concavely into the hill. Concave contours tend to gather water from a large area higher on the slope, and the heads of streams are localized downhill from hollows (Hack and Goodlett, 1960, p. 6; Reneau et al., 1990).

Profile curvature and contour curvature can be combined into a single diagrammatic classification of slopes (Figure 9-23). The horizontal axis of the diagram divides "water-gathering" slopes with concave contours (quadrants I and II) from "water-spreading" slopes with convex contours (quadrants III and IV). The vertical axis of the diagram separates slopes with convex profiles dominated by creep (quadrants II and III) from those with concave profiles dominated by wash (quadrants I and IV). Almost any segment of a land surface can be placed into one of the four quadrants of the diagrams (Troeh, 1965). The only exceptions are saddle-shaped surfaces, which require a higher order of mathematical analysis. Horizontal surfaces or planar slopes plot on the axes of the diagram.

Each of the slope elements of Figure 9-23 can be represented mathematically by a quadratic equation. Rotating a segment of a parabola around a vertical axis generates each surface. Uncertainty remains in deciding the appropriate sample spacing for fitting a parabolic surface to the ground. At each smaller interval, new roughness elements appear (p. 5). In perfecting his classification, Troeh surveyed agricultural lands around Cornell University and calculated the best-fitting paraboloid of rotation for each small area of surveyed land. In areas where the actual land surface could be represented by a single quadratic equation with vertical deviations no greater than 10 to 15 cm, a single soil type was represented. Where the land surface departed from the calculated surface by more than 15 cm, a new soil type appeared. Usually, the difference in soil was controlled by slightly better or worse drainage (Troeh, 1964).

Soils that vary as the result of slope and drainage are grouped in a **drainage catena,** named for the well-drained soil member in a group for which the other soil-forming factors (parent material, climate, organisms, and time) are similar (Brady and Weil, 1996, p. 633, Figure 19-3). The problems with the concept of a drainage catena is that the other soil-forming factors are rarely constant on a slope (p. 136). For example, thicker colluvium or regolith is likely on lower slopes; because of creep or erosion, the soil on the ridge crest may be much younger than that on the lower slopes.

FIGURE 9-23. Classification of slope elements by profile curvature and contour curvature (from Troeh, 1965).

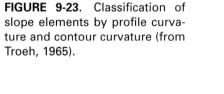

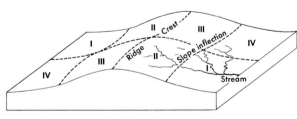

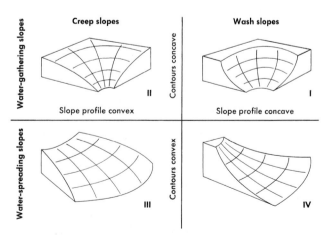

Mode and Rate of Slope Retreat

Most hillslopes are landforms that result when rivers cut valleys. Only tectonic scarps and sloping depositional surfaces (sand dunes, moraines, volcanoes, and so on) are not the secondary result of erosion. Ordinarily, a landscape comprises small slope elements, each reacting in a particular way to the local effectiveness of weathering, mass wasting, and erosion. All elements are related, however, because an accidental disequilibrium in any part of a slope affects adjacent segments above and below the site of the accident. Suppose that an animal burrowing on a hillside raises a mound of loose soil on the downhill slope. The slope is locally oversteepened, and the debris is rapidly spread downhill. Simultaneously, rain wash from uphill is trapped by the burrow, and creep is accelerated by the undermining. A dynamically stable, or graded, slope is an example of an open physical system (Chapter 2) through which both energy and matter move, a system that tends by self-regulating processes to maintain itself in the most efficient possible configuration. Slopes constantly change but tend toward some central graded state appropriate to the environment of the moment (Figure 1-6).

Questions that have stimulated generations of geomorphologists concern the retreat of slopes as river valleys deepen and widen. Are slope profiles stable, migrating backward parallel to themselves, or do slope angles decrease through time as summits are lowered? Are there climatic variations in the way slopes evolve? Do slopes persist through epochs of geologic time, or are present slopes the result of present-day processes? Can slope retreat be measured, or must it be deduced from indirect evidence? These questions are so deeply embedded in the fabric of geomorphic thought that they must be repeatedly reexamined in subsequent chapters. Only a brief preview is appropriate here while the processes that shape slopes are fresh in the reader's mind.

Some generalizations about slopes seem reasonably established. Straight valley-side slopes are associated with rapidly deepening valleys in a variety of tectonic and climatic settings and on a wide scale of sizes. In humid regions, slope profiles are commonly sigmoid, convex skyward near summits and concave on lower slopes. Creep is the dominant process on the upper convexity of the profile, and sheet wash or rill wash dominates on the concave lower elements. In plan view, well-drained noses alternate with water-gathering hollows in which tributary streams originate. Weathering is intensified along the axes of hollows. Hollows enlarge by erosion at the expense of intervening noses, but are

also filled with colluvium over intervals of 1000 to 10,000 years.

Creep and slope wash, which determine the relative proportion of convex and concave profile elements, are in part climate dependent. In some climates, convexities extend across low interfluves from one valley floor to the next (Figure 14-5). In other climates, concave slopes sweep upward from valley floors to knife-edge divides (Figure 15-1). Unless the climate changes, the regional dominance of one slope component or another will persist although declivities might change with time.

Cliffs and taluses are usually structurally determined (Chapter 12). Vertical cliffs can be cut by valley glaciers, waves, or rivers, but they are quickly veneered by their own taluses unless some erosional agent removes the debris from their bases. If this does not happen, the talus grows in height until the entire source of sliderock at the top of the cliff is buried. Thenceforth, the slope becomes a fossil or relict talus and gradually diffuses into a gently curved segment of a normal slope profile. If the sliderock is removed from the base of the talus as fast as it is supplied from the cliff at the top, the cliff and talus landforms migrate backward as long as the structural factors that define the cliff persist (Color Plate 10).

Colluvium of coarse rock debris on a slope may weather and change its angle of repose as it creeps downslope. The persistence of slope angles as valley sides retreat is thus related to the weathering character of the rock mass ("structure") and climate ("process"). If debris becomes finer as it moves downslope, eventually it is entrained in slope wash and transported over a concave profile to a stream. If, as is often the case with granitic rocks on desert slopes, large joint blocks "armor" slopes until they break down to coarse sandy grus, a sharp reentrant angle marks the foot of the hillslope and the head of an alluvial fan or pediment (Figure 13-5).

Much has been said here of the climatic and structural controls of hillslope development and retreat. Slope evolution through time is deferred until Chapter 15, and additional aspects of climatic geomorphology are reviewed in other chapters. In concluding this chapter, a few rates of slope retreat are quoted, only to suggest that such measurements can be made or deduced. Each of the five examples is from a region of dramatic slope development. Many more such examples have been compiled by Young (1974) and Saunders and Young (1983).

1. Cliffs of massive red or brown sandstone formations, generally of Mesozoic age, form much of the striking scenery of the Colorado Plateau (Color Plate

10). Schmidt (1989, Table 1) compiled scarp retreat rates for various cliff-forming strata in the Colorado Plateau that range from 0.5 to 6.7 km/10^6 yr (= mm/yr). Schumm and Chorley (1964, 1966, p. 30) estimated that one cliff had retreated 30 feet in the last 2500 years (3 to 4 mm/yr) by rock fall. However, the cliff was exposed to weathering for an indeterminate length of time before the measured blocks began to move. Where the Colorado River crosses the Colorado Plateau, its canyon is now an average of 12 km wide at the rim. By assuming that the Grand Canyon began to erode about 20 million years ago, these authors estimated the long-term widening rate of the canyon at about 0.6 mm/yr, or a rate of cliff retreat of 0.3 mm/yr on each canyon wall. Dohrenwend (1994, pp. 337–340) compiled numerous comparable rates of slope retreat from the Basin and Range Province.

2. LaMarche (1968) measured denudation rates in the White Mountains of California by the root exposure of bristlecone pines as much as 4000 years old (Figure 15-4). Steep hillslopes there are being lowered (parallel to the surface) at about 0.2 mm/yr; if the slope angle were 30°, horizontal recession at a rate of 0.4 mm/yr would result. These slopes are on heavily jointed dolomite in a montane climate of intense frost action, cloudbursts, and mudflows.

3. From a comprehensive study of slope development in a subarctic mountain valley in northern Sweden, Rapp (1960, p. 186) concluded that rock fall alone causes steep slopes there to retreat at about 0.04 to 0.15 mm/yr, or about 1 m in the 10,000 years of postglacial time. In the cold maritime climate, frost cracking and hydrofracturing are the most active processes. The bedrock is mica schist and limestone.

4. Mountain slopes under rainforest in Papua New Guinea are intensely scarred by active landslides, which on some slopes cover as much as 40 percent of the total area (Simonett, 1967, p. 73). Judging by the rates of postwar vegetation recovery, slide areas are not revegetated for 40 to 60 years. Therefore, an estimate of regional denudation rates can be made from the area and volume of landslide scars measured on aerial photographs. A combination of intense weathering, frequent and intense earthquakes, and steep slopes results in an estimated regional denudation rate of 1.0 to 1.4 mm/yr by landslides alone, on granitic terrane (Simonett, 1967, p. 83; Pain and Bowler, 1973). Assuming average forested slopes of clifflike 45° declivity, slope retreat here would approach 1 mm/yr. Similar rapid slope retreat and valley enlargement have been reported from Japan (Machida, 1966) and Panama (Garwood et al., 1979).

5. As a final example, the extreme deductions of King (1967, pp. 159, 163) concerning the South African land-scape can be cited. King asserted that tectonic movements initiate cycles of erosion by sending scarps retreating wavelike across the continent, each leaving behind a lower-level pediment (Chapter 13) while an older, higher land surface is consumed by the scarp. He believed that the Drakensberg scarp in Natal, 1200 m high, is one such scarp that was initiated at the coast in late Cretaceous time and has since migrated 250 km inland at a rate of about 3 mm/yr. King's hypothesis represents an extreme view of the importance of parallel scarp retreat and regional pedimentation, but the rates he cited are not excessive compared to the previously cited four examples, except for their vast spatial and temporal continuity.

One can conclude that steep valley walls and cliffs retreat by landsliding at rates of 0.5 to 5.0 mm/yr in a variety of climates on diverse lithologies. Other processes of slope retreat are slower and more variable; even on steep slopes and cliffs, recession rates range over two orders of magnitude from 0.01 to 1.0 mm/yr (Saunders and Young, 1983, p. 487).

REFERENCES

ALLISON, R. J., 1995, Slopes and slope processes: Prog. Phys. Geog., v. 19, pp. 265–279.

ANDERSSON, J. G., 1906, Solifluction, a component of subaerial denudation: Jour. Geology, v. 14, pp. 91–112.

BENTLEY, S. P., and SMALLEY, I. J., 1984, Landslips in sensitive clays, in Brunsden, D., and Prior, D. B., eds., Slope Instability: John Wiley & Sons Ltd., Chichester, UK, pp. 457–490.

BRABB, E. E., and HARROD, B. L., eds., 1989, Landslides: Extent and economic significance: A.A. Balkema, Rotterdam, 385 pp.

BRADY, N. C., and WEIL, R. R., 1996, Nature and properties of soils, 11th ed.: Prentice-Hall, Inc., Upper Saddle River, N.J., 740 pp.

BROOKS, G. R., Aylsworth, J.M., and 2 others, 1994, Lemieux landslide of June 20, 1993, South Nation Valley, Southeastern Ontario—a photographic record: Geol. Surv. Canada, Miscell. Rept. 56, 18 pp.

BRUNSDEN, D., 1984, Mudslides, in Brunsden, D., and Prior, D. B., eds., Slope Instability: John Wiley & Sons Ltd., Chichester, UK, pp. 363–418.

BUNTING, B. T., 1961, The role of seepage moisture in soil formation, slope development, and stream initiation: Am. Jour. Sci., v. 259, pp. 503–518.

CARSON, M. A., and KIRKBY, M. J., 1972, Hillslope form and process: Cambridge Univ. Press, Cambridge, UK, 475 pp.

CLAGUE, D. A., and DENLINGER, R. P., 1994, Role of olivine cumulates in destabilizing the flanks of Hawaiian volcanoes: Bull. Volcanol., v. 56, pp. 425–434.

CLARK, G. M., 1987, Debris slide and debris flow historical events in the Appalachians south of the glacial border, in Costa, J. E., and Wieczorek, G. F., eds., Debris flows/avalanches: Process, recognition, and mitigation: Geol. Soc. America Reviews in Engineering Geology, v. 7, pp. 125–138.

CLUFF, L. S., 1971, Peru earthquake of May 31, 1970; engineering geology observations: Seismol. Soc. America Bull., v. 61, pp. 511–521.

COLEMAN, J. M., PRIOR, D. B., and 2 others, 1993, Slope failures in an area of high sedimentation rate: Offshore Mississippi River delta, in Schwab, W. C., Lee, H. J., and Twichell, D. C., eds., Submarine landslides: Selected studies in the US Exclusive Economic Zone: U.S. Geol. Survey Bull. 2002, pp. 79–91.

COLMAN, S. M., and WATSON, K., 1983, Age estimated from a diffusion equation model for scarp degradation: Science, v. 221, pp. 263–265.

COTTON, D. S., and TE PUNGA, M. T., 1955, Fossil gullies in the Wellington landscape: New Zealand Geographer, v. 2, pp. 72–75.

CRANDELL, D. R., MILLER, C. D., and 3 others, 1984, Catastrophic debris avalanche from ancestral Mount Shasta volcano, California: Geology, v. 12, pp. 143–146.

CRUDEN, D. M., and VARNES, D. J., 1996, Landslide types and processes, in Turner, A. K., and Schuster, R. L., eds., Landslides: Investigation and mitigation: Transportation Research Board, Natl. Acad. Sci., Natl., Res. Council Spec. Rept. 247, pp. 36–75.

CULLING, W. E. H., and DATKO, M., 1987, The fractal geometry of the soil-covered landscape: Earth Surface Processes and Landforms, v. 12, pp. 369–385.

DALRYMPLE, J. B., BLONG, R. J., and CONACHER, A. J., 1968, An hypothetical nine-unit landsurface model: Zeitschr. für Geomorph., v. 12, pp. 60–76.

DOHRENWEND, J. C., 1994, Pediments in arid environments, in Abrahams, A. D., and Parsons, A. J., eds., Geomorphology of desert environments: Chapman & Hall, London, pp. 321–353.

EISBACHER, G. H., and CLAGUE, J. J., 1984, Destructive mass movements in high mountains: Hazard and management: Geol. Survey Canada, Paper 84–16, 230 pp.

von ENGELN, O. D., 1942, Geomorphology: The Macmillan Company, New York, 655 pp.

EVANS, S. G., and BROOKS, G. R., 1994, An earthflow in sensitive Champlain Sea sediments at Lemieux, Ontario, June 20, 1993, and its impact on the South Nation River: Canadian Geotech. Jour., v. 31, pp. 384–394.

FAIRCHILD, L. H., 1987, The importance of lahar initiation processes, in Costa, J. E., and Wieczorek, G. F., eds., Debris flows/avalanches: Process, recognition, and mitigation: Geol. Soc. America Reviews in Engineering Geology, v. 7, pp. 51–61.

FERGUSON, H. F., 1967, Valley stress release in the Allegheny Plateau: Eng. Geology, v. 4, no. 1, pp. 63–71.

de FRIETAS, M. H., and WATTERS, R. J., 1973, Some field examples of toppling failure: Géotechnique, v. 23, no. 4, pp. 495–513.

GARWOOD, N. C., JANOS, D. P., and BROKAW, N., 1979, Earthquake-caused landslides: A major disturbance to tropical forests: Science, v. 205, pp. 997–999.

GILBERT, G. K., 1909, The convexity of hilltops: Jour. Geology, v. 17, pp. 344–350.

GRYTA, J. J., and BARTHOLOMEW, M. J., 1989, Factors influencing the distribution of debris avalanches associated with the 1969 Hurricane Camille in Nelson county, Virginia, in Schultz, A. P., and Jibson, R. W., eds., Landslide processes of the eastern United States and Puerto Rico: Geol. Soc. America Spec. Paper 236, pp. 15–28.

HACK, J. T., and GOODLETT, J. C., 1960, Geomorphology and forest ecology of a mountain region in the central Appalachians: U.S. Geol. Survey Prof. Paper 347, 66 pp.

HAMPTON, M. A., LEE, H. J., and LOCAT, J., 1996, Submarine landslides: Rev. Geophysics, v. 34, pp. 33–59.

HANEBERG, W. C., and ANDERSON, S. A., eds., 1995, Clay and shale slope instability: Geol. Soc. America Reviews in Engineering Geology, v. 10, 153 pp.

HENDRON, A. J., JR., and PATTON, F. D., 1985, The Vaiont slide, a geotechnical analysis based on new geologic observations of the failure surface: U.S. Army Corps of Engineers Waterways Experiment Station, Geotechnical Laboratory Tech. Rept. GL–85–5, 144 pp. and Appendices.

HEWITT, K., 1988, Catastrophic landslide deposits in the Karakoram Himalaya: Science, v. 242, pp. 64–67.

HIGASHI, A., and CORTE, A. E., 1971, Solifluction: A model experiment: Science, v. 171, pp. 480–482.

HOLTZ, R. D., and KOVACS, W. D., 1981, An introduction to geotechnical engineering: Prentice-Hall, Inc., Englewood Cliffs, N.J., 733 pp.

HOOKE, R. L., 1987, Mass movement in semi-arid environments and the morphology of alluvial fans, in Anderson, M. G., and Richards, K. S., eds., Slope stability: Geotechnical engineering and geomorphology: John Wiley & Sons Ltd., Chichester, UK, pp. 505–529.

HOWARD, A. D., 1994, Detachment-limited model of drainage basin evolution: Water Resources Res., v. 30, pp. 2261–2285.

ISHII, R., and HIGASHI, M., 1997, Tree coexistence on a slope: An adaptive significance of trunk inclination: Proc., Royal Soc. London, Ser. B, v. 264, pp. 133–139.

JACOBSON, R. B., CRON, E. D., and McGEEHIN, J. P., 1989, Slope movements triggered by heavy rainfall, November 3–5, 1985, in Schultz, A. P., and Jibson, R. W., eds., Landslide processes of the eastern United States and Puerto Rico: Geol. Soc. America Spec. Paper 236, pp. 1–13.

JAHN, A., 1989, Soil creep of slopes in different altitudinal and ecological zones of Sudetes Mountains: Geografiska Annaler, v. 71A, pp. 161–170.

JOHNSON, A. M., and RODINE, J. R., 1984, Debris flow, in Brunsden, D., and Prior, D. B., eds., Slope instability: John Wiley & Sons Ltd., Chichester, UK, pp. 257–361.

JONES, F. O., EMBODY, D. R., and PETERSON, W. L., 1961, Landslides along the Columbia River valley, northeastern Washington: U.S. Geol. Survey Prof. Paper 367, 98 pp.

KIERSCH, G. A., 1964, Vaiont reservoir disaster: Civil Engineering, v. 34, pp. 32–39.

KING, L. C., 1967, Morphology of the earth, 2nd ed.: Oliver and Boyd Ltd., Edinburgh, 726 pp.

KIRKBY, M. J., 1967, Measurement and theory of soil creep: Jour. Geology, v. 75, pp. 360–374.

KOCHEL, R. C., 1987, Holocene debris flows in central Virginia, in Costa J. E., and Wieczorek, G. F., eds., Debris flows/avalanches: Process, recognition, and mitigation: Geol. Soc. America Reviews in Engineering Geology. v. 7, pp. 139–155.

KROHN, J. P., 1992, Landslide mitigation using horizontal drains, Pacific Palisades area, Los Angeles, California, in Slosson, J. E., Keene, A. G., and Johnson, J. A., eds., Landslides/landslide mitigation: Geol. Soc. America Reviews in Engineering Geology, v. 9, pp. 63–68.

LAMARCHE, V. C., 1968, Rates of slope degradation as determined from botanic evidence, White Mountains, California: U.S. Geol. Survey Prof. Paper 352-I, pp. 341–377.

LEE, F. T., 1989, Slope movements in the Cheshire Quartzite, southwest Vermont, in Schultz, A. P., and Jibson, R. W., eds., Landslide processes of the eastern United States and Puerto Rico: Geol. Soc. America Spec. Paper 236, pp. 89–102.

LEFEBVRE, G., 1996, Soft sensitive clays, in Turner, A. K., and Schuster, R. L., eds., Landslides: Investigation and mitigation: Transportation Research Board, Natl. Acad. Sci., Natl. Res. Council Spec. Rept. 247, pp. 607–619.

LEWIS, L. A., 1974, Slow movement of earth under tropical rain forest conditions: Geology, v. 2, pp. 9–10.

MACHIDA, H., 1966, Rapid erosional development of mountain slopes and valleys caused by large landslides in Japan: Tokyo Metropolitan Univ. Geog. Repts., no. 1, pp. 55–78.

MARANGUNIC, C., and BULL, C., 1968, The landslide on the Sherman Glacier, in The great Alaskan earthquake of 1964, v. 3, "Hydrology": Natl. Acad. Sci. Pub. 1603, Washington, D.C., pp. 383–394.

MASSON, D. G., 1996, Catastrophic collapse of the volcanic island of Hierro 15 ka ago and the history of landslides in the Canary Islands: Geology, v. 24, pp. 231–234.

MELOSH, H. J., 1983, Acoustic fluidization: American Scientist, v. 71, pp. 158–165.

———, 1987, The mechanics of large rock avalanches, in Costa, J. E., and Wieczorek, G. F., eds., Debris flows/avalanches: Process, recognition, and mitigation: Geol. Soc. America Reviews in Engineering Geology, v. 7, pp. 41–49.

MITCHELL, J. K., 1976, Fundamentals of soil behavior: John Wiley & Sons, Inc., New York, 422 pp.

MOON, B. P., 1986, Controls on the form and development of rock slopes in fold terrane, in Abrahams, A. D., ed., Hillslope processes: Allen & Unwin, Inc., Boston, pp. 225–243.

MOORE, G. W., and MOORE, J. B., 1988, Large-scale bedforms in boulder gravel produced by giant waves in Hawaii, in Clifton, H. E., ed., Sedimentologic consequences of convulsive geologic events: Geol. Soc. America Special Paper 229, pp. 101–110.

MOORE, J. G., NORMARK, W. R., and HOLCOMB, R. T., 1994, Giant Hawaiian underwater landslides: Science, v. 264, pp. 46–47.

NASH, D. B., 1984, Morphologic dating of fluvial terrace scarps and fault scarps near West Yellowstone, Montana: Geol. Soc. America Bull., v. 95, pp. 1413–1424.

NASH, D. F. T., 1987, Comparative review of limit equilibrium methods of stability analysis, in Anderson, M. G., and Richards, K. S., eds., Slope stability: Geotechnical engineering and geomorphology: John Wiley & Sons Ltd., Chichester, UK, pp. 11–75.

NEARY, D. G., and SWIFT, L. W., JR., 1987, Rainfall thresholds for triggering a debris avalanching event in the southern Appalachian Mountains, in Costa, J. E., and Wieczorek, G. F., eds., Debris flows/avalanches: Process, recognition, and mitigation: Geol. Soc. America Reviews in Engineering Geology, v. 7, pp. 81–92.

NORMAN, S. A., SCHAETZL, R. J., and SMALL, T. W., 1995, Effects of slope angle on mass movement by tree uprooting: Geomorphology, v. 14, pp. 19–27.

NORMARK, W. R., MOORE, J. B., and TORRESAN, M. E., 1993, Giant volcano-related landslides and the development of the Hawaiian Islands, in Schwab, W. C., Lee, H. J., and Twichell, D. C., eds., Submarine landslides: Selected studies in the U.S. Exclusive Economic Zone: U.S. Geol. Survey Bull. 2002, pp. 184–196.

PAIN, C. F., and BOWLER, J. M., 1973, Denudation following the November 1970 earthquake at Madang, Papua New Guinea: Zeitschr. für Geomorph., Supp. no. 18, pp. 92–104.

PHIPPS, R. L., 1974, The soil creep-curved tree fallacy: U.S. Geol. Survey Jour. Res., v. 2, pp. 371–377.

PIERCE, K. L., and COLMAN, S. M., 1986, Effect of height and orientation (microclimate) on geomorphic degradation rates and processes, late-glacial terrace scarps in central Idaho: Geol. Soc. America Bull., v. 97, pp. 869–885.

PIERSON, T. C., 1986, Flow behavior of channelized debris flows, Mount St. Helens, Washington, in Abrahams, A. D., ed., Hillslope processes: Allen & Unwin, Inc., Boston, pp. 269–296.

———, and COSTA, J. E., 1987, A rheologic classification of subaerial sediment-water flows, in Costa, J. E., and Wieczorek, G. F., eds., Debris flows/avalanches: Process, recognition, and mitigation: Geol. Soc. America Reviews in Engineering Geology, v. 7, pp. 1–12.

PIPER, D. J. W., SHOR, A. N., and HUGHES CLARK, J. E., 1988, The 1929 "Grand Banks" earthquake, slump, and turbidity current, in Clifton, H. E., ed., Sedimentologic consequences of convulsive geologic events: Geol. Soc. America Spec. Paper 229, pp. 77–92.

PLAFKER, G., and ERICKSEN, G. E., 1978, Nevados Huascarán avalanches, Peru, in Voight, B., ed., Rockslides and avalanches, I: Natural phenomena: Developments in geotechnical engineering 14A, Elsevier Scientific Publishing Company, Amsterdam, pp. 277–314.

———, and CONCHA, J. F., 1971, Geological aspects of the May 31, 1970, Peru earthquake: Seismol. Soc. America Bull., v. 61, pp. 543–578.

PORTER, S. C., 1970, Quaternary glacial record in Swat Kohistan, West Pakistan: Geol. Soc. America Bull., v. 81, pp. 1421–1446.

———, and OROMBELLI, G., 1981, Alpine rockfall hazards: American Scientist, v. 69, pp. 67–75.

PROFFER, K. A., 1992, Ground water in the Abalone Cove landslide, Palos Verdes Peninsula, southern California, in Slosson, J. E., Keene, A. G., and Johnson, J. A., eds., Landslides/landslide mitigation: Geol. Soc. America Reviews in Engineering Geology, v. 9, pp. 69–82.

RAPP, A., 1960, Recent development of mountain slopes in Kärkevagge and surroundings, northern Scandinavia: Geografiska Annaler, v. 42, pp. 65–200.

RENEAU, S. L., DIETRICH, W. E., and 3 others, 1990, Late Quaternary history of colluvial deposition and erosion in hollows, central California Coast Ranges: Geol. Soc. America Bull., v. 102, pp. 969–982.

ROBINSON, E. S., 1970, Mechanical disintegration of the Navajo Sandstone in Zion Canyon, Utah: Geol. Soc. America Bull., v. 81, pp. 2799–2805.

SASOWSKY, I. D., and WHITE, W. B., 1994, Role of stress release fracturing in the development of cavernous porosity in carbonate aquifers: Water Resources Res., v. 30, pp. 3523–3550.

SAUNDERS, I., and YOUNG, A., 1983, Rates of surface processes on slopes, slope retreat and denudation: Earth Surface Processes and Landforms, v. 8, pp. 473–501.

SCHAETZL, R. J., BURNS, S. F., and 2 others, 1990, Tree uprooting: Review of types and patterns of soil disturbance: Physical Geog., v. 11, pp. 277–291.

SCHMIDT, K.-H., 1989, Significance of scarp retreat for Cenozoic landform evolution on the Colorado Plateau, U.S.A.: Earth Surface Processes and Landforms, v. 14, pp. 93–105.

SCHUMM, S. A., 1967, Rates of surficial rock creep on hillslopes in western Colorado: Science, v. 155, pp. 560–561.

———, and CHORLEY, R. J., 1964, The fall of Threatening Rock: Am. Jour. Sci., v. 262, pp. 1041–1054.

———, and CHORLEY, R. J., 1966, Talus weathering and scarp recession in the Colorado Plateaus: Zeitschr. für Geomorph., v. 10, pp. 11–36.

SCHWAB, W. C., LEE, H. J., and TWICHELL, D. C., eds., 1993, Submarine landslides: Selected studies in the US Exclusive Economic Zone: U.S. Geol. Survey Bull. 2002, 204 pp.

SELBY, M. J., 1993, Hillslope materials and processes, 2nd ed.: Oxford Univ. Press, Oxford, UK 451 pp.

SHREVE, R. L., 1966, Sherman landslide, Alaska: Science, v. 154, pp. 1639–1643.

———, 1968, The Blackhawk landslide: Geol. Soc. America Spec. Paper 108, 47 pp.

SHRODER, J. F., JR., 1989, Hazards of the Himalaya: American Scientist, v. 77, pp. 564–573.

SIDLE, R. C., PEARCE, A. J., and O'LOUGHIN, C. L., 1985, Hillslope stability and land use: Am. Geophys. Union Monograph Series, v. 11, 140 pp.

SIMONETT, D. S., 1967, Landslide distribution and earthquakes in the Bewani and Torricelli Mountains, New Guinea, in Jennings, J. N., and Mabbutt, J. A., eds., Landform studies from Australia and New Guinea: Australian Natl. Univ. Press, Canberra, pp. 64–84.

SLOSSON, J. E., YOAKUM, D. D., and SHUIRMAN, G., 1992, Thistle landslide: Was mitigation possible? in Slosson, J. E., Keene, A. G., and Johnson, J. A., eds., Landslides/landslide mitigation: Geol. Soc. America Reviews in Engineering Geology, v. 9, pp. 83–93.

STATHAM, I., and FRANCIS, S. C., 1986, Influence of scree accumulation and weathering on the development of steep mountain slopes, in Abrahams, A. D., ed., Hillslope processes: Allen & Unwin, Inc., Boston, pp. 245–267.

STRAHLER, A. N., 1950, Equilibrium theory of erosional slopes approached by frequency distribution analysis: Am. Jour. Sci., v. 248, pp. 673–696, 800–814.

———, 1956, Quantitative slope analysis: Geol. Soc. America Bull., v. 67, pp. 571–596.

TAKAHASHI, T., 1991, Debris flow: A.A. Balkema, Rotterdam, 165 pp.

TAVENAS, F., CHAGNON, J.-Y., and LAROCHELLE, P., 1971, The Saint-Jean–Vianney landslide: Observations and eye-witness accounts: Canadian Geotech. Jour., v. 8, pp. 463–478.

TORRANCE, J. K., 1987, Slope stability, in Anderson, M. G., and Richards, K. S., eds., Slope stability: Geotechnical engineering and geomorphology: John Wiley & Sons Ltd., Chichester, UK, pp. 447–473.

TROEH, F. R., 1964, Landform parameters correlated to soil drainage: Soil Sci. Soc. America Proc., v. 28, pp. 808–812.

———, 1965, Landform equations fitted to contour maps: Am. Jour. Sci., v. 263, pp. 616–627.

TURNER, A. K., and SCHUSTER, R. L., eds., 1996, Landslides: Investigation and mitigation: Transportation Research Board, Natl. Acad. Sci., Natl. Res. Council Spec. Rept. 247, 673 pp.

VARNES, D. J., and SAVAGE, W. Z., eds., 1996, The Slumgullion earth flow: A large-scale natural laboratory: U.S. Geol. Survey Bull. 2130, 95 pp.

WASHBURN, A. L., 1967, Instrumental observations of mass-wasting in the Mesters Vig District, northeast Greenland: Meddelelser om Grønland, v. 166, no. 4, 296 pp.

WIECZOREK, G. F., 1996, Landslide triggering mechanisms, in Turner, A. K., and Schuster, R. L., eds., Landslides: Investigation and mitigation: Transportation Research Board, Natl. Acad. Sci., Natl. Res. Council Spec. Rept. 247, pp. 76–90.

WILLIAMS, G. P., and GUY, H. P., 1973, Erosional and depositional aspects of hurricane Camille in Virginia, 1969: U.S. Geol. Survey Prof. Paper 804, 80 pp.

YOUNG, A., 1974, The rate of slope retreat, in Brown, E. H., and Waters, R. S., eds., Progress in Geomorphology: Inst. British Geographers Special Pub. no. 7, pp. 65–77.

———, and SAUNDERS, I., 1986, Rates of surface processes and denudation, in Abrahams, A. D., ed., Hillslope processes: Allen & Unwin, Inc., Boston, pp. 3–27.

Chapter 10

The Fluvial Geomorphic System

Water, flowing down to the sea over and immediately beneath the land surface, is the dominant agent of landscape alteration. Near-surface weathering and groundwater solution provide a load for flowing streams, and mass wasting may dump great quantities of rock debris at the foot of slopes, but eventually rivers must carry all but a small fraction of the total rock waste from the lands to the sea. Estimates of the relative importance of various agents of continental denudation can hardly be better than orders of magnitude, but one careful evaluation (Garrels and Mackenzie, 1971, p. 114) credited rivers with 85 to 90 percent of the total present sediment transport to the sea, glaciers with about 7 percent, groundwater and waves with about 1 to 2 percent, and wind and volcanoes with less than 1 percent each. Therefore, to understand how landscapes evolve, one must understand how rivers do their work. The useful adjective **fluvial** (from L. *fluvius:* "river") pertains to the work of rivers, but in the geomorphic context, also includes the water-dominated preconditioning of rock debris by weathering and mass wasting before reaching a river channel (Leopold et al., 1964). Schumm (1977, p. vii) included in the **fluvial system** all the area and processes extending from drainage divides in the source areas of water and sediment, through the channels and valleys of the drainage basin, to depositional areas such as coasts. That inclusiveness is followed here.

The fluvial system is powered by the conversion of the potential energy of solar distillation and gravity to the kinetic energy of motion and heat (Figure 2-5). Most of the energy is lost to friction of internal turbulence in flowing water, but perhaps 2 to 4 percent (Rubey, 1938, p. 138) of the total potential energy of water flowing downhill over an erodible bed is converted to the mechanical work of erosion and transportation.

Fluvial processes obviously vary in intensity among climatic regions and along gradients of temperature, precipitation, altitude, and seasonality (Chapter 2). Oddly, on this hydrous planet, perhaps one-third of the land surface has no runoff to the ocean (Chapter 13). However, even arid regions with drainage into closed intermontane basins have landscapes of branching stream valleys. Infrequent or brief seasonal stream flow can shape otherwise dry landscapes. Furthermore, on the time scale in which landscapes evolve, climatic regions have shifted and changed in intensity, so that many regions now intensely arid show evidence of previous fluvial erosion and deposition (Chapter 4).

The fluvial system is so widespread and effective that it has been commonly regarded as the "normal" process of landscape evolution on earth (Cotton, 1948). W. M. Davis (1905a, reprinted 1954, p. 288; 1905b, reprinted 1954, p. 296) based his classic geomorphic cycle on the certainty that "the greater part of the land surface has been carved by . . . the familiar processes of rain and rivers, of weather and water." He defined a "normal" climatic region as "not so dry but that all parts of the surface have continuous drainage to the sea, nor so cold but that the snow of winter all disappears in summer." It is regrettable that Davis's definition betrays a certain provinciality that was common in eastern North America and western Europe at the time he wrote. Nevertheless, the subsequent recognition that vast areas of terrestrial landscape are shaped by processes not dominant in the cool, humid climates of Cambridge (Massachusetts), London, and Paris has not diminished the importance to geomorphology of the fluvial system.

OVERLAND FLOW

Infiltration, Throughflow, and Sheetflow

The proportion of surface and subsurface water that feeds a stream varies greatly with climate, soil type, bedrock, slope, vegetation, and many other factors. One estimate is that one-eighth of the annual runoff of the hydrologic cycle goes directly overland to the sea, and seven-eighths of the water goes underground at least briefly. The rapidity with which infiltrating water reacts with minerals (p. 128) explains why "pure" spring water closely reflects the chemistry of the local rocks. Even as river water begins its downhill flow, its chemical energy has been largely expended.

The *infiltration capacity* of soil is controlled by duration and intensity of precipitation, prior wetted condition of the soil, vegetation, soil mineralogy and texture, slope, and other factors (Horton, 1945, p. 307). In a humid climate during a light steady rain on a landscape with a continuous vegetation cover, infiltration may continue for a long time, the water moving downward through the zone of aeration toward the water table. However, eventually, the soil voids will fill with water, and *saturation overland flow* will begin (Montgomery and Dietrich, 1994, p. 230). If the rainfall is so intense that infiltration cannot keep pace, *Hortonian overland flow* may begin sooner, even if the soil is not saturated at depth. Subsequent developments are different if the deeper layers are saturated or unsaturated, but in either case, when the surface layer becomes saturated, overland flow begins, and soil particles loosened by raindrop impact or turbulence are entrained in the flow (Young and Wiersma, 1973; Wischmeier and Smith, 1978). The principle difference between Hortonian overland flow and saturation overland flow is that saturation adds seepage pressure (Figure 9-1) to the tendency for loose particles to be moved downslope. Direct runoff from well-vegetated slopes rarely exceeds a few percent of the precipitation, but on sun-dried, clay-rich surface crusts or bare rock, runoff may begin as soon as a thin surface layer is wetted by capillary absorption.

Seepage pressure is generated by infiltrated groundwater moving downslope within the soil profile. Variously defined as **throughflow** or **interflow** (McCaig, 1985, p. 123), this water moves relatively rapidly through interconnected cracks, animal burrows, decayed root channels and other soil voids at shallow depth, whether the deeper soil and parent material are saturated or not. Whereas groundwater flow is theoretically laminar, as defined by Darcy's law (p. 152), throughflow is within relatively large spaces in the soil (macropores) and flows with turbulence that can erode and further enlarge the soil voids (Beven and Germann, 1982). Where this water moves downslope but outward toward the soil surface, the seepage pressure reduces the effective particle weight and initiates sediment motion on a saturated surface.

In laboratory experiments with real soils or artificial sediment mixtures compacted into tilted boxes under simulated rainfall of varying intensity, the erosion and transport of soil particles can be studied. The impact energy of raindrops easily disturbs and entrains sand and small aggregates of sand, silt, and clay, and if the ground is saturated, water and sediment begin to move downslope. In experiments, a wire screen placed a few centimeters above the surface to break the raindrop impact without decreasing the precipitation intensity decreased soil losses by 90 percent or more (Young and Wiersma, 1973). On bare gentle slopes, the scattering effect of impact suppresses channel formation, and the muddy water flows as a thin, slow-moving surface layer called **sheetflow.** As slope increases, helical threads of secondary currents develop in the layer of water, which are capable of additional sediment scour (Moss et al., 1982).

Rills and Gullies

At some threshold of erosional and transportational energy that combines the factors of rainfall intensity, grain size, cohesion, infiltration capacity, slope, and surface microrelief, the threads of higher velocity and more turbulent current erode small channels, or **rills** (Dunne and Aubry, 1986, p. 35). The process quickly focuses at points of more rapid seepage discharge, undermining small pits or rill heads. The headward propagation of a rill is then limited only by the ability of conventional fluvial transport processes to remove the material provided by the seepage pressure. The process, called *piping,* is a common method of rill initiation (Howard and McLane, 1988). Piping in soil is similar to the larger-scale process of *sapping* that occurs where the land surface intersects the water table and springs emerge. There, erosion by groundwater flow undermines the overlying sediment and rock and cuts a steep-headed rill or channel that can expand headward and branch into a drainage network even without overland flow (Schumm et al., 1995). Sapping has been invoked to explain drainage networks on Mars as well as those on earth (Higgins, 1984; Howard and McLane, 1988).

When cut to a size large enough to survive the interval between rains (Figure 10-1), a rill collects surface runoff at its steep headward end, and the resulting waterfall migrates upstream, extending the rill into previously undissected upland and capturing other rills. On barren slopes, rills can be observed growing headward at a rate of centimeters per minute or faster. They tend to parallel each other, rather than to form dendritic networks, because each is growing headward on the steepest local gradient. By intrenching the surface a few centimeters deeper than the adjacent interfluve areas, each rill side and head becomes a local discharge area for the shallow throughflow. Water flow and piping continue at the rill heads even

after the surface soil has dried out, encouraging maintenance and growth of the evolving rill network. Although raindrop impact, creep, and many other processes can obliterate rills up to several centimeters in width and depth, rill development becomes dominant over creep flow and sheetflow on convex hillslopes a few hundred meters downslope from a ridge crest (Dunne and Aubry, 1986), and slope concavity begins (Figure 9-21).

A **gully** (Figure 10-2) is a stream channel with discrete cut banks, and often with a steep head, that distinguishes it from a rill (Higgins, 1990; Bocco, 1991; Dietrich and Dunne, 1993). Gullies are known to develop rapidly as a result of poor agricultural practices. One in Georgia, the size of which impressed the great British geologist Sir Charles Lyell in 1846, subsequently grew to a length of 230 m, a width of 150 m, and a maximum depth of 20 m, in little more than 100 years. It is eroded in the deeply weathered regolith on crystalline rocks of the Georgia piedmont and began to form after the area was deforested (Ireland, 1939). Gullies grow headward rapidly enough to capture adjacent rills and commonly develop a branching network of tributary channels, with dimensions scaled to their drainage area. They are the first step in the fluvial dissection of landscapes.

Because of their steep heads, gully floors are less steep than the intergully surface. Their steep, usually bare, banks are unstable, scarred by slumps and soil topples. Gullies in arid and semiarid regions are likely to be flat floored, wide (to carry massive bedloads), and steep walled (because of the strong vertical or columnar jointing typical of dry soils). Their longitudinal profiles are steep. When dry, which is most of the time, they are floored with coarse alluvium. In Spanish-speaking regions, these distinctive channels are called *arroyos*. The equivalent Arabic term, *wadi* (or *ouadi* in French), is widely used in North Africa and the Middle East.

Soil Erosion

The transition from overland flow to rills and gullies, and the resulting abrupt increase in soil erosion, seems to involve a number of *thresholds,* critical parameters at which processes abruptly change (Figure 1-7). For example, sheetflow over matted grass may be completely clear, even on steep slopes (Prosser et al., 1995). Yet, as soon as some threshold is exceeded, that water may abruptly converge on a depression or step in the hillside, gouge a gully head, and begin to erode vigorously (Montgomery and Dietrich, 1994; Dunne et al., 1995).

FIGURE 10-1. Rill erosion from less than 4000 m² of drainage on late-seeded winter wheat after summer fallow, near Mica, Washington. Rill in foreground is 45 cm deep (photo: U.S. Dept. Agriculture–Soil Conservation Service).

FIGURE 10-2. Gully formed in 26 years, Madison County, Georgia (photo: U.S. Dept. Agriculture–Soil Conservation Service).

Fluvial soil erosion is the collective result of creep, raindrop impact, sheetflow, throughflow, piping, rill wash, and erosion by larger channels, plus losses by groundwater solution (but excluding the work of wind). All the factors of soil formation (p. 135) contribute to the rate at which mass is removed from the landscape. Catastrophic events such as floods, droughts, fire, and land clearance for agriculture and construction will increase erosion rates by many orders of magnitude.

The amount of soil lost annually from agricultural land in the United States is estimated by the use of the **Revised Universal Soil Loss Equation** (RUSLE) (Renard et al., 1994; Brady and Weil, 1996, pp. 569–578).

$$A = RKLSCP$$

where A = predicted soil loss[1]
R = climatic erosivity (rainfall and runoff)
K = soil erodibility

___
[1]The equation was developed for use in English units, with soil annual loss in tons per acre. It can be converted to metric (SI) units of tonnes per hectare (Mg/ha) by multiplying by 2.24.

L = slope length
S = slope gradient or steepness
C = cover and management
P = erosion-control practice

The climatic erosivity parameter R is also called the rainfall erosion index, and measures the annual average regional soil loss from the erosive force of rainfall and runoff, including factors of total rainfall, intensity, and seasonal distribution. For a specific locality, climatic erosivity is estimated from regional maps of the United States that are contoured in values ranging from less than 10 in the arid regions of the west to more than 700 along the Gulf coast between Louisiana and Florida.

The soil erodibility factor K ranges from 0.0 to 1.0 based on the soil loss compared to that of a standard test plot 22 m long with a 9 percent slope kept continuously bare by tillage. It typically ranges from near 0 to 0.6, with a low value of about 0.2 for sandy soils with a high infiltration rate.

Parameters L and S together constitute the dimensionless topographic factor of the RUSLE. As with the erodibility factor, they are the ratio of soil loss from a plot compared to a standard plot 22 m long with a 9 percent slope, continuously fallow. The topographic factor increases with both slope length and steepness, obviously closely related to rill erosion. Typical numerical values range from 0.2 to nearly 3.0.

Crop cover and land management variables are included in the dimensionless factor C. It is very high for fields with row crops such as corn, using conventional plowing and cultivating techniques, less under closely spaced crops such as cereal grains, and still less under hay fields or meadows. Other subfactors such as prior land use, crop rotation, and tillage practice modify it. It has low values <0.1 under forest cover or heavy mulch, but approaches 1.0 under continuous row cropping with waste removed or plowed under.

The final dimensionless factor P concerns support practice, or special techniques such as contour plowing, strip cropping along hillside contours, grass-lined drainage ways, and terraces. If no support practice is in use, P = 1.0.

The product of the variables in the RUSLE is an estimate of the annual soil loss expressed in tons per acre (or Mg/ha) from farm land. Some skill and experience are necessary to apply the equation to farm lands, but local agricultural agencies can provide tables from which each of the factors can be estimated. The Revised Universal Soil Loss Equation is a computer software package that can be applied accurately to local conditions (Renard et al., 1994). Because the result is a product, if one or more of the variables can be reduced by intelligent farm practice, soil loss can be greatly reduced. For example, instead of plowing and planting bare fields straight up and down the slope (C $\simeq$ 1.0, P $\simeq$ 1.0), if cover crops (C $\simeq$ 0.2) are planted in contoured strips or terraces (P $\simeq$ 0.5), the soil loss can be reduced by 90 percent, without any change in the climate, soil, and slope parameters (Brady and Weil, 1996, p. 578). Annual soil loss of 5 tons per acre (11.2 Mg/ha) is regarded as the maximum loss allowable to maintain soil productivity (Brady and Weil, 1996, p. 578).

A continuing problem for geomorphologists is the observations that soil losses, whether measured from hillslopes or undissected landscapes, are always much higher than the sediment loads of the rivers that drain the sites would indicate (Figure 11-5). The scales of time and size are certainly part of the problem of the "missing" sediment in river systems; soil-loss measurements are restricted to relatively small plots, measured at most for a few years or decades. Another problem may be the multithousand-year storage time of colluvium in hillside hollows (p. 191). Sediment storage within the fluvial system is considered in greater detail in Chapter 11.

CHANNELED FLOW: CHANNEL PROCESSES AND GEOMETRY

Whether formed by chance or necessity, by headward erosion or downslope convergence, whether inherited or newly formed, systems of branching river channels dissect most of the subaerial landscape, each "in a valley proportioned to its size." (p. 232). Some channels are occupied by *permanent streams* that flow throughout the year, others have *intermittent streams,* and still others are used only briefly during and immediately after precipitation by *ephemeral streams.* Channel size is obviously related to river discharge, but it is common for rivers to overflow their channels at *flood stage.*

In humid regions, the mean annual discharge of rivers increases downstream unless structural controls such as karst divert some of the water. Most of the annual runoff from the land is poured into rivers at the heads of small tributaries although water can be added anywhere along a channel from surface runoff and from groundwater flow. Rivers in humid regions are called **effluent** because they receive contributions of groundwater. Rivers in arid regions generally lose water to the ground in addition to losing it by evaporation, and many of them dry up entirely without reaching the sea. These are called **influent** streams; their distinctive channel characteristics are described in Chapter 13.

Gaging Stations, Rating Curves, and Hydrographs[2]

For many years, agencies of governments have maintained *gaging stations* along rivers all over the world. At these stations, usually on segments, or *reaches,* that are straight and have carefully measured and relatively stable cross sections, the discharge, water-surface level, channel shape, stream velocity, amount of dissolved and suspended mineral sediment, and other variables are periodically or continuously recorded (Dingman, 1994, Appendix F).

Discharge (Q) is the amount of water passing through a channel cross section in unit time. It is defined as

$$Q = v\mathrm{A}$$

where v is the average flow velocity through a cross section of area A. In channel flow, area is the product

---

[2]In this section, all equations are expressed as in the original publications, most of which used the nonmetric U.S. units of measurement.

of channel width (w) and average depth (d), and the definition of discharge is commonly written as

$$Q = wdv$$

Q is expressed in units of volume per unit time, typically, m³/s or ft³/s. To determine Q with necessary precision, depth is measured at close intervals across a channel, and the velocity at several depths is measured at each interval. Then w, d, and v are calculated for the particular cross section.

Having determined Q for a variety of conditions, a *rating curve* or *rating table* for the station is established, from which Q can be directly determined from the water-surface height alone, which is read from a calibrated vertical staff or a float-operated water-level recorder installed at the station (Figure 10-3)

A graph of discharge plotted against time at a gaging station is called a **hydrograph.** Depending on vegetation, soil type, infiltration rate, and a host of other factors, the hydrograph reflects the precipitation in the watershed above the station. Stations progressively downstream have broader hydrographs with flatter crests at progressively later times, depending on the time required for the storm water to reach the station and various temporary storages within the basin (Figure 10-4). A line connecting two low points on a hydrograph separates the *base flow,* fed primarily by groundwater, from the *storm runoff,* the stream response to a particular precipitation event. Various graphic techniques are used to separate a hydrograph into the two components (Dunne and Leopold, 1978, pp. 287ff; Dingman, 1994, pp. 346ff). A widely used technique for predicting the response of a river to fluctuating precipitation is to synthesize a *unit hydrograph,* which generalizes or assumes many of the parameters that affect runoff and simply predicts the hydrograph for a station that would result from one unit of effective precipitation in the watershed (Dingman, 1994, pp. 402ff).

Bedrock Channels

It is obvious that soon after stream channels begin to incise, they should encounter the relatively unweathered rock beneath the soil and regolith. Many mountain streams have their channels entirely in rock, with only scattered patches of alluvium and colluvium on the channel floor and walls. Even in hilly regions of subdued relief and deep soil cover, rock outcrops are common in stream beds. Such exposures are typically the only places to collect rock specimens or map lithologic boundaries in otherwise soil- and vegetation-covered terranes.

Rivers incise channels into bedrock by a variety of processes, almost all of them strongly dependent on structural (*sensu lato*) factors. Solution weathering is a significant process in limestone and other soluble rocks. More generally important are *plucking* and *abrasion.* Plucking (or *quarrying*) is common in jointed rocks, where the hydraulic force of flowing water can push and pull detached joint blocks out of their initial position and roll or drag them downstream (book cover). The process is intensified in strongly seasonal climates, where wetting and drying or freezing and thawing can prepare blocks for removal and force

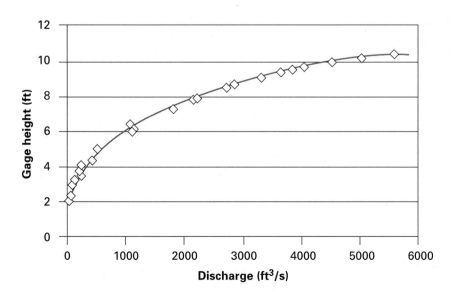

FIGURE 10-3. Rating curve for Corey Creek, near Mainesburg, Pa., defined by current-meter measurements below 500 ft³/s and by slope-area measurements at higher discharges (Station 108, Bailey et al., 1975, p. 120).

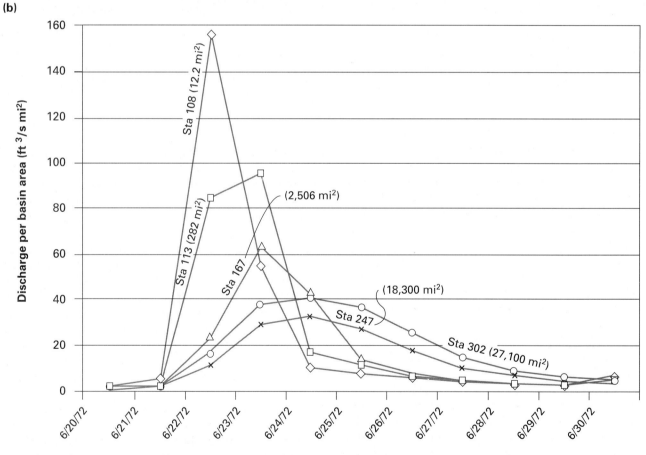

FIGURE 10-4. Hydrographs for five stations in the Susquehanna River basin during the hurricane Agnes flood of late June 1972. (a) Mean daily discharges. (b) Mean daily discharges divided by basin area upstream of the station. Station numbers increase downstream. Station 108, Corey Creek near Mainesburg, Pa.; Station 113, Tioga River at Tioga, Pa.; Station 167, Chemung River at Chemung, N.Y.; Station 247, Susquehanna River at Sunbury, Pa.; Station 302, Susquehanna River at Conowingo, Md. (data from Bailey et al., 1975).

open fractures large enough for water to enter them. During extreme flood events, joint blocks many meters on a side can be removed by violent whirlpool-like vertical macroturbulence. The process, especially effective on systematically jointed rock such as columnar basalt, is termed **kolk** (colk) or **kolking,** from a Germanic root referring to a hole or plunge pool scoured by violent flow as from a broken dike or dam. Baker (1973; Baker et al., 1987, p. 429) reintroduced the term to account for erosion of basalt blocks up to 36 ft (12 m) in diameter by the Lake Missoula paleoflood across the Columbia Plateau (p. 213). The hexagonal columnar jointing in the basalt (Figure 12-1) proved to be appropriately spaced for the spectacular dynamics of the paleoflood.

Abrasion is a common process, especially if a river is transporting sediment fragments that are abrasionally harder than the channel rocks. Because structural control is a fundamental property of landscapes (Chapter 12), many ridges and upland areas are made of rocks stronger than those that underlie adjacent lowlands, so their detritus can usually abrade channels downstream. In glaciated regions, the glacial drift usually includes mechanically fractured, sharp sand and gravel that can rapidly abrade shale and limestone channel beds. The impact of rolling and bouncing cobbles and boulders on the bed during extreme conditions is a special and powerful agent of channel erosion. Lahars and storm runoff that produce debris flows in headwater regions (Figure 9-16) also transport that debris through downstream channels, leaving many traces of plucking and abrasion.

Abrasion can be quite localized, especially in eddies or at the base of waterfalls. Sediment trapped in depressions under strong eddies can rapidly abrade **potholes,** hemispherical, cylindrical, or teardrop-shaped bedrock hollows (Wohl, 1993; Zen and Prestegaard, 1994) (Figure 10-5).

Streamlined potholes developed, grew, and merged into a narrow inner channel in Coy Glen, a small bedrock gorge near Ithaca, New York, in a few decades as a result of a gravel mining operation that allowed sharp quartz-rich glacial sand to spill into the gorge, which is eroded into Devonian shale and siltstone. A pothole approximately 1 m in length, 40 cm wide, and 20 cm deep in 1963 had grown to 1.5 m long, 80 cm wide, and 30 cm deep by 1983, by which time it had merged into a downstream line of similar potholes. Since the gravel mining has ended, the flat shale floor of Coy Glen has been swept clean of alluvium except for relict glacial boulders, and the potholes are no longer measurably growing. The well-known State Park at Watkins Glen, New York, as do many other gorges in the region, owes its striking

FIGURE 10-5. Pothole drilling at Glen of Pools, Watkins Glen State Park, N.Y. (photo: Tony Ingraham, New York State Office of Parks, Recreation and Historic Preservation).

scenery to a combination of "pothole drilling" or localized whirlpool abrasion and multiple sets of vertical joints. When an abraded pothole intersects a joint face, plucking can easily remove the corner fillets to convert the cylindrical pothole to a rectilinear pool (book cover).

In stratified rocks of contrasting lithology such as sandstone and shale or limestone and shale, each resistant layer forms a waterfall in a bedrock channel, usually with a plunge pool scoured in the underlying beds. Typically, the waterfall height equals the thickness of the resistant bed (Miller, 1990), and the waterfall retreats upstream by a combination of joint-controlled plucking of the resistant ledge-forming layer and abrasional undercutting of the underlying weak layer. Each resistant ledge becomes the controlling level for an upstream reach. The inflections in the long profile of a stream are called **nickpoints** (or knickpoints, from the German term *knickpunkte:* a bending point). Their subsequent evolution after an origin by processes other than local structural control, such as fault movement across the channel that creates a headward-migrating scarp, has been the source of

FIGURE 10-6. Hanging bedrock gorges and waterfalls from a 300-m slump headscarp, Kohala coast, Hawaii. Note initial volcanic dome surface that forms the skyline (photo: R. Craig Kochel).

much geomorphic speculation. In layered rocks, their evolution is controlled primarily by the dip of the strata relative to the channel gradient. As the waterfalls migrate headward, if they also migrate downdip they will pass below the channel bed; if they migrate updip, the waterfall may grow in height. If the strata dip transversely to the channel, the flow will be concentrated against the downdip side of the channel and the channel will migrate laterally (Figure 12-7).

Bedrock canyons in the Hawaiian Islands begin as shallow, straight valleys on the constructional slope of a volcanic dome (Figure 10-6). Later, headward erosion of nickpoints by spring sapping and headward erosion from the heads of great landslide scarps (Color Plate 6) creates *theater-headed valleys,* deep canyons that head in semicircular walls on which waterfalls abound (Baker, 1990). As the waterfalls retreat, they leave gorges with boulder-lined straight profile segments in which the boulders inhibit further intrenchment (Seidl et al., 1994).

During rapid channel incision in rock, sets of stress-relief joints may open parallel to the gorge walls and extend some meters back into the rock mass on each side of the gorge (p. 175). Such young joint sets add to the structural complexity of the rock mass and probably tend to steer the channel incision parallel to their strike. It is apparent that stream channels are subjected to a variety of structural controls as they erode into rock (Baker and Pickup, 1987). If a landscape is to be shaped by fluvial erosion, some of the work must to be done by direct channel erosion into rock, without intermediary weathering and mass wasting. The complexities of the processes are clearly worth much more study.

Alluvial Channels

All large rivers, and most small ones, have channels that are usually lined with *alluvium,* sediment that was carried to that channel reach by the river and that eventually will be carried farther downstream. Such *mobile channels,* as opposed to fixed channels in bedrock, can adjust to fluctuating stream flow in many ways. Sediment that reaches a channel during intense runoff continues to move downstream until the stream power decreases. There it stops, to line the channel until the next event with sufficient power moves it again. During extreme floods, the channel may be swept clean to the rock floor, but as the flood wanes, alluvium again buries the bedrock surface. Because of the great abundance of alluvium that lines the channels of rivers, its importance to the dynamics of fluvial geomorphology, and its importance to human activity, alluvium and alluvial channels are the subject of the remainder of this chapter as well as most of the following chapter.

Much of the alluvium in the lower reaches of large modern rivers is relict, deposited during the rise of sea level as the most recent ice sheets melted. In smaller rivers, former glacial or periglacial environments, more- or less-humid former climates, or tectonic movement have also created pockets of extensive alluvium that are not compatible with the dynamics of the modern river system. This chronic problem of relict alluvium further complicates the interpretation of river behavior and history.

Channel Width, Depth, and Current Velocity

The records of stream gaging stations are essential for the prediction of potential flood damage, stream pollution, and other disasters. These records also provide a voluminous history of stream flow. In 1953, L. B. Leopold and Thomas Maddock published an analysis of thousands of measurements from stream gaging stations all over the United States. They called their analysis of the relationships among stream discharge, channel shape, sediment load, and slope the **hydraulic geometry of stream channels** (Leopold, 1994, Chapter 10).

The first step in analyzing the hydraulic geometry of stream channels was to study the changes in channel width and depth, stream velocity, and suspended load at selected gaging stations during conditions ranging from low flow to bankfull discharge and flood (Leopold and Maddock, 1953, pp. 4–9). Over a wide range of conditions, they found that width, depth, velocity, and suspended load each increase as some small, positive power function of increased discharge. Some of the gratifyingly simple equations are

$$w = aQ^b, \quad d = cQ^f, \quad v = kQ^m$$

where Q = water discharge, w = water-surface width, d = mean water depth, and v = mean current velocity. The changes of load with discharge are described later in this chapter.

The numerical values of the arithmetic constants a, c, and k are not very significant for the hydraulic geometry of streams. They are simply constants of proportionality that convert given dimensions of width, depth, and velocity into the equivalent dimensions of discharge. The numerical values of the exponents b, f, and m, which describe the changes of width, depth, and velocity for a specified change of discharge, are very important. On log–log graphs, b, f, and m are the slopes of the linear regression lines of width, depth, and velocity plotted against discharge. Leopold and Maddock (1953, p. 9) found that the average of 20 representative gaging stations in central and southwestern United States gave values of the exponents b, f, and m as follows:

$$b = 0.26, \quad f = 0.40, \quad m = 0.34$$

These values signify that on the average for the 20 rivers, as the discharge of water past a gaging station increased, as after a heavy rain, the width of the channel (in feet) increased approximately as the fourth root of discharge (in ft^3/sec) ($w = aQ^{0.26}$), the mean depth increased approximately as the square root of discharge ($d = cQ^{0.40}$), and the mean velocity increased approximately as the cube root of discharge ($v = kQ^{0.34}$). Channel width, depth, and current velocity all increase at gaging stations during increased discharge, a conclusion of no surprise to anyone who has seen a river in flood. A larger sample of 315 at-a-station values of b, f, and m showed that certain types of rivers have diagnostic values of b, f, and m, and that the range of measured values is quite large. Their interpretation is deferred until the role of sediment transport by rivers has been reviewed (Figure 10-12).

Less intuitive are the results of comparing the changes in channel shape and stream velocity in a downstream direction (Leopold and Maddock, 1953, pp. 9–16). River discharge in humid areas increases downstream. When the mean annual discharge of many rivers past successive gaging stations was compared to the width, depth, and velocity at each station, the same equations were found to apply that had been derived for changes in flow past a single point. *As mean discharge of a river increases downstream, channel width, channel depth, and mean current velocity all increase.*

Everyone knows that effluent rivers get both wider and deeper as they grow larger downstream, but until Leopold and Maddock published their work, few had guessed that average current velocity also increases downstream. The conclusion violates our poetic impressions about wild, rapidly flowing mountain streams and deep, wide, placid rivers such as the Mississippi. To the average person, the fact that many rivers accelerate as they flow downstream during average discharge is a striking and sometimes surprising revelation. They do not immediately realize that much of the high-velocity current in a mountain torrent flows in circular eddies, with backward as well as forward motion. The mean current velocity that is calculated for a succession of gaging stations down a river is the average velocity throughout each channel cross section at the time when the discharge at each of the stations is equal to the calculated mean annual discharge for that station. Figure 10-7 reproduces an example of the evidence that mean current velocity increases downstream with mean annual discharge (Leopold, 1953). Both velocity and discharge are plotted on logarithmic scales to show as a straight line the power function relationship between the variables.

The numerical values of the three exponents b, f, and m that describe the variations in width, depth, and velocity with variable discharge at a gaging station are not the same as those that describe the downstream increases in width, depth, and mean velocity with progressively increasing mean annual discharge. In the downstream direction, at mean annual discharge,

FIGURE 10-7. Velocity and discharge along the Yellowstone-Missouri-Mississippi river system, demonstrating that mean velocity increases downstream with increasing mean discharge. Additional gaging stations, unnamed, are shown by x (data from Leopold and Maddock, 1953, Appendix A).

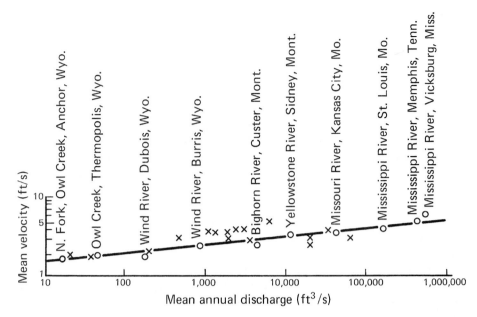

the average values for the exponents were found by Leopold and Maddock (1953, p. 16) to be

$$b = 0.5, \quad f = 0.4, \quad m = 0.1$$

These values demonstrate that in the downstream direction, channel width increases most rapidly with mean annual discharge (approximately as the square root of discharge), depth next most rapidly, and mean velocity increases only slightly, although it definitely increases. Leopold and Maddock (1953, p. 14) proposed that the increasing depth downstream permits more efficient flow in a river and overcompensates for the decreasing slope, thus providing a slight net increase in velocity at mean annual discharge.

Bathurst (1993, pp. 93ff, Table 4-3) and Park (1977, p. 140) confirmed the very narrow range of values for $b, f,$ and m downstream, suggesting that it represents a slight imbalance in the downstream direction between those factors that decrease mean velocity, such as decreased slope and increasing downstream flow resistance by channel bends and sediment bars, and those factors that increase mean velocity, such as increasing depth and decrease in grain size. The net result is a very slight increase in mean velocity downstream in rivers with a wide variety of discharge, load, and channel shapes.

A mathematical test of the hydraulic geometry equations suggests useful applications of the principles. Discharge is defined as cross-sectional area times mean velocity, or $Q = wdv$. If, as previously determined,

$$w = aQ^b, \quad d = cQ^f, \quad v = kQ^m$$

then by substitution,

$$Q = (aQ^b)(cQ^f)(kQ^m)$$

or

$$Q = ackQ^{b+f+m}$$

It follows that

$$a \times c \times k = 1.0$$

and

$$b + f + m = 1.0$$

The arithmetic constants $a, c,$ and k satisfy this test but are of no concern here. It is more interesting to verify that in the examples given for both single gaging stations and a downstream succession of gages, $b + f + m = 1.0$.

Carlston (1969a, p. 500) reviewed the statistical problems of evaluating these hydraulic geometry equations and confirmed their validity. He recalculated least-square solutions for the same data from 10 of the 20 river basins that were used to determine downstream changes by Leopold and Maddock. His more precise values of the exponents are

$$b = 0.461, \quad f = 0.383, \quad m = 0.155$$

For a larger set of 91 rivers, the downstream changes in width, depth, and velocity described by the values of the exponents b, f, and m are similar to those defined by Leopold and Maddock, and Carlston (Rhodes, 1987). The scatter of values in the larger sample shows a normal distribution around mean values that are similar to those calculated by Carlston. About 75 percent of the 91 rivers widen downstream faster than they deepen, as predicted by the mean values of b and f. Most of the 91 rivers also increase slightly in mean velocity downstream although the few that decrease cause a statistical skewing. Significantly, the most common values of b, f, and m are close to the values predicted by a theory of minimum variance. That is, they are close to the predicted optimal conditions of stream-power expenditure per unit of stream length, volume, bed area, and time (Rhodes, 1987, pp. 154–159). The significance of this example of the tendency toward least work in fluvial systems will be reviewed later in this chapter.

Longitudinal Profile

One of the interesting and significant aspects of stream flow is that as the quantity of water in the channel increases, the downstream slope of the water surface decreases. As an empirical rule, *slope is an inverse function of discharge* (Richards, 1982, p. 225). Water flows more efficiently in larger channels and therefore requires less slope to maintain its velocity. For effluent rivers, in which discharge increases downstream, the long profile should be concave to the sky, as a parabolic curve. For the Red River in Arkansas and Louisiana, a representative alluvial channel that is well adjusted to its discharge and sediment load, the relationship between slope (s) and mean annual discharge (Q_m) is $s \propto Q_m^{-0.55}$ (Figure 10-8). Carlston (1969b) measured the long profiles of numerous rivers in the Mississippi and Atlantic coast watershed in the eastern United States and confirmed that their longitudinal profiles are overall concave upward unless dominated by structural or tectonic controls (Figure 10-9). The average relationship of a group of midwestern and eastern U.S. rivers is $s \propto Q_m^{-0.65}$. However, he noted that many rivers have straight segments or even convex-skyward segments as part of the overall concavity. For a group of 79 rivers in Kansas, mostly with coarse sediment loads, $s \propto Q^{-0.25}$, and slope decreases only slightly as discharge increases (Osterkamp, 1978). Even more impressive, the Missouri River has a nearly constant slope despite increasing discharge ($s \propto Q_m^{-0.0}$) from the base of the Rocky Mountains nearly to the Mississippi River (Carlston, 1968, 1969b). Downstream from the confluence of the Missouri and the Mississippi Rivers, the relatively coarse sediment load of the Missouri steepens the slope of the Mississippi (Figure 10-9).

The inverse relationship between longitudinal river slope and average discharge can be illustrated by a practical application, if not explained by a general theory. In irrigation systems, each ditch must slope steeply enough to keep water moving and keep mud from settling in the channel, but not so steeply as to cause the water to erode the banks of the ditch. A delicate energy balance must be maintained. Through centuries of trial and error and experiment, farmers learned that successively smaller distributary irrigation ditches must be given successively steeper gradients to keep the water moving at the proper speed. The mathematical concept of a **regime canal,** which neither scours nor fills its channel, was first worked out for canals in India (Leopold and Maddock, 1953, pp. 43–48; Hey, 1988, p. 101), but the principles must date from the beginning of human history on the irrigated alluvial plains of the Tigris-Euphrates, Nile, Indus, and Ganges Rivers. In an irrigation system, the carefully controlled discharge decreases downstream, as the water from a single large feeder canal is divided and subdivided, but the inverse rule between slope and discharge holds, even in this reversal of the pattern of a normal river system. The added complexity of their highly variable discharge (as well as their complex history) makes rivers much harder to model than canals, however (Yu and Wolman, 1987).

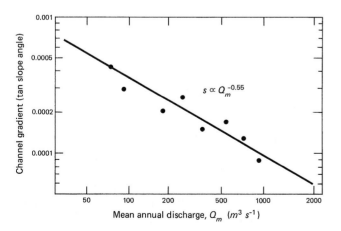

FIGURE 10-8. Slope-discharge relationship of the alluvial channel of the Red River in Arkansas and Louisiana (from Carlston, 1968).

FIGURE 10-9. Longitudinal profiles of seven rivers in the Mississippi drainage basin. Gradient units are slope tangents × 10⁵ (meters of drop per 100-km stream length) (from Carlston, 1969b, Figure 1).

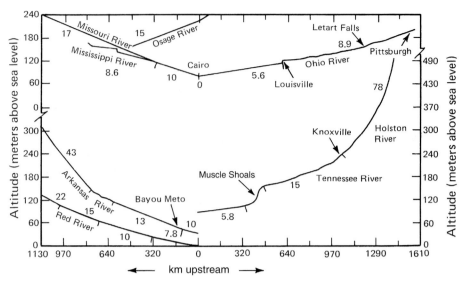

Base-Level Control

Where rivers enter the sea, the gravity potential of the falling water reaches zero. No further conversion of potential energy to stream work is possible, so sea level, and its projection under the land, is called the **ultimate base level** of stream erosion. Actually, most streams enter the sea with a considerable velocity and therefore have kinetic energy available to maintain their channel floors well below sea level, but this observation does not invalidate the use of sea level as the reference level for the limit of potential energy conversion. Within river systems, there are also nickpoints and other **local or temporary base levels** such as dams and lakes that delay upstream erosion but do not halt it.

Because rivers in humid regions have their greatest discharge and therefore their most gentle gradients near their mouths, they enter the sea at smooth, almost undetectable tangents. The Mississippi River at New Orleans is at sea level 170 km upstream from its multiple mouths in the Gulf of Mexico. The mighty Amazon River is only about 65 m above tidewater at Iquitos, Peru, 3600 km from its mouths, and at the mouth of tributary Rio Negro, still 1500 km from the sea, it is only about 15 m above sea level (Sioli, 1984, p. 129).

Repeated continental glaciations during the Quaternary Period introduced serious complications into the concept of base-level control. All large rivers now enter the sea either in *estuaries,* which are obviously the drowned segments of their former lower valleys, or across *deltas,* where borings demonstrate that 100 m or more of aggradation to a rising base level has filled former valleys (Figure 11-7). Sea level has fluctuated through a range of at least 100 m in harmony with glaciation, introducing cyclic perturbations in the hydrologic cycle on a time scale shorter than the age of most river valleys (Figure 4-7). Thus, when glaciers expanded, sea level fell and rivers eroded their valleys deeper. During the retreat of glaciers, sea level rose, and rivers with minor solid loads were drowned. Those with large loads, especially if glacier sediment was carried by the discharge, built deltas. It is not possible to demonstrate ultimate base-level control by modern sea level in any river system, yet the theoretical significance of base-level control is an important concept of geomorphology (for a recent review, see Schumm, 1993, pp. 279–280).

Dams on rivers impose an unusual form of local or temporary base-level control. The *trap efficiency* of a dam and a large reservoir can exceed 99 percent of the sediment load (Williams and Wolman, 1984, p. 9). Neither the Nile (Stanley and Warne, 1993) nor the Colorado now deliver sediment to the ocean; perhaps 13 percent of all fluvial discharge is now dammed, with corresponding sediment trapping (Milliman and Syvitski 1992, pp. 527, 540). The adjustment of a river's long profile to a dam is generally a flattening of the profile by deltaic deposition at the head of the reservoir and a steepening of the former profile by erosion immediately downstream of the dam (Williams and Wolman, 1984, p. 26). However, dams have many purposes. Some release water almost continuously for power generation. Others store water for most of the year and release it only during one season for irrigation. Still others are open most of the time and are closed only when a flood is anticipated. No single set of predicted responses is appropriate for all dams. In general, dams tend to reduce peak discharges and augment low flows so that the river discharge becomes more regular. The net loss of discharge can be very large: 64 percent of the Colorado River runoff is used for irrigation, and an additional 32 percent is lost by reservoir evaporation; almost none reaches the Gulf of California (Dynesius and Nilsson, 1994, p. 753). Landforms and channel shapes based on the episodicity of natural fluctuating discharge are most likely to be modified by dams.

OVERBANK FLOW: FLOODS

The water surface in a river channel rises during times of high discharge. The relationship is predictable enough to provide the data for calibrating a rating curve (Figure 10-3) by which discharge can be estimated directly from surface height alone. But at some particular discharge, depending on the channel geometry at the point of observation, a river will exceed the height of the channel banks, and overbank flow, or flooding, will occur. A **flood** is synonymous with discharge greater than bankfull channel capacity although many people carelessly refer to any high discharge as a flood.

Most floods have meteorologic causes occasioned by anomalies of atmosphere circulation that can vary greatly in temporal and spatial scales. An hour or less of intense thunderstorm precipitation can cause flooding in small stream valleys, or a decade can separate seasons of heavy precipitation and floods such as the El Niño floods in the Peruvian coastal desert. The spatial extent of flood precipitation can range from 1 km^2 for a local storm to 10^6 km^2 for a major tropical cyclone or hurricane (Hirschboeck, 1987, 1988).

Flood Frequency and Magnitude

Projects such as river basin planning studies and the design of bridges and road networks rely upon flood frequency analyses (U.S. Water Resources Council, 1981; Chow et al., 1988). Fundamentally, the goal of flood frequency analysis is to estimate the probability with which various discharge magnitudes will be exceeded in the future. In general, if the **exceedance probability** of a flood discharge of a specified magnitude is p, then on average that discharge level will be exceeded once every $T = 1/p$ yr. T is called the **recurrence interval** or **return period** of the flood, and is the inverse of the exceedance probability. For example, the flood discharge that has a 1 percent probability of being exceeded in any year is called the 100-year flood. However, this usage has the unfortunate connotation that only one T-year flood will occur in a T-year period, or that these floods occur regularly every T years. Neither connotation is correct; exceedances actually occur at irregular intervals.

The *Hurst effect* is a well-known feature of climatic and meteorologic data sequences such as floods; wet or dry (or warm or cold) years tend to occur in clusters (Kirkby, 1987). The phenomenon has also been called the "Joseph effect" with reference to Joseph's biblical prediction to the Egyptian pharaoh of seven fat years followed by seven lean ones. Because of the Hurst effect, it is not at all unusual to have several "hundred-year floods" in a single decade. Hurricanes in the Atlantic tropics have been shown to occur in multi-decadal clusters, correlated with wetter years in tropical Africa, for example (Figure 19-4). The Hurst effect is important to geomorphology because a sequence of closely spaced floods could have a greater geomorphic impact than a similar number of events of comparable magnitude but more uniform spacing. Instabilities generated by the earlier floods would not have time to equilibrate before the next flood struck.

To establish a relationship between the frequency and magnitude of floods, it is desirable to have a record at least as long as the return period of the events of interest. With such a record, analytical and graphical methods can be used to estimate the frequency with which various discharge magnitudes are exceeded. This is often done using an **annual flood series,** a list of the largest flood that occurred in each water year within the available record. A **water year** is a conveniently chosen 12-month period that need not begin on January 1. In the United States, water years generally start on October 1. The annual flood series is often sorted by magnitude so that one can refer to the ith largest. It is easy to show that, on average, the probability p_i that the ith largest flood will be exceeded in the future is (Loucks et al.,1981, pp. 108–109)

$$p_i = i/(n + 1)$$

Thus, the ith largest flood is often assigned a return period of

$$r_i = 1/p_i = (n + 1)/i$$

For example, the largest flood in a 25-year annual series would be assigned a return period of $(25 + 1)/1 = 26$

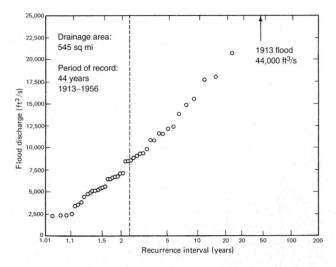

FIGURE 10-10. Gumbel graph of recurrence interval of annual floods of a specified magnitude, Miami River, Sydney, Ohio. Mean flood recurrence interval of 2.33 years is indicated (from Cross and Webber, 1959, Figure 337; later supplemental data do not significantly change the intervals).

years, and the fifth greatest annual discharge would be assigned a return period of $(25 + 1)/5 = 5.2$ years.

The magnitude and return period (or alternatively, the exceedance probability) of floods in an annual series can be plotted on probability paper that can be constructed for particular families of probability distributions, such as the normal, log-normal, or Gumbel EV-1 distributions (Gumbel, 1958). Each distribution describes a different relationship between the frequency and magnitude of flood events, and the interpretations can vary widely (Chow et al., 1988, Chapter 12; Malamud et al., 1996). From such a graph (Figure 10-10), the exceedance probability of a flood of any intermediate magnitude can be estimated by interpolation. The relative magnitudes of floods in an annual series should plot approximately as a straight line if the type of paper corresponds to the distribution from which the flood series was drawn (Loucks et al., 1981). However, sampling fluctuations often result in larger deviations at the low and high ends of the graph. A graphic solution is usually considered reliable for events with recurrence intervals of up to one-half the length of the annual series.

The *median flood* in an annual series has an exceedance probability of 50 percent, corresponding to a return period of 2 years. Thus, the sample median magnitude for the 44-year flood series in Figure 10-10 would be the average of the floods ranked nos. 22 and

23 in the series. For the Gumbel distribution, the *most probable annual flood* has an exceedance probability of 63 percent, corresponding to a return period of 1.58 years; for the normal distribution, the most probable flood is the same as the median flood.

An alternative analytical procedure is to rank all discharges above a selected magnitude or threshold. This approach can include more than one flood peak in some water years, and none in others. A minimum time duration for each event is commonly specified. Such a record is called a **base-stage series** or **partial-duration series.** When modeling flood frequency employing partial duration series, one can model both the frequency with which the specified threshold is exceeded and the likely magnitude of such exceedances. Alternatively, one can directly estimate the arrival rate (or frequency in events/yr) of all floods exceeding specified thresholds.

If the arrival rate of floods exceeding a threshold discharge is υ events per year, then for statistically independent events the annual probability p with which that discharge is exceeded is $p = 1 - \exp[-\upsilon]$. For υ on the order of 0.05 events/yr or less, the annual exceedance probability p essentially equals the exceedance frequency υ. The two approaches are often considered to be equivalent (Langbein, 1949).

It has been hypothesized, and demonstrated for several examples, that in well-adjusted alluvial rivers the most probable annual flood, with a return period of 1.58 years, is similar in magnitude to bankfull discharge (Dury et al., 1963; Leopold, 1994, pp. 134–145). This would imply that bankfull discharge corresponds to a flood level equaled or exceeded once per year in a partial duration series ($\upsilon = 1$ event/yr). Dury (1967) concluded that bankfull discharge is the dominant channel-shaping flow and is likely to be reached or exceeded once per year on average by streams in alluvial channels. Overall the adjustments of natural channels to both large floods and smaller discharges is a complex compromise between the frequency of different flow rates, their duration, the river's competence to transport sediment, the bed and bank material, and many other factors (Richards, 1982, pp. 122–144; Brakenridge, 1988).

Several of the variables of hydraulic geometry change abruptly when flood stage is reached. Most obvious are the changes of width and depth with discharge. Prior to overbank flooding, with increasing discharge at a station, depth typically increases more rapidly than width (p. 201). However, when overbank flooding begins, width abruptly increases manyfold, so that depth and velocity need not increase at all in order that $Q = wdv$. In the power functions for width, depth, and velocity at a station during flood,

$$w = aQ^b, \quad d = cQ^f, \quad v = kQ^m$$

the exponents $b \simeq 1.0, f \simeq 0.0$, and $m \simeq 0.0$. With those values of b, f, and m, the width of the flooded plain increases in linear proportion to discharge, and if the floodplain is very wide relative to the normal channel, depth and velocity need not increase at all. Actually, depth continues to increase slightly, but the mean velocity of a river in flood decreases to less than the mean velocity at bankfull stage because the broad, shallow sheet of water over the floodplain has a very low velocity that reduces the average of the entire cross section. Velocity remains essentially constant within the reach of maximum flooding as the flood crest migrates downstream along the length of a river. The downstream increase in flood discharge is accommodated not by faster flow, but by the slightly increased depth and the greatly increased width.

The slope along a meandering channel is always less than the down-valley slope of the floodplain because the length of the meandering channel is typically several times greater than the straight-line distance along the floodplain. Thus, rivers in flood have a down-valley surface gradient that is several times as steep as when the river is confined in a meandering channel. This increased slope enables the floodwaters to move down the valley efficiently even though the mean flood velocity need not be as high as in channeled flow—thus the peculiarity that rivers in flood flow on an increased average down-valley slope but at decreased mean velocity. The greater width of the floodplain more than compensates for the lack of increased velocity.

Catastrophic Floods and Paleofloods

Graphs of flood-frequency relationships may include one or more major floods of far greater discharge than would be predicted from the magnitude of lesser discharges. The 1913 flood in Figure 10-10 is such an example. Even larger catastrophic events can be due either to truly unusual and anomalous meteorologic events within the present climatic setting, to relict former climatic conditions such as glaciation, or to "accidents" such as the failure of a landslide or glacier dam that had temporarily impounded great volumes of water (Costa and Schuster, 1988; Mayer and Nash, 1987). Such an event, even if it occurred thousands of years earlier, may have been the principle event that shaped the present-day river channel and valley.

Paleoflood alluvium in the Grand Canyon provides evidence of at least 15 major floods in the past 4500 years, one of which had more than twice the discharge of any recorded flood (O'Connor et al., 1994). Since the closing of Glen Canyon dam in 1953, upstream from the Grand Canyon, floods have been suppressed and sediment has been trapped in the reservoir (Lake Powell). The sand bars in the canyon have been eroding. In a unique experiment, in March 1996, the dam gates were opened to create a modest flood of one-week duration in the Grand Canyon (Collier et al., 1997). Although results are as yet incompletely evaluated, some sand bars were replenished and two of the largest rapids in the canyon were widened. The cost of the experiment, including lost revenue from electric power generation, was substantial.

Many, perhaps most, rivers in glaciated regions have observable channel and valley forms that were shaped by catastrophic paleofloods from lakes that had been temporarily impounded by glaciers (Kehew and Lord, 1986; Teller and Kehew, 1994). The greatest floods known to have occurred on earth are the Pleistocene Lake Missoula paleofloods that repeatedly broke through a glacial lobe in eastern Washington and gouged thousands of square kilometers of anastomosing river channels across the Columbia Plateau to the Columbia River (Bretz, 1969; Baker, 1973; Waitt, 1985). Paleofloods in the Altay Mountains of Siberia may have been as large (Baker et al., 1993). Another great flood resulted from the overflow of pluvial (p. 409) Lake Bonneville into the Snake River (Malde, 1968; Jarrett and Malde, 1987; O'Connor, 1993). These great floods occurred more than 12,000 years ago, but their eroded channels and enormous alluvial deposits are the dominant features of the present landscape on the Columbia Plateau and the Snake River Plain (Baker et al., 1987, pp. 413–443). The maximum discharge of the Lake Missoula paleofloods (O'Connor and Baker, 1992) was at least $17 \pm 3 \times 10^6$ m³/s (= 600 $\times 10^6$ ft³/s; compare with Figure 10-7). Alluvium deposited by the Lake Bonneville paleoflood includes well-rounded basalt boulders more than 10 m in diameter (O'Connor, 1993). A considerable extrapolation of Hjulström's curve (p. 215) is required to estimate the flood velocity that could move such boulders. One estimate is that the peak discharge through a constricted reach of the Snake River canyon was about 1 million m³/s, for a duration of eight weeks (Jarrett and Malde, 1987). Midchannel alluvial bars as much as 100 m high, 2 km long, and 1 km wide were built of boulder gravel.

Postflood weathering and erosion have been insignificant in modifying the shape of the paleoflood deposits. Accounts of these great natural catastrophes should be read by all geomorphologists as a basis for speculation and contemplation on the role of infrequent but spectacular events in the shaping of "normal" landscapes.

SEDIMENT EROSION AND TRANSPORT IN CHANNELS

Solid and Dissolved Loads

Weathered rock, derived from the upland watershed or by plucking and abrasion in the channel, is carried by rivers in three forms. The compounds in solution or colloidal mixtures are the **dissolved load.** The solid matter is either fine-grained particles in suspension (**suspended load**) or coarse-grained particles that slide, roll, or bounce along the stream bed (**bedload, or traction load**). The division of the load varies greatly, controlled by climatic and structural factors. Table 15-1 gives some estimates for major rivers of the world. Rivers that are subject to large fluctuations in discharge, or that drain poorly vegetated regions, or that have generally large loads, tend to have high percentages of solid load (Laronne and Reid, 1993) and are likely to have braided channels (p. 222). Rivers that drain heavily forested regions, or overflow from lakes that act as settling basins, or drain karst regions, tend

to have mostly dissolved loads and are likely to meander (p. 221).

Both the dissolved and suspended loads in river water are routinely measured at gaging stations, and excellent data are available from many countries. Bedload has defied attempts to measure it accurately. Any device lowered to the bed of a stream to measure or collect the sediment in motion also deflects the boundary-flow conditions and distorts the measurement. Samplers designed since 1970 have nearly 100 percent efficiency for sediment in the coarse sand to fine gravel range (0.5–16 mm), but efficiency declines rapidly for both coarser and finer sizes. Obviously, trap efficiency is zero if the sampler orifice is smaller than the particles in motion (Emmett, 1980). Bedloads as great as 50 to 55 percent of the total solid load have been estimated for alluvial rivers on the Great Plains (Leopold and Maddock, 1953, pp. 29–30), but for most purposes bedload is simply assumed to be 10 percent of the total solid load, or about an additional 11 percent of the measured suspended load.

Dissolved load has no detectable effect on stream flow. The concentrations average only 130 ppm (Livingstone, 1963), so the solutions are too dilute to affect viscosity, turbulence, or the density of river water. Therefore, dissolved load, which can represent more than half the total work of fluvial denudation (Table 15-1), gets a "free ride" to the sea by rivers. No kinetic energy is required to move it.

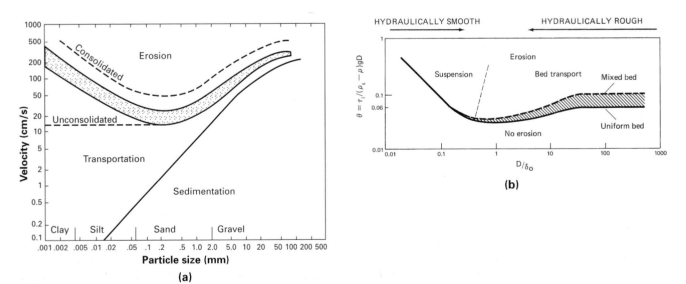

FIGURE 10-11. (a) Relation of mean current velocity in water at least 1 m deep to the size of mineral grains that can be eroded from a bed of similar-size grains and transported by the current. Dashed lines above and below upper curve are revisions proposed by Sundborg (1956). (b) The Shields curve, relating the Shields parameter Θ, or threshold of particle motion (erosion), to the grain Reynolds number D/δ_0, a measure of bed roughness.

Competence of Rivers

Competence is the measure of a stream's ability to transport a certain maximum grain size of sediment. Competence depends primarily on velocity although channel shape, shape and degree of sorting of the sediment particles, amount of suspended load, and water temperature can also affect competence. Figure 10-11a shows graphs of the relationship between mean velocity in a river and the size of particle that can be eroded or transported by it (Hjulström, 1935, 1939; Nevin, 1946; Sundborg, 1956, pp. 177–180). The assumptions in the figure caption should be noted because the mean velocity in a stream is not the velocity at the bank or bed of the stream, where much of the erosion and transport occur. Furthermore, the curves were experimentally derived from flume experiments on well-sorted sediments. If coarse and fine particles are mixed, the fine particles fill spaces between the coarse ones and are thereby protected, but they also help prevent the coarse grains from moving. Sediment transport by wind has similar competence limits (Figure 13-12).

Figure 10-11a demonstrates that the grain size of sediment that will stay in suspended transport is a power function of mean velocity (approximately a straight line on the log-log graph). A point on the lower graphed line, ("Stoke's law" of settling) between the transport and sedimentation fields, represents the mean velocity in a stream at least 1 m deep that will provide sufficient turbulence to keep a particle of a given size from settling out of suspension. The upper curve ("Hjulström's curve") is actually a field or zone of values for the velocity necessary to initially dislodge or erode particles. For pebbles, cobbles, and boulders, the erosion threshold is only slightly greater than the transportation velocity because they barely rise from the bottom as they bounce or roll as bedload. Actually, large stones on the bed of a stream are overturned or rolled by mean velocities somewhat less than those indicated by Hjulström's curve. The exact relationship between eroding velocity and large particle sizes has yet to be worked out (Novak, 1973). Probably 95 percent of the total work done by stream transport is attributable to the few percent of the load that is in the coarsest grain sizes (Bagnold, 1968, p. 53). The minimum threshold eroding velocity of 15 cm/s is for fine sand, but sand can be kept in transport by a mean velocity an order of magnitude less if some other process stirs it from the bottom. Silt and clay sizes (collectively called **mud** when wet and **dust** when dry) are kept in suspension by very slight currents, but the velocities necessary to erode a consolidated clay bank are comparable to those that move pebbles and cobbles because of the strong cohesiveness of clay-size particles and their relative smoothness.

Most formulas for the flow of water and sediment were experimentally determined in either closed pipes or open flumes with fixed walls. The geomorphic equivalent would be a bedrock channel with fixed sides and floor of specified hydraulic resistance, or *roughness.* Alluvial channels have mobile beds, so that width, depth, velocity, roughness, and other channel parameters constantly adjust to discharge and sediment load. Furthermore, whereas the concept of the regime canal (p. 209) applies to flows with almost constant discharge, rivers have extremely variable discharge, fluctuating through several orders of magnitude at a gaging station. The dynamic behavior of rivers and their alluvial channels may result in channel dimensions that are not appropriate for the discharge and sediment load at the time of measurement, the channel retaining a kind of "memory" of the most recent effective channel-shaping discharge (Yu and Wolman, 1987).

Despite the complexity of natural rivers and their channels, geomorphologists continue to rely heavily on hydraulic engineering formulas to explain how fluvial systems work. For example, the *Reynolds number* (*Re*) is a dimensionless number that distinguishes laminar from turbulent flow by the ratio of inertial and viscous forces. It is defined as

$$Re = \rho v R / \mu$$

where ρ is the fluid density, v is the mean velocity, R is the *hydraulic radius* of the channel (cross-sectional area divided by length of the wetted channel perimeter; approximated by stream depth), and μ is the fluid viscosity. When *Re* is <500 in water, viscous forces dominate and flow is *laminar,* with each element of fluid moving in parallel paths with uniform velocity. When *Re* is >2500 in water, inertial forces of mass and velocity dominate and flow is *turbulent.* Complex transitional flow occurs when 500<*Re*<2500 (Richards, 1982, p. 59; Knighton, 1984, p. 48). Water flow is turbulent in almost all natural channels.

The *Froude number* (*Fr*) is another dimensionless number that defines the type or degree of turbulence. It distinguishes *subcritical* or *tranquil* flow (*Fr* <1) from *supercritical* flow (*Fr* >1) with breaking surface waves and increased resistance to flow. Supercritical flow is uncommon in rivers with mobile-bed channels (Richards, 1982, p. 61). The formula for the Froude number is

$$Fr = v / \sqrt{gd}$$

where v is the mean velocity, g is the gravitational constant, and d is the depth. Note that both the Reynolds number and Froude number are velocity and depth dependent.

Several other hydraulic formulas relate the velocity of nonaccelerating channel flow to the total resistance of the bed so that the gravitational acceleration of water in an inclined channel is balanced by the surface area and roughness of the bed. One of them, the *Manning equation*, is

$$v = kR^{2/3}s^{1/2}/n$$

where v = mean velocity, k is a dimensionless constant (=1 in metric units and 1.486 in English units), R is the hydraulic radius (defined as the cross-sectional area divided by the wetted perimeter, but commonly approximated by mean channel depth), s is the longitude slope, and n is the *Manning roughness coefficient*, another dimensionless number that defines the flow resistance of a unit of bed surface. This resistance includes not only particle size, but also the shape of the bed and constructional bed forms such as ripples. A book of color photographs of stream reaches for which the Manning roughness coefficients have been computed has been published (Barnes, 1967). A reach being studied can be compared to one of the photographs and an appropriate value for n thereby selected. As in the formulas for the Reynolds number and Froude number, the Manning equation relates mean velocity to depth (as an approximation of R; but the approximation may be a poor one: Tinkler, 1981), as well as to surface slope and bed roughness (Richards, 1982, p. 63). It is apparent that channel depth is an especially critical dimension that affects velocity, turbulence, and bed shear stress.

For coarse sediment moving on the bed of a channel, the *boundary shear stress* may be a better measure of competence than the average water velocity (Baker and Ritter, 1975). The boundary shear stress τ_0 exerted on its bed by a stream is approximated by

$$\tau_0 = \gamma Rs$$

where γ is the specific weight of water (ρg), R is the hydraulic radius, and s is the slope. When $\tau_0 = \tau_i$, the threshold shear stress, particle motion on the bed will begin. The *Shields curve* (Figure 10-11b) relates two dimensionless variables, the *Shields parameter*, or threshold of particle motion,

$$\theta = \frac{\tau_i}{(\rho_s - \rho)gD}$$

to the *grain Reynolds number,*

$$D/\delta_0$$

where τ_i = threshold bed shear stress
  ρ_s = grain density
  ρ = fluid density
  g = acceleration of gravity
  D = grain diameter
  δ_0 = thickness of the laminar-flow boundary layer (a function of the threshold shear stress and the fluid viscosity)

The Shields curve relates θ, the threshold of sediment motion, which includes particle and fluid density, particle grain size, grain shape, and packing, to bed roughness, expressed by the ratio D/δ_0, which measures the extent that individual particles project above the laminar-flow boundary layer into the turbulent flow. The Shields curve is similar to Hulström's curve in that both define a threshold of sediment erosion. In the Shields curve, the threshold is at a minimum value of the Shields parameter corresponding to a certain bed roughness (Knighton, 1984, pp. 56–60; Komar, 1988). Fine sand and mud beds, in which the bed roughness is less than the thickness of the laminar-flow boundary layer ($D/\delta_0 < \delta_0$), are more difficult to erode than rougher beds. This effect reinforces the effect of cohesiveness in mud in resisting erosion as shown by Hjulström's curve (Figure 10-11a).

Recalling that the exponents $b + f + m = 1$ in the hydraulic geometry equations that describe width, depth, and velocity variations as power functions of discharge (p. 207), Rhodes (1977) used ratios derived from various hydraulic formulas to define various responses of channel cross sections to changing discharges (Figure 10-12, Table 10-1). On a triangular diagram, he plotted b, f, m values for 315 sets of single-station measurements so that at a labeled apex of the triangle, that exponent equals 1.0, and on the opposite side, the same exponent equals zero (Figure 10-12a). Rhodes divided the triangular diagram by five lines, defined as follows (Figure 10-12b):

1. $b = f$. If $b = f$, with variable discharge, the width-depth ratio stays constant. For all sets of values that plot in the left side of the diagram ($b < f$), channel width increases faster than depth with increasing discharge. Such streams typically carry mostly bed-load and commonly have a braided pattern. Streams on the right half of the diagram ($b < f$) deepen faster than they widen with increasing discharge, presumably because they have resistant banks and mobile

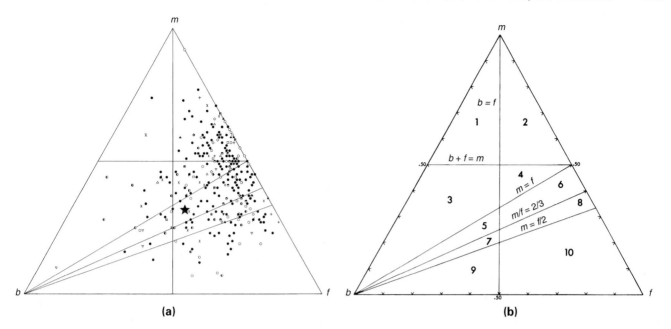

FIGURE 10-12.   (a) The *b-f-m* diagram showing plotting position of 315 sets of at-a-station hydraulic geometry exponents. Some points represent more than one station (Rhodes, 1977, Figure 1). (b) Divided *b-f-m* diagram showing numbered areas in Table 10-1 that delineate channel types (Rhodes, 1977, Figure 2).

beds. Most of the sampled channels show this behavior.

2. $m = f$. Changing competence with changing discharge is related to the increase in the velocity-depth ratio (m/f). For points that plot above the line $m = f$ ($m > f$), velocity increases faster than depth and competence should increase with increased discharge (at

constant width). This inference has not been proved by hydraulic theory, but is a reasonable hypothesis.

3. $m = f/2$. The significance of this line is derived from the formula for the Froude number. With increasing discharge, if velocity increases faster than the square root of depth (that is, if m, the exponent that describes the change in velocity, is greater than $f/2$, the expo-

TABLE 10-1

Expected Direction of Change of Morphologic and Hydrodynamic Parameters with Increasing Discharge for Different Channel Types

Channel Type	Width/depth Ratio (w/d)	Competence	Froude Number	Velocity/ Area Ratio (v/A)	Slope/ Roughness Ratio ($s^{1/2}/n$)
1	Increases	Increases	Increases	Increases	Increases
2	Decreases	Increases	Increases	Increases	Increases
3	Increases	Increases	Increases	Decreases	Increases
4	Decreases	Increases	Increases	Decreases	Increases
5	Increases		Increases	Decreases	Increases
6	Decreases		Increases	Decreases	Increases
7	Increases		Increases	Decreases	Decreases
8	Decreases		Increases	Decreases	Decreases
9	Increases		Decreases	Decreases	Decreases
10	Decreases		Decreases	Decreases	Decreases

Source: Rhodes (1977), Table 2.

nent that describes the change in the square root of depth, or $m > f/2$) the Froude number increases. Supercritical flow may result. If the Froude number decreases with increased discharge ($m < f/2$), supercritical flow will not be attained, and transport of particles larger than pebbles will be unlikely.

4. $m = b + f$. From the requirement that $b + f + m = 1$, for points plotting above this line, $m > 0.5$, and with increasing discharge the increase in velocity past a point is greater than the sum of the increase in width and depth ($b + f$). Such streams experience rapid increase in velocity relative to cross-sectional area, and probably flow in stable bedrock channels that exhibit decreased resistance with increasing discharge (Baker and Costa, 1987, p. 7).

5. $m/f = 2/3$. This division of the triangle diagram is derived from an approximation of the Manning equation in which depth is substituted for hydraulic radius. In the Manning equation, the power-function relationships of velocity with depth, slope, and the Manning roughness coefficient are such that if $m > 2/3f$, the ratio of slope to roughness ($s^{1/2}/n$) increases. Observations of slope demonstrate that it changes very little with increasing discharge, so for points that plot above this line, bed roughness probably decreases with increased discharge. Possible causes are greater depth, which reduces bottom shear stress, increased suspended load that reduces turbulence, or a more direct flow path in a larger channel.

The ten channel types marked out on the triangular diagram by the lines defined here are numbered on Figure 10-12b, and the predicted effect of increasing discharge for each of these is summarized in Table 10-1. Rhodes's compilation demonstrates the wide variety of morphologic and hydrodynamic changes that can be predicted in a single cross section of a channel because of the relative changes in only three variables: width, depth, and velocity (slope changes being negligible). The average values of b, f, and m calculated by Leopold and Maddock in 1953 ($b = 0.23$, $f = 0.40$, $m = 0.34$) happen to plot in Rhodes's channel type 6 (star, Figure 10-12a), but they are in no way representative of the 315 widely scattered sets of data complied by Rhodes. A similar conclusion was reached by Park (1977) using a somewhat smaller data set.

Bathurst (1993, pp. 90ff: Table 4-2) compiled typical values of b, f, and m at stations on a variety of bedrock and alluvial channels. He concluded that exponent f does not change much whether the bed is sand, cobble gravel, boulder gravel, or a steep series of waterfalls and pools, but that the exponent m

increases progressively with the grain size of bed material, mostly at the expense of a decrease in the exponent b. Velocity changes most rapidly with discharge (exponent m) in bedrock channels with pools and waterfalls because at low water, the pools may be almost still water, whereas at high discharges, the water roars through the pools and over waterfalls with a more nearly uniform slope, and velocity increases rapidly. Obviously, even at different gaging stations along a single river, the possible responses are enormously variable.

Capacity of Rivers; Variations of Suspended Load

Capacity is the theoretical maximum amount or mass of sediment load that a stream can transport. The grain size of the detritus may partly determine how much can be carried, but capacity is primarily a measure of the maximum amount, not grain size, of the load. It must be considered carefully because of frequent confusion of the idea of capacity with that of competence (Nevin, 1946). The maximum possible suspended load cannot be specified because at some arbitrary concentration of mud in water, a muddy river is simply called a mudflow instead (p. 176). One of the greatest reported suspended loads is that of a tributary to the Hwang He (Yellow River) of northern China. It dissects a great region of windblown silt (loess) and is reported to carry a suspended load of 48 percent silt by weight during high discharge (Eliassen, in Lane, 1937, p. 154).

Because dissolved load has no effect on the hydraulic geometry of rivers, and bedload defies accurate measurement, the load of a river that is usually measured is the amount of suspended load. The unit of measurement is the dry weight of sediment per volume of water or more commonly units such as tons per day of sediment for a specified discharge. Table 15-1 lists some annual river loads.

The hydraulic geometry of streams involves, in addition to discharge, slope, width, depth, and velocity, the amount of solid load as well as bed roughness and grain size of the load. The power-function equations relating these additional variables to discharge are similar in form to the equations for slope, width, depth, and velocity. For example, the power-function equation for the sediment load at a station is $L = pQ^j$, where L is suspended-sediment load, Q is discharge, and p and j are numerical constants (Leopold and Maddock, 1953, p. 21). Values for the exponent j range from 2.0 to 3.0. These large exponential values mean that as discharge at a station increases tenfold, the suspended load may increase a hundredfold to a thousandfold! Figure 10-13 is a typical graph of

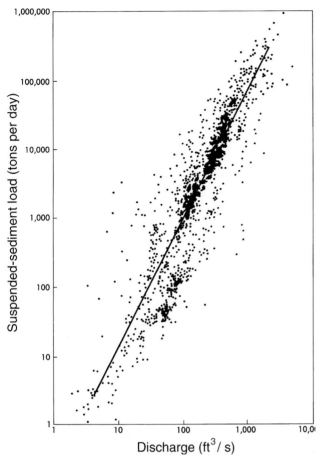

FIGURE 10-13. Relation of suspended-sediment load to discharge, Powder River at Arvada, Wyoming (Leopold and Maddock, 1953, Figure 13).

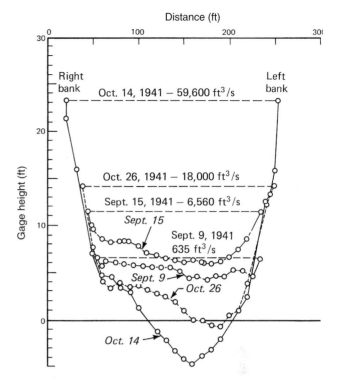

FIGURE 10-14. Channel cross sections during progress of a flood, September–October 1941, San Juan River near Bluff, Utah. Horizontal line is the water surface on the indicated date (Leopold and Maddock, 1953, Figure 22).

suspended-sediment load compared with discharge. The suspended-sediment load at a station increases much more rapidly with discharge than either channel width or depth; therefore, the enlargement of the channel by erosion cannot account for all the increased load. Most of the suspended sediment comes from the watershed upstream from the gaging station. The sediment is newly delivered to the stream by mass wasting and rill wash during the same rains or snowmelts that swell the discharge of the river.

Measurements of channel shape and suspended-sediment load confirm that streams move most of their loads during times of higher than average discharge. As the water rises, alluvial channels initially aggrade their bed (Figure 10-14), Presumably, the aggradation results from the sudden influx of new sediment from the uplands and the decreased roughness effect of the more dense water-sediment mixture. Local aggradation increases the bed slope in order to move that sediment downstream. During the peak discharge (Figure 10-14, October 14 cross section), the

channel may be scoured even deeper than during the preflood level, and then aggrade again during the later declining discharge. Thus competence is not to be simply correlated with velocity. At a given velocity during increasing discharge, bed aggradation occurs, but when that same velocity is reached during decreasing discharge, the bed may be eroded (Leopold and Maddock, 1953, pp. 30–35).

The sediment-transporting competence and capacity of a river also change abruptly during overbank flooding. Although an enormously greater sediment load is in transit (Figure 10-13) along the axis of the permanent channel, the shallow, low-velocity sheet of floodwater over the adjacent floodplain has little competence. As velocity is checked, sand, silt, and clay settle out of the floodwater, and the shallowness over the floodplain ensures that the fine sediment will reach the bottom of the water column before turbulence sweeps it downstream. People who live on floodplains know that a single flood may deposit a meter or more of muddy alluvium on their fields or in their houses. It may seem anomalous at first reading that a river in flood is flowing down a steeper gradient (p. 213) but may be less competent to transport sand and mud. Yet these interactions of the several semidependent variables of hydraulic geometry of rivers in flood illustrate well the fundamental role of channels and floodplains: to provide a low-gradient, efficient, sinuous or braided channel for average discharges and loads, and a steep, shallow, broad, constant-mean-velocity channel for overbank flood discharges. Truly, floodplains are for rivers. People occupy floodplains only at great risk, but judging from historical patterns, it is a risk that many find tolerable or necessary.

In some rivers, the net channel erosion of the peak runoff is gradually restored by deposition of alluvium during the low-water season. If channel shape were measured at the same location at the same season in successive years, one might believe that no permanent change had taken place in the river channel during a period of high discharge, unless one realizes that the mud, sand, and gravel that previously formed the stream bed at the location have been moved downstream toward the sea and have been replaced by new sediment from upstream. The geomorphology machine has surged ahead slightly.

The change of suspended-sediment load downstream with increasing discharge has not been measured directly, but it can be estimated indirectly from other parameters of downstream hydraulic geometry (Leopold and Maddock, 1953, pp. 21–26; Rhodes, 1987, p. 152). In the equation $L = pQ^j$, the value for exponent j in a downstream direction is 0.8, which by being less than 1 implies that the total suspended-sediment load increases downstream less rapidly than discharge, and, therefore the concentration of suspended sediment becomes more dilute toward the river mouth. Two factors are probably involved. First, few small tributaries and negligible overland flow enter a trunk stream directly, and these are the original source of most of the sediment load. Second, effluent groundwater that enters the trunk stream increases discharge but carries no suspended load, thus diluting the concentration.

In contrast to the decreasing sediment concentration downstream in through-flowing rivers of humid regions, the concentration of suspended load increases downstream in intermittent and ephemeral streams, and also in rivers of dry regions that decrease in discharge downstream as they lose water by infiltration and evaporation (Rhodes, 1987, p. 157). As emphasized in Chapter 13, the hydraulic geometry of fluvial systems in dry regions is very different from that of rivers in humid regions (Laronne and Reid, 1993).

Channel Shape, Habit, and Solid Load

The bed and banks of an alluvial channel are similar in grain size to the solid load of the river at that point, for if the stream is competent to move a fragment on the stream bed, it will do so, and that fragment becomes part of the bedload or suspended load. If the stream lacks competence to move fragments of a given size, they settle and become part of the alluvial bed. A constant exchange between bed and load takes place. Even though the channel is shaped primarily during high discharge (Figure 10-14), for every lesser discharge there is a combination of depth, velocity, and other variables that has a certain competence to erode or transport grains selectively up to a certain size limit (Figure 10-11). To the extent that channel roughness is determined by the grain size or bed forms of the alluvial bed, and roughness affects turbulence and competence, a complex readjustment between discharge, grain size of the load (and the bed), channel shape, velocity, and the amount of load is constantly in progress.

Water discharge (Q) determines the size, or cross-sectional area of an alluvial channel, but the amount and grain size of the sediment load determine both the channel cross-sectional shape (width-depth ratio) and **habit,** its planimetric shape or pattern (Schumm, 1977, 1985). Because channel shape varies so much, it is useful to refer to the **thalweg,** the line connecting points of maximum water depth in a general downstream direction along the channel, when describing channel habits. Most thalwegs pass through a succession of *pools* in the channel bed that are separated by *riffles*, which might be sedimentary bedforms or bedrock ledges. The pools and riffles of stream bed cause the thalweg to have an irregular slope, rising and falling in the downstream direction.

Some rivers have multiple channels that separate and rejoin over distances of many kilometers, with large islands separating several relatively stable channels (Nanson and Knighton, 1996). These are called **anastomosing,** or **anabranching,** channels. Single-channel alluvial rivers may have reaches that are straight, meandering, or braided. Six channel habits or patterns have been defined in terms of their relative stability, their sediment loads and grain size, ratio of suspended load and bedload, and other related variables (Figure 10-15). Changes in habit tend to be abrupt, when certain threshold conditions are reached or exceeded.

Straight Channels. Suspended-load channels (<3 percent bedload) with low gradients, low total load, and low velocity may be straight and narrow, with or

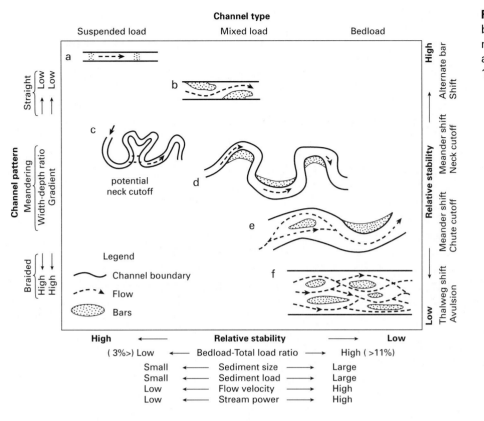

Channel type

Suspended load   Mixed load   Bedload

FIGURE 10-15. Channel patterns based on relative stability, sediment abundance and grain size, and other variables (Schumm, 1981, Figure 4).

High   ←   **Relative stability**   →   Low
(3%>) Low   ←   Bedload-Total load ratio   →   High (>11%)

Small	← Sediment size →	Large
Small	← Sediment load →	Large
Low	← Flow velocity →	High
Low	← Stream power →	High

without migrating submerged transverse bars. Such channels are rare, but if natural, may be quite stable (Figure 10-15a). With mixed suspended and bedloads, straight channels may develop alternate bars and a sinuous thalweg although the channel banks remain straight (Figure 10-15b). The alternate bars migrate downstream, remaining on the same side of the channel, thus exposing each bank to alternating intervals of erosion and protection.

When other types of channels are artificially straightened, they are almost always unstable. **Channelization,** artificial straightening of alluvial channels to decrease local flooding or to reclaim floodplains, greatly increases bankfull discharge, stream velocity, and down-valley gradient. The results are abrupt and dramatic (Emerson, 1971; Simon, 1994). The straightened reach usually deepens rapidly, followed by widening due to bank collapse. Aggradation downstream of the channelized reach may cause increased flooding there.

Meandering Channels. A single sinuous alluvial channel that carries only suspended load, with small total load and low velocity, is a simple example of *meandering* (Figure 10-15c). Channel width stays reasonably constant between stable banks, but as the curvature increases, a *neck cutoff* can occur. Otherwise,

these channels are relatively stable. With mixed suspended load and bedload, however, meandering channels begin to build *point bars* on the inside of meanders and undercut their outer banks as the thalweg alternately impinges on opposite banks (Figure 10-15d). The total load is usually larger in this type of meandering channel. Channels are wider at sharp bends than along regularly curving reaches. Such channels are quite unstable, with *chute cutoffs* across the back of the point bars adding to neck cutoffs as a process of shifting.

Alluvial meanders are remarkably regular, and their dimensions are proportional to channel width. The radius of curvature of meanders is usually between two and three times the channel width. The wavelength of most meanders varies between 10 and 14 times the channel width (Leopold, 1994, p. 58). Meandering involves inherent properties of flowing water, as well as size and shape of the channel, erodibility of the stream banks, proportion of suspended load and bedload, and probably other factors. In turn, meandering increases the channel length between two points and thus decreases the slope of the stream. Slope influences velocity and sediment-transporting capacity, so meanders not only are affected by other variables of stream flow but in turn affect those same variables.

Braided Channels.   As the total sediment load and the ratio of bedload to total load (more than 11 percent) increase, and the channel width-depth ratio and slope also increase, channels become unstable, and change from a meandering to a braided habit (Figure 10-15e). Point bars grow wide and are subject to frequent chute cutoffs, which isolate them as midchannel bars. Banks are also more erodible. Finally, the channel breaks into multiple thalwegs among gravel or sand bars, which may migrate or become vegetated as islands. The intertwining threads of channel are then *braided,* diverging (by "avulsion"), and converging among the bars and islands (Figures 10-15f, 11-8, and 17-13). At bankfull stage, a single, wide channel may be all that is seen; as the water level falls, the bars become progressively more exposed. The low-water channel floor is called a *braidplain,* which may be as wide as 20 km in major braided channels (Bristow and Best, 1993, p. 3). The bedload of braided streams may be sand or gravel, and there are behavioral variations of rivers that are primarily carrying one or the other (Germanoski and Schumm, 1993, p. 462). Most gravel-bed rivers also carry significant amounts of sand although sand-bed channels may have only minor amounts of gravel (Billi et al., 1992; Bristow and Best, 1993, p. 3). Fluctuating discharge and net aggradation both favor development of braided channels, both in natural channels and in flume experiments (Germanoski and Schumm, 1993; Warburton and Davies, 1994). This is probably related to the narrow range of velocity that separates deposition from erosion thresholds in bedload transport (Figure 10-11a). Transport rates are highly variable, primarily during high discharge, but also locally when chute cutoff occurs and intense scouring may begin at the confluent channels, or when channel avulsion abandons an old channel in favor of a new one (Bathurst, 1993, p. 80). Surges of bedload, especially gravel, migrate downstream (Hoey, 1992). Although it is conceivable that a braided channel could reach equilibrium among all the hydraulic geometry variables, the braided habit is most pronounced in channels that are **aggrading** or building up their beds. This is most commonly observed in glacial meltwater streams and on alluvial fans. Both kinds of rivers are notoriously variable in discharge, and both commonly show net aggradation.

Braided channels of bedload rivers tend to remain shallow at varying discharges, and they accommodate increased discharge by occupying additional channels. Therefore, the aggregate channel width across a braided reach can be calibrated to discharge and used in a manner similar to the use of a water-level recorder and rating curve to estimate discharge (Figure 10-3). The technique has promise for monitoring isolated, ungaged braided rivers by remote sensing techniques (Smith et al., 1996).

Channel Responses.   It could be visualized that a wide, shallow channel optimizes bed-surface area as a device to move bedload efficiently by shear. However, a more accurate statement is that because bedload is carried near the bottom of the stream, the velocity gradient increases rapidly upward in such a way that the relatively clear, faster-moving upper layer of water can erode banks made of grain sizes similar to those already in transport as bedload and thereby widen the channel. A related factor is that sand and gravel are noncohesive, and banks of coarse alluvium collapse readily, to build bars on the flat, shallow bed and create a braided pattern.

Rivers that carry mostly mud (silt and clay) as suspended load develop deeper, narrower channels with a catenary or trapezoid cross section. The wetted perimeter approaches the minimum length, that of a semicircle. A channel shape of this sort minimizes surface area and friction and therefore provides for the maximum transport of suspended load, which is carried by fluid turbulence, not bed shear. A deep, narrow channel has steep banks, which can be undercut, but the cohesive strength and smoothness of consolidated mud (Figure 10-11) resists bank erosion.

Schumm (1977, p. 109) found that at a given discharge, the width-to-depth ratio of rivers on the Great Plains is inversely proportional to the percentage of fine-grained alluvium in the eroding banks, and therefore also in the load. The relationship between channel shape and sediment grain size is given by the equation

$$F = 255 \, M^{-1.08}$$

where F is the width-depth ratio, and M is the weighted mean percent of silt and clay in the sediment. In many Great Plains rivers, the bedload may exceed half the total load, so the value for M is low and, inversely, F is high. The braided habit of these rivers is partly an adjustment among interdependent variables of width, depth, and sediment grain size, and partly a response to the external variables of discharge and load.

The meandering habit of many rivers, especially those that flow on fine-grained alluvium in humid regions, can also be related to the width-depth ratio of the channel and sediment grain size. As the suspended-sediment load (mud) increases in proportion to bedload, F decreases and the channel narrows and deepens. By these interrelated adjustments, more of the energy of the stream is expended against the banks and less against the deep bottom. The sinuosity of the thalweg and channel increases, and meanders form.

In flume experiments, Schumm and Khan (1972) attempted to produce meandering channels by adjusting slope, discharge, and sediment loads. The sediment in the flume was poorly sorted sand. They were able to produce channels with alternating bars and pools so that the thalweg meandered, but the channel banks remained essentially straight until the stream abruptly developed midchannel bars and became braided. A truly meandering channel could not be formed until 3 percent by weight of kaolinite clay was mixed into the water. The clay coated and stabilized the banks and bars and allowed the thalweg to deepen and expose the stabilized sandbars on alternating sides of the channel. True meanders formed. The experiment illustrates well the action of alluvial grain size as a factor in the hydraulic geometry of rivers.

It has been noted frequently that transitional forms are rare between wide, shallow channels with a braided habit and narrower, deeper channels that meander (Ikeda and Parker, 1989). Some reaches of alluvial rivers have changed historically from braiding to meandering or back (Nadler and Schumm, 1981), but the changes were abrupt. Perhaps the lack of intermediate forms is related to the fact that the minimum threshold of sediment erosion on both the Hjulström curve and the Shields curve (Figure 10-11) is in the fine sand range. If alluvium is noncohesive sand or coarser sediment, rising velocity drags it as bedload, and a wide, shallow braided channel develops. If the alluvium is cohesive and smooth, an equivalent velocity increase may erode it, but it is transported as suspended load in a deep, meandering channel.

THE CONCEPT OF GRADE IN FLUVIAL SYSTEMS

The concept of an open system that maintains itself in a steady state of most efficient configuration by internal self-regulation among variables was introduced in Chapter 2, by analogy to a rotary cement kiln. A river is an excellent example of an open system through which matter and energy flow, but within which are inherent tendencies toward self-regulation. Numerous examples of the interaction between discharge, load, channel shape, and other variables of hydraulic geometry have been cited. It is appropriate now to review the interaction of all the known variables of fluvial systems in the tendency to achieve **grade,** or long-term self-regulation, in a river channel. Although reduction of a regional landscape to low

relief near sea level might take millions of years, fluvial processes might achieve mutual adjustment in much shorter time, perhaps in centuries or millenniums (Schumm, 1977, p. 12).

Variables of Hydraulic Geometry

At least ten variables of hydraulic geometry are involved in the tendency for a river to achieve or maintain a graded state. Not all are of equal significance, and some are not within the self-regulatory ability of the river. Leopold and Maddock (1953) divided the variables of hydraulic geometry into three classes: *independent, semidependent,* and *dependent.*

Discharge, sediment load, and ultimate base level are the three *independent* variables in hydraulic geometry. The stream has little control over these factors; rather, it must adjust to them. Discharge is determined by precipitation and evaporation in the drainage basin, the permeability of the soil, the amount and type of vegetation, and the area of the drainage basin. Only the area of the drainage basin is affected by other changes in the river system. Headward erosion by tributaries can enlarge the drainage basin and thereby increase discharge, but even this process is limited because adjacent drainage nets are probably enlarging, too. Loss of part of a network through capture by a competing system can cause an abrupt decrease in discharge, but quite early in the erosional development of a landscape, the boundaries of each river's drainage basin become defined.

Sediment load is also nearly independent of the other variables of stream flow. Many of the same climatic, soil, and biologic factors that determine discharge also determine the amount of sediment that slopes deliver to streams. The type of rock mass is an additional powerful control for sediment load. Some rocks weather quickly to sand-size particles; others produce only silt and clay. Limestone weathering produces mostly a dissolved load, with little solid detritus. Streams erode their channels and thereby have some self-regulation of their load, but we have seen that most of the load reaches the river "ready-made" by weathering and mass wasting on the hillslopes of the drainage basins and in hollows at the heads of finger-tip tributaries. Some headwater terranes yield sediment so slowly that sediment load is said to be *supply-limited,* or *source-limited.* In most alluvial rivers, much more sediment is potentially available than can be moved, and the sediment load is then said to be *transport-limited.* When the river is competent to move the sediment, the sediment is there to be moved.

The ultimate base level of erosion is the third independent variable of stream flow. When a stream reaches the sea, it loses its identity. The potential energy of the stream is set by the altitude above sea level at which precipitation falls. Regardless of the discharge, load, or any other variable, a river that rises on a coastal plain only a few hundred feet above sea level will never be a mountain torrent.

The *semidependent* variables that interact to achieve the graded state include channel width, channel depth, bed roughness, grain size of the sediment load, velocity, and channel habit (the tendency for a stream to either meander or braid). These are semidependent inasmuch as they are partly determined by the three independent variables, but they are also partly capable of mutual self-regulation in a river. Width, depth, and mean velocity have been shown to be power functions of discharge. Competence, defined by the grain size of the load, is a function of depth and velocity. However, the mixture of grain sizes, especially the "mud ratio" M, has been shown to determine the width-depth ratio. Bed roughness is determined by the alluvial grain size, bed forms, and channel shape. The alluvial bed is built of grain sizes determined by the stream competence, yet that competence is in turn determined by such factors as velocity, channel shape and roughness, and the amount of load and discharge. In turn, bed roughness creates turbulence that affects the competence of the stream to entrain and move the load. Channel habits also interact with the other semidependent variables, as discussed in previous paragraphs. There is no simple way to diagram the many feedback loops and interactions among these semidependent variables.

Only one variable in the hydraulic geometry of a river, the downstream slope of the water surface, is regarded as being *dependent* on all other variables. Slope can be changed by building up one part of the channel, cutting down another, by changing channel length as by meandering or delta building (Schumm, 1993), or by regional tectonism. All these changes take time, so slope is usually the final adjustment the stream makes in becoming graded. If channel slope is changed suddenly, as by neotectonic movement (p. 45), it becomes mutually semidependent with the variables previously listed. Because it ordinarily cannot do so, it is subject to the influences of all the other known variables, and usually evolves on a time scale at least an order of magnitude longer than the mutual adjustments of the others.

Progressive Development of Grade

We can visualize the achievement of grade, or the graded state, in a river by first imagining an ungraded river that flows over a tectonic landscape, newly raised from the floor of the sea. Rainfall, and therefore discharge, varies over the new landscape. Water flows downhill along chance hollows and other water-collecting slopes. Lakes form in closed depressions. Drainage networks are eroded, and valleys are widened by mass wasting on slopes along the intrenching channels. Variables of the stream system are wildly out of equilibrium. Half the potential energy of such a river might be expended in one great vertical waterfall. The load of the stream is determined by landslides anywhere along the banks. Channel width and depth are restricted by the erodibility of the rocks that are crossed by the channel. The slope of the river approximates the initial slopes of the landscape.

Rather quickly, the semidependent variables of stream flow interact to form a channel system appropriate to the work at hand. Many of the hydraulic geometry equations for stream flow apply to both graded and ungraded streams. The independent variables of discharge, load, and base level may be the same for a graded and an ungraded river. The critical factor in the achievement of the graded condition is that the stream must flow on "adjustable" materials, so that changes of one variable can produce the appropriate changes in others. Alluvium is an excellent buffer for variations in stream energy because every fragment on the floodplain has some critical energy threshold of transportation. Having been once transported by the river, it will move again when the energy level is high enough. Thus, the establishment of an alluvium-lined channel on a continuous floodplain signals the achievement of grade in that part of a river.

Alluvium not only permits stream channels to adjust toward equilibrium; it is also a powerful absorbent for peak energy inputs into the river system. Discharge may increase by many orders of magnitude during a flood (Figure 10-13). If the excess energy of the flood is not fully absorbed by the increased load carried into the river from headwater slopes, alluvium is eroded until the ability of the river to do work is balanced by the work it is doing (Figure 10-14). Alluvium is analogous to a chemical buffering compound that is added in great excess to a solution to ensure that some property of the solution remains constant during a reaction.

Grade, as defined by an alluvial bed on an eroding *valley flat* (suballuvial rock floor, Figure 11-9), is theoretically established first in the downstream segments of eroding rivers and is gradually extended upstream (Davis, 1902). Larger rivers should attain grade earlier than smaller streams. A trunk stream may be at grade when its finger-tip tributaries are still gnawing headward into undissected slopes. These deductions have been greatly complicated by the impact of late Cenozoic sea-level fluctuations (Figure 4-7). A *graded reach,* or graded segment of a river, may form on easily eroded material upstream from a gorge across a resistant rock barrier. The resistant rock then forms the local or temporary base level for the graded reach. As the barrier is slowly eroded, the graded reach remains at grade because it can adjust progressively during the slow lowering of the temporary base level.

Grade as a Thermodynamic Equilibrium

Another approach to the concept of grade is to regard the graded condition of a river from the viewpoint of theoretical thermodynamics. In any steady-state physical system through which matter and energy move, we find a tendency for the least possible work to be done and also a tendency toward a uniform distribution of work. Nature is inherently conservative in these matters. In a river system, which derives its energy from water flowing downhill, the tendency for minimum work opposes the tendency for a uniform distribution of work in several different aspects of hydraulic geometry. Two examples follow.

The Long Profile as a Graded Form. If all the water of a river were added at the head of a single tributary, the minimum-work profile would be a waterfall

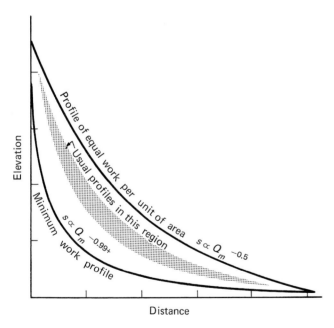

FIGURE 10-16. Schematic longitudinal profiles of rivers showing tendency to balance minimum total work with maximum distribution of work (from Langbein and Leopold, 1964, Figure 1).

straight down to sea level. In humid regions, where rivers gain water downstream, the minimum-work profile would be a curve in which the greatest loss of altitude takes place where the mean discharge (Q_m) is least, near the head of the river. The profile would be very steep near the head and almost horizontal near the mouth (Figure 10-16). Typically, in such a river, slope would be almost inversely proportional to discharge ($s \propto Q_m{}^{-1}$).

In contrast to the sharply concave theoretical profile of minimum work, the theoretical profile for uniformly distributed work would have a more nearly constant slope (Figure 10-16). The profile is only slightly concave skyward because as the rate of doing work increases downstream with discharge, the surface area of the stream bed also increases. The power expended per unit area of stream bed is constant on a river that widens downstream but decreases its slope only gradually. Typically, $s \propto Q_m{}^{-0.5}$ (Figure 10-8).

If two tendencies toward equilibrium oppose each other, the resulting equilibrium is most likely to be a predictable intermediate condition. For many rivers, the exponent for Q_m in the preceding equation ranges from -0.70 to -0.75 (Leopold, 1994, p. 276). Langbein and Leopold (1964) concluded from their studies of river dynamics that both the downstream profiles and channel cross sections of rivers approach the equilibrium form predicted from the principles of least work and uniform distribution of work, *provided that the channels are in material that is adjustable.* The presence of an alluvial floodplain is again implied as a condition of grade. The ideal long profile of a river flowing in an alluvial channel is a smooth concave-skyward profile that allows water to flow downhill to the sea with slight or negligible acceleration yet to expend only a few percent of the potential energy of the system on channel erosion or sediment transport (Bagnold, 1968).

Meanders in Graded Rivers. The free, regular meanders of suspended-load alluvial rivers have long aroused curiosity and artistic senses. It seems a waste of potential energy for a river to avoid the most direct path downhill, which in plan view would be a straight channel. However, meanders, like the long profile of alluvial rivers, have been demonstrated to be the result of balance between least work and uniform distribution of work (Langbein and Leopold, 1966; Leopold and Langbein, 1966). Meanders are *sine-generated curves* that have the statistical property of minimum variance. That is, the work done on each segment of stream bank is uniform, and a minimum amount of total work is done. Furthermore, the sine-generated curve is the shape of the average, or most probable, random walk between two points (Leopold, 1994, pp. 64ff). Thus meanders represent highly probable paths of fluid flow. Meltwater streams on glaciers meander though they may be sediment-free between ice walls. The Gulf Stream meanders up the east coast of the United States and eastward into the Atlantic, and high-altitude jet streams meander in the atmosphere.

Summary of Grade

The most concise definition of a graded stream was carefully phrased by Mackin (1948, pp. 471, 484). The hydraulic geometry of Leopold and Maddock required only a slight transposition of phrases in Mackin's definition. It is quoted here, as modified by Leopold and Maddock (1953, p. 51), as a summary of the concept of the graded river:

> A graded river is one in which, over a period of years, slope and channel characteristics are delicately adjusted to provide, with available discharge, just the velocity required for the transportation of the load supplied from the drainage basin. The graded stream is a system in equilibrium; its diagnostic characteristic is that any change in any of the controlling factors will cause a displacement of the equilibrium in a direction that will tend to absorb the effect of the change.

A few additional comments must be made about the concept of grade. Grade is a condition, not an altitude or a certain slope angle. It develops first near the mouths of rivers and gradually extends headward. Erosional lowering of the land continues for a long time after grade is achieved because as long as rivers carry sediment to the sea, they continue to lower the landscape over which they flow. A graded river is in a steady state only with regard to a certain time scale ("graded time," Figure 1-6). Over a time scale of millions of years, typical of the time intervals in which landscapes evolve ("cyclic time," Figure 1-6), the potential energy of an undisturbed river system gradually approaches zero, and the rate of change of the system also decreases. The river remains at grade, but the characteristics of the graded condition change with time. The significance of the concept of grade to the life history of regional landscapes is emphasized in Chapter 15.

REFERENCES

BAGNOLD, R. A., 1968, Deposition in the process of hydraulic transport: Sedimentology, v. 10, pp. 45–56.

BAILEY, J. F., PATTERSON, J. L., and PAULHUS, J. L. H., 1975, Hurricane Agnes rainfall and floods, June–July 1972: U.S. Geol. Survey Prof. Paper 924, 403 pp.

BAKER, V. R., 1973, Paleohydrology and sedimentology of Lake Missoula flooding in eastern Washington: Geol. Soc. America Spec. Paper 144, 79 pp.

———, 1990, Spring sapping and valley network development, with case studies by Kochel, R. C., Baker, V. R., Laity, J. E., and Howard A. D., in Higgins, C. G., and Coates, D. R., eds., Groundwater geomorphology: The role of subsurface water in earth-surface processes and landforms: Geol. Soc. America Spec. Paper 252, pp. 235–265.

———, BENITO, G., and RUDOY, A. N., 1993, Paleohydrology of late Pleistocene superflooding, Altay Mountains, Siberia: Science, v. 259, pp. 348–350.

———, and COSTA, J. E., 1987, Flood power, in Mayer, L., and Nash, D., eds., Catastrophic flooding: Allen & Unwin, Inc., Boston, pp. 1–21.

———, GREELEY, R., and 3 others, 1987, Columbia and Snake River plains, in Graf, W. L., ed., Geomorphic systems of North America: Geol. Soc. America, Centennial Special Vol. 2, pp. 403–468.

———, and PICKUP, G., 1987, Flood geomorphology of the Katherine Gorge, Northern Territory, Australia: Geol. Soc. America Bull., v. 98, pp. 635–646.

———, and RITTER, D. F., 1975, Competence of rivers to transport coarse bedload material: Geol. Soc. America Bull., v. 86, pp. 975–978.

BARNES, H. H., JR., 1967, Roughness characteristics of natural channels: U.S. Geol. Survey Water Supply Paper 1849, 213 pp.

BATHURST, J. C., 1993, Flow resistance through the channel network, in Beven, K., and Kirkby, M. J., eds., Channel network hydrology: John Wiley & Sons Ltd., Chichester, UK, pp. 69–98.

BEVEN, K., and GERMANN, P., 1982, Macropores and water flow in soils: Water Resources Res., v. 18, pp. 1311–1325.

BILLI, P., HEY, R. D., and 2 others, eds., 1992, Dynamics of gravel-bed rivers: John Wiley & Sons, Chichester, UK, 673 pp.

BOCCO, G., 1991, Gully erosion: Processes and models: Prog. in Phys. Geog., v. 15, pp. 392–406.

BRADY, N. C., and WEIL, R. R., 1996, Nature and properties of soils, 11th ed.: Prentice Hall, Upper Saddle River, N.J., 740 pp.

BRAKENRIDGE, G. R., 1988, River flood regime and floodplain stratigraphy, in Baker, V. R., Kochel, R. C., and Patton, P. C., eds., Flood geomorphology: John Wiley & Sons, Inc., New York, pp. 139–156.

BRETZ, J. H., 1969, The Lake Missoula floods and the channeled scablands: Jour. Geology, v. 77, pp. 503–543.

BRISTOW, C. S., and BEST, J. L., 1993, Braided rivers: Perspectives and problems, in Best, J. L., and Bristow, C. S., eds., Braided rivers: Geol. Soc. [London] Spec. Pub. no. 75, pp. 1–11.

CARLSTON, C. W., 1968, Slope-discharge relations for eight rivers in the United States: U.S. Geol. Survey Prof. Paper 600-D, pp. D45-D47.

———, 1969a, Downstream variations in the hydraulic geometry of streams: Special emphasis on mean velocity: Am. Jour. Sci., v. 267, pp. 499–509.

———, 1969b, Longitudinal slope characteristics of rivers of the midcontinent and the Atlantic East Gulf slopes: Internat. Assoc. Sci. Hydrol. Bull., v. 14, no. 4, pp. 21–31.

CHOW, V. T., MAIDMENT, D. R., and MAYS, L. W., 1988, Applied hydrology: McGraw-Hill, Inc., New York, 572 pp.

COLLIER, M. P., WEBB, R. H., and ANDREWS, C. D., 1997, Experimental flooding in Grand Canyon: Sci. American, v. 276, no. 1, pp. 82–89.

COSTA, J. E., and SCHUSTER, R. L., 1988, Formation and failure of natural dams: Geol. Soc. America Bull., v. 100, pp. 1054–1068.

COTTON, C. A., 1948, Landscape, aĺs developed by the processes of normal erosion, 2nd ed.: Whitcombe and Tombs Ltd., Wellington, N.Z., 509 pp.

CROSS, W. P., and WEBBER, E. E., 1959, Floods in Ohio, magnitude and frequency: Ohio Dept. Nat. Resources, Division of Water, Bull. 32, 325 pp.

DAVIS, W. M., 1902, Base-level, grade, and peneplain: Jour. Geology, v. 10, pp. 77–111 (reprinted 1954 in Geographical essays: Dover Publications, Inc., New York, pp. 249–278).

———, 1905a, Complications of the geographical cycle: Internat. Geog. Cong., 8th, Washington 1904, Repts., pp. 150–163 (reprinted 1954 in Geographical essays: Dover Publications, Inc., New York, pp. 279–295).

———, 1905b, Geographical cycle in an arid climate: Jour. Geology, v. 13, pp. 381–407 (Reprinted 1954 in Geographical essays: Dover Publications, Inc., New York, pp. 296–322).

DIETRICH, W.E., and DUNNE, T., 1993, The channel head, in Beven, K., and Kirkby, M. J., eds., Channel network hydrology: John Wiley & Sons Ltd., Chichester, UK, pp. 175–219.

DINGMAN, S. L., 1994, Physical hydrology: Macmillan Publishing Company, New York, 575 pp.

DUNNE, T., and AUBRY, B. F., 1986, Evaluation of Horton's theory of sheet wash and rill erosion on the basis of field experiments, in Abrahams, A. D., ed., Hillslope processes, Allen & Unwin, Inc., Boston, pp. 31–35.

DUNNE, T., and LEOPOLD, L. B., 1978, Water in environmental planning: W.H. Freeman and Company, San Francisco, 818 pp.

DUNNE, T., WHIPPLE, K. X., and AUBRY, B. F., 1995, Microtopography of hillslopes and initiation of channels by Horton overland flow, in Costa, J. E., Miller, A. J., and 2 others, eds., Natural and anthropogenic influences in fluvial geomorphology: The Wolman volume: Am. Geophys. Union, Geophys. Mono. 89, pp. 27–44.

DURY, G. H., 1967, Bankful discharge and the magnitude-frequency series: Aust. Jour. Sci., v. 30, p. 371.

———, HAILS, J. R., and ROBBIE, M. B., 1963, Bankfull discharge and the magnitude-frequency series: Aust. Jour. Sci., v. 26, pp. 123–124.

DYNESIUS, M., and NILSSON, C., 1994, Fragmentation and flow regulation of river systems in the northern third of the world: Science, v. 266, pp. 753–762.

EMERSON, J. W., 1971, Channelization: A case study: Science, v. 173, pp. 325–326.

EMMETT, W. W., 1980, A field calibration of the sediment-trapping characteristics of the Helley-Smith bedload sampler: U.S. Geol. Survey Prof. Paper 1139, 44 pp.

GARRELS, R. M., and MACKENZIE, F. T., 1971, Evolution of sedimentary rocks: W. W. Norton & Co., Inc., New York, 397 pp.

GERMANOSKI, D., and SCHUMM, S. A., 1993, Changes in braided river morphology resulting from aggradation and degradation: Jour. Geology. v. 101, pp. 451–466.

GUMBEL, E. J., 1958, Statistical theory of floods and droughts: Inst. Water Engineers Jour., v. 12, pp. 157–184.

HEY, R. D., 1988, Mathematical models of channel morphology, in Anderson , M. G., ed., Modelling geomorphological systems: John Wiley & Sons Ltd., Chichester, UK, pp. 99–125.

HIGGINS, C. G., 1984, Piping and sapping: Development of landforms by groundwater outflow, in LaFleur, R. G., ed., Groundwater as a geomorphic agent: Allen & Unwin, Inc., Boston, pp. 18–58.

———, 1990, Gully development, in Higgins, C. G., and Coates, D. R., eds., Groundwater geomorphology: The role of subsurface water in earth-surface processes and landforms: Geol. Soc. America Spec. Paper 252, pp. 139–155.

HIRSCHBOECK, K. K., 1987, Catastrophic flooding and atmospheric circulation anomalies, in Mayer, L., and Nash, D., eds., Catastrophic flooding: Allen & Unwin, Inc., Boston, pp. 25–56.

———, 1988, Flood hydroclimatology, in Baker, V. R., Kochel, R. C., and Patton, P. C., eds., Flood geomorphology: John Wiley & Sons, Inc., New York, pp. 27–49.

HJULSTRÖM, F., 1935, Studies of the morphological activity of rivers as illustrated by the River Fyris: Geol. Inst. Univ. Uppsala Bull., v. 25, pp. 221–527.

———, 1939, Transportation of detritus by moving water, in Trask, P. D., ed., Recent marine sediments: Am. Assoc. Petroleum Geologists, Tulsa, pp. 5–31.

HOEY, T., 1992, Temporal variations in bedload transport rates and sediment storage in gravel-bed rivers: Prog. in Phys. Geog., v. 16, pp. 319–338.

HORTON R. E., 1945, Erosional development of streams and their drainage basins: Geol. Soc. America Bull., v. 56, pp. 275–370.

HOWARD, A. D., and McLANE, C. F., III, 1988, Erosion of cohesionless sediment by groundwater seepage: Water Resources Res., v. 24, pp. 1659–1674.

IKEDA, S. and PARKER, G., eds., 1989, River meandering:

Am. Geophys. Union, Water Resources Mono. 12, 485 pp.

IRELAND, H. A., 1939, "Lyell" gully, a record of a century of erosion: Jour. Geology, v. 47 pp. 47–63.

JARRETT, R. D., and MALDE, H. E., 1987, Paleodischarge of the late Pleistocene Bonneville Flood, Snake River, Idaho, computed from new evidence: Geol. Soc. American Bull., v. 99, pp. 127–134.

KEHEW, A. E., and LORD, M. L., 1986, Origin and large-scale erosional features of glacial-lake spillways in the northern Great Plains: Geol. Soc. America Bull., v. 97, pp. 162–177.

KIRKBY, M. J. 1987, The Hurst effect and its implications for extrapolating process rates: Earth Surface Processes and Landforms, v. 12, pp. 57–67.

KNIGHTON, A. D., 1984, Fluvial forms and processes: Edward Arnold (Publishers) Ltd., London, 218 pp.

KOMAR, P. D., 1988, Sediment transport by floods, in Baker, V. R., Kochel, R. C., and Patton, P. C., eds., Flood geomorphology: John Wiley & Sons, Inc., New York, pp. 97–111.

LANE, E. W., 1937, Stable channels in erodible materials: Am. Soc. Civil Engineering Trans., v. 102, pp. 123–194 (including discussion).

LANGBEIN, W. B., 1949, Annual floods and the partial-duration series: Am. Geophys. Union, Trans., v. 30, pp. 879–881.

———, and LEOPOLD, L. B., 1964, Quasi-equilibrium states in channel morphology: Am. Jour. Sci., v. 262, pp. 782–794.

———, and LEOPOLD, L. B., 1966, River meanders-theory of minimum variance: U.S. Geol. Survey Prof. Paper 422-H, 15 pp.

LARONNE, J. B., and REID, I., 1993, Very high rates of bedload sediment transport by ephemeral desert rivers: Nature, v. 366, pp. 148–150.

LEOPOLD, L. B., 1953, Downstream change of velocity in rivers: Am. Jour. Sci., v. 251, pp. 606–624.

———, 1994, A view of the river: Harvard University Press, Cambridge, Mass., 298 pp.

———, and LANGBEIN, W. B., 1966, River meanders: Sci. American, v. 214, no. 6, pp. 60–70.

———, and MADDOCK, T., 1953, Hydraulic geometry of stream channels and some physiographic implications: U.S. Geol. Survey Prof. Paper 252, 57 pp.

———, WOLMAN, M. G., and MILLER, J. P., 1964, Fluvial processes in geomorphology: W.H. Freeman and Company, San Francisco, 522 pp.

LIVINGSTONE, D. A., 1963, Chemical composition of rivers and lakes: U.S. Geol. Survey Prof. Paper 440-G, 64 pp.

LOUCKS, D. P., STEDINGER, J. R., and HAITH, D. A., 1981, Water resources systems planning and analysis: Prentice-Hall, Inc., Englewood Cliffs, N.J., 559 pp.

MACKIN, J. H., 1948, Concept of the graded river: Geol. Soc. America Bull., v. 59, pp. 463–512.

MALAMUD, B. D., TURCOTTE, D. L., and BARTON, C. C., 1996, The 1993 Mississippi River flood: A one hundred or a one thousand year event?: Environmental & Engineering Geoscience, v. II, pp. 479–486.

MALDE, H. E., 1968, The catastrophic late Pleistocene Bonneville flood in the Snake River Plain, Idaho: U.S. Geol. Survey Prof. Paper 596, 52 pp.

MAYER, L., and NASH, D., eds., 1987, Catastrophic flooding: Allen and Unwin, Boston, 410 pp.

McCAIG, M., 1985, Soil properties and subsurface hydrology, in Richards, K. S., Arnett, R. R., and Ellis, S., eds., Geomorphology and soils: George Allen & Unwin (Publishers) Ltd., London, pp. 121–140.

MILLER, J. R., 1990, Influence of bedrock geology on knickpoint development and channel-bed degradation along downcutting streams in south-central Indiana: Jour. Geology, v. 99, pp. 591–605.

MILLIMAN, J. D., and SYVITSKI, J. P. M., 1992, Geomorphic/tectonic control of sediment discharge to the ocean: The importance of small mountainous rivers: Jour. Geology, v. 100, p. 525–544.

MONTGOMERY, D. R., and DIETRICH, W. E., 1994, Landscape dissection and drainage area-slope thresholds, in Kirkby, J. J., ed., Process models and theoretical geomorphology: John Wiley & Sons, Chichester, UK, pp. 221–246.

MOSS, A. J., GREEN, P., and HUTKA, J., 1982, Small channels: Their experimental formation, nature, and significance: Earth Surface Processes and Landforms, v. 7, pp. 401–415.

NADLER, C. T., and SCHUMM, S. A., 1981, Metamorphosis of South Platte and Arkansas Rivers, eastern Colorado: Phys. Geog., v. 2, pp. 95–115.

NANSON, G. C., and KNIGHTON, A. D., 1996, Anabranching rivers: Their cause, character and classification: Earth Surface Processes and Landforms, v. 21, pp. 217–239.

NEVIN, C., 1946, Competency of moving water to transport debris: Geol. Soc. America Bull., v. 57, pp. 651–674.

NOVAK, I. D., 1973, Predicting coarse sediment transport: The Hjulstrom curve revisited, in Morisawa, M., ed., Fluvial geomorphology: State Univ. of New York Publications in Geomorphology, Binghamton, New York, pp. 13–25.

O'CONNOR, J. E., 1993, Hydrology, hydraulics, and geomorphology of the Bonneville flood: Geol. Soc. America Spec. Paper 274, 83 pp.

———, and BAKER, V. R., 1992, Magnitudes and implications of peak discharges from glacial Lake Missoula: Geol. Soc. America Bull., v. 104, p. 267–279.

———, ELY, L. L., and 5 others, 1994, a 4500-year record of large floods on the Colorado River in the Grand Canyon, Arizona: Jour. Geology., v. 102, pp. 1–9.

OSTERKAMP, W. R., 1978, Gradient, discharge, and particle-size relations of alluvial channels in Kansas, with observations on braiding: Am. Jour. Sci., v. 278, pp. 1253–1268.

PARK, C. C., 1977, World-wide variations in hydraulic geometry exponents of stream channels: An analysis and some observations: Jour. Hydrology, v. 33, pp., 133–146.

PROSSER, I. P., DIETRICH, W. E., and STEVENSON, J., 1995, Flow resistance and sediment transport by concentrated overland flow in a grassland valley: Geomorphology, v. 13, pp. 71–86.

RENARD, K. G., FOSTER, G. R., and 2 others, 1994, RUSLE revisited: Status, questions, answers, and the future: Jour. Soil, Water Conserv., v. 49, pp. 213–220.

RHODES, D. D., 1977, The b-f-m diagram: Graphical representation and interpretation of at-a-station hydraulic geometry: Am. Jour. Sci., v. 277, pp. 73–96.

———, 1987, The b-f-m diagram for downstream hydraulic geometry: Geografiska Annaler, v. 69A, pp. 147–161.

RICHARDS, K. S., 1982, Rivers: Form and process in alluvial channels: Methuen & C., Ltd., London, 358 pp.

RUBEY, W. W., 1938, The force required to move particles on a stream bed: U.S. Geol. Survey Prof. Paper 189-E, pp. 120–141.

SCHUMM, S.A., 1977, The fluvial system: John Wiley & Sons, Inc., New York, 338 pp.

———, 1981, Evolution and response of the fluvial system, sedimentologic implications, in Ethridge, F. G., and Flores, R. M., eds., Recent and ancient nonmarine depositional environments: Models for exploration: Soc. Econ. Paleontologists and Mineralogists Spec. Pub. No. 31, pp. 19–29.

———, 1985, Pattern of alluvial rivers: Ann. Rev. Earth, Planet. Sci., v. 13, pp. 5–27.

———, 1993, River response to baselevel change: Implications for sequence stratigraphy: Jour. Geology, v. 101, pp. 279–294.

———, BOYD, K. F., and 2 others, 1995, A ground-water sapping landscape in the Florida panhandle: Geomorphology, v. 12, pp. 281–297.

———, and KHAN, H. R., 1972, Experimental study of channel patterns: Geol. Soc. America Bull., v. 83, pp. 1755–1770.

SEIDL, M. A., DIETRICH, W. E., and KIRCHNER, J. W., 1994, Longitudinal profile development into bedrock: An analysis of Hawaiian channels: Jour. Geology, v. 102, pp. 457–474.

SIMON, A., 1994, Gradation processes and channel evolution in modified west Tennessee streams: Process, response, and form: U.S. Geol. Survey Prof. Paper 1470, 84 pp.

SIOLI, H., 1984, The Amazon and its main affluents: Hydrography, morphology of the river courses, and river types, in Sioli, H., ed., The Amazon: Limnology and landscape ecology of a mighty tropical river and its basin: Dr. W. Junk Publishers, Dordrecht, pp. 127–165.

SMITH, L. C., ISACKS, B. L., and 2 others, 1996, Estimation of discharge from three braided rivers using synthetic aperture radar satellite imagery: Potential application to ungaged basins: Water Resources Res., v. 32, pp. 2021–2034.

STANLEY, D. J., and WARNE, A. G., 1993, Nile delta: Recent geological evolution and human impact: Science, v. 260, pp. 628–634.

SUNDBORG, A., 1956, The River Klarälven, a study of fluvial processes: Geografiska Annaler, v. 38, pp. 125–316.

TELLER, J. T., and KEHEW, A. E., eds., 1994, Late glacial history of large proglacial lakes and meltwater runoff along the Laurentide ice sheet: Quaternary Sci. Rev., v. 13, nos. 9–10, pp. 795–981.

TINKLER, K. J., 1981, Avoiding error when using the Manning equation: Jour. Geology, v. 90, pp. 326–328.

U.S. WATER RESOURCES COUNCIL (now called Interagency Advisory Committee on Water Data), 1981, Guidelines for determining flood frequency: U.S. Geol. Survey Office of Water Data Coordination Bull. 17B, 28 pp.

WAITT, R. B., JR., 1985, Case for periodic, colossal jökulhlaups from Pleistocene glacial Lake Missoula: Geol. Soc. America Bull., v. 96, pp. 1271–1286.

WARBURTON, J., and DAVIES, T., 1994, Variability of bedload transport and channel morphology in a braided river hydraulic model: Earth Surface Processes and Landforms, v. 19, pp. 403–421.

WILLIAMS, G. P., and WOLMAN, M. G., 1984, Downstream effects of dams on alluvial rivers: U.S. Geol. Survey Prof. Paper 1286, 83 pp.

WISCHMEIER, W. J., and SMITH, D. D., 1978, Predicting rainfall erosion loss: A guide to conservation planning: U.S. Dept. Agriculture, Agr. handbook No. 537, 58 pp.

WOHL, E. E., 1993, Bedrock channel incision along Piccaninny Creek, Australia: Jour. Geology, v. 101, pp. 749–761.

YOUNG, R. A., and WIERSMA, J. L., 1973, The role of rainfall impact in soil detachment and transport: Water Resources Res., v. 9, pp. 1629–1636.

YU, B., and WOLMAN, M. G., 1987, Some dynamic aspects of river geometry: Water Resources Res., v. 23, pp. 501–509.

ZEN, E., and PRESTEGAARD, K. L., 1994, Possible hydraulic significance of two kinds of potholes: Examples from the paleo-Potomac River: Geology, v. 22, pp. 47–50.

Chapter 11

Evolution of the Fluvial System

Schumm (1977) divided the entire fluvial system into three zones: zone 1, the drainage basin of water and sediment production; zone 2, a zone of river transport where the input of sediment can balance the output; and zone 3, regions of deposition on an alluvial plain, delta, or in deeper ocean water.

The previous chapter dealt at length with the processes by which water and sediment move overland in drainage basins to channel heads, and subsequently through the river channel system. This chapter considers the evolution of the fluvial landscape in response to the progressive organization of river networks. We review upland drainage basins first, with computer simulations that attempt to show how basins become dissected. Second, we analyze the progressive changes in channel network organization, with emphasis on the growth toward an equilibrium state or grade, especially by the development of an alluvium-lined valley floor. We also consider the role of overbank floods in shaping channels and floodplains. Finally, we discuss the subaerial depositional landscapes of alluvial terraces, fans, plains, and deltas in their tectonic and climatic settings. Here we return to a theme of earlier chapters: Many landscapes bear the imprint of prior conditions or unique paleoevents and are relict today.

DRAINAGE BASIN EVOLUTION

In nature and in laboratory experiments, rain falls over an entire surface, then infiltrates or flows overland under the influence of gravity until it collects in surface rills, gullies, and channel networks (Figure 11-1). All the runoff water, and the sediment that it has entrained, eventually leaves a unit of surface area through a single channel. The **drainage basin**, or **watershed**, of a river system is the surface upstream and uphill from a channel that sheds water and sediment into that channel. A drainage basin is an open system, into which and from which energy and matter flow, and its boundaries are normally well defined. Because so much of the terrestrial landscape is drained by rivers, drainage basins are a fundamental unit of geomorphic analysis. The network of branching stream channels includes only about 1 to 5 percent of the total basin area (Kirkby, 1993, p. 7), but its role in eroding and transporting sediment is critical to the entire basin's behavior and evolution. Drainage basins and their drainage networks are studied by a combination of the techniques used in measuring mass wasting (Chapter 9) and those used in hydrogeology.

Divides

In mountains, and in terranes with fine-textured badlands erosion (Figure 9-22), drainage basins are well defined by sharp ridges around their perimeters. Hiking trails in such landscapes are usually along or just below the ridge crests, uphill from the heads of stream channels, which may be rugged and impassable. The **divides** that define the boundary of adjacent basins are easy to see on topographic maps, aerial photos, and

(a)

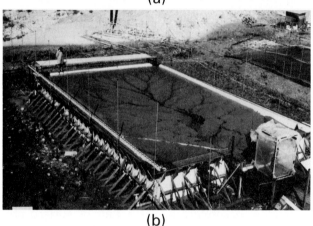

(b)

FIGURE 11-1. Rainfall-erosion facility (REF), Colorado State University. (a) Start of experiment. (b) Drainage network development (Schumm et al., 1987, Figure 2-9).

remotely sensed images as well as on the ground, over a wide range of size scales.

In regions of low relief, drainage-basin divides are much less obvious. On gentle, convex creep slopes (Figure 9-23), the actual surface divide may be indeterminate within hundreds or even thousands of meters. On such flat *interfluves* ("between rivers"), there is no overland flow except during precipitation, and at other times the divide must be visualized as a sinuous vertical surface that extends below the landscape into the zone of groundwater saturation, separating groundwater flow lines between adjacent basins (Figure 8-5). In smaller subparallel drainage basins along the flanks of a major divide, sloping ridges or noses (p. 192) mark the secondary divides.

It might seem that as fluvial erosion progresses, divides should become better defined. However, if upland slope processes such as creep, raindrop splash, and plant and animal ground disturbances inhibit rill formation, divides will remain vague, and at distances of hundreds of meters uphill from the head of rill incision (Dunne et al., 1995). Throughout the low-relief

areas of subtropical Africa, and probably in other tropical regions as well, drainage basins are marked by converging **dambos,** "tropical, seasonally waterlogged, predominantly grass covered, shallow linear depressions, [generally] without a marked stream channel" (Thomas and Goudie, 1985, p. v). Their floors are usually wet and are the sites of dry-season gardens and wells. They probably are important in tropical subsurface weathering and subsequent erosion. In Africa, the extensive regions of dambo landscape are also regions of known late Quaternary bioclimatic changes, which may have contributed to their formation (Thomas and Goudie, 1985). Possibly analogous features in temperate latitudes are the "seepage lines" of wetter soils that can be seen in freshly tilled fields, especially in the spring. They are difficult to trace on the ground, but from a low-flying airplane they are seen to form intricate networks of shallow linear depressions with darker, wetter soils. Bunting (1961) showed that the soil along the axes of seepage lines is deeper than between them because of the greater available moisture. They may mark the incipient courses of an eventual branching network of channels.

Channel Networks

Downhill from the level at which rill and gully incision begins (p. 199), water and sediment are carried primarily by a network of channels, defined by their steep eroded banks and eroded floors. The heads of some channels are vague; others are abrupt, headward-migrating waterfall nickpoints. At every scale, from that of a space photograph to an experimental facility (Figure 11-1), channel networks are the dominant feature of fluvial landscapes.

> Every river appears to consist of a main trunk, fed from a variety of branches, each running in a valley proportioned to its size, and all of them together forming a system of valleys, communicating with one another, and having such a nice adjustment of their declivities, that none of them join the principal valley, either on too high or too low a level; a circumstance which would be infinitely improbable, if each of these valleys were not the work of the stream that flows in it (Playfair, 1802, p. 102).

Playfair's law, quoted here, in addition to being written in a style only rarely achieved in scientific prose, illustrates the mathematician's appreciation of probability theory. In discussing the quotation, Playfair noted (1802, pp. 353, 355):

> The truth of the proposition . . . is demonstrated on a principle which has a close affinity to that on which chances are usually calculated . . . we must

conclude, that the probability of such a constitution having arisen from another cause, is, to the probability of its having arisen from the running of water, in such a proportion as unity bears to a number infinitely great.

For nearly 150 years, Playfair's law was quoted as a defensible and acceptable logical proposition, without any attempt to test it quantitatively. In 1945, Horton (1945, p. 280) demonstrated that enough hydrologic measurements were available to quantify the description and theories of developing drainage basins and river networks. "Horton's laws," a series of exponential or geometric equations, were expanded by Strahler (1952a, 1952b, 1954) and others, and the new subject of *quantitative fluvial geomorphology* was introduced (Morisawa, 1988). The goal was to establish quantitative, rather than qualitative, relationships between geomorphic processes and landforms. The technique is called **morphometry,** the measurement and mathematical analysis of various landform parameters.

One of the early accomplishments of fluvial morphometry was the analysis of branching drainage networks (Figure 11-2). An ingenious numbering technique was devised whereby the fingertip tributaries of streams, those that originate from overland or groundwater flow and flow to a junction without receiving any permanent tributaries themselves, are called **first-order** streams. The drainage basins of first-order streams can cover 50 percent of the area of higher-order basins (Marcus, 1980) and, as illustrated by the location of groundwater seepage and catastrophic debris slides (Figure 9-16), are the region in which mass-wasting merges into fluvial processes. A **second-order** stream begins at the junction of two first-order streams; it may receive additional first-order tributaries, but if it joins another second-order tributary, a **third-order** stream is formed, and so forth. In subsequent jargon (Shreve, 1966, 1967; James and Krumbein, 1969), first-order streams were called **external links** in topologic networks, and higher-order segments were called **internal links.**

The statistical analysis of parameters such as link lengths, bifurcation (branching) frequency, and bifurcation angles can become a highly practical subject. Similar botanical morphometric analysis has demonstrated that the growth forms of all species of trees can be classified into only 23 architectural modes or categories. Computer simulations prove that the pattern of branching that most closely approximates a real tree is one that maximizes the interception of light by the leaves and makes possible the most efficient photosynthesis (Honda and Fisher, 1978; Tomlinson, 1983). The implication is that the organization of tree branches, like drainage networks, is governed by physical laws of conservation of energy and energy

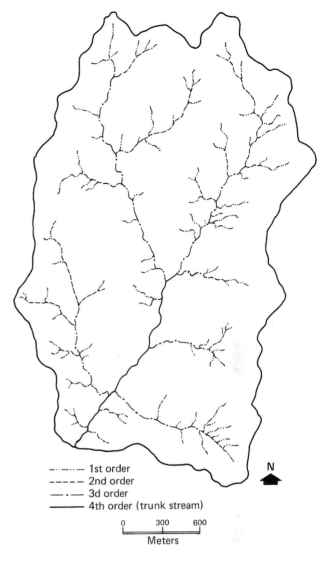

- –––··–– 1st order
- – – – – – 2nd order
- –– · –– 3d order
- ––––––– 4th order (trunk stream)

0   300   600
Meters

FIGURE 11-2.
Drainage network of Mill Creek, Ohio, from topographic map and field surveys (Morisawa, 1959). Symbols show system of stream ordering. Statistical data are listed in Table 11-1 and graphed in Figure 11-3.

distribution, and not by chance. To illustrate a practical problem, suppose that a cluster of houses is to be provided with water by a network of underground pipes. Each size of pipe has a stated cost per unit length, the cost of laying each size of pipe is specified, and each coupling costs a certain additional amount. What is the distribution array from a trunkline to individual houses that can be laid for the least cost? If we rephrase such a question in terms of natural drainage networks and make the important initial *assumption* that nature is inherently conservative in expending energy, we ask: What is the most efficient array of branching channels that can carry water and sediment from an area downhill to the sea?

Early in the studies of drainage networks, it was discovered that on homogeneous terranes they are highly ordered systems in which parameters such as mean link length, bifurcation ratio, downlink slope, drainage density (total length of links per unit area of drainage basin), stream azimuth, and many other dimensions and dimensionless ratios (Table 11-1; Figure 11-3) could be predicted from information about some segment of the system (Strahler, 1958). Hydraulics engineers and geomorphologists have a common interest in being able to predict such practical facts as the annual and peak discharges of a river, flood frequency, and duration of flood crest by combining a minimum number of observation stations within a river system with morphometric analysis.

Quantitative analysis of fluvial systems took a turn from the empirical (and tedious) analysis of real drainage basins to theoretical and statistical considerations of two-dimensional networks or systems in general. Leopold and Langbein (1962, p. 14) used a "random walk" technique to generate drainage networks that had dimensions and ratios similar to the natural networks analyzed by Horton, Strahler, and others. It is an instructive exercise to generate a simple river system by a "random walk" game. Start a number of markers at equal spaces along one edge of a piece of graph paper. Move the markers either ahead, left, or right, one space at a time, by the cast of a die. One restraint to random movement is that a marker cannot move backward. This is the game equivalent of gravity. If a marker intersects the path of another marker, it must thereafter follow that path. This rule of the game represents the surface tension and viscosity of water. The paths of the markers trace drainage nets in which first-order tributaries join to form second-order streams and so on until either all paths have merged into a single master stream or have diverged beyond any probable junction.

The significance of "random walk" models is that they seem to demonstrate the highly probable organization of drainage networks. High probability, in the concept of general systems theory (Chapter 2), represents the minimum energy expenditure to the system (maximum entropy) and the optimal distribution of energy expenditure within the system. The highly probable nature of drainage networks was repeatedly and emphatically demonstrated by Shreve (1966, 1967, 1969, 1975). He showed that all the empirical relationships of links and junctions that had been determined by earlier workers analyzing real drainage networks

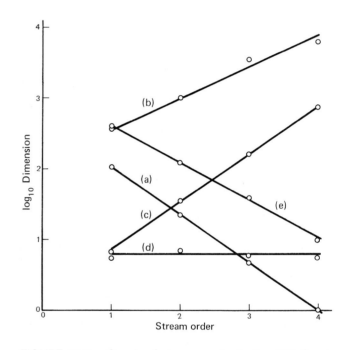

FIGURE 11-3. Graph of stream order in the Mill Creek drainage network compared to the channel and basin dimensions in Table 11-1.

Table 11-1

Analysis of Some Characteristics of the Mill Creek Drainage Network*

Dimension	(a)	(b)	(c)	(d)	(e)
Stream Order	Number of Streams	Average Length (ft)	Average Basin Area (10^5 ft^2)	Stream Density (mi/mi^2)	Average Channel Slope (tan < x 10^3)
1	104	364	6.97	5.45	396
2	22	993	33.73	7.02	123
3	5	3432	161.97	6.06	39
4	1	6283	747.14	5.66	10

Source: Morisawa (1962), Table 1.

*See Figures 11-2 and 11-3.

could also be demonstrated in "infinite topologically random channel networks." Although random processes do not perfectly generate all the details of branching drainage networks, models based on randomness became a standard against which natural networks could be compared (Abrahams, 1984, p. 185). If some property of a natural network deviates significantly from randomness, a search can be made for structural controls, or inheritance from a previous condition.

An intriguing recent development is that when three of Horton's laws, concerning bifurcation ratios, link lengths, and basin areas, were tested against a very large number of branching networks that were an unbiased statistical sample of the even much larger number of all possible branching networks given a certain number of first-order channels, the results justified neither the conclusion that networks are random nor that they represent maximum-entropy, least work systems (Kirchner, 1993). For example, in an unbiased sample of all possible networks that can be formed by merging N first-order channels, where $20 < N < 1000$, 96 percent of all bifurcation ratios, 95 percent of all length ratios, and 98 percent of all area ratios fell within the ranges considered typical of natural stream networks. Furthermore, when the sample set of all possible branching channels was divided in half by a shape factor so that the two subsets were decidedly nonrandom, the three Horton's law parameters showed no significant differences between the two subsets. Kirchner concluded that although channel networks may be either highly ordered by principles of maximum entropy and maximum efficiency, or random, Horton's laws and the many analyses of quantitative fluvial geomorphology that have followed from them are poor indicators of those states. He concluded: "Devising morphometric techniques to detect the characteristic structure of natural channel networks and explaining that structure in mechanistic terms remain central problems in quantitative fluvial geomorphology" (Kirchner, 1993, p. 594).

If properties such as the bifurcation ratio are similar for all parts of a network, so that, for example, each stream of order greater than 1 has, on average, the same number of streams of the next lower order joining it, the network is said to be *statistically self-similar*. Each part of such a network is a smaller version of the whole, and the network is scale-invariant, or *fractal* (p. 5) (Masek and Turcotte, 1994; Rodriguez-Iturbe et al., 1994). Branching networks based on the ordering system proposed by Strahler (Figure 11-2) are self-similar, as are, on average, the topologically random channel networks of Shreve (Peckham, 1995). One interesting property of self-similar networks is that their fractal dimension $D = 2$ produces a "space-filling" network,

meaning that if allowed to propagate to infinity, they would exactly fill a two-dimensional surface. Thus a drainage network, whether purposeful or random, attempts to fulfill its primary task of draining its entire basin (Masek and Turcotte, 1994; Peckham, 1995, p. 1027). However, most actual networks have a fractal dimension slightly less than 2, which is interpreted to mean that though they are potentially space filling, they do not continue to divide into smaller and smaller tributaries, but are truncated at some scale of branching by the transition from overland flow to gullies and rills.

A serious deficiency of fluvial morphometry for explanatory description is that even though real drainage networks can be demonstrated to be topologically random systems, there is no way to prove or disprove that real networks form in the same way that random models are generated (Leopold et al., 1964, p. 421). The techniques of quantitative fluvial geomorphology give excellent *descriptions* and *predictions* of drainage networks but no *explanation* for the manner in which they evolved.

The deficiency was illustrated by a morphometric study (Howard, 1971) in which stream networks were simulated by headward growth and branching. The technique is essentially the reverse of a random walk in which paths merge when they touch; at intervals specified by the rules of headward growth, each headward-expanding channel branches randomly. Howard's models, which are the equivalent of gully systems expanding into undissected uplands, satisfactorily simulate natural systems. His results were even better with models in which growth took place on a matrix that corresponded to an area of previously dissected upland. Howard (1971, pp. 48–49) concluded that it is dangerous to generalize about genetic processes from computer simulations. Many different simulation procedures generate similar topologically random networks. It is not possible to determine whether a given network was formed by (1) headward-branching growth of a gully system, or (2) progressive intersection and consolidation of small rills and channels as water flowed downhill, with or without preexisting irregularities (noses and hollows) that guided the evolution of the network. Leopold et al. (1964, p. 421) agreed that "a river or a drainage basin might best be considered to have a heritage, rather than an origin." The explanation of network evolution remains unsolved.

Computer-Simulated Basin Evolution

With the advent of powerful computers and software programs, physical stream-table experiments and primitive two-dimensional "random walks" have been largely replaced by sophisticated three-dimensional computer models of drainage basins simulating their evolution by eroding channel networks. Although none of the simulations can yet represent all the variables of real basin evolution, especially those of time- and spatially variable precipitation and runoff, they highlight some of the subtle factors that determine the course of basin and network evolution, and they allow long time series to simulate long-term (although unspecified in years) evolution.

Computer simulations combine basin evolution with the evolution of drainage networks by specifying a grided initial surface, usually gently sloping and with a certain random or fractal roughness that is some small fraction of the height of the surface. "Random walkers" are then placed on the surface by chance, and their subsequent paths are followed. These are the raindrops or the sedimentary particles loosened by impact. If they initially move to a depression, the depression is allowed to trap the sediment but lets water infiltrate or overflow. Overland flow rate is treated as *diffusional,* which, like soil creep, is a function of the slope curvature (p. 189). The computer program records the progressive movement downhill of water and sediment. At some irreversible threshold function based on upstream area and slope, water flow converges, becomes turbulent or *advective,* and merges into rills. If the water is carrying sediment to its capacity, it will not erode; if not, additional sediment is scoured and the rill grows. Some models are *detachment limited,* limiting the availability of sediment of a size that can be moved by both overland flow and channeled flow (Howard, 1994). Others are *diffusion limited,* measuring the wandering trajectories of the random walkers until they enter a network (Masek and Turcotte, 1993). Still others couple channel growth rates to diffusion in such a way that overland slopes steepen and converge toward gully heads (Figure 11-4) (Willgoose et al., 1991a, 1991b, 1991c).

As sediment is removed from the simulated land surface, branching networks evolve, and the entire surface becomes realistically dissected and is gradually lowered toward its "base level." Divides become defined and can be sharp or rounded depending on the relative effectiveness of sediment detachment and transport by overland and channeled flow. Uplands are generally convex; channels develop concave long profiles.

Some lessons learned from computer models confirm that many different programs produce similar, realistic-looking networks and basins. Even small variations of initial roughness can be extremely important in determining final conditions (Willgoose et al., 1991b, pp. 1694–1695). Drainage density evolves slowly at first, focused at the heads of initial gullies, then rapidly increases to near its final value (Willgoose et al., 1991b,

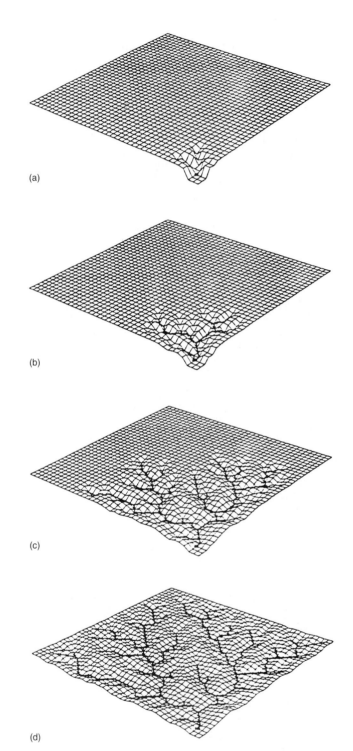

(a)

(b)

(c)

(d)

FIGURE 11-4. Isometric view of evolving channel network on a hillside at four successive times (Willgoose et al., 1991a, Figure 5).

p. 1690). Both *total* and *local relief* (the height difference between the highest and lowest points on all of or some part of the model) decrease with time. If base level is constantly lowered, the models reach a steady state of no change in gradients and only minor changes in flow direction, even after a lowering of the initial landscape by an amount equal to approximately three times the final steady-state relief (Howard, 1994, p. 2273).

Computer-simulated basins and networks provide many opportunities to test our inferences and assumptions about long-term landscape evolution. Even the choice of diffusion-limited or detachment-limited models give insight into the issues of supply-limited as opposed to transport-limited drainage networks (p. 223). If supply is limited, the growing networks simulate erosion into bedrock; if sediment supply is adequate, but transport capacity or competence is limited, the channels are entirely in alluvium. We can expect many further insights from future simulations of hillslope, drainage-basin, and river-network evolution.

SEDIMENT IN EVOLVING FLUVIAL SYSTEMS

Stream Power

In Chapter 2, the potential power (rate of doing work) of terrestrial runoff was calculated to be almost 12×10^9 kW, equivalent to the continuous expenditure of 1 horsepower on each 2.5 acres of land. Water entering a river has potential energy

$$PE = mgh$$

where m is the mass of water, g is the acceleration of gravity, and h is the height above sea level. As the water moves downstream, the potential energy is progressively converted to kinetic energy

$$KE = \frac{mv^2}{2}$$

The principle of conservation of energy requires that the loss of potential energy between two points on a river must be balanced by an equivalent increase in kinetic energy. However, all but a few percent of the converted energy is lost to heat by turbulent flow (p. 198). The rate of kinetic energy conversion (kinetic power,) per unit length of channel is

$$\Omega = \gamma Qs$$

where $\gamma = \rho g$, the specific weight of water, Q is discharge, and s is the slope (Bagnold, 1977; Knighton, 1984, p. 55). The power supply/unit bed area (ω) is then

$$\omega = \gamma Qs/w = \gamma dvs = \tau_0 v$$

because of the relationship $Q = wdv$, where v is the mean velocity, and depth is the approximate equivalent of hydraulic radius in the formula for boundary shear stress τ_0 (p. 216).

These equations relate stream power/bed area to the shear stress at the bed and determine whether the river will erode, transport its existing load, or deposit sediment on that portion of the bed. A **critical power** can be defined as that rate of useful energy conversion necessary to transport the existing load (Bull, 1979). If more power is being expended on a portion of river bed, erosion or *degradation* will result. If too much sediment is supplied, the river will *aggrade.* Continuous fluctuations by the variables of hydraulic geometry ensure that no simple conversion of stream power to conditions of erosion or deposition is possible, but over an appropriate interval of time, grade may be achieved or maintained. Bull (1979) emphasized that sections of rivers that are at or near the threshold of critical power are especially sensitive to changes of climate, base level, or human interference.

Sediment Sources, Storage, and Yields

Clearly, some drainage basins are detachment limited or supply limited, so they do not supply a sufficient load to absorb the stream power downstream. Rivers draining such basins are likely to be incising into bedrock channels. The hydraulic geometry parameters (Chapter 10) demonstrate that most sediment in a river has been derived from the drainage basin by processes such as creep, soil erosion, landslides, and rill incision. River incision also supplies sediment, but has the negative feedback effect of increasing the valley-side slopes, thereby increasing sediment supply and inhibiting the incision (Schumm and Rea, 1995). Probably most rivers are transport limited, with supply by weathering, mass wasting, and soil erosion exceeding the competence and capacity of the stream to move the sediment except during times of high discharge. At other times, the sediment is stored within the fluvial system as colluvium and alluvium. However, sediment output, as measured by the amount of sediment being carried out of a basin by its trunk river, is typically only about 5 to 7 percent of the sediment supplied from the drainage basin (Trimble, 1975, 1983, 1993). In the

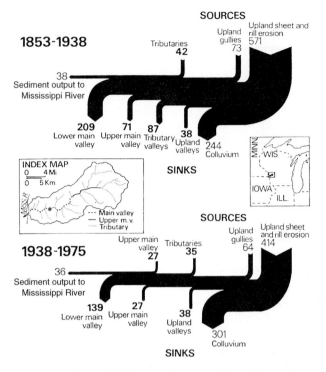

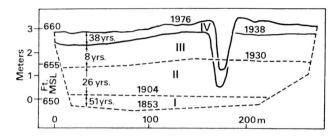

FIGURE 11-6. Cross section of Coon Creek, Wisconsin, valley floor at various dates. Location shown by the dot on Figure 11-5 index map (Trimble, 1983, Figure 5(A)).

FIGURE 11-5. Sediment budgets for Coon Creek, Wisconsin, 1853–1938 and 1938–1975. The basin is about 25 km southeast of La Crosse, Wisconsin, and has an area of 360 km². Numbers are annual averages for the periods in 10^3 Mg/yr and account for 3.6 x 10^7 Mg (tonnes) of sediment generated between 1853 and 1975. Bold numbers are measured, other numbers are estimated. Note that the upper main valley, which was a sink for sediment in the earlier period, had become a partial source of sediment in the later period (Trimble, 1983, Figure 2).

well-monitored basin of Coon Creek, in southwestern Wisconsin, almost 50 percent of the human-induced sediment loss from the basin upland is stored as colluvium and alluvium within the basin, and less than 7 percent has left the basin (Figure 11-5). The upland erosion was estimated using the Universal Soil Loss Equation (p. 201) for conditions at the time of settlement in 1853 and during the 1930s, the interval of peak agriculture and erosion loss. During eight years in the 1930s, the floodplain aggraded at almost 15 cm/yr, but from 1938 to 1976 aggradation (due to declining agricultural land use) decreased to only 1.6 cm/yr (Figure 11-6). Similar high denudation rates and massive alluvial storage have been documented for numerous river basins on the piedmont of both Maryland and the southeastern United States (Trimble, 1975; Jacobson and Coleman, 1986). These and similar studies remind us of the danger of using sediment output of rivers to estimate basin erosion, for most of the detritus has not left the basins even after long periods.

Downstream Changes in Alluvium Grain Size

Coarse alluvium decreases exponentially in mean grain size downstream by sorting and abrasion. The complex interplay among bed roughness, depth, velocity, slope, and grain size was discussed in Chapter 10. Size sorting by progressive loss of competence downstream is the most important of several causes for decreasing mean grain size in alluvium (Ferguson et al., 1996). A given grain size simply cannot be transported downstream beyond the point where the stream is competent to move it.

Abrasion also reduces the grain size of bedload alluvium in a downstream direction although abrasion is not a factor in reducing the grain size of the suspended load. In 260 km of transport down the Colorado River of Texas below Austin, granitic pebbles and cobbles from a known source area diminish in mean diameter by about 50 percent, quartz by about 30 percent, and chert by about 20 percent (Bradley, 1970, p. 68). The glacial outwash (p. 383) of the Knik River in Alaska shows a more dramatic decrease in grain size downstream (Bradley et al., 1972). Cobbles and boulders that are readily moved by periodic floods from an ice-dammed lake are reduced in size by 87 percent (to 13 percent of initial mean diameter) in only 26 km of river transport. This rapid reduction is consistent with many other observations of the rapid downstream change in grain size of glacial outwash.

An attempt was made to verify both sets of field measurements just cited by abrading gravel in laboratory conditions. Selected stones were dragged over the roughened bed of a toroidal abrasion tank by a current of water until they had traveled a distance equivalent to the downstream movement in the Colorado and Knik rivers. Fresh granitic pebbles from the Colorado River were reduced by only 10 percent of mean grain size in the tank in contrast to their 50 percent reduction in an equivalent field transport distance. Knik River gravel was reduced by only 8 percent in 26 km of laboratory transport in contrast to the 87 percent reduction in the field.

The authors of these studies concluded that sorting rather than abrasion was the major cause of the rapid downstream decrease in grain size in the aggrading glacial outwash stream bed. Sorting was also judged to be important to the downstream change of grain size in the Colorado River alluvium, but more important in that river was the abrasion of gravel that weathers slightly during temporary storage in the floodplain. A third factor in the downstream reduction of bedload alluvium could be the chipping and abrasion of cobbles as they shift and vibrate under current velocities that are not strong enough to transport them. Conceivably, cobbles could be abraded and rounded "in place" until they become small enough to be transported. The greater size reduction observed in actual rivers as opposed to flume experiments may be due in part to this kind of abrasion without net transport (Schumm and Stevens, 1972; Brewer et al., 1992).

It can be concluded that sorting, abrasion, and weathering during temporary storage all contribute to the observed decrease in mean grain size of bedload alluvium in a downstream direction. The abraded and weathered detritus is also transported, but in the form of dissolved and suspended load. As a very crude guess, the average alluvial fragment might take 1000 years to move downstream to the sea, most of that time being at rest (and weathering) in some part of the floodplain (Leopold et al., 1964, p. 328).

Active and Relict Alluvium

For many reasons, we consider alluvium to be in transit toward the sea, and therefore the landforms on alluvium must be ephemeral as compared to erosional landforms (back endpaper). Yet in terms of human lifetimes, or even the length of human history, landforms built of alluvium appear permanent. Written human history began more than 6000 years ago on the alluvial plains and deltas of the Middle East. Even today, an overwhelming majority of the human population lives on alluvium.

In the geologic record, alluvial conglomerates and sandstones are well known, but they comprise only a small fraction of the total volume of sedimentary rocks. Alluvial deposits, ancient and modern, are studied in greater proportion than their abundance would indicate because they may contain valuable placers of heavy ores, precious metals, or gems. Furthermore, terrestrial vertebrate fossils and the artifacts of ancient people are most commonly found in alluvium.

Quaternary tectonism and climatic fluctuations have made alluvium a complex material to interpret. The distinction between alluvium actively "in transit" and "relict" is especially critical in areas of former glaciation, where glacial outwash gravel or sand underlies or is adjacent to much finer-grained recent alluvium. The Mississippi River illustrates that interpretive problem well (Figures 11-7 and 11-8). In full-glacial time, the downstream part of the river entrenched a deep valley to the lowered sea level of the Gulf of Mexico. As the ice margin retreated northward from the upper midwestern states, sand and gravel alluvium of a braided Mississippi River filled the lower trench. Farther north, portions of the late-glacial braid plain still have morphologic expression (Figure 11-8). Later, when the ice margin was fringed by proglacial

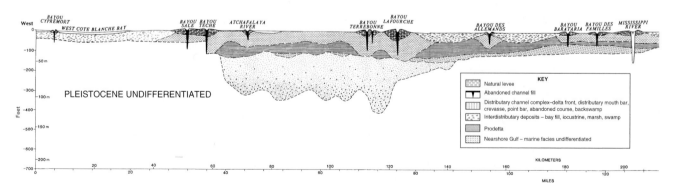

FIGURE 11-7. East-west cross section of the Mississippi delta plain south of New Orleans (from Autin et al., 1991, cross section G).

FIGURE 11-8. Contrast in stream patterns between relict late-glacial braid plain reach and adjacent Holocene meander belt of the Mississippi River alluvial plain near the Missouri-Arkansas state line (Autin et al., 1991, Figure 2).

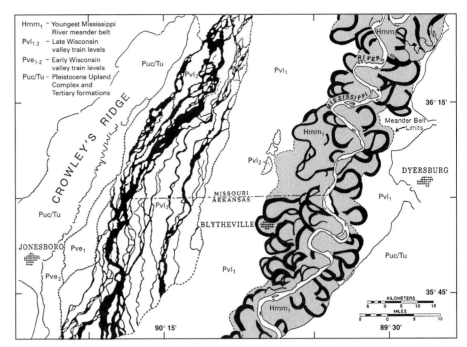

Hmm₁ – Youngest Mississippi River meander belt
Pvl₁₋₂ – Late Wisconsin valley train levels
Pve₁₋₂ – Early Wisconsin valley train levels
Puc/Tu – Pleistocene Upland Complex and Tertiary formations

lakes in the present Great Lakes region, the coarse alluvium was trapped in the lakes, and only mud and fine sand went down the Mississippi. By that stage, rising sea level was causing swampy deltaic deposition in Louisiana. In the last few thousand years, with sea level rising only slowly and with the modern delta prograding across the continental shelf, mud and organic deposition predominate on the floodplain. How much of the cross section in Figure 11-7 should be called active alluvium, or what qualifications to the term are required to explain the entire cross section?

As a working definition, we can restrict the definition of "active" alluvium to that sediment capable of being transported by the stream that now flows on it. Such a definition relates to both the depth of the alluvium and its grain size. The thickness of the modern alluvium is then defined by the flood-scour depth of the modern channel, and its maximum grain size by the maximum competence of the modern stream. Anything deeper or coarser is "relict" alluvium. The depth of modern channel scour should not be underestimated, however. Todd (1900) cited soundings of the bed of the Missouri River during high discharge that proved the channel to be 15 to 30 m deep in flood where it is only 3 to 15 m deep at low water (compare Figure 10-14). The flood-eroded channel was floored by bedrock, thus proving that the entire alluvial thickness is potentially in transit. The great vertical exaggeration of Figure 11-7 shows that the modern Mississippi channel is as deep as the uppermost stratigraphic units of prodelta, interdistributary, distributary channel, and natural levee deposits, which thereby become classed as the active alluvium.

Floodplain Morphology

Floodplains and their minor morphologic subdivisions are primarily depositional landforms. As such, their forms are genetically related to specific depositional processes (Figure 11-9; Table 11-2). Not all deposits create landforms, however. The lag deposits shown beneath the channel in Figure 11-9 might well be relict alluvium, for instance (compare Figure 11-7). Nanson and Croke (1992, p. 460) defined a *genetic floodplain* as "the largely horizontally bedded alluvial landform adjacent to a channel, separated from the channel by banks, and built of sediment transported by the present flow-regime."

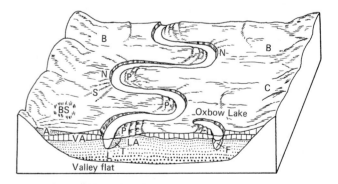

FIGURE 11-9. Typical associations of valley sediments. A-Alluvial fan; B-Backland; BS-Backswamp; C-Colluvium; F-Channel fill; L-Lag deposit; LA-Lateral accretion; N-Natural levee; P-Point bar; S-Splay; T-Transitory bar; VA-Vertical accretion. See Table 11-2 for sediment characteristics (S. C. Happ, in Vanoni, 1971, Figure 2-Q-1).

Table 11-2
Genetic Classifications of Valley Sediments*

Place of Deposition	Name	Characteristics
Channel	Transitory channel deposits	Primarily bedload temporarily at rest; part may be preserved in more durable channel fills or lateral accretions
	Lag deposits	Segregations of larger of heavier particles, more persistent than transitory channel deposits, and including heavy mineral placers
	Channel fills	Accumulations in abandoned or aggrading channel segments; ranging from relatively coarse bedload to find-grained oxbow lake deposits
Channel margin	Lateral accretion deposits	Point and marginal bars that may be preserved by channel shifting and added to overbank floodplain by vertical accretion deposits at top
Overbank floodplain	Vertical accretion deposits	Fine-grained sediment deposited from suspended load of overbank flood water, including natural levee and backland (backswamp) deposits
	Splays	Local accumulations of bedload materials, spread from channels onto adjacent floodplains
Valley margin	Colluvium	Deposits derived chiefly from unconcentrated slope wash and soil creep on adjacent valley sides
	Mass movement deposits	Earthflow, debris avalanche, and landslide deposits commonly intermix with marginal colluvium; mudflows usually follow channels but also spill overbank

Source: S. C. Happ, in Vanoni, 1971.
*Compare with Figure 11-9.

This definition precisely emphasizes the linkage between active alluvium, floodplains, and the hydraulic geometry of modern rivers (Chapter 10) but requires an elaborate set of additional terms for relict floodplains built by Pleistocene glacial meltwater or even during century- or decade-scale fluctuations in river regimens. It also raises questions about the definition of "the stream that [now] flows in it," in Playfair's law (p. 232, 327).

Lateral Accretion. Floodplains are built in two fundamental ways, by *lateral accretion* and by *vertical accretion*. In general, channel deposits that accumulate by lateral accretion are coarser than overbank flood deposits that cause vertical accretion. **Point bars** (Figure 11-9) are the most important component of lateral accretion. As a meandering channel migrates across the floodplain, the **cut bank** on the convex side is undercut and eroded. Piping by groundwater that flows out of the alluvium during falling stages of river discharge is an important factor in bank collapse and mobilization of alluvium. As banks collapse, the derived coarse bedload is carried a short distance downstream and deposited as a submerged bar, usually on the same side of the stream. The result is a cross-stratified deposit, with a subdued relief of low ridges and intervening swales, that may record many episodes of meandering channel migration (Figures 11-10 and 11-11). The crests of point bars approach the level of the former flood-plain on the cut-bank side unless the stream is actively entrenching or aggrading, so the level of the floodplain need not be changed significantly by lateral accretion. In streams that carry relatively coarse alluvium and lack natural levees, lateral accretion may account for 80 to 90 percent of the floodplain deposition (Wolman and Leopold, 1957, p. 96).

The age of point-bar deposits along the Little Missouri River was determined by Everitt (1968) by tree-ring chronology of cottonwood trees growing on various parts of the floodplain. An age map of the valley floor (Figure 11-11) shows that only relatively small areas of floodplain are older than 200 years. Everitt concluded that the reach of the river he studied redistributes every century a volume of sediment equal to the total present volume of alluvium along that reach, yet the floodplain has probably been within 1 m of its present elevation for at least 150 years. Even though a floodplain has retained a stable form for centuries, the alluvium that composes it is intermittently moving through the system, weathering while it is in storage and being abraded as it moves in the river. Sigafoos (1964) gave an excellent summary of the technique of using floodplain trees to document flooding and flood-plain deposition on a time scale of about a century.

Vertical Accretion. Large rivers with gentle gradients and a predominantly suspended load typically

(a)

(c)

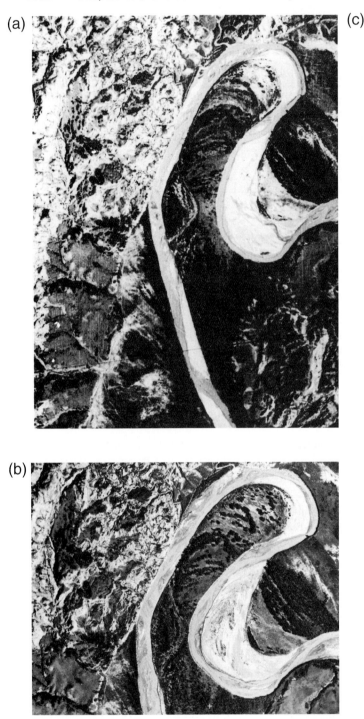

(b)

FIGURE 11-10. Floodplain and channel of the Little Missouri River near Watford City, North Dakota. For orientation and scale, see Figure 11-11. (a) 1939 vertical aerial photograph; (b) 1949 vertical aerial photograph; (c) 1958 vertical aerial photograph of the northernmost meander (photos: U.S. Department of Agriculture, from Everitt, 1968).

have meandering channels on broad floodplains. When they flood, the velocity in the overbank water may be very low, and the floodwaters may be weeks in receding into the channel. Under these conditions, floodplain deposition is largely by *vertical accretion.* Rates may exceed 1 cm/day during the annual flood on the Amazon River floodplain (Mertes, 1994). The deposits are usually fine sand or mud although splays

from breached channel banks may bring coarse sediment. Very large areas of the middle Mississippi River floodplain were submerged for weeks during the great flood of 1993. However, because of sediment dilution by the very large discharge and extended time of the flood, the typical floodplain accretion was relatively minor, on the order of a few millimeters except near levee breaks, where deposits 0.15 to 15 m in thickness

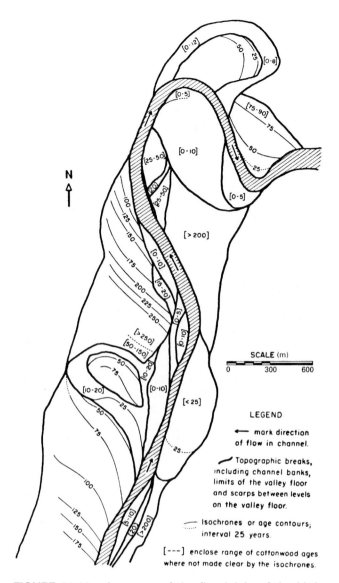

FIGURE 11-11. Age map of the floodplain of the Little Missouri River near Watford City, North Dakota, based on ages of cottonwood trees (Everitt, 1968).

adjacent backswamps. Under these conditions, a flood level high enough to overtop a segment of natural levee has a pronounced gradient into the backswamp, and floods are violent and severe. The current at a point of overflow erodes a gap or **crevasse** in the levee, and a **splay** spreads far across the backswamp (Figure 11-12). Sometimes the erosion is sufficient to begin an entirely new master channel, and a segment of the former channel is abandoned, to fill slowly with channel deposits and peat along a curved **oxbow lake.** If a meander recurves until it intersects an upstream portion of the channel, a **neck cutoff** (Figure 10-15) occurs, wherein the abrupt shortening of channel length locally increases the gradient and velocity in the cutoff, and eventually seals the truncated ends of the abandoned meander with bank deposits or **plugs.**

If a master stream flows between high natural levees, few tributaries can join it but must flow parallel to the trunk stream along the lowest axis of the backswamp. In older American terminology, such streams were called **Yazoo tributaries,** from the Yazoo River in Mississippi. A Yazoo tributary may be forced to extend downstream by the migrating meander belt of the trunk stream, but eventually it will be intersected by a meander and will be shortened again.

Vertical accretion cannot continue indefinitely unless the base level is rising. The postglacial rise of sea level has accentuated vertical accretion and the formation of natural levees and backswamps, as during the later history of the lower Mississippi River (Figure 11-7). The regular frequency of the most probable annual flood on most upland rivers suggests that vertical accretion is not a major contributor to floodplain development; otherwise, rivers with larger floodplains should flood progressively less frequently. On Coon Creek, however, the recent pulse of aggradation has raised the channel as well as the floodplain and need not alter the flood frequency (Figure 11-6). We may suspect that thick alluvium deposited by vertical accretion, such as the relict alluvium found beneath large rivers in their lower valleys (Figure 11-7), is the result of postglacial rise of sea level and would not be typical of hypothetical rivers graded to a stable sea level. Thin vertical-accretion alluvium may accumulate slowly on a floodplain if the river is constrained from migrating laterally (Ritter et al., 1973).

AGGRADATION IN THE FLUVIAL SYSTEM

Aggradation as a Normal Phase of Valley Evolution

The primary function of river systems is to carry away weathered debris fed into them in great quantity by mass wasting and overland flow. If any stream power

accumulated (Gomez et al., 1995). Earlier, from April to June 1973, the Mississippi River flooded large areas of its lower floodplain to depths of up to 4 m. During the two-month interval, vertical accretion averaged 53 cm along natural levees but decreased rapidly to less than 2 cm beyond 400 m from the natural levees and averaged only slightly more than 1 cm in the backswamps (Kesel et al., 1974). The abrupt loss of velocity at the edge of the flooded channel and the abundant sediment supply there commonly cause more and coarser deposition on the natural levees, which grade laterally into finer backswamp deposits. A result of vertically accreting natural levees may be that the river surface within the levees is many meters above the level of

FIGURE 11-12. Schematic reconstruction of Cenozoic Texas Coastal Plain depositional environments. Mixed-load sinuous to meandering channel with associated levees and crevasse splays forms relatively high, well-drained belt separating areas of swampy floodplain and interchannel lakes (Galloway, 1981, Figure 19).

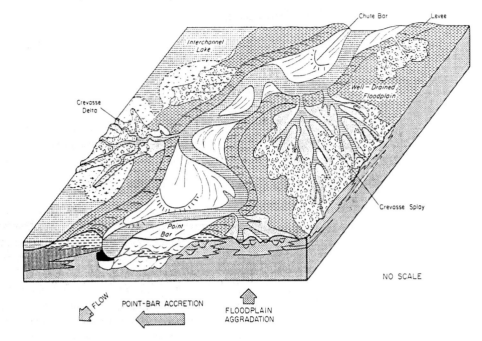

is available beyond that required to move the load, the river will erode its channel. The deduced normal cross section of a river valley thus should have an erosional rock surface, the **valley flat,** beneath alluvium in transit or temporarily stored in the floodplain. Examples of relict alluvium have previously been described in this chapter, where aggradation has been caused by rising base level, climatic change, or tectonism. All three of these factors have been so important in shaping the present landscape that we are hard pressed to find a river that has a "normal" floodplain underlain by alluvium in transit to the sea and that is slowly eroding a valley flat beneath the alluvium.

In view of the abundant climatic and tectonic events that can cause aggradation, it is appropriate to consider whether an aggradational phase can occur during the normal evolution of a fluvial landscape. A trunk stream might become graded early in the dissection history, while headward erosion is still in progress by first-order streams. Then, as the drainage network expands, greater discharge and load might require the longitudinal profile to be steepened by aggradation in the middle portion of the river system. When the uplands are totally dissected by competing drainage networks, no additional expansion can occur except by capture, and presumably the trunk stream will resume its slow downcutting (Davis, 1899, reprinted 1954, p. 260). Alluvial terraces might be left along the valley walls to record an aggradational phase that was a normal part of the continuous regional erosional evolution.

Cotton (1948, pp. 80–82) considered other possible causes of aggradation as a temporary phase of normal fluvial landscape development. Channel lengthening by both meandering and delta progradation causes a decrease of gradient, but the principles of hydraulic geometry suggest that aggradation need not result. Neither does it seem likely that the increased channel size of a graded river, or subsurface flow through alluvium, could cause a loss of competence and result in aggradation (Cotton, 1948, p. 81). The interaction of the many semidependent variables of the fluvial system is much too well balanced to permit such a disequilibrium to develop. Thus the only likely cause for an aggradational phase during fluvial dissection of a landscape, barring climatic and tectonic "accidents," seems to be the development of a graded trunk stream while drainage networks are still expanding. It should be recalled that most rivers are transport limited, rather than supply limited (p. 223), so an aggradational phase can be reasonably expected. The alluvial storage in Coon Creek (Figures 11-5 and 11-6) might be similar to a river's response to an expanding drainage-basin area.

Climatic Change and Aggradation

Rivers in alluvial channels on the Great Plains have been observed to change from a braiding to a meandering habit in response to decade-length historical climatic changes (p. 223). A generalization about the response to slight shifts in climate by fluvial systems in semiarid or subhumid regions that are near a threshold of change was set forth by Huntington (1907, p. 358) and is now called *Huntington's principle* (Fairbridge, 1968, p. 1125). Huntington postulated that increased aridity in areas of marginal rainfall leads to a loss of vegetative cover and increased mass wasting (Figure

Color Plate 1. "Gaussian hills that never were" by R.F. Voss. A perspective view of a computer-generated landscape made to look realistic by assigning a Gaussian probability distribution to the difference in altitude between any two prescribed points on a map, with suitable mathematical transformations. (Mandelbrot, 1983, plate C11.)

Color Plate 2. Complex structures in sedimentary rocks exposed in the headwall of "Cirque de Gavarnie" in the French Pyrenees. (photo: Takuma Arii.)

Color Plate 3. Tilted fault-block mountains of the Ambato Ranges, Argentina. City of Catamarca in southeast corner of 15x15 km image. (Landsat TM image 50213-1344.)

Color Plate 4. View southeastward toward the thrust-fault scarp of the Ancasti Range near Catamarca, Argentina.

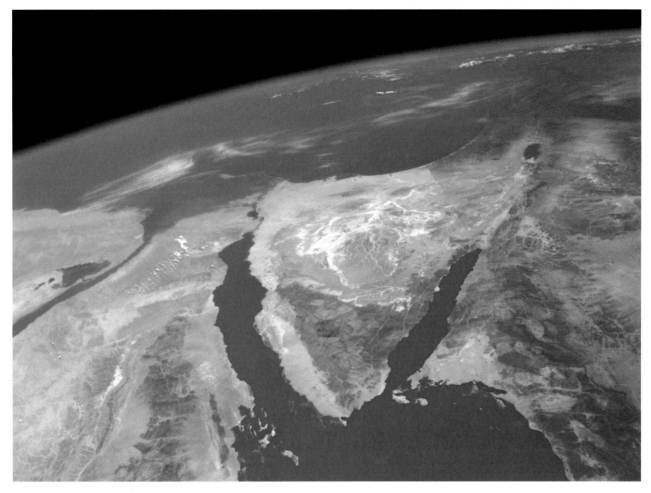

Color Plate 5. Space Shuttle oblique view toward the northwest. Sinai Peninsula and Nile River and delta (left); Gulf of Aqaba and Dead Sea rift (right). (NASA photograph S40-152-180.)

Color Plate 6. Kohala coast, north side of Hawaii, view west. Steep cliffs are the headscarp of a giant slump. Valley in foreground was cut to a glacially lowered sea level and aggraded during postglacial sea level rise. See also figure 10-6. (photo: E.F. Bloom.)

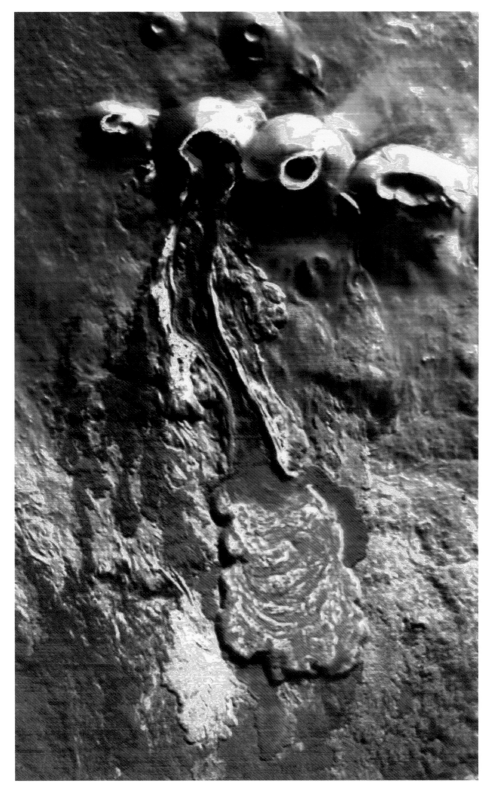

Color Plate 7. The "Great Tolbachik Fissure Eruption" (Fedotov and Markhinin, 1983) along a rift in Tolbachik Volcano, Kamchatka Peninsula. Thermal infrared multispectral scanner (TIMS) image made in September 1993 processed to enhance spectral contrast (Pieri et al., 1997). Line of scoria cones trends northward (left). White and red colors show the "Northern Breakout" basalt flows from a collapsed scoria cone that covered an area of 9km^2 during July to September, 1975. Barren tephra cover is imaged in blue; forested lower slopes of the volcano are brown. Main scoria cones have a basal diameter of about 1 km. (NASA-JPL image, courtesy D.C. Pieri. References listed in chapter 6.).

Color plate 8. Chao dacite plug dome, northern Chile, between snow-covered and glaciated volcanic peaks, Cerro Paniri (5960 m) to the west (left) and Cerro del León (5771 m) to the east (right). View is 15x15 km. (Landsat TM image 50522-1400.)

Color Plate 9. Cerro Galán, a caldera with a central resurgent dome in the Andes of northwestern Argentina (p. 112). Image is 60x60 km. (Landsat TM image 50451-1355.)

Color Plate 10. Structurally controlled benches and slopes along the Colorado River at Dead Horse Point, near Moab, Utah.

Color Plate 11. Cuestas on east flank of the Wind River Mountains. View north from near South Pass, Wyoming. (See figs. 12-7 and 12-13 and related text.)

Color Plate 12. Pilcomayo River flowing eastward across the Subandean Fold and Thrust Belt in southern Bolivia. River is believed to be antecedent to the deformation. (See figs. 12-18 and 12-21 and related text.)

Color Plate 13. View south from Fall River Pass (11,796 ft) in Rocky Mountain National Park, Colorado. Note contrast between smooth, convex upland and steep northeast - facing cirque headwall, which still contains snow in this August view. (Compare fig. 17-8.)

Color Plate 14. Fringing reef with low algal rim in the surf zone, Tongatapu, Kingdom of Tonga. Note solution notches on talus blocks and cliffs.

15-8). Stream channels become choked with the excessive loads supplied to them concurrent with a reduction in discharge. Aggradation results as rivers build up their alluvial beds until gradients are steep enough to move the greater load, which is likely to be of coarser grain size. Huntington's enunciation of this principle of fluvial response to increased aridity was conditioned by his observations in the interior of Asia, where decreased total rainfall typically coincides with increased seasonality of precipitation. This is often true, but seasonality need not be incorporated into the principle as a requirement. Clearly, climates with an intense dry season followed by concentrated precipitation are conducive to a maximum sediment load in streams (Chapter 13).

Huntington's principle further postulates that during times of greater precipitation, vegetation will stabilize landscapes, inhibit mass wasting, promote infiltration, and result in more sustained flow of rivers. Peak flood discharges will be minimized, and rivers will erode channels of more gentle longitudinal gradients. Most geomorphologists would mentally associate a braided habit with the aggradational phase and meandering with the more humid valley-deepening phase. Huntington suggested that the great gravel terraces that line the valleys of central Asia represent multiple cycles of arid or semiarid aggradation alternating with entrenchment during more humid intervals. The implied general correlation of humid and arid intervals with Quaternary glacial and interglacial episodes is obvious, but the current evidence suggests that in some climatic zones the wetter intervals were associated with high sea levels of interglacial ages, and the arid aggradational phases are recorded by coarse alluvium graded to low sea levels that are correlative with glacial ages (Chapter 18). In other regions, especially the presently arid southwestern United States, glacial ages were the more humid, or pluvial, ages as inferred by Huntington. Huntington's principle, with allowance for important regional variation and the possibility of significant lag times for some of the complex responses of the fluvial system, is an important key to relating Quaternary climatic change to landscape evolution (Chapter 18).

Depending on their prior condition, fluvial systems vary in their response to climate changes. Various lag times may be involved. For example, in southeastern Arizona and adjacent parts of Mexico, rivers responded to the onset of postglacial aridity and the replacement of forests by desert shrubs by a single pulse of aggradation about 8000 years ago. But in the lower Colorado River basin, there were two episodes of aggradation, the first inferred to be from the onset of postglacial monsoonal precipitation in the late Pleistocene and the second with the vegetation transition from forest to desert about 8000 years ago (Bull, 1991, p. 278). Rivers with glacierized headwaters generally aggraded massively during late-glacial ice-margin recession and intrenched in postglacial time. Thus two nearby river valleys in the American southwest might have aggraded in nearly opposite sequence: the one with glacierized headwaters aggraded in late-glacial time and then intrenched; the one with a nonglacierized basin intrenched during the cooler, wetter late Pleistocene and subsequently aggraded with the onset of postglacial aridity. Other rivers that drained into glacial-age lakes in arid regions may have aggraded to former lake levels and intrenched as the lakes evaporated. The combinations of processes and responses are nearly endless, and their study is a fascinating topic (Bull, 1991).

Fluvial Terraces

Most floodplains have a microrelief of a few meters or less, including low *floodplain terraces* deposited or eroded during floods of varying severity. The conditions of fluvial deposition during variable discharge, by either meandering or braided rivers, ensure such minor floodplain terraces. However, minor low terraces are not likely to be preserved for more than a few centuries before channel migration removes them (Figure 11-11).

If a river is slowly downcutting as its channel meanders laterally from one side of the valley to the other, portions of older floodplain or valley flat may be preserved as **unpaired terraces** (Figure 11-13), especially if they are *rock-defended* by a buttress of bedrock exposed in the valley wall. Unpaired terraces need not represent a change in the rate or nature of fluvial processes. They are little more than accidentally preserved fragments of former floodplains. Long ago, Playfair (1802, p. 103) cited such terraces as strong proof that valleys are indeed the work of the streams that flow in them.

Paired terraces, which occupy the same level on opposite valley walls, are significant evidence that a river has cut downward in an intermittent fashion. Like unpaired terraces, paired terraces may be remnants either of alluvium or of the former valley flat cut into rock. Rock-floored terraces, preserving part of the former valley flat, are called *strath terraces.* The distinction in the nature of terrace material is useful for descriptive purposes but does not have much genetic significance because most rock-cut terraces retain at least a veneer of alluvium. Paired alluvial terraces may represent climate-controlled aggradation alternating with downcutting, as inferred by Huntington and others, or simply pulses of intrenchment alternating with times when the valley flat was veneered with alluvium in

FIGURE 11-13. Terraces of the lower Rakaia River valley, Canterbury, New Zealand. Two high terraces in the distance are aggradational glacial outwash surfaces. Terraces in the foreground are unpaired erosional terraces, cut in glacial deposits (photo: National Publicity Studios, New Zealand).

transit. Even though a former floodplain was built by aggradation, it does not become a terrace except by subsequent erosion. It is important to remember that *fluvial terraces are cut, not built, by rivers* (Cotton, 1948, p. 187).

Flights of terraces on a weak-rock lowland may end downstream at a gorge through resistant rock that is the local and temporary base level for the upstream graded reach (p. 210). Because of slow downcutting at the resistant structure, the river upstream has time to open out a broad terraced valley. Such terraces are a special class of rock-defended terrace, where a single resistant obstruction will control and preserve multiple terraces (Cotton, 1948, p. 197). They are normally unpaired.

Thus far, only the cross-valley profile of fluvial terraces has been considered. The longitudinal profiles of terraces should also be studied for clues to river history. For instance, terraces may converge or diverge in the downstream direction. Convergence suggests progressive rejuvenation of the headwaters, perhaps by continuing tectonic uplift (Figure 5-28). Divergence suggests a progressive lowering of base level, either temporary or ultimate, more rapidly than the regional erosion rate. Differential erosion or tectonic subsidence in the downstream area could be invoked (Pierce and Morgan, 1992, pp. 26–27). Changes in the amount and grain size of the alluvium carried by a river when it flowed at each successive terrace level may also explain differences in longitudinal slope of terraces.

The lower reaches of many, perhaps most, river valleys have been profoundly altered by late Cenozoic sea-level fluctuations (pp. 210, 235). As noted for the lower Mississippi Valley, rivers intrenched their valleys to lower sea levels during glaciations and generally aggraded floodplains during interglacial high sea levels such as at present. Any cumulative tectonic movement superimposed on the sea-level oscillations is likely to generate sets of paired terraces extending at least tens of kilometers inland from the river mouths (Merritts et al., 1993). Separating the causes for the common phenomenon requires reconstructing sea-level history, climate-controlled hydraulic geometry, sediment supply, tectonism, vegetation history, and probably human interference.

Fluvial terraces on a single river system may be misinterpreted if the regional setting is ignored. Ritter (1967, 1972) described examples of rivers at the foot of the Beartooth Mountains in Montana. Rivers that rise in the mountain range have relatively steep gradients to carry coarse alluvium, whereas rivers that rise on the plains carry finer alluvium and have lower gradients. Therefore, even though two adjacent rivers drain to the same base level, near the mountain front the "mountain bred" stream is in a channel at a higher level than, and is likely to be captured by, the adjacent stream that heads at the mountain front. However, the low-gradient river cannot carry the coarse alluvium, and when capture occurs, an active aggradational phase begins, while the beheaded stream adjacent to it begins downcutting and forming terraces.

Valleys that now head in the mountains and also those that now originate at the mountain front have prominent terraces of coarse alluvium derived from the Beartooth Range, proving that capture has caused alternating episodes of aggradation and entrenchment, with terrace formation in each valley. Although a single valley would seem to record a complex history that might be given a tectonic or climatic explanation, the larger view demonstrates normal or "single-cycle" regional geomorphic development (see also Figure 13-7 and text discussion). The regional history of the Beartooth Range terraces is complicated further by outwash terraces from former valley glaciation.

DELTAS AND ALLUVIAL FANS

The third zone of the fluvial system is the region of sediment deposition, where water either enters the sea or a lake or evaporates on a desert floor. Two of the most common depositional landforms of alluvium result from the same cause: an abrupt loss of competence in a stream. *Deltas* are mostly subaqueous forms although their upper surfaces are usually slightly above water level. They result when a river enters a lake or the ocean and its velocity is abruptly reduced. *Alluvial fans* are subaerial forms, deposited at an abrupt decrease in competence where a stream channel loses water by infiltration, decreases its gradient, or widens abruptly. The two forms may merge; many deltas are capped by low-gradient alluvial fans. The "delta" of the Colorado River into the Gulf of California has separated the Salton Basin from the upper end of the Gulf, and its north side is the apex of a sea-level alluvial fan 70 m above the floor of the closed, below-sea-level Salton Basin.

Deltas

A level surface of water, whether a lake or the sea, has a powerful trapping effect on the suspended and bedloads of a river. A river enters standing water as a jet that maintains its central velocity for a distance equal to about four times the width of the river mouth. Then the flow begins to disperse by marginal turbulence and the mixing effects of wind, waves, and currents. Because of density differences produced by salinity, temperature, and sediment load, the river water and sediment may (1) sink to the bottom and flow downslope as a dense *turbidity current,* (2) flow directly forward into equally dense still water and mix turbulently within a distance equal to a few times the channel width, or (3) spread forward over the surface of the standing water (Bates, 1953). The third option is the most common where rivers enter the ocean because the suspended sediment load can only rarely increase the fluid density to equal the density of seawater. Eventually, velocity is sharply reduced and competence drops.

All the bedload and most of the suspended load is dropped close to the point where a stream enters a lake or the sea. The bedload usually forms a shallow arcuate **lunate bar** across the river mouth. To maintain cross-sectional area, the channel commonly breaks into multiple **distributary channels** across the accumulating alluvium and distributes new sediment over a broad arc concentric to the original river mouth. The resulting landform is a **delta.**

Deltas may be the subject of the oldest geomorphic description. Herodotus (fifth century B.C.) observed the growth of the Nile delta by the annual floods and correctly inferred that all of lower Egypt was the delta of the Nile (Stanley and Warne, 1993). The name *delta* comes from the Greek capital letter delta (Δ), which is the shape of the Nile delta (Figure 11-14, Color Plate 5).

Not all deltas have the same shape (Figure 11-15). The shape is influenced by river discharge and load, grain size, flood frequency and intensity, climate, vegetation, wave energy, tidal range, salinity, water temperature, and tectonism (Morgan, 1970). Some deltas have shifting sublobes that mark separate periods of growth (Figure 11-16). Others have been fairly stable in postglacial time, gradually *prograding* (building forward) along their entire seaward perimeter through a maze of distributaries. Marine deltas are almost entirely Holocene landforms, created near the end of the late glacial sea-level rise (Stanley and Warne, 1994).

The surface shape of deltas can be broadly classified into (1) fluvial- or river-dominated, (2) wave-dominated, and (3) tidal-dominated (Coleman et al., 1986; Galloway and Hobday, 1996, pp. 103–119). Very large rivers extending into sheltered or shallow gulfs or lakes build *digitate* or *bird-foot* deltas, with long, projecting natural levees little modified by waves or currents. The youngest Mississippi delta building across a very shallow continental shelf into the nearly tideless Gulf of Mexico is an exceptional example of a river-dominated delta (Figure 11-15a) (Coleman, 1988). If waves and currents are strong enough to erode the river-mouth bars and levees, the sand is carried laterally to form barrier beaches (Chapter 19), and the delta becomes cuspate or broadly rounded and smooth in a map view (Figure 11-15b). The Nile delta is wave-dominated, with barrier beaches enclosing a series of lagoons along its margin, and small cuspate deltas at the active Damietta and Rosetta distributary mouths (Figure 11-14). It is argued that the Nile delta is now a relict coastal plain landform because dams and irrigation networks have totally stopped the water and sediment delivery to the ocean (Stanley and Warne, 1993, p. 628). The Niger delta is also wave-dominated although tidal currents are significant. It has an arcuate shoreline of barrier beaches enclosing a maze of mangrove-covered dis-

FIGURE 11-14. The Nile delta and Suez Canal viewed toward the northeast (photo: NASA Shuttle Mission 41G-120-177).

tributary channels, none of which projects seaward (Figure 11-15c). In seas with high tidal range, the ebb and flow of the tides may distribute sediment rapidly offshore, and the distributary mouths become trumpet shaped, opening broadly to the sea between marsh islands or sand bars. Most of the nation of Bangladesh occupies the vast tidal-dominated merged deltas of the Ganges and Brahmaputra rivers (Coleman et al., 1986, pp. 326–329).

The longitudinal cross section of a simple or small delta into a lake or reservoir of water with density equal to the density of the river discharge shows three basic depositional units: *bottomset beds, foreset beds,* and *topset beds* (Gilbert, 1890). Only the third is subaerial. A surface gradient must be maintained to carry alluvium to the perimeter, and aggradation may extend upstream as a delta grows forward and reduces the river gradient. The *Gilbert-type* delta is relatively rare, restricted to freshwater lakes and reservoirs, although widely described in introductory texts as the most common type of delta.

The Mississippi River has one of the largest deltas and certainly the most studied one (Morgan, 1970; Coleman, 1988; Penland et al., 1988; Autin et al., 1991; Törnqvist et al., 1996). It has built forward to the edge of the continental shelf after having filled a deeply entrenched glacial-age valley (Figure 11-7). The entire delta complex is called a **delta plain.** The internal structure consists of a set of subdeltas, each an assemblage of linear, branching channel-sand units flanked by paired natural levees of poorly sorted sand and mud, which in turn grade laterally into highly organic backswamp mud. The delta complex resembles a pile of maple leaves, with the veins represented by the subdelta channel and levee deposits, and the intervein areas represented by backswamps. When a subdelta lobe was abandoned, compaction and subsidence lowered the area until only the drowned levee crests were exposed. Simultaneously, waves eroded the delta front and concentrated the sand as barrier spits and islands. These reworked shallow marine sediments of the erosion phases that follow the abandonment of a distributary system contribute up to 50 percent of the total sediment thickness in the Mississippi River delta plain. The geologic record of such shallow-water deltas could be largely of marine, rather than fluvial, sediments (Penland et al., 1988). Some later generations of distributary channels reoccupied segments of earlier drowned levee systems, in a few instances flowing in the opposite direction. At times, more than one subdelta were occupied simultaneously. Gradients are negligible throughout the modern delta plain.

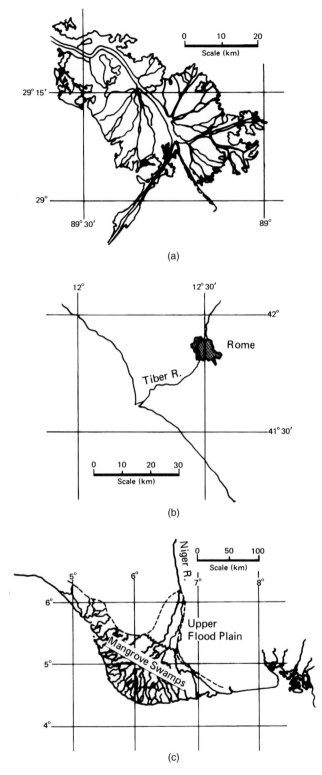

(a)

(b)

(c)

FIGURE 11-15. Some common delta forms. (a) Digitate ("birdfoot") delta of the Mississippi River. (b) Cuspate (concave seaward) Tiber delta. See also the Rosetta and Damietta mouths of the Nile River (Figure 11-14). (c) Lobate (convex seaward) Niger delta. Another common form, the bay-head delta, is shown in Figure 5-19.

Location of Major Deltas

On a relief map of the world, many of the largest rivers are seen to rise within orogenic belts but then flow across continents to the trailing margins where they build their deltas (Inman and Nordstrom, 1971; Audley-Charles et al., 1977; Potter, 1978). The Chang Jiang (Yangtze), Amazon, Mississippi, and Rhine rivers are representative of this pattern. Others, like the Ganges, Brahmaputra, Yukon, and Mackenzie rivers, flow parallel to the orogenic grain, either within the orogen or along one of its borders. A few rivers rise near a rifted continental margin, drain toward the interior of the continent, and then turn and find their way through the highlands of the rifted margin to the sea, often along ancient transverse fault valleys, as in the case of the Niger. It is uncommon to find large river deltas on a converging continental margin, for the local orogenic relief is too young and active for a large river system to have evolved. Because deltaic sediments and sedimentary rocks are important sources of petroleum, modern deltas are studied carefully to understand the depositional environments of ancient sediments (Galloway and Hobday, 1983; Whateley and Pickering, 1989).

Alluvial Fans

In its simplest classic form, an **alluvial fan** is a segment of a low cone with its apex at the mouth of a gorge or canyon (Figure 11-17). Contours are arcs of circles concentric on the head of the fan. In longitudinal or radial profile, a fan may have a straight slope, a series of radially flattening facets, or a continuous concave skyward slope. They are excellent examples of water-spreading wash slopes (Figure 9-23). Alluvial fans are especially abundant and significant landforms in arid intermontane regions (Figures 11-17, 13-2, and 13-3), but are common in humid regions as well.

An alluvial fan usually forms where a high-gradient stream leaves a confining gorge and discharges directly onto the floodplain of a trunk stream or the flat floor of an intermontane basin. A simple abrupt decrease in downstream gradient is not an adequate explanation for alluvial fans. More important is the abrupt widening and shoaling of the channel cross section. Deposition results from the abrupt loss of competence at the gorge mouth through great widening, infiltration, loss of depth, and in deserts, evaporation.

Fans are built of debris flows, colluvium, and bedload alluvium, especially gravel. Stratification may be sheetlike parallel to the fan surface or may contain many cut-and-fill structures of buried channels. Their surface morphology is characterized by radiating, branching distributary channels, most often braided

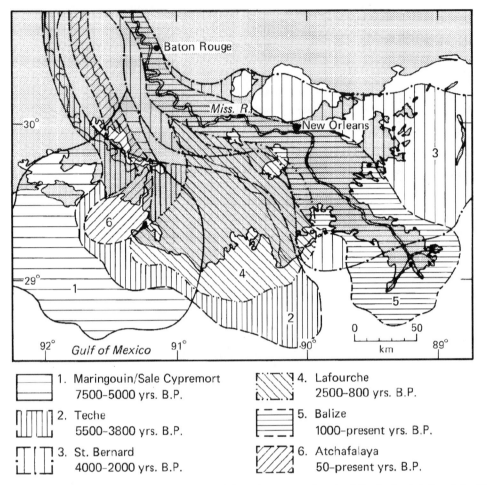

FIGURE 11-16. Chronology of the subdeltas that comprise the Mississippi delta plain (Coleman, 1988, Figure 2; chronology revised by Autin et al., 1991, p. 582, and Törnqvist et al., 1996). The modern digitate delta is shown in Figure 11-15(a).

rather than meandering. The permeable character of fan alluvium allows considerable infiltration so that fans may contain mudflow deposits recording the progressive loss of water after the suspended load became too thick to flow as a muddy stream.

The slope of alluvial fans increases with increased grain size and poorer sorting and increased concentration of the sediment load, and decreases with increased discharge (Hooke and Rohrer, 1979). Therefore, fans in arid regions, where the sediment load is coarse and may be delivered in the form of debris flows, tend to be steeper than fans in humid regions, where water discharge is larger and more continuous and sediment loads are constrained by vegetation. Some analyses distinguish between *fluvial fans,* or *"wet" fans,* comprised entirely of water-laid sediments, and *mud-flow fans* or

"dry" fans, more characteristic of arid regions where intermittent flash floods and debris flows add to fan volume (Schumm et al., 1987, pp. 285ff; Scott and Erskine, 1994). If the sediment load is less than the capacity and competence of a fan stream, the stream will entrench the upper fan and transport the sediment out to the margin of the fan, thereby reducing the stream gradient. A segmented fan will then develop, with younger lobes emerging from between areas of older alluvium (Figure 11-18). Stable channels are possible if the capacity and competence of the fluvial system are balanced by the sediment load. On actively growing fans, the sediment is dropped at the apex of the fan, filling preexisting channels and perhaps spreading over the fan surface until the steeper gradient permits transport. Fan-building streams typically have extreme variations in both

FIGURE 11-17. An alluvial fan in Death Valley, California (photo: John S. Shelton).

FIGURE 11-18. New fan forming on a braided floodplain in front of an erosionally truncated earlier fan at the mouth of Centre Creek, Avoka River valley, Canterbury, New Zealand (photo: C. O'Loughlin, Ministry of Forestry, New Zealand).

discharge and load, so fans commonly show complex surface morphology and internal structure (Blair and McPherson, 1992, 1994a, 1994b; Hooke, 1993).

Most fans are *aggradational,* or built-up, forms. One radial channel may carry most of the discharge for a time until its gradient becomes oversteep by deposition, when the stream abruptly shifts to a lower-gradient radius. The process ensures the uniform slope development along all radii and a resulting smooth conical form.

The alluvial fans of Death Valley (Figure 11-17) have been analyzed as equilibrium, or steady-state, landforms (Denny, 1965, 1967). Each fan has an area equal to one-third to one-half of the mountain watershed that feeds it. The surface areas of other alluvial fans are also roughly in linear proportion to their drainage areas, and their slopes generally decrease with increased basin area (Blair and McPherson, 1994a, Figures 14-22 and 14-23). However, there are wide variations in these dimensions. Fans dominated by debris flows are steeper, commonly having linear radial slopes in the range of 5°

to 15°. Fans dominated by flow, whether sheetflood or channeled flow, have lesser radial gradients in the range of 2° to 8° that decrease downstream. Some fans may be segmented, induced by continued basinward tectonic tilting (Hooke and Dorn, 1992).

A deduced sequence of fan formation, for example at the base of a fault scarp, might begin with the formation of a steep straight talus cone, followed by an accumulation of slightly lower-gradient debris flows and taluses around its periphery, followed still later by a broader, flatter, and more concave-upward alluvial fan dominated in its upper or proximal parts by debris flows and mudflows and in its outer, or distal, part by sandy or muddy sheetflood deposits (Blair and McPherson, 1994a, p. 395–396). Such a simple deductive sequence has many natural variants depending on the intensity of floods, the sediment supply and grain size, and the size of the basin into which the fan is being built.

The Kosi River of the Ganges plain in northern India has an exceptionally large, low-gradient alluvial

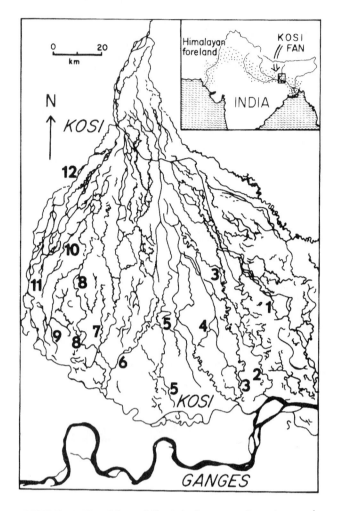

FIGURE 11-19. Map of Kosi drainage system, to resolution limit of about 80 m, constructed from March 1977 dry-season false-color satellite imagery. Dates for occupation of channels 1 to 12 are 1731, 1770–1775, 1807–1839, 1840–1873, 1873–1893, 1893–1921, 1921–1926, 1926–1930, 1930–1936, 1936–1942, 1942–1948?, and 1977, respectively (for sources, see Wells and Dorr, 1987, Figure 1).

fan at the base of the Himalaya Mountains. It is 154 km long and 143 km wide (Figure 11-19).

Repeated surveys in the past 228 years record a continuous westward shifting of the Kosi River for 113 km across its fan, in a series of episodic diversions of up to 19 km/yr (Wells and Dorr, 1987). Major westward diversions do not correlate with major earthquakes in the region, but seem to record a persistent tendency of the river when it abandons its channel to migrate to the previously unoccupied fan surface rather than shift eastward to a channel recently abandoned. Channel shifting occurs during monsoonal floods and commonly results in major human disasters (Nayak, 1996).

Many underfit trunk streams of glaciated valleys have low gradients that cannot transport the coarse alluvium delivered by valley-side tributaries, and the master stream is pushed laterally from one valley wall to the other by fans. Tributary fans may block the trunk stream and create a lake. Very steep alluvial fans are called **alluvial cones** and are comparable in form, if not in internal structure, to talus cones.

Piedmont Alluvial Plains

Large regional landforms develop from the lateral merging of adjacent low-gradient alluvial fans draining away from the base of dissected mountains. Most of the Great Plains Province of the central United States is veneered by an apron of late Cenozoic sand and gravel named the Ogallala Formation (Frye, 1971; Seni, 1980; Gustavson and Winkler, 1988), covering a region 1300 km long southward from South Dakota to Texas and 500 km wide (Figure 11-20). Most of the alluvium is sandy, but gravel channel deposits and fine-grained

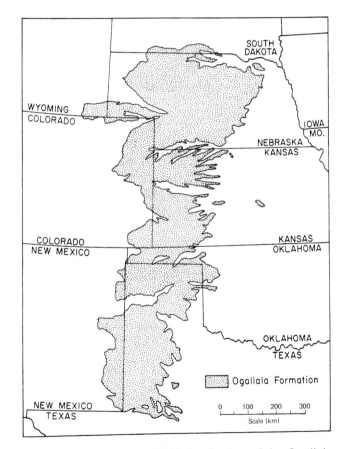

FIGURE 11-20. Generalized distribution of the Ogallala Formation in the Great Plains province of the United States (Frye, 1971, Figure 1).

vertical-accretion deposits are also known. The Ogallala Formation is as much as 150 m thick along the axes of buried valleys. Some combination of epeirogenic upwarp of the Rocky Mountains, and perhaps climatic change as well, caused the paleorivers of the Great Plains to aggrade until they entirely filled their valleys and coalesced across former interfluves. Eolian sand and loess (Chapter 13) blew off the braided floodplains and blanketed adjacent uplands, indicating that the climate was semiarid, at least in the southern part of the Ogallala depositional region (Gustavson and Winkler, 1988). The modern Kosi fan of India (Figure 11-19) has been suggested as a model for the environment of the Ogallala Formation deposition. Pleistocene climatic changes and renewed tectonism probably ended the aggradation phase (Stanley and Wayne, 1972). The present relief of the unglaciated Great Plains is essentially the depositional surface of the Ogallala Formation and similar, but slightly younger, Pleistocene alluvium. The rivers that now flow east to the Mississippi are at grade on this ancient alluvial surface. The regional slope is only about 0.1 percent.

Similar piedmont alluvial plains form the Argentine Patagonian plains on the eastern side of the Andes Mountains, the Indo-Gangetic Plain on the southern foreland of the Himalaya Mountains, and the Canterbury Plains of New Zealand. Late Cenozoic tectonism has played an important role in initiating these and other great alluvial plains in foreland basins along active orogens (Chapter 3). However, climatic changes during the Quaternary Period, including glaciation in the mountains, repeatedly caused the rivers to change their hydraulic geometry, and much of the alluvium is relict.

REFERENCES

ABRAHAMS, A. D., 1984, Channel networks: A geomorphological perspective: Water Resources Res., v. 20, pp. 161–168.

AUDLEY-CHARLES, M. G., CURRAY, J. R., and EVANS, G., 1977, Location of major deltas: Geology, v. 5, pp. 341–344.

AUTIN, W. J., BURNS, S. F., and 3 others, 1991, Quaternary geology of the lower Mississippi Valley, in Morrison, R. B., ed., Quaternary nonglacial geology: Conterminous U.S.: Geol. Soc. America, The Geology of North America, v. K-2, pp. 547–581.

BAGNOLD, R. A., 1977, Bed load transport by natural rivers: Water Resources Res., v. 13, pp. 303–312.

BATES, C. C., 1953, Rational theory of delta formation: Am. Assoc. Petroleum Geologists Bull., v. 37, pp. 2119–2162.

BLAIR, T. C., and McPHERSON, J. G., 1992, The Trollheim alluvial fan and facies model revisited: Geol. Soc. America Bull. v. 104, pp. 762–769. (Reply: Geol. Soc. America Bull., v. 105, pp. 564–567).

———, 1994a, Alluvial fan processes and forms, in Abrahams, A. D. and Parsons, A. J., eds., Geomorphology of desert environments: Chapman & Hall, London, pp. 354–402.

———, 1994b, Alluvial fans and their natural distinctions from rivers based on morphology, hydraulic processes, sedimentary processes, and facies assemblages: Jour. Sedimentary Res., v. A64, pp. 450–489. (Discussion and Reply: Jour. Sedimentary Res., v. A65, pp. 581–586, pp. 706–711).

BRADLEY, W. C., 1970, Effect of weathering on abrasion of granitic gravel, Colorado River (Texas): Geol. Soc. America Bull., v. 81, pp. 61–80.

———, FAHNESTOCK, R. K., and ROWEKAMP, E. T., 1972, Coarse sediment transport by flood flows on Knik River, Alaska: Geol. Soc. America Bull., v. 83, pp. 1261–1284.

BREWER, P. A., LEEKS, G. J. L., and LEWIN, J., 1992, Direct measurement of in-channel abrasion processes, in Bogen, J., Walling, D. E., and Day, T. J., eds., Erosion and sediment transport monitoring programmes in river basins: Internat. Assoc. Hydrol. Sciences (IAHS) Pub. no. 210, pp. 21–29.

BULL, W. B., 1979, Threshold of critical power in streams: Geol. Soc. America Bull., v. 90, pp. 453–464.

———, 1991, Geomorphic response to climatic change: Oxford University Press, Inc., New York, 326 pp.

BUNTING, B. T., 1961, The role of seepage moisture in soil formation, slope development, and stream initiation: Am. Jour. Sci., v. 259, pp. 503–518.

COLEMAN, J. M., 1988, Dynamic changes and processes in the Mississippi delta: Geol. Soc. America Bull., v. 100, pp. 999–1015.

———, ROBERTS, H. H., and HUH, O. K., 1986, Deltaic landforms, in Short, N. M., and Blair, R. W., JR., eds., Geomorphology from space: A global overview of regional landforms: NASA publication SP-486, pp. 317–352.

COTTON, C. A., 1948, Landscape as developed by the processes of normal erosion: Whitcombe and Tombs, Ltd., Wellington, New Zealand, 509 pp.

DAVIS, W. M., 1899, The geographical cycle: Geog. Jour., v. 14, pp. 481–504. (Reprinted 1954 in Geographical essays: Dover Publications, Inc., New York, pp. 249–278.)

DENNY, C. S., 1965, Alluvial fans of the Death Valley region, California and Nevada: U.S. Geol. Survey Prof. Paper 466, 62 pp.

———, 1967, Fans and pediments: Am. Jour. Sci., v. 265, pp. 81–105.

DUNNE, T., WHIPPLE, K. X., and AUBRY, B. F., 1995, Microtopography of hillslopes and initiation of channels by Horton overland flow, in Costa, J. E., Miller, A. J., and 2 others, eds., Natural and anthropogenic influences in fluvial geomorphology: The Wolman volume: Am. Geophys. Union, Geophys. Mono. 89, pp. 27–44.

EVERITT, B. L., 1968, Use of the cottonwood in an investigation of the recent history of a flood plain: Am. Jour. Sci., v. 266, pp. 417–439.

FAIRBRIDGE, R. W., 1968, Terraces, fluvial-environmental controls, in Fairbridge, R. W., ed., Encyclopedia of geomorphology: Reinhold Book Corp., New York, pp. 1124–1138.

FERGUSON, R., HOEY, T., and 2 others, 1996, Field evidence for rapid downstream fining of river gravels through selective transport: Geology, v. 24, pp. 179–182.

FRYE, J. C., 1971, The Ogallala Formation-a review, in Ogallala Aquifer Symposium: Texas Tech. Univ. Internat. Center for Arid and Semi-arid Land Studies Spec. Rept. no. 39, pp. 5–14a.

GALLOWAY, W. E., 1981, Depositional architecture of Cenozoic Gulf Coastal Plain fluvial systems, in Ethridge F. G., and Flores, R. M., Recent and ancient nonmarine depositional environments: Models for exploration: Soc. Econ. Paleontol. Mineral. Spec. Pub. No. 31, pp. 127–155.

———, and HOBDAY, D. K., 1996, Terrigenous clastic depositional systems, second edition: Springer-Verleg Berlin Heidelberg New York, 489 pp.

GILBERT, G. K., 1890, Lake Bonneville: U.S. Geol. Survey Mon. 1, 438 pp.

GOMEZ, B., MERTES, L. A. K., and 3 others, 1995, Sediment characteristics of an extreme flood: 1993 upper Mississippi River valley: Geology, v. 23, pp. 963–966.

GUSTAVSON, T. C., and WINKLER, D. A., 1988, Depositional facies of the Miocene-Pliocene Ogallala Formation, northwestern Texas and eastern New Mexico: Geology, v. 16, pp. 203–206.

HONDA, H., and FISHER, J. B., 1978, Tree branching angle: Maximizing effective leaf area: Science, v. 199, pp. 888–890.

HOOKE, R. L., 1993, The Trollheim alluvial fan and facies model revisited: Discussion: Geol. Soc. America Bull., v. 105, pp. 563–564.

———, and DORN, R. I., 1992, Segmentation of alluvial fans in Death Valley, California: New insights from surface exposure dating and laboratory modelling: Earth Surface Processes and Landforms, v. 17, pp. 557–574.

———, and ROHRER, W. L., 1979, Geometry of alluvial fans: Effect of discharge and sediment size: Earth Surface Processes, v. 4, pp. 147–166.

HORTON, R. E., 1945, Erosional development of streams and their drainage basins: Geol. Soc. America Bull., v. 56, pp. 275–370.

HOWARD, A. D., 1971, Simulation of stream networks by headward growth and branching: Geog. Analysis, v. 3, pp. 29–50.

———, 1994, A detachment-limited model of drainage basin evolution: Water Resources Res., v. 30, pp. 2261–2285.

HUNTINGTON, E., 1907, Some characteristics of the glacial period in non-glaciated regions: Geol. Soc. America Bull., v. 18, pp. 351–388.

INMAN, D. L., and NORDSTROM, C. E., 1971, On the tectonic and morphologic classification of coasts: Jour. Geology, v., 79, pp. 1–21.

JACOBSON, R. B. and COLEMAN, D. J., 1986, Stratigraphy and recent evolution of Maryland piedmont flood plains: Am. Jour. Sci., v. 286, pp. 617–637.

JAMES, W. R., and KRUMBEIN, W. C., 1969, Frequency distributions of stream link lengths: Jour. Geology, v. 77, pp. 544–565.

KESEL, R. H., DUNNE, K. C., and 3 others, 1974, Lateral erosion and overbank deposition on the Mississippi River in Louisiana caused by 1973 flooding: Geology, v. 2, pp. 461–464.

KIRCHNER, J. W., 1993, Statistical inevitability of Horton's laws and the apparent randomness of stream channel networks: Geology, v. 21, pp. 591–594 (Reply: Geology, v. 22, pp. 380–381).

KIRKBY, M. J., 1993, Network hydrology and geomorphology, in Beven, K., and Kirkby, M. J., eds., Channel network hydrology: John Wiley & Sons Ltd., Chichester, UK, pp. 1–11.

KNIGHTON, D., 1984, Fluvial forms and processes: Edward Arnold (Publishers) Ltd, London, 218 pp.

LEOPOLD, L. B., and LANGBEIN, W. B., 1962, Concept of entropy in landscape evolution: U.S. Geol. Survey Prof. Paper 500-A, 20 pp.

LEOPOLD, L. B., WOLMAN, M. G., and MILLER, J. P., 1964, Fluvial processes in geomorphology: W. H. Freeman and Company, San Francisco, 522 pp.

MARCUS, A., 1980, First order drainage basin morphology-definition and distribution: Earth surface processes, v. 5, pp. 389–398.

MASEK, J. G., and TURCOTTE, D. L., 1993, A diffusion-limited aggregation model for the evolution of drainage networks: Earth, Planet. Sci. Ltrs., v. 119, pp. 379–386.

———, 1994, Statistical inevitability of Horton's laws and the apparent randomness of stream channel networks: Comment: Geology, v. 22, p. 380.

MERRITTS, D. J., VINCENT, K. R., and WOHL, E. E., 1993, Long river profiles, tectonism, and eustasy: A guide to interpreting fluvial terraces; Jour. Geophys. Res., v. 99, pp. 14,031–14,050.

MERTES, L. A. K., 1994, Rates of flood-plain sedimentation on the central Amazon River: Geology, v. 22, pp. 171–174.

MORGAN, J. P., 1970, Depositional processes and products in the deltaic environment, in Morgan, J. P., ed., Deltaic sedimentation, modern and ancient: Soc. Econ. Paleontol. Mineral. Spec. Pub. No. 15, pp. 31–47.

MORISAWA, M. E., 1959, Relation of quantitative geomorphology to stream flow in representative watersheds of the Appalachian Plateau province: Office of Naval Research Project NR 389-042, Tech. Rept. 20, 94 pp.

———, 1962, Quantitative geomorphology of some watersheds in the Appalachian Plateau: Geol. Soc. American Bull., v. 73, pp. 1025–1046.

———, 1988, The Geological Society of America Bulletin and the development of quantitative geomorphology: Geol. Soc. America Bull. v. 100, pp. 1016–1022.

NANSON, G. C., and CROKE, J. C., 1992, A genetic classification of floodplains: Geomorphology, v. 4, pp. 459–486.

NAYAK, J. N., 1996, Sediment management of the Kosi River basin in Nepal, in Walling, D. E., and Webb, B. W., eds., Erosion and sediment yield: Global and regional perspectives: Internat. Assoc. Hydrol. Sciences (IHAS) Pub. no. 236, pp. 583-586.

PECKHAM, S. D., 1995, New results for self-similar trees with applications to river networks: Water Resources Res., v. 31, pp. 1023–1029.

PENLAND, S., BOYD, R., and SUTER, J. R., 1988, Transgressive depositional systems of the Mississippi delta plain: A model for barrier shoreline and shelf sand development: Jour. Sedimentary Petrol., v. 58, pp. 932–949.

PIERCE, K. L., and MORGAN, L. A., 1992, Track of the Yellowstone hotspot: Volcanism, faulting, and uplift, in Link, P. K., Kuntz, M. A., and Platt, L. B., eds., Regional geology of eastern Idaho and western Wyoming: Geol. Soc. America Mem. 179, pp. 1–53.

PLAYFAIR, J., 1802, Illustrations of the Huttonian theory of the earth: Dover Publications, Inc., New York, 528 pp. (facsimile reprint, 1964).

POTTER, P. E., 1978, Significance and origin of big rivers: Jour. Geology, v. 86, pp. 13–33.

RITTER, D. F., 1967, Terrace development along the front of the Beartooth Mountains, southern Montana: Geol. Soc. America Bull., v. 78, pp. 467–483.

———, 1972, Significance of stream capture in the evolution of a piedmont region, southern Montana: Zeitschr. für Geomorph., v. 16, pp. 83–92.

———, KINSEY, W. F., and KAUFFMAN, M. E., 1973, Overbank sedimentation in the Delaware River Valley during the last 6000 years: Science, v. 179, pp. 374–375.

RODRIGUEZ-ITURBE, I., MARANI, M., and 2 others, 1994, Self-organized river basin landscapes: Fractal and multifractal characteristics: Water Resources Res., v. 30, pp. 3531–3539.

SCHUMM, S. A., 1977, The fluvial system: John Wiley & Sons, Inc., New York, 338 pp.

———, MOSELEY, M. P., and WEAVER, W. E., 1987, Experimental fluvial geomorphology: John Wiley & Sons, Inc., New York, 413 pp.

———, and REA, D. K., 1995, Sediment yield from disturbed earth systems: Geology, v. 23, pp. 391–394.

———, and STEVENS, M. A., 1972, Abrasion in place: A mechanism for rounding and size reduction of coarse sediment in rivers: Geology, v. 1, pp. 37–40.

SCOTT, P. F., and ERSKINE, W. D., 1994, Geomorphic effects of a large flood on fluvial fans: Earth Surface Processes and Landforms, v. 19, pp. 95–108.

SENI, S. J., 1980, Sand-body geometry and depositional systems, Ogallala Formation, Texas: Texas Bureau of Economic Geology, Rpt. of Invest. no. 105, 36 pp.

SHREVE, R. L., 1966, Statistical law of stream numbers: Jour. Geology, v. 74, pp. 17–37.

———, 1967, Infinite topologically random channel networks: Jour. Geology, v. 75, pp. 178–186.

———, 1969, Stream lengths and basin areas in topologically random channel networks: Jour. Geology, v. 77, pp. 397–414.

———, 1975, Probabilistic-topologic approach to drainage-basin geomorphology: Geology, v. 3, pp. 527–529.

SIGAFOOS, R. S., 1964, Botanical evidence of floods and floodplain deposition: U.S. Geol. Survey Prof. Paper 485-A, 35 pp.

STANLEY, D. J., and WARNE, A. G., 1993, Nile delta: Recent geological evolution and human impact: Science, v. 260, pp. 628–634.

———, 1994, Worldwide initiation of Holocene marine deltas by deceleration of sea-level rise: Science, v. 265, pp. 228–231.

STANLEY, K. G., and WAYNE, W. J., 1972, Epeirogenic and climatic controls of early Pleistocene fluvial sediment dispersal in Nebraska: Geol. Soc. America Bull., v. 83, pp. 3675–3690.

STRAHLER, A. N., 1952a, Dynamic basis of geomorphology: Geol. Soc. America Bull., v. 63, pp. 923–938.

———, 1952b, Hypsometric (area-altitude) analysis of erosional topography: Geol. Soc. America Bull., v. 63, pp. 1117–1142.

———, 1954, Quantitative geomorphology of erosional landscapes: Internat. Geol. Cong., 19th, Algiers 1952, Comptes Rendus, sect. 13, fasc. 15, pp. 341–354.

———, 1958, Dimensional analysis applied to fluvially eroded landforms: Geol. Soc. America Bull., v. 69, pp. 279–300.

THOMAS, M. F., and GOUDIE, A. S., eds., 1985, Dambos: small channelless valleys in the tropics: Characteristics, formation, and utilization: Zeitschr. für Geomorph., Supp. no. 52, 222 pp.

TODD, J. E., 1900, River action phenomena: Geol. Soc. America Bull., v. 12, pp. 486–490.

TOMLINSON, P. B., 1983, Tree architecture: American Scientist, v. 17, pp. 141–149.

TÖRNQVIST, T. E., KIDDER, T. R., and 8 others, 1996, A revised chronology for Mississippi River subdeltas: Science, v. 273, pp. 1693–1696.

TRIMBLE, S. W., 1975, Denudation studies: Can we assume stream steady state?: Science, v. 188, pp. 1207–1208.

———, 1983, Sediment budget for Coon Creek basin in the Driftless Area, Wisconsin, 1853–1977: Am. Jour. Sci., v. 283, pp. 454–474.

———, 1993, The distributed sediment budget model and watershed management in the Paleozoic Plateau of the Upper midwestern United States: Physical Geog., v. 14, pp. 285–303.

VANONI, V. A., 1971, Sediment transportation mechanics: Q. Genetic classification of valley sediment deposits: Hydraulics Div. Jour., Am. Soc. Civil Engineers Proc., Paper 7815, no. HYl, pp. 43–53.

WELLS, N. A., and DORR, J. A., JR., 1987, Shifting of the Kosi River, northern India: Geology, v. 15, pp. 204–207.

WHATELEY, M. K. G., and PICKERING, K. T., 1989, Deltas: Sites and traps for fossil fuels: Geological Society (London) Spec. Pub. no. 41, 360 pp.

WILLGOOSE, G., BRAS, R. L., and RODRIGUEZ-ITURBE, I., 1991a, A coupled channel network growth and hillslope evolution model: 1. Theory: Water Resources Res., v. 27, pp. 1671–1684.

———, 1991b, A coupled channel network growth and hillslope evolution model: 2. Nondimensionalization and applications: Water Resources Res., v. 27, pp. 1685–1696.

———, 1991c, A physical explanation of an observed link area-slope relationship: Water Resources Res., v. 27, pp. 1697–1702.

WOLMAN, M. G., and LEOPOLD, L. B., 1957, River flood plains: Some observations on their formation: U.S. Geol. Surv. Prof. Paper 282-C, pp. 86–109.

Chapter 12

Structural Control of Fluvial Erosion

INTRODUCTION

No geomorphologist doubts the importance of rock structure to the shape of erosional landforms. *Structure*, along with *process* and *time*, is a cornerstone of explanatory description (Chapter 1). Structural factors of rock weathering are discussed in Chapters 7 and 8, and the role of rock structure in mass wasting is frequently cited in Chapter 9. This chapter is concerned with the effects on fluvial erosion of the rock mass, including all its discontinuities such as stratification and other primary layering, foliation and similar metamorphic fabrics, joints, faults, and folds.

The preceding two chapters emphasized fluvial erosional, transportational, and depositional processes and forms, and it was not necessary to consider structural control of fluvial processes except incidentally. Drainage networks were assumed to develop over homogeneous terranes in order to simplify the study of fluvial processes. However, on the real earth, rivers cross belts or zones of contrasting structures, and their hydraulic geometry changes in response to the erodibility of each structure and the type of load it delivers to the river.

The present subaerial landscape is nothing more than an instantaneous and arbitrary interface between atmosphere and lithosphere, marking a transitory phase of the erosion process. Erosion, including tectonic denudation (p. 78), has probably removed more than 10 km from many continental landscapes and as much as 30 km from some mountain ranges (p. 121). During such tremendous amounts of erosion, whole structural terranes can be removed. For instance, coastal plain sediments can be stripped off as streams intrench into the older rocks beneath. Folded and faulted sedimentary structures can be stripped from basement complexes of igneous and metamorphic rocks. The relief that had been buried under unconformities and thrust sheets, perhaps part of ancient landscapes, can be exhumed.

During the progress of erosion, it is inevitable that some structures will be more resistant to destruction than others. Relative erodibility, or rock-mass resistance, is a subjective property, not simply defined. For instance, gravel alluvium may survive on very old river terraces, not because it became strongly cemented or lithified but because it is so permeable that rainfall percolates rapidly through it and no surface streams can form to erode it (Rich, 1911). Emerged coral reefs on tropical coasts (Figure 3-5) may have their relief preserved or even enhanced because the permeable reef rock is lowered only slightly while solution lowers the adjacent less permeable lagoon-floor carbonate sand and mud. In both examples, passive resistance is more effective than strength.

A loose distinction can be made between the sedimentary *cover strata* of continents and the igneous and metamorphic *basement* rocks that underlie them. Some cover strata, for example, thick-bedded quartz-rich sandstones, are significantly more resistant to erosion than many igneous and metamorphic rocks. More commonly, the cover strata are more erodible and are easily stripped off the basement rocks by fluvial, glacial, and eolian processes.

Such an erodibility contrast is to be seen in the eastern United States along the **Fall Zone,** a seaward-dipping exhumed erosion surface between crystalline rocks of the Piedmont and New England Provinces and the overlying coastal plain sediments of Cretaceous and younger ages (Figure 1-5). Postorogenic erosion created this surface of low relief before it was buried by onlapping marine sediments. Subsequent gentle epeirogenic warping has uplifted the landward region and depressed the seaward part so that a narrow bevel or facet of older metamorphic rocks has been exposed on the inland edge of the coastal plain by erosion of the former cover strata. The steepened gradients and many waterfalls on rivers where they cross the Fall Zone gave the region its name. Its location at the head of tidewater from New England southward to Washington D.C., and from there southward along the inner edge of the Coastal Plain Province, with local water power available at the many waterfalls, made it the site of colonial American settlements and subsequently of most of the eastern cities of the United States.

FRACTURES AS STRUCTURAL CONTROLS

Joints

Systematic or nonsystematic joint sets cut most indurated rock masses. Joints are probably the most common rock structures, breaking igneous, metamorphic, and sedimentary rocks in all kinds of orogenic and epeirogenic settings (Pollard and Aydin, 1988). They control weathering, mass wasting, and erosion on a wide range of scales, from microcracks (Figure 7-2) to regional lineations that help characterize entire geomorphic provinces. Some are the result of contraction on cooling (Figure 12-1). Others are sheeting joints, the result of pressure release (Figure 7-3). Still others are due to regional tectonic stresses (cover photo) (Engelder and Geiser, 1980). Whatever their origin, joints provide openings for groundwater movement, and weathering and mass wasting are intensified. Joints that open parallel to gorges and steep-sided valleys probably aid in enlarging and perpetuating the valleys (p. 175).

Masses of jointed bare rock, stacked like well-worn childrens' blocks, are called **tors** (Figure 12-2). They are not transported, but result from differential weathering and erosion along joints, at least one set of which is nearly horizontal. Tors may form at depth by joint-con-

trolled weathering within the regolith (Figure 7-13), or they may form subaerially by differential weathering and erosion of structures within the rock mass (Wahrhaftig, 1965; Eggler et al., 1969). The tropical savanna climate (Chapter 13) has been cited as especially favorable for tors (Thomas, 1994, pp. 325-327). Tors in tropical regions probably develop by differential weathering along fractures in the deep regolith of the region, by the same processes that round originally angular blocks to create corestones (Figures 7-1 and 7-13). When exposed by subsequent erosion, they produce a landscape of monumentlike towers (Figures 12-2 and 12-3).

An extensive literature concerns the origin of tors in midlatitudes. Some argue that they are relics of mid-Tertiary hot, humid climate weathering that extended much farther poleward (Chapter 4). The extensive tor landscape on the Haast Schist in southern New Zealand (lat. 41°S) has been partly exhumed from a former cover of mid-Tertiary continental sediments, and could be relict (Stirling, 1990). However, the region had a cold, wet periglacial climate during Pleistocene ice ages, and the tors might have been exposed by solifluction during such times. If so, they are polygenetic (Fahey, 1981). The extensive tor landscapes of central and southern Great Britain, such as on foliated granite in Dartmoor (Figure 12-3), are also regarded as polygenetic, formed during former times of deep weathering but exposed by intense Pleistocene periglacial solifluction (Gerrard, 1978).

Thick-bedded sandstone formations in the central Appalachian Mountains south of the limit of glaciation commonly form flattened ridge crests strewn with tors. These upland surfaces are part of the evidence cited for uplifted Tertiary peneplain remnants in the Appalachian Mountains (Chapter 15), but also were the sites of periglacial solifluction that could have exposed the tors and aided in developing the broad upland summits (Clark and Hedges, 1992; Clark et al., 1992, p. 63). It is curious that climates as unlike as the tropical savanna and the periglacial tundra are credited with producing tor landscapes, and even more intriguing that a succession of these contrasting climatic types may be responsible for producing tors. They are among the most fascinating and enigmatic landforms known, especially relating to the concept of equifinality (p. 14).

Faults

The important distinction between tectonic fault scarps and structurally controlled fault-line (erosional) scarps is illustrated by Figure 5-11. Where unlike rock types abut each other at a fault, a variety of scarp landforms can develop by differential erosion. However, erosion

FIGURE 12-1. Downstream from Palouse Falls, Washington. Cliff-and-talus canyon walls are controlled primarily by vertical columnar joints in multiple lava flows (photo: F. O. Jones no. 29, U.S. Geological Survey).

along a fault does not necessarily produce a scarp. In a faulted terrane of massive igneous or metamorphic rock, only the *crushed zone* along the faults may be subject to more rapid weathering and erosion. On aerial photographs, crystalline terranes show a characteristic network of differentially eroded zones on both fault lines and joints (Color Plate 3).

A confused or chaotic landscape is likely to result from structurally controlled erosion of tectonic **mélanges.** In many regions of the Mediterranean and circum-Pacific tectonic belts, great masses or blocks of rigid igneous, metamorphic, and sedimentary rocks, some several kilometers in length, are embedded in a more plastic matrix of shale or low-grade, fine-grained metamorphic rock (Hsü, 1968). The tectonic interpretation of these mélanges is a matter of debate, but they probably are formed by thrust faulting and submarine slumping in an orogenic marine sedimentational envi-

ronment such as an accretionary prism (Figure 3-2) (Raymond, 1984; Cowan, 1985). Little has been written about the geomorphology of mélange terranes. Hsu (1971) described a mélange landscape in Cyprus and compared it with parts of eastern Taiwan. In both regions, resistant blocks of graywacke, chert, ultrabasic igneous rock, or limestone form hills or boulder-strewn ridges surrounded by claystone lowlands. Maxwell (1974, p. 1201) described the mélange topography of northern California as "accidental, characterized by scattered blocks jutting out of a rounded or land-slid surface." The mélange landscape in the Napa Valley, California, is

. . . hummocky topography, with resistant blocks protruding above grassy swales developed on the matrix. Hundreds of blocks protrude above the surface throughout the terrane. Where streams

FIGURE 12-2. Tors in a tropical savanna landscape, Townsville, northern Queensland, Australia. Minor lithologic variations extend through the stacked blocks without offset. See also Figure 12-3.

FIGURE 12-3. Ridge-crest tors of foliated granite, Dartmoor, Devonshire, England.

mélange hills do not seem to fit into any other available descriptive category.

DIFFERENTIAL FLUVIAL EROSION ON LAYERED ROCKS

Most sedimentary rocks, and many igneous and metamorphic rocks, have layers of contrasting erodibility. The layering may be the result of sedimentary stratification, igneous intrusion, metamorphic differentiation, or a host of other causes. Weathering emphasizes even slight contrasts in lithology, but on the larger scale, belts of distinctive erosional topography form on the exposed edges of rock layers. Where layers are horizontal, their exposed edges, or *outcrops,* are parallel to contours and form altitudinally zoned structural landforms. Where layers are tilted, much more complex geometry develops. Much of the geomorphic diversity of landscapes is due to the structural control of layered rocks, and an extensive descriptive and explanatory terminology has evolved. Much of the terminology is shared with the field of structural geology because geomorphic form is one important way of recognizing structures in the field.

Layered Igneous and Metamorphic Rocks

Volcanic plateaus are built of piles of lava and tephra sheets and intrusive sills crossed by numerous dikes (p. 107). Either a single flow or a series of rapidly sequential eruptions may cover a very large area with fresh flows of lava or ignimbrite, which then cool as a unit.

have cut deeply into it, blocks in profusion crown knobs high above the stream valleys, jut out from valley walls, and stick up from stream bottoms, commonly deflecting the streams around them. Blocks range from pebble-sized or even smaller to mountain-sized (Phipps, 1984, p. 107).

The mélange structure may extend to a depth of 10 km. A unique geomorphic aspect of mélange terrane is that the hills of resistant rock have no "roots," and no prediction of sequential landscape evolution can be made. Although not strictly an example of fault-controlled scarps, the abrupt structural boundaries of

The interior of the cooling unit looses heat both upward and downward and as it cools, it contracts and fractures into a crudely hexagonal pattern of columnar joints (Long and Wood, 1986; DeGraff and Aydin, 1987). However, the upper and lower marginal zones usually cool more rapidly and shatter into unsystematic blocks of glassy or vesicular rock. Columnar jointing provides high-angle sets of closely spaced joints that promote rock falls, cliffs, and taluses. The glassy or shattered marginal zones commonly form taluses separating joint-controlled cliffs (Figure 12-1). The spectacular effects of the Lake Missoula and Lake Bonneville floods (p. 213) were in large part structurally controlled by the plateau basalts of the region.

Differential fluvial erosion on volcanic structures other than basalt or ignimbrite plateaus also produce distinctive structurally controlled landforms although they are rarely of regional extent. The complex internal structure of a composite cone (Figure 6-15) may, as a result of deep erosion, be expressed as radiating, concentric, or arcuate ridges of more resistant rock units. The central conduit or **neck** of a volcano is especially likely to survive erosion if it has been filled with solidified lava and the surrounding conical mound of tephra and lava is removed.

Metamorphic terranes commonly have zones or layers of more resistant rock alternating with more erodible belts. A typical example would be quartz-rich gneiss alternating with biotite schist. Weathering and erosion more rapidly lower the schistose belts, and the resistant layers are etched out. The scale may be measured in meters, or as in the coastal region of Maine, in strike ridges of resistant gneiss 100 m or more in height and as much as 50 km in length (Figure 12-4). Metamorphic terranes have a dominant regional structural grain. Hills have their long axes on strike. Streams follow erodible belts or cross ridges at fracture zones. Entire regional landscapes such as the New England geomorphic province of the northeastern United States become a series of structurally controlled subparallel ridges and valleys (Flint, 1963). The Precambrian Grenville metamorphic terrane of eastern North America provides some especially impressive examples of differential erosion. Marble belts between schists and gneisses have formed weak-rock lowlands that curl with the metamorphic grain (Figure 12-5).

Folded Sedimentary Rocks

Most of the present-day subaerial landscape is eroded on stratified sedimentary rocks. In the United States, for instance, only 8.4 percent of the area is of Precambrian terrane, but strata of the Carboniferous, Creta-

FIGURE 12-4. Satellite image of Casco Bay, Maine. North is to the top; water is black on this image. Gneiss ridges form long peninsulas and islands projecting into the Gulf of Maine (NASA ERTS E-1472-14530).

ceous, and Quaternary Systems each cover about 14 percent of the area (Gilluly, 1949, p. 576). Sedimentary rocks cover 66 percent of the total continental area (excluding ice-covered Antarctica) although the coverage varies greatly from 87 percent of Europe to only 52 percent of North America (Blatt and Jones, 1975). The sedimentary cover strata range from easily eroded, unconsolidated Cenozoic glacial drift or coastal-plain sediments to stronger, better lithified, and possibly folded and faulted Mesozoic and Palaeozoic rocks. Thicknesses range from a few meters to more than 10 km. Usually the thickest sections of cover strata are now contorted in orogenic belts. Epeirogeny has, however, uplifted large continental areas with only slight warping of the cover strata, which are only 1 to 2 km in thickness (p. 44). Thus epeirogeny and orogeny have created regionally extensive subaerial landscapes that are largely eroded from sedimentary structures. The differential erodibility of shale, limestone, and sandstone is the dominant structural control of landscapes over two-thirds of the continental area.

Flat-lying sedimentary rocks of contrasting erodibility form *structural benches* on valley walls (Color Plate 10). Many plateau surfaces are structurally controlled by resistant *caprock* that persists at a high elevation, while the terrain around it is being lowered. The caprock is usually a sedimentary rock, but it can also be a lava flow or a concretionary weathered layer such as laterite (p. 141) or calcrete (p. 142).

Tilted or folded strata provide more complex structural controls to erosion. Generally, the outcropping edge of a resistant stratum will form an escarpment or

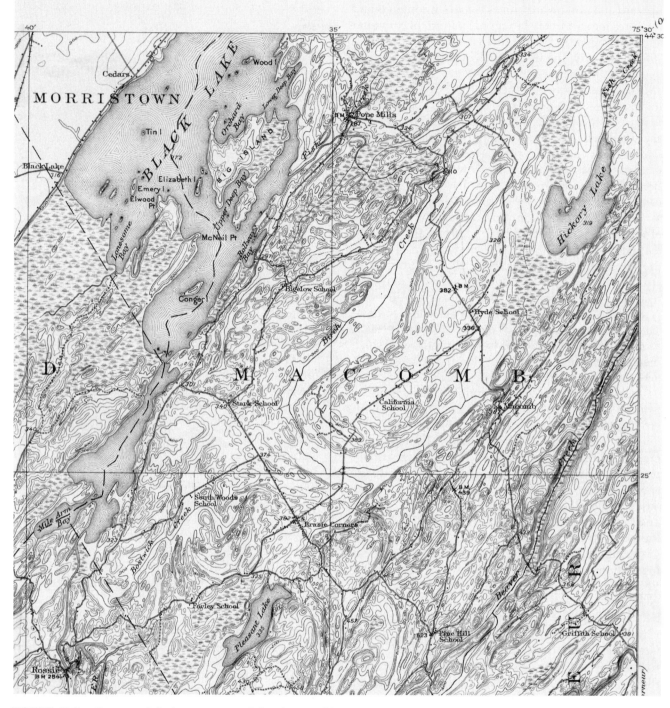

FIGURE 12-5. Contorted drainage pattern following marble and calc-silicate lowlands among gneiss hills, northwestern Adirondack Mountains, New York (part of Hammond, N.Y. 15′ topographic map quadrangle, 1935 edition).

cliff, commonly joint controlled. The *dip slope,* away from the crest of the escarpment, is parallel to the stratification and tends to resist dissection as weaker overlying strata are eroded. A single, thin layer of sandstone or limestone may armor or protect an extensive shale dip slope, for instance (Color Plate 11).

Popular terminology has evolved to differentiate structurally controlled ridges on tilted resistant layers (Figure 12-6). If the more resistant unit is nearly horizontal and caps a broad flat-topped hill, the hill is called a **mesa.** If the diameter of the cap rock is less than the height of the hill above the surrounding terrain, the term

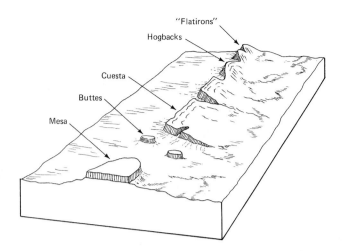

FIGURE 12-6. Terminology of structurally controlled scarps and ridges.

butte is more often applied. If the resistant bed dips gently, an asymmetric **cuesta** results, with a steep escarpment and a gentle dip slope. When the dip of the layer is comparable to the angle of repose of the sliderock on the escarpment face, the ridge becomes approximately symmetric and is known as a **hogback.** "Flatirons" are interfluve remnants of dissected hogbacks.

The structural control of tilted strata imposes a powerful asymmetry on drainage networks. Escarpment streams are steep, short, and have high gradients. Dip-slope streams are likely to have more gentle gradients, larger watersheds, more tributaries, and more sustained flow (Figure 12-7). As erosion progresses, a trunk stream flowing along a *weak-rock lowland* between two

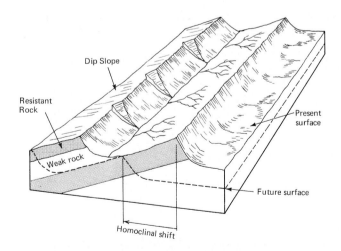

FIGURE 12-7. Asymmetric drainage networks on homoclinal ridges. Steep-gradient escarpment streams erode faster than gentle-gradient dip slope streams. Ridge crests and valleys will migrate laterally downdip as the landscape is lowered by erosion.

cuestas will migrate laterally in the down-dip direction, as it strips away the less resistant strata. The entire ridge and valley system migrates laterally as well as downward with time in a process termed **homoclinal shifting.** By this process, structurally controlled ridges and valleys must migrate laterally many times their own width as a region of dipping strata is lowered by erosion.

Beyond homoclinal dips, the next degree of structural complexity in stratified rocks is a series of parallel anticlines and synclines with horizontal fold axes (Figure 12-8). Tectonic ridges (Chapter 5) are normally anticlines or domes, but in structurally controlled eroded landscapes, hills and ridges are just as common on synclinal as on anticlinal structures. An early observation of this fact led to the concept of **topographic inversion,** whereby structural depressions such as synclinal axes become ridges or hills (Figure 12-8). An impressive amount of erosion is implied, but as demonstrated earlier (p. 121), the amount of rock eroded from the continental masses can be measured in tens of kilometers, and multiple topographic inversions must have resulted during the dissection of fold mountains such as the central Appalachians. Therefore, it is doubtful whether synclinal mountains are any more a measure of the depth of erosion than anticlinal mountains.

One may safely infer that erosion is in progress while an orogeny is in progress. The rising anticlines of tectonic landscapes (Figure 12-9) are cut by gullies as soon as runoff begins to drain from them. When a resistant stratum becomes exposed in a rising anticline, it is still deeply buried in the axes of adjacent synclines. By the time the regional landscape has been eroded sufficiently to expose the resistant stratum at the level of the synclinal axes, the anticlines have been breached, and the synclinal floors become flat or gently basined summits, flanked by escarpments that face outward. If only a single resistant unit is embedded in a thick sequence of erodible rocks such as shale, the resistant stratum is very likely to cap synclinal ridges as erosion progresses. If resistant and erodible strata are interlayered at close intervals, the landscape will have numerous synclinal and anticlinal ridges formed on the various resistant units (Color Plate 12). Where folding becomes extremely complex, dissection favors neither synclines nor anticlines (Color Plate 2).

The abundance of synclinal mountains has led some authors to suggest that unusually intense strain during folding encourages the subsequent erosional breaching of anticlines. However, breached anticlines are not so much the result of the tensional fracture pattern at the crests of folds as they are simply a consequence of Powell's dictum (p. 118) that early-formed topographic prominences, most commonly anticlines, are the first structures to be attacked by erosional agents.

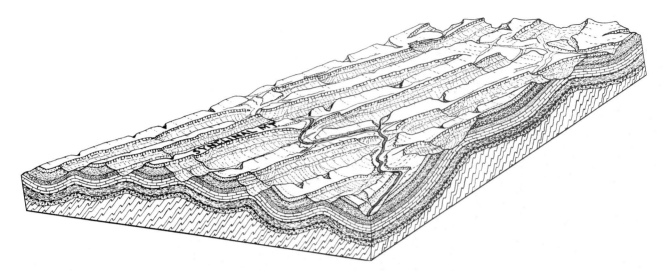

FIGURE 12-8.   Block diagram of dominantly synclinal ridges in a terrane of folded sedimentary rocks (from von Engeln, 1942, Figure 183). It is unlikely that basement rocks would be folded, as suggested in this figure. More likely a *décollement,* or detachment zone, would separate basement blocks from the folded cover strata.

FIGURE 12-9.
Doubly plunging anticlines and domes of limestone east of Shiraz, Iran, in the Zagros Range. Some domes are breached, others show only consequent valleys down their flanks (NASA ERTS E-1221-06293).

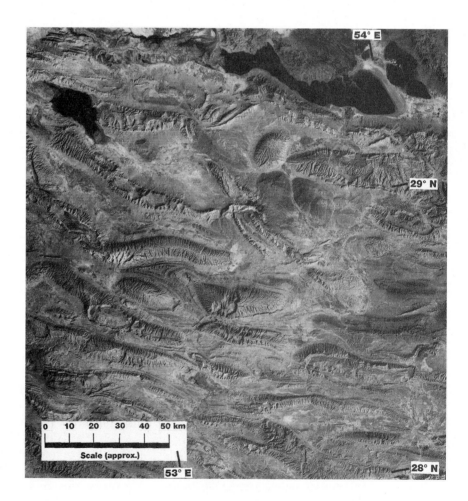

FIGURE 12-10.
Dissected plunging anticline, Zagros Mountains, western Iran. Note that the limestone ridges, resistant in this arid climate, are dissected into flatirons on the flank of the structure but wrap continuously around the structural nose in the distance, where their longer, gentler dip slopes protect the interlayered shale (photo: Hunting Surveys Ltd. associated with Aerofilms Ltd. Copyright reserved).

Plunging structures introduce a still higher level of complexity to structurally controlled landforms. Cuesta- and hogback-forming resistant strata not only dip laterally from anticlinal toward adjacent synclinal axes, but the structures plunge along strike. The elliptical form of doubly plunging anticlines is well shown in Figure 12-9. In closer view (Figure 12-10), the resistant beds are seen to dip radially from an elongate anticlinal axis forming steep hogbacks on the flank of the structure and more gently dipping, inward-facing escarpments around the nose of the fold. Because the resistant strata typically dip more steeply on the sides of a plunging anticline than around its ends, their outcrop width is narrower on the flanks, and erosional *breaching* of the structure is more likely to occur there than on the end or nose of the fold. In the arid Zagros Mountains of Iran, multiple limestone strata are strong ridge formers, separated by more erodible shale layers that form subsequent valleys along strike (Figure 12-10).

If the resistant strata are more gently but repeatedly folded, and are separated by thicker sections of erodible rock, individual structures may not form closed ellipses, but form continuous, sinuous ridges. Such subparallel belts of *zigzag ridges* characterize the folded central Appalachians (Figure 12-11a). The same

resistant rock, such as the Tuscarora quartzite of the Appalachians, may form ridge after ridge as it is repeatedly intersected by the land surface (Figure 12-11b). Faults that cause strata to be offset laterally also displace ridge crests controlled by the resistant layers.

A discovery of great tectonic as well as geomorphic significance is that in the southern and central Appalachians, the Piedmont, Blue Ridge, and Valley and Ridge Geomorphic Provinces (Figure 1-5) were all displaced westward by tens to hundreds of kilometers during the late Palaeozoic Alleghanian Orogeny. They are all **allochthonous terranes** formed far to the east of their present position and transported westward along a series of layer-parallel thrust faults in lower Palaeozoic sedimentary strata (Figure 12-12). Predicting the effects of deep erosion on such terranes is complicated because with an additional 3 to 5 km of erosion on the overlying allochthonous thrust plate, a new landscape would be exposed beneath it, in many places on a terrane of the same rocks that have been removed from the upper plate. A comparison of the cross section of Figure 12-12 drawn in 1989, with the front face of the block diagram of Figure 12-8, drawn in 1942, illustrates vividly how plate tectonic theory and resulting research discoveries have revised our understanding of

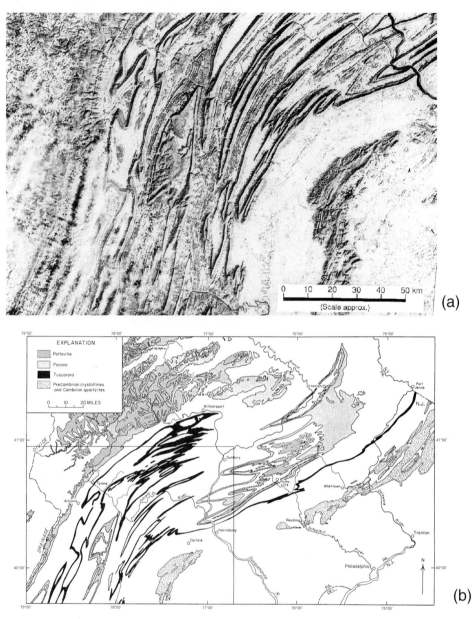

FIGURE 12-11. (a) Zigzag ridges in the Valley and Ridge Province of the Appalachian Mountains in Pennsylvania. For location, see Figure 12-11(b). Low-angle lighting from the southeast may cause the ridges to appear inverted. Most of the ridges in the north-central part of the image are breached anticlines (NASA ERTS E-1513-15220). (b) Simplified geologic map of the major ridge-forming strata in the Appalachian Mountains of Pennsylvania. The area of Figure 12-11(a) is outlined. Note that only a few resistant ridge formers are repeatedly exposed across this thin-skinned (p. 40) fold belt (after Thompson, 1949, Plate I).

folded structures. It is suspected, but not yet proved, that the higher-grade metamorphism responsible for the anthracite coal in the tight syncline around Scranton, Pennsylvania, (Figure 12-11b) was caused by the complete removal by erosion of a former overlying thrust sheet 7 to 9 km in thickness (Sevon, 1986). One can only wonder how many other landscapes on complexly folded rocks might have been once covered by many kilometers of an overthrust sheet of which no remnants remain.

Structurally controlled *domes* and *basins* are the result of erosion on double-plunging folds. Although common in fold belts of orogenic regions (Figure 12-9), broad domes and basins are important features of epeirogeny as well. Epeirogenic activity contempora-

neous with deposition creates belts or rings of distinctive lithologic facies that intensify the erodibility contrast of the concentrically disposed strata (Miller et al., 1990; Driese et al., 1994, Figure 13). In the Central Lowland of the United States, the Michigan Basin has a central structural depression of more than 4000 m, although it is now a low plateau between adjacent shale lowlands occupied by Lakes Michigan, Huron, and Erie. To the south, the Nashville Dome is 80 km wide and 200 km long but has only a gentle central uplift. Resistant strata form cuestas encircling the cores of many epeirogenic structures, facing inward onto an eroded lowland on the domes and facing outward from the basins. The Paris Basin of northern France is ringed by outward-facing escarpments, especially in the

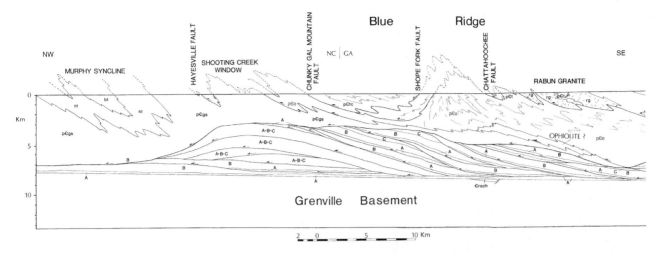

FIGURE 12-12. Seismic interpretation of the Blue Ridge region of the southern Appalachian Mountains. A sheet of Precambrian metamorphic rock (units labeled p-Є) has been thrust many tens of kilometers northwestward over lower Paleozoic sedimentary rocks, which are also stacked and repeated in thrust sheets. A = Cambrian rocks, B = Cambrian/Ordovician rocks, C = middle Ordovician and younger rocks. The Precambrian rocks are not part of the Grenville basement, but are in an allochthonous sheet (from Costain et al., 1989).

northeast quadrant (Joly, 1984, Figures 8-19 and 8-20). The low wooded ridges and intercuesta lowlands between Paris and the Rhine graben were the scene of bitter fighting during World Wars I and II. Control of the cuesta ridges meant tactical advantage to one of the opposing armies. The English Channel nearly transects a broad structural dome that is ringed by inward-facing cuestas of chalk, which are called **downs** in England. The core of the dome is deeply eroded in Jurassic claystones that underlie the chalk, and Cenozoic sediments were subsequently deposited on the eroded basin floor.

In most of the examples just cited, topographic inversion has produced a topographic basin on erodible shale in the center of an epeirogenic dome or a low plateau centered on a former epeirogenic basin. However, in the Black Hills of South Dakota, the Henry Mountains of Utah, and elsewhere, sharp domal uplift over a core of strong crystalline rock has created a central mountain range with ringing hogbacks and cuestas. Similar rows of upturned strata are common along the fronts of many mountain ranges (Figure 12-13).

The most extreme orogenic deformation of layered rocks consists of great overturned, recumbent, often thrust-faulted folds known as **nappes** from their type locality in the Alps. Some include cores of mobilized basement rock; others involve only cover strata (Figure 12-14). Alpine mountain sides may show strata more than 1000 m in thickness, all overturned. Detached pieces of nappes or thrust plates may cap ridges underlain by much younger rocks. Nappes, like mélanges, produce complex structurally controlled landforms,

usually well exposed because they are within great mountain ranges.

STRUCTURALLY CONTROLLED DRAINAGE PATTERNS

In regions of folded sedimentary rocks, structurally controlled strike ridges form most of the higher terrain. As with all upland landforms, these are best regarded as residuals of less-rapid fluvial erosion, with the intervening valleys the more rapidly lowered parts of the landscape. Systems of valleys eroded on contrasting structures conform to those structures, giving to both valley profiles and drainage networks distinctive, structurally controlled geometry. Both empirical and genetic classifications of structurally controlled drainage networks have evolved and are among the more useful and important tools of geomorphic analysis.

Empirical Classification by Form

Arthur D. Howard (1967) compiled and classified drainage patterns into categories of basic patterns, modifications to the basic patterns, and varieties of the modified patterns. His very thorough classification is summarized in Table 12-1 and Figures 12-15, 12-16, and 12-17. The classification refers to the regional patterns of an aggregate of valleys and gullies, not to the patterns of single stream channels, which are described by

FIGURE 12-13. View north along the Rocky Mountain Front Range near Morrison, Colorado. Rounded summits to the west are on Precambrian schist, flanked by hog-backs of light-colored Carboniferous sandstone. The prominent central hogback is on eastward-dipping sandstone of the Cretaceous Dakota Formation (Lovering and Goddard, 1950, Figure 6) (photo: T. S. Lovering 5a, U. S. Geological Survey).

such terms as crooked, straight, meandering, and braided (Chapter 10).

Drainage patterns follow regional structural trends and cover a wide range of scales. They are subjectively defined and transitional from one to another. Nevertheless, the basic and modified terms in Table 12-1 are part of the working vocabulary of geomorphologists and aerial photographic interpreters. They can best be studied by comparing Figures 12-15 through 12-17 with a selection of topographic maps of various scales.

Genetic Classification of Drainage Patterns

Consequent and Subsequent Rivers and Their Valleys. River systems that develop as a direct result of, and in harmony with, a preexisting land surface are said to be **consequent.** One may consider that the flow of water was a direct consequence of whatever slopes were initially present. (A less common term for small valleys or gullies is **insequent,** shortened from "initial consequent.") Even newly emerged seafloors or fresh basalt plateaus are not likely to be devoid of initial relief, and runoff will fill initial hollows to the lowest point on the rim and then drain down to the next hollow (Figure 15-9). Only in arid regions will the depressions on an undissected landscape remain free of permanent lakes. Each basin joins the next lower one until an entire drainage network is established, as in computer simulations (Figure 11-4).

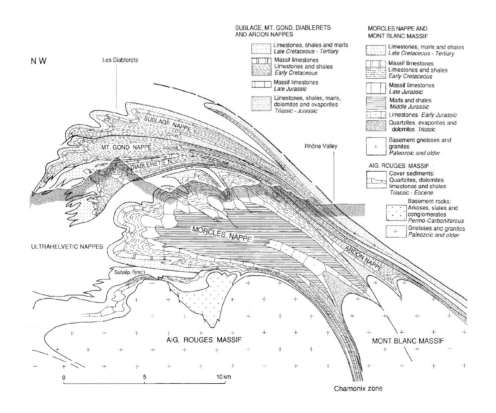

FIGURE 12-14. Nappe geometry of the western Helvetic Alps, Switzerland. Present topography is traced across the inferred structures. The cover nappes have been partially stripped off and piled to the northwest above the deeper basement-cored nappes (from Escher et al., 1993, Figure 3).

Table 12-1

Significance of Basic and Modified Basic Drainage Patterns

Basic	Significance	Modified Basic	Added Significance or Locale
Dendritic	Horizontal sediments or beveled, uniformly resistant, crystalline rocks. Gentle regional slope at present or at time of drainage inception. Type pattern resembles spreading oak or chestnut tree.	Subdendritic Pinnate Anastomotic Distributary (Dichotomic)	Minor secondary control, generally structural. Fine-textured, easily erodible materials. Floodplains, deltas, and tidal marshes. Alluvial fans and deltas.
Parallel	Generally indicates moderate to steep slopes but also found in areas of parallel, elongate landforms. All transitions possible between this pattern and type dendritic and trellis.	Subparallel Collinear	Intermediate slopes or control by subparallel landforms. Between linear loess and sand ridges.
Trellis	Dipping or folded sedimentary, volcanic, or low-grade metasedimentary rocks; areas of parallel fractures; exposed lake or seafloors ribbed by beach ridges. All transitions to parallel pattern. Type pattern is regarded here as one in which small tributaries are essentially same size on opposite sides of long parallel subsequent streams.	Subtrellis Directional trellis Recurved trellis Fault trellis Joint trellis	Parallel elongate landforms. Gentle homoclines. Gentle slopes with beach ridges. Plunging folds. Branching, converging, diverging, roughly parallel faults. Straight parallel faults and/or joints.
Rectangular	Joints and/or faults at right angles. Lacks orderly repetitive quality of trellis pattern; streams and divides lack regional continuity.	Angulate	Joints and/or faults at other than right angles. A compound rectangular-angulate pattern is common.
Radial	Volcanoes, domes, and erosion residuals. A complex of radial patterns in a volcanic field might be called multiradial.	Centripetal	Craters, calderas, and other depressions. A complex of centripetal patterns in area of multiple depressions might be called multicentripetal.
Annular	Structural domes and basins, diatremes, and possibility stocks.		Longer tributaries to annular subsequent streams generally indicate direction of dip and permit distinction between dome and basin.
Multibasinal	Hummocky surficial deposits; differentially scoured or deflated bedrock: areas of recent volcanism, limestone solution, and permafrost. This descriptive term is suggested for all multiple-depression patterns whose exact origins are unknown.	Glacially disturbed Karst Themokarst Elongate bay	Glacial erosion and/or deposition. Limestone. Permafrost. Coastal plains and deltas.
Contorted	Contorted, coarsely layered metamorphic rocks. Dikes, veins, and migmatized bands provide the resistant layers in some areas. Pattern differs from recurved trellis [Figure 12-16(H)] in lack of regional orderliness, discontinuity of ridges and valleys, and generally smaller scale.		The longer tributaries to curved subsequent streams generally indicate dip of metamorphic layers and permit distinction between plunging anticlines and synclines.

Source: Howard (1967).

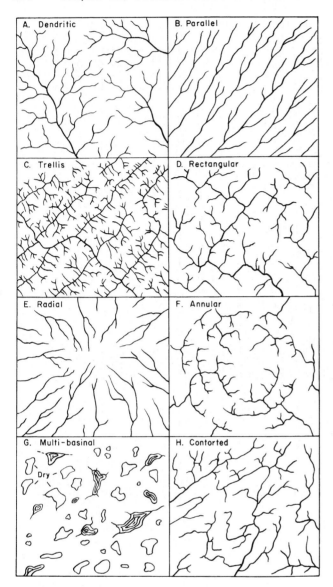

FIGURE 12-15.   Basic drainage pattern. See Table 12-1 (Howard, 1967, Figure 1).

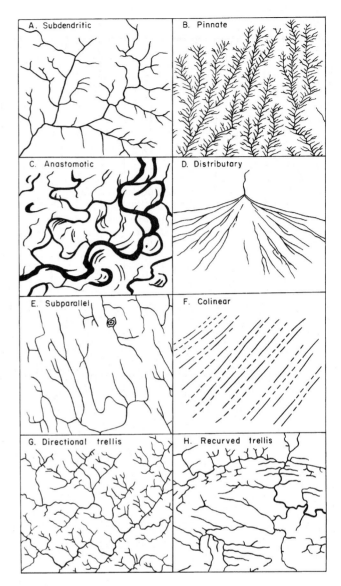

FIGURE 12-16.   Modified basic patterns. Each pattern occurs on a wide range of scales. See Table 12-1 (Howard, 1967, Figure 2).

Initially, all runoff is consequent. As valleys are eroded, however, structural control may be exerted. Rivers obliquely crossing belts of rocks with contrasting erodibility tend to follow the zone of easy erosion before crossing the next resistant structure (Figure 12-18). Tributaries that expand into uplands underlain by belts of erodible rocks expand their drainage basins more rapidly than those that erode into resistant rocks. Thus structurally controlled, or **subsequent,** valleys evolve, which develop independent of and subsequent to the primary, or consequent, drainage (Figure 12-19). Of the eight basic drainage patterns in Table 12-1, only dendritic, parallel, and radial are likely to be consequent. The others are all linked closely to control by structures. The contorted pattern of Figure 12-5 is a good example

of subsequent drainage on a complex terrane although deranged somewhat by glacial erosion and deposition.

Antecedent, Superposed, and Captured Drainage. In orogenic regions, the valleys that are first eroded over a landscape may later have new tectonic ridges rise across their trends. The rivers may be *defeated* (forced to abandon their valleys), or they may be able to maintain gorges across the landmass rising in their paths. Streams that have maintained their valleys across tectonic ridges are said to be **antecedent** because the river is older than or antecedent to the deformation (p. 85)

Yet another class of rivers and valleys are called **superposed.** The word is a contraction of "superim-

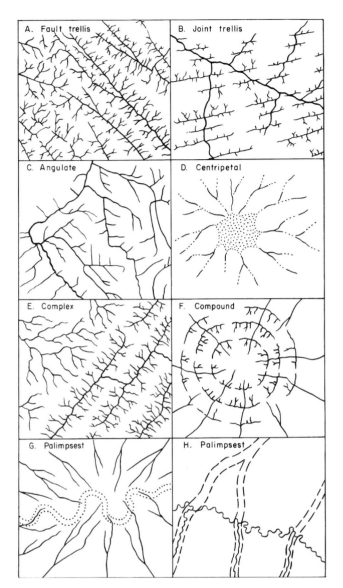

FIGURE 12-17. Other modified basic patterns, (A)-(D). See Table 12-1. Complex, compound, and palimpsest patterns, (E)-(H), are combinations or superpositions of basic or modified basic patterns (Howard, 1967, Figure 3).

FIGURE 12-18. Structural control of the Pilcomayo River in southern Bolivia across the Subandean Fold and Thrust Belt of the eastern Andes. The general southeasterly consequent course of the river has evolved into a series of short reaches parallel to resistant sandstone hogbacks, separated by shorter, high-gradient reaches across the resistant strata. View approximately 25 km wide. See Color Plate 12 for the regional view (Landsat Thematic Map image no. 50213-1342).

posed" and implies that a river system has been let down or laid over a landscape. Suppose that a land newly emerged from the sea has a thin cover of sediments that bury an older terrane of complexly folded rocks. On emergence, a consequent drainage system, perhaps with a dendritic pattern, might form. As the valleys deepen, they are eventually eroded into the old land under the covering strata. The dendritic pattern becomes superposed on the older rocks without regard to structural control. When the covering beds have been entirely removed by erosion, the only clue to superposition is the anomalous positions of the river valleys. An even more complex case of superposition

would be if the former cover were a deformed, folded thrust sheet or nappe (Figure 12-12). One can only guess at the possibilities for drainage anomalies superposed on the rocks below the thrust.

Superposition is extremely common in glaciated regions, where interglacial and preglacial drainage networks are filled with glacial drift, and the present streams almost randomly occupy alternating segments of old reexcavated valleys and new, postglacial valleys (Figure 4-8). Short gorgelike segments superposed on bedrock form scenic attractions in the Finger Lakes region of central New York (book cover) and in the "Dells" region of southern Wisconsin, among many other places.

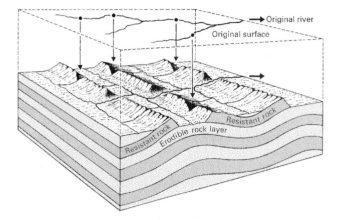

FIGURE 12-19. Expansion of a consequent valley system over an initial landscape, with a later stage of erosion when subsequent tributaries have become dominant. Arrows project downward from valley segments of the consequent system that were defeated by resistant layers and stream capture to become wind gaps.

A common problem in geomorphology is distinguishing between antecedent and superposed rivers. In the Middle Rocky Mountains, for instance, major rivers cross mountain ranges in narrow gorges although easy alternative routes are available around the ends of the ranges a few miles away (Figure 12-20). Early geologists thought the rivers were antecedent, implying that the mountain blocks had risen very recently, after the rivers had become established. Later work proved that a mid-Cenozoic interval of deep lacustrine and alluvial filling buried large parts of the mountains. Rivers such as the

Yellowstone, Bighorn, and Laramie formerly flowed over alluvium and were superposed on the resistant rocks of the buried ranges as the area was epeirogenically uplifted and dissected. The evidence for superposition includes (1) remnants of former alluvium now on the mountainsides at suitable altitudes, and (2) the similarity of the alluvium upstream and downstream of the gorges, both of which demonstrate a former continuous cover.

The eastern Subandean Belt of the Andes Mountains is a thin-skinned foreland fold and thrust belt (Figure 3-3, Color Plate 12), where Palaeozoic to Cenozoic strata are being folded and thrust eastward onto the Amazon shield by active orogeny that began in the Miocene Epoch (Jordan et al., 1983, pp. 346-347). The major rivers draining eastward from the higher interior of the Andes are clearly antecedent to the folds of the Subandean Belt although subsequent tributaries are rapidly excavating strike valleys along erodible shale beds (Figures 12-18 and 12-21). The folds of the Subandean Belt are propagating eastward, and a difficult semantic problem arises when a live anticline has risen across the valley of a river that was formerly flowing on a thin alluvial cover. Is the river then superposed, or antecedent? The usually obvious distinction between previously folded rocks at depth and overlying undeformed cover strata, diagnostic of superposition, is blurred if the cover strata are also involved in the deformation, in which case the river is clearly antecedent.

Stream capture (or "piracy") has been a popular concept of geomorphologists for generations. Either by chance or aided by a structural advantage, one stream may erode into the drainage basin of another and either

FIGURE 12-20. Superposed gorge of the Wind River across the Owl Creek Mountains in central Wyoming (photo: William W. Locke).

FIGURE 12-21. Antecedent Pilcomayo River flowing eastward across active folds of the Subandean Belt in southern Bolivia. The youngest folds are to the east, and some that are known at depth from petroleum exploration have not yet propagated through surficial alluvium. View approximately 25 km wide. See Color Plate 12 for the regional view (Landsat Thematic Map image no. 50213-1342).

progressively or abruptly divert, or *capture,* new tributaries. The trunk stream that is thereby deprived of part or all of its former watershed is said to have been **beheaded.**

Quite commonly, subsequent drainage systems capture segments of earlier consequent drainage. Figure 12-19 suggests how the process could occur, especially if the original drainage on cover strata had been superposed on a folded belt. As subsequent valleys expand along weak-rock belts, they progressively intersect the earlier consequent drainage, which by that stage would be a series of graded reaches alternating with gorges or **water gaps** across the resistant strata. After capture, the former course across the resistant ridges might be marked by **wind gaps,** notches that mark the former drainage line. Figure 12-19 shows only

a single trunk stream that continues to drain the same region during the change from consequent to subsequent pattern. It is easy to visualize however, that in the competition among regional trunk streams, substantial areas of watershed might be permanently gained or lost by capture (Clark, 1989).

The effects of a capture on the hydraulic geometry of both the capturing and the beheaded system can be dramatic. Most commonly, the capturing stream has a relatively steep longitudinal gradient in a low-order tributary at the point of capture, whereas the defeated stream is likely to have a lower gradient. The sudden diversion of a greatly increased discharge through the capturing system causes rapid intrenchment, perhaps with a waterfall or rapids that migrate rapidly up the captured segment from the point of initial capture. Any sharp change in direction of a stream, downstream from an anomalously steep section, suggests an **elbow of capture.**

Another common mechanism of capture is provided by the unequal gradients of two parallel streams emerging from a mountain front. The one carrying the greater bedload may aggrade its channel above the level of the adjacent stream into which it then easily diverts. The terraces on rivers at the foot of the Beartooth Mountains in Montana, described in the preceding chapter (p. 246), exemplify this process.

Cenozoic orogeny has created many mountain belts that generally parallel the converging margins of continents. During deformation, antecedent streams may maintain their valleys across foreland fold belts (p. 39), or cross-axial drainage may be superposed on deeper structures from thin cover strata. In either case, drainage networks are likely to develop with their long axes roughly transverse across orogenic belts (Oberlander, 1965, 1985). Later, as subsequent valleys deepen and lengthen, regional drainage networks are likely to develop trellis patterns, parallel to the orogen. Their great length makes the long subsequent tributaries of a trellis system susceptible to capture by a short, high-gradient stream valley expanding into the orogenic belt from the adjacent coast or foreland basin, if it happens to find a weak or narrow zone in the resistant ridge that separates it from the subsequent valley inland behind the ridge. If a capture occurs, the length of flow of the captured tributaries is shortened, and as the nickpoint (p. 205) segment migrates upstream from the elbow of capture and smooths out, stream terraces may be left behind, graded toward the beheaded trunk stream rather than toward the capturing river.

Gilbert (1877) proposed a "law of equal declivities," which states that if two streams of unequal longitudinal gradient drain the opposite sides of a ridge, the divide between them will shift laterally, or *migrate,* until their gradients are equal. If one of the streams is deepening its valley more rapidly than the other, by virtue of a structural advantage, the divide may shift or "creep" laterally until the expanding stream entirely captures the drainage of the lesser competing stream. Possibly, the migrating divide may intersect a trunk stream, suddenly diverting a larger headwater region into the capturing stream. At that instant, the divide "leaps" abruptly to a new regional position. Creeping and leaping divides are part of the regional adjustments among drainage systems that result from progressive adjustments to structure as a landscape evolves by fluvial erosion.

REFERENCES

BLATT, H., and JONES, R. L., 1975, Proportions of exposed igneous, metamorphic, and sedimentary rocks: Geol. Soc. America. Bull., v. 86, pp. 1085–1088.

CLARK, G. M., 1989, Central and southern Appalachian water and wind gap origins: Review and new data, in Gardner, T. W., and Sevon, W. D., eds., Appalachian geomorphology: Geomorphology, v. 2, pp. 209–232.

———, BEHLING, R. E., and 4 others, 1992, Central Appalachian periglacial geomorphology: A field excursion guidebook, 27th Internat. Geog. Congress: The Pennsylvania State University Agronomy Department, Agronomy Series Number 120, 248 pp.

———, and HEDGES, J., 1992, Origin of certain high-elevation local broad uplands in the central Appalachians south of the glacial border, U.S.A.-A paleoperiglacial hypothesis, in Dixon, J. C. and Abrahams, A. D., eds., Periglacial geomorphology: The Binghamton Symposia in Geomorphology: International Series, No. 22, John Wiley & Sons Ltd., Chichester, UK, pp. 31–61.

COSTAIN, J. K., HATCHER, R. D., JR., and ÇORUH, C., 1989, Appalachian ultradeep core hole (ADCOH) project site investigation. Regional seismic lines and geologic interpretation, in Hatcher, R. D., Jr., Thomas, W. A., and Viele, G. W., eds., The Appalachian-Ouachita Orogen in the United States: Geol. Soc. America, The Geology of North America, v. F-2, Plate 12.

COWAN, D. S., 1985, Structural styles in Mesozoic and Cenozoic mélanges in the western cordillera of North America: Geol. Soc. America Bull., v. 96, pp. 451–462.

DEGRAFF, J. M., and AYDIN, A., 1987, Surface morphology of columnar joints and its significance to mechanics and direction of joint growth: Geol. Soc. America Bull., v. 99, pp. 605–617.

DRIESE, S. G., SRINAVASAN, K., and 2 others, 1994, Paleoweathering of Mississippian Monteagle Limestone preceding development of a lower Chesterian transgressive systems tract and sequence boundary, middle Tennessee and northern Alabama: Geol. Soc. America Bull., v. 106, pp. 866–878.

EGGLER, D. H., LARSON, E. E., and BRADLEY, W. C., 1969, Granites, grusses, and the Sherman erosion surface, southern Laramie Range, Wyoming: Am. Jour. Sci., v. 267, pp. 510–522.

ENGELDER, T., and GEISER, P., 1980, On the use of regional joint sets as trajectories of paleostress fields during the development of the Appalachian Plateau, New York: Jour. Geophys. Res., v. 85, pp. 6319–6341.

ESCHER, A., MASSON, H., and STECK, A., 1993, Nappe geometry in the western Swiss Alps: Jour. Structural Geol., v. 15, pp. 501–509.

von ENGELN, O. D., 1942, Geomorphology: The Macmillan Company, New York, 655 pp.

FAHEY, B. D., 1981, Origin and age of upland schist tors in central Otago, New Zealand: New Zealand Jour. Geol. Geophys. v. 24, pp. 399–413.

FLINT, R. F., 1963, Altitude, lithology, and the Fall Zone in Connecticut: Jour. Geology, v. 71, pp. 683–697.

GERRARD, A. J. W. 1978, Tors and granite landforms of Dartmoor and eastern Bodmin Moor: Proc. of the Ussher Society, v. 4, pp. 204–210.

GILBERT, G. K., 1877, Report on the geology of the Henry Mountains (Utah): U.S. Geog. Geol. Survey Rocky Mtn. Region (Powell), 160 pp.

GILLULY, J., 1949, Distribution of mountain building in geologic time: Geol. Soc. America Bull., v. 60, pp. 561–590.

HOWARD, A. D., 1967, Drainage analysis in geologic interpretation: A summation: Am. Assoc. Petroleum Geologists Bull., v. 51, pp. 2246–2259.

HSÜ K. J., 1968, Principles of mélanges and their bearing on the Franciscan-Knoxville paradox: Geol. Soc. America Bull., v. 79, pp. 1063–1074.

HSU T. L., 1971, Mélange occurrence in Cyprus: Geol. Soc. China Proc. for 1970, no. 14, pp. 155–164.

JOLY, F., 1984, Paris Basin, in Embleton, C., ed., Geomorphology of Europe: John Wiley & Sons, Inc., pp. 154–161.

JORDAN, T. E., ISACKS, B. L., and 4 others, 1983, Andean tectonics related to geometry of subducted Nazca plate: Geol. Soc. America Bull., v. 94, pp. 341–361.

LONG, P. E., and WOOD, B. J., 1986, Structures, textures, and cooling histories of Columbia River basalt flows: Geol. Soc. America Bull., v. 97, pp. 1144–1155.

LOVERING, T. S., and GODDARD, E. N., 1950, Geology and ore deposits of the Front Range, Colorado: U. S. Geol. Survey Prof. Paper 223, 319 pp.

MAXWELL, J. C., 1974, Anatomy of an orogen: Geol. Soc. America Bull., v. 85, pp. 1195–1204.

MILLER, J. R., RITTER, D. F., and KOCHEL, R. C., 1990, Morphometric assessment of lithologic controls on drainage basin evolution in the Crawford Upland, south-central Indiana: Am. Jour. Sci., v. 290, pp. 569–599.

OBERLANDER, T., 1965, The Zagros streams: A new interpretation of transverse drainage in an orogenic zone: Syracuse Geog. Ser., no. 1, Syracuse Univ. Press, Syracuse, N.Y., 168 pp.

———, 1985, Origin of drainage transverse to structures in orogens, in Morisawa, M.E. and Hack, J. T., eds., Tectonic geomorphology: Allen & Unwin, Inc., Boston, pp. 155–182.

PHIPPS, S. P., 1984, Ophiolitic olistostromes in the basal Great Valley sequence, Napa County, northern California

Coast Ranges, in Raymond, L. A., ed., Mélanges: Their nature, origin and significance: Geol. Soc. America Spec. Paper 198, pp. 103–125.

POLLARD, D. D., and AYDIN, A., 1988, Progress in understanding jointing over the past century: Geol. Soc. America Bull., v. 100, pp. 1181–1204.

RAYMOND, L. A., ed., 1984, Mélanges: Their nature, origin and significance: Geol. Soc. America Spec. Paper 198, 170 pp.

RICH, J. L., 1911, Gravel as a resistant rock: Jour. Geology, v. 19, pp. 492–506.

SEVON, W. D., 1986, Susquehanna River water gaps: Many years of speculation: Pennsylvania Geology, v. 17, no. 3, pp. 4–7.

STIRLING, M. W., 1990, The Old Man Range and Garvie Mountains: Tectonic geomorphology of the central Otago peneplain, New Zealand: New Zealand Jour. Geol. Geophys., v. 33, pp. 233–243.

THOMAS, M. F., 1994, Geomorphology in the tropics: A study of weathering and denudation in low latitudes: John Wiley & Sons, Chichester, UK, 460 pp.

THOMPSON, H. D., 1949, Drainage evolution in the Appalachians of Pennsylvania: Annals, N.Y. Acad. Sci., v. 52, pp. 31–62.

WAHRHAFTIG, C., 1965, Stepped topography of the southern Sierra Nevada, California: Geol. Soc. America Bull., v. 76, pp. 1165–1189.

Chapter 13

Arid and Savanna Landscapes; Eolian Processes and Landforms

The processes of landscape development in dry regions differ only in frequency and intensity, rather than in kind, from those in humid regions. Most of the processes, including weathering, mass wasting, and fluvial erosion and deposition, have been reviewed in prior chapters. In this chapter, the landforms of perennially dry and seasonally dry climatic zones are described with minimal further reference to the details of the shaping processes, with the exception of the work of wind. Although wind is a global phenomenon, it is an effective geomorphic agent only in those places with dry soil and incomplete vegetative cover, at least seasonally.

DRY CLIMATES

The largest single identifiable climatic region on earth is the dry region, where either seasonal or annual precipitation is insufficient to maintain vegetational cover and permit perennial streams to flow. Climatologists have devised various empirical formulas to define dry climates in terms of the ratio of precipitation to evapotranspiration. The temperature of a region must be incorporated into such formulas because it is a major factor in evapotranspiration, but otherwise temperature is not a diagnostic parameter of arid climates. For instance, the Köppen system (Köppen and Geiger, 1936) defined the boundary between semiarid climates and humid climates with winter rains by the relationship

$$P \leq 20T$$

where P is the annual precipitation in millimeters, and T is the mean annual temperature in Celsius degrees. The truly arid (desert) climate was defined as one in which precipitation is only one-half as great as the amount that locally separates the semiarid from the humid climate, or

$$P \leq 10T$$

Köppen and his assistants calculated other similar empirical formulas for climates in which precipitation is nonseasonal or is concentrated in the summer. Although apparently related only to mean annual precipitation and evaporation, Köppen's choice of boundaries was based primarily on vegetation. The parameters that he chose to define aridity also define areas in which insufficient moisture is available to support continuous vegetational cover of the ground. By Köppen's definitions, 26 percent of the land area is dry.

In another climatic classification, Thornthwaite (1948) attempted to correlate potential evapotranspiration to other climatic variables, such as temperature and length of day, and thereby to define more precisely the limits of dry climates as expressed in plant growth. His formulas apply primarily to the United States, where the minimum annual precipitation required for a humid climate increases southward from about 500 mm in North Dakota to about 750 mm in Texas. Less than 250 mm of annual precipitation produces a desert

FIGURE 13-1.   Dry regions of the world (after Meigs, 1953). Areas with mean annual temperature < 10°C are excluded. Definitions of degrees of aridity are discussed in the text.

in almost any temperature range. By Thornthwaite's definitions, fully 36 percent of the land area is dry, even excluding the high-latitude frozen deserts.

Meigs (1952, 1953) used Thornthwaite's definitions but remapped areas based on better climatic data. Meigs's map (Figure 13-1) shows 34 percent of the land area to be dry, including 4 percent defined as extremely arid, where no precipitation has fallen for at least 12 consecutive months at least once in the period of record and where rainfall lacks a seasonal rhythm.

In the most recent attempt to delimit and classify arid regions (UNESCO, 1977), an *aridity index* for 1600 weather stations was used in combination with information about soils and vegetation to create a map of bioclimatic aridity defined by the ratio of annual precipitation (P) to annual potential evapotranspiration (ETP). Four zones were defined:

Hyperarid: $P/ETP < 0.03$
Arid: $0.03 < P/ETP < 0.20$
Semiarid: $0.20 < P/ETP < 0.50$
Subhumid: $0.50 < P/ETP < 0.70$

The UNESCO classification of aridity encompasses about 43 percent of the earth's land area, significantly larger than the area included in previous analyses, but

includes large areas of semiarid or subhumid grasslands that would not be considered "deserts" except in drought years or when they have been degraded by overgrazing or other human impact (Thomas and Middleton, 1994). Nevertheless, by any of the classifications, between 25 and 40 percent of the earth's land area is presently arid, even excluding the high latitude deserts that are either ice- or snow-covered, or underlain by perennially frozen ground (Chapter 14). That area, at least in Africa, increased significantly after the mid-Cenozoic "greenhouse" climate (Chapter 4) that made tropical Africa warmer and wetter than now between about 8 and 3 Ma (deMenocal, 1995).

In low latitudes, two belts of aridity coincide with the subtropical anticyclonic belts of high atmospheric pressure about 15° to 30° north and south of the equator (pp. 26, 53). They are fringed by relatively narrow belts of semiarid transitional climates (Figure 13-1). In middle latitudes, dry climates are localized in the interior of the large continents. Temperature and evaporation are not as high in these regions as in the subtropics, and some precipitation may fall as snow. Midlatitude dry climates are characterized by relatively wide belts of semiaridity around smaller cores of true desert. Freezing temperatures during winter months are a common feature of midlatitude semiarid climates.

Small areas of dry climate also are on the west sides of the continents, where cold ocean currents flow offshore (Figure 4-2), and on the downwind, or rain shadow, side of mountain ranges.

It needs to be emphasized that when climatologists define dry climates, they commonly include the criterion of discontinuous vegetative cover. In whatever way aridity is defined, it includes the concept that erosion is not inhibited by a continuous ground cover of plants. Furthermore, rivers flow only seasonally, and even when flowing, they decrease in discharge downstream, in contrast to the streams in humid regions. Moisture deficiency causes groundwater to move very slowly. Secondary minerals accumulate in soil profiles, mostly from dust fall, and soil structure is strongly affected. For the geomorphologist, discontinuous flowing rivers, lack of continuous ground cover, and accumulation of secondary minerals in soil profiles define dry regions; each has a significant effect on landscape evolution.

Semiarid climates usually support a sparse grassland or **steppe** vegetation. In the United States, the boundary between humid and semiarid climates is approximated by the transition westward from medium-height grasses with a continuous turf or sod in the humid regions to short, shallow-rooted bunch grasses on otherwise bare ground in the semiarid regions. In arid regions, even the bunch grasses disappear, and the vegetation is, at best, widely spaced shrubs and salt-tolerant bushes. The maximum sediment yields have been reported from the arid-semiarid transition (Figure 15-8), especially if the natural vegetation and soil have been disturbed by the grazing of domestic cattle or by cultivated agriculture.

LANDFORMS IN DRY REGIONS

Steppes and Prairies

The semiarid and subhumid *steppes, prairies, veld,* and *pampas* of the world are characterized by (1) grass cover; (2) annual precipitation up to twice the amount that locally defines the desert boundary; and (3) rivers that are graded externally to the region, usually to the sea. Slopes and drainage networks are integrated regionally, and erosional landforms evolve toward sea level rather than to the rising base level of a closed basin.

Steppe regions are generally plains or low dissected plateaus. Large areas of grassland, such as the Great Plains of North America, are the constructional surfaces of huge piedmont alluvial plains (Figure 11-20). The Pampas region of Argentina is a monotonous plain of windblown silt (**loess**) and sand that covers an area of more than 1 million km^2. The dissected plains

of north central China and the interior plains of central Asia are even larger semiarid steppes on loess. In other regions, especially Australia and southern Africa, grassy plains are either erosional or depositional surfaces. Most have now been converted to agricultural use because cereal grains have soil and climatic requirements very similar to those of the grasses from which they evolved (p. 51).

Desert Plains and Plateaus

By far the greatest proportion of the world's arid regions are monotonous plains and plateaus in the interiors of large continents (Figure 13-1). There is no compelling reason why dry regions should be nonorogenic, but nonorogenic regions are likely to form the large interior parts of continents where moisture-bearing winds cannot penetrate, and nonorogenic regions also lack the relief necessary for orographic precipitation. In addition, internal drainage, characteristic or even diagnostic of deserts, causes alluviation that buries minor relief. Wind has been suggested as a leveling agent of desert plains as well, but evidence presented later in this chapter suggests that wind is more likely to increase desert relief by excavating basins and constructing major dune landforms.

Where alluvial fans merge in a desert basin, a saline lake (**playa, salar,** or **sabkha**) may form. Here, mud and salts accumulate to form the most level topographic surfaces developed on earth. Water less than a meter in depth may briefly accumulate, but when the water has evaporated, a great flat of polygonal-cracked mud or salt is all that remains. Residual brines may ooze from beneath the encrusted surface of the playas, too concentrated to evaporate further but liquid enough to engulf motor vehicles that are driven onto the surface. On other playas, hard crusts of salt and mud provide nearly perfect surfaces for racing high-speed cars and landing airplanes and spacecraft.

A variety of desert processes produce a surface concentrate of gravel called a **reg,** or stony desert (known in China as a *gobi,* or a *gobi desert*)(Walker, 1982, p. 368). A **desert pavement** may form a stable flat surface mosaic of closely fitted cobbles and pebbles in a layer only one or two rocks deep, each polished by sand blast and often faceted as well. If wind abrasion is not severe, *rock varnish* may coat the exposed surface of the pavement. If the rocks have been moved, their varnish-free undersurfaces betray the movement, so animal trails and vehicle tracks across reg remain visible for decades or centuries. However, if a meter-size area of desert pavement is removed, adjacent sediment creeps toward the resulting depression and largely eliminates it within a few decades (Haff and Werner, 1996).

Desert pavements have traditionally been attributed to deflation (p. 291) of finer particles, leaving behind a lag of wind-abraded pebbles and cobbles (Dixon, 1994, pp. 74–77). Many desert pavements have been produced by deflation, but other processes can also make them. For example, sheetflood (p. 282) and channeled flow during and after storms can scour away finer sediment (Williams and Zimbelman, 1994). Freezing and thawing or wetting and drying cycles can move gravel-size particles upward to accumulate as a surface layer above a finer substrate. As eolian dust accumulates on a desert landscape (p. 143), it can filter down through gravel and form pedogenic horizons beneath the desert pavement. Cosmogenic dating of some clasts prove that they have been "floating" at the surface for as long as 85,000 years, even while a layer of eolian dust 60 cm or more in thickness has accumulated beneath them (Wells et al., 1995).

Approximately one-fourth of the Sahara and the Arabian desert are sandy areas, or *erg* (p. 300). The remainder is either **hammada** (barren rock) or reg. Rock ledges are pitted and polished by abrasive wind-borne sand, but the dominant relief is fluvial. Even in the driest desert plains, drainage networks cover the landscape. Some are used by present runoff, even if rain falls less than once in 12 months, but an unknown portion of the desert dissection is probably due to formerly greater runoff. The issue of climatic change (Chapters 4 and 18) is of major significance to theories of landscape evolution in arid regions. We do not yet know enough about either present processes or late Cenozoic climatic change in the dry regions to distinguish confidently between active and relict landforms. It has been argued that all but the most arid landscapes are essentially "fluvial" (Graf, 1988, p. 10).

Mountainous Deserts

The desert Basin and Range Province of the southwestern United States and Mexico, the Pacific coastal desert of Chile and Peru, and the thin-skinned fold mountains adjacent to the Persian Gulf (Figure 12-9) are each examples of desert mountains (Figure 13-2). Their areas are small compared to the continent-size deserts of Australia and the Sahara and the great desert plains of Arabia, trans-Caspian former USSR, and western China. However, the American Basin and Range Province has been particularly significant in the development of the science of geomorphology, for it was the excellent analytic and descriptive literature about this area written in the late nineteenth century that became the basis for W. M. Davis's deductions about of landscape evolution (Chapter 15). Europeans of that time also followed the exploration of the American west with great interest, so for a time the arid and semiarid territories of America became the leading topic of geomorphic thought. As noted in Chapter 5 (p. 78) the nearly unconscious association of fault-block tectonic relief with arid and semiarid landscapes has permanently biased geomorphic thought, especially in the United States (Graf, 1988, pp. 20–21).

FIGURE 13-2. Death Valley, California, northwest from Dante's View. Compare Figure 13-3. Note saline playa at the base of massive alluvial fans and the exposed regolith in the foreground (photo: U.S. Borax Co.).

Mountains in deserts are likely to be islands of non-desert climate rather than part of the desert. Orographic precipitation and sometimes snowmelt feed streams that dissect mountains in deserts much as they dissect mountains in other climatic zones. The contrast in morphogenesis is most notable at the base of desert mountains, where the streams discharge from the mountains onto the surrounding dry *piedmont* ("foot-of-the-mountain") slopes (Cooke et al., 1993, p. 19).

A common landform of intermontane deserts is the alluvial fan (p. 249). Influent rivers decrease in discharge as they enter desert basins and deposit their entire sediment load as they spread laterally, evaporate, or infiltrate. It is no exaggeration to say that desert mountains are progressively buried in their own waste. The great apron of coalescing fans at a mountain front is called a **bajada** (or **bahada**). Within the Great Basin section of the Basin and Range Province, where the drainage is entirely internal, mountains compose about 38 percent of the landscape area, desert flats about 20 percent, and fans and bajadas 31 percent (Figure 13-3). Minor areas of playas, arroyos, badlands, dunes, and volcanic fields make up the remainder of the area (Cooke et al., 1993, p. 20).

An especially interesting mountainous desert is the Sonoran region of the Basin and Range Province in southwestern United States and adjacent Mexico.

Much of the area is graded to sea level by intermittent streams and by tributaries to the lower Colorado River and Gila River. Although the region appears to consist of mountains rising above broad alluvial plains or fans, much of the intermontane surface is of planed rock, similar to that which composes the mountains. An estimated 20 percent of the region consists of mountains, 40 percent of barren or thinly veneered bedrock plains, and 40 percent of thick alluvium (McGee, 1897, p. 91). Early explorers were surprised to find that the gently sloping piedmont surfaces 5 to 8 km from the mountain front were cut across hard granitic rocks, with only a thin layer of alluvium on the surface: "so thin that it may be shifted by a single great storm" (McGee, 1897, p. 91). The wide extent of barren piedmont plains in the Sonoran district is probably permitted by the external drainage, which exports alluvium to the sea rather than permitting it to accumulate in closed intermontane basins.

The significance and extent of bare or thinly veneered rock surfaces in the Sonoran Desert and elsewhere has led to extensive further study. Bryan (1922, p. 88) formally adopted the elegant term **mountain pediment** for

> a plain which lies at the foot of mountains in an arid region or in headwater basins within a mountain mass. The name is applied because the plain appears to be a pediment upon which the mountain stands. A mountain pediment is formed by erosion and deposition of streams, usually of the ephemeral type, and is covered with a veneer of gravel in transit from higher to lower levels. It simulates the form of an alluvial fan.

In classic architecture, a pediment is the triangular end of a low-pitched gable roof, on which a frieze is often carved. The term *mountain pediment* was chosen because in profile, when viewed from a distance, each mountain peak seems to stand at the crest of a low conical plain, triangular in profile (Figure 13-4). Today, the landform is referred to simply as a **pediment**. It has no relation to the term *pedistal,* with which it is frequently confused, or to the very general descriptive term *piedmont,* which simply means "at the foot of the mountain." The true statement that pediments are piedmont landforms has confused generations of beginning geomorphology students.

As first used by Gilbert (1882, p. 183), the term *pediment* was casually applied to the surfaces of the alluvial fans that encircle desert mountains. Bryan's definitive study, following McGee's observations in the same general area, emphasized that pediments are slopes of transportation cut on bedrock, usually covered with a veneer of alluvium in transit from high to

FIGURE 13-3. Bajada and saline playa of Death Valley, California. View approx. 20 km wide; north to the top. Rock varnish (p. 279) of variable darkness indicates the relative age of the fan surfaces (satellite image: SPOT Image Corp., Reston, VA © CNES, 1990, courtesy T. G. Farr).

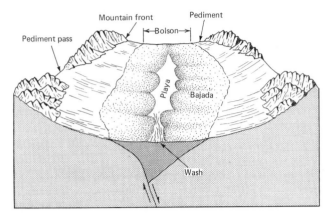

FIGURE 13-4. Typical assemblage of mountainous desert landforms.

lower levels. In form and function, a pediment is similar to an alluvial fan, the difference being that a pediment is an erosional landform and a fan is constructional.

The slope on pediments, as on alluvial fans, increases toward the mountain front. Typical gradients range from 1° to 7°. Headward portions of pediments may penetrate into a mountain mass, isolating segments of the range and possibly intersecting with pediments from the other side of the range at *pediment passes* (Howard, 1942). The pediment form, like that of an alluvial fan, is an excellent example of a water-spreading wash slope (Figure 9-23). Functionally, it is an ideal surface to distribute and dissipate a mass of water and sediment introduced at its apex. There are no significant differences in slope, roughness, or size between alluvial fans and pediments.

By original definition and conventional use, the pediment is a landform of dry regions. However, by extension, the word has been applied to any gently sloping piedmont erosional surface of low relief in climates ranging from subpolar to the humid tropics, wherever tectonic stability has permitted erosion and transport to become balanced for a long enough time to develop an extensive surface (Dohrenwend, 1994, p. 321).

Theories about the origin of pediments are closely related to geomorphic processes of dry regions. McGee's dramatic eye-witness account of a **sheetflood** in the Sonoran Desert (McGee, 1897, pp. 100–101) was especially influential in establishing the sheetflood as an important geomorphic process.

The shower passed in a few minutes and the sun reappeared, rapidly drying the ground to whiteness. Within half an hour a roar was heard in the foothills, rapidly increasing in volume; the teamster was startled, and set out along the road up the valley at best speed; but before he had gone 100 yards the flood was about him. The water was thick with mud, slimy with foam and loaded with twigs, dead leaflets and other flotsam; it was seen up and down the road several hundred yards in either direction or fully half a mile in all, covering the entire surface on both sides of the road, save a few islands protected by exceptionally large mesquite clumps at their upper ends. The torrent advanced at race-horse speed at first, but, slowing rapidly, died out in irregular lobes not more than a quarter of a mile below the road; yet, though so broad and tumultuous, it was nowhere more than about 18 inches and generally only 8–12 inches in depth. . . . For perhaps five minutes the sheetflood maintained its vigor, and even seemed to augment in volume; the next five minutes it held its own in the interior, though the advance of the frontal wave slackened and at length ceased; then the torrent began to disappear at the margin, the flow grew feeble in the interior, the water shrank and vanished from the margin up the slope nearly as rapidly as it had advanced, and in half an hour from the advent of the flood the ground was again whitening in the sun, save in a few depressions where muddy puddles still lingered.

McGee's description led to a general assumption that sheetfloods erode weathered bedrock surfaces to the perfection of pediments. Other geomorphologists proposed that lateral corrasion by streams (or **streamfloods,** episodic high-intensity discharges confined to channels) is the dominant process of pediment formation (Johnson, 1932). The controversies between proponents of these and other hypotheses of pediment evolution were summarized by Hadley (1967). Cooke et al. (1993, p. 199) noted that sheetflow cannot be the cause of a low-relief surface because the low-relief surface is a prerequisite for sheetflow.

Field observations of desert storms are sufficiently rare that processes of erosion and deposition are poorly known. Rahn (1967) described examples of flood activity on piedmont slopes in Arizona and concluded that the floods on pediments were streamfloods, confined to channels but exhibiting high velocity and capable of impressive erosion and deposition. Only when rain fell directly onto bajadas were sheetfloods produced. Rahn concluded that pediments are likely to be "born dissected" by channeled flow near mountain fronts and later smoothed by weathering and sheetfloods as the mountain front retreats.

Apparently, pediments widen by the lateral migration of rills and ephemeral channels that periodically impinge on the mountain front and undercut taluses

FIGURE 13-5. Mountains near Maricopa, Arizona, surmounting a pediment. The boulder-controlled mountain front has an abrupt junction angle with the pediment (photo: R. E. Moeller).

and cliffs. They also expand by the normal mass-wasting retreat of cliffs and taluses and by abrasion of the rock-cut pediment surface by sheetfloods (Figure 13-5). The relative importance of the three processes seems to vary from region to region. Analysis of these details is difficult, because all erosion is slow in dry regions, and some pediments probably formed under a different climate and are now being slowly dissected. A strong contemporary opinion is that many, if not all, pediments were first formed in a more humid climate, were buried, and are now being exhumed (Tuan, 1959; Oberlander, 1974). An even stronger view is that the deep regolith found on pediments and mountain sides beneath lava flows older than 8 million years in the Mojave Desert not only proves the great age of the pediments and mountain sides, but also demonstrates that the landscape, at least in part, dates from the time of regolith formation, perhaps during a Miocene "greenhouse" climate (Oberlander, 1972, 1989; Dohrenwend, 1994, p. 340).

As pediments expand into the mountain slopes, the lower part of the pediments in turn may be progressively buried by the aggrading bajada alluvium (Figure 13-4). For this reason, pediments in truly arid regions, where the local base level is a playa, are generally only narrow rock-cut fringes between the mountain fronts and the bajada. The pediments plunge downslope beneath the thickening alluvial fill toward the center of the **bolson,** or intermontane basin. By those who insist on precise definitions, the smoothly graded wash slope at such a mountain front would be called a pediment in its upper part, where the entire thickness of alluvium is in transit, and a bajada in its lower part, where the alluvium has permanently come

to rest in the bolson. The boundary between pediment and bajada is then defined by the thickness of alluvium that can be episodically reworked by contemporary fluvial processes. This definition is analogous to the definition of active and relict floodplain alluvium in humid regions (p. 240).

The geometry of desert basins and mountains is such that as a basin is filled, its surface area increases, whereas the area of the adjacent mountain decreases by erosion (if it is not tectonically active). Therefore, successive increments of mass eroded from the mountain cause a decreasing rate of vertical accretion in the receiving basin and the development of a convex-skyward buried bedrock surface, even though the piedmont surface is always concave-upward (Figure 13-6). Lawson (1915) referred to the convex buried rock surface as the **suballuvial bench** and to the concave exposed portion as the *subaerial bench.* The latter term is synonymous with pediment. Wells on the margins of many bolsons in the Great Basin demonstrate that their subsurface geometry is as Lawson inferred. The original fault scarp of the basin-range structure is now far out under the bajada, and large portions of pediments truncate older sedimentary formations or indurated fan gravels. Figure 13-6 illustrates that a slight lowering of base level by tectonism, stream capture, or climatic change could extend and regrade large areas of a pediment across former bajada sediments.

In plan view, pediments, like alluvial fans, range from semicircular flat cones, to nearly flat, planar surfaces, to a series of shallow, gently concave valleys. At the base of a fault scarp or at a sharp erodibility contrast between mountain rocks and those of the adjacent basin, the head of a pediment may be quite linear. On

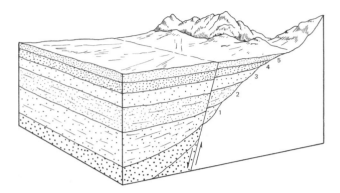

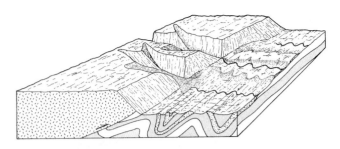

FIGURE 13-7. Evolution of topography where a stream flows from resistant rocks across deformed weak rocks (adapted from Rich, 1935, Figure 3).

FIGURE 13-6. Lawson's theory for the origin of a convex-upward suballuvial rock bench. Equal volumes of alluvium cause successively smaller increments of aggradation in the widening bolson. Minor continued faulting demonstrates that the original fault-block boundary is now well away from the mountain front. Vertical scale is exaggerated.

other piedmonts, the pediments extend well into the mountain mass as broad embayments, leaving an extremely sinuous mountain front. Some embayed mountain fronts may result from the lateral migration of rivers as they emerge into the basin. The suballuvial rock floors in these embayments are essentially rock-defended, or strath (p. 245) terraces and may be in the form of nested steps, the lower ones younger. Other pediments seem to have resulted from intermittent sheetfloods separated by intervals of surface weathering and rillwork. These pediments are smoother and more nearly planer. Many pediments show more bare rock and extensive dissection in their upper, proximal regions. This could be the result of the gradual reduction of the mountain mass in the drainage basin so that sediment supply decreases with time and the pediment is regraded to a reduced gradient (Dohrenwend, 1994, p. 350).

Pediments that form on weak-rock lowlands at the base of mountains carved from resistant rock are especially likely to be complex, multiple surfaces. Weak-rock lowlands are rapidly eroded by rivers from the mountain front, leaving broad, fan-shaped pediments heading at major canyon mouths. These are cut across erodible rocks, but are protected from deep dissection by an alluvial veneer of coarse gravel transported from the mountain canyons (Figure 13-7). Their coarse alluvial bedload requires the pediment streams to maintain steep gradients to the local basin floor. If, however, the alluvial cover is breached, gullies form in the underlying weak rocks and erode rapidly headward. These gullies carry primarily silt and clay from the erodible substrata, and the pediment cover gravel is not abun-

dant enough to choke the gullies. Therefore, they grow headward with steep headwalls and the pediment cover gravel slumps down their flanks from interfluve remnants. Lateral shifting of the gullies in this manner can introduce multiple levels of pediment remnants without any tectonic movement or climate change.

Tectonic uplift and climatic change can compound the development of multiple, "nested" pediment remnants (Oberlander, 1989, p. 57). In the Sierras Pampeanas of northwestern Argentina, fission-track dating of volcanic ash interbedded in thin pediment alluvium has documented at least five pediment levels that range in age from 2.5 to 0.3 million years. Although some may be the result of multiple dissection eposides on the piedmont slopes, active tectonism is certainly involved in the older, most dissected pediments (Strecker et al., 1989).

The most extensive pediments are in regions that have sufficient precipitation to allow at least occasional direct surface runoff to the sea. Under these conditions, which are technically semiarid rather than arid, weathered sediment can be transported to the sea instead of filling bolsons and progressively burying the lower edges of developing pediments. The Sonoran Desert in Arizona and in adjacent Mexico, the classic region of pediment development in North America, has pediments, alluvium-floored plains, and river channels all graded to sea level in the Gulf of California. Now, water rarely flows in the channels. The entire Sonoran piedmont landscape that so impressed early explorers may be relict from former times of greater precipitation.

The interior plateaus of large parts of Africa are great, gently sloping pediments, and large parts of the dry continent, Australia, are also pediment-prone. The arid interior of Asia is less well known than other dry regions, but there, too, pediments can be predicted to be among the dominant landforms. Considering that 26 to 43 percent of the land area is classified as arid or semiarid, the pediment may be the most common erosional landform on earth.

If there is a landscape characteristic of arid regions, it should consist of barren rock or gravel-armored plains scarred by braided ephemeral channels and arroyos. Eolian processes produce deflation basins and dune fields. Mountains are fringed with cliffs, taluses, pediments, bajadas, and playas. Large areas of the earth do have such landform assemblages. Even so, the degree to which the forms were shaped by processes that were formerly more intense has yet to be resolved.

THE TROPICAL SAVANNA

The Savanna Climate

A **savanna** is a tropical region of sparse open woodlands and thorny shrubs, with grass ground cover. The tropical savanna climatic type was included with the tropical rain forest as the "tropical rain climates" by Köppen (Köppen and Geiger, 1936) and the "tropical humid climates" by Thornthwaite (1948). It belongs neither to the arid climates nor to the constantly wet tropical rain forest. In addition to being hot, the tropical savanna climate is dry for a significant part of each year, even though total rainfall exceeds the definitions used to define the dry regions by climatologists. In the savanna climate, which covers about 15 percent of the earth's land area, precipitation is strongly seasonal, usually coming during the hottest season. Toward the equator, the savanna grades into tropical rain forest, and poleward it grades into the subtropical dry belts (Figure 7-14). Most photographs of savanna landscapes convey the impression of aridity because most are taken during the dry season when the grass is brown and many trees have shed their leaves. During the wet months, however, very heavy rainfalls are common, and vast areas of savanna plains are under standing water. The year-round high temperatures and seasonally abundant water promote intense chemical weathering, even though rivers flow only during the rainy season and for a brief interval thereafter.

In savanna regions, fires are common during the dry season; some are started by natural causes, but many are started by people in order to clear the land or to drive wild game. Some biogeographers suspect that savanna grassland has greatly expanded because of human activity. If so, this important geomorphic region has become much more important in the later Cenozoic Era. At the earliest, savannas could not have existed prior to the early Tertiary evolution of grasses (Chapter 4). Grass pollen is common in Miocene continental sediments, and savannas probably had became widespread by then (p. 51).

Savanna Landforms

The savanna landscape of Africa was first made known through the work of German geomorphologists early in this century (Bornhardt, 1900; Passarge, 1928). They called it an *Inselberglandschaft,* or **inselberg landscape,** with reference to the steep-sided, isolated hills and mountains of barren rock that rise like islands from nearby flat plains (Figure 13-8). During the wet months, wide areas of the plains are flooded, and some of the hills are truly islands.

The term *inselberg* subsequently was extended to the residual knobs on pediments in arid and semiarid regions, and the distinction between geomorphic

FIGURE 13-8. Inselberg near Einasleigh, Queensland, Australia (photo: Ian Douglas).

processes in truly dry regions and the savanna wet-and-dry region became blurred. In an effort to restore precision, Willis (1934; King, 1948; Howard and Selby, 1994, pp. 127–130) proposed that the barren, domelike "sugar-loaf" hills without taluses so common in savanna regions should be called *bornhardts,* after the pioneer geomorphologist. Twidale (1982, pp. 15, 124) maintained this usage by distinguishing the most common granite inselbergs from unspecified other types, and designating the most abundant of the granite inselbergs as bornhardts. He defined a **bornhardt** as a domical inselberg developed on a single massive joint block. He designated two other minor types of inselbergs on granite rocks: **castle koppies,** angular and castellated hills (like the small turrets and slotted walls of a medieval castle), and **nubbins,** knolls of block or boulder-strewn bedrock exposure. In his usage, bornhardts, castle koppies, and nubbins are structurally controlled landforms related to corestones and tors (Figures 7-13, 12-2, and 12-3). As such, they are asserted to be azonal or nonclimate-controlled landforms shaped by the structural properties of granitic rocks (including such factors as mineralogy and fractures) (Twidale, 1982, p. 24). Nevertheless, the unique regional landscape of inselbergs, especially the most common variety of granite bornhardts, rising like islands above surrounding flat plains, requires a special set of conditions that are most often found in the tropical savanna climate. These include warm temperatures, at least seasonally humid conditions, active subsurface weathering by groundwater, and slow or only episodic fluvial incision (Twidale, 1982, p. 136). Tectonically stable cratonic plains also seem to be required, with their tendency to remain near sea level and undergo very slow denudation (p. 44).

An alternative hypothesis is that, considering the major global changes of climate during the Cenozoic area, many stable cratonic regions, including mid- or high-latitude regions such as the Canadian shield, have experienced "greenhouse" climates comparable to the subtropical savanna of today (Chapter 4). Because large inselbergs are exceptionally durable landforms, it is possible that they have survived for many millions of years in nonglaciated shield regions of low relief and low erosional energy (Thomas, 1994, p. 342ff). Similarly, they are found surrounded by rainforest in the continuously wet equatorial tropics, and might be forming there today or might be relicts of a former seasonally drier climate. Thomas (1994, p. 343) noted that even though they were originally described as landforms unique to savanna climate and are striking features of that region: "In fact, their distribution appears not to be controlled by climate zone." The role of climate change in the present distribution of inselbergs and related landforms remains an intriguing problem of geomorphic research (Büdel, 1982).

In the savanna climate, with high year-round temperatures and heavy seasonal rainfall, chemical weathering extends to great depths (Figure 7-13; Thomas, 1994, pp. 19ff). With only seasonal runoff, however, the low gradient cratonic savanna-zone rivers cannot remove the massive sediment loads furnished by weathering, which is primarily bedload of quartz-rich sandy grus. Rather than entrenching their valleys, savanna rivers braid or flood across extremely flat **wash plains.** Runoff that does not evaporate infiltrates to the water table. The result of rapid chemical weathering by groundwater and inhibited stream erosion is that wash plains are underlain by deep zones of altered rock (Büdel, 1957). Although the ground surface is a monotonous plain of alluvium and saprolite (p. 141), at a depth of perhaps 100 m an irregular surface is in the process of formation at a "weathering front" in the zone of seasonally fluctuating water table. Small masses of fresh rock that are bypassed by the subsurface weathering are eventually exhumed as corestones and tors (Figures 7-13, 12-2, and 12-3); such small forms are commonplace on jointed granitic and similar rocks in all humid and formerly humid climates. Under the optimal climate conditions of the tropical savanna, and usually in a nonorogenic tectonic setting of cratonic crystalline rocks with irregular joint spacing that includes some very large unfractured monoliths, the inselberg landscape evolves. When the weathering front has bypassed a large monolith, its preservation and eventual expression as an inselberg seem assured. The huge size of some savanna inselbergs leads to the conviction that they are of considerable age. Denudation rates based on cosmogenic dating techniques imply that inselbergs in South Australia have lost only a meter or so of rock from their summits in the entire Pleistocene (Bierman and Turner, 1995 p. 382). Thus structural control and climatic factors combine to produce one of the more striking landscapes of the earth. The process has been referred to as a "double surface of leveling" ("*doppelten Einebnungsflächen*") by Büdel (1957). Figure 13-9 summarizes Büdel's hypothesis of inselberg development.

It is also possible that purely chance events at the surface, such as a landslide, might strip the regolith off an emerging rock mass, allow it to be washed clean, and thereby preserve it from future deep chemical weathering. The development of a benched landscape on the Sierra Nevada without structural control has been attributed to such randomness of exposure (Wahrhaftig, 1965).

Some inselbergs are encircled by shallow moats at their bases, probably due to concentrated subsurface

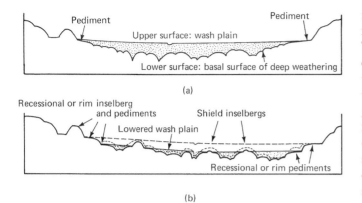

(a)

(b)

FIGURE 13-9. Hypothesis of a "double surface of leveling" in the tropical savanna. (a) Wash plain of seasonal flooding is as much as 100 m above the weathering front. Pediments fringe the wash plain. (b) Wash plain is lowered by rejuvenation or climatic change. Inselbergs and marginal pediments are exhumed or regraded to the lowered wash plain (from Büdel, 1957, Figures 5 and 7).

weathering by runoff and infiltration at the base of the inselberg (Twidale, 1968, pp. 105–124; 1976, pp. 7–17; Twidale and Bourne, 1975). As wash plains are lowered, these moats and weathered notches in the flanks of steep-sided inselbergs (Figure 13-10) are exhumed to record former ground levels. Other inselbergs have narrow pediments around their bases (Figure 13-9), which led to the confusing extension of the word *inselberg* to residual hills that are surrounded by much larger pediments in other types of dry regions. The savanna pediments are narrow and of low gradient, 3° being the maximum reported in Ghana (Clayton, 1956). The minor development of pediments and the great extent and flatness of wash plains distinguish the savanna inselberg landscape from the pedimented landscapes of dry regions. The contrast is due to the intensity of chemical weathering in the savanna climate, by which exposed rock is so weakened that it is easily eroded, and to the heavy but seasonal precipitation, by which the detritus is distributed across the wash plains.

There seems to be adequate evidence to postulate that the inselberg landscape defined by Bornhardt is a landscape assemblage especially well displayed in the savanna climate. However, just because they are found in that climate, it does not follow that they are only formed there. On the wetter, equatorial side of the savanna regions, the wash plains become forest-covered, stabilized, and eroded by perennial streams. On the dry, poleward side of the savanna belt, desert landforms replace the inselbergs and wash plains because chemical weathering becomes less effective. However, both climatic boundaries of the savanna zone have shifted repeatedly during the late Cenozoic Era, and some undetermined amount of the savanna landscape, especially in adjacent climatic regions, is almost certainly relict. Deep saprolitic weathering remnants in places as diverse as the Mojave Desert, the Canadian Shield, and the New England states may indicate that

FIGURE 13-10. Steepened margin of a granite inselberg caused by multiple episodes of subsurface weathering and subsequent exposure of the weathering front, Pildappa Hill, South Australia (photo: C. R. Twidale).

those regions experienced savannalike climates well into the late Cenozoic Era (pp. 58, 283).

EOLIAN PROCESSES AND LANDFORMS

Dynamics of Eolian Transport and Deposition

Wind is always turbulent. All the equations for turbulent flow in water (Chapter 10) also apply to air though the values of the parameters are very different. As in turbulent flow in water or any other fluid, above a thin boundary layer where the velocity is zero, the velocity initially increases rapidly, then less rapidly, with height (Figure 13-11). In air, the thickness of the stagnant boundary layer over a flat granular bed is equal to about 1/30 of the grain diameter. Above the stagnant layer, wind velocity is proportional to the logarithm of the height, from which the important parameter V_*, the **drag velocity,** was defined as being proportional to the rate of increase of wind velocity with the logarithm of height (Bagnold, 1941, p. 51). By definition,

$$V_* = \sqrt{\tau/\rho}$$

where τ is the surface shear stress, and ρ is the density of air. Although V_* has the dimensions of a velocity, it really represents the vertical velocity gradient near a rough surface. When it exceeds a threshold level (V_{*t}) equal to the vertical settling velocity of a particle of a certain size, that particle will begin to oscillate at the surface beneath the wind, then suddenly leave the sur-

face in a nearly vertical direction (Greeley and Iverson, 1985, p. 70; Pye and Tsoar, 1990, p. 90ff; Lancaster and Nickling, 1994). Bagnold (1941) demonstrated that the threshold velocity is proportional to the square root of grain diameter for grains larger than about 0.1 mm (Figure 13-12a). The relationship is similar to the threshold of erosion in water defined by Hjulström's curve (Figure 10-11a, redrawn as Figure 13-12b). Others have shown subsequently that in air as in water, grains smaller than 0.06 mm (silt and clay) are transported primarily in suspension because their settling velocity is less than the vertical component of turbulent flow. However, in air much of the entrainment of fine particles is caused by the impact of larger grains (Pye and Tsoar, 1990, p. 95). Figure 13-12b is simplified from Figure 10-11a and redrawn with a linear vertical axis for a direct comparison of the threshold of erosion in air and water. However, Figure 13-12a uses the calculated values of the threshold velocity V_{*t} in air as a vertical axis, whereas Figure 13-12b uses Hjulström's measured average velocity in flowing water at least 1 m in depth, so the curves are not exactly analogous. In both curves, the threshold of erosion is linear for larger grains, but below a minimum grain size of about 0.08 mm in air and about 0.2 mm in water, the threshold velocity for erosion rises steeply. In water, this is explained in part by cohesion and in part by the relative thickness of the laminar-flow boundary layer (Figure 10-11b). In air, fine particles on a flat surface lie entirely within the stagnant boundary layer, and are not easily mobilized. Therefore, the Shields curve for air is similar to that for water for sand and larger grains, but rises much more

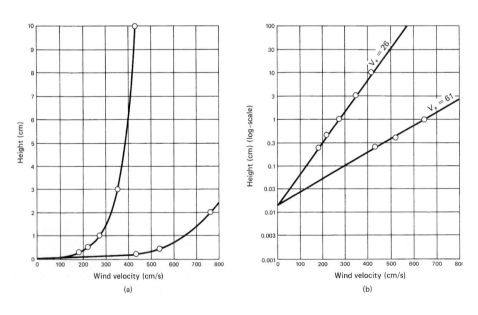

(a)

(b)

FIGURE 13-11.
Velocity distribution for two different wind strengths on (a) linear and (b) logarithmic height scales. The slope of the lines in Figure 14-4b is V_*, the velocity gradient near the surface. (Bagnold, 1941, Figure 15 a and b, © Chapman and Hall Ltd, London, 1973).

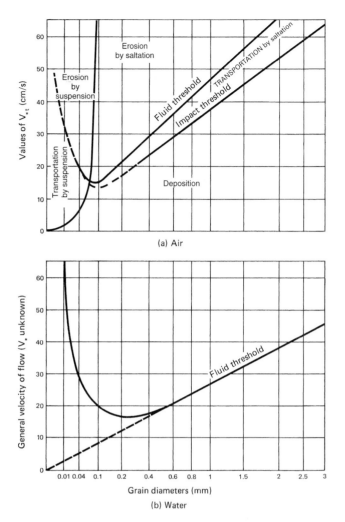

FIGURE 13-12. Variation of the threshold velocity with grain size for (a) air and (b) water. Drawn on a square root scale to show the relation $V \propto \sqrt{d}$ for larger grains (a, from Pye and Tsoar, 1990, Figure 4-4; b, Bagnold, 1941, Figure 29, © Chapman and Hall, Ltd., London, 1973).

grains have already been set in motion, their impacts will cause larger grains to move at velocities only about 80 percent of the threshold velocity predicted for the particular grain size (Figure 13-12a). This effect is not obvious in water, where the viscosity of water absorbs the momentum of the impacting grains.

Because of the great density contrast between air and rock fragments, only very fine detritus, with an effective diameter of less than 0.2 mm and usually less than the minimum size of very fine sand (0.0625 mm), can be maintained in suspension by the upward component of atmospheric turbulence. This is *dust* that grades into extremely fine particulate matter called *haze, smoke,* or *aerosols,* which can be carried entirely around the world by planetary winds before settling out or being washed out of the air.

Loose sand particles with a diameter of 0.2 mm or larger can be stirred by a steady light wind of 5 m/s but do not remain in suspension. They roll or slide along the ground or bounce, transferring momentum to other grains by collisions. Bagnold (1941) made a thorough laboratory and field study of the physics of windblown sand, and by ingenious experiments first demonstrated that most sand grains transported by wind do so in short asymmetric trajectories, never far from the ground. The process is termed **saltation.** A falling grain strikes one or more other grains and makes a visible dimple or crater in the sand surface. Although the disturbed particles are scattered in various directions, as they rise into the airstream and fall again, they are swept forward. Just prior to impact, they are usually approaching the ground at an angle of only 10° to 16° (Bagnold, 1941, p. 19). Above some threshold velocity, the momentum transferred from the wind to saltating grains is sufficient to maintain continuous forward transport of saltating sand. A lesser amount of sand is driven forward over the ground surface without saltating. This component Bagnold called **surface creep.** It is analogous to the traction load or bedload transported by rivers.

The effect of increased wind speed is to impart a higher forward velocity to saltating grains, which in turn produces a rapid increase in the amount of sand in transport (Figure 13-13). Most of the sand saltating over a surface of similar-size grains does not rise higher than 1 or 2 m, regardless of wind speed. Because of the logarithmic decrease in wind velocity close to the ground, the median height of the trajectories of all saltating sand grains is only about 1 cm above the ground.

The impact of a saltating sand grain can move a much larger grain forward by surface creep. Bagnold (1941, pp. 33–37) estimated that significant creep occurs

steeply for silt and clay-size particles. For example, a layer of dry portland cement powder will not move on the floor of a wind tunnel even when V_{*t} exceeds 100 cm/s, a wind strong enough to move fine pebbles with diameters of 4.6 mm (Bagnold, 1941, p. 90). This phenomenon should be obvious to anyone who has observed the very localized plume of dust that trails a moving vehicle on a dry dusty road on a windy day. Immediately behind the disturbance, the dust is again immobile.

For sand particles exposed to wind, also there is a lower *impact threshold* of motion, at a velocity less than the fluid threshold, caused by other sand grains already in motion striking particles at rest. If smaller

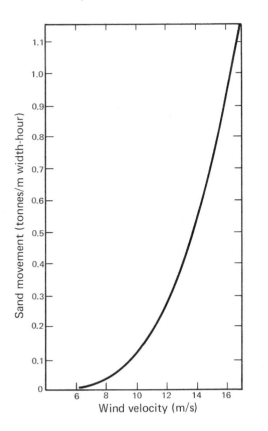

FIGURE 13-13. Relation between the flow of average dune sand and the wind velocity as measured at a standard height of one meter (Bagnold, 1941, Figure 22).

in grains six times the diameter of saltating grains. However, creeping grains move only a few millimeters with each impact, whereas saltating grains have characteristic trajectories a meter or more in length. Here is a potent method of size sorting, in which the size fraction moving by saltation greatly outdistances coarser grains that move only by surface creep. During strong winds over poorly sorted alluvium, the finest fraction is completely exported from the region as dust; a selected size fraction (perhaps including three-fourths of the total volume moved) is driven downwind by saltation; a comparatively small amount of coarse sand (perhaps one-fourth of the total sand volume) is driven downwind by creep; and a gravel lag is left behind. Wind passing from a sheet of moving sand onto a barren rock surface is no longer slowed by the drag of loose, saltating sand. The abrupt increase in surface wind velocity sweeps the hard surface clean, but when any swept-up grains again pass onto a sand surface, the wind velocity is checked, particle momentum is

absorbed by impacted grains, and deposition results. Deserts thus consist of various combinations of bare rock *hammada,* gravel or stony *reg* (or *gobi*), and sandy *erg* (p. 300). The boundaries between these various terranes are abrupt.

Geomorphic Significance of Eolian Processes

As much as 0.7 percent of the solar energy received at the earth's surface is converted into thermal and density contrasts in the atmosphere, which produce winds (Figure 2-2). Considering the enormous amount of solar energy available for geomorphic processes, eolian processes and landforms should have a significant place in geomorphology. Yet the direct effectiveness of eolian processes is rated quite low (p. 198), less than rivers, glaciers, and wind-driven waves. The present general opinion of the minor role of eolian geomorphic processes is especially surprising in view of the very substantial proportion of earth's land area that is arid and semiarid, where vegetation is restricted in its ability to cover protectively the weathered detritus at the ground surface.

Wind is considered to be a comparatively minor agent of geomorphic change primarily because of the low density of air as compared to rock and water. The relatively low ratio of the densities of immersed rock and water [(2.65 − 1.0)/1.0 ≃ 1.6:1] means that most of the fluvial sediment load can be carried in suspension, buoyed by the turbulence of the moving and often muddy water. However, rock fragments are 2000 times heavier than the air they displace and require an air velocity for their transport that is 29 times greater than the velocity of water required to transport the same-size particle (Pye and Tsoar, 1990, p. 103). Only very fine dust particles are moved in suspension except at extremely high wind velocities. Also because of its low density, moving air exerts a relatively low pressure against obstructions, whereas water flowing downhill can accomplish significant erosion by simple hydraulic force against submerged rock ledges.

Eolian processes and landforms are most obvious in arid regions, but there are many other climatic regions where sediment supply and exposure to winds also permit eolian landforms to develop. For example, belts of coastal dunes are on sandy coastlines in all climatic regions although their frequency is less in the humid tropics (p. 297). The braided channels made by glacial meltwater are notable sources for windblown dust and sand owing to their variable discharges and their extensive nonvegetated exposed bars and abandoned channels. The massive sheets of loess (p. 301) downwind from rivers in glaciated areas attest to the effectiveness of meltwater streams to provide a renewable supply of fine-grained alluvium for subsequent wind transport. Dry periglacial regions are also susceptible to eolian processes (Figure 14-8). Fine-grained soils in any climatic region are likely to have minor eolian landforms such as blowouts and dunes, especially if they have been disturbed by inappropriate agricultural practices.

Wind Erosion

Wind erosion occurs either by **deflation,** the process of removing loose sand and dust, or by **abrasion,** using the entrained sand grains as tools against rock surfaces or other grains (Laity, 1994, p. 506). The widespread industrial use of sandblasting by an air jet with entrained fine sand illustrates the potent abrasive capacity of windblown sand (Suzuki and Takahashi, 1981). It has been shown experimentally that abrasion by impact, in contrast to abrasion by rubbing, can be accomplished by particles that are softer than the surface they strike. Even ice crystals at $-10°C$ to $-25°C$, with a hardness of between 2 and 3.5 on the Mohs scale of hardness used by geologists, have been observed to break edges and corners from cleavage rhombs of minerals with hardness of 3 to 5.5 (Dietrich, 1977). Rock surfaces are fluted, grooved, pitted, or polished within decades after exposure (Hickox, 1959; Whitney and Dietrich, 1973, p. 2571). Sand-size angular quartz grains become rapidly rounded by impact with each other in experimental apparatus and produce large quantities of silt-size fragments in the process. Further, salt weathering has been shown to be very effective, in combination with wetting and drying, in breaking down sand-size quartz grains to silt sizes (p. 302). The implication is that significant amounts of dust can be derived from sand by abrasion even though desert sand is notable for its excellent sorting and general lack of silt and clay particles. The finer fractions are transported far from their sources in sandy deserts.

Cobbles and pebbles that show wind abrasion are called **ventifacts** (Laity, 1995) (Figure 13-14). If undisturbed, a ventifact develops a facet facing the effective wind direction and sloping down to windward. Most ventifacts have multiple facets that intersect along sharp keel-like ridges, testifying to more than one direction of effective abrasion. Large wind-eroded boulders and talus blocks, unlikely to have moved, may be used to infer one or more effective wind directions. Smaller cobbles and pebbles are likely to be moved by a great variety of processes, so their facets have little significance for determining the directions of

effective winds (Sharp, 1949). Bricks and blocks of easily abraded gypsum cement placed on a plot in a California desert were undermined by creep and wind scour and tilted, rotated, and overturned, sometimes within a few months (Sharp, 1980). The detailed fluting, pitting, and rill marks on some ventifacts have been attributed to abrasion by dust rather than by sand (Whitney and Dietrich, 1973).

Undercut, mushroom-shaped pedestal rocks in desert areas are commonly attributed to wind abrasion. However, experiments (Figure 13-13) prove that the median height of windblown sand transport is only about 1 cm over a layer of similar-size grains and little more than 10 cm over hard-rock surfaces. A variety of additional experiments and field observations agree that there should be a zone of maximum abrasion between 10 and 40 cm above ground level, with little abrasion above a height of 2 m (Cooke et al., 1993, p. 292). Most authorities now doubt whether wind abrasion could cut the notches or overhangs that are 1 to 2 m in height on desert cliffs or large boulders. More likely, weathering processes (p. 125) involving exfoliation or spalling on sheltered surfaces are the cause of the undercut surfaces. Some contain rock paintings or are encrusted with gypsum crystals, neither of which could survive if abrasion were an active process.

Minor eolian abrasional landforms include the chutes or flumes that notch low escarpments around playas and other desert plains. They are aligned with the locally effective winds and are probably scoured by windborne sand being swept through small gullies. They are usually cut in weakly indurated sediments and are separated by tonguelike or streamlined teardrop-shaped, wind-abraded ridges called **yardangs** (Blackwelder, 1934; Ward and Greeley, 1984; Laity, 1994, pp. 523–531). Deflation of weakly indurated sediment is at least as important as is abrasion in shaping yardangs (McCauley et al., 1977). The shape of yardangs probably originates from initial surface irregularities such as gullies and their interfluves at a scarp edge of a deflation basin, and evolve toward relatively stable streamlined bodies with a width-length ratio of 1/4 and a tapered downwind end, considered to be the shape of least resistance of a solid body immersed in a moving fluid (Ward and Greeley, 1984, p. 832). However, yardangs are dynamic, small landforms that are probably removed as fast as new ones form by differential erosion.

On the high, cold, dry Altiplano Plateau of the central Andes in Argentina, Chile, and Bolivia, volcanic ignimbrite sheets that cover thousands of square kilometers are systematically eroded by southeast-trending parallel *quebradas* (gullies) that are aligned with the

FIGURE 13-14. Ventifacts on lag gravel between lake-shore dunes, Sleeping Bear, Leelanau County, Michigan. North is at the bottom. P = prevailing winds; H = high-velocity winds; unlabeled arrow at bottom indicates other effective winds (photo: M. I. Whitney).

powerful prevailing winds from the northwest (Color Plate 9, Figure 13-15). They are probably eroded in part by fluvial action along cooling joints in the ignimbrite plateaus, with wind abrasion preferentially elongating and extending those open joints that are aligned with the wind, and perhaps with drifting sand filling those joints that trend obliquely to the wind. Aligned quebradas in one area in Chile are typically spaced 30 to 200 m apart and range in depth from a few meters to 30 m (Guest, 1969). They may be several kilometers in length and commonly trend oblique to the local slope. They are a characteristic feature of many ignimbrite sheets of the Altiplano and are a useful criterion for

identifying areas of ignimbrite on space images (Fielding et al., 1986).

Deflation hollows of various sizes are known to form quickly by wind erosion on exposed erodible terranes. When the vegetation on stabilized dunes is destroyed by fire or overgrazing, *blowouts* develop within months. When excavated below root depth, a blowout can grow as deep as the effective winds can lift sand (Figure 13-16). Dry creep keeps the adjacent slopes at the appropriate angle of repose. The limiting depth for blowouts is usually the water table because moist sand resists deflation, and plants may grow on the floor of the basin as well.

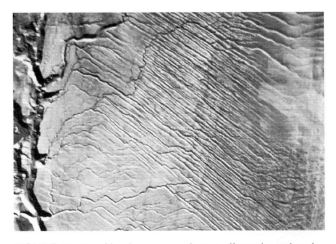

FIGURE 13-15. Northwest-southeast aligned *quebradas* (arroyos) in an ignimbrite sheet, northeastern Chile (Landsat Thematic Mapper 4, image y5050614004X0, approx. 15 km wide).

Over the more arid sections of the American Great Plains, thousands of small basins mark sandy districts. They were termed "buffalo wallows" by early explorers because animals frequented the ponds found in them. Although primarily the result of deflation during dry years, buffalo and domestic cattle surely aid deflation by loosening the sun-dried soil in the floors of the blowouts as they seek water. After being farmed for decades, some of the Great Plains blowouts were observed to deflate actively during the dry, fallow years of the mid-1930s.

Larger closed basins in arid regions, some of them many kilometers in diameter and more than 100 m in depth, have variously been attributed to solution, differential compaction, faulting, and deflation. Tectonic causes can usually be affirmed or eliminated by geologic mapping or drilling, but the other possibilities are difficult to separate. Big Hollow, near Laramie, Wyoming, is about 5 km wide, 15 km long, and 100 m deep. It is considered an inactive blowout, developed on fine-grained Mesozoic sedimentary rocks. The enormous Qattara Depression (Figure 13-17) in the western Egyptian Sahara is estimated to have an excavated volume of 3200 km^3 (Peel, 1966, p. 21). Its floor reaches 134 m below sea level, less than 100 km inland from the Mediterranean coast. Annual precipitation ranges from 25 to 50 mm on the northern rim to less than 25 mm in the western part of the basin although torrential winter rains have been reported (Albritton et al., 1990). The Qattara Depression may have originated as a Miocene river valley that was abandoned and segmented by deep karst solution during the late Miocene "Messenian crisis" when the Mediterranean basin became a desert at least 2 km below sea level (p. 59). A cavern in Eocene limestone was penetrated by a water well 228 to 246 m below the western floor of the depression, strengthening the evidence that paleokarst processes may have helped deepen the basin (Albritton et al., 1990, p. 959). Structural control is also a possibility (Gindy, 1991). When the Mediterranean basin refilled in Pliocene time, the sea again transgressed over western Egypt, depositing thin limestone. Subsequently, the Plio-Pleistocene and Miocene limestone caprock was breached, and a combination of mass wasting and slope retreat on easily eroded mudstones that underlie the surrounding limestone escarpments and wind

FIGURE 13-16. Active blowout in coastal dunes on Cape Cod, Massachusetts. Vegetated blowout in background is at the water table.

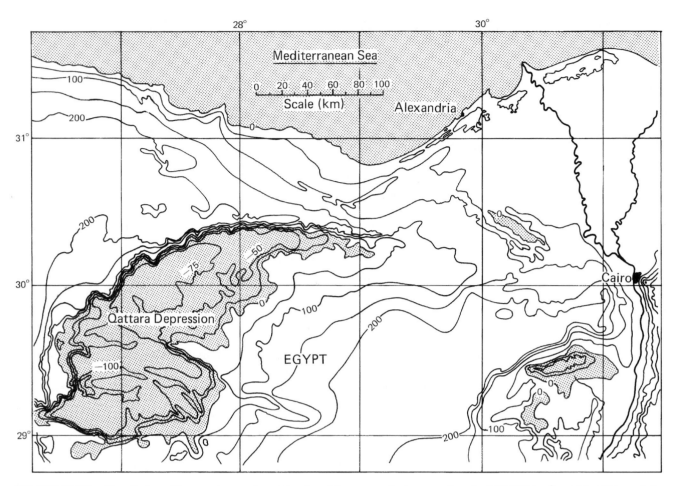

FIGURE 13-17.   The Qattara Depression, a desert basin with an area below sea level of 19,500 km², within 56 km of the Mediterranean coast. Contours in meters. The depression has been suggested as a site for hydroelectric power generation by siphoning water from the Mediterranean Sea and letting it fall through turbines to the floor of the basin, where it would evaporate (redrawn from Ball, 1933).

deflation of the resulting sediment from the basin floor have been the primary processes involved in deepening the basin (Ball, 1927, 1933). If the Qattara Depression is assumed to have formed by deflation in the last 2 million years, the eolian denudation rate is 9 cm/1000 yr, impressively comparable to reported rates of fluvial denudation (Chapter 15). The possible role of limestone solution and exhumed paleotopography in the origin of the basin await further research.

Numerous very large depressions in Pliocene continental sediments of the Mongolian Desert, the **Pang Kiang Hollows** described by Berkey and Morris (1927, pp. 336–337, 347), also are confidently attributed to a combination of Pleistocene deflation and arid scarp retreat. In a general review of desert landscapes, Peel (1966) urged renewed attention to the origin of large desert basins, suggesting that wind erosion, rather than tending to reduce the relief of desert landscapes, might actually increase it. Only the water table can limit the

depth of deflation if structures do not produce a gravel lag and wind strength is adequate. The water table might actually be lowered as basins develop (Murray, 1952). It is notable that erosional desert landscapes, unlike fluvial landscapes, do not evolve toward the ultimate base level of the sea. Although tectonic desert basins seem to be sediment traps (Figures 13-4 and 13-6), it is possible that some basins could deepen almost without limit by deflation alone.

Depositional Landforms of Eolian Sand

A patch of sand on a desert plain has a curiously definitive size limitation. Under gentle winds, barely strong enough to stir the sand grains, the patch is spread downwind and grows thinner or disperses. No new sand is delivered to the patch from upwind because the available sand there is sheltered by bedrock or gravel irregularities. Under strong winds, however, a preexisting sand patch larger than a critical size of a few meters grows in thickness and in size as its upwind border extends still farther upwind. The change in behavior is related to wind-transport dynamics described in preceding paragraphs. Strong winds create surface velocities over hard surfaces that can transport sand grains downwind. Over sand, however, the wind energy near the surface is absorbed by saltating sand, and the new sand derived from upwind is deposited in increasing amounts until an equilibrium is reached with the amount leaving the patch. For net accumulation to occur on an exposed sand patch, it must have a minimum initial downwind length of 4 to 6 m, several times the length of the saltation trajectory (Bagnold, 1941, p. 183; Lancaster, 1996).

A smooth sheet of sand is inherently unstable under gentle wind. The saltation method of sand movement results in most grains of a given size following trajectories of similar length. At the upwind edge of a sand patch, where transport first begins, a lag of coarser grains is left behind. The saltating grains moving downwind from that lag area impact at the mean distance of saltation and produce another lag area. By this process, **wind ripples** develop, transverse to the wind direction. Their typical wavelength is 1 m or less, very similar to the characteristic flight length of saltating grains, but probably more influenced by statistical fluctuations in the scatter of the impacted grains (Lancaster, 1994, pp. 475ff). Sand ripples are characterized by a lag concentrate of coarse grains at their crests where the fine grains have been removed, and finer grains in their troughs. Because of the rather low angle of impact of saltating grains, each ripple crest receives more than the average number of saltation impacts, and creates a shadow zone of fewer impacts in its leeside trough.

Under stronger winds, the surface of a sand patch loses its rippled form because less size sorting takes place. Then the entire surface layer is sheared forward by saltation and surface creep. However, under such conditions, new sand is also trapped from upwind sources, and the sand sheet is not eliminated but increases in area and thickness as long as an upwind sand supply is available.

Large accumulation landforms of windblown sand are called **dunes**. The dynamics just described make a sharp distinction between ephemeral ripples on a sand surface, which form under gentle winds, and dunes, which are distinctly larger and stable or growing landforms. Even larger-scale complex and compound dunes are called **draas** (Wilson, 1972; Lancaster, 1994). To survive and grow, dunes must be at least 4 to 6 m in downwind length and are commonly much larger. It is common for the sand on dunes to be rippled, just as large

swells on the ocean surface have small wind waves on their surfaces in turn. Although ripples have the coarsest sand on their crests, on dunes the coarse grains creep or roll to the troughs and the finest sand remains on the crests (Lancaster, 1989, p. 75; 1995, p. 106).

Bagnold (1941, p. 188) classified sand accumulations under five headings: sand shadows and sand drifts, dunes, whalebacks or zibar dunes, undulating fixed sand sheets, and sand seas or ergs. That classification is followed here, with modifications and additions noted by later authors (Lancaster, 1994; 1995).

Sand Shadows and Sand Drifts. Wherever wind velocity is checked by a fixed obstruction, sand may strike the obstruction and then fall at its windward base or be swept into the lee of the obstruction and accumulate as a streamlined mound. Both landforms are classed as **sand shadows.** They are fixed in form and size by the size and aerodynamic shape of the obstruction. For instance, a small tree raised above the ground on a narrow trunk will not accumulate a shadow, whereas a tuft of grass or a ground-hugging shrub may produce a substantial lee shadow. Some vegetation will trap sand and grow upward within the accumulating mound, forming a crude hemispherical mound of sand held in a mesh of roots.

Sand drifts are the accumulations in the lee of a local high-velocity zone between obstacles. Most commonly they accumulate downwind from a place where wind is funneled through a gap at a higher-than-average velocity. Another important site of sand drifts is at the foot of a downwind cliff or escarpment. Sand blown off the plateau surface sinks into the still air in the lee of the cliff. When such sand drifts grow large enough, they begin to trail off downwind from the base of the cliff. On images of the Martian surface, and on photos and electronic images of terrestrial deserts, sand shadows and drifts are collectively called *wind streaks.*

Dunes. True dunes can best be regarded as deformable obstructions to airflow. They are free to move, divide, grow, and shrink, but they are not dependent on a fixed obstruction for their maintenance. Two principles that have been previously emphasized are critical to dune formation. First is the principle that a free sand patch larger than 4 to 6 m is likely to increase in size and thickness and thus become a mound or hill. The second principle is that, like a cliff, a dune develops a lee-side pocket of low-velocity air that promotes deposition there. As soon as a mound of sand grows to moderate size, it develops the classic dune asymmetry of a gentle windward slope and a steep, angle-of-repose, *slip face* to leeward (Figures 13-18 and 13-19). The sand on the windward slope is firm and compact, but as it blows

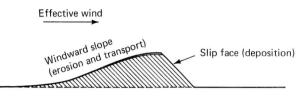

FIGURE 13-18. Cross section of an active dune. Much of the slip-face deposition finally results from sand flow and slumping.

over the crest and cascades down the slip face, it may become loose enough to make walking difficult. Note that the internal structure of a dune is fixed by the original deposition and slight compaction; whereas the dune *form* may move with or without changing shape, the internal *structure* is immobile. Eroded edges of slip-face bedding are commonly exposed on windward slopes (Figure 13-19). It is less common for gently dipping sheetlike accumulations of windward-slope deposits to be preserved in dune structures.

Dunes have a wide range of forms, some of which were unknown until aerial photography and space images became available for large areas of hitherto unexplored sand regions in the Sahara and Arabia (Figure 13-20).

Star dunes may be several hundred meters in height and several kilometers in diameter, with radiating sinuous ridges culminating at a common crest. They accumulate under effective winds that blow from

FIGURE 13-19. Dipping strata of an earlier dune cycle truncated by cross winds. Barchan dunes migrated from left to right. White Sands, New Mexico (photo: Tad Nichols).

several different directions (Breed and Grow, 1979, p. 275; Lancaster, 1994, pp. 481–482; 1995, pp. 71–76). At various seasons, they may show slip faces on various sides of their multiple radiating arms. They seem to grow in height rather than migrate, as though they are at some focus of depositional winds.

Longitudinal or **linear dunes** may be huge landforms extending hundreds of kilometers in length, a kilometer or more in width, and several hundred meters in height. They are especially well developed in the heart of the Trade Wind deserts, where the wind is either from a constant direction or varies seasonally. Longitudinal dunes cover approximately 30 percent of the total area of eolian deposition (Tsoar and Møller, 1986). Their great size and abundance, and the fact that many of them are now relict, have lead to much speculation about their origin. Bagnold (1941) suggested that they were built by seasonal shifting winds that preferentially elongated one arm of a large barchan dune until it became a wavy, elongate longitudinal dune. Livingstone (1986) and Tsoar (1989) demonstrated that longitudinal dunes extend parallel to the dominant wind direction, but that other effective winds, at an angle to the dominant flow, change direction as they cross the dune crest, to blow nearly parallel to the dune crest. These winds may either erode or deposit, because of velocity changes, but always with a longitudinal transport component. Great systems of longitudinal dunes cross the arid interiors of Australia and South Africa (Figure 13-21), but most are now covered by sparse vegetation and soil and seem to be relict from one or more earlier periods of greater aridity. Active longitudinal dunes are found in areas nearly devoid of vegetation, where winds are persistent for many months and sand supply is irregular. The dune ridges are separated by gravel-armored reg or hammada.

Transverse dunes (Figure 13-22) are associated with massive quantities of sand and relatively ineffective winds. They are common along coastlines (Cooper, 1958, 1967; Inman et al., 1966) and on zones of erodible sandy bedrock or alluvium. Although their crests are ridges roughly perpendicular to the effective wind, in detail the crests have sinuous or "fish-scale" outlines or break up into fields of isolated barchan or longitudinal dunes. The *barchanoid ridge* is a common transitional type (Figure 13-20).

Barchan dunes (Figure 13-22) are classic eolian landforms. Their crescentic form consists of a gently inclined windward slope and a steep lee side around which the horns or cusps of the dune project downwind, making the slip face concave to the downwind direction. Barchans are isolated dunes that migrate freely across rock or gravel desert plains, usually downwind from some other dune form. Although they

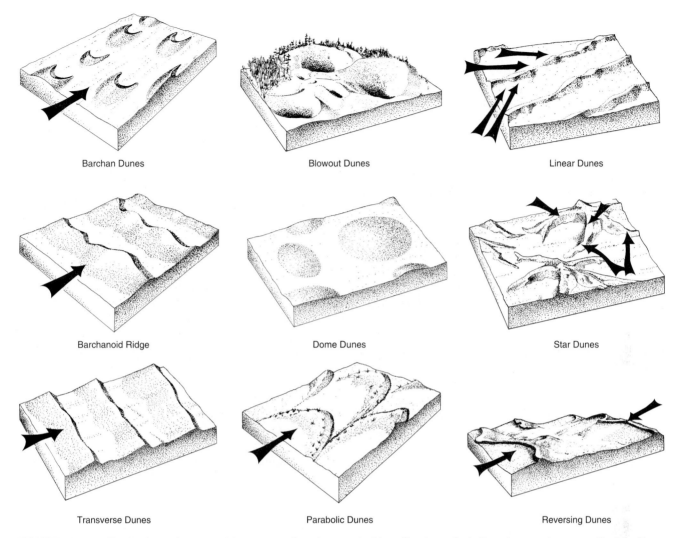

Barchan Dunes

Blowout Dunes

Linear Dunes

Barchanoid Ridge

Dome Dunes

Star Dunes

Transverse Dunes

Parabolic Dunes

Reversing Dunes

FIGURE 13-20. Basic dune forms, with arrows showing probable effective wind directions, where applicable (from McKee, 1979, pp. 11–13). Sizes range from 10 m to more than 100 km (see text).

have a nearly constant sand mass, they are large enough to maintain themselves as they migrate and represent a remarkable balance between accumulation, transportation, and erosion. Barchans in the Salton Sea basin in California have been photographed and measured repeatedly over the past 50 years as they migrate eastward at 15 to 25 m/yr (Haff and Presti, 1995).

Parabolic dunes and irregular **blowout dunes** are characteristic of partially stabilized sandy terranes that develop blowouts (Figures 13-16 and 13-23). The crescentic shape of parabolic dunes has a very different orientation from that of barchan dunes, likely to confuse a novice. On the lee side of an elongate blowout a dune may accumulate, its windward slope rising from the blowout and the slip face *convex* downwind in plan view. A parabolic dune may become elongated (a "hairpin" dune) by downwind migration of a blowout until

it is split into two minor longitudinal ridges paralleling the long axis of the blowout. These dunes are usually much longer and narrower than barchan dunes and are always associated with a central blowout. They should not be confused with barchans, especially on aerial photographs, because the resulting interpretation of effective wind direction will be reversed.

Sand dunes on coasts, except in deserts, commonly are parabolic and blowout dunes because, even though the shorezone may provide a copious supply of sand, and onshore winds may be an effective transport medium, the water table is likely to be near the surface, which limits the depth to which deflation can work. Further, coastal vegetation will quickly colonize and stabilize dunes as they migrate inland. Kocurek and Havholm (1993) defined *dry, wet,* and *stabilized* eolian systems, which are distinguished by the relative role of

FIGURE 13-21. Dunes of the northern Namib Sand Sea, Namibia. Coastal ocean currents transport sand northward, where it is blown inland and northward. Dune patterns include barchans and transverse dunes near the coast, and complex linear (longitudinal) dunes that break into fields of star dunes near ephemeral river channels in the east (Landsat MSS 40394-08264: courtesy N. Lancaster).

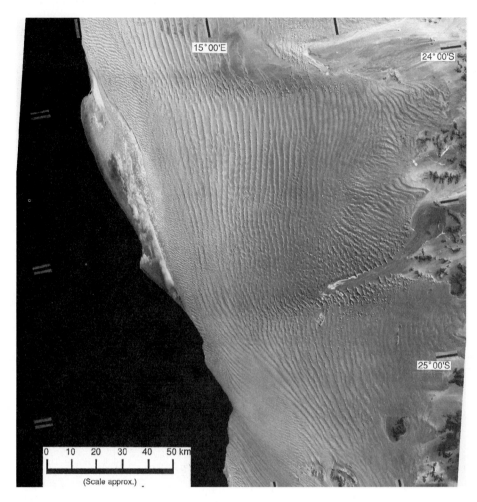

FIGURE 13-22. Transverse or barchanoid coastal dunes breaking inland into barchans. Guerro Negro dune field, Baja California, Mexico. View northward, Laguna Manuela in background. Dunes in foreground are about 6 m high (photo: D. L. Inman).

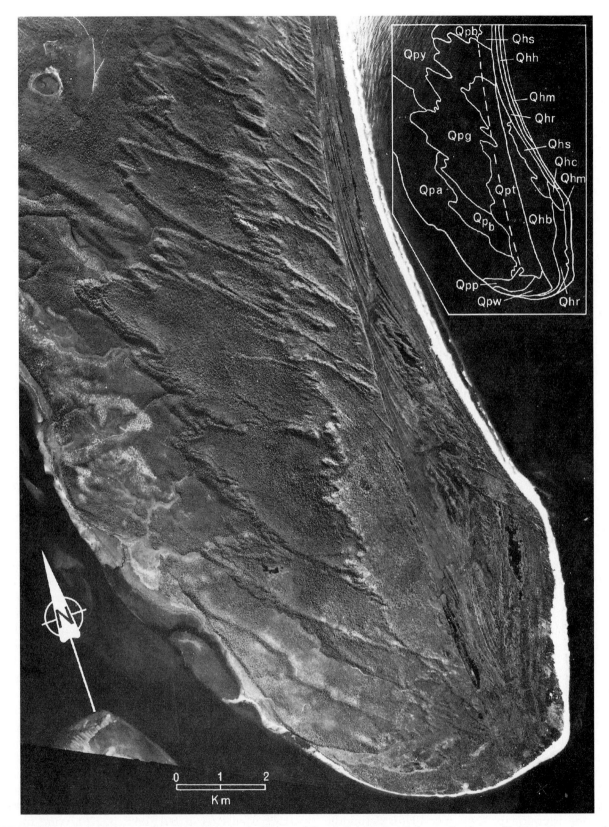

FIGURE 13-23.   Parabolic dunes at the south end of Fraser Island, Queensland, Australia. Large Pleistocene dunes (symbols Qp), stabilized by vegetation, were blown inland by the southeasterly Trade Wind, and are separated from younger Holocene dunes (symbols Qh) and beach ridges by a prominent former shoreline. A buried older shoreline is indicated by the dashed line. (Ward, 1977, Plate I: Crown Copyright, Australian Department of Minerals and Energy, 1977.)

the water table and surface stabilizing factors such as vegetation in controlling sand movement. Dry systems have no water table or stabilization limitations; in wet systems, erosion is limited downward by the water table or a capillary fringe that traps sand blowing over the surface; in stabilized systems, vegetation, cementation, or mud coatings prevent the free movement of sand. Coastal sand dunes are usually constrained to the wet or stabilized eolian systems. Compare the Mexican coastal dunes in a hyperarid climate (Figure 13-22), which evolve inland into barchans or barchanoid ridges migrating inland across a hard, salt-cemented but barren plain, with eastern Australian parabolic dunes (Figure 13-23), which are largely vegetated but intermittently active, stabilized by both the water table and vegetation (Ward and Grimes, 1987).

Coastal sand dunes commonly include a large proportion of biogenetic calcareous sand that readily recements as *eolianite* (Pye and Tsoar, 1990, pp. 274–284). Most of the surface relief of the island of Bermuda is of eolianite dunes. It is common on many other biologically productive coasts as well. When freshly quarried, eolianite can be soft enough to be cut with hand tools and makes an excellent building material when it hardens on exposure.

Whalebacks and Zibar Dunes. Where longitudinal dunes have migrated downwind, the coarser sand is left behind as a prominent low, rolling ridge of coarse sand, lacking slip faces. These **whalebacks** or **zibar dunes** (Pye and Tsoar, 1990, p. 215) attest to the stability of longitudinal dunes in form and spacing over long periods. Either very large longitudinal dunes migrated a long distance downwind or longitudinal dunes repeatedly followed the same tracks across the desert in order to leave these large residual ridges. Either case implies former wind effectiveness, sand supply, or both, far in excess of present conditions. The whalebacks described by Bagnold (1941, p. 230) in the Egyptian desert are 1 to 3 km in width and 50 m high and extend for 300 km. The modern longitudinal dunes that may surmount these great ridges are much smaller and less continuous. Sometimes as many as four chains of modern parallel longitudinal dunes surmount a single whaleback (for similar Arabian forms, see Holm, 1960). These forms represent one of the more intriguing aspects of the effect of climatic change on eolian landforms.

Undulating Fixed Sand Sheets. Large areas of semiarid steppe or prairie (p. 279), especially in the middle latitude interiors of the large continents, are underlain by sandy sediments, have a sparse grass cover, and have intermittent rain or winter snow. As the result of either drought or fire, local areas of shallow deflation basins develop, alternating with subdued parabolic dune forms (Madole, 1995; Muhs et al., 1996). Winds are not sufficiently effective to create dunes with active slip faces, but the combination of deflation and deposition makes a landscape of undulating hills, similar in appearance to the top of a cloud cover as viewed from an airplane (Figure 13-24). During wet years or seasons, ponds accumulate in hollows between the hills. High permeability and low precipitation prevent stream dissection even when the water table is near the surface. The Sand Hills region of Nebraska (Ahlbrandt and Fryberger, 1980) is a typical example. Large areas of the Russian steppe, the Argentine pampa, and Iran are also of this origin. As a transitional landscape between arid and subhumid climates, these large regions can be expected to show many relict landforms of conditions both more moist and more arid than now experienced. Buried soil profiles, sand-filled drainage systems (Loope et al., 1995), polycyclic drainage networks, and buried sites of human occupancy attest to the mobility of sandy terranes under marginal precipitation.

Sand Seas or Ergs. In contrast to the undulating fixed sand sheets of grassy steppes, from 20 to 45 percent of arid regions of the old world consists of vast "sand seas" or *ergs* (Lancaster, 1994). No large sand seas are in North or South America. Their surfaces consist of a variety of dune forms, ranging from low mounds to huge star dunes. These largest of all the sandy landscapes are fixed primarily by lithologic factors. Sand is thick and continuous enough for the surface to include the full size range of eolian landforms: ripples, dunes, and draas (defined by Wilson, 1973, as ridges about 1 to 3 km apart). There may be barren rock areas and reg; Wilson (1973, p. 78) suggested a preliminary definition of an **erg** as "an area where wind-laid sand deposits cover at least 20 percent of the ground, and which is large enough to contain draas." The minimum size ranges from 1 to 40 km^2. At least 30 named ergs exceed 120,000 km^2 (Wilson, 1973, Table 1). Those in Africa, Arabia, and Australia are on or downwind from easily weathered, sandy sedimentary rocks, probably deposited in Cenozoic or Mesozoic marine embayments or freshwater lakes. They are more nearly "timeless" residuals than evolving landforms, either constructional or erosional, although sand is known to both enter and leave even the largest ergs (Lancaster, 1989; 1995). Some sand seas show evidence of Quaternary episodic activity induced by climatic and sea-level fluctuations, and tectonism (Lancaster, 1994, p. 501; 1995, p. 226). Wind transport and deposition constantly shift the surface layers, but the regional landscape is one of awesome monotony (Figure 13-25).

FIGURE 13-24. The Sand Hills region of Nebraska. Stabilized dunes along Nebraska Highway 97, south of the Dismal River (photo: W. J. Wayne).

Dust and Loess

Loose silt- and clay-size particles ("dust") can be extremely resistant to wind shear and entrainment (p. 288). However, if stirred into suspension by vehicles, animals, impacting sand grains, or vortical winds ("dust devils," "whirlwinds"), dust can be lifted hundreds or thousands of meters into the air and carried long distances (Middleton, 1989). Dust that has been transported for distances of less than 100 km has grain sizes in the silt range (approximately 5 to 50 μm). Finer dust stays in suspension as an aerosol and is precipitated only by rainfall; it is mostly clay size (2 to 10 μm) (Péwé et al., 1981, p. 3). It can travel around the earth on high-altitude winds. Fine windblown dust is the major component of deep-sea sediments; the Sahara is the major source for the equatorial Atlantic Ocean (deMenocal, 1995), and central Asia contributes most of the dust to the North Pacific (Schneider et al., 1990). Dust from central Asia is seasonally collected in Hawaii (Whalley et al., 1987; Schneider et al., 1990), and Saharan dust is a major component of the soils on lime-

FIGURE 13-25. Dunes in the Rub' al Khali, an erg covering 600,000 km² of southeastern Saudi Arabia (photo: Arabian American Oil Company).

stones in the Caribbean region (Muhs et al., 1990). During Pleistocene cold intervals, dust was more abundant in the atmosphere (Chapter 18), reaching three to five times more abundant than in interglacial times in the northern hemisphere (Rea, 1994; Yung et al., 1996).

In the arid southwestern United States, dust is the major source of soil-forming parent material (Reheis et al., 1995). Most of the fine-grained sediment on alluvial fans and reg plains is wind-blown, as demonstrated by the content of CaCO3 and salts in the fine fraction of the soil that is greatly in excess of that which could be derived from the substrate, as well as by other mineralogic criteria. As noted in the discussion of desert soils (p. 143), calcic horizons in desert soils are now attributed largely to calcareous dust rather than to *in situ* weathering. Current rates of dust deposition as measured by sediment traps at a variety of sites in California and Nevada are more than adequate to account for the silt and clay in the modern soils of the region; the modern rates are probably excessive because of human activity. Dust deposition rates in the American Great Basin were probably low during the glacial intervals because many desert basins contained lakes and were not dust sources. Further, vegetation cover and downward translocation of clay and carbonate in the Great Basin were more effective during cooler and wetter late Pleistocene time (Chapter 18). Dust deposition was rapid during the early Holocene as desert playas became exposed by desiccation (Figure 18-10).

Thick deposits of homogeneous unstratified silt are draped over or bury other landforms in the middle latitudes. The extremely well-sorted, fine-grained sediment is generally known by the German term **loess,** meaning simply "loose," or unconsolidated. Loess and similar deposits form a cover 1 to 100 m in thickness over an estimated 10 percent of the earth's land area (Pécsi, 1968) and 17 percent of the United States (Higgins and Modeer, 1996, p. 585). Loess has an eolian origin, much of it derived from glacial outwash and other alluvium (Figure 14-8) or from deserts, although minor water transport, mass wasting, and pedogenic processes are accepted as part of the overall process of loess accumulation (Pye, 1987, pp. 198–200; 1995). Deposits similar to loess but showing evidence of downslope or water retransport are often referred to as "loesslike" or "secondary loess." The extremely good size sorting during long-distance transport of windblown silt accounts for most of the peculiar and significant properties of loess. It is a material, not a kind of landscape, but where thick enough to bury the underlying landscape, its surface deserves consideration as a constructional landform. Furthermore, the soil-forming, slope stability, and erosional properties of loess terranes also give them special geomorphic interest. The engineer-

ing properties of loess are well reviewed by Higgins and Modeer (1996).

Most of the world's loess is dominantly silt-size quartz (mostly 20 to 60 μm; Smalley and Smalley, 1983), but reflects the lithology of the source terranes and may be highly calcareous. The abundance of quartz indicates a source-area weathering environment in which quartz was stable but minerals such as felspars, mica, and ferromagnesian minerals were chemically weathered. Because detrital quartz is usually sand-size, the abundance of quartz in loess is puzzling (Smalley, 1995). Glacial abrasion (Chapter 17) to silt size is clearly indicated by the angular, fractured grains of quartz in the loess of formerly glaciated areas. Experimental evidence of eolian abrasion and salt weathering of quartz to silt-size fragments seem to have resolved some former concerns about the origin of the silt-size quartz in nonglacial loess (Whalley et al., 1987; Pye, 1987, p. 18).

A large region of loess and loesslike sediments is in north central China, where they cover at least 635,000 km^2 (Liu, 1988, p. 13). The Chinese loess is probably not glacially derived. It becomes coarser toward the deserts of northwestern China and grades northward into sandy loess and the sand in the Gobi desert of Mongolia (Derbyshire, 1983; Pye, 1987, pp. 240–242). It locally exceeds 400 m in thickness (Zhang et al., 1991, Preface). Loess deposition began at about 2.4 Ma, according to paleomagnetic correlations and other dating techniques (Kukla et al., 1988; Liu, 1988, p. 52). Deposition continues today at a rate of several millimeters per year (Derbyshire, 1983). It has numerous soil profiles developed within it, and its history is intimately involved with late Cenozoic climate change (Chapters 4 and 18).

In detailed studies, loess sheets in North America have been shown to consist of multiple depositional units, generally correlative with glacial or stadial conditions and separated by unconformities with soil profiles representing interglacial or interstadial (p. 404) conditions (Forman et al., 1992; Maat and Johnson, 1996). Loess deposits thin exponentially southeastward away from source areas, usually outwash plains (Winspear and Pye, 1995) (Figure 13-26).

Buried soils demonstrate that deposition was intermittent, probably controlled by the availability of a renewable source such as the braided floodplain of a glacial meltwater river. Fossil snail shells in the Mississippi Valley loess are of species that inhabit moist woodlands. Probably the thick loess along the downwind sides of glacial outwash rivers was trapped by trees growing adjacent to the floodplains. Grass is also an effective dust trap, from which periodic rains wash the accumulated dust down to the soil surface. Winter dust blowing from barren fallow fields commonly discolors

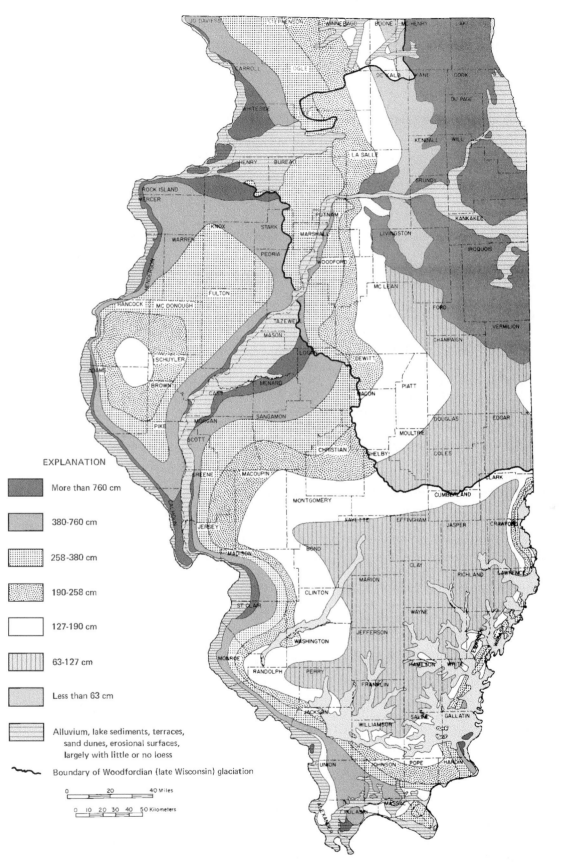

FIGURE 13-26. Loess thickness in Illinois (compare with Figure 14-8) (from Willman and Frye, 1970, Plate 3).

the snow downwind of the source areas. When the snow melts, the resulting loess would show no stratification, even though it was a seasonal accumulation.

The loess of the Argentine Pampas is derived from weathered volcanic rocks and contains a high percentage of pyroclastic glass shards and plagioclase indicative of an Andean source (Teruggi, 1957; Sayago, 1995). Its origin and age must be related to the chronology of eolian deflation in the high volcanic regions of the Andes (Figure 13-15). Deposition began in the late Pliocene and continues today.

The region along the southern periphery of the European ice sheets and across central Asia are historically famous for their loess soils. In a broad band from western France across southern Germany, Austria, Hungary, Poland, Rumania, Bulgaria, Ukrania, Russian, and the southern states of the former Soviet Union, loess forms the parent material for the soils that produce cereal grains and pastures. It is closely associated with periglacial phenomena (p. 406).

Because of its fineness, clay and carbonate content, and homogeneity, loess consolidates readily and forms stable steep cliffs, often with a "case-hardened" surface of secondary calcium carbonate. Most of the calcium carbonate in loess is derived from weathering in the source area, not from weathering or groundwater transport after deposition (p. 143). No more than 3 to 10 percent of the carbonate content of the Chinese loess is estimated to have been derived from local weathering by atmosphere CO_2 or by groundwater (Liu, 1988, p. 4).

Loess erodes readily in humid climates and forms a characteristic high-density network of gullies with steep or vertical banks controlled by the columnar soil structure typical of loess. Parabolic dunes of loesslike silt and clay (**lunettes**) accumulate on the lee sides of dry lake beds in southern Australia (Hills, 1940; Stephens and Crocker, 1946) and in Texas and New Mexico (Sabin and Holliday, 1995; Holliday, 1997). The loess sheets associated with the last glaciation are relatively undissected and unweathered, but beyond the drift border of the last ice sheets, older loesses have much more developed erosional landscapes. Quaternary loess sheets can be dated by stratigraphic correlation with tephras and glacial drift and by paleomagnetic correlation with other Quaternary deposits (Chapter 18).

REFERENCES

AHLBRANDT, T. S., and FRYBERGER, S. G., 1980, Eolian deposits in the Nebraska Sand Hills: U. S. Geol. Survey Prof. Paper 1120-A, 24 pp.

ALBRITTON, C. C., JR., BROOKS, J. E., and 2 others, 1990, Origin of the Qattara Depression, Egypt: Geol. Soc. America Bull., v. 102, pp. 952-960. (Reply: v. 103, pp. 1375–1376.)

BAGNOLD, R. A., 1941, The physics of blown sand and desert dunes: Methuen & Co. Ltd., London, 265 pp.

BALL, J., 1927, Problems of the Libyan desert: Geog. Jour., v. 70, pp. 21–38, 105–128, 209–222.

——, 1933, Qattara depression of the Libyan desert and the possibilities of its utilization for power-production: Geog. Jour., v. 82, pp. 289–314.

BERKEY, C. P., and MORRIS, F. K., 1927, Geology of Mongolia: Natural history of central Asia, v. 2, Am. Museum Nat. Hist., New York, 475 pp.

BIERMAN, P., and TURNER, J., 1995, [10]Be and [26]Al evidence for exceptionally low rates of Australian bedrock erosion and the likely existence of pre-Pleistocene landscapes: Quaternary Res., v. 44, pp. 378–382.

BLACKWELDER, E., 1934, Yardangs: Geol. Soc. America Bull., v. 45, pp. 159–166.

BORNHARDT, W., 1900, Zur Oberflächengestaltung und Geologie Deutsch-Ostafrikas: D. Reimer, Berlin, 595 pp.

BREED, C. S., and GROW, T., 1979, Morphology and distribution of dunes in sand seas observed by remote sensing, in McKee, E. D., ed., A study of global sand seas: U.S. Geol. Survey Prof. Paper 1052, pp. 253–302.

BRYAN, K., 1922, Erosion and sedimentation in the Papago country, Arizona: U.S. Geol. Survey Bull. 730-B, pp. 19–90.

BÜDEL, J., 1957, Die "doppelten Einebnungsflächen" in den feuchten Tropen: Zeitschr. für Geomorph., v. 1, pp. 201–228.

——, 1982, Climatic geomorphology (English translation by L. Fischer and D. Busche): Princeton University Press, NJ, 443 pp.

CLAYTON, R. W., 1956, Linear depressions (Bergfüssniederungen) in savanna landscapes: Geog. Studies, v. 3, pp. 102–126.

COOKE, R., WARREN, A., and GOUDIE, A., 1993, Desert geomorphology: UCL Press Limited, London, 526 pp.

COOPER, W. S., 1958, Coastal sand dunes of Oregon and Washington: Geol. Soc. America Mem. 72, 169 pp.

——, 1967, Coastal dunes of California: Geol. Soc. America Mem. 104, 131 pp.

DERBYSHIRE, E., 1983, Origin and characteristics of some Chinese loess at two locations in China, in Brookfield, M. E., and Ahlbrandt, T. S., eds., Eolian sediments and processes: Developments in sedimentology 38, Elsevier Science Publishers, Amsterdam, pp. 69–90.

DIETRICH, R. V., 1977, Impact abrasion of harder by softer materials: Jour. Geology, v. 85, pp. 242–246.

DIXON, J. C., 1994, Aridic soils, patterned ground, and desert pavements, in Abrahams, A. D., and Parsons, A. J., eds., Geomorphology of desert environments: Chapman & Hall, London, pp. 64–81.

DOHRENWEND, J. C., 1994, Pediments in arid environments, in Abrahams, A. D., and Parsons, A. J., eds., Geomorphology of desert environments: Chapman & Hall, London, pp. 321–353.

FIELDING, E. J., KNOX, W. J., JR., and BLOOM, A. L., 1986, SIR-B radar imaging of volcanic deposits in the Andes: IIEE Trans. on Geoscience and Remote Sensing, v. GE-24, pp. 582–589.

FORMAN, S. L., BETTIS, E. A., III, and 2 others, 1992, Chronologic evidence for multiple periods of loess deposition during the Late Pleistocene in the Missouri and Mississippi River Valley, United States: Implications for the activity of the Laurentide ice sheet: Palaeogeog., Palaeoclimatol., Palaeoecol., v. 93, pp. 71–83.

GILBERT, G. K., 1882, Contributions to the history of Lake Bonneville: U.S. Geol. Surv. 2nd Ann. Rept., pp. 167–200.

GINDY, A. R., 1991, Origin of the Qattara Depression, Egypt: Discussion: Geol. Soc. America Bull., v. 103, pp. 1374–1375.

GRAF, W. L., 1988, Fluvial processes in dryland rivers: Springer-Verlag, Berlin, 346 pp.

GREELEY, R., and IVERSON, J. D., 1985, Wind as a geological process on Earth, Mars, Venus and Titan: Cambridge Univ. Press, Cambridge, UK, 333 pp.

GUEST, J. E., 1969, Upper Tertiary ignimbrites in the Andean Cordillera of part of the Antofagasta Province, northern Chile: Geol. Soc. America Bull., v. 80, pp. 337–362.

HADLEY, R. F., 1967, Pediments and pediment-forming processes: Jour. Geol. Education, v. 15, pp. 83–89.

HAFF, P. K., and PRESTI, D. E., 1995, Barchan dunes of the Salton Sea region, California, in Tchakerian, V. P., ed., Desert aeolian processes: Chapman & Hall, London, pp. 153–177.

HAFF, P. K., and WERNER, B. T., 1996, Dynamical processes on desert pavements and the healing of surface disturbances: Quaternary Res., v. 45, p. 38–46.

HICKOX, C. F., JR., 1959, Formation of ventifacts in a moist, temperate climate: Geol. Soc. America Bull., v. 70, pp. 1489–1490.

HIGGINS, J. D., and MODEER, V. A. JR., 1996, Loess, in Turner, A. K., and Schuster, R. L., eds., Landslides: Investigation and mitigation: Transportation Research Boards, Natl. Acad. Sci., Natl. Res. Council Spec. Rept. 247, pp. 585–606.

HILLS, E. S., 1940, The lunette, a new landform of eolian origin: Aust. Geographer, v. 3, pp. 15–21.

HOLLIDAY, V. T., 1997, Origin and evolution of lunettes on the High Plains of Texas and New Mexico: Quaternary Res., v. 47, pp. 54–69.

HOLM, D. A., 1960, Desert geomorphology in the Arabian peninsula: Science, v. 132, pp. 1369–1379.

HOWARD, ALAN D., and SELBY, M. J., 1994, Rock slopes, in Abrahams, A. D., and Parsons, A. J., eds., Geomorphology of desert environments: Chapman & Hall, London, pp. 123–172

HOWARD, A. D., 1942, Pediment passes and the pediment problem: Jour. Geomorphology, v. 5, pp. 2–31, 95–136.

INMAN, D. L., EWING, G. C., and CORLISS, J. B., 1966, Coastal sand dunes of Guerrero Negro, Baja California, Mexico: Geol. Soc. America Bull., v. 77, pp. 787–802.

JOHNSON, D. W., 1932, Rock fans of arid regions: Am. Jour. Sci., v. 223, pp. 389–420.

KING, L. C., 1948, A theory of bornhardts: Geog. Jour., v. 112, pp. 83–87.

KOCUREK, G., and HAVHOLM, K. G. 1993, Eolian sequence stratigraphy-A conceptual framework, in Weimer, P., and Posamentier, H., eds., Siliclastic sequence stratigraphy: Recent developments and applications: Amer. Assoc. Petrol. Geol. Mem. 58, pp. 393–409.

KÖPPEN, W., and GEIGER, R., 1936, Handbuch der Klimatologie, v. 1, part C: Gebrüder Borntraeger, Berlin, 44 pp.

KUKLA, G., HELLER, F., and 4 others, 1988, Pleistocene climates in China dated by magnetic susceptibility: Geology, v. 16, pp. 811–814.

LAITY, J. E., 1994, Landforms of aeolian erosion, in Abrahams, A. D., and Parson, A. J., eds., Geomorphology of desert environments: Chapman & Hall, London, pp. 506–535.

———, 1995, Wind abrasion and ventifact formation in California, in Tchakerian, V.P., ed., Desert aeolian processes: Chapman & Hall, London, pp. 295–321.

LANCASTER, N., 1989, The Namib Sand Sea: Dune forms, processes and sediments: A. A. Balkema, Rotterdam, 180 pp.

———, 1994, Dune morphology and dynamics, in Abrahams, A. D., and Parsons, A. J., eds., Geomorphology of desert environments: Chapman & Hall, London, pp. 474–505.

———, 1995, Geomorphology of desert dunes: Routledge, London and New York, 290 pp.

———, 1996, Field studies of sand patch initiation processes on the northern margin of the Namib Sand Sea: Earth Surface Processes and Landforms, v. 21, pp. 947–954.

———, and NICKLING, W. G., 1994, Aeolian sediment transport, in Abrahams, A. D., and Parsons, A. J., eds., Geomorphology of desert environments: Chapman & Hall, London, pp. 447–473.

LAWSON, A. C., 1915, The epigene profiles of the desert: Calif. Univ. Pub. Geol. Sci., v. 9, pp. 23–48.

LIU, T. S., 1988, Loess in China, 2nd ed.: Springer-Verlag Berlin, 224 pp.

LIVINGSTONE, I., 1986, Geomorphological significance of wind flow patterns over a Namib linear dune, in Nickling, W. G., ed., Aeolian geomorphology: Allen & Unwin, Inc., Boston, pp. 97–112.

LOOPE, D. B., SWINHART, J. B., and MASON, J. P., 1995, Dune-dammed paleovalleys of the Nebraska Sand Hills: Intrinsic versus climatic controls on the accumulation of the lake and marsh sediments: Geol. Soc. America Bull., v. 107, pp. 396–406.

MAAT, P. B., and JOHNSON, W. D., 1996, Thermoluminescence and new ^{14}C age estimates for Late Quaternary loesses in southwestern Nebraska: Geomorphology, v. 17, pp. 115–128.

MADOLE, R. F., 1995, Spatial and temporal patterns of late Quaternary eolian deposition, eastern Colorado, U.S.A.: Quaternary Sci. Rev., v. 14, pp. 155–177.

MCCAULEY, J. F., GROLIER, M. J., and BREED, C. S., 1977, Yardangs, in Doehring, D. O., ed., Geomorphology in arid regions: Publications in Geomorphology, State University of New York in Binghamton, pp. 233–269.

McGEE, W J, 1897, Sheetflood erosion: Geol. Soc. America Bull., v. 8, pp. 87–112.

MCKEE, E. D., ed., 1979, A study of global sand seas: U.S. Geol. Survey Prof. Paper 1052, 429 pp.

MEIGS, P., 1952, Arid and semiarid climatic types of the world: Internat. Geog. Cong., 17th, Washington 1952, Proc., pp. 135–138.

———, 1953, World distribution of arid and semiarid homoclimates, in Reviews of research in arid zone hydrology. UNESCO, Paris, pp. 203–210 and maps.

deMENOCAL, P. B., 1995, Plio-Pleistocene African climate: Science, v. 270, pp. 53–59.

MIDDLETON, N. J., 1989, Desert dust, in Thomas, D. S. G., ed., Arid zone geomorphology: Halsted Press: a division of John Wiley & Sons, New York, pp. 262–283.

MUHS, D. R., BUSH, C. A., and 3 others, 1990, Geochemical evidence of Saharan dust parent material for soils developed on Quaternary limestones of Caribbean and western Atlantic islands: Quaternary Res., v. 33, pp. 157–177.

MUHS, D. R., STAFFORD, T. W., JR., and 6 others, 1996, Origin of the late Quaternary dune fields of northeastern Colorado: Geomorphology, v. 17, pp. 129–149.

MURRAY, G. W., 1952, The water beneath the Egyptian western desert: Geog. Jour., v. 118, pp. 443–452.

OBERLANDER, T. M., 1972, Morphogenesis of granitic boulder slopes in the Mojave Desert, California: Jour. Geology, v. 80, pp. 1–20.

———, 1974, Landscape inheritance and the pediment problem in the Mohave Desert of southern California: Am. Jour. Sci., v. 274, pp. 849–875.

———, 1989, Slope and pediment systems, in Thomas, D. S. G., ed., Arid zone geomorphology: John Wiley & Sons, Inc., New York (A Halsted Press Book), pp. 56–84.

PASSARGE, S., 1928, Panoramen afrikanischer- Inselberglandschaften: Dietrich Reimer, Berlin, 15 pp., 25 plates.

PÉCSI, M., 1968, Loess, in Fairbridge, R. W., ed., Encyclopedia of geomorphology: Reinhold Book Corp., New York, pp. 674–678.

PEEL, R. F., 1966, The landscape in aridity: Inst. Brit. Geographers Trans., no. 38, pp. 1–23.

PÉWÉ, T. L., PÉWÉ, E. A., and 2 others, 1981, Desert dust: Characteristics and rates of deposition in central Arizona, in Péwé, T. L., ed., Desert dust: Origin, characteristics, and effects on man: Geol. Soc. America Spec. Paper 186, pp. 169–190.

PYE, K., 1987, Aeolian dust and dust deposits: Academic Press Inc. (London) Ltd., London, 334 pp.

———, 1995, Nature, origin and accumulation of loess: Quaternary Sci. Rev., v. 14, pp. 653–667.

———, and TSOAR, H., 1990, Aeolian sand and sand dunes: Unwin Hyman Ltd, London, 396 pp.

RAHN, P. H., 1967, Sheetfloods, streamfloods, and the formation of pediments: Assoc. Am. Geographers Annals, v. 57, pp. 593–604.

REA, D. K., 1994, Paleoclimatic record provided by eolian deposition in the deep sea: The geologic history of wind: Rev. Geophysics, v. 32, pp. 159–195.

REHEIS, M. C., GOODMACHER, J. C., and 6 others, 1995, Quaternary soils and dust deposition in southern Nevada and California: Geol. Soc. America Bull. v. 107, pp. 1003–1022.

RICH, J. L., 1935, Origin and evolution of rock fans and pediments: Geol. Soc. America Bull., v. 46, pp. 999–1024.

SABIN, T. J., and HOLLIDAY, V. T., 1995, Playas and lunettes on the southern High Plains: Morphometric and spatial relationships: Assoc. Am. Geographers Annals., v. 85, pp. 286–305.

SAYAGO, J. M., 1995, The Argentine neotropical loess: An overview: Quaternary Sci. Rev., v. 14, pp. 755–766.

SCHNEIDER, B., TINDALE, N. W., and DUCE, R. A., 1990, Dry deposition of Asian mineral dust over the central North Pacific: Jour. Geophys. Res., v. 95, no. D7, pp. 9873–9878.

SHARP, R. P., 1949, Pleistocene ventifacts east of the Big Horn Mountains, Wyoming: Jour. Geology, v. 57, pp. 175–195.

———, 1980, Wind-driven sand in Coachella Valley, California: Further data: Geol. Soc. America Bull. v. 91, pp. 724–730.

SMALLEY, I. J., 1995, Making the material: The formation of silt-sized primary mineral particles for loess deposits: Quaternary Sci. Rev., v. 14, pp. 645–651.

———, and SMALLEY, V., 1983, Loess material and loess deposits: Formation, distribution and consequences, in Brookfield, M. E., and Ahlbrandt, T. S., eds., Eolian sediments and processes: Developments in sedimentology 38, Elsevier Science Publishers, Amsterdam, pp. 51–68.

STEPHENS, C. G., and CROCKER, R. L., 1946, Composition and genesis of lunettes: Royal Soc. S. Australia Trans., v. 70, pp. 302–312.

STRECKER, M. R., CERVENY, P., and 2 others, 1989, Late Cenozoic tectonism and landscape development in the foreland of the Andes: northern Sierras Pampeanas (26°-28°S), Argentina: Tectonics, v. 8, pp. 517–534.

SUZUKI, T., and TAKAHASHI, K., 1981, An experimental study of wind abrasion: Jour. Geology, v. 89, pp. 23–36.

TERUGGI, M. E., 1957, The nature and origin of Argentine loess: Jour. Sed. Petrology, v. 27, pp. 322–332.

THOMAS, D. S. G., and MIDDLETON, N. J., 1994, Desertification: Exploding the myth: John Wiley & Sons Ltd, Chichester, UK, 194 pp.

THOMAS, M. F., 1994, Geomorphology in the tropics: A study of weathering and denudation in the low latitudes: John Wiley & Sons Ltd, Chichester, UK, 460 pp.

THORNTHWAITE, W., 1948, An approach toward a rational classification of climate: Geog. Rev., v. 38, pp. 55–94.

TSOAR, H., 1989, Linear dunes-forms and formation: Prog. in Phys. Geog., v. 13, pp. 507–528.

———, and MØLLER, J. T., 1986, The role of vegetation in the formation of linear sand dunes, in Nickling, W. G., ed., Aeolian geomorphology: Allen & Unwin Inc., Boston, pp. 75–95.

TUAN, YI-FU, 1959, Pediments in southeastern Arizona: Univ. Calif. Pub. Geog., no. 13, 140 pp.

TWIDALE, C. R., 1968, Geomorphology, with special reference to Australia: Thomas Nelson (Australia) Ltd., Melbourne, 406 pp.

———, 1976, Analysis of landforms: John Wiley & Sons Australasia Pty. Ltd., Sydney, Australia, 572 pp.

———, 1982, Granite landforms: Elsevier Scientific Publishing Co., Amsterdam, 372 pp.

———, and BOURNE, J. A., 1975, Episodic exposure of inselbergs: Geol. Soc. America Bull., v. 86, pp. 1473–1481.

UNESCO, 1977, World distribution of arid regions [Map]: Laboratoire de Cartographie thématique du CERCG, CNRS, Paris. (Explanatory note, UNESCO,1979, 54 pp.)

WAHRHAFTIG, C., 1965, Stepped topography of the southern Sierra Nevada, California: Geol. Soc. America Bull., v. 76, pp. 1165–1189.

WALKER, A. S., 1982, Deserts of China: American Scientist, v. 70, pp. 366–376.

WARD, A. W., and GREELEY, R., 1984, Evolution of yardangs at Rogers Lake, California: Geol. Soc. America Bull., v. 95, pp. 829–837.

WARD, W. T., 1977, Sand movement on Fraser Island: A reponse to changing climates, in Lauer, P. K., ed., Occasional papers in anthropology no 8: Anthropology Museum, Univ. Queensland, Aust., pp. 113–126.

———, and GRIMES, K. G., 1987, History of coastal dunes at Triangle Cliff, Fraser Island, Queensland: Aust. Jour. Earth Sci., v. 34, pp. 325–333.

WELLS, S. G., McFADDEN, L. D., and 2 others, 1995, Cosmogenic ^{3}He surface-exposure dating of stone pavements: Implications for landscape evolution in deserts: Geology, v. 23, pp. 613–616.

WHALLEY, W. B., SMITH, B. J., and 2 others, 1987, Aeolian abrasion of quartz particles and the production of silt-size fragments: Preliminary results, in Frostick, L. E., and Reid, I., eds., Desert sediments: Ancient and modern: Geol. Soc. [London] Spec. Pub. 35, pp. 129–138.

WHITNEY, M. I., and DIETRICH, R. V., 1973, Ventifact sculpture by windblown dust: Geol. Soc. America Bull., v. 84, pp. 2561–2581.

WILLIAMS, S. H., and ZIMBELMAN, J. R., 1994, Desert pavement evolution: An example of the role of sheetflood: Jour. Geology, v. 102, pp. 243–248.

WILLIS, B., 1934, Inselbergs: Assoc. Am. Geographers Annals, v. 26, pp. 123–129.

WILLMAN, H. B., and FRYE, J. C. 1990, Pleistocene stratigraphy of Illinois: Illinois Geol. Survey, Bull. 94, 204 pp.

WILSON, I. G., 1972, Aeolian bedforms-their development and origins: Sedimentology, v. 19, pp. 173–210.

———, 1973, Ergs: Sedimentary Geol., v. 10, pp. 77–106.

WINSPEAR, N. R., and PYE, K., 1995, Textural, geochemical and mineralogical evidence for the origin of Peoria Loess in central and southern Nebraska USA: Earth Surface Processes and Landforms, v. 20, pp. 735–745.

YUNG, Y. L., LEE, T., and 2 others, 1996, Dust: A diagnostic of the hydrologic cycle during the last glacial maximum: Science, v. 271, pp. 962–963.

ZHANG, Ze., ZANG, Zh., and WANG, V., 1991, Loess deposit in China: Geological Publishing House, Beijing, 202 pp.

Chapter 14

Periglacial Geomorphology

THE PERIGLACIAL ENVIRONMENT

Definition and Extent

As originally defined and generally used early in this century, the term **periglacial** referred to the region in Europe that was peripheral to the margins of the Pleistocene glaciers where frost action was inferred to have been particularly active in the past. The traditional emphasis was on past climatic conditions now more equable, but the term applies equally well to high-latitude and high-altitude regions of today, primarily in the northern hemisphere. The current tendency is to consider the periglacial environment as near-glacial in the sense of either location or intensity of conditions. No single climate definition of the periglacial region has been formulated, but essential to all definitions is the concept of intense frost action and bare ground for at least part of the year (thus excluding glacier-covered regions, which are a separate topic). In some periglacial regions, the mean annual temperature is close to 0°C, and the annual range varies only slightly because of maritime influences. In other regions, a short summer interval may be distinctly hot, with 20 hours or more of sunshine, but winters are dark and intensely cold and dry.

Many temperate and high-altitude regions experience seasonal or occasional frosts that primarily influence plant growth, but to be classed as periglacial, the effects of freezing and thawing must "radically affect the nature of the ground surface" (Williams and Smith, 1989, p. 2). Criteria include displacement of soil material, migration of groundwater, unique landforms, and distinctive vegetation, all related to the intensity of freezing and thawing. About 38 percent of the earth's land area is encompassed by this broad definition, including tundra, boreal forests, cold steppes, and highlands in any latitude.

Permafrost

Permafrost is soil or rock that remains frozen for more than one year. Permafrost is not required by most definitions of the periglacial environment (Washburn, 1980; Péwé, 1983; Thorn, 1992). Features described on subsequent pages, including gelifluction and patterned ground, are commonly regarded as periglacial phenomena, but are known to form in areas that lack permafrost. Nevertheless, many periglacial regions are underlain by permafrost, and permafrost can be regarded as sufficient, if not necessary, evidence of the periglacial environment (Thorn, 1992, p. 17). For practical reasons, the modern periglacial zone is regarded here as coincident with the region of permafrost (Figure 14-1), but excluding the glacier-covered areas. Synonyms include *cryergic* or *cryogenic* processes and regions.

By restricting the definition of "periglacial" by the criterion of perennially frozen subsoil (permafrost), the modern periglacial region becomes much more coherent and identifiable (Péwé, 1983; Nelson and Outcalt, 1987). Further, perennially frozen ground produces soil and sediment structures and landforms that remain after the climate changes, so relict periglacial conditions can be identified, and the formerly affected areas can be shown on maps.

Permafrost now underlies 22 to 25 percent of the world's non-glacierized (pp. 353–354) land area, including about 82 percent of Alaska, at least 50 percent of Canada and Russia, and 20 percent of China (Figure 14-1). All that is required to produce permafrost is a mean annual ground surface temperature below 0°C (Figure 14-2). Only 1 or 2 m at the surface is subject to seasonal thaw. This is the **active layer.** Because ground temperature at a depth of a meter or two closely approximates mean annual air temperature, the latter measure is often used to define the permafrost region (Nelson and Outcalt, 1987). However, differences in precipitation, soil moisture, porosity, vegetation cover, dissolved components, and groundwater pressure have marked effects on the thickness of the frozen layer, and subcategories of *alpine, sporadic, discontinuous,* and *continuous* permafrost are used. In general, the southern limit of continuous permafrost in Canada is approximated by the mean annual surface air isotherm of −8°C, and the southern limit of discontinuous permafrost follows the −1°C air isotherm (French and Slaymaker, 1993, p. 22). Some permafrost may have persisted since the last glaciation; in other areas, the permafrost has formed in postglacial time. For this reason, permafrost is not defined as *permanently* frozen, but only *perennially* frozen. As with glaciers, permafrost thickness and limits have been shrinking during the last century (Burn, 1992). The maximum known thickness of permafrost is 1600 m, but over wide areas of the arctic, a thickness of several hundred meters is commonplace (French, 1996, p. 56).

The periglacial region is impressive in size and geomorphic significance. Some 38 percent of the earth's land area is now under either a cold, treeless, snow climate (tundra) or a boreal forest and snow climate (taiga), within which most localities would be considered periglacial. At least half of the combined areas of these two large, cold regions is underlain by permafrost. At various times during the Pleistocene Epoch, recently enough so that some relict landforms record periglacial processes, perhaps as much as 40 percent of the land area has been subjected to periglacial conditions (Tricart, 1970, p. 28). The geomorphic significance of relict periglacial conditions is reviewed at the end of this chapter.

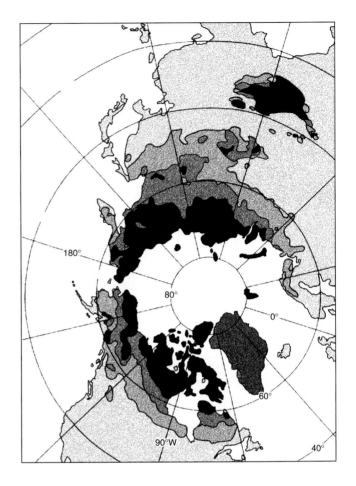

FIGURE 14-1. Contemporaneous distribution of permafrost in the Northern Hemisphere, derived from the "surface frost index" and long-term climate data from several thousand stations. The frost index, computed from mean monthly temperature and precipitation normals, is based on the ratio of freezing to thawing degree-days (Nelson and Outcalt, 1987). Computed values were reclassified into three categories of permafrost: sporadic (light pattern), widespread discontinuous (intermediate pattern), and continuous (dark pattern). Glacierized areas such as Greenland are excluded. Isolated areas of alpine permafrost on high mountains are not shown (from Anisimov and Nelson, 1996, Figure 4).

PERIGLACIAL PROCESSES

Periglacial processes are characterized more by the criterion of degree or intensity than of kind. Weathering and soil formation are controlled by the intensity and frequency of freezing and thawing and by the generally

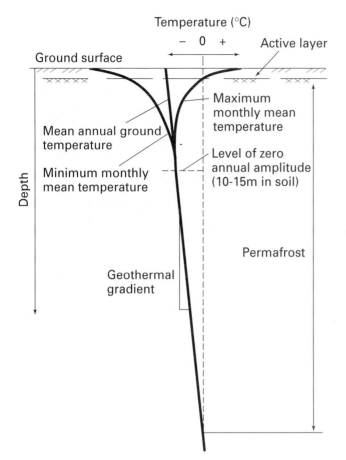

FIGURE 14-2. Typical temperature regime in permafrost (after French, 1996, Figure 5-2).

frozen subsoil. Mass wasting is dominated by gelifluction in the active layer over permafrost (p. 176). Rock falls are common in mountainous regions. Fluvial erosion is inhibited by the brevity of seasonal stream flow, bank erosion by ice, and the massive loads of sediment fed into streams by gelifluction. Wind becomes a relatively important agent of transport and deposition, with sand dunes and loess sheets creating major downwind constructional landforms. In the review that follows, most of the processes have been described in previous chapters and will be only briefly noted. The emphasis in this chapter is on ways the processes interact to produce a landscape recognizable as the result of periglacial activity.

Weathering

Freezing and Frost Cracking. Despite continued uncertainty about the exact process by which rocks

are fractured by freezing and thawing of interstitial water (p. 126), there is no doubt that regions now in the periglacial environment are characterized by great quantities of angular, fractured rock detritus (Figure 14-3). The size of the fragments is largely a function of rock-mass properties such as lithology, pore size, permeability, degree of microcracking, freezing intensity, and water content (Lautridou, 1988). *Frost scaling* causes thin flakes to break from rock surfaces; *frost splitting* is thought to be a true breaking of previously uncracked rock; and *frost wedging* exploits preexisting fractures. Laboratory experiments suggest that repeated freezing and thawing can disaggregate mineral grains as small as 1 μm in diameter (McDowall, 1960; Lautridou, 1988, p. 43). Water is a necessary but little-evaluated component in the process, but precipitation records for periglacial regions do not give much useful information about the availability of water within the ground. Freezing must be slow enough to permit pore water to migrate toward the freezing front. Certainly, dry rocks will not be disrupted by repeated temperature fluctuations through 0°C. Even for saturated rocks, the critical temperature for fracturing is probably well below 0°C. A range of -4°C to -15°C, reached by continuous slow cooling at a rate of less than -0.1°C/hr to -0.5°C/hr, is calculated to be optimal for crack growth in granite and marble (Walder and Hallet, 1985). Rocks with medium-size pores and moderate permeability show an increase in shattering down to temperatures of -8°C to -10°C, but little change at colder temperatures. However, a temperature below -10°C has been reported to be more effective in fracturing chalk (Lautridou, 1988). Nevertheless, with all these qualifications and uncertainties, alternate freezing and thawing is generally considered the dominant weathering process in the periglacial zone.

Chemical and Biologic Weathering. Freezing and thawing have been so closely associated with periglacial weathering that other processes have been slighted. Hydration, hydrolysis, and oxidation will occur whenever aerated water contacts minerals, but at low temperatures and with few microorganisms, chemical processes proceed only very slowly. Crusts of hydrated salts form on rocks at McMurdo Sound, Antarctica, and might be important to rock weathering there. In other coastal periglacial regions, where sea salts collect by evaporation under low atmospheric humidity, similar weathering will also be expected (Hall, 1992, pp. 113–115; French, 1993, pp. 148–150).

FIGURE 14-3. Wright Valley, South Victoria Land, Antarctica. The mean annual temperature is about −20°C. No soil develops on the steep taluses (photo: C. Bull, Institute of Polar Studies, Ohio State University).

An instructive illustration of weathering in a polar desert was provided by Kelly and Zumberge (1961), who analyzed a sequence of quartz diorite specimens from a single exposure at Marble Point, on McMurdo Sound, Antarctica. The samples ranged in degree of weathering from fresh rock to completely disaggregated, iron-stained grus. Although the authors assumed that a sequence of chemical changes would be illustrated by the samples, in reality no chemical weathering was found, except for oxidation of iron-rich biotite and accessory minerals. Bulk chemical and mineralogic composition did not change. No clay minerals formed. The disintegration was caused not by oxidation or hydration but by frost cracking and the subsequent crystallization of salts in pores and fractures. The site is exposed to sea spray during the brief summer, but no rain falls. This study provided substantial quantitative support for the general but previously unsubstantiated opinion that mechanical weathering predominates over chemical weathering in cold, dry climates. Even in the maritime Antarctic islands, mechanical weathering greatly dominates, and weathering in general is slow (Hall, 1992, p. 119). The question of limestone solution in cold regions was discussed in Chapter 8. As yet, no weathering effects have been attributed to the endolithic cyanobacteria found in Antarctic rocks (p. 133).

Ground Ice

Frost heaving is the process of lifting surface particles by the formation of ice in soil. As ice nuclei form in soil, free water migrates toward these nuclei, which grow as veins or lenses of ice while adjacent sediments are dewatered (Everett, 1961; French, 1993, pp. 145–148). The grain size and pore size of sediments determine the rate of ice-crystal segregation and water movement. Silt is the optimal grain size for ice-crystal growth, and frozen silt may contain 50 to 150 percent ice by weight (French, 1996, p. 80). The volume of ice in permafrost sediment can vary from near zero (dry permafrost) to greater than 90 percent (Harry, 1988, p. 121).

Needle ice consists of clusters of slender ice crystals up to several centimeters in length that form overnight in frosty wet soils. Needle ice is a common phenomenon in seasonal frost climates well beyond the climatic range of permafrost. Although the phenomenon is much less impressive than the formation of larger segregated ice masses in perennially frozen ground, needle ice plays an important role in loosening surface soils and in lifting small stones to the surface of fine sediments. On hill slopes, frost heave and needle ice promote soil creep by lifting particles at right angles to the sloping surface, then permitting them to settle vertically on remelting.

Segregated ice masses grow within permafrost by *diffusion* under temperature and pressure gradients, by *gravity transfer* of thaw water from the active layer, and by *intrusion* of groundwater under pressure during freezing (Harry, 1988, p. 120). Most segregated ice is at a shallow depth just below the active layer, reflecting the seasonal availability of thaw water adjacent to the still-frozen permafrost. Furthermore, the growth of segregated ice is inhibited by the confining pressure of overburden weight, which increases with depth.

Ice wedges (Figure 14-4) are downward-tapering masses of foliated ice and fine sediment that penetrate permafrost to depths of more than 10 m, and may be 2 or 3 m wide at the base of the active layer. Most authorities accept the contraction theory of ice wedge formation (Leffingwell, 1915; Lachenbruch, 1962) as the best explanation. As outlined by Lachenbruch (Figure 14-4), the contraction theory is based on seasonal cracks that develop in totally frozen ground during winter. Ice and soil, like other solids, contract on cooling, and may explosively develop vertical cracks several millimeters wide and a meter or more in depth. In the spring, water from the active layer seeps down the crack and refreezes in the cold permafrost. During the following summer, warming and expansion (but not thawing) of the frozen layer causes sediments to be upturned and deformed against the ice wedge. Each winter the freeze-cracking is repeated along the initial plane of weakness, and each summer a new lamella of ice is introduced. After several centuries, large ice wedges have formed, their spacing and intersections outlining ice-wedge polygons. **Sand wedges** can form by a similar process in arid cold regions, where blowing sand fills surface cracks that open during the coldest season. Figure 14-4 should be studied carefully to note that ice wedges form below the active layer, due to the thermal contraction of solid soil and ice. The stress generated by the 9 percent expansion of water on freezing is an entirely different process, unrelated to ice-wedge formation (Worsley, 1985). Because they can be replaced by later sediments as *ice-wedge casts*, ice wedges are among the features that are preserved as valid criteria of former permafrost in regions not now periglacial. However, the identification of fossil ice wedges is difficult and controversial (Black, 1976; Péwé, 1983; Johnson, 1990).

Permafrost is not completely impermeable, and within sediments of suitable permeability, groundwater may flow under gravity. Where it refreezes, **intrusive ice** in the form of veins, sills, and laccoliths (p. 85) accumulates. Some intrusive ice masses arch the land surface upward to form distinctive permafrost landforms (p. 320).

Mass Wasting

Because of the perennially frozen substrate, mass wasting is not a deep phenomenon in permafrost regions. **Gelifluction** and **frost creep** (Figure 9-7) occur only in the active layer, rarely more than 2 m thick (Figure 14-2). Resistant rock ledges are exposed to freeze and thaw as the active layer flows downhill, and they stand out as tors (p. 252) and low structurally controlled ridges on debris-covered slopes. Gelifluction occurs even on very low slope angles, perhaps as low as 1° on rocks that break down to clay-rich debris (note background landscape, Figure 14-5). On bouldery debris, gelifluction lobes may have slopes as steep as 20° (Tricart, 1970, pp. 105–106) (Figure 14-6).

Along with the specific processes of gelifluction and frost creep, a variety of other mass movements are common features of the periglacial environment. Snow and rock avalanches are frequent, especially during the springtime. Mud flows and debris flows break out of gelifluction lobes and sheets, and are really only an extension of the gelifluction process. On clay-rich soils,

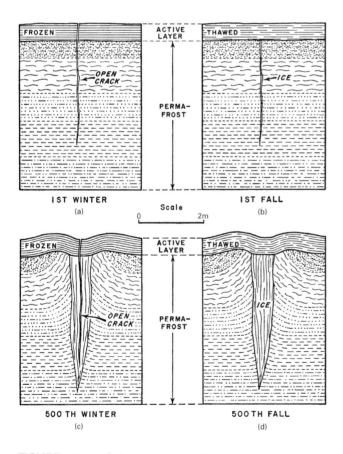

FIGURE 14-4. Schematic evolution of an ice wedge according to the contraction-crack theory (from Lachenbruch, 1962, Figure 1).

FIGURE 14-5. Tundra landscape of gentle convex slopes above river-bank cliffs on the Colville River near Umiat, Alaska. The bentonitic shale is prone to gelifluction and other mass-wasting processes (photo: T. L. Marlar, U.S. Army Cold Regions Research and Engineering Lab).

debris flows in channels with levees have been described (foreground, Figure 14-5) (Anderson et al., 1969; Harris, 1987; French, 1988).

The annual cycle of freeze and thaw in the active layer over permafrost produces large lateral stresses in the mixture of rock fragments, ice, and water. On level ground, slabs and cobbles may be lifted by frost heaving, but fine sediment fills in beneath them so that the

FIGURE 14-6. Large gelifluction lobe, Mesters Vig district, northeast Greenland (photo: A. L. Washburn).

layer becomes sorted, with the coarsest fragments at the surface.

Erosion and Transport Processes

Nivation. A subtle progression of processes can be observed beginning with weathering at the margins of a snow bank, to meltwater erosion from beneath the snow, to the downhill erosive creep of water-saturated or refrozen snow. The three processes are grouped under the heading of **nivation,** the intensification of weathering and erosion associated with patches of snow that persist into the summer season in the periglacial environment. Nivation and gelifluction have been ranked, along with freezing and thawing, as the three groups of processes that characterize the periglacial environment (St. Onge, 1969). The definition of nivation is complex because it includes both weathering and erosion, and describes only the intensification of processes that are also active in nonperiglacial regions (Thorn, 1988; 1992).

Snow patches that persist in sheltered positions on hillslopes below the altitude of permanent snowfields have been reported to produce **nivation hollows** in erodible sediments, even within a single season (Flint, 1971, pp. 135–136). Weathering is intensified in the saturated ground beneath the snow. The saturated snowmass produces a flow of water toward its edges, especially on the downhill side. Water percolating to the base of the snowbank carries away fine particles from the rock or soil beneath. Freezing and thawing may be intensified at the edges of the shrinking snowbank as well, although the snow insulates the rock and

FIGURE 14-7. Nivation hollows on cryoplanation terraces cut into a greenstone ridge on Indian Mountain near Hughes, Alaska (Reger and Péwé, 1976, Figure 1) (photo: T. L. Péwé).

soil beneath it. Periglacial hill slopes commonly are lined with shallow nivation hollows, especially on the shaded and downwind side of ridge crests where snow accumulates and is shaded from sun for the longest time (Figure 14-7). Permafrost promotes widening rather than deepening of the hollows because only the active layer is removed by nivation processes. In the conterminous United States, the favored location for nivation hollows is on the northeastern side of ridges, where snow is blown by westerly winds and is sheltered from sunlight in early summer. In the western Great Basin, nearly 80 percent of the nivation hollows face toward the northeast (Dohrenwend, 1984).

There has been a tendency to equate nivation hollows with incipient glacial cirques (p. 378) and to imply a genetic progression from residual snow patches in sheltered hollows, to permanent snowfields in basins, to cirques from which alpine glaciers flow (Chapter 17). However, most current authors express doubt about the ability of nivation to create depressions large enough to become cirques. Most mountain regions that currently support late-season snowbanks were glaciated in the late Pleistocene, and it is possible that the hollows in which snowbanks persist might be relict glacial forms. Thorn (1988, p. 22) calculated that contemporary nivation rates in the Rocky Mountains are too slow by orders of magnitude to create even small cirques within postglacial time. It follows that even if nivation hollows might eventually grow into cirques, the process is so slow that they are likely to be glaciated one or more times in the process. Landforms formerly attributed to nivation in Great Britain are reinterpreted as immature glacial cirques (Ballantyne and Harris, 1994, p. 246). Nearly all the 10,000 nivation landforms surveyed in the western Great Basin are relict even though they are on erodible rocks or sediments. They are most abundant just below the altitude of late Pleistocene glaciation but are uncommon within the glaciated area, implying that they formed under a formerly colder climate, perhaps contemporaneously with glaciation at higher altitudes (Dohrenwend, 1984).

The originator of the term *nivation* (Matthes, 1900) found no evidence that seasonal snow patches moved or abraded their beds. Subsequently, study of the nivation process in the Snowy Mountains of Australia (Costin et al., 1964) proved conclusively that the downhill creep and slide of snow banks dragged basal rock fragments over bedrock and produced visible, though minor, abrasion marks on both the transported fragments and the subsurface. Steel rods notched to various depths were embedded in rock in nivation hollows, and their deformation or breakage under the force of moving snow was calibrated by laboratory tests (Costin et al., 1973). The pressure generated ranged from 160 to 3790 kPa (1.6 to 37.9 bar), equal to or greater than the shear stress calculated for the base of glaciers (p. 362). Clearly, seasonal snow patches in nivation hollows, with floors that slope at angles of 10° to 27°, do move downhill and abrade their floors with bedload fragments. Because one of the definitive features of glaciers is evidence of motion (Chapter 16), proof of snow-patch abrasion makes the distinction between glacial erosion and nivation even less clear. In areas of possible former periglacial or glacial conditions, abraded surfaces have been cited as proof of glaciation. Such evidence may need to be reevaluated. At least it should be used with great caution.

Fluvial Erosion. Rivers in the periglacial zone have extremely complex hydraulic geometry. Many periglacial regions would be classed as arid or semi-arid, based on their total annual precipitation. Runoff, controlled by snow melting, is temperature-dependent rather than precipitation-dependent. Infiltration is limited over permafrost. If there is rainfall, the direct runoff is increased even more by additional snowmelt. Seasonal and diurnal fluctuations are great. High-gradient mountain streams and large rivers on plains may continue to flow under ice cover throughout the winter. Even so, discharge is greatly decreased because of the lack of overland flow in winter. Small, low-gradient rivers may freeze solid and cease to flow entirely. Because the periglacial zone in general is semiarid or arid, river discharges are rarely high, except during the melting season or high-precipitation events.

Springtime is the season of violent fluvial action. When meltwater begins to flow into headwater streams, the streams swell and burst their ice cover or flow over frozen river beds, progressively developing a flood wave that moves downstream. If the thaw season begins in the headwaters and progresses toward the river mouth, as it generally does in the larger south-to-north-flowing rivers of arctic North America and Siberia, huge ice jams and floods are seasonal events that are part of the streams' regimen. Briefly, the rivers move much larger loads and much coarser sediment than their mean annual discharges would suggest. Some cited values illustrate the point: Eurasian Arctic rivers discharge 60 percent of their total annual discharge during May, June, and July (Gordeev et al., 1996, p. 671); The Colville River in northern Alaska carries 43 percent of its annual discharge and 73 percent of its total inorganic suspended load in a three-week spring period; 60 percent of the Yukon River discharge is during June, July, and August (Telang et al., 1991, p. 78). Up to 50 percent of the runoff is from melting snow. Initially, the floods flow over basal ice on those rivers that were frozen solid, so bank erosion is much more extensive than channel deepening. Bank erosion is also intensified by floating ice.

The discharge of small periglacial streams, especially mountain streams, follows a distinct daily cycle during the summer. Shortly after the hottest part of the day, the rivers reach peak discharge as overland runoff reaches the channels. Before dawn after a freezing night (if "night" is an appropriate term for extended twilight), runoff will have nearly ceased. Sediment transport and deposition goes through a daily cycle under such conditions, with alluvium deposited among shifting, braided channels. The sediment load delivered to streams also fluctuates with the daily cycle of gelifluction and other forms of mass wasting. Most periglacial streams, other than mountain torrents, flow on aggraded beds. Mass wasting seems to deliver sediment faster than periglacial rivers can move it to the sea. Aggradation results, but it is not yet clear how much of the aggradation is modern and how much is relict from Pleistocene times. The thick gold-bearing alluvium of the Yukon Lowland contains many skeletons of extinct Pleistocene mammals, confirming its relict character.

Eolian Transport. Low precipitation, diurnal fluctuations in discharge, freeze-drying of exposed sediment, and strong winds combine to make the periglacial zone a region of strong eolian action. Clouds of sand and dust can be seen rising from the flood plains of rivers as the spring flood subsides (Figure 14-8). Dunes are common features on the downwind banks of rivers. Sand sheets, less well sorted than dunes and sometimes stratified, are widespread in northern Europe and are attributed to windblown sediment settling on snow and being prevented from forming dunes (Dijkmans and Mücher, 1989). Very large areas of present and former periglacial regions are mantled with loess (p. 301), but loess is not of uniquely periglacial origin, and is a poor diagnostic criterion for a former periglacial environment (Thorn, 1992, p. 21). Eolian sediment in cold climates may be limited more by the available supply of sediment than by the transportation ability of the wind (Neuman, 1993). The great increase in atmospheric dust content during glaciations (p. 302) can be attributed to both an increase in wind intensity and the exposure of extensive sediment source areas on meltwater floodplains, drained ice-marginal lakebeds, and continental shelves exposed by lowered sea level.

PERIGLACIAL LANDFORMS

Major Landforms

The combination of intense frost action, permafrost, gelifluction, nivation, strong winds, incomplete vegetation cover, and seasonal flowing rivers imparts a strong geomorphic imprint on landscapes. Periglacial plains are wastelands of patterned ground, rubble-strewn hillslopes, gelifluction lobes and sheets, and

FIGURE 14-8. Clouds of silt transported by wind from the Delta River floodplain, central Alaska. Compare with Figure 13-26 (photo: U.S. Navy, courtesy T. L. Péwé).

seasonal bogs or shallow lakes of surface water slow to infiltrate because of the permafrost layer. The background of Figure 14-5 gives a good impression of the relief of periglacial plains on nonresistant rocks. Gelifluction causes mass wasting on very low slopes, producing broad convex summits. Stream action can be so brief and ineffective that small valleys cannot entrench but are choked by gelifluction debris. Mountainous periglacial regions add block fields, rock glaciers, massive taluses, and a variety of terraced or benched slopes shaped by nivation and gelifluction. Many well-illustrated descriptions of periglacial scenery have been published (Muller, 1947; Hopkins et al., 1955; Wahrhaftig, 1965; Ferrians et al., 1969; Péwé, 1975). Only the major landforms and some of the more spectacular minor forms are noted here. For more detailed descriptions of periglacial phenomena, books by Tricart (1970), Washburn (1980), Giardino et al. (1987), Clark (1988), Williams and Smith (1989), French and Slaymaker (1993), and French (1996) are recommended, especially to American readers. An enormous literature in Russian, Polish, French, and the Scandinavian languages is readily available, as well.

Asymmetric Valleys.   Some periglacial valleys show notable asymmetry in cross profile, even where struc-

tural control can be eliminated. The most typical condition is for north-facing slopes (in the northern hemisphere) to be steeper than the south-facing slopes (French, 1996, p. 202). Where valleys trend north or south, the asymmetry disappears. The theory is that gelifluction is more intense on the valley side facing the sun for the longest time each day, and the shaded side stays colder, less prone to gelifluction, and steeper. Furthermore, the copious colluvium from the south-facing slopes forces rivers to preferentially erode and steepen their north-facing valley walls. Relict valleys with such asymmetry are common in southern Indiana (latitude 38°N) adjacent to the Wisconsin glacial margin (Wayne, 1967). By contrast, Tricart (1970, p. 137) suggested that near the southern limit of the arctic periglacial regions, permafrost, gelifluction, and gentle slopes might persist on cold, north-facing valley walls even as drier, warmer, nonpermafrost slopes develop steeper gradients on the south-facing valley sides.

Cryoplanation Terraces and Cryopediments.   Flattened summit areas and benched hillslopes in Alaska were attributed to an assemblage of periglacial weathering, erosional, and depositional processes collectively termed **altiplanation** (Eakin, 1916, pp. 78-82). Current authors prefer the equivalent term, **cryoplanation**

(Priesnitz, 1988). In Alaska, they cluster in a narrow vertical range that implies an altitude-controlled zone optimal for terrace development, near the climatic snowline. Their distribution reinforces the hypotheses that nivation is important to their development (Nelson, 1989). By implication, the composite of cryoplanation processes could result in a regional periglacial surface of low local relief, not controlled by any base level. The deduced or implied penultimate landform of periglacial regions, the "cryoplain" or "altiplain," has never been seriously proposed as a climatic analog of a peneplain or a "pediplain" (pp. 345 and 348). However, **cryoplanation terrace** (or the Russian equivalent, **goletz terrace**) is a widely accepted term for the typical benchlike landforms (Figure 14-7).

Cryopediments were first recognized as relict periglacial landforms in eastern Europe in the 1950s (for English-language reviews, see French, 1996, p. 175; Washburn, 1980, p. 237). The term has been extended to present-day periglacial regions. In shape and size, they are similar to cryoplanation terraces, but are on lower piedmont slopes backed by steep mountain slopes, analogous to pediments in arid and semiarid regions (p. 281). Both cryoplanation terraces and cryopediments begin as aligned nivation hollows or a nivation bench on a periglacial hillside, after which frost weathering maintains a steep slope or cliff at the back of the bench while gelifluction maintains a smooth, gentle "slope of transportation" that, like a semiarid pediment, is primarily an erosional surface even when covered by gelifluction debris in transit across it (Te Punga, 1956; Reger and Péwé, 1976). At one locality in the Canadian Yukon, which was never glaciated, the well-developed pediments are regarded as relicts of a former, even colder, periglacial climate (Fried et al., 1993).

Block Fields, Rock Streams, and Rock Glaciers. Masses of angular rock rubble on summits (**block fields** or **felsenmeer**) or streaming down mountainsides (**rock streams**) are striking features of modern periglacial landscapes, and their relict equivalents are also well known. Relict rock streams in the Appalachian Mountains consist of large, subangular boulders and are now devoid of vegetation. They extend as much as 1000 m downhill from a cliff or exposed ledge (Figure 14-9). Although not now in motion, the Appalachian rock streams were almost certainly formed by periglacial mass wasting (Potter and Moss, 1968; Clark and Ciolkosz, 1988; Clark et al., 1992). Gelifluction and frost creep are the likely processes because, although the rock streams now lack the water-retaining fine sediments in the upper few meters that would be required to maintain motion, the large blocks have interstitial finer sediments among them at depth. Hickory Run boulder field, one of the larger of the relict Appalachian rock streams, has a surface gradient of only 1° (Smith, 1953).

Rock glaciers were once thought to be moribund valley glaciers but are now recognized as distinct landforms, transitional between periglacial and glacial in origin (Giardino et al., 1987). White (1971a, p. 44) defined rock glaciers as "accumulations of unsorted, till-like, coarse to fine rock debris, with an ice core and/or interstitial ice, having a glacier shape and spreading downvalley." Active rock glaciers are either tongue-shaped along valley axes or lobes on valley walls. They have steep frontal slopes, often with active

FIGURE 14-9. Hickory Run boulder field, Carbon County, Pennsylvania. Boulders reach 2 m in length (photo: W. Bolles, Pennsylvania Dept. of Education).

FIGURE 14-10. Sulphur Creek rock glacier, northern Absaroka Mountains, Wyoming (photo: Noel Potter, Jr).

taluses, and arcuate pressure ridges on their surfaces (Figure 14-10). All known rock glaciers are derived either from glacial debris downvalley from an active or inactive glacier, or from avalanche debris, sliderock, or colluvium on a valley wall (Johnson, 1987, p. 191; Humlum, 1988; Whalley and Martin, 1992; Barsch, 1996, p. 11).

If a mass of poorly sorted colluvium has sufficient ice packed within its pore spaces, the plastic deformation of the ice can produce downhill motion similar to but slower than glacier flow. The arbitrary difference between a rock glacier and a glacier is that the former is mostly rock debris with interstitial ice, whereas the latter is mostly ice with included rock fragments. Barsch (1996, pp. 4–5) emphasized that rock glaciers are "supersaturated" with ice, i.e., they contain a volume of ice greater than the normal pore volume of the unfrozen debris. Several rock glaciers in the central Rocky Mountains are ice-cored and either are transitional upslope to true valley glaciers or are degenerating former glaciers (White, 1971a, 1971b, 1976; Potter, 1972; Benedict et al., 1986). Larger rock glaciers in Alaska are of ice-cemented rock fragments, similar to glaciers in all but the relative proportion of ice and rock (Wahrhaftig and Cox, 1959). Rock glaciers move forward by internal deformation

and debris slides at their termini, resulting in a rolling or continuous-tread form of transport. Although they have not been proved to erode the rock surfaces over which they move, their movement is clearly transitional to glacial processes. A rock glacier derived from a degenerating glacier might be used to infer climate warming; one derived from snow and ice that filled sliderock on a talus might imply climatic cooling; obviously, interpretations should be cautious.

Minor Landforms

The landscape of periglacial regions is marked by a variety of patterns, many of which barely deserve the name "landform." Much of the terrain is depositional rather than erosional because of the contrast between intense mass wasting and inefficient stream erosion. On the alluvial plains and other surfaces of low relief, permafrost overlain by the active layer contorts, erupts, or collapses the land surface into a maze of polygonal patterns, bogs or shallow lakes, ice-cored domes, and pits. Only the more striking forms are illustrated here.

Patterned Ground. The relationship of ice wedges to permafrost has been described previously (p. 313). Polygons, outlined by open cracks or shallow depression but not underlain by ice wedges, are called **frost-crack polygons** (Washburn et al., 1963). They are not diagnostic of permafrost and can form in any seasonally cold region if snow cover is thin during a cold interval. Some frost cracks may be preserved by windblown silt or sand packing into them, but for the most part, the patterns are visible only during cold weather and are not preserved.

More prominent polygonal patterns are in areas of continuous permafrost, with extremely low mean annual temperatures, probably averaging −6 °C to −8 °C. Permafrost is normally several hundred meters thick. In such regions, networks of **ice-wedge polygons** cover many thousands of square kilometers (Figure 14-11). Each polygon, from 3 to 30 m in diameter or more, is outlined by shallow troughs, beneath each of which is a wedge-shaped foliated mass of ice, a meter or more broad at the top and extending downward to 10 m or more.

Polygons may be made of sorted or unsorted sediment and may have raised centers or raised rims (Figure 14-11) (Washburn, 1980; Krantz et al., 1988, and references therein). Sorted polygons involve the most severe rearrangement of rock fragments, with coarser particles lifted to the surface or thrust outward to the edges of the polygons. A higher proportion of fine-grained sediment probably promotes sorting because water is held longer without refreezing in fine sediments, and ice masses may segregate from the freezing

FIGURE 14-11.
Ice-wedge polygons, northeastern Mackenzie River delta, Canada. Dark high-centered polygons are 10 to 20 m in diameter (photo: J. R. Mackay).

soil to disrupt and heave the original stratification, if any. Sorted polygons may be preserved because seasonal-flowing water permeates through the coarse network of polygonal boundaries without eroding the fine sediments from the polygon cores. Sorted polygons are not diagnostic of the permafrost zone although they are found there. They seem to be more abundant on wet, cold mountain plateaus and other places with intense freeze-thaw activity.

During thawing, which begins at the surface and migrates downward into the active layer, dewatering increases the bulk density. The density contrast is believed to be adequate to promote free convection of sediment in the active layer, with the formation of polygonal patterns (Hallet and Prestrud, 1986; Hallet and Waddington, 1992). A related process of convective

circulation of water within the active layer, caused by the anomalous greater density of water at 4°C at the surface of the active layer compared to the less dense water at 0°C at the permafrost contact beneath the active layer, may also promote free convection and polygonal patterns (Gleason et al., 1986; Krantz et al., 1988). Although water convection in the density gradient of 0°C to 4°C is far too weak to transport sediment, it can transfer heat to the base of the active layer, creating a corrugated relief that then promotes differential sorting in the overlying active layer.

Sorted polygons, when occurring on slopes, are distorted laterally by gravity to **sorted stripes** (Figure 14-12), parallel segregations of coarse and fine sediment trending downhill. Like sorted polygons, sorted stripes require active freeze-thaw processes but are not restricted to regions of permafrost. Once formed, the coarse stripes will channel runoff from the hillside. The percolating water probably erodes fine sediment from beneath the coarse debris and aids in lowering the stripes of coarse debris into trenches down hillsides. The excellent drainage provided by sorted stripes enables hillslopes to remain stable at unusually steep angles, perhaps to 30°, without slumping or sliding.

Frost Mounds: Palsas and Pingos. A variety of *frost mounds* form from periglacial groundwater processes. Some form within a single winter season by refreezing of the active layer (Pollard, 1988; Seppälä, 1988). A **palsa** in the traditional Lappland terminology was a peat hummock with a core of ice and frozen peat. In wide usage, it now included mounds with cores of mineral sediment as well (Nelson et al., 1992). Many of these minor landforms are associated with the discontinuous permafrost region (Figure 14-1) where winter freezing is severe but groundwater is mobile for an extended summer season and peat bogs are abundant.

Under certain hydrologic conditions, lenticular intrusions of ice arch the overlying ground into a fractured dome or blister (Figure 14-13) called a **pingo**. Pingos occur at the edges of alluvial plains or old lakebeds and at the bases of long slopes. Probably the hydrostatic head in groundwater is intensified by pressure generated at the base of the permafrost or in unfrozen layers within permafrost. At some low point on an aquifer, pressure builds up sufficiently to arch the overlying permafrost and build up a mass of ice in the core of the mound. These are called "open system" pingos because groundwater continues to flow toward them and they can grow. Other pingos are thought to be of a "closed-system" type, wherein the progressive inward and downward freezing of an old lake or bog generates pressure in the enclosed saturated area sufficient to

FIGURE 14-12. Relict patterned ground and stone stripes on Snake River basalt, north of Glenns Ferry, Idaho (Malde, 1964). Plateau gradient is 2 percent southwestward; colluvium gradient is 20 percent southeastward. Light-colored areas are grassy; dark areas are basalt fragments (vertical aerial photo: U.S. Dept. of Agriculture).

drive pore water upward, where it freezes as a segregated mass in the core of a mound (Pissart, 1988; French, 1996, Figure 6-20). Fossil pingos, low circular mounds enclosing poorly drained basins, are reported from many areas of western Europe (Ballantyne and Harris, 1994, pp. 74–83 and references therein). A few similar forms in Illinois (Flemal et al., 1973) and Pennsylvania (Marsh, 1987) have been described, but they seem less common in North America than in Europe (Flemal, 1976). Like fossil ice wedges, fossil pingos are excellent evidence of former permafrost, and their presence near the southern limits of former glaciers gives important information about the former extent of the periglacial zone.

Thermokarst and Relict Periglacial Landforms

Thermokarst. The global warming trend of the last two centuries has caused a reduction in area of permafrost regions. Since most permafrost temperatures are only 5°C to 10°C below freezing, they are inherently unstable in a time of general warming. Some areas that were formerly underlain by continuous permafrost now have discontinuous or only sporadic occurrences, and the temperature of permafrost has increased by as much as 2°C to 4°C within this century (Lachenbruch et al., 1988; Burn, 1992; Burn and Smith, 1993). A global warming of 2°C would reduce the area of permafrost by 25 to 44 percent (Anisimov and Nelson, 1996). As ice-rich permafrost thaws, the reduction of total volume and the collapse of ground over former segregated ice masses produces a pitted karstlike terrain, to which the term **thermokarst** has been applied (Figure 14-14) (Czudek and Demek, 1970). Many areas of permafrost are now preserved only because of the insulating effects of forests or tundra vegetation. If land is cleared for roads, railways, airports, or pipelines, the collapse of permafrost can lead to destruction of the installations. An enormous literature of permafrost engineering has grown in recent years as arctic oil fields

FIGURE 14-13. Pingo, Wollaston Peninsula, Victoria Island, Northwest Territories, Canada (photo: A. L. Washburn).

FIGURE 14-14. Thermokarst in a gravel road near Prudhoe Bay, Alaska. Failure to use sufficient insulation during construction resulted in thawing of underlying ice wedges and differential subsidence (photo: F. E. Nelson).

and pipelines have been developed and built (Walker et al., 1987).

Relict Periglacial Landforms. Former periglacial regions peripheral to the Pleistocene ice sheets cannot be compared directly with the arctic tundras of today. Relict low-altitude periglacial soil structures have been discovered at latitudes at least as low as 40°N in the central United States (Johnson, 1990, Figure 1). Even if the periglacial interval was brief, its record at a latitude more than halfway toward the equator from the north pole should remind us not to extrapolate modern climatic conditions to the Pleistocene record. In the central plains of the United States at the edge of a Pleistocene ice sheet, the summer sun at noon was within 16° of the zenith. Days and nights must have been of nearly the same length as present ones and not at all like the seasonal alternation of continuous daylight and darkness that characterizes the tundra regions north of the Arctic Circle. Former low-latitude periglacial regions are outstanding examples of "no-analog" climates, in which geomorphic process and biota were unlike any region on earth today (p. 62).

Büdel (1982, pp. 109, 119) argued that fully 95 percent of the landforms of the midlatitude northern hemisphere are relict from former periglacial conditions. He regarded subtropical savanna and periglacial geomorphic processes as the two most important sets of landscape-forming processes on earth. In reviewing Büdel's viewpoint, Cotton (1963, p. 769) acknowledged that it is possible that most of the landforms in midlatitude regions, regarded by W. M. Davis and others of his time as the result of the "normal" processes of fluvial erosion (pp. 198–199), may be relict from past periglacial conditions. It would indeed be an odd trick of fate if the landforms of the northeastern United States and western Europe, which provided the observational evidence for the deduced "normal cycle of erosion" (p. 346), were actually relict periglacial landforms. Geomorphologists with field experience in present-day periglacial regions are steadily demonstrating the impact of former periglacial processes in regions of presently temperate climates. Both for its present and past importance, periglacial processes and landforms are an important current topic of geomorphology despite continuing difficulty with precise definitions (Thorn, 1992).

REFERENCES

ANDERSON, D. M., REYNOLDS, R. C., and BROWN, J., 1969, Bentonite debris flows in northern Alaska: Science, v. 164, pp. 173–174.

ANISIMOV, O. A., and NELSON, F. E., 1996, Permafrost distribution in the northern hemisphere under scenarios of climatic change: Global and Planet. Change, v. 14, pp. 59–72.

BALLANTYNE, C. K. and HARRIS, C., 1994, Periglaciation of Great Britain: Cambridge University Press, Cambridge, UK, 330 pp.

BARSCH, D., 1996, Rockglaciers: Indicators for the present and former geoecology in high mountain environments: Springer-Verlag Berlin Heidelberg, 331 pp.

BENEDICT, J. B., BENEDICT, R. J., and SANVILLE, D., 1986, Arapaho rock glacier, Front Range, Colorado, U.S.A.: A 25-year resurvey: Arctic and Alpine Res., v. 18, pp. 349–352.

BLACK, R. F., 1976, Periglacial features indicative of permafrost: Ice and soil wedges: Quaternary Res., v. 6, pp. 3–26.

BÜDEL, J., 1982, Climatic geomorphology (transl. by L. Fischer and D. Busche): Princeton Univ. Press, Princeton, New Jersey, 443 pp.

BURN, C. R., 1992, Recent ground warming inferred from the temperature in permafrost near Mayo, Yukon Territory, in Dixon, J. C. and Abrahams, A. D., eds., Periglacial geomorphology: John Wiley & Sons Ltd., Chichester, UK, pp. 327–350.

———, and SMITH, M. W., 1993, Issues in Canadian permafrost research: Prog. in Phys. Geog., v. 17, p. 156–172.

CLARK, G. M., BEHLING, R. D., and 4 others, 1992, Central Appalachian periglacial geomorphology: Guidebook, Field excursion C.20c, 27th Internat. Geog. Cong.: Pennsylvania State Univ. College of Agriculture, Agronomy Series Number 120, 248 pp.

CLARK, G. M., and CIOLKOSZ, E. J., 1988, Periglacial geomorphology of the Appalachian highlands and interior highlands south of the glacial border-a review: Geomorphology, v. 1, pp. 191–220.

CLARK, M. J., ed., 1988, Advances in periglacial geomorphology: John Wiley & Sons Ltd., Chichester, UK, 481 pp.

COSTIN, A. B., JENNINGS, J. N., and 2 others, 1973, Forces developed by snowpatch action, Mt. Twynam, Snowy Mountains, Australia: Arctic and Alpine Res., v. 5, pp. 121–126.

———, and 2 others, 1964, Snow action on Mt. Twynam, Snowy Mountains, Australia: Jour. Glaciology, v. 5, pp. 219–228.

COTTON, C. A., 1963, A new theory of the sculpture of middle-latitude landscapes: New Zealand Jour. Geol. Geophys., v. 6, pp. 769–774.

CZUDEK, T., and DEMEK, J., 1970, Thermokarst in Siberia and its influence on the development of lowland relief: Quaternary Res., v. 1, pp. 103–120.

DIJKMANS, J. W. A., and MÜCHER, H. J. 1989, Niveo-aeolian sedimentation of loess and sand: An experimental and micromorphological approach: Earth Surface Processes and Landforms, v. 14, pp. 303–315.

DOHRENWEND, J. C., 1984, Nivation landforms in the western Great Basin and their paleoclimatic significance: Quaternary Res., v. 22, pp. 275–288.

EAKIN, H. M., 1916, The Yukon-Koyukuk region, Alaska: U.S. Geol. Survey Bull. 631, 88 pp.

EVERETT, D. H., 1961, Thermodynamics of frost damage to porous solids: Faraday Soc. Trans., v. 57, pp. 1541–1551.

FERRIANS, O. J., KACHADOORIAN, R., and GREENE, G. W., 1969, Permafrost and related engineering problems in Alaska: U.S. Geol. Survey Prof. Paper 678, 37 pp.

FLEMAL, R. C., 1976, Pingos and pingo scars: Their characteristics, distribution, and utility in reconstructing former permafrost environments: Quaternary Res., v. 6, pp. 37–53.

———, HINKLEY, K. C., and HESLER, J. L., 1973, DeKalb mounds: A possible Pleistocene (Woodfordian) pingo field in north-central Illinois, in Black, R. F., Goldthwait, R. P., and Willman, H. B., eds., The Wisconsinan Stage: Geol. Soc. America Mem. 136, pp. 229–250.

FLINT, R. F., 1971, Glacial and Quaternary geology: John Wiley & Sons, Inc. New York, 892 pp.

FRENCH, H. M., 1988, Active layer process, in Clark, M. J., ed., Advances in periglacial geomorphology: John Wiley & Sons Ltd., Chichester, UK, pp. 151–198.

———, 1993, Cold-climate processes and landforms, in French, H. M., and Slaymaker, O., eds., Canada's cold environments: McGill-Queens' University Press, Montreal and Kingston, Canada, pp. 143–167.

———, 1996, The periglacial environment, 2nd ed.: Addison Wesley Limited, Essex, UK, 341 pp.

———, and SLAYMAKER, O., 1993, Canada's cold land mass, in French, H. M., and Slaymaker, O., eds., Canada's cold environments: McGill-Queens' University Press, Montreal and Kingston, Canada, pp. 3–27.

FRIED, G., HEINRICH, J., and 2 others, 1993, Periglacial denudation in formerly unglaciated areas of the Richardson Mountains (NW Canada), in Barsch, D., and Mäusbacher, R., eds., Some contributions to the study of landforms and geomorphic processes: Zeit. für Geomorph., Supp. no. 92, pp. 35–59.

GIARDINO, J. R., SHRODER, J. F., JR, and VITEK, J. D., eds., 1987, Rock glaciers: Allen & Unwin, Inc., Winchester, Massachusetts, 355 pp.

GLEASON, K. I., KRANTZ, W. B., and 3 others, 1986, Geometrical aspects of sorted patterned ground in recurrently frozen soil: Science, v. 232, pp. 216–220.

GORDEEV, V.V., MARTIN, J.M., and 2 others, 1996, A reassessment of the Eurasian river input of water, sediment, major elements, and nutrients to the Arctic Ocean: Am. Jour. Sci., v. 296, pp. 664–691.

HALL, K. J., 1992, Mechanical weathering in the Antarctic: A maritime perspective, in Dixon, J. C. and Abrahams, A. D., eds., Periglacial geomorphology: John Wiley & Sons Ltd., Chichester, UK, pp. 103–123.

HALLET, B., and PRESTRUD, S., 1986, Dynamics of periglacial sorted circles in western Spitsbergen: Quaternary Res., v. 26, pp. 81–99.

HALLET, B., and WADDINGTON, E. D., 1992, Buoyancy forces induced by freeze-thaw in the active layer: Implications for diapirism and soil circulation, in Dixon, J. C., and Abrahams, A. D., eds., Periglacial geomorphology: John Wiley & Sons Ltd., Chichester, UK, pp. 251–279.

HARRIS, C., 1987, Mechanisms of mass movement in periglacial environments, in Anderson, M. G., and Richards, K. S., eds., Slope stability: John Wiley & Sons Ltd., Chichester, UK, pp. 531–559.

HARRY, D. G., 1988, Ground ice and permafrost, in Clark, M. J., ed., Advances in periglacial geomorphology, John Wiley & Sons Ltd., Chichester, UK, pp. 113–149.

HOPKINS, D. M., KARLSTROM, T. N. V. and 5 others, 1955, Permafrost and ground water in Alaska: U.S. Geol. Survey Prof. Paper 264-F, pp. 113–146.

HUMLUM, O., 1988, Rock glacier appearance level and rock glacier initiation line altitude: A methodological approach to the study of rock glaciers: Arctic and Alpine Res., v. 20, pp. 160–178.

JOHNSON, P. G., 1987, Rock glacier: Glacier debris systems or high-magnitude low-frequency flows, in Giardino, J. R., Shroder, J. F., Jr., and Vitek, J. D., eds., Rock glaciers: Allen & Unwin, Inc., Winchester, Massachusetts, pp. 175–192.

JOHNSON, W. H., 1990, Ice-wedge casts and relict patterned ground in central Illinois and their environmental significance: Quaternary Res., v. 33, pp. 51–72.

KELLY, W. C., and ZUMBERGE, J. H., 1961, Weathering of a quartz diorite at Marble Point, McMurdo Sound, Antarctica: Jour. Geology, v. 69, pp. 433–446.

KRANTZ, W. B., GLEASON, K. J., and CAINE, N., 1988, Patterned ground: Sci. American, v. 259, no. 6, pp. 68–76.

LACHENBRUCH, A. H., 1962, Mechanics of thermal contraction cracks and ice-wedge polygons in permafrost: Geol. Soc. America Spec. Paper 70, 69 pp.

———, CLADOUHOS, T. T., and SALTUS, R. W., 1988, Permafrost temperature and the changing climate, in Senneset, K., ed., Permafrost, v. 3, Fifth Internat. Conference on Permafrost: Tapir Publishers, Trondheim, Norway, pp. 9–17.

LAUTRIDOU, J.-P., 1988, Recent advances in cryogenic geomorphology, in Clark, M. J., ed., Advances in periglacial weathering: John Wiley & Sons Ltd., Chichester, UK, pp. 33–47.

LEFFINGWELL, E. deK., 1915, Ground-ice wedges: The dominant form of ground-ice on the north coast of Alaska: Jour. Geology, v. 23, pp. 635–654.

MALDE, H. E., 1964, Patterned ground in the western Snake River Plain, Idaho, and its possible cold-climate origin: Geol. Soc. America Bull., v. 75, pp. 191–207.

MARSH, B., 1987, Pleistocene pingo scars in Pennsylvania: Geology, v. 15, pp. 945–947.

MATTHES, F. E., 1900, Glacial sculpture of the Bighorn Mountains, Wyoming: U.S. Geol. Survey, 21st Ann. Rept., pt. 2, pp. 167–190.

McDOWALL, I. C., 1960, Particle size reduction of clay minerals by freezing and thawing: New Zealand Jour. Geol. Geophys., v. 3, pp. 337–343.

MULLER, S. W., 1947, Permafrost or permanently frozen ground and related engineering problems: Edwards Bros., Ann Arbor, Michigan, 231 pp.

NELSON, F. E., 1989, Cryplanation terraces: periglacial cirque analogs: Geografiska Annaler, v. 71A, pp. 31–41.

———, HINKEL, K. M., and OUTCALT, S. I., 1992, Palsa-scale frost mounds, in Dixon, J. D., and Abrahams, A. D., eds., Periglacial geomorphology: John Wiley & Sons Ltd., Chichester, UK, pp. 305–325.

———, and OUTCALT, S. I., 1987, A computational method for prediction and regionalization of permafrost: Arctic and Alpine Res., v. 19, pp. 279–288.

NEUMAN, C. M., 1993, A review of aeolian transport processes in cold environments: Prog. in Phys. Geog., v. 17, p. 137–155.

PÉWÉ, T. L., 1975, Quaternary geology of Alaska: U.S. Geol. Survey Prof. Paper 835, 145 pp.

———, 1983, The periglacial environment in North America during Wisconsin time, in Porter, S. C., ed., Late-Quaternary environments of the United States, v. 1, The Late Pleistocene: Univ. of Minnesota Press, Minneapolis, pp. 157–189.

PISSART, A., 1988, Pingos: An overview of the present state of knowledge, in Clark, M. J., ed., Advances in periglacial geomorphology: John Wiley & Sons Ltd., Chichester, UK, pp. 279–297.

POLLARD, W. H., 1988, Seasonal frost mounds, in Clark, M. J., ed., Advances in periglacial geomorphology: John Wiley & Sons Ltd., Chichester, UK, pp. 201–229.

POTTER, N., JR., 1972, Ice-cored rock glacier, Galena Creek, northern Absaroka Mountains, Wyoming: Geol. Soc. America Bull., v. 83, pp. 3025–3057.

———, and MOSS, J. H., 1968, Origin of the Blue Rocks block field and adjacent deposits, Berks County, Pennsylvania: Geol. Soc. America Bull., v. 79, pp. 255–262.

PRIESNITZ, K., 1988, Cryoplanation, in Clark, M. J., ed., Advances in periglacial geomorphology: John Wiley & Sons Ltd., Chichester, UK, pp. 49–67.

REGER, R. D., and PÉWÉ, T. L., 1976, Cryoplanation terraces: Indicators of a permafrost environment: Quaternary Res., v. 6, pp. 99–109.

SEPPÄLÄ, M., 1988, Palsas and related forms, in Clark, M. J., ed., Advances in periglacial geomorphology: John Wiley & Sons Ltd., Chichester, UK, pp. 247–278.

SMITH, H. T. U., 1953, Hickory Run boulder field, Carbon County, Pennsylvania: Am. Jour. Sci., v. 251, pp. 625–642.

ST. ONGE, D. A., 1969, Nivation landforms: Canada Geol. Survey Paper 69-30, 12 pp.

TELANG, S. A., POCKLINGTON, R., and 3 others, 1991, Carbon and mineral transport in major North American, Russian arctic, and Siberian Rivers, in Degans, E. T., Kempe, S., and Richey, J. E., eds., Biogeochemistry of major world rivers: John Wiley & Sons Ltd., Chichester, UK, pp. 75–104.

TE PUNGA, M. T., 1956, Altiplanation terraces in southern England: Biul. Peryglac., v. 4, pp. 331–338.

THORN, C. E., 1988, Nivation: A geomorphic chimera, in Clark, M. J., ed., Advances in periglacial geomorphology: John Wiley & Sons Ltd., Chichester, UK, pp. 3–31.

———, 1992, Periglacial geomorphology: What, where, when?, in Dixon, J. C. and Abrahams, A. D., eds., Periglacial geomorphology: John Wiley & Sons Ltd., Chichester, UK, pp. 1–30.

TRICART, J., 1970, Geomorphology of cold environments (transl. by E. Watson): Macmillan and Co. Ltd., London, 320 pp.

WAHRHAFTIG, C., 1965, Physiographic divisions of Alaska: U.S. Geol. Survey Prof. Paper 482, 52 pp.

———, and COX, A., 1959, Rock glaciers in the Alaska Range: Geol. Soc. America Bull., v. 70, pp. 383–436.

WALDER, J., and HALLET, B., 1985, A theoretical model of the fracture of rock during freezing: Geol. Soc. America Bull., v. 96, pp. 336–346.

WALKER, D. A., WEBBER, P. J., and 5 others, 1987, Cumulative impacts of oil fields on northern Alaskan landscapes: Science, v. 238, pp. 757–761.

WASHBURN, A. L., 1980, Geocryology: A survey of periglacial processes and environments: John Wiley & Sons, Inc., New York, 406 pp.

———, SMITH, D. D., and GODDARD, R. H., 1963, Frost cracking in a middle-latitude climate: Biul. Peryglac., v. 12, pp. 175–189.

WAYNE, W. J., 1967, Periglacial features and climatic gradient in Illinois, Indiana, and western Ohio, east-central United States, in Cushing, E. J., and Wright, H. E., Jr., eds., Quaternary paleoecology: Yale Univ. Press, New Haven, Connecticut, pp. 393–414.

WHALLEY, W. B., and MARTIN, H. E., 1992, Rock glaciers: II Models and mechanisms: Prog. in Phys. Geog., v. 16, pp. 127–186.

WHITE, S. E., 1971a, Rock glacier studies in the Colorado Front Range, 1961 to 1968: Arctic and Alpine Res., v. 3, pp. 43–64.

———, 1971b, Debris falls at the front of Arapaho rock glacier, Colorado Front Range, U.S.A: Geografiska Annaler, v. 53, ser. A, pp. 86–91.

———, 1976, Rock glaciers and block fields, review and new data: Quaternary Res., v. 6, pp. 77–97.

WILLIAMS, P. J., and SMITH, M. W., 1989, The frozen earth: Fundamentals of geocryology: Cambridge University Press, Cambridge, UK, 306 pp.

WORSLEY, P., 1985, Periglacial environment: Prog. in Phys. Geog., v. 9, pp. 391–401.

Chapter 15

Landscape Evolution

DEDUCTIVE GEOMORPHOLOGY

What happens to a terrane that lies exposed to atmosphere, rain, and life for millions of years? Are the changes predictable, sequential, and evolutionary? These are the questions that carry geomorphologists beyond the routine of observing and describing contemporary processes into the speculative or deductive realms. The time scale becomes that of the geologists, and many geographers, especially those who are interested in human interactions with the landscape, turn away. Time has always been a difficult concept for geomorphologists, as for all scientists. Until recent decades, time scales longer than a few centuries or millenniums were known only with poor precision, and relative terms such as landscape "youth," "maturity," and "old age" were used with only the poorest relation to numerical years. Newer methods of radiometric dating, especially those that measure exposure ages of rock surfaces by their accumulation of cosmogonic nuclides, promise a revolution in geomorphology comparable to the upsetting introduction of inexpensive wrist watches into a primitive society: Time then demands attention; structures cannot be described without specifying their age in radiometric years; processes are necessarily integrated over time to define their rates.

Chapters 7 through 14 have been concerned with the destructive and depositional processes that change landscapes, the rates at which those processes work, and the observational evidence that landscapes differ from place to place because of geologic structure and the nature, intensity, and duration of the processes of change. The reasoning has been largely from specific examples or experiments toward statements of general principles. This is the *inductive* approach, and it is a vital part of the system of logical thinking we call science.

The other side of the logical coin of science is the *deductive* approach. Here we reason from general principles toward analysis of specific problems. Chapters 7 through 14 have also provided numerous examples of this approach. For example, the hydraulic geometry of stream channels, the Manning equation, Hjulström's curve, and the Shields curve (Chapter 10), though compiled from extensive laboratory and field measurements, are well established and now serve as generalizations for deducing future responses and events at a particular place, given a particular set of conditions or changes. Geomorphologists, like all scientists, constantly use both inductive and deductive reasoning in solving problems, usually without consciously considering the distinction. The deductive approach to geomorphology primarily concerns the changes of landscapes through time. We cannot watch a landscape evolve even though we have abundant reasons to think that it does. We wish to make a motion picture of a landscape's life history, but our source material is a series of still photographs of many different landscapes. We must reason from general principles if we are to deduce the history of any landscape.

Deductions about landscape evolution have been severely criticized for having exceeded the restraints of observational evidence. It is true that in the first decades of this century the deductive system of "explanatory description" was pushed well beyond what had been proved experimentally. A great danger of deductive reasoning is that if the general principles are wrong, inadequate, or incorrectly applied, even the most careful logical procedures will inevitably lead to erroneous final deductions. Suppose, for instance, that a general principle had been established through repeated observations in humid regions (it has not) to the effect that hillside slopes decrease in steepness through time. If this principle were incorrectly used to predict the future shape of a desert slope, the deduced shape would have little relation to real landforms because the various processes of mass-wasting operate at different relative rates in humid and in arid climates.

Figure 2-5 gives adequate evidence that more water annually falls on the land than evaporates from it. At least 60 percent of all landscapes evolve under conditions of excess water runoff. In this chapter, the sequential evolution of landscapes under conditions of humid climate and abundant flowing rivers is given first importance. However, the very large areas of present aridity (approximately 30 percent of the land area) and permafrost (perhaps 20 percent of the land area) impose specific subaerial processes on rock structures. The results are expressed as landforms that would evolve along climate-specific paths, if given enough time. Other climatic types, especially the tropical wet and dry savanna environment, also may produce distinctive suites of evolving landforms. However, the issue of *equifinality* (p. 14), and the probability of major climate changes within the time span of the evolution of modern landscapes, makes evolutionary deductions based on climate-controlled processes a risky activity.

Glaciation imposes a unique set of subice processes on landscapes, and it is considered separately in two subsequent chapters because the resulting landforms are so different from those that take shape subaerially. Coastal and submarine processes are also so different from subaerial ones that they require separate consideration (Chapters 19 and 20).

PROOFS THAT FLUVIAL LANDSCAPES EVOLVE SEQUENTIALLY

No one who has seen a muddy river or a gullied hillside doubts that erosional processes consume the subaerial landscape. Nevertheless, the burden of proof rests on geomorphologists to demonstrate that landforms are produced, or that fluvial landscapes evolve, in a systematic fashion through time. As noted in Chapter 1, one of the fundamental justifications for geomorphology is that changes can be predicted. There are at least six experimental or observational proofs that landscapes evolve in predictable sequences of forms. Some are more rigorous than others, but all are useful.

Laboratory Experiments and Computer Simulations

Experiments on landscape evolution can be conducted on sand tables or other models. A groove molded in a sloping surface conducts water and, as it does so, changes its shape to accommodate the discharge. A mound of fine sediment under a spray mist develops drainage networks down its slopes. By careful choice of materials and experimental conditions, many landforms can be reproduced in miniature in a laboratory, and their evolution observed.

The difficulty with all scale models is that changes of dimensions in length, mass, and time do not alter the intrinsic properties of matter. Further, physical constants such as gravitation cannot be scaled down. For example, water in a model channel a few centimeters deep has the same density and viscosity as water in a real river; therefore, turbulence in the water due to these properties is grossly out of scale in the model. Dimensionless parameters such as D/δ_0, the ratio of bed grain size to the thickness of the surface laminar-flow layer (Figure 10-11b), cannot be scaled. Furthermore, as the linear dimension of an object is decreased, its volume decreases as the cube of the length, but its surface area decreases only as the square of the length. Thus, small particles have much larger surface areas in proportion to their masses than large particles of the same shape. Surface effects cause very fine-grained wetted particles in a model landscape to cohere in a fashion totally inappropriate to the non-cohesive sand or gravel of the real landscape that they are meant to represent in scale. An experimental river channel on a sand table proves how channels form on sand tables but very little more.

One constructive approach to scaling problems is to build large enough models so that they approximate small samples of real systems (Figure 11-1). Another approach is to use mathematical rather than "hardware" models such as sand tables. Geometric, dynamic, and kinematic variables can be programmed in accordance with known or assumed laws in deterministic models; probabilities can be stated and evaluated in probabilistic models (as in the case of random walk drainage networks, p. 234); or optimization criteria can be stipulated (such as the principles of least work or uniform distribution of work in river systems, p. 218) (Woldenberg, 1985, p. xv). Predicting the progressive degradation of initially steep scarps by the diffusion equation (Figure 9-20) is a good example of simple mathematical modeling in two dimensions. More elaborate two- and three-dimensional computer models of landscape evolution include numerous variables of weathering, mass wasting, and erosion, and feedback loops among them (p. 236). Modeling aided by computers is a growing subject of geomorphology, as well as of related engineering fields of hydrology, rock mechanics, and soil mechanics (Anderson, 1988).

However, the inherent problems of interpreting topologically random networks should be recalled (p. 235). An integrated system of river valleys can be equally well explained by either headward growth and branching of a stream network or progressive integration of small gullies and rills down an initial hillslope. Many probabilistic simulations generate branching systems that look like drainage networks and evolve in an intuitively acceptable way, yet the programs do not tell us how real networks have evolved or how they might change in the future.

Real Landscapes Evolving Under Accelerated Conditions

When a former equilibrium is disturbed by natural or man-made disaster, new landforms may develop quickly. By observing the changes, we can infer the sequences to be expected when changes are much slower. Complex drainage networks may evolve on tidal mud flats during the few hours of each low tide. After volcanic tephra falls, new drainage systems develop within a period of months or years. Large gullies, 10 m or more in depth and several hundred meters in length, have been known to develop in less than a century after land was cleared for agriculture (Figure 10-2).

Accelerated erosion due to natural or man-made catastrophe is good proof that landscapes evolve because the changes are full-scale except in time. Unfortunately, the "accident" that induces accelerated change commonly affects only one or a few processes of change. Accelerated erosion due to forest clearing and farming is caused by more rapid and concentrated runoff, but soil formation is not speeded up correspondingly, and mass wasting may change from soil creep to slumping and earthflow. Thus the resulting landforms are not those that would have formed if all processes of change had been accelerated proportionally. Accelerated erosion cannot be directly extrapolated to interpret landscape evolution on the scale of geologic time.

Playfair's Law of Accordant Junctions, and Other Probabilistic Deductions

A third defense for the proposition that fluvial landscapes evolve sequentially is embodied in Playfair's law (p. 232) and more modern statistical studies of drainage networks (Chapter 11). Playfair's statement is not a rigorous proof of a natural law but only an observation about a highly probable condition. He argued that if valleys did not evolve by the work of the streams that flow in them, the "nice adjustment of their declivities" would be a highly improbable state. Having considered the hydraulic geometry of streams and the concept of graded rivers, the average reader of this book is better equipped to appreciate the significance of Playfair's words than the greatest scholars of 1802.

With our increasing awareness of late Cenozoic climate changes, we are confronted with defining what Playfair meant by "the stream that [now] flows" in a valley. Is it the stream defined by the hydraulic geometry during the latest water year, or by the "100-year flood," or by the 10,000 years of Holocene time? Or did another kind of river cut the valley in which the stream now flows, perhaps one that carried a load of glacial outwash or periglacial solifluction debris (pp. 240, 322)? The rivers that now flow in valleys with "such a nice adjustment of their declivities," may have little resemblance to the kinds of rivers that did most of the valley erosion.

With a significant change in climate, such as those that are known during late Cenozoic time, various geomorphic thresholds are exceeded, processes change, and the landscape responds in complex ways, (Rinaldo et al., 1995) (Figure 1-7). Aridity might lead to a loss of vegetational cover and accelerated overland sediment transport, possibly causing aggradation in upland channel networks according to Huntington's principle (p. 244). Later as the upland adjusts to the new fluvial regime, gully incision might resume, but the sediment eroded from them might cause a wave of aggradation downstream as an intrinsic change not related to any extrinsic cause. It has even been proposed that the landforms explained by the deduced "normal" system of landscape evolution for midlatitude cool-humid climates, are largely relict from a time of much colder climate with perennially frozen ground and intense frost action (p. 322).

Progressive Lose of Potential Energy in Fluvial Systems

We must learn to recognize the influence of climatic change on landforms, without losing sight of the fundamental observation that rivers carry by far the bulk of the eroded debris from the lands to the sea except under the most extreme climatic disruptions. As long as a river is carrying sediment to the sea, it is lowering the landscape that provides the gravity potential for flow. There can be no long-term steady state in a physical system of declining energy supply ("cyclic time," Figure 1-6). We observe sediment in transport to the sea by fluvial systems and can prove that the sediment is derived from erosion of the landscape. These facts are powerful support for the assertion that landscapes evolve with time and, furthermore, that the rate of evolutionary change is likely to decrease with time as the energy of the system declines in the absence of tectonic uplift.

Just as the sediment in transport by rivers is a powerful proof that landscapes evolve, so is the alluvium that is temporarily left behind as river terraces during valley deepening (Chapter 11). Alluvial terraces have surface features and internal structure that prove they were once part of a river's floodplain. If they are not now reached by floods, either the valley has been deepened since they formed or there has been some diversion of water from the river. This proof of valley erosion is another of Playfair's defenses of the theories of his deceased colleague, James Hutton (Playfair, 1802, p. 103).

LANDSCAPES CAN BE ARRANGED IN SERIES

The fifth method of proving that landscapes evolve is not based on rigorous logic but on pragmatic observation. It is still the best practical demonstration of landscape evolution. At the beginning of this chapter, we noted that the conceptual goal is a motion picture of a single landscape evolving, but the source material is a series of still photographs of many landscapes. We can assemble any group of photographs of nontectonic eroding landscapes into progressive or transitional sequences, making due allowance for regional differences of climates and rock types, and always being alert for relict or exhumed forms. If we had a large enough collection of such photographs so that we could say with confidence that a certain percentage of all landscapes are in a certain state at one moment of time, we could apply the **ergodic theorem** from statistical mechanics and deduce that any single landscape is statistically likely to be in that same state for the same percentage of its life history (Paine, 1985). The ergodic theorem asserts that sampling an assemblage of systems at one time is equivalent to sampling one system at a succession of times. Unfortunately, rather than having a statistically valid sample of all landscapes, geomorphologists must attempt to deduce a life history for a landscape from only a limited number of case studies. Nevertheless, ergodic reasoning is a powerful tool for geomorphic deduction even though the open systems of the land surface and the processes of change are too complex for rigorous statistical prediction.

The ability to arrange landscapes in some sequential order does not tell us in which direction the order proceeds. Even if we make our motion picture, we do not know which way to run the film. For the establishment of a unidirectional series, we return to the preceding proof of landscape evolution, that rivers carry sediment. If eroding landscapes form a developmental series, it must be in the direction of larger valleys and smaller residual hills.

The practical fact that landscapes can be sequentially arranged has a theoretical basis in thermodynamic principles. As noted in Chapter 10, the graded state results from a tendency toward equilibrium between opposing principles of least work and uniform distribution of work. As a landscape erodes, progressively better adjustment between opposing tendencies must result. In their discussion of river-channel morphology, Langbein and Leopold (1964, p. 793) made the significant inference that as rivers tend toward equilibrium between the two opposing principles, "deviations in time will approximate those which can be observed in an ensemble of places." Some conditions of landscapes are more probable than others, and nature favors tendencies toward the probable. The two most probable changes in nontectonic fluvial landscapes are, first, that the rivers will become graded and, second, that their potential energy will decline through time. Both changes permit predictions to be made about the changes in appearance of individual landscapes with the passage of time. A difficulty remains in specifying whether the changes of the landscape lie within "graded time" or "cyclic time" (p. 15). On the shorter time scale of "graded time," fluvial systems evolve from the ungraded to the graded condition, then remain in the graded condition thereafter. The ergodic theorem cannot be applied to graded time because the condition of grade is essentially timeless. On the longer time scale of cyclic time, possibly involving many millions of years, landscape change is progressive, and the concepts of the ergodic theorem can be applied although not with the necessary statistical rigor.

It may seem a waste of time and intelligence to outline these first five proofs of landscape evolution because they are all quite apparent. Yet only 200 years ago, the concept of slow, orderly development of landscapes was a challenge to the established religious and philosophical order. The acceptance of Hutton's concept of uniformity and continuity of process, and the corollary concept of the enormousness of geologic time, set the intellectual stage for the theory of organic evolution in the mid-nineteenth century.

The Sixth Proof: Radiometrically Dated Landscapes

Some landscapes, such as those eroded from volcanic rocks or emerged coral reefs, can have their structure dated by radiometric means. Knowing the length of time that has elapsed since inception, a rate of denudation can be determined and used to estimate the survival time of the subaerial landscape. Knowing that a landscape has a finite life does not aid in deducing its appearance at various stages of the aging process, but by dating the rocks in landscapes of similar structure but of different ages, a precise chronology of the erosional stages can be made.

A Hawaiian example of the geomorphic perspective provided by radiometric dates is the lowland on the northeastern side of Oahu (Figure 15-1). The great scarp of the Nuuanu Pali truncates basalt flows of the Koolau Range that are about 1.8 to 2.7 million years old (Clague and Dalrymple, 1987, p. 51). It may be part of the caldera rim that was breached by a great submarine landslide more than 1.4 million years ago (p. 187). Subsequently, in the past 0.9 million years, small scoria cones and flows have covered part of the eroded caldera floor. Thus the lowland relief shown by Figure 15-1 is the result of mass-wasting and erosion between about 1.4 and 0.9 million years ago, slightly modified by later events.

Other datable erosional landscapes, such as dissected till plains of the last glaciation, surfaces covered by tephra of a known age, or coral reefs elevated above sea level (Chappell, 1974), are available to measure the actual progress of erosional dissection. Radiometric dating methods such as fission-track dating can measure the time that has elapsed during several kilometers of erosion. *Cosmogenic radiation* dating methods now being developed should be able to define the time interval during which surface rocks and soil, perhaps to a depth of about 1 m, have been exposed to cosmic radiation by measuring the accumulation of a variety of cosmic-ray-produced nuclides (Dorn and Phillips, 1991; Bierman, 1994; Bierman and Turner, 1995; Anderson et al., 1996). These methods offer a proof of landscape evolution that was not available to the great geomorphologists of earlier generations. Several examples are described in subsequent paragraphs.

RATES OF LANDSCAPE EVOLUTION

The progress of sequential evolution of landscapes traditionally has been described in relative, not absolute, terms. The words *youth, maturity, old age,* and *senility* were applied to landforms by analogy to organisms, noting that the absolute interval of time required to attain maturity may vary by orders of magnitude among the various forms of life. The stages were defined only by observed criteria, not by time in years.

FIGURE 15-1. View northwest along the headscarp of the Nuuanu Pali, Oahu, Hawaii. Erosion on the northeastern side of Oahu is rapid because of high rainfall brought by the Trade Winds.

There are a number of ways, however, by which **denudation rates,** or the depth of rock removed from an area in a specified time interval, can be measured. By applying the appropriate rate to the total volume of a landscape, quantified estimates of the durability of subaerial landforms can be made.[1]

Methods of Measuring Denudation Rates

Degradation and Exposure of Historical Monuments.

The stone monuments and walls of civilizations more than a few centuries old commonly show loss of surface detail and mass through weathering and erosion (Figure 7-6). Typical surface loss and progressive inscription illegibility by solution on limestone and marble tombstones in the eastern United States is in the range of 0.5 to 1.5 mm/century (0.5 to 1.5 cm/1000 yr) (Meierding, 1993). Arkosic sandstone monuments in cemeteries near Middletown, Connecticut, are weathering at a rate of about 6 mm/century (6 cm/1000 yr) (Matthias, 1967; Rahn, 1971), almost the same rate as the present average rate of landscape lowering for the entire United States (Table 15-1). Carved and polished tombstones may not weather in the same way as naturally exposed rock, but they are exposed to equivalent weathering conditions and provide a good index of relative weathering by various lithologies.

The U.S. Department of Energy is studying the design of an underground repository for civilian nuclear waste. The design life for the repository is to be at least 10,000 years, and consideration must be given to designing a surface monument that will warn future generations of the dangers of site disturbance. The oldest surviving human structures are the pyramids of Egypt, which are less than half as old as the proposed life of the warning monument at the nuclear waste repository. They and other ancient structures have been seriously degraded, but the primary cause for their damage has been human, not natural. Ancient monuments have been quarried for reused building stone, used as military practice targets, and otherwise despoiled. But even without human defacing, an inscription carved in stone at the nuclear waste site would have to be protected from weathering to survive for 10,000 years. No precedent exists.

In addition to losses of their own surface detail, ancient monuments and walls may mark the former ground level, so that regional denudation can be measured. Neolithic earthworks on the chalk hills of southern England have preserved the underlying chalk from solution, while around the structures the land surface has been lowered at a rate of 10 to 12.5 cm/1000 yr (Pitty, 1971, p. 191). Soil erosion has exposed the foundations of Roman ruins and lowered the land surface adjacent to paved Roman roads in Italy at a rate of 20 to 50 cm/1000 yr in about the last 2000 years (Judson, 1968a). The post-Roman rates are high, probably as the direct consequence of human activity.

Sediment Loads of Rivers.

The most obvious way to measure denudation rates by fluvial processes is to monitor the amount of load a river carries from its watershed. Although simple in principle, the actual measurements are difficult. Only the suspended load of rivers is routinely measured. Dissolved load is less frequently recorded, and bedload has generally defied direct measurement (p. 214). Corrections for volumetric changes during weathering must always be kept in mind because, as noted in Chapter 7, chemical weathering produces mostly dissolved bicarbonate compounds, but most of the bicarbonate ions are derived from the atmosphere, not from the weathered minerals.

Careful estimates of river loads and regional denudation for the entire United States were first attempted by Dole and Stabler (1909). They concluded that the United States was being lowered at a rate of 1 inch in 760 years (3.3 cm/1000 yr). A later study (Judson and Ritter, 1964) had much more data available, especially for dissolved loads. The solid load of Table 15-1 includes an assumed bedload equal to 10 percent of the suspended load. The regional denudation rates in Table 15-1 vary from 17 cm/1000 yr lowering of the Colorado River Basin to 4 or 5 cm/1000 yr for Atlantic and Gulf Coast drainage basins. Rates of denudation by limestone solution are comparable (Figure 8-7). Judson (1968b) estimated that owing to erosion induced by agriculture, the present average denudation rate for the United States of 6 cm/1000 yr is approximately double the precolonial rate.

Total sediment loads for most of the world's rivers are poorly known, but some major rivers at least have had their average annual load and **yield** (t/km^2) of suspended sediment estimated (Milliman and Meade, 1983; Milliman and Syvitski, 1992; Walling and Webb, 1996). The variation of yield, which is the annual suspended load per unit drainage area, for major rivers in

---

[1] Throughout this section of the chapter, denudation rates are expressed as centimeters per 1000 years of land-surface lowering even though the duration of the measured intervals vary from decades to millions of years. The dangers inherent in such comparison of process rates over various intervals of time are discussed on p. 6.

Table 15-1
Rates of Regional Denudation in the United States[a]

Drainage Region	Drainage Area ($10^3 km^2$)	Runoff (m^3/sec)	Annual Load						Denudation Rate (cm/1000 yr)
			Dissolved (tonnes/ km^2)	(%)	Solid [b] (tonnes/ km^2)	(%)	Total (tonnes/ km^2)		
Colorado	629	0.6	23	5.2	417	94.8	440		17
Pacific Slopes, California	303	2.3	36	14.7	209	85.3	245		9
Western Gulf	829	1.6	41	28.9	101	71.1	142		5
Mississippi	3238	17.5	39	29.2	94	70.8	133		5
S. Atlantic and E. Gulf	736	9.2	61	55.5	48	44.5	109		4
N. Atlantic	383	5.9	57	45.2	69	54.8	126		5
Columbia	679	9.8	57	56.4	44	43.6	101		4
Totals	6797	46.9	43	26.5	119	73.5	162		6

Sources: Judson and Ritter (1964) and Judson (1968b).
[a]Great Basin, St. Lawrence, and Hudson Bay drainage are not included.
[b]Solid load includes assumed bed load equal to 10 percent of suspended load.

various geographic regions shows certain regional trends (Figure 15-2). The tonnages of suspended load were converted to cm/1000 yr of denudation over the drainage area on the assumption that the density of the denuded rock is 2.65 g/cm^3, a typical value for silicate and carbonate rocks. Non-Arctic Asian rivers and rivers draining glaciated mountains have the highest denudation rates. Africa, Arctic drainage, and Australia have generally low rates. The research on which Figure 15-2 was based should be carefully read as an illustration of the assumptions and extrapolations that are required for such summaries. For example, the Colorado River basin was estimated in 1964 to have a denudation rate of 17 cm/1000 yr (Table 15-1). In 1968, another authority estimated the Colorado River denudation rate at 14 cm/1000 yr (Holeman, 1968). By 1983, the series of dams on the Colorado River had reduced the load that it carries to the sea to less than 0.1 percent of the previous measurements (Milliman and Meade, 1983, p. 5). Since the construction of the Aswan dam on the Nile River, it too carries essentially no suspended load to the Mediterranean Sea, and the delta coast is actively eroding. Some of the gaging stations used for compiling Figure 15-2 are upstream of river mouths, so additional sediment load may be added or removed farther downstream. The best documented example is the Huang He (Yellow River), where 33 percent of the suspended load that leaves the loess-covered hills of northern China is deposited on the lower alluvial plain, and another 43 percent is dropped in the delta region. Only 24 percent of the upstream river load actually reaches the ocean (Milliman and Meade, 1983, p. 3). With all of their cautious assumptions, Milliman

and Meade (1983, p. 18) concluded that the average annual sediment yield from the continents is about 180 t/km^2, which is equivalent to a bedrock denudation rate of 6.8 cm/1000 yr. Milliman and Syvitski (1992) increased the estimate of total suspended load by about 50 percent by including many additional small, high gradient rivers that drain from orogenic belts directly into the nearby ocean. Although the additional drainage area of these small rivers offsets to a large extent any increase in average sediment yield, such small rivers that drain actively tectonic regions have very high denudation rates. An extremely high contemporary erosion rate of 5 to 6 m/1000 yr has been reported for the eastern mountains of Taiwan, by which those mountains may have reached an equilibrium altitude at which rapid erosion balances rapid tectonic uplift (Dahlen and Suppe, 1988, p. 169) (p. 118).

Abundant evidence has been presented in previous chapters that most of the sediment transported by rivers is derived from the upland slopes adjacent to first-order tributaries. The smaller and more headward the drainage basin, the larger the proportion of total stream load it supplies. Whereas soil erosion in the conterminous United States is estimated at more than 5×10^9 t/yr, the suspended sediment load of rivers draining the same area is less than 0.4×10^9 t/yr, only 8 percent of the reliably estimated soil loss (Milliman and Meade, 1983, p. 18). In fact, the sediment load of a trunk stream is never as great as the sum of the loads from its tributaries (Figure 11-5). This anomaly may be explained as a phenomenon of either climatic change or the explosive impact of human beings on landscapes during the last few centuries and millenniums (Trimble,

FIGURE **15-2.** Variation of annual suspended sediment load with drainage area. Yields for Asian and glacial rivers are large, but desert rivers (Australia, SW Africa) have small yields (Milliman and Meade, 1983, Figure 3).

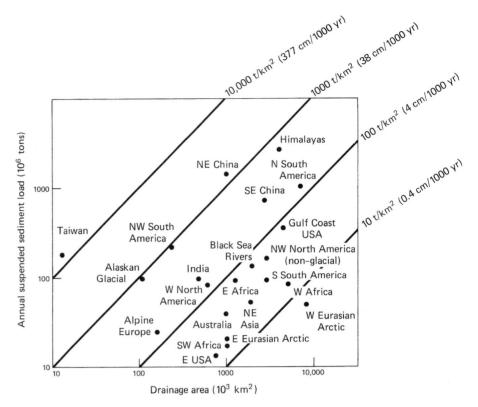

1975, 1977). Alternatively, floodplains and hollows on hillslopes may naturally store alluvium much longer than has been suspected (p. 191).

The rates of slope retreat by mass wasting (p. 193), when converted to approximate denudation rates, imply regional lowering at a rate an order of magnitude more rapid than the estimates based on suspended-sediment loads in rivers. The contradiction could be explained by hypothesizing that only about 10 percent of the area of a fluvial eroding landscape actually furnishes the stream load, and the other 90 percent is nearly immobile under colluvial or alluvial cover. Some evidence supports that hypothesis. In the Amazon basin, for instance, 84 percent of the total sediment load is furnished by the 12 percent of the drainage area that is high in the Andean headwaters (Gibbs, 1967). River-water chemistry and submarine cores off the mouths of the Amazon provide no hint that the largest river on earth traverses an enormous rainforest lowland for most of its length (Kronberg et al., 1986).

Submarine Sedimentation Rates. Very little fluvial sediment ever reaches the abyssal plains of the ocean floor. Nearly all of it is deposited on the continental margins. Therefore, geophysical studies of the sedimentary thickness on the continental margins can be used to derive longer-term rates of denudation. More

assumptions are involved, including major uncertainties about the exact age of the sediments in the wedge under the continental margins, the area of continent that furnished the sediment, the contribution of marine biogenic sediments, volume and density changes during diagenesis, loss by uplift and recycling of previously deposited sediments, and many other factors. Representative papers on the subject are those by Curray and Moore (1971), Mathews (1975), Bell and Laine (1985), and Poag and Sevon (1989).

Mathews (1975) proposed that the volume of the Cenozoic wedge of sediment in the western Atlantic Ocean, if derived from the Appalachian region of eastern North America, would require an average of about 2 km of regional denudation in the last 60 million years, giving a Cenozoic denudation rate of about 3.3 cm/1000 yr, very similar to the rates calculated for the eastern United States from modern river sediment loads (Table 15-1). However, he calculated that the denudation rate for the total Cenozoic Era was much less in the northern part of eastern North America, which includes the crystalline terranes of the Adirondack Mountains, New England, and eastern Canada; in those areas, the Cenozoic rate has been only about 0.5 cm/1000 years.

The sediment distribution and lithology along the eastern U.S. continental margin demonstrates that the present river systems were in place across the Appalachian Mountains (Figure 12-11) by early Mesozoic time, not long after the opening of the Atlantic basin. At least three phases of rapid sediment accumulation, correlating with times of source-area tectonic uplift, alternated with two longer phases of slower sedimentation, when the Appalachian region was presumably worn to low relief. The most recent interval of Appalachian uplift seems to have been from mid-Miocene (38 million years ago) to the present (Poag and Sevon, 1989, p. 128).

Depth of Erosion over a Known Interval of Time.
In the arid to semiarid White Mountains of California, basalt lava flows with an age of 10.8 million years were erupted onto a dissected landscape of low relief. Later, the area was block faulted and eroded. The lava flows now form ridges that are 50 m or more above most of the surrounding landscape. By drawing topographic profiles of the present landscape and projecting the base of the former lava flows above the profiles, cross-sectional areas can be obtained that when divided by profile lengths, give the mean lowering of the landscape since the flow was removed (Figure 15-3). The average thickness of the basalt, 37 m, must be added to the average amount of lowering to give the total denudation since the flows stopped and dissection began. By this technique, the most probable denudation rate during the past 10.8 million years was about 1.6 to 2.0 cm/1000 yr for the region (Marchand, 1971). The rates were probably greater on granitic igneous rocks than on basalt and dolomite in this cold, dry upland climate. Another study of the same region (LaMarche, 1968) concluded that the denudation rate during the last few thousand years has been at least an order of magnitude more rapid because the roots of bristlecone pine trees as old as 4000 years have been progressively exposed by mass wasting and erosion at rates of 15 to 20 cm/1000 yr (Figure 15-4). Holocene climatic changes may have intensified the denudation rates. Possibly the rapid recent rates need correction for their relatively brief duration (p. 6).

FIGURE 15-4. Bristlecone pines, White Mountains, Inyo County, California. Centers of roots extend about 60 cm above ground level, a measure of degradation while the tree has been growing (photo: V. C. LaMarche Jr., U.S. Geological Survey).

The foregoing two studies, from the same region but varying greatly in the scales of time that were studied and the evidence that was used, have a common basis in that a reference horizon, either the base of a lava flow or the former ground level on roots of long-lived trees, was used to measure regional denudation over an extended time interval. In Chapter 7, several other studies of reconstructed fold geometry, metamorphic facies, and geothermal heat flow were cited as proof that many landscapes are now developed on rocks that were formerly buried to the depths of 20 km or more (p. 121). The 24-km thickness of rock that has been removed from the Adirondack Mountains was mostly eroded between 1.1 billion and 500 million years ago, because Cambrian and Ordovician marine sediments transgressed onto a low-relief terrain eroded across the Grenville age (1.1 billion years) metasedimentary rocks. The average denudation rate for the entire erosional interval would have been 2 cm/1000 yr. During subsequent uplift in early Cretaceous time, the ancient rocks were again unroofed, at a rate of about 4 cm/1000 yr (Roden-Tice and Tice, 1996). The presently exposed Sierra Nevada granodiorite batholiths were emplaced at depths of less than

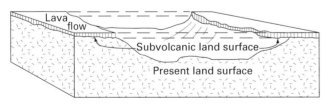

FIGURE 15-3. Former landscape in the White Mountains, California, buried by a basalt flow averaging 37 m in thickness, now preserved only on ridges above valleys 50 m or more in depth (from Marchand, 1971, Figure 3).

10 km over an interval of Cretaceous time until about 80 million years ago, implying an average denudation rate thereafter of as much as 12 cm/1000 yr. However, most of that erosion had been accomplished by middle to late Eocene time, after which Oligocene volcanic rocks erupted over the exposed granodiorite terrane on a surface of low relief. If 10-km erosion had been accomplished between 80 and 38 million years ago, the average rate would have been almost 24 cm/1000 yr (Christiansen and Yeats, 1992, p. 288). For the Swiss Alps, 30 km of denudation during the past 30 million years (p. 121) gives an average rate of 100 cm/1000 yr.

Fission-track ages in the Appalachian region demonstrate that the thickness of former rock that has been removed since late Paleozoic time ranges from 3 km in the Appalachian Plateau to more than 9 km in the eastern part of the Valley and Ridge Province, which includes the anthracite coal district (Roden and Miller, 1989, Figure 3). About 6 km of the maximum 9 km was removed by erosion during and soon after the compressional phase of the Appalachian Orogeny when the Atlantic Ocean basin closed. Much of the erosional detritus of the late Paleozoic mountains was shed to the west, into the shallow epicontinental sea that covered the midcontinent region. The denudation rate during the 25–30 million year interval from the late Paleozoic to early Jurassic time might have been as high as 20 to 38 cm/1000 years from an Andean-type landscape with peaks about 4 km in height (Slingerland and Furlong, 1989, p. 35). Thereafter, the remaining 3 km of measured denudation has been removed sometime during the last 230 million years. Although the rate of erosion was certainly not constant over this long interval of Mesozoic and Cenozoic time, the average rate would have been only 1.3 cm/1000 yr.

For the Nanga Parbat massif of the western Himalaya Mountains in Pakistan, fission-track ages require regional erosion rates as high as 3 to 4 mm/yr (300 to 400 cm/1000 yr) during the last 1 to 2 million years (Burbank and Beck, 1991). For fission-track studies, the erosion rate is derived from a cooling model that requires rapid erosion to strip off surface rocks and steepen the geothermal gradient; uplift rates are assumed to be equal to the denudation rates on the assumption that such extreme erosion rates could not persist without continuing orogeny. The inferred denudation rates for the Nanga Parbat region are many times faster than the denudation rates derived from sediment loads of Himalayan rivers (Figure 15-2) and are among the most rapid inferred by any technique. Their significance to continental erosion is discussed in later sections of this chapter.

Papuan Erosion: A Mathematical Exercise

Volcanoes offer a unique opportunity to measure rates of erosional landscape evolution. Their constructional forms are typically symmetrical and predictable. Their initial size and shape are nearly independent of climate-controlled erosional processes, and they are found in all climates. Their eruptive rocks can be dated by a variety of radiometric methods, and they evolve from active, growing landforms to skeletal remnants over a time span of 10 million years or more (Ruxton and McDougall, 1967; Karátson, 1996). One study of volcanic erosional evolution and some extended conclusions from it are reviewed here as an example of the approach.

The Hydrographers Range is a dissected andesitic composite cone in northeastern Papua New Guinea. Annual rainfall in the region probably ranges from 2250 mm at sea level to more than 3000 mm above an altitude of 1000 m. Most of the mountains are covered by tropical rain forest. Potassium-argon dates establish that the volcano ceased eruption about 650,000 years ago, when it began its erosional history. Most of the central summit portion is now dissected, and its initial shape is not known, but it probably was at an altitude of about 2000 m and had a considerable amount of constructional relief around multiple eruptive centers. Remnants of the youngest lava flows form the interfluves below an altitude of 975 m, and the predissection shape of the volcano can be reconstructed with confidence below that altitude. The dissecting valleys deepen and steepen rapidly toward the interior of the dissected mountain. By measuring the areas of numerous valley cross sections at selected altitudes, the volume of rock eroded in 650,000 years was calculated. Using assumptions similar to those used in the previous example of the White Mountains in California (Figure 15-3), denudation rates were tabulated for each of a succession of reconstructed contour lines between 60 and 975 m above sea level (Figure 15-5). Above 1000 m, in the summit area, only rough estimates of the amount of erosion could be made. The reported denudation rates range from only 8 cm/1000 yr at 60 m above sea level to 52 cm/1000 yr above 900 m altitude (Ruxton and McDougall, 1967). The best-fitting equation for the relationship between denudation rate and initial altitude is shown in Figure 15-5.

Because denudation rate is so strongly correlated with altitude in the range of 60 to 1000 m, it is reasonable to extrapolate the equation given in Figure 15-5 to 2000 m, the probable initial constructional height of the Hydrographers Range. By such extrapolation, the denudation rate near the original summit must have

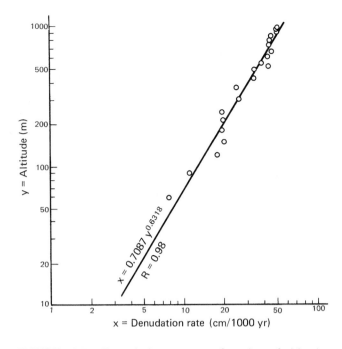

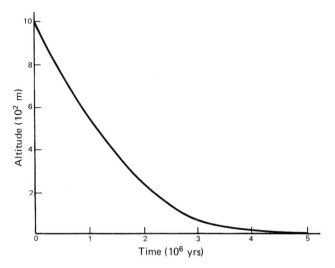

FIGURE 15-6. Time required to lower the summit of the Hydrographers Range, eastern Papua New Guinea, from an initial assumed altitude of 1000 m. Rates are shown on Figure 15-5.

FIGURE 15-5. Denudation rate as a function of altitude, eastern Papua New Guinea (from Ruxton and McDougall, 1967).

been about 86 cm/1000 yr, sufficient to lower the central region about 560 m since it was built. This extrapolation is close to the 490 m of summit-area lowering that can be roughly estimated from reconstructing the original cone-shaped profile. Both estimates imply that the central part of the range has been lowered by fully one-fourth of its initial altitude in only 650,000 years.

The equation (Figure 15-5) that links denudation rate to altitude can be integrated between any successive altitudes to give the amount of time necessary to lower the range the specified amount. Figure 15-6 graphs the future denudation history of the region but assumes only the present altitude of 1000 m to avoid the uncertainties of the unknown configuration of the actual summit region. At the denudation rates reported by Ruxton and McDougall (1967), the mountain range could be lowered from 1000 m to 500 m in 1.1 million years, to 100 m in 2.8 million years, to 10 m in 4.0 million years, and to sea level in 4.9 million years. Although the denudation rate approaches zero as altitude approaches zero, the rate remains finite, and the time for total erosion of the volcanic terrane to sea level is also finite.

The denudation rates for the Hydrographers Range are not excessive compared to other regions (compare the rates shown in Figure 15-5 with Tables 15-1 and 15-2 and Figure 15-2), but the integrated time for

denuding the entire region is unusually short. The apparent rapid time for total removal of the Hydrographers Range might be caused by the initial assumption that denudation rate is only a function of altitude and not of some other variable such as area of watershed above a certain altitude. As a mountainous area is reduced, the total orographic rainfall might decrease, even at low elevations.

Figure 15-7 gives additional evidence for inferring the sequential evolution of a mountainous terrain during denudation. In the Hydrographers Range, **local relief** (vertical difference between ridge crests and valley floors), slope gradients, and average slope length all decrease with decreasing denudation rates, and therefore also with altitude. Extrapolating in time, Figure 15-7 predicts that as the Hydrographers Range is lowered, its valleys will become broader and shallower and will develop decreased side-slope gradients. Highest areas are eroding most rapidly, so the original conic form should become lower and more dissected, and the interstream ridges should become lower and less steep-sided. After only a few million years, this constructional volcanic cone will become a low range of radiating ridges, separated by broad, shallow valleys. No mention is made in the original description of structural controls being exerted on the denudation rate by massive lava flows, dikes, or sills (Figure 6-15), but such controls could convert the initial radial pattern to an annular or centripetal pattern (Table 12-1).

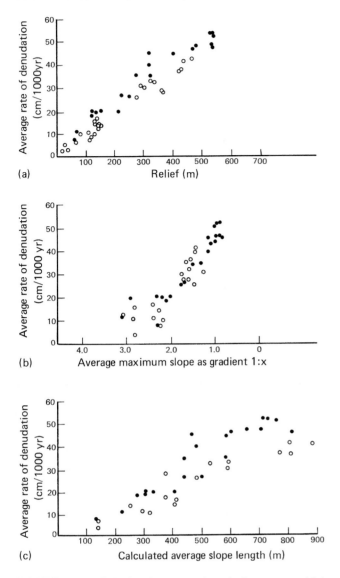

FIGURE 15-7. Relation between denudation rate and (a) local relief, (b) average maximum slope, and (c) average slope length in the Hydrographers Range. Data from more dissected sectors are shown as closed circles; data from less dissected sectors are shown as open circles (from Ruxton and McDougall, 1967, Figure 5).

Climate and Relief Factors in Denudation Rates

Table 15-2 gives some estimates of regional denudation rates characteristic of various climates in areas of steep and "normal" lowland relief. Although chemical weathering rates show no correlation with relief (Lasaga et al., 1994, p. 2379), the much more effective physical weathering processes (p. 127) and mass wasting are obviously accelerated by steep relief. As expected from the preceding data compilations, mountainous regions erode at rates of at least 10 to 100

cm/1000 yr, an order of magnitude more rapid, regardless of climate, than lowlands with typical rates of 1 to 10 cm/1000 yr. The figures offer impressive support for Powell's dictum (p. 118) that mountains are ephemeral landforms. Mountains with valley glaciers seem to erode most rapidly of all, probably because freezing water and glacier erosion add to the erosional work of rivers. Studies later than those used for the compilation of Table 15-2 give even higher denudation rates for glacierized (p. 353) mountains, ranging up to 50 m/1000 yr for the tectonically active southeastern Alaskan coastal ranges (p. 381). Research on denudation rates in other cold climates are complicated by the fact that many of the regions were only recently deglaciated, and landforms shaped primarily by glaciers are now out of equilibrium with fluvial processes. Under such conditions, rapid geomorphic change is inevitable (Harbor and Warburton, 1993; Hallet et. al., 1996). Temperate and tropical humid climates have ranges of rates that are comparable although some of the most extreme denudation rates are recorded in southeast Asia and the tropical islands of Taiwan, Philippines, and Indonesia, where a combination of small drainage basins with high relief, seismicity, intense weathering, and human interference of the natural ecosystems all combine (Figure 15-2).

Another way of comparing the climatic control of fluvial denudation rates is to compare the sediment yield of drainage basins of comparable size but located in a variety of climates (Langbein and Schumm, 1958; Wilson, 1973; Douglas, 1967). Figure 15-8 (upper curve) shows the sediment yields of 94 small watersheds in the United States, grouped into 6 categories of similar runoff (Langbein and Schumm, 1958). The lower curve of Figure 15-8 is based on a compilation of data from a wide range of published sources, calculated on the same basis as the upper curve (Douglas, 1967). Both curves show that the sediment yields, converted to denudation rates, have maximum values at a mean annual runoff of about 30 mm, which, at a mean annual temperature of 10°C, requires an *effective precipitation* (the amount of precipitation required to produce the observed runoff) of about 250 to 300 mm/yr. The lower graph shows low denudation rates for runoff of 200 to 600 mm/yr, or an effective precipitation of 760 to 1270 mm/yr. At still higher mean annual runoff, the denudation rate rises again (Douglas, 1967, Figure 2).

The disparity in absolute values of the denudation rates shown by the two graphs has been attributed to the fact that most of the 94 U.S. sites used to compile the upper curve were in agricultural areas, whereas the lower curve was compiled from drainage basins in Australia and elsewhere that were in natural vegetation. Agricultural activity is known to at least double

Table 15-2

Denudation Rates for Various Climate and Relief Conditions

Climate	Relief	Typical Range for Rate of denudation, (cm/1000yr)
Glacial	Normal (= ice sheets)	5–20
	Steep (= valley glaciers)	100–500
Polar/montane	Mostly steep	1–100
Temperate maritime	Mostly normal	0.5–10
Temperate continental	Normal	1–10
	Steep	10–20+
Mediterranean	—	1–?
Semiarid	Normal	10–100
Arid	—	1–?
Subtropical	—	1?–100?
Savanna	—	10–50
Rainforest	Normal	1–10
	Steep	10–100
Any climate	Badlands	100–100,000

Source: Saunders and Young (1983, p. 497).

the denudation rate, so the two curves may be in reasonable agreement. Conversely, Australia seems to have exceptionally low denudation rates (pp. 58, 347).

For temperatures typical of the United States, the peak sediment yield corresponds to the vegetational boundary between desert shrubs and grasslands. Several other researchers have drawn similar conclusions (summarized in Yair and Enzel, 1987, Figure 1). In general, the denudation rate increases with precipitation and decreases with vegetation cover. As mean annual runoff increases from zero in deserts to about 30 mm in

semiarid regions, fluvial denudation increases because, although there is more runoff, the precipitation is inadequate to maintain continuous vegetation cover. As precipitation further increases, grass and forest become more effective in stabilizing the soil, and the denudation rate decreases. At precipitation levels in excess of the amount required to maintain a closed forest cover, the denudation rate begins to rise again although not to the levels measured in semiarid to subhumid climates (Douglas, 1967; Kirkby, 1980, pp. 3-5). The amount of denudation by chemical solution would also increase in

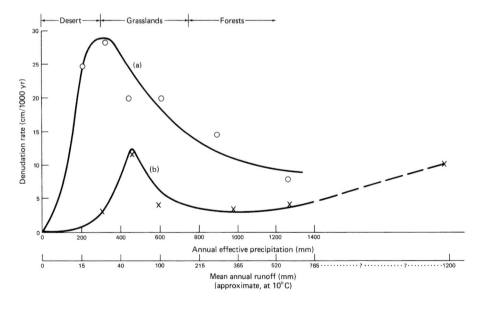

FIGURE 15-8. Variation of denudation rates with annual effective precipitation or runoff. Each symbol represents the arithmetic average of a group of similar stations. Effective precipitation is the precipitation necessary to produce a given amount of runoff at a mean annual temperature of 10°C. The vegetational boundaries apply primarily to temperature conditions in the United States. The conversion graph does not extend beyond 1400 mm of effective precipitation. (a) 94 stations in the United States in 6 groups (after Langbein and Schumm, 1958); (b) 90 stations in Australia and elsewhere (derived from sediment yield) in 6 groups (after Douglas, 1967).

very wet regions although this was not considered in these denudation studies. These inferences are crudely supported by the denudation rates listed in Table 15-2, with lowlands in semiarid, subtropical, and savanna climates yielding a range of denudation rates approximately ten times greater than lowlands in either arid or temperate and tropical humid climates. Clearly, type and extent of vegetation cover is at least as important as precipitation in determining denudation rates, and temperature is a minor factor. The importance of vegetation cover is emphasized by the contrasting denudation rates of the two curves in Figure 15-8, with agricultural disturbance clearly being a major factor in the denudation rates in the United States (Larson et al, 1983).

Although opinions may conflict concerning denudation rates under moderate to high precipitation, all authors agree that as precipitation approaches zero, the fluvial denudation rate must also approach zero. With no net runoff from the land, no net lowering of the landscape can occur unless by mass wasting and wind. The lowering of mountains by denudation is neutralized or even exceeded by aggradation of desert basins with sediments that have bulk density lower than that of the eroded rock mass. Arid mountains are literally buried in their own detritus. Erosion to a rising base level of aggradation is the most important factor in the deduced erosional history of mountainous desert landscapes (Chapter 13).

Conclusions from Figure 15-8 invite comparison with Huntington's principle (p. 244). In semiarid shrub or grassland environments with high denudation rates, an increase in precipitation should increase the vegetation cover and reduce erosion, as deduced by Huntington. In an extremely arid region, however, increased precipitation might actually increase the denudation rate, if it does not exceed the amount required to stabilize the landscape with grass.

Structural and Tectonic Factors in Denudation Rates

It is impossible to quantify the relative importance of climate-controlled processes and structure in the denudation of various tectonic terranes. Active orogenic belts can develop in arid or humid, cold or warm climatic belts. High mountains are always cold, and many are glaciated if sufficient precipitation is available. Platforms and shields tend to have regionally monotonous climates, but they range from deserts in Australia, Africa, and Arabia, to rain forests in Brazil, to periglacial plains in Siberia and Canada.

Active Orogens.   The local relief, or height difference between ridge crests and valley floors, may be the most important single factor in determining sediment yield or denudation rate. Given the extremely rapid denudation rates reported for active orogenic regions such as the Himalaya Mountains or Taiwan, local relief could not be maintained for more than a few million years unless active orogenic uplift maintained it. Many studies use fission-track dates, which are a measure of the rate of unroofing of a mass of cooling rock, as a direct proxy for uplift rate, by the logic that rapid erosion must be sustained by equally rapid uplift.

Shields or Cratons.   As noted in Chapter 3, these large areas of ancient continental crust have been tectonically inactive for hundreds of millions of years, experiencing at most broad epeirogenic warping of less than a few hundred meters. Their low relief discourages orographic precipitation, which along with their large area, usually results in dry or subhumid climates. Low precipitation plus low relief equates to ineffective fluvial erosion. In warm, humid climates, a deep regolith reduces the rate of weathering. Erosion is typically supply-limited rather than transport-limited although neither weathering, mass-wasting, nor erosion are very effective. Large areas of Brazil, Africa, Australia, and India fall within this category of structural terrane. The Canadian and Siberian shields differ from similar terranes in low-latitude regions in that Quaternary glaciation has removed the former saprolite (p. 141) and exposed the "bare bones" of shield structures to contemporary weathering, but glaciation and periglacial conditions have also deranged the fluvial drainage network and choked it with solifluction debris so that denudation rates are also low, even if transport-limited rather than supply-limited.

Sedimentary Platforms.   Large areas of "layer cake" stratigraphy with gently warped domes and basins are typical of continental regions. As noted in Chapter 12, these regions develop a characteristic cuestaform topography with only moderate to low regional relief, usually less than 600 m. Large, structurally adjusted, well-integrated drainage networks such as the Mississippi, Paraná, or Volga rivers carry eroded sediment to trailing continental margins. Denudation rates are typically low, in the range of 1 to 10 cm/1000 yr, although many of these regions are now experiencing increased erosion caused by intensive agriculture.

Significant areas of sedimentary platforms are underlain by karst-prone rocks, mostly limestone (Chapter 8). Typical denudation rates of karst terrains (Figure 8-7) are similar to those of shale and sandstone terrains eroded by surface fluvial processes. Since most limestone caverns are integrated with surface fluvial systems, cave deposits offer useful information for measuring regional denudation rates. Paleomagnetic stratigraphy in caves on the Appalachian Plateau records downcutting of adjacent river valleys at rates of between 5 and 6 cm/1000 yr (Sasowsky et al., 1995; Springer et al., 1997). Very similar rates were determined for the valley floors near caves in northern England (Gascoyne et al., 1983) and in the Southeastern Highlands of Victoria, Australia (Webb et al., 1991, p. 230).

Rates of Denudation and Tectonism Compared

In Chapters 3 and 5, numerous examples were cited of orogenic and epeirogenic uplift increasing the height of landscapes by as much as a few millimeters per year, rarely even 1 cm/yr (10 m/1000 yr) or more. The denudation rates reported in the preceding sections of this chapter should be compared with constructional rates in order to determine whether newly rising landscapes can outpace denudation and increase regional relief or whether landscapes are doomed to be kept low by erosion despite tectonic uplift. Most denudation rates for mountainous regions in various climates range from 10 to 100 cm/1000 yr, roughly 10 percent of the reported orogenic uplift rates. Denudation rates for large watersheds such as the Amazon or Mississippi are generally between 1 and 10 cm/1000 yr, again roughly an order of magnitude less than long-term epeirogenic uplift rates. It would appear that only extremely high mountain ranges, or mountains of easily eroded sediments in regions of high rainfall (Taiwan), might suffer denudation rates so great that their absolute elevation could not increase despite strong uplift. For example, an intriguing but unwarranted extrapolation of the best-fitted equation for Figure 15-5 would give a denudation rate of 233 cm/1000 yr for a hypothetical 10,000-m mountain range in Papua New Guinea. The rate is consistent with the regional Himalayan denudation rate of 70 cm/1000 yr inferred from submarine sedimentation rates (Curray and Moore, 1971), the denudation rate of more than 100 cm/1000 yr for the Kosi River, which drains the highest parts of the Himalaya Mountains, and the remarkable denudation rates of 300 to 400 cm/1000 yr reported for the Nanga Parbat Range (p. 336). Because of the power function increase of denudation rate with relief and altitude (Figure 15-5), perhaps the Himalayas are as high as any mountain range on earth can be, given the present effectiveness of destructive geomorphic processes.

All these estimates and extrapolations lead to the conclusion that if tectonic movements raise new land above the sea, the land is likely to increase in area and altitude much faster than erosion can destroy it. However, denudation rates as well as the strength and rheidity of rock masses set an upper limit to the height of subaerial relief on this planet at about 10 km (p. 13).

At least two end members of the contest between tectonic uplift and erosion can be envisioned. In one extreme, as exemplified by the intense erosion of eastern Taiwan (Figure 15-2), easily eroded mélange and accretionary-prism mudstones (Figure 3-2) are rapidly uplifted in a humid, tropical, seismically active region, and erosion limits this steady-state mountain range to altitudes no higher than 1 to 2 km despite intense tectonism (p. 118). At the other extreme, the Altiplano Plateau of the central Andes in Bolivia and adjacent countries rose very rapidly in the last 12 million years, in a chronically arid region. Once it rose to heights of 4 to 6 km, the Altiplano Plateau became even more cold and arid, and except for wind erosion, there is little or no denudation of the vast upland. Such a tectonic mass might become self-limiting by isostasy, for if nothing is removed, the orogen cannot continue to be uplifted indefinitely. Instead, it may grow wider, as in the case of the Andean Altiplano or Tibet. It is interesting to postulate that the relief of all orogenic regions is bounded by these end members of erosion-limited and tectonic-limited conditions (Dahlen and Suppe, 1988).

SEQUENTIAL DEVELOPMENT OF FLUVIAL LANDSCAPES

The necessary foundation has now been laid for deducing the sequential development of a nontectonic regional landscape under a climate that ensures year-round runoff by rivers that increase their discharge as they flow to the sea. In such a fluvial climate, continuous vegetational cover by grass and trees is assumed. It is necessary to deduce the development of individual river valleys and drainage basins prior to considering the regional landscape because in fluvial landscapes networks of valleys are the basic landform units (Schumm and Ethridge, 1994).

Sequential Development of River Valleys

As drainage networks expand over a subaerial landscape, most of the erosion is accomplished by the first-order rills and gullies (Figures 10-1 and 10-2). Gradients are steep and as streams branch and grow headward and organize their subsidiary drainage basins, the landscape changes rapidly. New slopes shed waste directly into the first-order channels, which can deepen only if their streams can first move out the detritus supplied by mass wasting from their own valley walls. A wavelike front of intense erosion sweeps inland across the old landscape as new drainage networks expand.

If we watch a computer simulation of such an outward-expanding drainage network, we see the initial slopes steepen toward the encroaching tributaries (Figure 11-4). The area becomes a complex of small interfluves, side slopes, and gully floors. Then, as the network expands headward, the minor dissection is consolidated into larger and less complex landforms, perhaps a single valley as a former first-order stream becomes a higher-order trunk stream. At this stage, regional dissection becomes less rapid although the local relief between interfluve and valley floor continues to increase.

As a stream network expands, trunk streams and major tributaries change their functions from dissection to transportation. Downcutting may be deferred by an aggradation phase, as a former V-shaped gully is choked by sediment from its own expanding watershed. Quite early in the deduced sequence of events, we can expect the beginning of floodplains because alluvium is temporarily or seasonally stored and weathered before it is removed downstream. Dams of alluvium may divert the streams against their banks to begin the process of spur trimming and valley-floor widening. By the time a stream has developed its own tributary system and is draining a watershed of 1 km^2, it is likely to be in a well-formed valley approximately 1 to 10 m in depth, its low-water channel meandering or braiding around masses of poorly sorted colluvium and alluvium. Valley slopes will be at the angle of repose of the local rock mass or weathered debris and will extend down either to the channel banks or to fragments of floodplains.

Eventually, valley-side slopes will be trimmed back by stream erosion, and alluvium will be deposited on the valley floor so that the river flows between alluvial banks except at peak discharge. Midchannel bars may divert the main current to one or the other valley wall so that for a time the stream has one bank of alluvium and one of bedrock. This process aids in valley widening and the development of a floodplain.

A significant event in the deduced sequential evolution of a stream valley is the development of a continuous floodplain. Depending on the grain size of the load and the complex interplay of hydraulic geometry variables, the stream channel will meander or braid in a layer of its own alluvium. At this stage, the mutual adjustment of channel shape, discharge, load, and slope becomes possible, and the stream is graded in an alluvial channel (p. 224). Erosion is not completed, but from this time onward, the river must mutually or sequentially readjust the entire complex of hydraulic geometry variables in response to changing conditions.

Local perturbations such as landslides are less likely to affect a graded river because any debris from the valley wall will accumulate as a cone or sheet on the edges of the floodplain and may weather there for centuries before the stream begins to rework it. As a consequence, valley-side profiles will change from initially straight slopes directly to the stream banks to compound curves, blending above into convex summits of divides and below into concave wash slopes that in turn blend into floodplains (Figure 9-21).

When streams carry increased discharge, their longitudinal gradients decrease (Chapter 10). Therefore, in an expanding fluvial network such as that being used as the model for this deduction, an initial high-gradient first-order tributary or gully will be replaced in time by successively lower-gradient valley segments. A net regional lowering of altitude must result. If the example of the Hydrographers Range is valid, with lower altitude and decreasing denudation rate, local slope gradient, slope length, and local relief will also decrease (Figure 15-7). Slower mass wasting on more gentle slopes should permit more intense chemical weathering and thicker soils, with a corresponding reduction in the amount and grain size of sediment delivered to the river.

In general, then, the changes we observe in a downstream direction along a present river valley are those we can deduce for the temporal changes at any point along the river valley: a reduction in down-valley gradient, progressive growth of a continuous flood plain, removal of waterfalls and rapids except those on resistant structures, progressive widening of the cross-valley profile by spur trimming and adjacent valley-side retreat, and progressive loss of local first-order tributaries draining adjacent slopes.

Sequential Evolution of Regional Fluvial Landscapes

A regional fluvial landscape is not simply an assemblage of isolated individual drainage systems. Whereas the deduced development of a single river valley can assume a continued headward growth of the drainage network, on the regional scale there must come a time when independent drainage systems intersect, interfere, or compete. Deductions about the sequential evolution of regional fluvial landscapes must be strongly conditioned by the restraint of a finite limit for basin expansion.

No generalization is possible about the initial appearance of a regional landscape just beginning to be dissected by rivers. The favorite initial stage postulated by traditional deductive geomorphologists was the former floor of a shallow sea, newly emerged above sea level. The assumptions were that the sea floor is

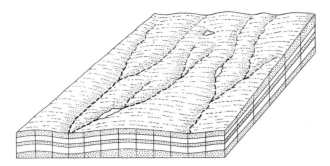

FIGURE 15-9. Hypothetical initial landscape, newly emerged from the sea. The initial surface is conformable to the sedimentary beds that underlie it. Drainage is consequent on the surface irregularities (based on drawings by W. M. Davis and C. A. Cotton).

smooth and featureless and that it emerges quickly relative to erosional processes. Figure 15-9 suggests such a landscape, with scale omitted. The figure could represent 1 or 100 km². Any other constructional landscape—volcanic, tectonic, or depositional—could as well be taken as the initial form. Equally, or perhaps more probably, a landscape already undergoing fluvial dissection becomes the initial form for a new sequence of changes. It is well to recall the quotation (p. 235) that a landscape is best considered to have a heritage rather than an origin. Even computer simulations are strongly controlled by minor initial relief (p. 236).

Regional landscapes require additional deductions about the sequential development of divide areas between adjacent or opposed drainage networks. Initially, the divide area is large relative to the expanding fluvial basins and networks. As the new generation of drainage basins intersect, however, the old land is converted to a series of fragments isolated between vigorous headward-eroding streams. Drainage divides become more sharply defined at this stage, but whether the divides become knife-edged ridges between adjacent steep valleys or smooth, convex hilltops depends on complex interrelations between structure, climate, and stream gradients.

As river valleys dissect a landscape, they are influenced by many subtleties of structural control. Weak-rock belts develop subsequent stream valleys. Divides become localized on the most resistant rocks although superposition from cover strata onto a resistant structure may introduce complexities into the deduced evolutionary patterns.

Eventually, an evolving region will consist of a series of divides, valley sides, and valley floors, all mutually adjusted and graded with respect to each other. The drainage density is determined by structure and climate; drainage patterns reflect structural control

or lack of it; divides are on the most resistant rock type between adjacent trunk streams; and the highest terrain is the region farthest inland, at the headwaters of the largest streams. A critical stage in regional landscape evolution is the final elimination of whatever initial forms were present. In significance, this stage of complete upland dissection is analogous to the development of a continuous floodplain along a segment of stream valley. In no way can the two events be considered simultaneous, however, because floodplains are constantly under development along a valley segment somewhere along each river.

Local, Available, and Critical Relief

Deductions concerning the development of regional fluvial landscapes subsequent to complete upland dissection depend on the interaction between mean regional elevation above sea level and the drainage density. With high initial elevation or closely spaced streams, valley-side slopes are likely to intersect as narrow divides while the valleys are still being deepened. With low regional elevation or widely spaced streams, individual rivers will have completed their initial phase of rapid deepening and will have become the trunk streams of expanding networks while large interfluve areas are as yet unaffected by the new dissection. Glock (1932) defined **available relief** as the vertical distance between the initial upland flats and graded valley floors of dominant rivers and argued that if the available relief exceeded the local relief (the vertical distance from ridge crests to adjacent stream channels), total regional dissection of the old land would be complete before flood plains began to accumulate on the local valley floors (Figure 15-10a). If the available relief were less than the local relief, downcutting by trunk streams would be inhibited, because they would become lined with alluvium and graded, and upland denudation would be very slow (Figure 15-10b). Glock envisioned a special case of **critical relief** such that regional disappearance of relict terrain on the interfluves would exactly coincide with the development of graded segments in the adjacent valleys. The concepts of available and critical relief need redefinition to incorporate the factor of drainage density, but the basic concepts are fruitful nonetheless. With high available relief (or a high drainage density), regional landscapes will quickly be completely dissected to the stage called *ridge-and-ravine topography* by Hack (1960, p. 89). Subsequently, until alluvial floodplains extend upstream into the region, a long interval of time might pass, and a great amount of regional denudation might occur, without any significant sequential changes in the

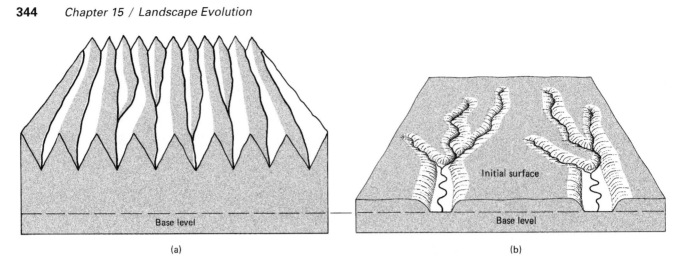

FIGURE 15-10. Complexities in comparing the deduced development of individual valleys with the development of the regional landscape. (a) With high available relief, upland is totally dissected, but streams are vigorously downcutting. (b) With low available relief, trunk streams are not downcutting, and upland dissection is very slow.

appearance of the landscape (Figure 15-11). Competing streams might capture or be captured, resistant structures might develop as regional divides, and initially buried structures might be exposed and exert new controls on sequential development. But unless climate or structure change radically, the drainage pattern and local relief will seem "timeless" appropriate to the concept of graded time (p. 15). Isostatic uplift in response to erosional unloading will extend this stage of development for an even longer time.

The Later Stages of Sequential Development

Accepting the premise that a landscape is at a finite altitude when tectonic and isostatic uplift become negligible, we can deduce the later stages of sequential landscape development. The deduction incorporates the concept of *cyclic time*, measured in millions of years, during which the potential energy of the fluvial system gradually declines. Within the long-term span of the cycle, on the shorter time scale of *graded time* (hundreds or thousands of years) the system will appear to be in equilibrium (Figure 1-6). When graded, a river floodplain widens only slowly because the meandering channel only rarely encounters the bedrock of the valley wall. The loss of local relief in the headwater regions will decrease the sediment load and probably the grain size of the load, so all the hydraulic variables of the stream system must continue to change although in mutually compatible ways. The denudation rate will probably decline from 1 to 10 cm/1000 yr to 0.1 to 1 cm/1000 yr. The net effect must be a gradual lowering of the longitudinal valley profile as the river slowly shifts laterally, trimming the valley flat beneath the

floodplain. As relief is lowered, the slopes in interstream areas become more gentle. As floodplains become significant parts of the total landscape, the proportion of stream-dissected upland terrain must necessarily decrease. The observations that large quantities of the upland soil and slope-derived sediment that enter fluvial systems are not delivered to the sea are important to deductions about the later stages of landscape development. If sediment storage within fluvial systems is a normal part of their behavior and not the result of Holocene climate change or human activity (Figure 11-5), we must deduce that an old landscape will be largely covered with colluvium and alluvium that low-gradient streams are relatively incompetent to transport. As the fluvial denudation rate declines, wind erosion and transport may become significant, especially on moisture-deficient landscapes where agricultural practices cause seasonally exposed bare soil (Kirkby, 1980; Trimble, 1988).

When summit convexities are graded to lower slope concavities, which in turn are graded to floodplains, which are graded down-valley to the sea, what else can happen? Surely, as long as solar energy is supplied and sediment is exported, the landscape must continue to evolve. Local relief will decrease, slope angles will decrease, chemical weathering will increasingly dominate mechanical weathering, mass wasting will become slower and slower, and each increment of change will take a longer time. The extrapolations provided by the denudation rates in the Hydrographers Range (Figure 15-6) offer some clues to the later stages of landscape development. The exponential loss of relief with time is such that of the nearly 5 million years of subaerial denudation that is predicted for an initial elevation of 1000 m, nearly one-half of the time is

FIGURE 15-11. Complete regional fluvial dissection of loess, Shanxi Province, China (NASA ERTS E-1525-02455).

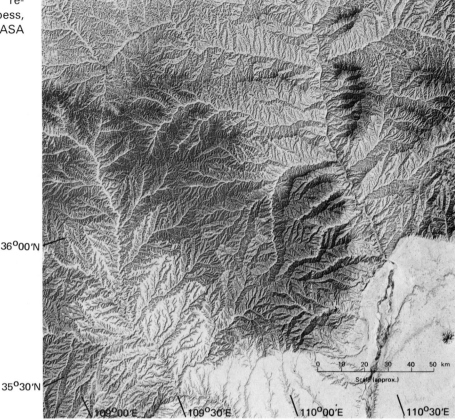

required to remove the final 10 m of relief. Unless tectonic uplift or volcanism renew the available relief, fluvial landscapes seem doomed to a condition of low elevation above sea level and low local relief for most of their life histories. A corollary to Powell's dictum is that if mountains cannot long remain mountains, they are very likely to become low rounded hills, deeply mantled with weathered residuum and drained by low-gradient river networks that are thoroughly adjusted to the slight but still declining available energy. Given the extreme range of denudation rates cited in preceding pages, the history of most landscapes should be an interval of several million years or more during active orogeny when erosion is very rapid and as much as 10 km of overlying rock mass is removed, followed by a time measured in tens of millions of years of low relief and extremely slow denudation. Without tectonism, the loss of subaerial mass might follow a curve similar to that in Figure 15-6, but much steeper in the early time and very flat for an extremely long later time interval. Significant to the study of modern landforms is the fact that this deduced long time interval included major global climatic

changes (Chapter 4). Few modern landscapes of low relief can be assumed to have been shaped by the processes now operating on them, or at least not by the modern intensity of those processes.

PENEPLAINS AND OTHER DEDUCED ENDFORMS OF EROSION

The History of an Idea

During the nineteenth century, geomorphologists gradually accepted the ideas of Hutton and Playfair that vast amounts of subaerial denudation had preceded the shaping of the modern landscape. The enormousness of geologic time was demonstrated by stratigraphy and faunal succession. The erosion of valleys by rivers and glaciers became an accepted uniformitarian concept, and the progressive changes of landscapes through extended time became the subject for geomorphic speculation and debate. In 1874, G. K. Gilbert demonstrated that the basin-range fault blocks were much younger than the folding and metamorphism of the rocks within the ranges (p. 76), and in 1876, J. W. Powell used the phrase "the great denudation" for this erosion interval. Powell's dictum (p. 118), which has been noted repeatedly in this chapter, illustrates the attitudes of the time.

By 1883, W. M. Davis had begun development of his *geographical cycle*, or *cycle of erosion*, inductively generalizing from his own field work and that of his contemporaries that fluvial landscapes evolve in a predictable genetic sequence, which he characterized by analogy with the biologic terms *youth, maturity,* and *old age* (Davis, 1885, 1899, 1902). Present readers should note that Davis wrote during the peak of intellectual enthusiasm for Darwinian evolution. Determinism was a popular philosophical scheme, and Davis's use of the term *cycle* implied not so much a steady-state condition but a directed sequence of events from an assumed initial state to a deduced endform. It was cyclic only by the assumption that the initial state could be recreated by tectonism or volcanism.

Davis assumed that rapid uplift initiated a "normal cycle of erosion" (Figure 15-12). The prevailing tectonic theory of his time called for worldwide pulses of orogeny followed by stillstand, so his assumption was readily accepted. (It is not so readily accepted today, with evidence that the Himalaya Mountains have been rising for at least 40 million years, and the central Andes for at least 12 million years.) After uplift, he deduced a brief stage of youth, when relief rapidly increases as rivers deepen their valleys and extend their drainage basins (Figure 15-12, interval 1-2) and regional landscapes are dissected. He chose as the criterion of maturity in valleys the development of free-swinging meanders in trunk streams, presumably on alluvial floodplains. For the regional landscape, his primary criterion of maturity was the total removal of the former land surface from interfluves. This includes the time of maximum relief, which then begins a slow decline (Figure 15-2, interval 2-3). Other criteria, such

as total elimination of waterfalls and lakes from river valleys, development of sharp-crested divides, and maximum drainage density, were also deduced by Davis but are probably better omitted because they are usually determined by structure or climate.

Davis emphasized that old age is not separated from maturity by any specific criteria. As in organisms, where maturity is defined by the specific condition of reproductive ability, in geomorphic analysis the twin criteria of meandering rivers and complete upland dissection are specific features that can be observed. Old age is less easily defined. In organisms, it is only an extension of maturity, with a decrease in metabolic activity, loss of facilities, and an intangible decline of ability. In landscapes, this long interval is marked by continued slow loss of relief (Figure 15-12, intervals 3-4 and beyond 4). In Davis's concept, the additional time required to reduce the relief JK in Figure 15-12 to half its value might require as much time as all that had preceded. This suggestion of a "half-life," or exponential decay of landscape relief, is well supported by the projected destruction of the Hydrographers Range in Papua New Guinea (Figure 15-6). As one criterion of old age in landscapes, Davis suggested that floodplains would become several times wider than the meander belt. Recent studies of fluvial mechanics give no evidence that meander belts develop such higher-order "supermeanders" or that floodplains increase their width through time beyond the amplitude of the meanders. Further, Davis's old landscapes were deduced to be covered with broad sheets of alluvium over which the rivers sluggishly meandered. A more likely deduction is that low hills of deeply weathered rock and colluvium would form more of the regional terrain than Davis implied.

The common but regrettable use of maps of the lower Mississippi River meander belt to illustrate an old-age landscape has confused deductions about the later stages of erosion. Most introductory students of geology and geography in the United States are falsely taught that the very young depositional landscape of the aggraded Mississippi Valley (Figure 11-8) is representative of very old erosional landscapes!

In an attempt to describe the form of landscapes that have undergone long-continued weathering and erosion in humid climates, Davis (1889) introduced the elegant word **peneplain**. He used as a root the word *plain*, in the geographical context of a regional surface of very low relief near sea level. Realizing that the ultimate base level is the limit of subaerial erosion, which like a mathematical limit may be approached but might not be achieved, he prefixed to the word *plain* the Latin derivation "pene," meaning almost. Thus "peneplain" was introduced to the scientific literature as *a surface of*

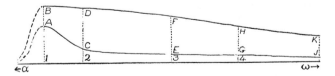

FIGURE 15-12. W. M. Davis's concept of the erosion cycle. Time (nonlinear) is plotted from α to ω on the horizontal axis. Upper line (BDFHK) represents the height of upland interfluves; lower line (ACEGJ) represents the height of valley floors; AB, CD, etc. are a measure of the local relief. Time intervals are marked by 1, 2, 3, 4, and beyond (simplified from Davis, 1899, Figure 4).

regional extent, low local relief and low absolute altitude, produced by long-continued fluvial erosion.

How can we define a landscape that is "almost" gone? Apparently, the peneplain that W. M. Davis had in mind was far from a mathematical plane, or even a "peneplane," as some have proposed to spell the word. He envisioned an erosionally dissected landscape, one that includes the drainage networks of several master streams that enter the sea, in which the total relief is no more than about 100 m at the regional divides. Cotton (1948, p. 272) offered an important clue to the concept:

In 1914 the late Professor Davis for the author's benefit defined the limiting steepness of the slopes of a peneplain as such as one might lay out a straight road on in any direction "and trot on it". In these days this would be a high-gear road.

In our modern era of excessive horsepower, it is hard to envision such a gently sloping erosional landscape, but it would not be planar.

Recent measurements of denudation rates provide quantitative support for the peneplain concept. Even at very low denudation rates of 1 cm/1000 yr or less, fluvial landscapes could be lowered as much as 10 m/million years. From the rates listed in Table 15-1, Judson and Ritter (1964) concluded that the entire conterminous United States could be removed to sea level in 11 to 12 million years if no isostatic or tectonic uplift were to occur. Schumm (1963) concluded that regional peneplanation might require 15 to 110 million years, allowing for isostatic compensation. Ahnert (1970) suggested that 11 to 18.5 million years would be required to reduce the relief of a large midlatitude drainage basin to 10 percent of its original relief. Ohmori (1985) calculated that with tectonic uplift and denudation rates typical for Japan, the mean altitude of any region will decrease to about 100 m above sea level in 10 million years, regardless of its initial altitude. Clearly, the concept of cyclic time fits comfortably within the late Cenozoic Era (back endpaper) (Young and Saunders, 1986, p. 15) and large areas of anorogenic continental landscapes have had time to be reduced to low altitude and low local relief. A complication that is yet to be evaluated is the importance of epeirogenic vertical movements (p. 45) with neotectonic rates well in excess of probable denudation rates.

Ancient continental shields have persisted within a few hundred meters above and below sea level for substantial periods of geologic time. The relief on these cratonic areas has probably been regraded repeatedly during the last billion years. The present low relief of the Canadian shield, for example, attributed by some to intense and repeated continental glaciation, is replicated by the sub-Paleozoic and sub-Cretaceous unconformities where those rocks lap onto the shield (p. 381). Apparently, the shield has not had significant relief for at least the last 400 million years.

The exhumed erosion surface that truncates Precambrian basement rocks of the Argentine Pampean Ranges (Color Plate 3) is another example of an ancient erosion surface of very low relief. It shows comparable monotony whether it has been exhumed from beneath late Paleozoic, Mesozoic, or late Cenozoic cover strata (p. 78). However, because the cover strata are not marine, the Pampean surface cannot be proved to have formed near sea level. Fairbridge and Finkl (1980) have proposed that ancient cratonic shields are repeatedly exhumed and modified, and are maintained at low relief throughout geologic time by a variety of erosional process. They propose that the term *peneplain* be applied to these polygenetic surfaces, broading its definition to remove the restrictive condition of cyclic fluvial erosion.

The continent nation of Australia is emerging as a unique laboratory for the study of very ancient landforms. Long-term denudation rates as low as 0 to 2 m/million years (less than 2 mm/1000 yr; compare Tables 15-1 and 15-2) throughout the Mesozoic and Cenozoic Eras have resulted in a landscape whose main features are 10^7 to 10^8 years old (Ollier, 1991; Gale, 1992). Landscapes that had been interpreted as resulting from multiple cycles of Cenozoic erosion have been demonstrated to be exhumed from beneath Cretaceous cover strata with little subsequent alteration (Nott, 1994). An early Miocene valley system dated by basalt lavas that filled some of the valleys in New South Wales is still well preserved (Young and McDougall, 1993). A belt of late Eocene fossil sand dunes marks the inland limit of an early Tertiary marine transgression in South Australia (Benbow, 1990). Inselbergs near Adelaide have been exposed to cosmic rays for at least 0.5 million years and are eroding at a rate of less than 0.7 ± 0.1 m/million years (0.7 ± 0.1 mm/1000 yr) (Bierman and Turner, 1995). Much of the continent's landscape is relict from pre-Pleistocene time. It seems clear that in Australia at least, landscapes with modest relief have survived for 10^7 years or more under very slow denudation rates. The modern aridity of the continent is a relatively recent late Cenozoic phenomenon (Nanson et al., 1992), so something more than climate is required to explain the Australian anomalous landscape. Perhaps its origin as part of the ancient Gondwana shield or craton offers a structural explanation for the exceptional antiquity of the Australian landscape. Although much of the continent has low relief, most of the landscape would not fit any current concept of a peneplain.

Other Concepts of Sequential Development and Endforms

W. Penck: Tectonics and Landscape Evolution.

Davis's cycle of erosion made only passing reference to tectonic movements other than initial uplift followed by stillstand and erosion. He did allow for interruptions to the cycle by renewed intermittent uplift, but these were treated only as temporary disturbances that would initiate new downcutting and develop river terraces and compound valley forms. Radiometric dates from orogenic belts now clearly prove that tectonic uplift, metamorphism, and volcanism can continue for several tens of millions of years during which there is much evidence, especially from fission-track dating, for rapid denudation rates. The most notable attempt at formulating a systematic geomorphology that related tectonic movement to erosion can be found in the writings of Walther Penck (Penck, 1924; English transl., 1953). His major works, written during a terminal illness and published posthumously in fragmentary form, are rightfully known for their complex and obscure German prose. Nevertheless, they attracted much attention among geomorphologists as an alternative or even a challenge to Davis's cyclic concept. Davis himself wrote a critical review of Penck's proposals (Davis, 1932) that became the primary source for most English-speaking geomorphologists. Unfortunately, certain errors of translation and interpretation in Davis's review have been misattributed to Penck ever since (Simons, 1962). The impact of Walther Penck's contributions on American geomorphology can be best judged by the published papers and discussions of a 1939 symposium on the subject (von Engeln, 1940). One of the better reviews of Penck's theories was by O. D. von Engeln (1942, pp. 256–268).

Penck accepted the idea that landscapes could be reduced to endforms of low relief but maintained that these endforms never became the initial forms for new episodes of dissection. By his theory, uplift was not rapid and then zero, as Davis supposed, but always began slowly, waxed to a maximum, and then waned gradually to a stop. During the long stage of initial slow uplift, all prior forms were destroyed, and a new surface of low relief, adjusted to a balance between uplift and degradation, developed. This initial surface of low relief was called **Primärrumpf** by Penck and was the basis for his deductions. Instead of hypothesizing sequential forms, he maintained that slopes, once developed, retreat parallel to themselves and persist until they intersect. During downcutting by rivers, such as might be caused by accelerating uplift, the initial slopes would be broad and gentle, but as the rate of uplift accelerated, inner valley slopes would be steeper.

Because each increment of slope profile retreated from the valley axis and was followed by a lower slope element of steeper gradient, convex skyward slopes would be diagnostic of accelerated uplift. Similarly, straight slopes would indicate a constant rate of stream downcutting, and concave slopes would indicate a waning rate of entrenchment. Thus, as uplift died away, landscapes were theorized to be an assemblage of concave forms, to be converted to a Primärrumpf during the next episode of initial slow uplift. Few modern geomorphologists accept Penck's deductions about the persistence of slopes through time.

No erosional sequence of forms was allowed by Penck's scheme because each morphologic assemblage was related to a certain tectonic condition. His commendable primary goal was to use geomorphology to determine regional tectonic history. Penck denied any climatic control of geomorphic processes other than glaciation, believing that tectonics alone determined landform assemblages. Yet it is a curious fact that Penck's ideas of slope retreat have been found most analogous to processes that shape mountain fronts and pediments in semiarid climates (Chapter 13).

Regional Pedimentation.

A later generation of geomorphologists have used the pediment landform as the basis for deductions about sequential evolution of landscapes. Maintaining that hillslopes retreat parallel to themselves and leave a residual low slope of transportation (the pediment) at their base, these authors propose that huge areas can be reduced to a series of coalescing pediments capped by residual rock knobs and that successive generations of coalescing pediments (or *pediplains*, by analogy with peneplains) can migrate across a landscape, each one consuming higher and older surfaces by parallel scarp retreat, while in turn being consumed by the scarp retreat of a lower and younger generation of pediments (p. 194). No sequence of landforms is admitted by this theory either, because once formed, the pediments are asserted to be essentially immutable, stable landforms (King, 1953). The origin and significance of pediments and related landforms were considered in Chapter 13, but it is worth noting again that very large areas of the earth's surface are pediment-prone, and in the geologic past, before the evolution of vascular plants, the entire subaerial landscape may have been shaped by pediment-forming processes that we now associate with dry climates.

Etchplains.

Another climate-based theory of landscape evolution deduces that very stable, extremely ancient (early Mesozoic or Cenozoic) landscapes have evolved slowly in the tropical seasonally wet and dry climate, or savanna. As noted in Chapter 13, this extensive climatic type includes large areas of crystalline continental shields with low relief and little or no epeirogenic movement. The combination of structural factors such as crustal stability, crystalline rocks, and low tectonic relief, with the climatic factors of seasonal precipitation, deep chemical weathering, and low-gradient rivers that cease to flow in the dry season, create the *inselberg* and *washplain* landscape characterizing by a double front of weathering (Figure 13-9). Even slight epeirogenic uplift, or a climate change that reduces the length or severity of the dry season, could initiate fluvial dissection of the deep regolith and expose a landscape that displays the relief of the former weathering front, etched from essentially unweathered rock. The resulting **etchplain** may be rugged or smooth depending on the weathering behavior of the rock mass, but the metamorphic grade of cratonic terranes tends to homogenize weathering resistance, and many areas in central Africa, Australia, India, Sri Lanka, and northern South America have monotonous expanses of low-relief plains with scattered inselbergs and extensive areas of exposed rock. The relief on etchplains is not controlled by fluvial base level, so an etchplain need not meet the criterion of peneplains that they form near sea level. However, as cratonic regions inherently have low "continental freeboard" (p. 44), etchplains are generally at low altitude even if they are not graded to sea level.

The concept of the etchplain, although introduced through studies in Uganda by the English geomorphologist E. J. Wayland (Thomas, 1994, p. 287), has been largely developed by Germans, especially Büdel (1982) and Bremer (1971, 1981, 1993), and Australians (Mabbutt, 1965; Finkl, 1979; Twidale, 1987). An excellent chapter-length review of the subject by Thomas (1994) includes many of his own observations of etchplains in Nigeria and Sierra Leone.

Because etchplains form beneath regolith, they are inherently rejuvenated or exhumed landforms wherever they are found. Thus, their terminology is complex, including *incipient, dissected, semi-striped, incised,* and *buried* subcategories (Thomas, 1994, Table 9-1). Furthermore, their low-latitude locations have experienced multiple Cenozoic episodes of wetter climates, with deep weathering, alternating with drier climate and active pediment formation. Probably all etchplains are polygenetic, and even high-latitude cratons such as the Canadian and Scandinavian shields may owe their present low relief to early Cenozoic or older episodes of deep weathering, subsequently stripped of regolith by fluvial and glacial erosion (pp. 340, 382). Fairbridge and Finkl (1980, pp. 78, 82) advocated that the importance of the subregolith etching process is so important

in creating the low relief on all cratons that the definition of peneplain (p. 346) should be relaxed to include etchplains and their inherently polygenetic origin.

"Dynamic Equilibrium" Landscapes. The emphasis on internal self-adjustment among complex variables in open physical systems, such as characterizes modern stream hydrology (Chapter 10), has led to a hypothesis of "steady-state" landscapes (Hack, 1960). Critics of Davis's cycle of erosion argue that when alluvial floodplains develop, even in small streams, the landscape is essentially graded and therefore deny the progressive development of graded master streams as a criterion of the stage of regional evolution. The only sequential change they envision is an early one from initial disequilibrium to subsequent equilibrium. Thereafter, as long as the energy in the open system remains constant, the landforms also remain constant even though mass is removed from the eroding terrane. Hack (1960, p. 85) applied the name "dynamic equilibrium" to this concept of a time-independent, open-system landscape in a steady state. He regarded most landscapes that had previously been called "maturely dissected peneplains" as nothing more than erosionally graded steady-state surfaces that carry no connotation of sequential development. Only when the energy of the whole system ultimately declines, as when the available relief is consumed, will the landscape evolve to one of reduced local relief, decreased slope gradients, and blunted divides.

Hack's landscape in dynamic equilibrium could be accommodated in Davis's cyclic theory by assuming extreme available relief, during the reduction of which great volumes of rock must be removed by rivers whose spacing and gradient are climate- and structure-dependent and, therefore, time-independent. However, his avowed purpose was to explain landscapes in an essentially noncyclic conceptual framework, and Hack rightfully challenged some of the hypotheses of multiple residual peneplains that have been inferred from accordance of summits in the Appalachian region. Summit accordance and many other landscape features, such as gravel-covered interfluves and summit regions, can well be explained by long-continued erosion and progressive adjustment to structure. Multiple cycles of erosion, each one terminating in a peneplain that was then uplifted and dissected, are not required (Morisawa, 1989).

One can very well agree with Hack that most landscapes are in an uneasy dynamic equilibrium between the energy available for work and the work being done, during any particular interval of "graded" time. However, if regional elevation, and therefore the energy available for geomorphic change, continues to decline at a measurable rate, no landscape can be said to represent an open system in a steady state on the scale of "cyclic" time. As in the analogy of a rotary cement kiln (Figure 2-1), the landscape must be constantly evolving toward equilibrium with the declining energy. The major contrast between Hack's "steady-state" landscape and Davis's cycle is that Hack deduced total regional dissection and graded rivers as a nearly permanent condition rather than as the mature stage of an evolving series. The choice between these models depends on the relative rates of regional denudation and late Cenozoic tectonism. If, as in the example of the Hydrographers Range, total denudation in 5 million years can be inferred, evolving landforms can be deduced. If, however, tectonics and climatic change invalidate the assumption of initial uplift or other constructional process followed by still-stand and landscape evolution, the dynamic equilibrium model, changing only from disequilibrium to equilibrium, is most suitable as a basis for interpreting the present landscape. Until late Cenozoic climate fluctuations and tectonic history, especially the role of neotectonic epeirogeny, are better known, there seems to be room for multiple hypotheses in the interpretation of low-relief erosional landscapes (Figure 1-6).

REFERENCES

AHNERT, F. 1970, Functional relationships between denudation, relief, and uplift in large mid-latitude drainage basins: Am. Jour. Sci., v. 268, pp. 243–263.

ANDERSON, M. G., ed., 1988, Modelling geomorphologic systems: John Wiley & Sons Ltd., Chichester, UK, 458 pp.

ANDERSON, R. S., REPKA, J. L., and DICK, G. S., 1996, Explicit treatment of inheritance in dating depositional surfaces using in situ [10]Be and [26]Al: Geology, v. 24, pp. 47–51.

BELL, M. and LAINE, E. P., 1985, Erosion of the Laurentide region of North America by glacial and glaciofluvial processes: Quaternary Res., v. 23, pp. 154–174.

BENBOW, M. C., 1990, Tertiary coastal dunes of the Eucla Basin, Australia: Geomorphology, v. 3, pp. 9–29.

BIERMAN, P. R. 1994, Using in-situ produced cosmogenic isotopes to estimate rates of landscape evolution: A review from the geomorphic perspective: Jour. Geophys. Res., v. 99, pp. 13,885–13,896.

———, and TURNER, J., 1995, [10]Be and [26]Al evidence of exceptionally low rates of Australian bedrock erosion and the likely existence of pre-Pleistocene landscapes: Quaternary Res., v. 44, pp. 378–382.

BREMER, H., 1971, Flüsse, Flächen, und Stufenbildung in den feuchten Tropen: Wurzburger Geographische Arbeiten, v. 35, 194 pp.

———, 1981, Reliefformen und reliefbildende Prozesse in Sri Lanka, in Bremer, H., Schnütgen, A., and Späth, H., eds., Zur Morphogenese in den feuchten Tropen: Verwitterung und Reliefbildung am Beispiel von Sri Lanka, Relief Boden Paläoklima, v. 1, pp. 7–183.

———, 1993, Etchplanation, review and comments of Büdel's model: Zeitschr. für Geomorph., Supp. no. 92, pp. 189–200.

BÜDEL, J., 1982 Climatic geomorphology (transl. by L. Fischer and D. Busche): Princeton Univ. Press, Princeton Univ. Press, Princeton, New Jersey, 443 pp.

BURBANK, D. W., and BECK, R. A., 1991, Rapid, long-term rates of denudation: Geology, v. 19, pp. 1169–1172.

CHAPPELL, J., 1974, Geomorphology and evolution of small valleys in dated coral reef terraces, New Guinea: Jour. Geology, v. 82, pp. 795–812.

CHRISTIANSEN, R. L., and YEATS, R. S., 1992, Post-Laramide geology of the U.S. Cordilleran region, in Burchfiel, B. C., Lipman, P. W., and Zoback, M. L., eds., The Cordilleran Orogen: Conterminous U.S.: Geol. Soc. America, The geology of North America, v. G–3, pp. 261–406.

CLAGUE, D. A., and DALRYMPLE, G. B., 1987, The Hawaiian-Emperor volcanic chain, Part I, Geologic evolution, in Decker, R. W., Wright, T. L., and Stauffer, P. H., eds., Volcanism in Hawaii: U.S. Geol. Survey Prof. Paper 1350, v. 1, pp. 5–54.

COTTON, C. A., 1948, Landscape as developed by the processes of normal erosion, second ed.: Whitcombe and Tombs Limited, Christchurch, NZ, 509 pp.

CURRAY, J. R., and MOORE, D. G., 1971, Growth of the Bengal deep-sea fan and denudation in the Himalayas: Geol. Soc. America Bull., v. 82, pp. 563–572.

DAHLEN, F. A., and SUPPE, J., 1988, Mechanics, growth, and erosion of mountain belts, in Clark, S. P., Jr., Burchfiel, B. C., and Suppe, J., eds., Processes in continental lithospheric deformation: Geol. Soc. America Spec. Paper 218, pp. 161–178.

DAVIS, W. M., 1885, Geographic classification, illustrated by a study of plains, plateaus, and their derivatives [abs.]: Am. Assoc. Adv. Sci. Proc., v. 33, pp. 428–432.

———, 1889, Topographical development of the Triassic formation of the Connecticut Valley: Am. Jour. Sci., v. 37, pp. 423–434.

———, 1899, The geographical cycle: Geog. Jour., v. 14, pp. 481–504 (reprinted 1954 in Geographical essays: Dover Publications, Inc., New York, pp. 249–278).

———, 1902, Base-level, grade, and peneplain; Jour. Geology, v. 10, pp. 77–111 (reprinted 1954 in Geographical essays: Dover Publications, Inc., New York, pp. 381–412).

———, 1932, Piedmont benchlands and Primärrumpfe: Geol. Soc. America Bull., v. 43, pp. 399–440.

DOLE, R. B., and STABLER, H., 1909, Denudation: U.S. Geol. Survey Water-Supply Paper 234, pp. 78–93.

DORN, R. I., and PHILLIPS, F. M., 1991, Surface exposure dating: Review and critical evaluation: Phys. Geog., v. 12, pp. 303–333.

DOUGLAS, I., 1967, Man, vegetation, and the sediment yield of rivers: Nature, v. 215, pp. 925–928.

von ENGELN, O. D., ed., 1940, Symposium: Walther Penck's contribution to geomorphology: Assoc. Am. Geographers Annals, v. 30, no. 4, pp. 219–284.

———, 1942, Geomorphology: The Macmillan Company, New York, 655 pp.

FAIRBRIDGE, R. W., and FINKL, C. W., JR., 1980, Cratonic erosional unconformities and peneplains: Jour. Geology, v. 88, pp. 69–86.

FINKL, C. W., JR., 1979, Stripped (etched) landsurfaces in southern Western Australia: Australian Geog. Studies, v. 17, pp. 33–52.

GALE, S. J., 1992, Long-term landscape evolution in Australia: Earth Surface Processes and Landforms, v. 17, pp. 323–343.

GASCOYNE, M., FORD, D. C., and SCHWARCZ, H. P., 1983, Rate of cave and landform development in the Yorkshire Dales from speleothem age data: Earth Surface Processes and Landforms, v. 8, pp. 557–568.

GIBBS, R. J., 1967, Geochemistry of the Amazon River system: Part I: Geol. Soc. America Bull., v. 78, pp. 1203–1232.

GLOCK, W. S., 1932, Available relief as a factor of control in the profile of a landform: Jour. Geology, v. 40, pp. 74–83.

HACK, J. T., 1960, Interpretation of erosional topography in humid temperate regions: Am. Jour. Sci., v. 258–A, pp. 80–97.

HALLET, B., HUNTER, J., and BOGEN, J., 1996, Rates of erosion and sediment evacuation by glaciers: A review of field data and their implications: Global and Planet. Change, v. 12, pp. 213–235.

HARBOR, J., and WARBURTON, J., 1993, Relative rates of glacial and nonglacial erosion in alpine environments: Arctic and Alpine Res., v. 25, pp. 1–7.

HOLEMAN, J. N., 1968, Sediment yield of major rivers of the world: Water Resources Res., v. 4, pp. 787–797.

JUDSON, S., 1968a, Erosion rates near Rome, Italy: Science, v. 160, pp. 1444–1446.

———, 1968b, Erosion of the land, or what's happening to our continents?: Am. Scientist, v. 56, pp. 356–374.

———, and RITTER, D. F., 1964, Rates of regional denudation in the United States: Jour. Geophys. Res., v. 69, pp. 3395–3401.

KARÁTSON, D., 1996, Rates and factors of stratovolcano degradation in a continental climate: A complex morphometric analysis for nineteen Neogene/Quaternary crater remnants in the Carpathians: Jour. Volcanol. and Geothermal Res., v. 73, pp. 65–78.

KING, L. C., 1953, Canons of landscape evolution: Geol. Soc. America Bull., v. 64, pp. 721–752 (discussion and replies, v. 66, pp. 1205–1214).

KIRKBY, M. J., 1980, The problem, in Kirkby, M. J., and Morgan, R. P. C., eds., Soil erosion: John Wiley & Sons Ltd., Chichester, UK, pp. 1–16.

KRONBERG, B. I., NESBITT, H. W., and LAM, W. W., 1986, Upper Pleistocene Amazon deep-sea fan muds reflect intense chemical weathering of their mountainous source lands: Chem. Geology, v. 54, pp. 283–294.

LAMARCHE, V. C., JR., 1968, Rates of slope degradation as determined from botanic evidence, White Mountains, California: U.S. Geol. Survey Prof. Paper 352–I, pp. 341–377.

LANGBEIN, W. B., and LEOPOLD, L. B., 1964, Quasi-equilibrium states in channel morphology: Am. Jour. Sci., v. 262, pp. 782–794.

LANGBEIN, W. B., and SCHUMM, S. A., 1958, Yield of sediment in relation to mean annual precipitation: Am. Geophys. Union, Trans., v. 39, pp. 1076–1084.

LARSON, W. E., PIERCE, F. J., and DOWDY, R. H., 1983, The threat of soil erosion to long-term crop production: Science, v. 219, pp. 458–465.

LASAGA, A. C., SOLER, J. M., and 3 others, 1994, Chemical weathering rate laws and global geochemical cycles: Geochimica et Cosmochimica Acta, v. 58, pp. 2361–2386.

MABBUTT, J. A., 1965, The weathered landsurface of central Australia: Zeitschr. für Geomorph., v. 9, pp. 82–114.

MARCHAND, D. E., 1971, Rates and modes of denudation, White Mountains, eastern California: Am. Jour. Sci., v. 270, pp. 109–135.

MATHEWS, W. H., 1975, Cenozoic erosion and erosion surfaces of eastern North America: Am. Jour. Sci., v. 275, pp. 818–824.

MATTHIAS, G. F., 1967, Weathering rates of Portland Arkose tombstones: Jour. Geol. Education, v. 15, pp. 140–144.

MEIERDING, T. C., 1993, Inscription legibility method for estimating rock weathering rate: Geomorphology, v. 6, pp. 273–286.

MILLIMAN, J. D., and MEADE, R. H., 1983, World-wide delivery of river sediment to the oceans: Jour. Geology, v. 91, pp. 1–21.

MILLIMAN, J. D., and SYVITSKI J. P. M., 1992, Geomorphic/tectonic control of sediment discharge to the ocean: The importance of small mountainous rivers: Jour. Geology. v. 100, pp. 525–544.

MORISAWA, M., 1989, Rivers and valleys of Pennsylvania, revisited, in Gardner, T. W., and Sevon, W. D., eds., Appalachian geomorphology: Geomorphology, v. 2, pp. 1–22.

NANSON, G. C., PRICE, D. M., and SHORT, S. A., 1992, Wetting and drying of Australia over the past 300 ka: Geology, v. 20, pp. 791–794.

NOTT, J., 1995, The antiquity of landscapes on the North Australia Craton and the implications for theories of long-term landscape evolution: Jour. Geology, v. 103, pp. 19–32.

OHMORI, H., 1985, A comparison between the Davisian scheme and landform development by concurrent tectonics and denudation: Univ. Tokyo Dept. Geography Bull. no. 17, pp. 19–28.

OLLIER, C. D., 1991, Ancient landforms: Belhaven Press, London, 233 pp.

PAINE, A. D. M., 1985, 'Ergodic' reasoning in geomorphology: Time for a review of the term?: Prog. in Phys. Geog., v. 9, pp. 1–15.

PENCK, W., 1924, Die Morphologische Analyse: J. Engelhorn's Nachfolger, Stuttgart, 283 pp.

———, 1953, Morphological analysis of landforms (transl. by H. Czech and K. C. Boswell): St. Martin's Press, New York, 429 pp.

PITTY, A. F., 1971, Introduction to geomorphology: Methuen & Co. Ltd., London, 526 pp.

PLAYFAIR, J., 1802, Illustrations of the Huttonian theory of the earth: Dover Publications, Inc., New York, 528 pp. (facsimile reprint, 1964).

POAG, C. W., and SEVON, W. D., 1989, A record of Appalachian denudation in postrift Mesozoic and Cenozoic sedimentary deposits of the U.S. Middle Atlantic continental margin, in Gardner, T. W., and Sevon, W. D., eds., Appalachian geomorphology: Geomorphology, v. 2, pp. 119–157.

RAHN, P. H., 1971, Weathering of tombstones and its relationship to the topography of New Uk: Jour. Geol. Education, v. 19, pp. 112–118.

RINALDO, A., DIETRICH, W. E., and 3 others, 1995, Geomorphological signatures of varying climate: Nature, v. 374, pp. 632–635.

RODEN, M. K., and MILLER, D. S., 1989, Apatite fission-track thermochronology of the Pennsylvania Appalachian basin, in Gardner, T. W., and Sevon, W. D., eds., Appalachian geomorphology: Geomorphology, v. 2, pp. 39–51.

RODEN-TICE, M., and TICE, S. J., 1996, Early Cretaceous unroofing history of the Adirondack Mountains, New York State through apatite fission-track thermochronology [abs.]: Internat. workshop on fission-track dating, Gent, Belgium, Aug. 26–30, 1996, p. 93.

RUXTON, B. P., and McDOUGALL, I., 1967, Denudation rates in NE Papua from K-Ar dating of lavas: Am. Jour. Sci., v. 265, pp. 545–561.

SASOWSKY, I. D., WHITE, W. B., and SCHMIDT, V. A., 1995, Determination of stream-incision rate in the Appalachian plateaus by using cave-sediment magnetostratigraphy: Geology, v. 23, pp. 415–418.

SAUNDERS, I., and YOUNG, A., 1983, Rates of surface processes on slopes, slope retreat and denudation: Earth Surface Processes and Landforms, v. 8, p. 473–501.

SCHUMM, S. A., 1963, The disparity between present rates of denudation and orogeny: U.S. Geol. Survey Prof. Paper 454–H, 13 pp.

———, and ETHRIDGE, F. G., 1994, Origin, evolution and morphology of fluvial valleys, in Dalrymple, R. W., Boyd, R., and Zaitlin, B. A., eds., Incised-valley systems: Origin and sedimentary sequences: SEPM (Soc. Sedimentary Geol.) Spec. Pub. 51, pp. 11–27.

SIMONS, M., 1962, The morphological analysis of landforms: A new review of the work of Walter Penck: Inst. Brit. Geographers Trans., no. 31, pp. 1–14.

SLINGERLAND, R., and FURLONG, K. P., 1989, Geodynamic and geomorphic evolution of the Permo-Triassic Appalachian Mountains, in Gardner, T. W., and Sevon, W. D., eds., Appalachian geomorphology: Geomorphology, v. 2, pp. 23–37.

SPRINGER, G. S., KITE, J. S., and SCHMIDT, V. A., 1997, Cave sedimentation, genesis, and erosional history in the Cheat River canyon, West Virginia: Geol. Soc. America Bull., v. 109, pp. 524–532.

THOMAS, M. F., 1994, Geomorphology in the tropics: A study of weathering and denudation in low latitudes: John Wiley & Sons Ltd., Chichester, UK, 460 pp.

TRIMBLE, S. W., 1975, Denudation studies: Can we assume stream steady state?: Science, v. 188, pp. 1207–1208.

———, 1977, The fallacy of stream equilibrium in contemporary denudation studies: Am. Jour. Sci., v. 277, pp. 876–887.

———, 1988, The impact of organisms on overall erosion rates within catchments in temperate regions, in Viles, H. A., ed., Biogeomorphology: Basil Blackwell Ltd., Oxford, UK, pp. 83–142.

TWIDALE, C. R., 1987, Etch and intracutaneous landforms and their implications: Australian Jour. Earth Sci., v. 34, pp. 367–386.

WALLING, D. E., and WEBB, B. W., 1996, Erosion and sediment yield: A global overview, in Walling, D. E., and Webb, B. W., eds., Erosion and sediment yield: Global and regional perspectives: Internat. Assoc. Hydrol. Sciences (IAHS) Pub. no. 236, pp. 3–19.

WEBB, J. A., FINLAYSON, B. L., and 2 others, 1991, The geomorphology of the Buchan karst—implications for the landscape history of the Southeastern Highlands of Australia, in Williams, M. A. J., DeDekker, P., and Kershaw, A. P., eds., The Cainozoic in Australia: A re-appraisal of the evidence: Geol. Soc. Australia Spec. Pub. 18, pp. 210–234.

WILSON, L., 1973, Variations in mean annual sediment yield as a function of mean annual precipitation: Am. Jour. Sci., v. 273, pp. 335–349.

WOLDENBERG, M. J., ed., 1985, Models in geomorphology: Allen & Unwin, Inc., Boston, 434 pp.

YAIR, A., and ENZEL, Y., 1987, The relationship between annual rainfall and sediment yield in arid and semi-arid areas. The case of the northern Negev, in Ahnert, F., ed., Geomorphological models: Theoretical and empirical aspects: Catena Supp. 10, pp. 121–135.

YOUNG, A., and SAUNDERS, I., 1986, Rates of surface processes and denudation, in Abrahams, A. D., ed., Hillslope processes: Allen & Unwin, Inc., Boston, pp. 3–27.

YOUNG, R., and MCDOUGALL, I., 1993, Long-term landscape evolution: Early Miocene and modern rivers in southern New South Wales, Australia: Jour. Geology, v. 101, pp. 35–49.

Part V

Glaciers and Glaciation

Glaciers are masses of ice (with included rock debris, water, and air) on land, which show evidence of present or past internal deformation and motion. Water in the solid state has thermal and physical properties very different from those in the liquid state. The transition from one state to the other is abrupt. Therefore, landscape development by glaciers is very different from the processes and forms of the fluvial system that are dominated by liquid water, and the boundaries between glacial landforms and fluvial landforms are unusually sharp.

Glaciers, especially continent-size ice sheets, pose unique problems for geomorphic analysis. In one sense, the surface of an ice sheet is a constructional landform, built of layers of snow compressed to ice. In another sense, glacier ice is an easily deformed plastic solid and assumes a domal or ellipsoidal shape that can be defined as being structurally controlled. Erosion plays only a minor role in shaping the surface of a glacier, yet glaciers are powerful eroding agents of other rocks. An estimated 7 percent of all modern erosion is being done by glaciers (p. 198), which makes them second only to rivers (although a poor second) in total erosional effectiveness. Nevertheless, considering their limited area today, they make a respectable contribution to continental denudation. Glaciation increases the erosion rate on a mountain by at least an order of magnitude over the rate on a comparable, but unglaciated, mountain (Table 15-2).

We study glaciers in three ways: first, as mobile, diverse, dynamic landforms shaped by unusual geomorphic process such as rheid flow, brittle fracture, and melting; second, as peculiar but abundant monomineralic rock masses with unusual thermal and rheologic properties; and third, as powerful geomorphic agents that are shaping and have shaped many terrains. Three distinct scientific disciplines are involved in studying glaciers. The first is glacier morphology, the description and explanation of ice surfaces or "ice-scapes." Next is glaciology, the science of ice. These two subjects are covered in Chapter 16. The third scientific discipline is glacial geomorphology, which describes and explains the geomorphic work of glaciers, past and present. It is considered separately in Chapter 17. A useful distinction is to use "glacierized" to describe a landscape partly or entirely covered by glaciers, and distinguish that term from "glaciated," meaning a landscape now ice-free, but with the diagnostic landforms and sediments of former glaciers.

Although glaciers are powerful agents of geomorphic change, no attempt has ever been made to deduce a "cycle of erosion" for glacierized landscapes. For one reason, glacial erosion operates largely independent of base-level control by sea level, therefore there is no progressive reduction of potential energy in the glacial system that would support the concept of cyclic time (Figure 1-6). In addition, widespread glaciation has been only episodic in geologic history, at times when plate motions, ocean circulation, and atmospheric conditions have combined to permit ice accumulation on land. Even though the earth is now in an "icehouse" climate and has experience multiple, extensive glaciations in the past few million years (Chapter 4), the primary result has been modification of preexisting, largely fluvial, landscapes. Especially in mountains, those modifications have been awesome. Some details of the geomorphic impact of glaciers and associated phenomena during the past 130,000 years are provided in Chapter 18.

No geomorphologist can appreciate the work of ancient glaciers in shaping some of the world's most dramatic landscapes unless he or she understands the physical and thermal properties of ice. But all the theoretical studies in the world cannot replace a walk onto the surface of a modern glacier for comprehending their geomorphic significance.

Chapter 16

Glaciers as Landforms: Glaciology

Of all the water near the surface of the earth, only about 2 percent is on the land, in the solid state as glacier ice (Figure 2-5). Even this small fraction is sufficient to cover entirely one continent (Antarctica) and most of the largest island (Greenland) with ice to an average thickness of 2.2 km over Antarctica and 1.5 km over Greenland. At the present time, about 10 percent of the earth's land area is ice covered. An additional 20 percent has been ice covered repeatedly during the glaciations of the Pleistocene Epoch, much of it as recently as 15,000 to 20,000 years ago. Glaciation has been the dominant factor in shaping the present landscape of North America northward of the Ohio and Missouri rivers and of Eurasia northward of a line from Dublin eastward through Berlin to Moscow and beyond the Urals (Figure 16-1). In addition, mountains and plateaus in all latitudes have been glaciated to an altitude 1000 to 1500 m lower than their present snowlines.

GEOMORPHOLOGY OF GLACIER SURFACES

The simplest morphologic subdivision of glaciers is to distinguish those that flow between confining rock walls, or **valley glaciers,** and those that bury the rocky landscape and flow unconfined by virtue of their great thickness, the **ice caps** and **ice sheets** (Sharp, 1988). Although the morphology of all glaciers is controlled by the rheidity of ice (p. 13), important differences are to be noted whether flow is confined or unconfined.

Glaciers originate in regions where snow accumulation exceeds loss, and they flow outward or downward to regions where losses exceed accumulation (Figure 16-2). The *terminus,* or downglacier extremity, represents the line where losses by all causes (melting, sublimation, erosion, and calving into water: collectively termed **ablation**) equal the rate at which ice can be supplied by accumulation and forward motion. Some valley glaciers have steep ice faces at their termini, but others end deeply buried in transported rock debris, so that the exact location of the terminus is uncertain. In the *zone of accumulation,* vectors of particle motion are downward into the ice mass as each year's snowfall adds a new surface layer to the glacier. Erosion at the glacier bed is intensified. In the *zone of ablation,* vectors of particle motions point toward the ice surface, because a surface layer is annually removed, exposing progressively deeper ice (Figure 16-2). Net sediment deposition is probable. This is the predominant method of bringing sediment to the surface of a glacier, more important than thrust faulting (p. 366). At the cross section of the glacier where upglacier net accumulation and downglacier net ablation are in exact balance, internal flow vectors are parallel to the glacier surface. This equilibrium cross section of a glacier is located beneath the **equilibrium line,** the downglacier edge of net annual accumulation.

FIGURE 16-1. Northern hemisphere ice sheets and other glaciers during the last ice age. Possible floating ice shelves not shown (Denton and Hughes, 1981, p. viii, reprinted by permission. Copyright © 1981, John Wiley & Sons, Inc.).

Valley Glaciers

In a mountainous terrain, the colder temperatures and usually higher precipitation at higher altitudes produce ice fields at the heads of glaciers. This is the zone of accumulation (Figure 16-2), and as a rule of thumb, about 65 percent of the surface area of a valley glacier is in this zone. The annual discharge of ice through the cross section beneath the equilibrium line is an important measure of the glacier's regimen. The equilibrium line is approximated by the annual **snow line** on the glacier and adjacent mountain slopes.

Downglacier from the equilibrium line, a glacier has net annual loss of ice. Melting is the major factor in valley glaciers, but sublimation, wind ablation, and iceberg calving into lakes or the ocean also contribute. A glacier is in equilibrium only if annual discharge through the equilibrium-line cross section is equal to the net annual accumulation upglacier and the net annual ablation downglacier. Otherwise, it will grow or shrink.

Many large valley glaciers are fed by one or more *tributary glaciers* (Figure 16-3). Although tributary glaciers do not erode their floors as deeply as does the trunk

FIGURE 16-2. Trajectories of particle motion in a valley glacier. The large area of the zone of accumulation is not well shown in this longitudinal cross section. Sections labeled (+) and (−) are local areas of compressing and extending flow.

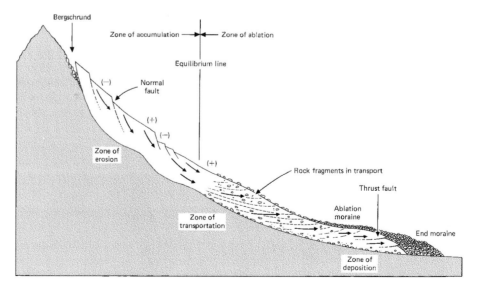

FIGURE 16-3. Yentna Glacier, Alaska Range. Late summer view; equilibrium line is visible toward the upper left. Rhythmic ridges on the leftmost tributary are *ogives,* formed by seasonal melting as the glacier descends a buried ridge. Lateral moraines of tributary glaciers become medial moraines in foreground, but maintain their individual identity (photo: Austin Post, U.S. Geological Survey).

glacier, the surfaces of merging glaciers join at a common level because ice is easily deformed and responds to the lateral stresses imposed by an entering tributary. Tributary glaciers do not turbulently mix with the trunk glacier, as rivers do. Because of the nature of flow in ice, the tributary glacier retains its identity as a separate stream or surface ribbon of ice (Figure 16-3) although it may become progressively attenuated or enfolded into the trunk glacier.

Among the oldest observations of glacier flow were reports that lines of stakes emplaced straight across valley glaciers would bulge progressively downvalley year after year, proving that the center part of the ice surface moved faster than the margins. Many subsequent experiments have confirmed the fact (Figure 16-4). Because most of the differential shear is very near the base and lateral margins of a glacier, the mean velocity of the surface ice is within a few percent of the mean velocity of the entire glacier cross section (Paterson, 1994, p. 270). As a useful result of this approximation, glaciologists can estimate the annual discharge of ice through the equilibrium-line cross section of a glacier by measuring surface velocities only. It is not necessary to drill boreholes and measure velocities within the ice. In this respect especially,

the laminar flow of a glacier is much simpler than the turbulent flow of a river, for which an entire velocity cross section must be measured to determine mean discharge (p. 203).

Valley glaciers have much higher surface gradients than rivers, commonly averaging 10 percent (6°). On ice cliffs or ice falls steeper than 45°, ice avalanches are nearly continuous. Ledges buried beneath the ice are reflected by *crevasse fields* and ice pinnacles (*séracs*) on the ice surface (Figures 16-3 and 17-4). The long profile of a glacier surface is likely to be marked by level areas and steep drops, much like the surface gradient of a river through rapids and across pools.

Within the zone of accumulation of a valley glacier, velocity generally increases downglacier. This is called *extending flow* because an imaginary mass within the glacier becomes longer and thinner as it moves downvalley (assuming that the valley sides are parallel). Within the zone of ablation, longitudinal velocity generally decreases, perhaps to zero at the toe of the glacier. This is called *compressing flow* because an imaginary mass within the glacier becomes shorter and thicker as it flows (again assuming no change of the valley width). The equilibrium cross section is therefore the place where the flow rate is at the maximum.

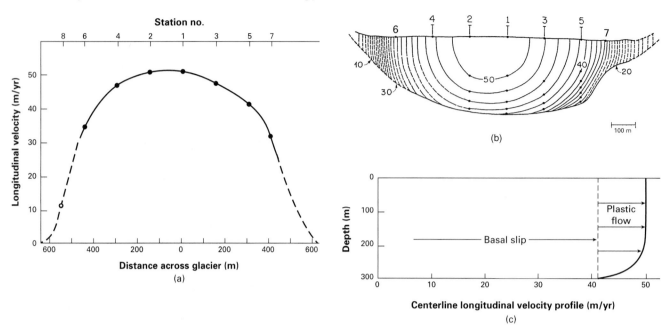

FIGURE 16-4. Flow in the Athabaska Glacier, Alberta, Canada. (a) Lateral variation in surface velocity. Solid circles, surveyed markers; open circle, estimated from adjacent measured velocities. (b) Distribution of longitudinal velocity in cross section from boreholes at stations 1-5, with extrapolated marginal velocities. (c) Inferred longitudinal velocity profile at the centerline of flow half way between boreholes 1 and 2. Basal slip accounts for 81 percent of the surface velocity in this profile, and more than 70 percent of the surface velocity across half the glacier width (data from cross section A, Raymond, 1971).

Locally within a glacier, other areas of extending and compressing flow also can be identified (Figure 16-2). For example, flow may be compressive upstream from a bed obstruction, with the ice thickening and the glacier surface bowing upward as a reflection of the buried obstacle. Extending flow is common where ice passes over a buried sill and drops rapidly. At such places, the ice may fracture into a crevasse field or simply thin rapidly by extending flow. In plan view, compressing and extending flow can be expected, respectively, on the inside and outside of curves, where shear rates are high. Similar surface fracture patterns on floating ice shelves show regions of extending and compressing flow (Figure 16-8). Beneath the ice, regions of extending and compressing flow can control basal slip (p. 366), and determine whether the glacier erodes its bed or deposits sediment (Chapter 17).

On the surface of most valley glaciers in the ablation zone, a variety of ephemeral landforms develop due to melting, collapse of ice tunnels, and erosion by streams flowing on, in, or along the glacier. Many features are analogous with karst features and may be collectively called **glacier karst**. *Thaw lakes* and swallow holes (*moulins*) mark the ice surface. Segregations of debris-rich ice stand as ice-cored mounds or moraine ridges above the debris-covered surface (Figure 16-3).

Lakes may drain and then abruptly form again when a crevasse or moulin opens at depth, and then deformation or clay-rich sediments seal it closed again. Ice tunnels within or at the base of the glacier discharge sediment-laden water from the terminus.

Ice Caps and Ice Sheets

Unconfined glaciers may be so large that they form their own orographic precipitation pattern. The accumulation zone is in the central region of the dome, and the ablation zone is on the periphery, below the equilibrium line. The regimen of such an ice cap or ice sheet is very sensitive to slight climatic changes. Ice caps and ice sheets, unlike valley glaciers, are examples of *positive feedback systems*. For example, any increase in accumulation will increase the height and therefore the area of the accumulation zone relative to the ablation zone, resulting in further accelerated growth. Conversely, a decrease in accumulation will lower the central dome, further decrease the relative area of accumulation, and accelerate shrinkage. Because the climate has warmed since they formed, numerous small or moderate-size ice caps on Iceland, Scandinavia, and the Canadian Arctic islands could not re-form today if their summits ever lowered to near the equilibrium line; the present sustaining orographic precipitation is a consequence of the ice caps themselves, not of any mountain mass beneath them.

The radial profiles of continent-size ice sheets and the somewhat smaller, but otherwise similar, ice caps differ very little from each other (Figure 16-5). Theories of radial flow predict parabolic or elliptical radial profiles very similar to those actually observed (Figures 16-5 and 16-6). The ice must spread radially in plan view from the area of accumulation, as well as outward along any cross section. Except for the complexity of outlet glaciers and ice streams (p. 360–361), the general flow lines (Figure 16-7) are probably much as inferred by early workers: downward beneath the zone of accumulation near the center and then outward to the zone of ablation. Obstructions such as enclosing mountain ranges obviously complicate the flow patterns. Note the considerable vertical exaggeration on Figures 16-5, 16-6, and 16-7. At true scale, the thickness of the ice could barely be shown by a slight thickening of the horizontal base lines. Published maps and radar profiles (Walker et al., 1968; Drewry, 1983; Bogorodsky

et al., 1985) give an excellent idea of the surface relief, bottom relief, and thickness of large ice sheets.

Most of the Greenland ice sheet perimeter is a series of broad, gently sloping ramps. Along some parts of the perimeter, where the terminus is on land, the glacier ends as a vertical ice cliff as much as 100 m in height. The cliffs may be due to a combination of factors, such as low temperatures, compass orientation relative to a low sun angle that causes sublimation directly from the cliff face, and cold, brittle ice almost free of enclosed sediment except at the base. The Greenland ice sheet (known as the *Inland Ice*) rests on a basin-shaped interior plateau surrounded by mountain ranges (Figure 16-6) (Reeh, 1989). Around much of its margin, the ice moves between exposed or slightly buried mountain peaks as **outlet glaciers.** These have many of the geomorphic properties of valley glaciers. At the head of each outlet glacier is an area of shallow concavity sloping toward the outlet, with many wide, deep crevasses. This is the "drawdown" area, or "iceshed" of that particular outlet glacier.

At least two discrete outlet glaciers drain significant portions of the Greenland ice sheet at rates much more rapid than the average radial spreading rates. Jakobshavns Isbrae reaches the west coast of Greenland as a high gradient, thick (2800 m) outlet glacier that may be the world's fastest moving nonsurging glacier (Echelmeyer et al., 1991, 1992). The continuous forward motion is 7 km/yr near the floating terminus, and it produces many of the icebergs that float south through Davis Strait. This single outlet glacier drains 6.5 percent of the total area of the Greenland ice sheet although it is only 6 km wide and 85 to 90 km in length. Spectacular surface zones of marginal shearing separate the ice stream from adjacent slow-moving ice. A similar tongue of fast-moving ice in northeastern Greenland was identified from repeated satellite radar images (Fahnestock et al., 1993). It drains at least twice the area of Jakobshavns Isbrae. The factors that control such rapidly moving portions of an ice sheet have yet to be determined, but they violate generalizations about uniform trajectories of flow radially from the center of an ice sheet as implied by Figure 16-7.

Much of the perimeter of the Antarctic ice sheet is also a great cliff, but that cliff ends at a *grounding line* in the ocean (Figure 16-7). A few **nunataks,** peaks surrounded by ice, rise above the surface (Figure 16-8). The West Antarctic sector, in the longitudinal zone that faces northward toward the Pacific Ocean between New Zealand and South America, is grounded 1 to 2 km below sea level in two large tectonic basins from which ice flows into the floating *ice shelves* of the Ross Sea and Weddell Sea and into numerous other smaller ice shelves (Figure 16-8). Episodically, slabs of ice with

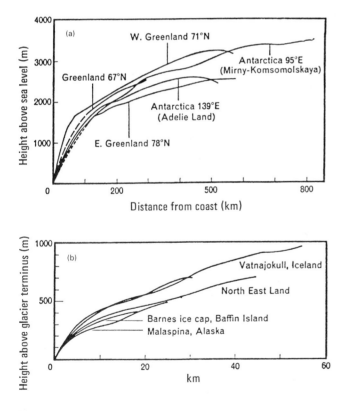

FIGURE 16-5. Surface profiles of (a) the ice sheets of Greenland and Antarctica, and (b) smaller ice sheets and ice caps. In both figures, bedrock relief is slight relative to ice thickness (Robin, 1964, Figure 5).

FIGURE 16-6. Profiles and estimated temperatures of the Greenland ice sheet (the Inland Ice). Vertical exaggeration 50:1 (Weidick, 1975, Figure 4).

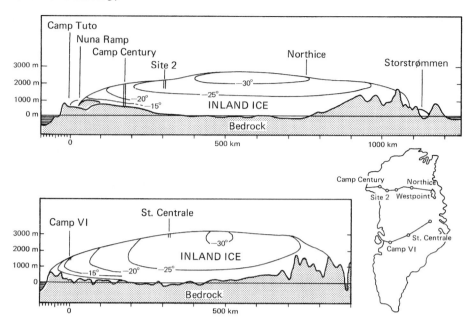

areas measured in thousands of square kilometers separate from the ice shelves and drift into the southern ocean (Swithinbank, 1988; Rott et al., 1996). Large outlet glaciers are held back by *ice rises,* buried hills over which the ice rides in the transitional zone where outlet glaciers become floating ice shelves (Figure 16-8). Separated by the ice rises are several rapidly flowing *ice streams,* moving over water-saturated sediments at

rates of up to several meters per day (Engelhardt et al., 1990; Alley and Whillans, 1991).

The surface of the East Antarctic ice sheet is a monotonous dome with only minor areas of exposed bedrock, mostly near the coast (Drewry, 1983; Radok, 1985). Flow lines define ice divides and drainage basins, but surface gradients are very low, and little relief is apparent. Much of the East Antarctic ice sheet

FIGURE 16-7. Particle paths and calculated ages of ice in a cross section of Law Dome, a small independent ice dome on the edge of the Antarctic ice sheet in Wilkes Land (111°E longitude) (from Budd and Morgan, 1973, Figure 4).

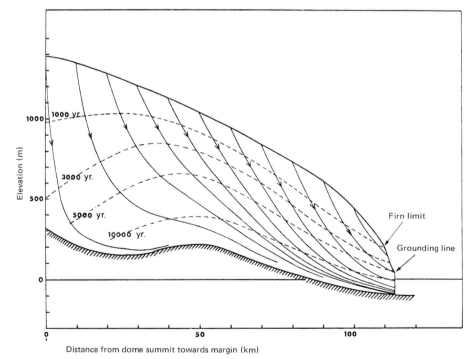

has precipitation of less than 50 mm/yr of water equivalent, making it one of the driest regions on earth. Surface temperatures are the coldest known on earth.

GLACIOLOGY

Glaciology is the scientific study of all ice, not just of glaciers. It includes the study of ice crystals in high clouds, hail, and snow; frozen lake, river, and ocean water; and glacier ice. The Voyager and Galileo spacecraft exploration of Jupiter, Saturn, and Uranus and their moons has broadened the interest of glaciology to include the ices of ammonia (NH_3) and methane (CH_4), for these along with H_2O ice are probably the materials of which many of the moons of the outer planets are composed (Klinger et al., 1985; Jankowski

and Squyres, 1988). Glaciologists can be metallurgists, meteorologists, physicists, geographers, or geologists. What we know about glacier ice has been learned from a wide field of experiment, observation, and theory (Lliboutry, 1965; Hutter, 1983; Weertman, 1983; Paterson, 1994; Hooke, 1998).

Formation of Glacier Ice

The conversion of snow to glacier ice follows a regular sequence. Fresh snow is full of entrapped air and may have a bulk specific gravity even lower than 0.1. Snowflakes readily *sublime* (vaporize directly from the solid state without melting), so aging snowflakes lose their frilly margins and become more globular. Meltwater often freezes onto snowflake nuclei, so aging also tends to increase the size of the ice granules.

FIGURE 16-8. Oblique aerial photograph up the Fleming outlet glacier, Graham Land, Antarctica. The grounding line is somewhere in midphoto. Buffer Ice Rise (foreground) deflects and fractures the floating Wordie Ice Shelf (photo: U.S. Navy for U.S. Geological Survey, TMA 1835F33, courtesy R. S. Williams, Jr.).

Old snow, such as remains on sheltered mountain slopes in early summer, has the texture of very coarse sand. This is the late-season *corn snow* or *buckshot snow*, familiar to avid skiers. The loose granular mass is about one-half ice and one-half entrapped air and has a bulk specific gravity of about 0.5.

If snow has survived a summer melting season, it is called **firn** in German or **névé** in French. Both terms are widely used in English. Firn, or névé, is an intermediate step in the conversion of snow to glacier ice, with a bulk specific gravity between 0.5 and 0.8. It is granular and loose unless it has formed a crust. It represents the net positive balance between winter accumulation and summer loss.

As successive annual layers accumulate, the deep firn is compacted. Individual ice grains freeze together. By definition, when the grains are frozen together so that the mass is impermeable to air, firn becomes glacier ice. The bulk specific gravity is usually about 0.8 by this stage of consolidation. Glacier ice is a polycrystalline mass that includes a variable amount of entrapped air, as well as dust and rock fragments that have fallen, washed, or been blown onto the ice surface, and rock that has been eroded from beneath the glacier. The bulk specific gravity of glacier ice ranges from about 0.8 to 0.9, close to the specific gravity of pure, gas-free ice (0.916).

Ice Deformation

The only form of ice that occurs naturally on planet Earth crystallizes in the hexagonal crystal system, and is characterized by three intermolecular glide axes 120° apart on a plane perpendicular to the *c* optical axis (Figure 16-9). The glide axes define a *basal glide plane* on which single ice crystals easily deform under very low shear stress. The basal plane is the orientation of the incomplete hexagonal skeletal ice crystals we call snowflakes. Under a low sustained shear (directional) stress of 0.3 MPa (3 bar), an ice crystal can be drawn out into a thin ribbon in only a few days (Figure 16-9) (Glen, 1952).

When an ice specimen is subjected to a slight shear stress (pressure or tension) and then released, it elastically recovers its original shape. When the shear stress is increased to above a certain threshold called the *elastic limit* or *yield strength*, irreversible deformation or *strain* results (Figure 16-10a). If shear stress is maintained, the sample deforms by continuous plastic strain or *creep*.

Polycrystalline ice under shear stress deforms very much as iron deforms when it is heated to a bright red color. Either in single crystals or in polycrystalline

masses, at only slightly increased stresses above the yield stress the strain rate of ice becomes very rapid. By the simplifying assumption that ice is a perfect plastic, which does not deform at all at stresses less than 0.1 MPa (1 bar) but deforms infinitely rapidly at a slightly larger stress, theoretical models of ice flow have been developed that closely approximate the flow of glaciers. The approximation is sometimes referred to as the "one-bar flow law." It is plotted on Figure 16-10b as a horizontal line at a shear stress of 0.1 MPa (1 bar). Several lines of evidence suggest that shear stresses at the base of glaciers are seldom far from the range 0.05 to 0.15 MPa (0.5 to 1.5 bar), so the simplifying assumption of the one-bar flow law has proved useful (Paterson, 1994, p. 240).

More precisely, the plastic deformation of ice compiled from laboratory experiments and measurements in ice tunnels and boreholes (Figure 16-11) fits a power-function equation of the form

$$\dot{\varepsilon} = A\tau^n$$

where $\dot{\varepsilon}$ is the creep rate, or rate of permanent deformation measured in percent change of length per time (which simplifies to reciprocal times; see p. 12), τ is the shear stress, and A and n are constants. This equation is called "Glen's flow law" in recognition of the original experimenter (Glen, 1952; Paterson, 1994, pp. 85; Menzies, 1995, pp. 139, 150; Hooke, 1998, pp. 41ff). Experimental values of n for both single crystals and polycrystalline ice generally have a mean of about 3 (Figure 16-10b). When plotted on a log-log graph, the value of n of about 3 is illustrated by a thousandfold increase in creep rate for each tenfold unit increase in shear stress (Figure 16-11).

If an ice grain is constrained from easy deformation on its basal glide plane, as for example, if it is completely surrounded by other grains of ice and held in an unfavorable orientation relative to the applied shear stress, it will still deform, but much more slowly. The creep rate is 500 times more rapid for a given shear stress in the "easy glide" basal plane of a single crystal than in various "hard glide" crystal orientations or in polycrystalline aggregates (Figure 16-11) (Weertman, 1983, p. 224; Hooke, 1998, p. 34). The cluster of experimental and field-determined strain-stress relationships in the lower central part of Figure 16-11 are regarded as the best ones to use in calculations or modeling of glacier flow. They demonstrate that, in general, a mass of polycrystalline ice subjected to a shear stress of 0.3 MPa (3 bar) will deform by an amount equal to its original length in about 1 year.

Although absolute creep rates vary greatly between single crystals in the easy glide orientation and polycrystalline glacier ice with varying amounts of

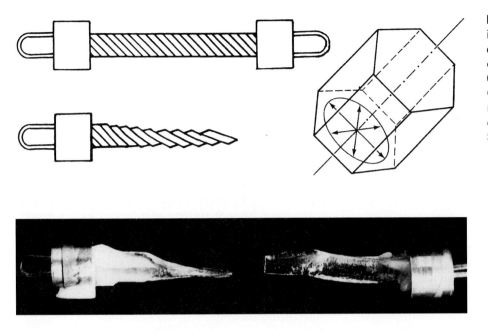

FIGURE 16-9. Single crystal of ice, 13 mm² cross section, *c* optical axis 45° to specimen axis, deformed by sustained tension of 0.4 kg (= 0.3 MPa) (photo: J. W. Glen, reproduced from the Journal of Glaciology by permission of the International Glaciological Society).

impurities, the power function (*n*) that describes the increase in creep rate with increased shear stress is similar in most experiments. The constant *A*, but not *n*, is strongly temperature-dependent, so that for a given shear stress, the strain rate is only one-fifth as great in ice at −25°C as in ice at −10°C (Paterson, 1994, p. 86).

It is important to note that confining, or cryostatic, pressure does not directly affect the strain rate of ice. Only shear, or directional, stress causes plastic deformation. As with rock and other solids, confining pressure only prevents brittle fracture and thereby permits creep to be measured. Confining pressure does not make the ice at the bottom of a glacier any more "plastic" than the ice at the surface. Many serious errors in interpreting glacier flow have been made because of this misconception.

Thermal Classification of Glaciers

Confining, or cryostatic, pressure does affect glacier deformation indirectly. Pressure depresses the freezing temperature of water almost linearly from 0°C at 0.1 MPA (1 bar) to −22°C at a pressure of 2070 MPa (2070 bar), beyond which other forms of ice crystallize. Given

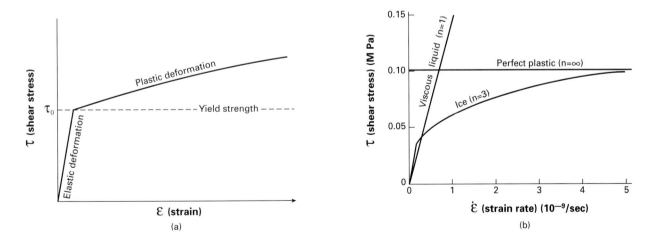

FIGURE 16-10. (a) A schematic typical stress-strain curve. τ_0 is the yield strength, marking the onset of plastic creep, ε is the strain, or deformation. (b) Shear stress τ versus strain rate $\dot{\varepsilon}$ for ice, with $n = 3$ (p. 362). For comparison, the strain rates of a "perfect" plastic ($n = \infty$) and a viscous liquid ($n = 1$) are shown.

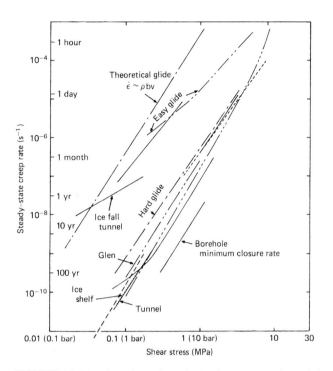

FIGURE 16-11. Log-log plot of steady-state or the minimum creep rate versus stress. Creep rate is normalized to −10°C. Creep rate and stress are for, or have been changed to, the equivalent of a uniaxial tension or compression test. Axes are reversed from Figure 16-10b. Data from numerous sources listed in Weertman (1983) (reproduced with permission from the Annual Review of Earth and Planetary Sciences, v. 11, © 1983 by Annual Reviews Inc.).

the nearly uniform specific gravity of glacier ice, the pressure-melting temperature of ice (the temperature at which ice and water phases coexist) decreases at a rate of about 0.7°C/1000 m of ice thickness. Thus, if water is encountered at any depth within a glacier, the temperature at that depth can be precisely calculated (Figure 16-12). From the foregoing discussion, two categories of glaciers can be visualized. One category is of glaciers that are cold from top to bottom, with the entire thickness of ice well below the pressure-melting temperature. The other category is of glaciers that are at or very close to the pressure-melting temperature so that small pressure variations can produce melting. Actually, this thermal classification is better applied to parts of glaciers rather than to entire valley glaciers or ice sheets because the thermal status of one part of a glacier may be different from that of another part. A discussion of the thermal status of glaciers and parts of glaciers uses terminology first proposed more than 60 years ago (Lagally, 1932; Ahlmann, 1933, 1935), but the examples and conclusions are much more recent. The thermal condition of glaciers has a very important

bearing on both their form and their ability to shape landscapes.

Cold, Polar, or Dry-base Glaciers. A 1387-m borehole through the Greenland ice sheet at Camp Century, about 100 km from the west edge of the ice sheet, was completed in 1966. At the time, it was the deepest borehole made in ice and the first to penetrate entirely through an ice sheet. The temperature profile of the Camp Century borehole (Figure 16-12) illustrates well the characteristics of a **cold**, or **polar**, glacier. The ice temperature at a depth of 10 m was −24°C, very close to the mean annual air temperature at Camp Century. A minimum temperature of −24.6°C was reached at a depth of 154 m; then the temperature rose steadily to −13.0°C at the base of the ice. The entire thickness of ice at the Camp Century borehole is far below the pressure-melting temperature. No water phase can exist, and the rather slow surface motion of only 3.3 m/yr must be due entirely to plastic creep and minor brittle fracturing near the surface.

Cold, or polar, glaciers are also called **dry-base** glaciers, because the basal ice must be frozen firmly to the rock beneath. Experiments have demonstrated that the surface bond between rock-forming minerals and ice is considerably stronger than the yield strength of ice, so that any motion of a cold glacier must be by internal plastic deformation. Just one consequence of this fact is that cold glaciers cannot be expected to erode by sliding over bedrock. Most of the motion is concentrated in the basal, relatively "warm" ice, however, and differential shear in the basal zone might break off rock projections or fracture larger rock fragments during transport.

Temperate Glaciers. As the confining pressure due to the weight of the overlying ice increases downward in a glacier, the melting temperature is depressed. If the mean annual surface temperature of the accumulating snow and ice is close to 0°C and melting occurs, then with progressive burial, pressure melts additional ice, and the ice and water phases coexist at progressively lower temperatures with depth. This is the definition of a **temperate** glacier.

An important characteristic of a temperate glacier derives from the negative slope of the pressure-melting curve (Figure 16-12). Because the temperature at the base is less than at the surface, no geothermal heat can flow upward as it does in a polar glacier. The geothermal heat that reaches the base of a temperate glacier cannot be conducted upward but must melt basal ice. This geothermal heat, plus an increment of heat generated by sliding friction that is also trapped at the base of the ice, assures that a temperate glacier must move on a wet base.

FIGURE   16-12. Temperature profiles of boreholes at Camp Century, Greenland (Weertman, 1968), Byrd Station, Antarctica (Gow, et al., 1968), and a hypothetical temperate glacier.

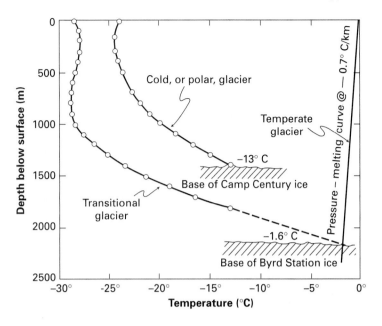

FIGURE   16-12. Temperature profiles of boreholes at Camp Century, Greenland (Weertman, 1968), Byrd Station, Antarctica (Gow, et al., 1968), and a hypothetical temperate glacier.

Flow rates in temperate glaciers are likely to be at least twice as fast as in polar glaciers. Striated, grooved basal rock surfaces are possible, as are numerous other significant forms of glacier transport and erosion (Chapter 17) because the ice slides over wet rock or saturated sediment. In addition, the subice water adds its erosional, transportational, and depositional work to that being done by the ice (Chapter 17).

Intermediate Categories. Ahlmann (1933, 1935) subdivided polar glaciers (*arctic* glaciers in his original terminology) into two classes, *sub-polar* and *high-polar.* Although both are cold at depth, summer melting on the sub-polar glacier might produce a surface layer 10 to 20 m deep, which is at the melting temperature. The various zones of dry snow, wet snow, and water percolating into surface ice and refreezing are well shown by satellite radar images (Fahnestock et al., 1993). Ahlmann's high-polar glacier would be typified by the Greenland ice sheet at Camp Century (Figures 16-6 and 16-12) in which not even a surficial layer of snow melts in the summer.

Lagally (1932), in a classification otherwise almost identical to Ahlmann's polar and temperate glaciers, proposed a *transitional* type that would be cold or polar at the top but be at the pressure-melting temperature near the base. The transitional category of Lagally was accepted as an interesting theoretical possibility but was generally overlooked until 1968, when a borehole was drilled at Byrd Station in Antarctica (Figure 16-12) (Gow et al., 1968). At a depth of 2164 m, the bottom of the ice was reached, but at the basal contact was a film of water judged to be at least 1 mm in thickness, under

enough pressure to rise 50 m up the drill hole and displace the antifreeze solution that had been used to keep the hole open. At the base of the ice sheet, the calculated melting temperature is −1.6°C. The surface temperature is −28.8°C. The steeper-than-average thermal gradient in the lower half of the Byrd Station borehole suggests a geothermal heat flow about 50 percent greater than the world average. Because West Antarctica is an ice-covered volcanic archipelago, the higher heat flow is reasonable. The effectiveness of the wet-base condition is demonstrated by the fact that the Byrd Station drill bit was unable to obtain a sample of the subice material because the ice moved laterally several centimeters per day and constantly pinched the drilling tool in the borehole. The substrate may be either rock or saturated, clay-rich basal sediment.

Subsequent radar depth soundings of the entire Antarctic ice sheet have indicated that large areas of the basal ice are wet and have a basal shear stress of only about 20 kPA (0.2 bar) (Oswald and Robin, 1973; Drewry, 1983; Radok, 1985). Surprisingly then, much of the Antarctic ice sheet is wet based in spite of the extremely cold surface temperatures. Even more surprising is the identification by radar and seismic exploration of a large lake, 14,000 km² in area and as much as 500 m in depth, 4 km below the central Antarctic ice (Kapitsa, et al., 1996; Ellis-Evans and Wynn-Williams, 1996). Since liquids have no yield strength (Figure 16-10), this and other large areas at the base of the Antarctic ice sheet are floating with zero basal shear stress. The implications for ice-sheet stability are the subject of intense debate.

GLACIER MOVEMENT

Plastic Flow

A slab of ice frozen to an inclined rock surface will deform downslope as a solid, without any melting and refreezing. The rate of movement at the ice surface is in nonlinear proportion to the thickness and surface slope of the ice (Paterson, 1994, p. 240). Theory and observation of actual glaciers agree that the motion is most rapid at the surface and progressively less rapid with depth (Figures 16-4 and 16-13). Theory and observation also agree that the direction of the flow is controlled by the surface slope of the glacier, not by the slope of the rock floor beneath it, as long as the ice thickness is significantly greater than the bedrock relief. It is quite possible for basal ice to move into and out of an ice-filled basin or over an obstruction, controlled only by the slope on the surface of the ice above the basin or obstruction. Thus ice can flow "uphill" over and around obstacles on its rock floor.

Theoretical models that assume ice to be a plastic solid with rheidity defined by Glen's flow law give excellent predictions of the shape, thickness, and rate of flow both of glaciers and of floating ice shelves, but the results are not perfect. An obvious reason is that real glaciers are complex masses of ice, with internal temperature gradients and impurities, that move over rock surfaces along discontinuities that cannot be modeled theoretically. Also, during movement, deforming ice crystals in a polycrystalline mass grow and reorient themselves in the stress field by a process of solid-state recrystallization. Thus the grain size and orientation, which greatly affect the rate of deformation, change during deformation (Hooke, 1998, pp. 36ff). Ice under high shear stress may fracture or form small, interlocking grains, but if the stress is low and uniform, these small grains "anneal" to form oriented grains as wide as 10 cm. Ice recovered from deep boreholes in glaciers usually has crystals several cm across, oriented with their *c* optical axes parallel and vertical; in this configuration, horizontal "easy glide" flow is facilitated, and the deformation rate increases by a factor of 4 over the rate for randomly oriented grains (Figure 16-11) (Weertman, 1983, p. 229).

Brittle Fracture: Faults and Crevasses

In their thin downstream or marginal regions, glaciers also "flow" by movement along discrete shear planes such as thrust faults. Pressure of moving ice against marginal ice that is either frozen fast to rock, or too full of rock debris to move, or blocked by a large obstruction, causes a thrust fault to form, dipping upglacier.

Actively moving ice shears forward and over the stagnant basal portion. Thrust faults of this sort may reach entirely through the glaciers and are one means by which rock debris from the base of the ice is brought upward to the surface.

Brittle failure also contributes to surface topography. Wherever a glacier moves over an obstruction at shallow depths or drops down a buried cliff, the surface ice is fractured by a maze of crevasses, mostly vertical and extending tens of meters in depth. These are tension cracks that heal at the depth where the tensional stresses can be compensated by plastic deformation. The deepest known crevasses, in the cold brittle ice of Antarctica, are about 50 m deep. Crevasses often become bridged by drifting snow and are one of the major dangers for explorers. A prominent arcuate crevasse, called the **bergschrund**, is at the extreme head of most valley glaciers (Figure 17-3). It is caused by snow compaction and ice motion away from the rock face. Deep crevasses in valley glaciers provide a route for surface rock debris to be incorporated into deeper ice.

Basal Slip: Enhanced Plastic Flow and Regelation Slip

The strain rate of glacier ice is typically proportional to the third power of the shear stress (Figure 16-11). Near

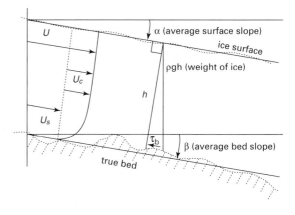

FIGURE 16-13. Simplified cross section through a glacier moving at a constant velocity on a sloping bed. Surface velocity *U* is the sum of plastic creep (U_c) and basal slip (U_s). Forces at the bed show that the resisting component τ_b (basal shear stress) balances the *driving stress* parallel to the bed, which is the weight of the overlying ice (ρgh) multiplied by the sine of the surface slope: $\tau_b = \rho gh \sin \alpha$. For typical gradients < 10°, bed slope (β) is irrelevant.

the base of a glacier, where bedrock irregularities protrude into the ice, local shear stresses increase, and local ice deformation is correspondingly very greatly increased. This **enhanced plastic flow** is proportional to the size of the obstruction, so the larger the obstruction, the more important is the enhanced plastic flow. Although strictly speaking the process is one of plastic flow, it is restricted to a zone near the ice-rock interface, and is usually regarded as a form of basal slip.

If ice is at or very close to its melting temperature, an additional form of basal movement can occur, called **regelation slip.** This is not a solid-state phenomenon but involves pressure-dependent melting and freezing. If the ice is at the melting temperature for the appropriate confining pressure, on the upglacier side of an obstruction the ice presses against rock, locally increasing pressure, depressing the melting temperature, and producing water *at the same ambient temperature as the surrounding ice.* The water migrates along grain boundaries within the ice or over the rock surface to regions of less pressure, especially on the downglacier side of the obstruction where ice is being drawn away from the rock. Here the water refreezes as bubble-free clear blue *regelation ice* (Figure 16-14).

Regelation is essentially an isothermal process. The only thermal gradients are those produced by the pressure gradients around the obstruction. The heat released by refreezing on the downstream side of the obstruction migrates upstream, primarily through the obstruction, to be absorbed by the melting at the upstream side. The heat transfer is faster through small obstacles, so the smaller the obstruction the more important is regelation slip.

Mathematical models of basal slip are imperfect because it is essentially the result of boundary conditions between unlike materials and phases. The controlling factor, assuming that the temperature is at or very near the melting point, is the bed roughness, or size of the limiting obstruction (Drewry, 1986, p. 10; Paterson, 1994, p. 139). Large obstructions are more efficiently passed by enhanced plastic flow, but smaller obstructions are more efficiently passed by regelation slip. As shown by borehole measurements (Figure 16-4), basal movement (including regelation slip) can account for 50 percent or more of the total observed surface motion of valley glaciers. Theory suggests that the controlling wavelength of bedrock obstruction, which would permit the optimal contributions by both plastic deformation and regelation slip, is about 0.5 m (Paterson, 1994, p. 142). However, obstructions of this size are not common on glaciated rock surfaces. Streamlined glaciated landforms about ten times as large are common and are thought to be stable forms under moving ice, somewhat like ripple marks and dunes under moving water and air. The lack of agreement between theory and observation illustrates the imperfections of present understanding of basal sliding.

Water at the Base of Glaciers: Surges

Glacier Hydrology. Water occurs on (*supraglacial*), within (*englacial*), and beneath (*subglacial*) temperate glaciers, at the base of transitional glaciers, and seasonally on and in some cold glaciers as well. It flows as rivers or collects in meltwater lakes on the surface of glaciers, flows along pressure gradients through karst-like enlarged fractures within the ice (Figure 16-15), and flows along the base of the ice over impermeable beds or permeates into suitable substrates. Many analogies with karst hydrology (Chapter 8) can be found in glaciers, although the opening of conduits in glaciers is by melting rather than solution, and the flow of ice may close established conduits and open new ones (Menzies, 1996; Hooke, 1998, Chapter 8).

Moulins serve as conduits for surface meltwater to flow down into the glacier, eventually joining subglacial channel systems. They are inclined as much as 45° toward the glacier margin because the water tries to flow along the internal water pressure gradients, perpendicular to the equipotential lines (Figure 16-15) in the saturated zone of the glacier (Holmlund, 1988).

Enhanced plastic flow and regelation slip are typically associated with a film of water at the ice-rock interface, but the most important effect of the water layer is to reduce the resistance to basal sliding. Bed roughness is an important factor in controlling glacier flow (Figure 16-13). Roughness typically includes a wide range of dimensions, as shown by fractal analysis (Figure 1-2). If the water film covers and negates the smaller part of the roughness spectrum, the flowing ice "sees" only the fewer, larger irregularities, and will slide much faster (Weertman and Birchfield, 1983, p. 22). By carefully mapping the glacial erosional features on a recently deglaciated limestone surface, Walder and Hallet (1979) concluded that at least 20 percent of the surface area had been separated from the base of the glacier by water-filled channels, with the remainder covered by a thin water film.

Cavities with dimensions of several meters may be opened on the downglacier side of rock knobs that protrude into temperate glaciers. These become water filled, especially during the summer melt seasons, and may coalesce into a system of water-filled channels with water pressure almost equal to the cryostatic pressure. Channels may be closed by ice flow during the winter season, and reopen or form in new areas during subsequent meltwater seasons.

FIGURE 16-14.
Experimental demonstration of the formation of regelation ice beneath a temperate glacier (from Peterson, 1970).

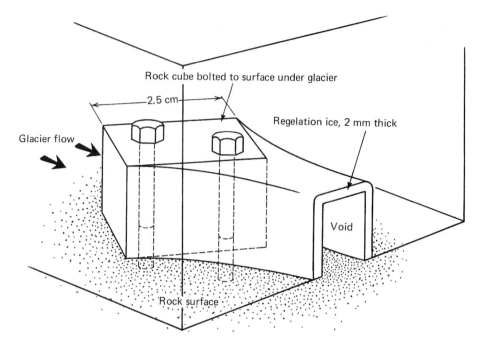

Complex sets of subice tunnels have been observed under an outlet glacier from a small Norwegian ice cap that is used for hydroelectric power generation (Hooke et al, 1985), as well as under other temperate glaciers (Hock and Hooke, 1993).

Deformable, or Soft, Beds. Where wet-base glaciers move over a layer of saturated sediment rather than hard rock, an additional important mode of glacier movement has been demonstrated. The sediment layer become a **deformable bed,** dragged in the direction of glacier movement by the transfer of the glacier's basal shear stress downward into the sediment layer (Boulton, 1996). As shown by the Coulomb equation that governs mass movement (Figure 9-1), pore pressure within the sediment layer reduces the overburden pressure, and the *effective* shear stress on a glacier bed of saturated sediment can be reduced to near zero (Iverson et al., 1995). If the sediment is fine grained, especially silt-size, drainage is inhibited and pore pressure is maximized. Sand and gravel beds may be well drained and have much lower pore pressure. Clay-rich beds inhibit infiltration from the overlying ice, and by decreasing the effective shear stress at the base of the overlying ice, can produce rapid basal slip without much deformation of the underlying sediment.

Surging Glaciers. In a small percentage of glaciers, the ice has been observed to move suddenly forward, advancing at a rate of kilometers per year instead of

the average rate of a few hundred meters per year (Paterson, 1994, Chapter 14). These **surging glaciers** typically flow normally for 10 to 100 years, then suddenly surge forward. The zone of rapid motion begins high in the glacier, and advances wavelike to the terminus. The surging part of a glacier is dominated by extending flow, but the lower end is in compressing flow until the surge reaches it, when it rapidly thins and surges forward (McMeeking and Johnson, 1986).

Surges are unknown in many glaciated regions, yet over 6 percent of the 2356 glaciers in the St. Elias Mountains on the Alaska-Canada border have surged, and other concentrations are in specific regions in Asia, the Andes, Iceland, eastern Greenland, and Svalbard

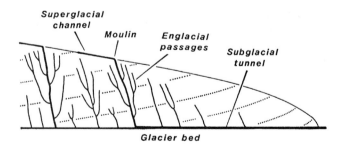

FIGURE 16-15. Equipotential lines (dotted) and internal drainage in an ideal cross section of a temperate glacier on a horizontal, impermeable bed. Flow is perpendicular to the equipotential lines and toward the outlet of the subglacial tunnel (Shreve, 1985, Figure 4).

(Clarke et al., 1986). Surges rarely last for more than a few months to a year, but they create a chaotic pattern of crevasses on the glaciers (Figure 16-16) (Kamb et al., 1985). They were first reported on a group of Alaskan glaciers early in this century (Tarr, 1909) and were attributed to an excessive accumulation of snow from avalanches produced by an earthquake in 1899. This idea has not been supported by subsequent observations, but after the major Alaskan earthquake of 1964, a close watch was kept on the glaciers of the Alaska Range in hopes of observing the initial phases of surging.

Events leading up to the well-documented 1982-1983 surge episode of Variegated Glacier in Alaska had been monitored for a decade (Kamb et al., 1985; Raymond and Harrison, 1988). The ice in the middle and upper parts of the glacier had been thickening since 1973, and the velocity of flow had been increasing, even though the terminus remained stagnant. Until 1978, most of the increased rate of flow was during the summer, but subsequently the ice began to move more rapidly in winter as well. This indicated that basal sliding was accelerating even in winter. In 1982, the surface of the glacier in the accumulation zone began to move forward at rates of 2 to 15 m/day. The surge then propagated rapidly downglacier, increasing from a presurge flow rate of 0.2 m/day or less to 40 to 60 m/day. The surface of the glacier was intensely crevassed by the rapid motion. A borehole to the base of the glacier confirmed that 95 percent of the surging motion was by basal sliding. The zone of water saturation in the ice had risen to within 90 percent of the ice thickness. A basal layer of impermeable, shearing sediment was detected by seismic reflection techniques beneath Variegated Glacier in Alaska prior to the 1982-1983 surge. It may have helped confine water near the bed and build the water pressure to a level at which the glacier was nearly floating during the surge (Richards, 1988). When the water in the ice migrated down to the terminus and drained away in a spectacular flood, the surge stopped. The excess water that accumulated within the glacier was only indirectly correlated with air temperature and precipitation. Apparently, it is a characteristic of Variegated Glacier to accumulate water over an interval of about 20 years, then rapidly drain during a surge (Kamb et al., 1985).

Flow in Ice Sheets

For various reasons excluding hydrostatic pressure, shear strain (the vertical gradient of velocity) in an ice sheet, as in a valley glacier (Figure 16-13), should be most rapid near its base. Ice at higher temperatures deforms more rapidly, and the highest temperatures are near the base of an ice sheet because of the poor thermal conductivity of ice and the geothermal heat flow (Figure 16-12). The increased grain size of the ice crystals and their progressively better orientation with each other in the horizontal plane also facilitate plastic flow. A contribution of frictional (deformational) heat must also be considered. Further, the greatest shear stress is at the base of the ice although this factor does not vary greatly. If any regelation slip occurs, it too will be at the base. Therefore, reasonable models of flow in ice caps are obtained by assuming that all the shear takes place in a basal layer only a few meters to tens of meters in thickness, and the overlying ice is passively transported. Under such assumptions, flow rate is proportional only to the surface slope of the ice, which develops an elliptical or parabolic profile (Figures 16-5 and 16-6). An excellent approximation of the thickness of many ice sheets and ice caps within a few hundred kilometers of their margins is given by the equation

$$h = 3.4\sqrt{d}$$

where h is the thickness of the ice (in meters) on a horizontal base at distance d in from the margin of the ice sheet (in meters). This equation is derived from Glen's flow law (p. 362) with an assumed basal shear stress of 50 kPa (0.5 bar) (Paterson, 1994, p. 242). A few minutes with a calculator will demonstrate how well this equation approximates actual surface profiles of ice sheets and ice caps (Figure 16-5). However, many of the western and southern lobes of the Laurentide ice sheet during the last ice age (Figure 16-1) had surface slopes considerably flatter than would be predicted by the equation, implying basal shear stresses as low as 1 to 30 kPa (0.01 to 0.3 bars) for ice that flowed northwest across the Canadian arctic near the Alaskan border (Beget, 1987), and only 0.7 to 22 kPa (0.007 to 0.22 bars) for ice lobes in the Great Plains and Great Lakes lowlands (Clark, 1992, 1994). Such shear stresses are much too low to be the result of ice deformation. Rather, they have been attributed to shearing in a deformable bed of water-saturated sediment at the base of the ice.

The pattern of flowlines in the former North American Laurentide ice sheet (Chapter 17) suggests that where the ice was moving primarily over resistant rock such as the metamorphic and igneous terranes of the Canadian Shield it moved primarily by plastic deformation, enhanced basal plastic flow, and regelation slip. However, when it expanded onto the more easily erodible sedimentary-rock terranes of the northern Great Plains and the Interior Lowland around the Great Lakes (Figure 18-7), a deforming bed of fine-grain sediment greatly accelerated movement (Clark, 1994). The implications of such a deforming bed, with resulting

FIGURE 16-16. Views upglacier of the Variegated Glacier, Alaska (a) before a surge, July 1982, and (b) during a surge, 4 July 1983 (Kamb et al., 1985, photos courtesy B. Kamb).

(a)

(b)

very flat surface gradients on the ice, much thinner ice than previously estimated, highly lobate and rapidly fluctuating ice margins, and less than expected post-glacial isostatic recovery, are considered further in Chapters 17 and 18.

REFERENCES

AHLMANN, H. W., 1933, Scientific results of the Swedish-Norwegian Arctic Expedition in the summer of 1931, Part VIII, Glaciology: Geografiska Annaler, v. 15, pp. 161–216, 261–295.

———, 1935, Contribution to the physics of glaciers: Geog. Jour., v. 86, pp. 97–113.

ALLEY, R. B., and WHILLANS, I. M., 1991, Changes in the West Antarctic ice sheet: Science, v. 254, pp. 959–963.

BEGET, J., 1987, Low profile of the northwest Laurentide ice sheet: Arctic and Alpine Res., v. 19, pp. 81–88.

BOGORODSKY, V. V., BENTLEY, C. R., and GUDMAND-SEN, P. E., 1985, Radioglaciology: D. Reidel Publishing Company, Dordrecht, The Netherlands, 254 pp.

BOULTON, G. S., 1996, Theory of glacial erosion, transport and deposition as a consequence of subglacial sediment deformation: Jour. Glaciology, v. 42, pp. 43–62.

BUDD, W. F., and MORGAN, V. I., 1973, Isotope measurements as indicators of ice flow and paleo-climates, in van Zinderen Bakker, E. M., ed., Paleoecology of Africa and the surrounding islands and Antarctica: A.A. Balkema, Cape Town, South Africa, v. 8, pp. 5–22.

CLARKE, G. K. C., SCHMOK, and 2 others, 1986, Characteristics of surge-type glaciers: Jour. Geophys. Res., v. 91, pp. 7165–7180.

CLARK, P. U., 1992, Surface form of the southern Laurentide ice sheet and its implications to ice-sheet dynamics: Geol. Soc. America Bull., v. 104, pp. 595–605.

———, 1994, Unstable behavior of the Laurentide ice sheet over deforming sediment and its implications for climate change: Quaternary Res., v. 41, pp. 19–25.

DENTON, G. H., and HUGHES, T. J., eds., 1981, The last great ice sheets: John Wiley & Sons, Inc., New York, 484 pp.

DREWRY, D. J., ed., 1983, Antarctica: Glaciological and geophysical folio: Scott Polar Research Institute, Cambridge, UK, 9 map sheets with text.

———, 1986, Glacial geologic processes: Edward Arnold (Publishers) Ltd., London, 276 pp.

ECHELMEYER, K., CLARKE, T. S., and HARRISON, W. D., 1991, Surficial glaciology of Jakobshavns Isbrae, west

Greenland: Part I. Surface morphology: Jour. Glaciology, v. 37, pp. 368–382.

ECHELMEYER, K., HARRISON, W. D., and 2 others, 1992, Surficial glaciology of Jakobshavns Isbrae, west Greenland: Part II. Ablation, accumulation and temperature: Jour. Glaciology, v. 38, pp. 169–181.

ELLIS-EVANS, J. C., and WYNN-WILLIAMS, D., 1996, A great lake under the ice: Nature, v. 381, pp. 644–646.

ENGELHARDT, H., HUMPHREY, N., and 2 others, 1990, Physical conditions at the base of a fast moving Antarctic ice stream: Science, v. 248, pp. 57–59.

FAHNESTOCK, M., BINDSCHADLER, R., and 2 others, 1993, Greenland ice sheet surface properties and ice dynamics from ERS-1 SAR imagery: Science, v. 262, pp. 1530–1534.

GLEN, J. W., 1952, Experiments on the deformation of ice: Jour. Glaciology, v. 2, pp. 111–114.

GOW, A. J., UEDA, H. T., and GARFIELD, D. E., 1968, Antarctic ice sheet: Preliminary results of first core hole to bedrock: Science, v. 161, pp. 1011–1013.

HOCK, R., and HOOKE, R. L., 1993, Evolution of the internal drainage system in the lower part of the ablation area of Storglaciären, Sweden: Geol. Soc. America Bull., v. 105, pp. 537–546.

HOLMLUND, P., 1988, Internal geometry and evolution of moulins, Storglaciären, Sweden: Jour. Glaciology, v. 34, pp. 242–248.

HOOKE, R. L., 1998, Principles of glacier mechanics: Prentice Hall, Upper Saddle River, New Jersey, 248 pp.

————, WOLD, B., and HAGEN, J. O., 1985, Subglacial hydrology and sediment transport at Bondhusbreen, southwest Norway: Geol. Soc. America Bull., v. 96, pp. 388–397.

HUTTER, K., 1983, Theoretical glaciology: D. Reidel Publishing Company, Dordrecht, The Netherlands, 510 pp.

IVERSON, N. R., HANSON, B., and 2 others, 1995, Flow mechanism of glaciers on soft beds: Science, v. 267, pp. 80–81.

JANKOWSKI, D. G., and SQUYRES, S. W., 1988, Solid-state ice volcanism on the satellites of Uranus: Science, v. 241, pp. 1322–1325.

KAMB, B., RAYMOND, C. F., and 6 others, 1985, Glacier surge mechanism; 1982-1983 surge of Variegated Glacier, Alaska: Science, v. 227, pp. 469–479.

KAPITSA, A. P., RIDLEY, J. K., and 3 others, 1996, A large deep freshwater lake beneath the ice of central East Antarctica: Nature, v. 381, pp. 684–686.

KLINGER, J. D., BENEST, A., and 2 others, eds., 1985, Ices in the solar system: D. Reidel Publishing Company, Dordrecht, The Netherlands, 954 pp.

LAGALLY, M., 1932, Zur Thermodynamik der Gletscher: Zeitschr. für Gletscherkunde, v. 20, pp. 199–214.

LLIBOUTRY, L., 1965, Traité de glaciology (2 vols.): Masson et Cie, Paris, 1040 pp.

McMEEKING, R. M., and JOHNSON, R. E., 1986, On the mechanics of surging glaciers: Jour. Glaciology, v. 32, pp. 120–122.

MENZIES, J., 1995, Dynamics of ice flow, in Menzies, J., ed., Modern glacial environments: Processes, dynamics, and sediments: Butterworth-Heinemann Ltd., Oxford, UK, pp. 139–196.

————, 1996, Hydrology of glaciers, in Menzies, J., ed., Modern glacial environments: Processes, dynamics, and sediments: Butterworth-Heinemann Ltd., Oxford, UK, pp. 197–239.

OSWALD, G. K. A., and ROBIN, G. D., 1973, Lakes beneath the Antarctic ice sheet: Nature, v. 245, pp. 251–254.

PATERSON, W. S. B., 1994, The physics of glaciers, 3d ed.: Pergamon Press, Tarrytown, New York, 480 pp.

PETERSON, D. N., 1970, Glaciological investigations on the Casement Glacier, southeast Alaska: Ohio State Univ. Inst. of Polar Studies, Rept. no. 36, 161 pp.

RADOK, U., 1985, The Antarctic ice: Sci. American, v. 253, no. 2, pp. 98–105.

RAYMOND, C. F., 1971, Flow in a transverse section of Athabaska Glacier, Alberta, Canada: Jour. Glaciology, v. 10, pp. 55–84.

————, and HARRISON, W. D., 1988, Evolution of Variegated Glacier, Alaska, U.S.A., prior to its surge: Jour. Glaciology, v. 34, pp. 154–169.

REEH, N., 1989, Dynamic and climatic history of the Greenland ice sheet, in Fulton, R. J., ed., The Quaternary geology of Canada and Greenland: Geol. Soc. America, The geology of North America, v. K-1, pp. 795–822.

RICHARDS, M. A., 1988, Seismic evidence for a weak basal layer during the 1982 surge of Variegated Glacier, Alaska, U.S.A.: Jour. Glaciology. v. 34, pp. 111–120.

ROTT, H., SKVARCA, P., and NAGLER, T., 1996, Rapid collapse of northern Larsen ice shelf, Antarctica: Science, v. 271, pp. 788–792.

SHARP, R. P., 1988, Living ice: Understanding glaciers and glaciation: Cambridge University Press, Cambridge, UK, 225 pp.

SHREVE, R. L., 1985, Esker characteristics in terms of glacier physics, Katahdin esker system, Maine: Geol. Soc. America, v. 96, pp. 639–646.

SWITHINBANK, C., 1988, Satellite image atlas of glaciers of the world: Antarctica: U.S. Geol. Survey Prof. Paper 1386-B, 278 pp.

TARR, R. S., 1909, The Yakutat Bay region, Alaska: U.S. Geol. Survey Prof. Paper 64, 183 pp.

WALDER, J., and HALLET, B., 1979, Geometry of former subglacial water channels and cavities: Jour. Glaciology, v. 23, pp. 335–362.

WALKER, J. W., PEARCE, D. C., and ZANELLA, A. H., 1968, Airborne radar sounding of the Greenland ice cap: Flight 1: Geol. Soc. America Bull., v. 79, pp. 1639–1646.

WEERTMAN, J., 1968, Comparison between measured and theoretical temperature profiles of the Camp Century, Greenland, borehole: Jour. Geophys. Res., v. 73, pp. 2691–2700.

————, 1983, Creep deformation of ice: Ann. Rev. Earth Planet. Sci., v. 11, pp. 215–240.

————, and BIRCHFIELD, G. E., 1983, Basal water film, basal water pressure, and velocity of traveling waves on glaciers: Jour. Glaciology, v. 29, pp. 20–27.

WEIDICK, A., 1975, Review of Quaternary investigations in Greenland: Grønlands Geologiske Undersøgelse Miscell. Papers, no. 145; distributed as Ohio State Univ. Inst. of Polar Studies, Rept. no. 55, 161 pp.

Chapter 17

Glacial Geomorphology

Glacial geomorphology concerns the processes by which glaciers shape landscapes and analyzes the resulting glaciated landforms. The information in Chapter 16 about the deformation of ice and the flow of glaciers at various temperatures, with or without associated meltwater, is a necessary prerequisite to understand this chapter, but here the emphasis is on glaciers and meltwater as geomorphic agents that erode the rock mass over which they flow and transport or deposit the resulting debris.

Although the glacier-related geomorphic work of meltwater and wind extend many kilometers beyond a glacier margin, the direct work of glacier ice ends at an abrupt, although shifting, boundary. This is one of the few examples that can be cited where geomorphic processes and landforms change at a sharp boundary. Only at coasts is there such an equally abrupt process-controlled change of landforms.

A problem of glacial geomorphology is that most of the processes work beneath ice and are exceptionally difficult to observe. Glaciated terrain is not a landscape until it is exposed by the disappearance of the glacier. In that sense, all glaciated landscapes are relict even if they were exposed only by the previous year's retreat of a glacier terminus.

Valley glaciers create a unique, distinctive landscape by their vigorous erosional shaping of mountain ranges. As noted later in the chapter, the assemblage of sharp pyramidal peaks, knife-edged ridges, and U-shaped troughs with hanging tributaries is so much a part of our mental image of glacierized and glaciated mountains that we are hard pressed to deduce what the landforms were on a nonglaciated high mountain range during the "greenhouse" climate of the middle Cenozoic Era (Chapter 4). Even though valley glaciers only occupy and modify preexisting river valleys, those modifications can be extreme.

The Pleistocene continental ice sheets pose another set of challenges to geomorphologists. Their erosional work may have been primarily to strip away a former regolith and modify former drainage systems, and their deposits may be measured only in tens of meters of thickness, but they created the landscapes and soils of northern Europe and North America, where the industrial revolution began and agriculture has become a business instead of a means of individual survival. Soils on the vast glaciated and loess plains are young but fertile, formed in transported sediments rather than in an *in situ* regolith. A large percentage of the readers of this book live on a landscape of relict glacial landforms, little modified by 10,000 to 20,000 years of postglacial time.

GLACIAL EROSION AND SEDIMENT TRANSPORT

Valley glaciers probably derive most of their sediment loads from intense weathering and mass wasting on adjacent slopes. Rock debris that composes lateral and medial moraines seems to stay fairly high in the glacier cross section. An unknown amount is actually eroded by the moving glacier.

Ice sheets that bury landscapes cannot gain sediment by mass-wasting because nothing is exposed above the ice. The Antarctic and Greenland ice sheets are notably free of any sediment other than minute amounts of windblown dust, except very near their bases. Neither ice sheet appears to be transporting much rock debris, but they may not be good models for the former ice sheets of Europe and North America, which expanded and shrank repeatedly during the Pleistocene Epoch.

Fluvial sediment transport beneath wet-base glaciers has largely replaced other mechanisms as the process most favored to deliver sediment to the margins of glaciers (Evenson and Clinch, 1985; Gustavson and Boothroyd, 1987). More and more, glacial and fluvial processes are being studied in conjunction with each other as a "glaciofluvial system" analogous to the fluvial system introduced in Chapter 10 (Ashley et al., 1985; Drewry, 1986; Gurnell and Clark, 1987; Lawson, 1993). Even more complex are the glaciolacustrine and glaciomarine environments, where a calving ice margin stands in fresh or salt water. Extensive segments of the former ice sheets were grounded in water, and the resulting sediments are the subject of many sedimentologic studies (Dowdeswell and Scourse, 1990; Anderson and Ashley, 1991).

Processes and Bedforms

Rock masses that are eroded by overriding glacier ice produce detritus with a wide range of grain sizes that must then be transported away. Left behind are rock surfaces with diagnostic minor landforms and surface markings, collectively called **bedforms.** Unlike water and wind, glacial transport has no competence limit (p. 215), and all grain sizes from clay to giant boulders are transported, often thoroughly mixed. Some of the debris is entrained into the basal ice. A larger fraction is probably transported by water from beneath wet-base glaciers. Erosion, entrainment, and transport form a continuum that ends only when the particles reach a final resting place.

Glaciotectonism. Both the vertical (static) load and the shear stress of the forward motion of glacier ice contribute to deforming or moving underlying rock masses or sediments. Pore water pressure and the resulting *effective stress* (p. 170) applied to the substrate by the weight of overlying ice are critical factors in determining the response of glacially overridden materials (van der Wateren, 1995). Shallow slabs of frozen sediment overlying a deeper horizon saturated with confined water, for example, might be easily thrust forward even by very low basal shear stress beneath over-

riding ice. Ice-push ridges of shale and mudstone, some stacked as imbricate thrust sheets, are well known on the Canadian prairies (Aber et al., 1989, p. 37). Similar forms are widely distributed in Denmark, Iceland, the Netherlands, and elsewhere. Their arcuate map patterns and deformation vectors that are clearly related to former ice motion, and their shallow detachment zones, are criteria that demonstrate a glacial origin.

Crushing and Fracture. The crushing strength of most rocks greatly exceeds any combination of normal and shear stresses that can be applied to the bed by overriding glacier ice. However, at a point contact between a rock particle embedded in the ice and the rock substrate, pressure is adequate to chip or flake both the particle and the substrate, just as mild finger pressure on the broad head of a thumbtack can drive the point into hard wood. A wide range of distinctive fractures can be seen on glaciated rock surfaces, including *chattermarks* and *crescentic fractures* (Drewry, 1986, p. 34). These are made as hard particles rolling at the ice-rock interface elastically compress the rock surface in front of them and stretch the rock behind them to create rhythmic fractures. Similar forms are reproduced by rolling a hard steel ball over glass under a heavy load. Large rock obstructions that protrude upward into ice can also be broken by the excess ice pressure on their up-glacier sides, even though temperate ice will attempt to flow around such an obstruction by enhanced plastic flow and regelation slip (p. 366). Because near-surface rock masses are usually weathered and full of microfractures and pores, fractures occur at stresses lower than the predicted strength of the rocks.

Plucking and Hydraulic Jacking. During regelation slip, water migrates to the downglacier side of a subice obstruction and refreezes (p. 366). Rock fragments, usually small, can be incorporated into the regelation layer and carried away by the glacier. This process is called glacial **plucking.** It is not clear whether regelation ice has enough tensile strength to actually pull away lee-side pieces of rock or whether pieces already loosened by the stress of the moving ice are simply frozen in and carried away (Boulton, 1974, pp. 69–75).

Plucking is most effective on fractured rocks. Jahns (1943) reconstructed the sheeting joints on crystalline metamorphosed rocks near Boston, Massachusetts (p. 121) to show that plucking had removed several times the volume of rock from the downglacier, or **lee** sides of hills than abrasion had removed from the upglacier, or **stoss** (German for *push*) sides.

A related process involving rapid increases of water pressure in subice cavities causes basal ice to move suddenly, pushing joint-bounded blocks of rock forward.

Blocks up to 1 m in length have been observed. The process is called **hydraulic jacking** (Drewry, 1986, pp. 46–47).

Abrasion. The abraded scratches and grooves that characterize glaciated rock surfaces are undoubtedly caused by rock fragments in the ice rather than by the ice itself (Boulton, 1974, p. 43). If the glacier is sliding over its base on a water film, any fragment larger than the thickness of the water film will be rolled or dragged against the rock surface. Glaciated surfaces show a range of abraded microrelief forms from hairline *striations* to *grooves* a meter or more in depth and width. Careful study of glacial striations and grooves can sometimes reveal the direction of ice motion, as when a piece of rock is chipped off and dragged downglacier over the surface, or when a resistant projection protects less resistant rock in its lee from abrasion (Figure 17-1). Regelation slip must be an effective process of abrasion, because if a sand grain is embedded in basal ice, it will move into contact with the rock surface as the enclosing ice melts. Abrasion is ineffective beneath polar glaciers, where there is no sliding at the ice-rock interface (p. 364). The strongest proofs that a region has been glaciated by a wet-base glacier is a regional pattern of striated and grooved bedrock surfaces.

Not only bedrock surfaces are abraded; the rock fragments carried in glaciers also are subjected to abrasion. A significant proportion of glacially transported pebbles and cobbles have striated or polished planocurved facets on one or more sides, similar in shape to the sole of an old shoe, showing that they were held in contact with other rocks as they ground past each other.

Abrasional effectiveness depends on the relative hardnesses of the "tool" held by the ice and the rock substrate, the flux of abrasive particles in contact with the bed, and the effective force with which the particles are held against the bed (Boulton, 1974; Hallet, 1979). At the base of a temperate glacier, the normal stress (ice thickness) is a factor in determining basal shear stress (Figure 16-13), which in turn affects the basal ice velocity. The flux of abrasive particles in contact with the bed is reduced by a debris concentration above a certain relatively low value. Glaciers with significant basal sliding will abrade their beds until the basal load becomes large and the sliding rate decreases, when deposition will begin (Hallet, 1981).

Some polished and striated rock surfaces show patterns of flow that diverge from the regional direction in minute, complex ways (Figure 17-2). Local flow directions show eddylike patterns, helical flow along shallow troughs, and potholes that suggest a highly mobile abrasive substance (Gjessing, 1966). Possibly, the base of some temperate and intermediate glaciers is underlain by water-saturated, fine-grained sediment under high water pressure and therefore low effective stress, that deforms under the shearing stress applied by the overriding ice (Alley, 1989; Alley et al., 1989). The *deformable bed* (p. 368) moves generally in the same direction as the overlying ice but sensitively responds to the locally intensified stress gradients around obstructions. Whether it is capable of abrading the rock surface beneath it, or whether it absorbs all the shearing motion of the glacier within itself, is not known. A deformable bed of saturated, shearing sediment could deform, erode, or accumulate, depending on the stress field within it and within the glacier ice immediately above it (Boulton, 1996). Such saturated basal sediment has been exposed by rapid retreat of some glacier termini and has been observed in tunnels dug along the base of valley glaciers. It also makes constructional landforms when it squeezes up into crevasses near the margins of glaciers.

FIGURE 17-1. Glacially abraded surface of limestone, with a protruding chert concretion, a streamlined tail of limestone, and a shallow marginal moat (photo: A. Meike).

FIGURE 17-2. Glacially scoured metamorphic rock suggesting a highly plastic abrasive medium, near Portør, Norway, on the Skagerrak coast (photo: Just Gjessing).

The combination of crushing, plucking, and abrasion exploits structural details of rock masses, especially fractures. A variety of streamlined bedforms, such as *stoss-and-lee topography, roches moutonnées,* and *crag-and-tail* hills are variants of the combined abrasion and plucking process (Figure 17-3). They range in length from a few meters to about 1 km. They may form equally well under ice sheets or valley glaciers, but a wet-base status is necessary. All are small hills with abraded upglacier sides, but the crag-and-tail variant has a streamlined tail of compact sediment or erodible bedrock sheltered behind a more resistant crag (Figure 17-1), whereas the other two forms have steepened, plucked lee faces. Many of these bedforms may be only slightly modified by glaciation and are primarily inherited from the preglacial relief (Lindström, 1988).

Fluvial Erosion Beneath Glaciers. With the recognition of the widespread occurrence of wet-base glaciers, both at the present time and during former ice ages, the erosional role of flowing water beneath glaciers has received new emphasis. The basal film of water observed beneath temperate and intermediate glaciers is not capable of mechanical erosion, although by virtue of its high confined pressure, high dissolved CO_2 content, and low temperature it can chemically corrode carbonate rocks (Hallet, 1976; Drewry, 1986, pp. 77–82). However, when it is flowing in subglacial channels, the relatively warm and often muddy meltwater can erode both the base of the ice and the bedrock or sediment beneath the ice. *Potholes* in bedrock eroded by vortices in subglacial rivers are well-known glacial erosional landforms. Very large

FIGURE 17-3. Intertidal rock knob on Kennebunk Beach, Maine, showing stoss-and-lee topography. Ice moved from left to right, abrading the upglacier (stoss) side and plucking the downglacier (lee) side along joints.

tunnel valleys were eroded into unconsolidated sediments by subglacial meltwater along some margins of the late Pleistocene ice sheets. Those in central Minnesota are up to 1 km in width and 30 km in length (Mooers, 1989). Whether the tunnels entrench into the substrate or melt upward into the ice, and whether they close by ice deformation or become filled with sediment is a complex function of the amount of meltwater, the basal water pressure, ice plasticity, sediment load, and many other variables.

Erosion of subglacial sediment and its transportation to the ice margin by subglacial and englacial streams is claimed to be the source of almost all the outwash in front of the Malaspina Glacier in Alaska (Gustavson and Boothroyd, 1987). An extreme view is that episodic subglacial floods from beneath the southern part of the North American Laurentide ice sheet were responsible for most of the glacial erosional landforms on both bedrock and sediment, including polished and abraded streamlined forms (Sharpe and Shaw, 1989; Shaw, 1989). A continuing debate concerns the relative importance of subglacial water and highly deformed basal sediment in producing the polished, striated, and fluted surfaces on hard rocks so diagnostic of glacial erosion (Figures 17-1 and 17-2).

Erosional and Residual Landforms

Troughs. The erosional landforms of valley glaciers rank with the most spectacular scenery on earth. The European Alps are so thoroughly dissected by valley glaciers that the entire assemblage is sometimes simply referred to as *alpine topography* (Figures 17-4, 17-5, and 17-6). The fundamental landform component is the **glacial trough,** the channel of a former valley glacier. The cross profile of a glacial trough is a catenary curve, not unlike the cross profile of a river channel but quite different from the profile of an entire river valley. Numerical modeling of glacial troughs, based on physical concepts of least work and optimal erosional effectiveness, confirm that their generally parabolic cross profiles are highly efficient for ice flow (Hirano and Aniya, 1988, 1989; Harbor, 1992). With reasonable assumptions of erosional efficiency, a glacier could convert a V-shaped valley into a glacial trough in as little as 10^4 years (Harbor, 1992). Obviously, rock-mass strength (Chapter 7) is a very important factor in the rate and degree of trough formation (Augustinus, 1992, 1995). Computer simulations of trough erosion along a narrow belt of highly erodible rock predict a narrower, deeper trough, as expected (Harbor, 1995).

Troughs are on the order of 1 km or more in width, and commonly hundreds of meters or more in depth. Concave slopes dominate, rising from the valley floor to steep cliffs or stepped-back, ice-abraded shoulders

(Figures 16-3, 17-6, and 18-9). In long profile, glacial troughs may be a series of basins separated by cross-valley ledges of more resistant rocks (Figure 17-5) (Gutenberg et al., 1956). The long profile is structurally controlled by intense glacial plucking and abrasion on more highly fractured or easily eroded rock masses to create the basins. Alternating segments of compressing and extending flow may also differentially erode the glacier bed and enhance initial structurally controlled irregularities, or possibly create harmonic changes in glacier dynamics that cause regularly spaced basins without structural control. Typically, deep-basin erosion is localized by valley-wall constrictions, and sills are associated with widening of the valley walls (Shoemaker, 1986). If not filled with sediment, the basins form chains of lakes, each with a waterfall over the downstream ledge to the next lake. These delightful chains are called **paternoster lakes,** from the resemblance to shining beads on a rosary. It is helpful to visualize a glacial trough as the analog to a river channel. Just as a river channel thalweg may have alternating pools and riffles (p. 220), the basal ice can flow up and down as long as the surface gradient provides the necessary downvalley gravitational potential.

Trough Lakes and Fiords. In a glacial trough that has either a rock sill or a massive end moraine blocking its lower end, a **trough lake** will drown the valley floor. Trough lakes are usually long, steep-sided, deep, and cold, with a minimal inflow of sediment because of the clean abraded valley walls and the upstream settling basins in paternoster lakes on most tributary streams. The Swiss lakes, the Como region of northern Italy, the Lake District of England, the Finger Lakes of central New York (Figure 17-7), and many other regions of large trough lakes have become justly famous as scenic tourist areas. Hydroelectric projects make use of the large storage volumes and abrupt gradients associated with trough lakes in mountainous regions.

The side walls of glacial troughs are straightened by abrasion and plucking. The interlocking spurs that interrupt views along a river valley are cut away, giving a series of **ice-faceted spurs** that align between lateral valleys. Where tributary glaciers joined the trunk ice stream, smaller glacial troughs form, but because the smaller glaciers were less erosive than the trunk glacier, their floors are usually well above the floor of the main valley. Waterfalls cascade down the main valley wall from these **hanging troughs.** The accordance of valley floors, so important to Playfair's law (p. 232), is absent in alpine glacial scenery. However, if we think of glacial troughs as analogous with river channels rather than valleys the comparison is better because tributary river beds are commonly shallower than their trunk channel.

GLACIERS IN AN ALPINE REGION

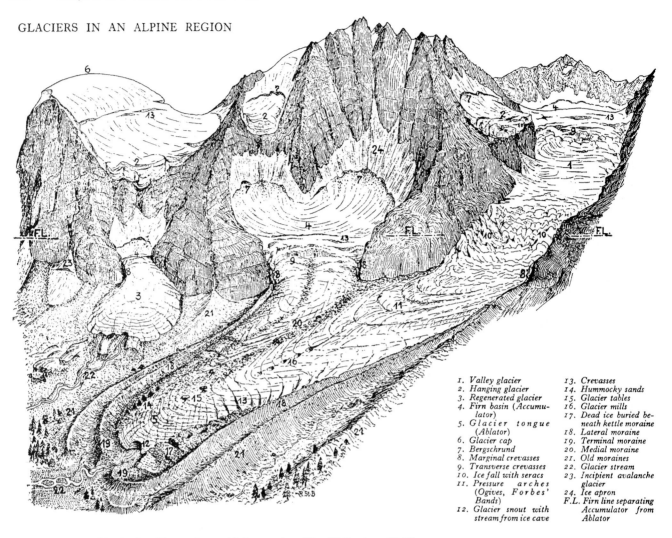

FIGURE 17-4.   Glacier landforms in an Alpine region (Streiff-Becker, 1947).

1. *Valley glacier*
2. *Hanging glacier*
3. *Regenerated glacier*
4. *Firn basin (Accumulator)*
5. *Glacier tongue (Ablator)*
6. *Glacier cap*
7. *Bergschrund*
8. *Marginal crevasses*
9. *Transverse crevasses*
10. *Ice fall with seracs*
11. *Pressure arches (Ogives, Forbes' Bands)*
12. *Glacier snout with stream from ice cave*
13. *Crevasses*
14. *Hummocky sands*
15. *Glacier tables*
16. *Glacier mills*
17. *Dead ice buried beneath kettle moraine*
18. *Lateral moraine*
19. *Terminal moraine*
20. *Medial moraine*
21. *Old moraines*
22. *Glacier stream*
23. *Incipient avalanche glacier*
24. *Ice apron*
F.L. *Firn line separating Accumulator from Ablator*

Fiords are glacial troughs that have been drowned by the sea (Figure 20-5). Their form differs from that of other troughs in that near the outer coast, many fiords shoal rapidly to only a fraction of the depth that they attain between more restricting walls farther inland, where multiple basins may be 1 km or more in depth. In some fiords, the sill at the mouth is an end moraine, but usually it is bedrock and is attributed to a reduction in intensity of vertical erosion as the valley glacier either began to float (Crary, 1966) or spread laterally as a piedmont lobe. The analysis of extending and compressing flow in glaciers (p. 357) assumes that longitudinal flow is confined between parallel walls so that no lateral expansion or contraction occurs. Where their valley walls widen, as on a piedmont plain, glaciers spread laterally and thin abruptly. Basal erosion decreases dramatically and the trough floor rapidly rises to a terminal sill (Shoemaker, 1986).

Cirques.   At the head of each trough, in the accumulation area of the glacier that eroded the valley, are steep-headed, semicircular basins, or **cirques** (Figure 7-4; Color Plate 13). The bowl shape is a consequence of intense frost action and nivation processes adjacent to the ice field plus the flow lines in the zone of accumulation (Figure 16-2) that carry surface rock debris downward toward the base of the glacier, where the bed is heavily abraded. A cirque has much the shape of a human heel print in sand and is often "down-at-the-heel," that is, more deeply eroded as a basin behind an abraded rock sill at the lip of the cirque. A cirque lake, or **tarn,** dammed by the rock sill, occupies the basin in many cirques.

Cirques are a unique and diagnostic feature of alpine glaciation. The association of active cirques with the accumulation areas of modern valley glaciers led to the recognition of similar relict forms very early in the

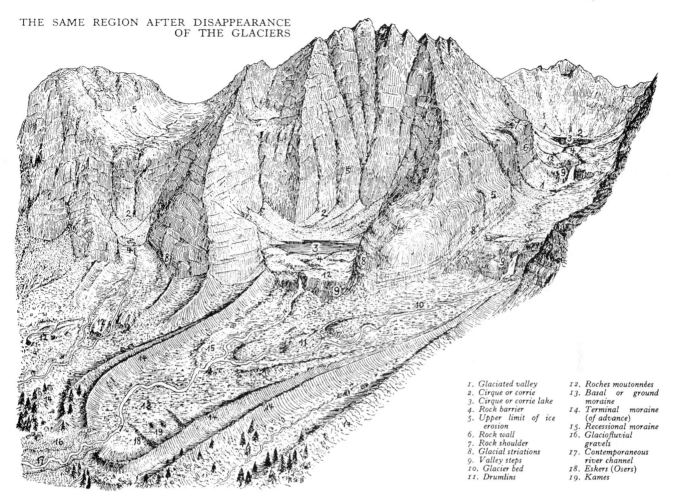

THE SAME REGION AFTER DISAPPEARANCE
OF THE GLACIERS

1. Glaciated valley
2. Cirque or corrie
3. Cirque or corrie lake
4. Rock barrier
5. Upper limit of ice erosion
6. Rock wall
7. Rock shoulder
8. Glacial striations
9. Valley steps
10. Glacier bed
11. Drumlins
12. Roches moutonnées
13. Basal or ground moraine
14. Terminal moraine (of advance)
15. Recessional moraine
16. Glaciofluvial gravels
17. Contemporaneous river channel
18. Eskers (Osers)
19. Kames

FIGURE 17-5. Glacial landforms following deglaciation of an Alpine region (Streiff-Becker, 1947).

FIGURE 17-6. View northeastward over Mount Cook (3754 m) up the Tasman Glacier, New Zealand. Horns and arêtes are prominent (photo: New Zealand National Publicity Studios).

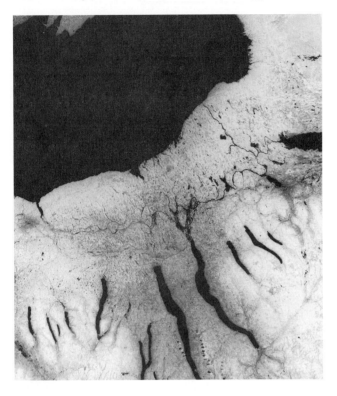

FIGURE 17-7. The Finger Lakes of central New York. All the lakes drain north to Lake Ontario (top) via the Oswego River. Cayuga Lake, the easternmost of the two longest lakes, is 61 km long. A light snow cover enhances the relief (NASA ERTS E-1243-15244).

history of geomorphology. By nature of their position high on mountain slopes, many cirques that fed Pleistocene valley glaciers still carry small cirque glaciers or greatly reduced valley glaciers in them. In North America, they commonly are oriented toward the northeast, most sheltered from the afternoon sun and in a favorable orientation for windblown snow to accumulate (Color Plate 13). Many cirques are compound, with a nested series of progressively smaller and higher forms representing stages in the reduction of icefield size in late-glacial and postglacial time. The altitude of a cirque floor, tarn, or a rock sill is a convenient approximation of the former equilibrium line altitude.

Cols, Arêtes, and Horns. If a plateau or mountain range is lightly but incompletely dissected by glacial troughs and cirques, the former convex upland surface remains as remnants between the steep, concave glaciated forms (Figure 17-8; Color Plate 13). This condition is well shown in the central Rocky Mountains and the Big Horn Mountains, where it has been described as a *scalloped upland*, or *biscuit-board* topography. The latter term describes the landscape well, by analogy to a thick layer of biscuit dough on a cutting board, with arcuate pieces removed by cirque erosion.

Progressive glacier erosion of cirques and troughs leads to the development of sharp-edged divides between them. As two steep-headed cirques intersect, U-shaped notches cut the divide between their head-

FIGURE 17-8. A scalloped upland on a broad, gentle divide (foreground) evolving by intersection of opposing cirques into a *fretted upland* of arêtes, horns, and cols at higher altitudes (background). Compare Color Plate 13.

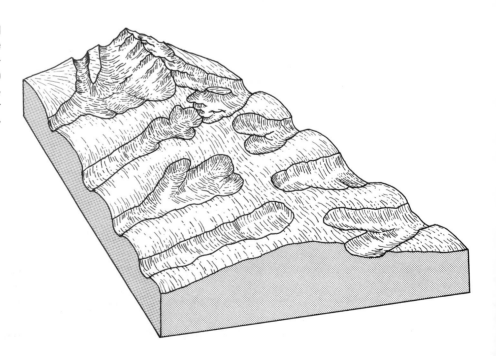

walls. Such a gap is a **col** in alpine terminology. A saw-toothed divide between opposing or parallel troughs, consisting of a series of cols and intervening ridge segments, is an **arête.** Where three or more cirques intersect, a higher pyramidal or triangular-faceted peak may rise above the general level of the arête. These peaks are the **horns** (or *Matterhorn peaks,* after the type example) of alpine scenery. A summit area of arêtes, cols, and horns represents the complete destruction of the preglacial surface (Figure 17-6), and is analogous to the complete dissection of a regional fluvial landscape. The most prominent horns are at the intersection of three or more radiating arêtes.

Through Valleys and Breached Divides. If mountain glaciation increases in intensity until a mountain ice cap develops, former regional divides may be breached by glacial erosion, and major diversions of preglacial stream flow result. The Brooks Range of Alaska was glaciated more intensely on the southern, snowier side of the regional divide, whereas only short valley or piedmont glaciers formed on the arid northern Arctic Slope. As the ice divide was south of the buried bedrock divide, erosion by the north-flowing glaciers breached the preglacial regional divide in several places, and major ice streams flowed northward down the steep troughs. Breached divides of this sort become strongly abraded and low. In the Brooks Range, the divide at Anaktuvuk Pass, between the Yukon and Arctic drainage, is only slightly over 600 m above sea level and is on a swampy valley floor among moraines (Porter, 1966). In Patagonia, glacial erosion has captured many rivers on the more arid Argentine (eastern) side of the Andes Mountains and diverted them to the wetter Chilean (western) coast.

Similar breached divides in the glaciated northern Appalachian Mountains were produced by multiple advances of the Pleistocene ice sheets that overrode the preglacial divides between the St. Lawrence-Great Lakes drainage and the southerly draining Delaware, Susquehanna, and Ohio Rivers. Powerful erosion through constricted valleys at the base of the ice sheets produced the Finger Lakes troughs (Figure 17-7) on the north side of the divide but left flat-gradient, swampy **through valleys** between north- and south-flowing tributaries (Tarr, 1905). The present divide is inferred to be close to the preglacial position in most areas, but valley forms have been modified into a series of troughs "like the intersecting passages of some unroofed ants' nest" (Clayton, 1965, p. 51; see also Linton, 1963; Coates, 1966). Undoubtedly, much of the erosion of the through valleys can be attributed to meltwater overflowing into the southerly drainage from large ice-

dammed lakes north of the divide at times when the ice margin blocked the St. Lawrence lowland. Each ice advance must have been preceded by an intense period of fluvial erosion in meltwater channels across the divides, and each retreat was accompanied by a similar interval of overflowing ice-marginal lakes.

Regional Denudation

Valley Glaciers. Table 15-2 and the related text present some evidence that the regional denudation rate in mountains with glaciers exceeds 1 m/1000 yr, the most rapid of any climatic region. Even higher rates, up to 50 m/1000 yr, have been calculated for large, fast-moving temperature valley glaciers in southeastern Alaska (Hallet et al., 1996). The combination of high relief, steep slopes, intense nivation, and temperate glaciers must indeed have a powerful impact on landscapes. The share of the work borne by glaciers alone is hard to evaluate because glaciers are just one aspect of geomorphic processes in high snowy mountains (Harbor and Warburton, 1993). It seems certain that alpine glaciation, in which most of the upland is buried under a mountain ice cap from which valley glaciers radiate, must be one of the most rapid means of destroying mountains. The **gipfelflur,** or "sea of peaks" all at about the same elevation, is a well-known feature of alpine scenery (Figure 17-6), but whether the horns that rise above mountain ice fields represent remnants of an ancient erosion surface or are nothing more than the mean level of intersection of adjacent cirques and troughs cannot be determined. Further, in the Holocene Epoch (a postglacial or interglacial interval), glaciers all over the earth have greatly receded, and the high modern denudation rates of mountainous regions are at least partly due to the many oversteepened slopes relict from glacial erosion and deposition but now exposed to mass-wasting and fluvial erosion (Church and Ryder, 1972).

Ice Caps and Ice Sheets. An extreme range of views can be found concerning the effectiveness of large ice sheets in denuding landscapes. White (1972) argued that the multiple Laurentide ice sheets of eastern North America completely stripped up to 1 km of Paleozoic and younger sedimentary rocks off large areas of the Canadian shield. Other authors (Ambrose, 1964) have noted that the buried relief of the crystalline shield, where it is unconformably overlain by lower Paleozoic or Cretaceous sedimentary rocks, is very similar to the local relief on the shield today. By this line of reasoning, multiple Pleistocene glaciations have done little more than to remove the weathered regolith (Dyke et al., 1989, p. 182). The oldest known glacial deposits in the central United States contain igneous and metamorphic rocks derived from the Canadian shield, proving that the shield was exposed at least in part prior to the onset of Quaternary glaciation (Gravenor, 1975). The view of minimal regional denudation by ice sheets is supported by the rather small volume of glacial deposits on the periphery of the Laurentide ice sheet and by the fact that most of the glacial sediment was derived from relatively near the ice margin, especially in the southern quadrant in the United States. From the volume of glacial sediment on the North American continent, many authors have argued that a total of no more than 20 m was eroded from the Canadian shield during the Quaternary ice ages. Only 1 to 2 m of resistant shield rock was removed by the latest glaciation although as much as 6 m might have been removed by the earliest ice sheets, which eroded the preglacial regolith (Kaszycki and Shilts, 1987, p. 158). Such rates are of the same order as the rates of fluvial erosion expected for shield terranes (Figure 15-2). They imply that ice sheet erosion is not intense, and that ice sheets could even be regarded as "protection" from fluvial erosion. A less confident estimate is that a total of 120 m has been removed from the glaciated portion of North America by Pleistocene glacial erosion, and that most of the volume is now in the deep sea, not on land. This estimate is based on many assumptions of deep-sea sediment volumes, grain-size distribution, and age (Bell and Laine, 1985). Ice sheets do not seem to be dramatic erosive agents.

Discoveries of bauxite pebbles in glacial deposits on Martha's Vineyard, Massachusetts (Kaye, 1967); deep weathering on Long Island, New York, that is characteristic of the humid tropics (Liebling, 1973; Blank, 1978); and remnants of saprolite in New England and Quebec (LaSalle et al., 1985) all support the inference that in preglacial times, weathering typical of the present tropics or subtropics affected northeastern North America (Chapter 4). Such weathering is known to affect rocks to depths of 100 m or more (Figure 7-14), so it is possible that glacial erosion has accomplished little more than to remove the former regolith from northeastern North America. The very large crystalline **erratics** derived from the Precambrian shield, but widely scattered in southern Canada and the glaciated United States east of the Rocky Mountains, need not represent erosion into fresh, unweathered rock; they are more likely corestones (Figures 7-1 and 7-13) from the preglacial weathered layer (Feininger, 1971). Minimal glacial erosion into resistant rock masses is also inferred from the conformity of roches moutonnées around Boston, Massachusetts, to sheeting joints that are almost certainly preglacial (Jahns, 1943) and from

similar hills in Sweden that have analogs in the relief on sub-Cretaceous and sub-Cambrian unconformities (Lindström, 1988). The nearly total lack of rock debris in the tabular icebergs derived from Antarctica implies that glacial erosion is now inactive there (Warnke, 1970). However, the thick deposits of glaciomarine sediments around Antarctica prove that at some time in the Cenozoic Era, the Antarctic ice sheet eroded a considerable mass off the continent, including a deep regolith (p. 58).

GLACIAL DEPOSITION

Several traditional terms have been used to describe rock debris that is both in transport by modern glaciers and is inferred to have been deposited by former glaciers. The oldest word is *moraine*, used to describe the hummocky or ridged depositional landscapes beyond the margins of alpine glaciers and sometimes also to describe the rock debris in and on glaciers today (Figures 17-4 and 17-5). Some confusion results if the word *moraine* is used to describe both the material and the landforms, so only the latter meaning is used here. **Moraine,** then, is simply a collective term for all depositional landforms of direct glacial origin. One possible classification of moraine and related glaciofluvial forms is outlined on the following pages.

For centuries before the glacial theory of Agassiz (1840), the poorly sorted, unconsolidated surficial deposits of northern Europe were supposed to have been the result of the biblical flood of Noah's time. In Germany, the term *diluvium* was widely use with reference to the "great flood." In the British Isles, the same material was for the same reason called *drift*, a term also used to describe marine sediment in transport along coasts. With the acceptance of the glacial theory to explain these deposits, **glacial drift** (or simply **drift** if no ambiguity is possible) has come to be the general English term for all sediment transported and deposited by glaciers and associated fluvial activity. It is important not to confuse landforms and materials when studying glacial geomorphology, so *drift* is used consistently to describe glacier-transported sediment, and *moraine* is used as a general term for glacier-depositional landforms.[1] The value of the distinction should become apparent in the following pages.

---

1. See the discussion on p. 12 concerning the analogous use of the terms *terrane* and *terrain*.

Glacial Drift

The following brief glossary defines some of the more common terms for categories of glacial drift and associated sediments. Details of composition, sorting, stratification, and genesis are left to specialized texts (Evenson et al., 1983; Eyles, 1983; Drewry, 1986; Goldthwait and Matsch, 1989; Menzies, 1995, 1996), but this vocabulary is necessary to appreciate the genesis and morphology of morainic and related glaciofluvial landscapes.

1. **Till.** Nonsorted, nonstratified glacial drift, deposited directly from ice without reworking by meltwater. The nongenetic term for any mixed sediment, **diamicton,** is recommended for poorly sorted debris that cannot be proved to be glacier-derived. Some authors object so much to the unique glacial origin implied by the term *till* that they use only the term *diamicton* for all poorly sorted till and till-like sediments, including the following subcategories and many others (Dreimanis, 1989).
   a. *Lodgment* till—Compact, tough, dense till, often clay-rich, deposited at the base of a flowing temperate or wet-base glacier. Stones may show surface abrasion and preferred orientation parallel to former flow.
   b. *Melt-out* till—Sediment released from ice by melting. *Basal melt-out* till can be deposited wherever the heat generated by friction, deformation, or geothermal heat flow cannot be conducted rapidly enough through overlying ice. It is compact and nonsorted, but is not sheared by ice flow. *Surface melt-out* till can be derived from either superglacial, englacial, or subglacial debris that has arrived at the glacier surface by ablation. It is most common in polar glaciers, where ablation is primarily by sublimation. On temperature glaciers, surface sediments are usually reworked by water.
   c. *Resedimented* till—Surface melt-out till commonly slides, slumps, or flow as it accumulates on ablating ice, especially if it is water saturated. While retaining most properties of till, resedimented till may show crude layering or deformation structure. It is transitional to the following categories of glaciofluvial, or water-laid, drift.
2. **Ice-contact stratified drift.** Drift modified by meltwater during or after deposition. Considerable stratification and sorting is usual. Although water-modified, the large range of sizes, chaotic sedimentary structures, slump structures, faults, surface topography, and inclusions of till all confirm an origin in contact with, or in close proximity to, melting ice.
3. **Outwash.** Alluvium deposited by meltwater not in close proximity to melting ice. Sorting and stratification are more regular than in ice-contact stratified drift. Similar to other coarse alluvium, except as influenced by the extreme variability of discharge from glacial sources.

In addition to these three categories of drift, the following sediment types are also commonly associated with glaciation:

4. **Glaciolacustrine sediment.** Deltaic and lake-bottom deposits, usually fine grained, formed in settling basins on or near melting ice. Commonly forms silt and clay plains of broad extent, with distinctive erosional and soil-forming properties. Sometimes *varved*, or composed of rhythmic silt-clay couplets each representing an annual cycle of deposition in a seasonally frozen lake.
5. **Eolian sediment.** Sand dunes, sand sheets, and loess derived from other types of glacial drift. Genetically undifferentiated from other eolian sediments except by regional and temporal association with other drift (Derbyshire and Owen, 1996).
6. **Glaciomarine sediment.** "Glacially eroded, terrestrially derived sediment within a variety of facies in the marine environment. Sediment may be introduced into the depositional environment by fluvial transport, by ice-rafting, as an ice-contact deposit, or rarely by eolian transport" (Molnia, 1983, p. 136).

Moraines and other Constructional Drift Landforms

Some moraines and constructional glaciofluvial landforms are genetically related to a specific kind of drift. Examples are *till plains,* which by definition are low-relief areas of eroded or deposited till; *eskers,* which by origin in an ice tunnel or channel must be made of ice-contact stratified drift; or *outwash plains,* which are depositional plains of outwash. Other terms for constructional glacial landforms do not indicate the nature of the drift from which they are built. For example, *end moraines* range in composition from entirely till to entirely ice-contact stratified drift: Almost all moraines contain some mixture of these materials.

In an attempt to sort out the nomenclature of moraines, Prest (1968) proposed a classification based on the orientation of morainic landforms with respect to ice movement (Table 17-1). The classification is

Table 17-1
Classification of Moraine[a]

Transverse to Ice Flow (Controlled Deposition)	Parallel to Ice Flow (Controlled Deposition)	Nonoriented (Uncontrolled Deposition)
Ground moraine—corrugated. End moraine; includes terminal, recessional, and push moraine. Ice-thrust moraine. Ribbed moraine. De Geer moraine; includes cross-valley moraine. Interlobate and kame moraine. Linear ice-block ridge; includes ice-pressed and ice-slumped ridge, and crevasse filling.	Ground moraine—fluted and drumlinized; includes drumlin, drumlinoid ridge, crag-and-tail hill. Marginal and medial moraine. Interlobate and kame moraine. Linear ice-block ridge; includes ice-pressed and ice-slump ridge, and crevasse filling.	Ground moraine—hummocky (low relief); includes some ablation moraine. Disintegration moraine—hummocky and/or pitted (mainly high relief); includes dead ice, stagnation, collapse, and ablation moraine. Interlobate and kame moraine. Irregular ice-block ridge and rim ridge; includes ice-pressed and ice-slump ridge, and ablation slide moraine.

Source: Prest (1968).

[a]Included are landforms constructed from till or other drift that was deposited directly from, or in association with, glacier ice.

intended for landforms built by continental ice sheets but includes forms built by valley glaciers as well. The scheme is vaguely similar to the classification of sand dunes (Figure 13-20) in that longitudinal (parallel to ice flow), transverse (perpendicular to ice flow), and nonoriented forms are included. The transverse forms are mostly ice-marginal deposits, built during the last recession of the ice margin, but include some forms that are shaped beneath ice during ice advance. The forms that parallel ice flow, by contrast, are nearly all controlled by actively moving ice, although some accumulate in stagnant, melting ice along channels and crevasses parallel to former flow directions. The nonoriented forms are commonly built of a mixture of melt-out till and ice-contact stratified drift, during a time of regional melting with insignificant control by ice motion. Longitudinal forms are likely to predate transverse and nonoriented forms in the same region. For instance, younger *end moraines* or *kame moraines* may be banked against, or draped over, *drumlins* and fluted (corrugated) ground moraine. Some of the morainic terms apply equally well to transverse, longitudinal, and nonoriented forms. For example, melt-out till pressed into the base of a crevassed and melting ice sheet may produce ridges oriented in several directions or produce a nonoriented maze of ridges and mounds. Drift is a notoriously mixed sediment; within a single moraine (Figure 17-9), deltaic structures, lodgment till, and deformed ice-contact stratified drift may be found close together.

Glacial landforms tend to occur in characteristic regional patterns because the terrane over which the

FIGURE 17-9. The concept of a glacial series or morpho-sequence. A group of morainic and related glaciofluvial landforms are either synchronous or sequential in time and are genetically related.

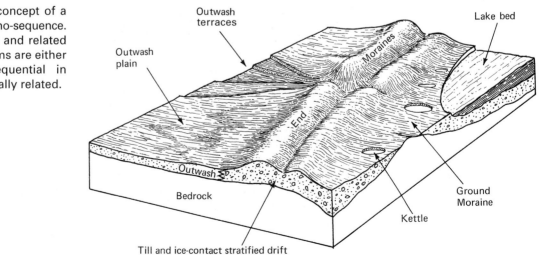

glacier moved and the ablation history largely determine the regional character of both the drift and the landforms. Many morainic forms can be assembled in a genetic sequence that was called a *glacial series* (Figure 17-9) by early workers (Penck and Brückner, 1909; von Engeln, 1921), and later incorporated into the *morphosequence* concept (Koteff and Pessl, 1981, p. 5). Each morphosequence refers to a continuum of landforms comprised of glaciofluvial sediments, grading from ice-contact environments to distant outwash. All the forms in a morphosequence are regarded as contemporary and a chronology can be built from successive morphosequences. The principle is a good one. At any one time during the uncovering of a landscape by ice, an end moraine may be forming at the ice margin, fronted by an outwash plain. Ice tunnels and crevasses behind the moraine may be the sites of eskers, crevasse fillings, and kames. Meltwater may be traced through an eroded channel across the end moraine to a fan or delta that coalesces with the outwash plain in front of it. The ice-contact, or *proximal*, part of the end moraine is of till or ice-contact stratified drift. On the *distal* or outwash side of the end moraine, the drift is mostly outwash, with inclusions of ice-contact stratified drift or resedimented till. The entire assemblage represents contemporary events and, when understood as a unit,

gives a good concept of the activity along a receding glacier terminus.

Morainic Landforms Constructed Primarily of Till.
Many of the ice-parallel morainic landforms are streamlined, and when developed in clusters give a vivid impression of ice motion. Similar erosional forms, such as roches moutonnées and crag-and-tail hills, have been described previously (p. 376). Best known of the streamline ice-molded forms are **drumlins** (Muller, 1974). These are elliptical hills, blunt on the upglacier end, with an elongate downglacier tail (Figure 17-10). They occur in clusters, most frequently in areas of fine-grained lodgment till. Internal composition is not a criterion for calling a streamlined hill a drumlin, and known examples range from those that are nearly all rock with a thin veneer of till to those that are entirely till with no rock core. Classic examples are to be seen in northern Ireland, New Brunswick, eastern Massachusetts, southern Ontario, central New York, and southern Wisconsin. In each region, hundreds or thousands of drumlins give the landscape a distinct "grain," or linear trend. The drumlins tend to be of comparable internal structure, size, and shape within each region. The similar streamline shape of drumlins of diverse internal structure illustrates that they are stable forms

FIGURE 17-10. Drumlin field southeast of Lake Athabaska, Saskatchewan, Canada. Ice flowed from the northeast (upper right). Also shown are numerous kettle lakes and a system of anastomosing eskers. Lakes are approximately 1 km in diameter. Low-angle illumination is from the southeast. If the drumlins appear inverted, view the picture upside down (photo: Surveys and Mapping Branch, Dept. of Energy, Mines, and Resources, Canada).

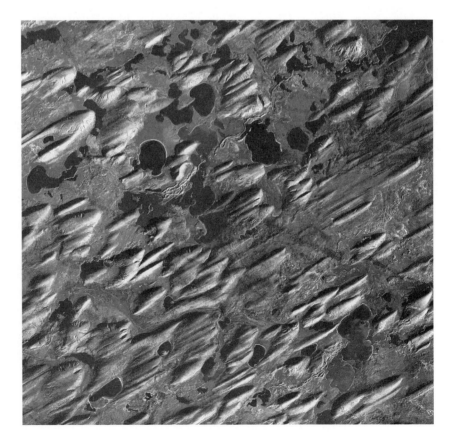

in equilibrium with local conditions at the base of moving ice. The concept of *equifinality,* the tendency of diverse processes to produce similar landforms (p. 14), is well illustrated by drumlins and related streamline erosional forms (compare Figures 17-1 and 17-10). Table 17-1 lists as variants such forms as fluted (ridged) ground moraine (flutes) and elongate "drumlinoid" ridges.

A hypothesis of "dilatant flow" within water-saturated basal till has been proposed to explain drumlin formation (Smalley and Unwin, 1968; Smalley, 1981). A mixture of rock debris, ice, and water, similar to the deformable sediment bed beneath some wet-base glaciers (p. 368), could flow as a viscous fluid as long as it retains water. If, however, an increased pressure region upglacier of a bedrock obstruction or even of a large stone within the mass were to cause local dewatering, the mixture would stiffen and lodge. A similar effect could result if a mass of coarser or better-sorted sediment were dewatered by shear within enclosing viscous sediment (Boulton and Hindmarsh, 1987). The initial obstruction would be enlarged by the mass of rigidified till and continue to grow to some equilibrium streamlined form or until another obstruction in the vicinity caused a similar response. The hypothesis has the advantage of explaining the size, shape, compositional variation, and spatial distribution of drumlins. The process requires a wet-base glacier and a deformable substrate. Local pressure and temperature gradients at the base of the ice, associated with extending and compressing flow, may localize drumlin forma-

tion in certain zones broadly parallel to the ice margin. A stimulating alternative hypothesis is that drumlins are the streamlined remnants of former drift eroded by subglacial meltwater floods (p. 377).

Landforms Constructed Primarily of Ice-Contact Stratifed Drift. A great variety of distinctive landforms are made of ice-contact stratified drift. Oriented and nonoriented forms are both common. The terminology is complicated because various terms can be applied to either single landforms or groups of forms.

Kame is representative of the complexities of this category of landforms. A kame is an isolated hill or mound of stratified drift deposited in an opening within or between ice blocks or in a moulin (Figure 17-11). If the opening was between ice and a valley wall, the resulting form is called a **kame terrace. Kame deltas** are flat-topped hills with an ice-contact proximal face and a semicircular plan view, deposited in standing water at an ice margin. A **moulin kame** is believed to form within stagnant ice at the base of a *moulin* or glacier-karst swallow hole (p. 358). It is clear that a variety of environments in or adjacent to melting glaciers can produce kames.

Another important landform of ice-contact stratified drift is the **esker** (Figure 17-10). An esker is a sinuous or meandering ridge of stratified sand and gravel, commonly with very well-rounded pebbles and cobbles. Eskers form in open channels on glaciers and in tunnels within or beneath them. If large quantities of

FIGURE 17-11. Internal structure of a kame near West Durham, Maine. Coarse gravel foreset beds at bottom were deposited by a prograding delta among ice blocks. Overlying glaciofluvial sand and gravel were probably deposited by an aggrading meltwater river.

sediment-laden meltwater are discharging through a karstlike hydrologic system within a glacier, the sediment may be deposited on the floors of various tunnels or channels while the ice walls or roofs are widened or raised by melting. Eskers show sinuosities comparable to the dimensions of rivers and clearly show by internal structures that they are water-laid. However, they cross bedrock ridges, rising and falling as they are traced across moraine landscapes. Confined flow driven by the pressure gradient within the overlying ice must be invoked to permit them to be deposited on uphill gradients. Most trend broadly perpendicular to major end moraines and parallel to former ice flow and ice surface gradient because water flowed under hydrostatic head toward the ice margin (Shreve, 1985) (Figure 16-15). Eskers seem to be systematically related to regions of crystalline bedrock with discontinuous, permeable drift cover, rather than with regions of fine-grained till (Clark and Walder, 1994). The esker systems of northeastern United States, eastern Canada, and southern Finland and Sweden are noteworthy.

A **kettle** is a basin of nondeposition within morainic terrain, formed where an ice block was buried by till or ice-contact stratified drift and then melted. Many areas of moraine are chaotic mazes of kettles among kames and **crevasse fillings,** the latter representing sediment castings made in ice molds that then melted. Extremely complex internal structures, representing multiple episodes of melting and drift deposition, characterize many areas of kames and kettles (Figure 17-12). Small kettles develop angle-of-repose side slopes and circular outlines very similar to those of collapse dolines (Figure 8-10). Larger ones are complexly segmented by crevasse fillings (Figure 17-10). Kettle lakes are favorite scenic attractions and a fisherman's paradise. Thoreau's Walden Pond may be the best known.

Landforms Constructed Primarily of Outwash. Meltwater alluvium aggrades the floors of glaciated valleys downstream from melting glaciers. Meltwater streams have steep gradients and braided patterns appropriate for their massive bedloads and fluctuating discharges. Where outwash is confined between valley walls downstream from a glacier, the aggraded surface is called a **valley train** (Figure 17-13). On the margin of an ice sheet in areas of low relief, broad **outwash plains** form from confluent alluvial fans that emerge from multiple meltwater valleys through end moraines. The Icelandic term **sandur** (plural: sandar) is sometimes applied to outwash plains. The Icelandic sandar are notable for major episodic floods that inundate hundreds of square kilometers during events called *jökulhlaups* ("glacier bursts"), when ice dams are floated by subglacial water pressure and release masses of impounded water from the margins of

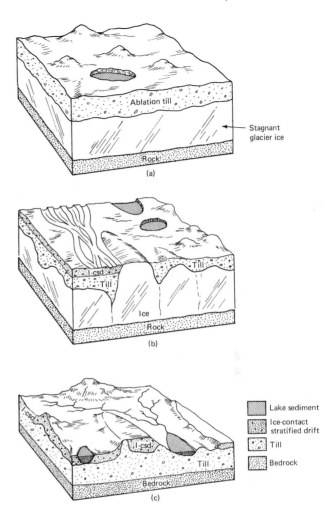

FIGURE 17-12. Hypothetical sequence of events in the evolution of a flat-topped kame. (a) Cover of melt-out till over stagnant ice. (b) Till slumps and washes into deepening crevasses, exposing ice on a chaotic ablation surface. (c) Glaciofluvial sediment washed into a depression when ice melted, later became the crest of a kame, surrounded by a rim of till. Other ice-contact stratified drift deposits are slumped and tilted by final melting.

mountain ice caps. Active valley trains and outwash plains are excellent source areas for dune sand and loess (Figure 13-26), because they carry massive sediment loads, but may nearly dry up at night or during winter, when melting is slower. Eolian deflation may then remove sand and silt from the dry river bed, but the sediments are renewed with each runoff event.

Landforms Constructed of Glacolacustrine, Eolian, and Glaciomarine Sediments. When ice-marginal lakes drain, a variety of minor depositional landforms remain. The deeper parts of the lake will typically become a lacustrine plain of relatively fine-grained sediment. Slumps and mudflows may accompany the

FIGURE 17-13. View upstream over the braided valley train of the Tasman River, South Island, New Zealand. Tasman Glacier, mostly covered with melt-out till, in right distance. Mount Cook is high peak on the left (photo: V. C. Brown & Son, Photographers).

draining of the lake. Beach ridges and deltas may define former shorelines. Eolian landforms include dune fields, sand plains, and loess sheets that drape over or bury prior topography (Chapter 13). Because of isostatic uplift, extensive areas of emerged glaciomarine sediments are common near coasts. They create a complex set of landforms that include submarine end moraines, ice-contact deltas, eskers, and emerged beach ridges (Figure 17-14). The quick-clay terranes of glaciomarine sediments are noteworthy for their slope stability problems (Figure 9-9).

INDIRECT GEOMORPHIC EFFECTS OF GLACIATION

Glacial Isostasy

A special form of epeirogenic movement of particular significance in the Pleistocene Epoch is **glacial isostasy.** The theory of isostasy holds that the pressure at the base of all vertical sections of the earth, above some deep level of compensation, is equal. Nature's "great experiments in isostasy" (Daly, 1940, p. 309) were the alternate

loading and unloading of large continental areas with ice sheets during late Cenozoic time (Figure 17-15). Seawater must evaporate and accumulate on land as ice in order for ice sheets to grow. Thus each glaciation involved the mass transfer of a layer of water perhaps 120 m in thickness from the 71 percent of the earth's surface that is ocean, and the concentration of that water mass, as glacier ice as much as 3 to 4.5 km in thickness, on about 5 percent of the earth's surface. We infer that the ocean floor responded to the removal of water load by isostatically rising (Tushingham and Peltier, 1991). It is much easier to prove that the glaciated areas larger than about 50 km in diameter were basined beneath their loads of ice because when the ice was melting, lakes and seas often formed shorelines near the glacier margins. The shorelines of these former ice-marginal water bodies show regular tilting, always concentric upward toward the center of former maximum ice thickness. Uplift of 120 m is known from areas on the eastern side of Hudson Bay, Canada, in a total time of less than 6000 years (Andrews, 1991, Figure 10).

The rate of postglacial uplift is rapid at first, then progressively slower. It approximates an exponential decay curve with a *relaxation time* (the time required for the deformation to decrease to $1/e$, or 37 percent, of

FIGURE 17-14. Emerged Holocene beach ridges on Ostergarnsholm, eastern Gotland, Sweden. The contemporary uplift rate is about 2 mm/yr (Figure 17-16) (photographer and copyright holder: Arne Philip, Visby, Sweden. Courtesy IGCP Project Ecostratigraphy).

the original value) of from 1000 to 5000 years. Initial uplift rates of 9 to 10 cm/yr have been reported for Greenland (Washburn and Stuiver, 1962; Ten Brink, 1974) and for arctic Canada (Andrews, 1991). The maximum known current uplift rate is 9 mm/yr, recorded near the north end of the Baltic Sea (Figure 17-16) (Kakkuri and Chen, 1992).

Where uplift rates are so rapid, landforms such as storm beaches are raised in a few decades sufficiently high to be unreached by the next major storm, and wide, descending flights of beach ridges mark the margins of seas and lakes (Figure 17-14). River gradients must also be affected by the tilting associated with isostatic uplift, and lakes or swamps may have formed or drained, but the extreme deranged drainage patterns of glaciated regions make the subtleties of tilting hard to recognize. Rivers flowing away from the center of uplift should show simultaneous rejuvenation as their gradients are increased by tilting; rivers flowing toward the center of maximum uplift, as do many of the rivers that drain into Hudson Bay, should have their gradients reduced by the greater differential uplift of their lower reaches; this contributes to the abundance of lakes and swamps in glaciated regions. The differential tilt on postglacial shorelines is typically of the order of 1 m/km, or 0.1 percent, which is comparable to the gradient of many river segments, so the effect on stream hydrology cannot be neglected.

The Great Lakes of North America form a natural laboratory for studying postglacial uplift and tilting (Teller, 1987). Abandoned shorelines and contemporary water-level gages prove that the lakes are deepening on their southern shores and shoaling on the northern. For example, Lake Michigan and Lake Huron form a continuous water level through the very deep Straits of Mackinac. The northeastern parts of the combined lakes are now rising about 30 cm/100 yr relative to their southwestern parts (Figure 3-7). Furthermore, the outlet of the combined lakes near Detroit is rising at

FIGURE 17-15. Estimated surface profile of the Fennoscandian ice sheet along a line eastward from Trondheim, Norway, to Helsinki, Finland. Note the very large area of ice that was grounded below sea level in the Baltic basin (Robin, 1964, Figure 7).

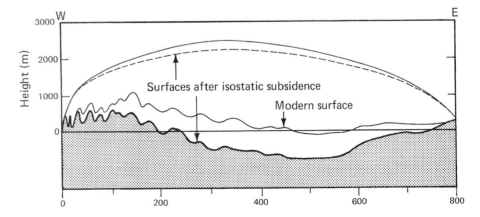

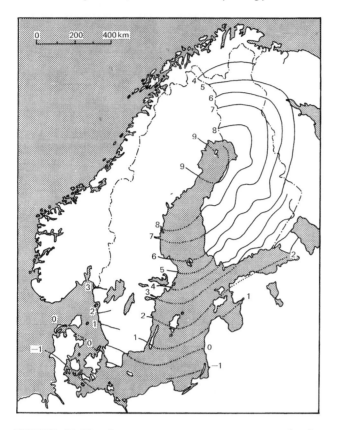

FIGURE 17-16. Contemporary emergence rates in the Baltic basin and vicinity (mm/yr). Emergence is relative to sea level, which is also rising 1 mm/yr or more (Flint, 1971, Figure 13-2).

about 7.5 cm/100 yr relative to a former glacial overflow at Chicago. The discharge of Lake Michigan and Lake Huron will be diverted naturally from the St. Lawrence system to the Mississippi in the next 3200 years if control dams in the Chicago Sanitary and Ship Canal do not prevent the changes (Clark and Persoage, 1970, p. 632). The magnitude of the potential change suggests that widespread regional drainage diversions are possible by postglacial uplift. However, it should be emphasized that the uplift is a *recovery,* or a return to a previous condition, and not the equivalent of net tectonic uplift.

Glacial Eustatic Fluctuations of Sea Level

The volume of water abstracted from the sea at times of ice-sheet expansion and returned to the sea during interglacial intervals of the Pleistocene Epoch has had a worldwide effect on landscape evolution. Global changes of sea level are called **eustatic** changes, meaning that the entire ocean is affected; if they are due to fluctuating ice volume, they are called **glacial eustatic**

sea-level changes. Estimates of the maximum glacial eustatic lowering of sea level during the latest full glacial interval, as compared to today, vary from 80 to 140 m, with the best estimate being about 120 m (Figure 18-4). The lowering was enough to lay bare about one-half of the continental shelf area so that all rivers had to extend and deepen their valleys to the lower base level. At present, all rivers enter the sea either in estuaries or across thick deltaic fills. Both features are a direct consequence of glacial eustatic sea-level fluctuations but are so much a part of the late Cenozoic fluvial geomorphic system that we find it hard to deduce what river mouths would be like if sea level had been stable for the last few million years (Chapter 11).

Glacial eustatic fluctuations of sea level have affected all the coastal regions of the earth, with much greater geomorphic impact than postglacial uplift, which principally affected only the deglaciated areas. The rising sea level of the last 20,000 years has submerged all coasts except those with rapid glacial-isostatic or tectonic uplift. During the most rapid phases of deglaciation, from about 15,000 to 7000 years ago, sea level may have risen at times at a rate of 10 mm/yr. The consequences of this submergence to coastal morphology are considered at length in Chapters 19 and 20.

REFERENCES

ABER, J. S., CROOT, D. G., and FENTON, M. M., 1989, Glaciotectonic landforms and structures: Kluwer Academic Publishers, Dordrecht, The Netherlands, 200 pp.

AGASSIZ, L., 1840, Études sur les glaciers: privately published, Neuchâtel, Switzerland, 346 pp.

ALLEY, R. B., 1989, Water-pressure coupling of sliding and bed deformation: II. Velocity-depth profiles: Jour. Glaciology, v. 35, pp. 119–129.

———, BLANKENSHIP, and 2 others, 1989, Water-pressure coupling of sliding and bed deformation: III. Application to Ice Stream B, Antarctica: Jour. Glaciology, v. 35, pp. 130–139.

AMBROSE, J. W., 1964, Exhumed paleoplains of the Precambrian shield of North America: Am. Jour. Sci., v. 262, pp. 817–857.

ANDERSON, J. B., and ASHLEY, G. M., eds., 1991, Glacial marine sedimentation; Paleoclimatic significance: Geol. Soc. America Spec. Paper 261, 232 pp.

ANDREWS, J. T., 1991, Late Quaternary glacial isostatic recovery of North American, Greenland, and Iceland: A neotectonics perspective, in Slemmans, D. B., Engdahl, E. R., and 2 others, eds., Neotectonics of North America: Geol. Soc. America Decade Map Vol. 1, pp. 473–486.

ASHLEY, G. M., SHAW, J., and SMITH, N. D., eds., 1985, Glacial sedimentary environments: Soc. Econ. Paleontologists and Mineralogists Short Course No. 16, 246 pp.

AUGUSTINUS, P. C., 1992, The influence of rock mass strength on glacial valley cross-profile morphometry: A case study from the southern Alps, New Zealand: Earth Surface Processes and Landforms, vol. 17, pp. 39–51.

———, 1995, Glacial valley cross-profile development: The influence of in situ rock stress and rock mass strength, with examples from the Southern Alps, New Zealand: Geomorphology, v. 14, pp. 87–97.

BELL, M., and LAINE, E. P., 1985, Erosion of the Laurentide region of North America by glacial and glaciofluvial processes: Quaternary Res., v. 23, pp. 154–174.

BLANK, H. R., 1978, Fossil laterite on bedrock in Brooklyn, New York: Geology, v. 6, pp. 21–24.

BOULTON, G. S., 1974, Processes and patterns of glacial erosion, in Coates, D. R., ed., Glacial geomorphology: State Univ. of New York Publications in Geomorphology, Binghamton, New York, pp. 41–87.

———, 1996, Theory of glacial erosion, transport and deposition as a consequence of subglacial sediment deformation: Jour. Glaciology, v. 42, pp. 43–62.

———, and Hindmarsh, R. C. A., 1987, Sediment deformation beneath glaciers: Rheology and geological consequences: Jour. Geophys. Res., v. 92, pp. 9059–9082.

CHURCH, M., and RYDER, J. M., 1972, Paraglacial sedimentation: A consideration of fluvial processes conditioned by glaciation: Geol. Soc. America Bull., v. 83, pp. 3059–3072.

CLARK, P. U., and WALDER, J. S., 1994, Subglacial drainage, eskers, and deforming beds beneath the Laurentide and Eurasian ice sheets: Geol. Soc. America Bull., v. 106, pp. 304–314.

CLARK, R. H., and PERSOAGE, N. P., 1970, Some implications of crustal movement in engineering planning: Canadian Jour. Earth Sci., v. 70, pp. 628–633.

CLAYTON, K. M., 1965, Glacial erosion in the Finger Lakes region (New York State, U.S.A.): Zeitschr. für Geomorph., v. 9, pp. 50–62 (discussion, v. 10, pp. 475–477).

COATES, D. R., 1966, Discussion of K. M. Clayton, "Glacial erosion in the Finger Lakes region (New York State, U.S.A.)": Zeitschr. für Geomorph., v. 10, pp. 469–474.

CRARY, A. P., 1966, Mechanism for fiord formation indicated by studies of an ice-covered inlet: Geol. Soc. America Bull., v. 77, pp. 911–929.

DALY, R. A., 1940, Strength and structure of the earth: Prentice-Hall, Inc., Englewood Cliffs, New Jersey, 434 pp.

DERBYSHIRE, E., and OWEN, L. A., 1996, Glacioaeolian processes, sediments and landforms, in Menzies, J., ed., Past glacial environments: Sediments, forms and techniques: Butterworth-Heinemann Ltd., Oxford, UK, pp. 213–237.

DOWDESWELL, J. A., and SCOURSE, J. D., eds., 1990, Glacimarine environments: Processes and sediments: Geol. Soc. [London], Spec. Pub. 53, 423 pp.

DREIMANIS, A., 1988, Tills: Their genetic terminology and classification, in Goldthwait, R. P., and Matsch, C. L., eds., Genetic classification of glacigenic deposits: A. A. Balkema, Rotterdam, pp. 17–83.

DREWRY, D., 1986, Glacial geologic processes: Edward Arnold (Publishers) Ltd., London, 276 pp.

DYKE, A. S., VINCENT, J. S., and 3 others, 1989, The Laurentide ice sheet and an introduction to the Quaternary geology of the Canadian shield, in Fulton, R. J., ed., Quaternary Geology of Canada and Greenland: Geol. Soc. America, The Geology of North America, v. K-1, pp. 178–189.

von ENGELN, O. D., 1921, The Tully glacial series: N.Y. State Museum Bull. 227-228, pp. 39–62.

EVENSON, E. B., and CLINCH, J. M., 1985, Debris transport mechanisms at active alpine glacier margins: Alaskan case studies, in Kujansuu, R., and Saarnisto, M., eds., INQUA till symposium, Finland 1985: Geol. Survey Finland, Spec. Paper 3, pp. 111–136.

EVENSON, E. B., SCHLÜCHTER, C. H., and RABASSA, A., eds., 1983, Tills and related deposits: A.A. Balkema, Rotterdam, 454 pp.

EYLES, N., ed., 1983, Glacial geology: An introduction for engineers and earth scientists: Pergamon Press, Inc., Elmsford, New York, 409 pp.

FEININGER, T., 1971, Chemical weathering and glacial erosion of crystalline rocks and the origin of till: U.S. Geol. Survey Prof. Paper 750-C, pp. 65–81.

FLINT, R. F., 1971, Glacial and Quaternary geology: John Wiley & Sons, Inc., New York, 892 pp.

GJESSING, J., 1966, On "plastic scouring" and "subglacial erosion": Norsk Geografisk Tidsskrift, v. 20, pp. 1–37.

GOLDTHWAIT, R. P., and MATSCH, C. L., eds., 1988, Genetic classification of glacigenic deposits: A. A. Balkema, Rotterdam, 294 pp.

GRAVENOR, C. P., 1975, Erosion by continental ice sheets: Am. Jour. Sci., v. 275, pp. 594–604.

GURNELL, A. M., and CLARK, M. J., eds., 1987, Glacio-fluvial sediment transfer: An alpine perspective: John Wiley & Sons Ltd., Chichester, UK, 524 pp.

GUSTAVSON, T. C., and BOOTHROYD, J. C., 1987, A depositional model for outwash, sediment sources, and hydrologic characteristics, Malaspina Glacier, Alaska: A modern analog of the southeastern margin of the Laurentide Ice Sheet: Geol. Soc. America, v. 99, pp. 187–200.

GUTENBERG, B., BUWALDA, J. P., and SHARP, R. P., 1956, Seismic explorations on the floor of Yosemite Valley, California: Geol. Soc. America Bull., v. 67, pp. 1051–1078.

HALLET, B., 1976, Deposits formed by subglacial precipitation of $CaCO_3$: Geol. Soc. America Bull., v. 87, pp. 1003–1015.

———, 1979, A theoretical model of glacial abrasion: Jour. Glaciology, v. 23, pp. 39–50.

———, 1981, Glacial abrasion and sliding: Their dependence on the debris concentration in basal ice: Ann. Glaciology, v. 2, pp. 23–28.

———, HUNTER, L., and BOGEN, J., 1996, Rates of erosion and sediment evacuation by glaciers: A review of field data and their implications: Global and Planet. Change, v. 12, pp. 213–235.

HARBOR, J. M., 1992, Numerical modeling of the development of U-shaped valleys by glacial erosion: Geol. Soc. America Bull., v. 104, pp. 1364–1375.

————, 1995, Development of glacial-valley cross sections under conditions of spatially variable resistance to erosion: Geomorphology v. 14, pp. 99–107.

————, and WARBURTON, J., 1993, Relative rates of glacial and nonglacial erosion in alpine environments: Arctic and Alpine Res., v. 25, pp. 1–7.

HIRANO, M., and ANIYA, M., 1988, A rational explanation of cross-profile morphology for glacial valleys and of glacial valley development. Earth Surface Processes and Landforms, vol. 13, pp. 707–716.

————, 1989, A rational explanation of cross-profile morphology for glacial valleys and of glacial valley development: A further note: Earth Surface Processes and Landforms, vol. 14, pp. 173–174.

JAHNS, R. H., 1943, Sheet structure in granites: Its origin and use as a measure of glacial erosion in New England: Jour. Geology, v. 51, pp. 71–98.

KAKKURI, J., and CHEN, R., 1992, On horizontal crustal strain in Finland: Bull. Géodésique, v. 66, pp. 12–20.

KASZYCKI, C. A., and SHILTS, W. W., 1987, Glacial erosion of the Canadian shield, in Graf, W. L., ed., Geomorphic systems of North America: Geol. Soc. America, Centennial Special Vol. 2, pp. 155–161.

KAYE, C. A., 1967, Fossiliferous bauxite in glacial drift, Martha's Vineyard, Massachusetts: Science, v. 157, pp. 1035–1037.

KOTEFF, C., and PESSL, F., JR., 1981, Systematic ice retreat in New England: U.S. Geol. Survey Prof. Paper 1179, 20 pp.

LASALLE, P., DeKIMPE, C. R., and LAVERDIERE, M. R., 1985, Sub-till saprolites in southeastern Quebec and adjacent New England: Erosional, stratigraphic, and climatic significance, in Borns, H. W., Jr., LaSalle, P., and Thompson, W. B., eds., Late Pleistocene history of northeastern New England and adjacent Quebec: Geol. Soc. America Spec. Paper 197, pp. 13–20.

LAWSON, D. E., 1993, Glaciohydrologic and glaciohydraulic effects on runoff and sediment yield in glacierized basins: US Army Corps of Engineers Cold Regions Res. and Engin. Lab. Mono. 93–2, 108 pp.

LIEBLING, R. S., 1973, Clay minerals of the weathered bedrock underlying coastal New York: Geol. Soc. America Bull., v. 84, pp. 1813–1816.

LINDSTRÖM, E., 1988, Are roches moutonnées mainly preglacial forms?: Geografiska Annaler, v. 70a, pp. 323–331.

LINTON, D. L., 1963, The forms of glacial erosion: Inst. Brit. Geographers Trans., v. 33, pp. 1–28.

MENZIES, J., ed., 1995, Modern glacial environments: Processes, dynamics and sediments: Butterworth-Heinemann Ltd., Oxford, UK, 621 pp.

————, 1996, Past glacial environments: Sediments, forms and techniques: Butterworth-Heinemann Ltd., Oxford, UK, 598 pp.

MOLNIA, B. F., 1983, Subarctic glacial-marine sedimentation: A model, in Molnia, B.F., ed., Glacial-marine sedimentation: Plenum Press, New York, pp. 95–144.

MOOERS, H. D., 1989, On the formation of the tunnel valleys of the Superior Lobe, central Minnesota: Quaternary Res., v. 32, pp. 24–35.

MULLER, E. H., 1974, Origins of drumlins, in Coates, D. R., ed., Glacial geomorphology: State Univ. of New York Publications in Geomorphology, Binghamton, New York, pp. 187–204.

PENCK, A., and BRÜCKNER, E., 1909, Die Alpen im Eiszeitalter: Tauchnitz, Leipzig, 3 v., 1199 pp.

PORTER, S. C., 1966, Pleistocene geology of Anaktuvuk Pass, Central Brooks Range: Arctic Inst. of North America Tech. Paper 18, 100 pp.

PREST, V. K., 1968, Nomenclature of moraines and ice-flow features as applied to the glacial map of Canada: Canada Geol. Survey Paper 67–57, 32 pp.

ROBIN, G. D., 1964, Glaciology: Endeavour, v. 23, pp. 102–107.

SHARPE, D. R., and SHAW, J., 1989, Erosion of bedrock by subglacial meltwater, Cantley, Quebec: Geol. Soc. America Bull., v. 101, pp. 1011–1020.

SHAW, J., 1989, Drumlins, proglacial meltwater floods, and ocean responses: Geology, v. 17, pp. 853–856.

SHOEMAKER, E. M., 1986, The formation of fjord thresholds: Jour. Glaciology, v. 32, pp. 65–71.

SHREVE, R. L., 1985, Esker characteristics in terms of glacier physics, Katahdin esker system, Maine: Geol. Soc. America Bull., v. 96, pp. 639–646.

SMALLEY, I. J., 1981, Conjectures, hypotheses, and theories of drumlin formation: Jour. Glaciology, v. 27, pp. 503–505.

————, and UNWIN, D. J., 1968, Formation and shape of drumlins and their distribution and orientation in drumlin fields: Jour. Glaciology, v. 7, pp. 377–390.

STREIFF-BECKER, R., 1947, Glacierization and glaciation: Jour. Glaciology, v. 1, pp. 63–65.

TARR, R. S., 1905, Drainage features of central New York: Geol. Soc. America Bull., v. 16, pp. 229–242.

TELLER, J. T., 1987, Proglacial lakes and the southern limit of the Laurentide Ice Sheet, in Ruddiman, W. F., and Wright, H. E., JR., eds., North America and adjacent oceans during the last deglaciation: Geol. Soc. America, The geology of North America, v. K-3, pp. 39–69.

TEN BRINK, N. W., 1974, Glacio-isostasy: New data from West Greenland and geophysical implications: Geol. Soc. America Bull., v. 85, pp. 219–228.

TUSHINGHAM, A. M., and PELTIER, W. R., 1989, Ice/3G: A new global model of late Pleistocene deglaciation based upon geophysical predictions of post-glacial relative sea level change: Jour. Geophys. Res., v. 96, pp. 4497–4523.

WARNKE, D. A., 1970, Glacial erosion, ice rafting, and glacial-marine sediments: Antarctica and the Southern Ocean: Am. Jour. Sci., v. 269, pp. 276–294.

WASHBURN, A. L., and STUIVER, M., 1962, Radiocarbon-dated postglacial delevelling in northeast Greenland and its implications: Arctic, v. 15, pp. 66–73.

van der WATEREN, F, M., 1995, Processes of glaciotectonism, in Menzies, J., ed., Modern glacial environments: Processes, dynamics and sediments: Butterworth-Heinemann Ltd., Oxford, UK, pp. 309–335.

WHITE, W. A., 1972, Deep erosion by continental ice sheets: Geol. Soc. America Bull., v. 83, pp. 1037–1056.

Chapter 18

Late Quaternary Climatic Geomorphology

Late Quaternary is an informal term, generally used to include Late Pleistocene and Holocene time, or about the last 130,000 years (back endpaper). The interval includes oxygen isotope stages 1 to 5. It includes the last interglaciation, the most recent "ice age," and the entire "postglacial" or Holocene Epoch of the past 10,000 years. The sequence of interglacial-glacial-interglacial climates is only the latest of such fluctuations that have been operating for at least the past 2.5 million years. It is important not for its role in shaping major landforms, which were more or less as they are today at the beginning of late Quaternary time, nor for its uniqueness, although the amplitude of late Quaternary climatic swings has been greater than many previous cycles (Figure 4-7). Rather, late Quaternary events are important because they have shaped the details of the landscape and soils on which we walk today, and have seen the rise of human civilizations that have modified that landscape with increasing effectiveness.

Chapter 4 could be usefully read or reviewed in preparation for what follows here. Intervening Chapters 7 to 17 provide systematic overviews of the processes and landforms of the fluvial, arid, periglacial, and glacial geomorphic systems that are in this chapter considered in the context of episodes of climate change, relatively brief compared to the age of most landforms, but powerful enough so that their relics can be seen in the present palimpsest landscape (p. 35).

The late Quaternary interval has the most reliable chronology of all geologic history. Numerous radiometric and cosmogenic isotopes can be measured with a precision that permits ages to be determined to within 1 or 2 percent of the total elapsed time. Other techniques such as tree-ring and ice-layer counting give annual resolution of events, at least during the Holocene Epoch. Thus, within late Quaternary time, the rates of geomorphic processes can be determined with precision, and these rates can then be applied to older, undated, or poorly dated landscapes and events. If for no other reason, the ability to date late Quaternary landscapes and measure their rates of geomorphic change justifies detailed study.

LATE QUATERNARY CLIMATES

The literature of late Quaternary climate studies contains numerous graphs wherein some climate parameter such as mean annual temperature, annual or seasonal precipitation, or length of growing season is plotted against a suitable time scale. Even more common are *proxy records*, in which the temporal record of some measured parameter such as oxygen-isotope ratios, widths of annual growth rings in trees, or relative abundances of certain pollen grains, is interpreted as a record of specific climatic conditions. Proxy records are subject to multiple interpretations, as the following examples illustrate, and it would be rare if a phenomenon as complex as climate could ever be interpreted from a single variable.

Benthic δ¹⁸O Records

Figure 18-1 illustrates the $\delta^{18}O$ record in four representative deep-sea cores that cover the last 150,000 years. They are similar to the inset of Figure 4-7. In each graph, abrupt decreases of $\delta^{18}O$ in benthic foraminifera of almost 2‰ were in progress at about 130,000 years ago and again at about 10,000 years ago. These rapid decreases are inferred to be the result of some combination of ocean warming and the return of isotopically

light glacial meltwater to the world ocean (pp. 54–56). They are called *Terminations* I and II, and are typical of the terminations that mark the end of each cold, glacial interval in the longer isotopic record (Figure 4-7). The four curves of Figure 18-1 were not dated by radiometric methods, but were confidently correlated by calculating an average sedimentation rate that placed the center of these two terminations at 11,000 and 127,000 years. (Various authors use slightly different times for correlating the terminations.) Although estimates of the relative importance of ocean warming and ice-sheet melting range from 30 to 70 percent for each variable (p. 56), the benthic foraminifera that were analyzed in these cores live in very cold deep water, and their oxygen-isotope fluctuations are inferred to be primarily a

FIGURE 18-1. $\delta^{18}O$ values measured on benthic foraminifera in deep-sea cores from four widely separated regions. Curves are inferred to be proxy records of ice volume and therefore sea level changes for about the past 150,000 years. Terminations I and II and conventional oxygen-isotopic stage designations are shown with estimated ages (based on Ruddiman and McIntyre, 1981, Figure 2).

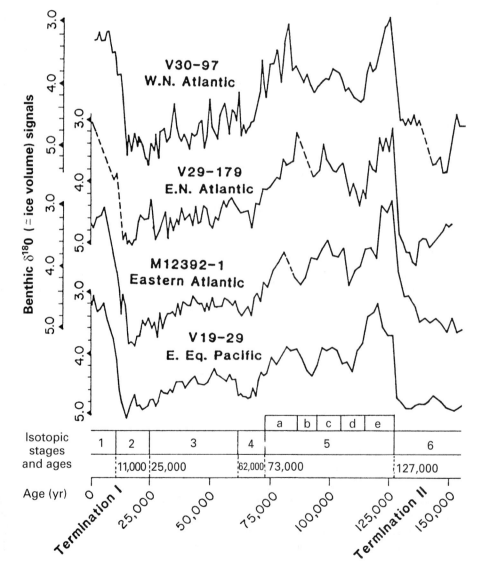

record of ice-volume, and therefore sea-level, changes (Ruddiman and McIntyre, 1981; Shackleton, 1987). The terminations clearly show that each ice age ended abruptly, requiring little more than 10,000 years to change from full-glacial to full-interglacial conditions.

Vostok, Antarctica, Ice cores

The Russian-maintained Vostok Station is on the eastern Antarctic ice sheet (78° 27′S; 106° 51′E) at an altitude of 3,490 m. The mean annual temperature there is −55.5°C. A series of bore holes in the ice sheet at Vostok have collected continuous ice cores to a maximum depth of 2755 m, from which a remarkable climatic record has been constructed (Figure 18-2) (Barnola et al., 1987; Genthon et al., 1987; Jouzel et al., 1987; Lorius et al., 1988; Jouzel et al., 1989; Jouzel et al., 1993; Vostok Project Members, 1995).

The δD (deuterium, or ^{2}H) record (curve *a*) is the equivalent of a δ^{18}O record, and is the proxy for the condensation temperature of snow in the troposphere above Vostok (Figure 4-4). With suitable corrections for changes in isotopic content of the oceanic sources of the moisture, changes in ice-cap altitude, and other factors, the temperature deviations from the current surface temperature at Vostok for the past 150,000 years have been calculated (Figure 18-2, curve *b*). Both Terminations I and II show an abrupt 6°C temperature rise from full-glacial to interglacial conditions at Vostok (Jouzel et al., 1993, p. 408). In Figure 18-2, the terminations are estimated to be centered at 13,000 and 140,000 years ago. Snowfall at Vostok correlates linearly with air temperature and is inferred to have been only half of the present low accumulation rate during the coldest parts of the isotopic record (intervals B, D, F, and H, Figure 18-2, curve *a*).

The dust content of the Vostok ice cores increased by at least a factor of 10 whenever the inferred temperature dropped by more than 4°C below its modern value during isotope stages 2, 4, and 6. Thus, glacial intervals were not only colder, but were times either of more extensive deserts, more intensive winds in the desert source regions for the Antarctic dust, or more efficient meridional dust transport (Basile et al., 1997). Patagonia was the main contributor to dust fallout in eastern Antarctica, but other minor sources were Australia and South Africa (Basile et al., 1997). The extensive Atlantic continental shelf of South America, which was broadly exposed during glacial times of low sea level, was not a source of the dust, based on trace-element analyses.

The time scale for the Vostok ice-core record was produced by "curve matching" the isotopic proxy temperature record in the ice cores with comparable deep-sea δ^{18}O records such as those in Figure 18-1. Terminations I and II are especially obvious in both the ice-core and marine records and provide good references for curve matching (Figure 18-2, curves *b* and *c*). The curve matching is supported by models of ice flow and also by the content of cosmogenic ^{10}Be in the ice that helps calibrate accumulation rates (Jouzel et al., 1993, p. 408).

The most exciting aspect of the Vostok record is the measured CO_2 content of the air trapped by the conversion of firn to glacier ice (Figure 18-2, curve *d*). The CO_2 content of the entrapped air rose abruptly, from about 200 ppmV (parts per million volume) in full-glacial times to about 280 ppmV in interglacials. The Vostok ice cores offered the first evidence that glaciations were times of low atmospheric CO_2 and interglaciations were times of higher CO_2. CH_4, another important greenhouse gas, shows a similar trend. Could the reduction in greenhouse gases cause ice ages? Perhaps, but it is also possible that cold glacial oceans cause the reduced greenhouse CO_2 because colder water absorbs more of it from the atmosphere. The time resolution of the Vostok cores is too coarse to tell whether the changes in atmospheric CO_2 lead or follow the changes in temperature, so cause and effect remain the subject of debate. But it is clear that atmospheric greenhouse gases played an important role in late Quaternary climate changes, if only by amplifying the effects of other causes such as the orbital variations of incoming solar radiation (pp. 56–58).

Coral Reef Chronology

A few ppm of the radioactive isotopes of uranium occur in a precisely measured ratio in ocean water (Chen et al., 1986). As do other trace elements such as Ba, Mg, and Sr, uranium substitutes for calcium in the $CaCO_3$ of coral limestone. ^{238}U and ^{234}U decay radioactively at different rates, and their conversion to ^{230}Th as part of the uranium series of radioactive decay provides a tool for dating coral reefs from the entire late Quaternary with analytical errors of less than 1 percent of the age of the sample (Edwards et al., 1987).

In many tropical regions of tectonic stability, such as oceanic islands distant from both spreading ridges and subduction zones, a coral reef about 6 m above present sea level forms a coastal terrace. It typically is dated by the uranium-series method at about 125,000 years old (Szabo et al., 1994). A prominent terrace of similar age is common on tectonically uplifted coral-capped tropical islands in orogenic belts, with a sequence of younger emerged terraces below it (Figure 3-5). By correlating

FIGURE 18-2. Vostok, Antarctica, ice core records. Depth scale at top, estimated age scale at bottom. Curve **a:** δD (deuterium) record; less negative values represent warmer air temperatures. Climate stages A to H are separated by boundaries with estimated ages in thousands of years. Curve **b:** Smoothed surface temperature record at Vostok derived from δD and δ¹⁸O measurements, expressed as deviations from the current temperature. (Jouzel et al., 1993, p. 408, revised the maximum range of temperature deviation to 6°C rather than the estimated 9°C shown here.) Curve **c:** Synthetic δ¹⁸O oceanic record derived from averaging sets of benthic foraminifera analyses such as those in Figure 18-1 (Martinson et al., 1986). Curve **d:** CO₂ concentrations of air trapped the ice: heavy line is the best estimate with uncertainty bands on either side (Lorius et al., 1988, Figure 1).

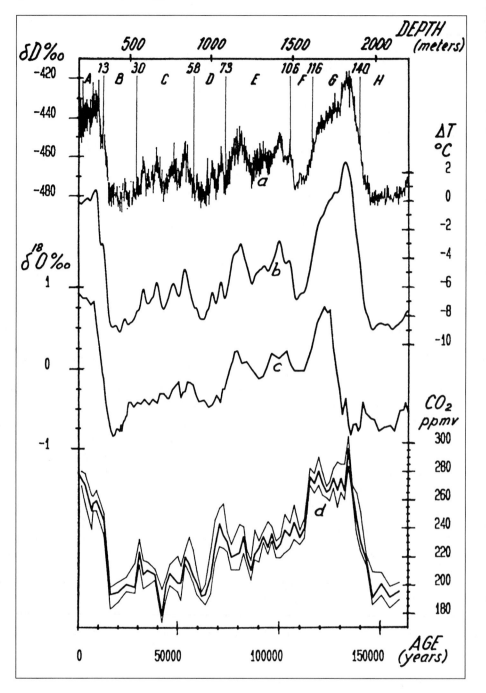

the 125,000-year-old terrace with oxygen-isotope stage 5e of the deep-sea record, a convincing radiometrically dated chronology for late Quaternary sea-level changes (Figure 18-3), oceanic temperature fluctuations, and ice-core stratigraphy, has been established. The deep-sea oxygen-isotope records and the ice-core records provide continuous time series, but the high-precision uranium-series dates on coral reefs establish the ages of critical points in those time series (Chappell, et al., 1996,

p. 235). Because reef corals grow only in shallow water, the uranium-series ages of emerged corals are also useful to establish sea-level changes and tectonic uplift rates (p. 41).

The high precision that has been achieved by the uranium-series dating method has been applied with particular success in dating Termination I in coral reefs (Figure 18-4). A series of boreholes in coral reefs in Barbados, Papua New Guinea, Tahiti, and elsewhere have

FIGURE 18-3. Late Quaternary sea-level history based on a reconciliation of coral-reef terrace heights and ages in Papua New Guinea (black circles) with deep sea oxygen-isotope ratios (open squares). Undated but stratigraphically correlated low sea-level deposits between reef terraces (black triangles) are also shown. Sea-level positions based on oxygen isotope ratios have error estimates of ± 5000 years and ± 7 m vertical range (Chappell et al., 1996, Figure 1, reprinted with kind permission from Elsevier Science–NL, Amsterdam).

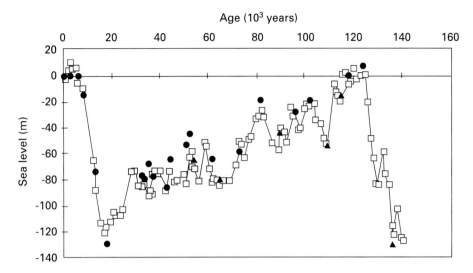

provided samples or reef growth in stratigraphic succession during the rapid rise of sea level during Termination I (Fairbanks, 1989). By dating each sample by both the radiocarbon and uranium-series methods, not only has the history of Termination I been detailed, but the radiocarbon time scale has been calibrated for the past 20,000 years and more (Figure 18-4). The substantial errors in the radiocarbon time scale have previously been corrected by comparison with tree-ring chronologies back to about 11,400 years before present (Kromer and Becker, 1993), but for older samples, the necessary corrections were unknown. Paired dates on Barbados coral samples showed that radiocarbon dates in the range of 20,000 years are about 3500 years too young (Bard et al., 1990). These results have been confirmed for the past 13,000 years by similar paired dates on corals from Papua New Guinea (Edwards et al., 1993) and Tahiti (Bard et al., 1996).

Precise uranium-series dates on corals show that reefs grew upward rapidly during the rising sea level of Termination I, with two intervals of especially rapid sea-level rise at about 14,000 and 11,000 years ago separated by an interval of slower rise, called a "period of reduced melting" (Figure 18-4). The period of reduced melting correlates approximately to the Younger Dryas cold interval in the European pollen record, which is described in more detail in a later part this chapter (p. 403). Correlative fluctuations in the $\delta^{18}O$ values of pelagic foraminifera (Bard et al., 1987) demonstrate that Termination I involved important fluctuations in meltwater production by receding ice sheets, sharp changes in sea-surface temperatures, and perhaps pulses of rapid CO_2 release from the ocean to the atmosphere (Edwards et al., 1993).

Pollen Record at Grande Pile, Vosges, France

Very few terrestrial sites have recorded continuous sedimentation through the late Quaternary interval. One such site known as Grande Pile is on the Vosges Plateau of northern France. In a basin among older moraines and glaciated rock knobs, 19 m of finely laminated, pollen-rich sediment accumulated during the last 140,000 years (Woillard, 1978; Woillard and Mook, 1982; Seret et al., 1990; de Beaulieu and Reille, 1992). Figure 18-5 is a simplified pollen diagram in which the percentage abundance of all tree and shrub pollen ("Trees") is compared to the percentage abundance of all nonarboreal pollen and fern spores ("Herbs"). An increasing percentage of tree pollen, which ranges from 5 to 95 percent of the total pollen, is the proxy record of a warming temperature. Abundant herb pollen implies cold steppe or tundra conditions. Figure 18-5 is drawn with a uniform depth scale (except between 2.5 and 4.5 m) but because the rate of organic sedimentation was higher during forested times, relatively short time intervals of warmth are represented by each unit of sediment thickness. The time scale for colder intervals is compressed because of slow sedimentation rates during those times.

A series of radiocarbon dates of decreasing precision document the age of the Grande Pile pollen succession back to a minimum of 70,000 years ago. The entire record has been correlated to deep-sea oxygen isotope stages 1 to 6, including the five subdivisions of stage 5, lettered as 5a to 5e (compare Figures 18-1 and 18-5). Pollen zones numbered 1 to 21 in Figure 18-5 identify times of specific pollen assemblages; for example, Zone 1 has grass and herb pollen typical of a

FIGURE 18-4. Barbados sea-level curves based on radiocarbon-dated and uranium-series-dated shallow water corals. The Younger Dryas cold interval of slower sea-level rise was traditionally dated at 10,000 to 11,000 years age by radiocarbon, but actually lasted for 1300 years, between 11,700 and 13,000 years ago (Fairbanks, 1990, Figure 2).

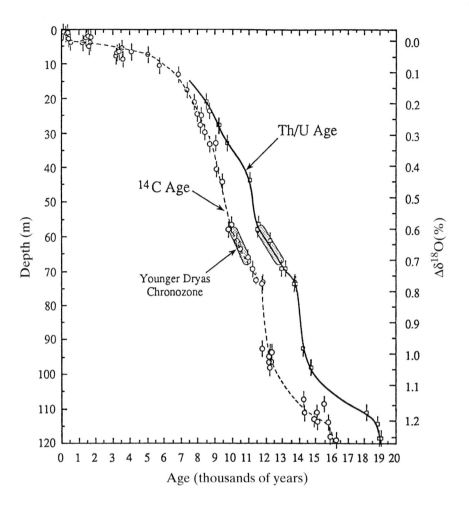

periglacial steppe beyond the margin of the Saale glaciation (oxygen isotope stage 6). Zone 2, correlated with the Eem interglaciation and oxygen isotope stage 5e, shows an upward succession progressing from tundra with some birch and juniper, through boreal birch-pine forest, to hazel-oak-alder forest at the warmest part of stage 5e (Woillard, 1978, p. 11), which was warmer than the present (de Beaulieu and Reille, 1993, p. 431). The subsequent cooling into Zone 3 is recorded by a progressive change to a cold temperate forest of fir and spruce, to a boreal forest (taiga) of pine, spruce, and birch, and then to a cold grass steppe. The completeness of the vegetational record from cold to warm to cold argues for the completeness of the sedimentation history through the entire Eem interglaciation. The subsequent two intervals of forest growth at Grande Pile (Zones 4, 5, 6, and 8) also were of warm temperate forest, separated by an interval of cold steppe (Zone 7). They are correlated with oxygen isotope stage 5c and 5a (Figure 18-1), and are regarded as part of the early Weichsel (Wisconsin) glaciation (Woillard and Mook, 1982, p. 160; Mangerud, 1991). Pollen Zones 12 and 18

define periglacial steppe grassland and are correlated with oxygen isotope stages 4 and 2, separated by a series of fluctuations during oxygen isotope stage 3, with radiocarbon ages ranging from 62,000 to 29,000 years ago, when the vegetation at Grande Pile fluctuated between cold steppe grassland and boreal forest. The abrupt reforestation between Zones 18 and 19 is clearly correlative with Termination I.

Summit, Greenland, Ice cores

Two ice cores, 28 km apart, have been drilled through the Greenland ice sheet near its highest point, known as Summit (72° 34′N; 37° 37′W, altitude 3238 m). The mean annual temperature at Summit is −32°C. The core drilled by a European team (GRIP: European Greenland Ice-core Project) was 3029 m long. The basal 6 m was of silty ice with pebbles. The adjacent core drilled by a U.S. Team (GISP2: US Greenland Ice-sheet Project 2) reached bedrock at a depth of 3053 m. The basal 13 m of the core was of banded brown silty ice with rock inclusions. The two cores show excellent cor-

FIGURE 18-5. Diagram of total tree and shrub pollen versus herb pollen from Grande Pile, France. Depth scale (m). Radiocarbon dates at various levels are shown, with suggested correlations with the deep-sea oxygen-isotope record (simplified from Woillard and Mook, 1981, Figure 1).

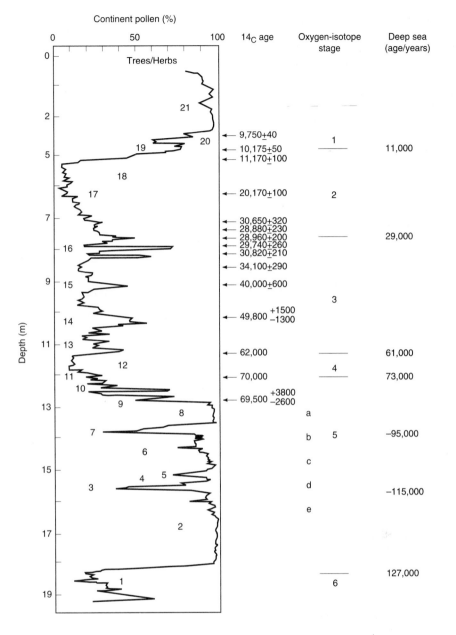

relation of $\delta^{18}O$ ratios, dust layers, and other marker horizons except for their basal 10 percent, judged to be more than 87,000 years old (Grootes et al., 1993). The GISP2 core stratigraphy is regarded as reliable for the past 110,000 years (Brook et al., 1996). The basal parts of both cores are probably deformed by flow and do not provide reliable climate records. However, because of the high snow accumulation rate at Summit, strong seasonal temperature and precipitation contrasts, and proximity to dust and tephra sources (pp. 95), the Summit cores provide a continuous, extremely detailed record of climate-proxy parameters for most of late Quaternary time (Taylor et al., 1993), especially relevant because of the proximity of the

North American and European ice sheets and the North Atlantic Ocean (Figure 18-6).

Of major significance to interpreting late Quaternary climates, the Summit ice cores show numerous strong $\delta^{18}O$ fluctuations during the last glaciation, equivalent to temperature fluctuations of 6 to 9°C (0.63‰ $\delta^{18}O = 1$°C, Grootes et al., 1993, p. 553). These fluctuations, known as *Dansgaard-Oescher cycles*, occurred at intervals of 2000 to 3000 years. Their cold minima correlate with layers of sand-size ice rafted debris in North Atlantic deep sea cores, indicative of massive iceberg discharges from the Laurentide ice sheet, but including sediments from Iceland and the Scandinavian ice sheet as well (Bond and Lotti, 1995).

FIGURE 18-6. $\delta^{18}O$ record from the GRIP ice core, Summit, Greenland, between depths of 1500 to 2675 m, covering a time span from 10,000 to 87,000 years ago. Linear depth scale; time scale established by counting annual layers back to 14,500 years; beyond that by ice flow modeling. In the upper 1500 m of the ice core that covers 10,000 years of Holocene time $\delta^{18}O$ values are nearly constant at $-35 \pm 1‰$. Warm peaks of Dansgaard-Oescher cycles 1 to 21 are numbered for reference. The late glacial cold interval known as the Younger Dryas followed warm peak no. 1 (modified from Dansgaard et al., 1993, Figure 1).

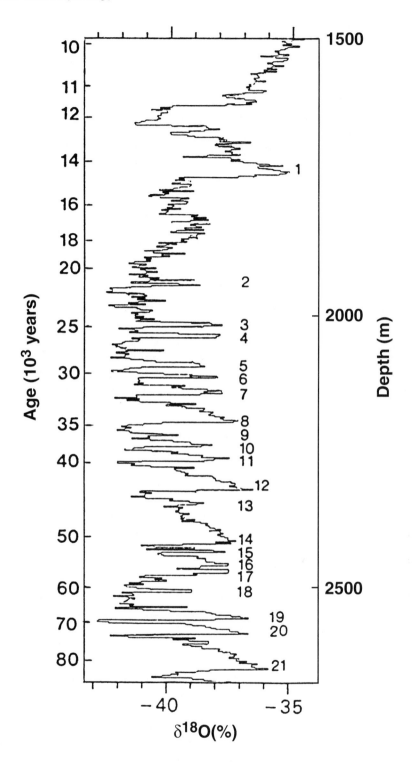

The larger ice-rafting episodes between 10,000 and 40,000 years ago, called *Heinrich events*, record massive surges of ice out of Hudson Strait from the heart of the Laurentide ice sheet. The cold minima of the Dansgaard-Oescher cycles also correlate with increased terrestrial dust (p. 304) and sea-salt content in the Summit ice cores that indicate more intense storminess in the region (Mayewski, et al., 1994). Taken together, the Summit ice cores and North Atlantic sediment cores record extreme instability of atmospheric and oceanic circulation and of the large ice sheets adjacent to the North Atlantic Ocean during the last ice age, in sharp contrast to the relative tranquility of the Holocene Epoch (Ditlevsen et al., 1996).

In particular, near the end of the last glaciation as the climate was warming, during a brief 1300-year interval known as the *Younger Dryas* the climate over and around the North Atlantic Ocean returned to near-glacial conditions. The name refers to a previously established pollen zone in peat bogs and lakes in northern Europe, in which pollen of the Arctic tundra plant *Dryas* is abundant and tree pollen sharply decreases. The increased dust content of the Greenland atmosphere was rich in Ca^{2+} and Mg^+ ions representative of a continental source area rather than sea salt, so a continental-scale reduction of vegetation cover and stronger winds is inferred (Mayewski et al., 1993).

During the Younger Dryas interval, sea surface temperatures in the Atlantic Ocean near Portugal dropped by 7°C (Bard et al., 1987). Correlative sharp cooling and glacier advances have been reported from New Zealand and western North America (Denton and Hendy, 1994). The event is also very likely recorded in submarine sediments near the coast of Venezuela (Hughen et al., 1996). Even corals growing on the coast of Papua New Guinea recorded a slowing of sea level rise (p. 399) and related changes in oceanic isotope chemistry, but with a time lag of about 600 years (Edwards et al., 1993). The biologic and geomorphic impact of the Younger Dryas event must have been severe as glaciers and permafrost regions abruptly expanded. Then, this major reorganization of circum-arctic atmospheric circulation, a disruption of the North Atlantic Ocean thermohaline circulation, and probably global circulation changes, ceased abruptly. The Younger Dryas event ended within a transitional intervals of no more than 10 to 20 years (Alley et al. 1993), yet represented a climate fluctuation one-third or more of the full glacial-interglacial range. The general warming trend of the early Holocene resumed.

The Younger Dryas interval is attributed to a sudden flood of fresh glacial meltwater into the North Atlantic from the St. Lawrence River estuary. The thermohaline origin of NADW is very sensitive to any surface freshwater dilution (pp. 53ff). Computer simulations demonstrate that even modest freshening of the North Atlantic could stop NADW production, perhaps within a few decades (Broecker, 1995). It seems likely that the Younger Dryas interval, and perhaps the other Dansgaard-Oescher and Heinrich cold events as well, involved highly complex interactions among atmospheric circulation, iceberg discharge from unstable ice-sheet margins, and North Atlantic thermohaline circulation changes. The Summit ice cores seem to have recorded the changes faithfully although interpreting those records is still in progress.

Interpreting the Proxy Records

The overall similarity of the dated coral reef sea-level chronology, the $\delta^{18}O$ marine records, the proxy climate records from ice cores, and the pollen records, provide great confidence that the general climate history of the late Quaternary has been established (Figures 18-1, 18-2, 18-3, 18-5, and 18-6). A cold glacial maximum about 140,000 years ago was followed by rapid deglaciation and warming at Termination II, culminating in an interglaciation that was globally warmer than the present, with sea level a few meters higher than now and with atmospheric CO_2 levels comparable to those at present. Then, a series of temperature, sea-level, and atmospheric CO_2 fluctuations were superimposed on a cooling trend that culminated in the most recent glacial maximum approximately 20,000 years ago. Again, within about 15,000 years at Termination I, the climate warmed, the large ice sheets melted, sea level rose about 120 m, and the CO_2 content of the atmosphere increased to its historical level.

The chronology established by uranium-series dates on coral reefs, and used to calibrate the deep-sea-sediment and ice-core records, fits the Milankovitch orbital chronology (Figure 4-6) very well. With the late Quaternary record of climate change well founded, the orbital chronology has been extrapolated to a paleomagnetic reversal 788,000 years ago and beyond with confidence (Figure 4-7).

The major glaciation of oxygen isotope stage 6 just prior to Termination II, in many regions the most extensive of all, is called the Illinoian in North America and the Saale or Riss in Europe. It began less than 0.5 million years ago and consisted of several advances separated by local retreats but is not yet adequately dated.

The next youngest recognized event in the glacial-interglacial sequence was an interglaciation, called the Sangamon in North America, the Eem or Riss-Würm in Europe, and oxygen isotope stage 5e in the deep-sea record. Sea level was 5 to 6 meters above present, spanning or recurring during the interval from about 130,000 to 120,000 years ago. Numerous coral reefs grew above present sea level at that time, and the eroded bases of these reefs are the foundation of modern reefs (Chapter 19). The continents were somewhat warmer than present. Deciduous forests extended into what is now the conifer zone north of the Great Lakes, for instance. The warm interglacial interval persisted for only about 12,000 years, however (Mangerud, 1991, p. 46).

Sometime slightly less than 120,000 years ago, the earth's climate cooled sharply, and the most recent major glaciation began. Sea level fell in a series of pulses, separated by brief intervals of high sea level that were nevertheless below present sea level (Figure 18-3). The entire interval from about 115,000 to about 10,000 years ago is now called the Wisconsin glaciation in North America and the Weichsel or Würm glaciation in Europe. It includes oxygen isotope stages 2, 3, 4, and substages 5a to 5d. It consisted of intervals of ice advance, called **stadials,** separated by intervals of ice retreat that were colder than an interglacial, called **interstadials.** The later stadials seem to have been more severe than the early ones, culminating in the maximum advance of the northern hemisphere ice sheets only about 20,000 years ago. Most of the outermost moraines of the last glaciation around the periphery of the Laurentide and Scandinavian ice sheets are dated close to that time.

The late-glacial marginal retreat of the Laurentide and Scandinavian ice sheets was dramatic. By 11,000 years ago, the ice margins were fluctuating but receding north of the Great Lakes (Teller and Kehew, 1994) and in the Baltic Sea. Sea level was rising rapidly (Figure 18-4). The Gulf Stream had diverted back toward northern Europe from its glacial maximum track toward Portugal. The warming trend was sharply reversed during the Younger Dryas interval, but by 9000 years ago, the Scandinavian ice sheet was separating into small highland residuals in northern Sweden, and the sea was entering Hudson Bay, calving the thickest part of the complex Laurentide ice sheet and radically changing the flow directions of the ice that remained on the highlands east and west of the Bay (Fulton, 1989). By 6500 years ago, the remnant Laurentide ice sheets were gone, sea level was within a few meters of its present level, and the climate was approximately as today.

In the past 6000 years, several climatic oscillations have affected geomorphic and pedologic processes, but they are not comparable to the glacial-interglacial or even stadial-interstadial changes. Perhaps they have involved a temperature range of 1°C in the middle latitudes, in contrast to the 8°C to 10°C difference between glacial and postglacial climates in the same regions. Between 6000 and 3000 years ago, most areas were warmer and drier than today, although both the timing and the relative intensity of warming and drying varied widely in various localities (Webb et al., 1987, p. 46). Most of the cirque glaciers in the United States and Canadian Rockies and the European Alps were gone, as were most of the small mountain ice caps in Norway. The valley glaciers along the Alaskan coast retreated far up their valleys. Then, in a cooling trend with a series of minor oscillations, the climate approached present conditions, and valley glaciers expanded somewhat. This medieval cold interval of a few centuries has

become known at the "Little Ice Age" (Grove, 1987). The twentieth century has seen a dramatic retreat of valley glaciers from the moraines built during the last times of expansion between A.D. 1500 and 1750.

LATE QUATERNARY GEOMORPHOLOGY

Relict Glacial and Periglacial Landscapes

The most obvious geomorphic result of late Quaternary climate fluctuations has been the extensive region of glacial erosion and deposition in middle latitudes, mostly in North America and Europe. Mountains, even in low latitudes, were also extensively glaciated. Glacial drift is widespread within the glaciated regions of Figure 16-1. Drift-free landscapes show the diverse results of glacial erosion. It is hardly surprising that glaciated landforms are abundant, given that the Wisconsin glaciation lasted for at least 100,000 years as opposed to the mere 10,000 years of "postglacial" time. The examples that follow are selected from a much longer list of landscapes that are relict from late Quaternary glaciations.

Northern United States and Southern Canada. The southern quadrant of the North American Laurentide ice sheet was temperate and wet-based (p. 369). In the midwestern United States, the Laurentide ice sheet reached southward almost to 39°N latitude, well over half way from the north pole to the equator. Meltwater was abundant even during the Wisconsin glacial maximum. Glaciofluvial and glaciolocustrine sediments and landforms abound north of the terminal moraines. The ice margin was highly lobate, with relatively minor preglacial or earlier glacial landforms influencing the flow of the ice (Figure 18-7). Earlier glaciations had obliterated old drainage networks and excavated the Great Lake basins (p. 61). In the glaciated Interior Lowland Province and the Great Plains (Figure 1-5) the Wisconsin glaciation spread a thin sheet of drift across older drifts and loess. The end moraines and ground moraine of the Wisconsin glaciation include complex assemblages of glacial landforms (Chapter 17) with moderate to steep local relief. The end moraines range in height from a few tens to a few hundred meters, and in width from a few hundred meters to a few kilometers. They determine the drainage divides of low-order drainage basins, and between them are extensive flat plains, commonly called "prairies" that are underlain variously by till, outwash, glaciolacustrine sediment, or loess. In many regions, the moraine arcs are the locus of farmhouses and villages because in their natural state they were forested and provided a ready source of construc-

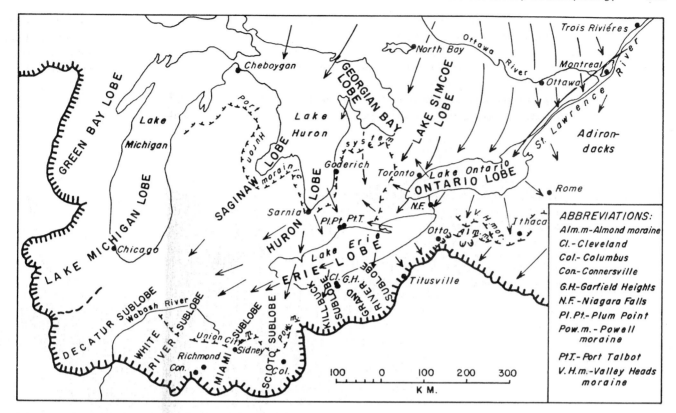

FIGURE 18-7. Distribution of glacial lobes and sublobes in the region of the Great Lakes during the Wisconsin glaciation. Arrows indicate glacial movements inferred from striations and drumlin orientations. The Almond and Valley Heads moraines in western New York mark the approximate southern boundary of the Ontario lobe; the area to the south was traversed by an older glacial flow from the northeast (Dreimanis and Goldthwait, 1973, Figure 1).

tion lumber. By contrast, the intervening plains are relatively treeless and become progressively grassier in a westerly direction until they merge into the open grasslands of the northern Great Plains. Between about 9000 and 7000 years ago, a grassland prairie "peninsula" extended farther east and north into Wisconsin, Illinois, and Minnesota. Subsequently, the forest-prairie boundary migrated westward again (Webb et al., 1987, Jacobson et al., 1987, p. 282).

In the northern Great Plains, the Laurentide ice moved southward and westward, up the regional slope. Large lakes formed at the ice margin by the obstructed drainage (Teller, 1987). The combination of an increasing dry climate to the west, clay-rich soils, and ice-marginal lakes created an extremely thin, dynamic ice sheet (pp. 369ff) that left complex moraine systems and innumerable shallow kettle lakes. Throughout the glaciated Interior Lowland and northern Great Plains, where strata are nearly horizontal and rock exposures are rare because of the veneer of glacial drift, end moraines form the only significant local relief.

It is not generally appreciated that the slowing but continuing postglacial isostatic uplift in the Great Lakes region is projected to divert Lakes Superior, Michigan, and Huron (Figure 3-7) back into the Mississippi drainage basin within about 3200 years (p. 390). Isostatic recovery is a return to prior crustal equilibrium, so it is likely that the present large drainage basin of the St. Lawrence River has been only an ephemeral Holocene phenomenon and that the St. Lawrence-Mississippi divide will return to the vicinity of Detroit, Michigan, unless prevented from doing so by human intervention. Calculations of long-term denudation rates and sediment yields such as those reviewed in Chapter 15 must always consider the impact of such relatively recent effects of glaciation.

Eastward from the east end of Lake Erie, in the more rugged topography of the northern Appalachian Mountains, the lobate pattern of the midwestern ice margin broke down. Local relief is greater, rock-mass strength is greater, and the present climate is more maritime. Laurentide ice was thicker and moved over either crystaline bedrock or a sandy substrate rather than over a deformable shale-rich bed as in the Great Lakes region (p. 369). In the glaciated northern Appalachian Mountains of the eastern United States and Canada, much of the relief is relict from the middle Cenozoic "greenhouse" climate, with very old bedrock landforms stripped of their former regolith but retaining their structurally controlled forms (p. 382). End moraines are rare north of the southern New England coastal plain. At the end of the Wisconsin glaciation, the ice sheet broke up into isolated valley-filling masses of dead ice that melted to produce diverse assemblages of landforms typical of ice-contact stratified drift, such as eskers and kames.

Southward from the maximum advances of the Wisconsin ice margins, tundra vegetation extended at least 300 km on the Appalachian Plateau (Jacobson et al., 1987; Clark and Ciolkosz, 1988). Periglacial patterned ground is known from many regions of central and east-central United States (Chapter 14). Periglacial processes must have influenced the development of the Mississippi Basin, but the effects have not been separated from the massive effects of glaciation. Although tundralike, the late-glacial periglacial zone of the United States has no analog in the present high-latitude tundra, for the sun remained well above the horizon throughout winter days and was high overhead during the summer (p. 61). Loess deposition was widespread, especially downwind from large rivers (Figure 13-26). Soils in regions of former glaciation and permafrost have developed under a postglacial succession of tundra, evergreen, and deciduous forests or grasslands, and are unlikely to be the simple result of contemporary soil-forming climatic variables (Chapter 7).

Northern Europe. The present climate of northern Europe grades from cool maritime forests on the Atlantic seaboard in two directions: (1) eastward to continental steppe grasslands in Poland, Ukrania, and Russia, and (2) southward to Mediterranean subhumid conditions. It approximates a mirror image of North American climatic zonation. During the Weichsel glaciation, nonglaciated Europe north of the Alps became a periglacial region. The present maritime influence was greatly decreased by the width of the exposed continental shelf and by the southward diversion of the Gulf Stream, which then did not affect the European climate north of Portugal (Ruddiman and McIntyre, 1973). The Alpine ice cap that formed synchronously with the expanding Scandinavian ice sheet blocked warm air and moisture from the south.

The Weichselian-age end moraines across the northern European plains define a sharp geomorphic boundary. North of the moraines the landscape has abundant kettle lakes, eskers and kames, outwash plains, and deranged drainage patterns typical of recently deglaciated regions (Mangerud, 1991; Böse, 1994). South of the Weichsel outermost end moraines,

the local relief has been greatly reduced by intense periglacial solifluction and loess deposition. Lakes are rare because basins were filled with solifluction debris.

Parallel to the moraines across Germany and Poland are a series of ice-marginal meltwater channels called *Urstromtäler*. These broad, shallow channels of periglacial braided rivers carried ice-marginal drainage from as far east as Warsaw, past Berlin, and northward to the North Sea (Figure 18-8). Postglacial rivers such as the Weser, Elbe, Oder, and Vistula are displaced from their filled former valleys and make a series of sharp bends as they alternately follow the floors of Urstromtäler and cross end moraines. The gentle gradients and wide beds of the Urstromtäler have offered many opportunities for canals, railroads, and highways to cross the northern European plains.

South from the Weichsel moraines to the Alps, Europe in glacial times was a cold, treeless periglacial terrain. During cold stadials, mean annual temperatures may have reached $-7°C$ to $-12°C$ even though the pollen of steppe and tundra herbs indicate that summer temperatures rose to $7°C$ to $8°C$ (Vandenberghe, 1992, p. 72). Correspondingly, winter temperatures must have been well below $-20°C$ during the coldest stadials. The extreme seasonal temperature range intensified periglacial solifluction activities. Runoff and river discharge were strongly seasonal, with a massive bedload carried in braided channels during summer floods, preceded by ice jams that cut back channel banks and widened the periglacial flood plains (p. 308). The entire set of periglacial processes are an example of "no analog" conditions (p. 61).

Strong winds, regional low relief, and the seasonally dry meltwater channels of periglacial northern Europe promoted intense eolian activity (Kozarski, 1991). In addition to thick loess, extensive areas of sand dunes and *cover sand* blanket the landscape. The cover sand is a relatively thin sheet of sand with some silt, horizontally laminated but lacking the form or internal stratification of dune sand. It is inferred to have been deposited on snow or in wet depressions (Vandenberghe, 1991).

As the Weichsel ice margin retreated, the climate warmed and became more humid, but with several sharp fluctuations. Birch forests first replaced steppe grasslands, and by about 13,000 years age pine and birch forests were widespread in northwestern Europe. Then, abruptly (Figure 18-6) the climate cooled sharply, the forests died and were replaced by the Younger Dryas tundra, and the Scandinavian ice sheet margin expanded 50 km or more to build a prominent arc of end moraines from the west coast of Norway across southern Sweden and southern and eastern Finland (Lundqvist, 1986; Donner, 1995). These *Fennoscandian* moraines mark the final major ice margin of the reced-

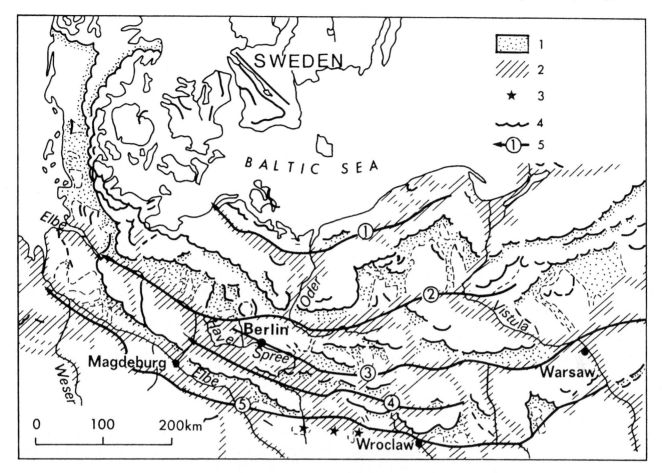

FIGURE 18-8. End moraines and associated meltwater features of the Weichsel glaciation on the north European plain: (1) outwash gravel; (2) Urstromtäler; (3) major alluvial cones from the Mittelgebirge; (4) major ice margins; (5) principal routes of ice-marginal meltwater drainage to the west (Duphorn et al., 1984, Figure 8-3).

ing Scandinavian ice sheet, which at the end of the Younger Dryas interval began to break up into masses of stagnant ice and was entirely gone within 1400 more years. The final phase of deglaciation was characterized by eskers that are a characteristic feature of the Scandinavian and Finnish landscapes (Donner, 1995, p. 106).

About 5000 years ago, the climate became maritime and moist and perhaps 2°C warmer than the average of today. Dense deciduous forests covered northwestern Europe, and peat bogs were widespread on uplands and lowlands. About 3000 years ago, the forests began to degenerate, perhaps owing to a slightly cooler, drier climate but more likely owing to the beginning of Neolithic agriculture. People have strongly affected subsequent geomorphic evolution in western Europe by cutting the forests and planting crops.

The highlands of Scandinavia and the Alps support small ice caps and valley glaciers today although all but a possible few have reformed after the mid-Holocene and are not the remnants of the last glaciation.

As far south as the Mediterranean coast, western European landscapes are dominated by the multiple episodes of wet and dry periglacial processes. Slopes are mantled with thick gelifluction sheets. Valleys are heavily aggraded. Sand plains with stabilized dunes are extensive. Cave stratigraphy and rock paintings prove that Paleolithic people hunted the tundra animals during the Weichsel glaciation and witnessed the periglacial processes (Valladas et al., 1986).

High-altitude New Guinea. One of the most striking results of glaciation is the landscape of high-altitude equatorial regions. On the island of New Guinea, 3° to 6° south of the equator, only three of the highest mountain peaks, all above 4600 m in height, now support icefields and glaciers (Allison and Peterson, 1989; Brown, 1990). The largest of the ice fields has an area of only about 7 km². From an undetermined time prior to 22,000 years ago until as recently as 9000 years ago, mountain ice caps or valley glaciers shaped at least 20 mountains, covering a total area of 2000 to 2200 km² (Hope and Peterson, 1976, p. 174). The firn limit was about 1000 m lower than present, and the treeline, as well as vegetational zones in the lower rainforests, were comparably lowered (Galloway et al., 1973; Hope and Peterson, 1975; Bowler et al., 1976). The result is that areas now in high-altitude grasslands, just above the treeline, have U-shaped glaciated troughs, paternoster lakes, hanging valleys, and other typically alpine landforms (Figure 18-9).

Moraine ridges cross valley floors and dam lakes. Soils are shallow. Weakly developed cirques dissect mountain ridges into arêtes. Areas between 3200 and 2200 m, now in montane rainforest, were covered by alpine grasslands until about 10,000 years ago. Valley forms change from broad, open troughs in the deglaciated highlands to steep, deep gorges across the forested lower mountain ridges. Entire assemblages of landforms, each appropriate to a modern climatic and vegetational zone on the mountain sides, are found relict about 1000 m lower. The entire shift is consistent with a climatic cooling of about 5° to 6°C, with no sig-

nificant change in precipitation (Löffler, 1972, 1977, pp. 63–68). The climatic interpretation is complicated by the influence of human agriculture for the last several thousand years but is nevertheless very similar to the history of montane equatorial regions in eastern Africa and western South America.

Relict Periglacial Landforms in Korea. Superimposed on the colder and probably drier glacial-age climates, increased continentality due to lowered sea level probably contributed to the geomorphic contrasts of some regions. For example, southern Korea today has a humid subtropical climate, with temperature extremes greatly reduced by the moderating effect of the seas around the peninsula. During full-glacial lowering of sea level, the entire Yellow Sea was dry land and the Korean peninsula was incorporated into the Asian mainland. Massive periglacial gelifluction deposits on the south coast of Korea (35°N), presumably dating from the most recent glacial age, overlie a deep red saprolite horizon. These may be the most equatorward evidence of low altitude periglacial

FIGURE 18-9. Pindaunde Valley, a glacial trough with rock-basin lakes on the flank of Mount Wilhelm, Papua New Guinea. The building at the lower end of the lakes is at an altitude of 3500 m. Glaciated landforms extend down to 3200 m, below which steep V-shaped valleys are covered with tropical rainforest (photo: G. S. Hope).

processes. They testify to the extreme geomorphic changes that have occurred there, mostly as a result of the change from a maritime to a continental climate (Guilcher, 1976, pp. 655–658).

Alternating Late Quaternary Pluvial and Nonpluvial Processes

The term **pluvial** ("rainy") implies greater total precipitation but in fact is more accurately defined as greater effective precipitation, the excess of precipitation over evapotranspiration (p. 338). The climatic causes can be greater precipitation, reduced evaporation, more effective seasonal distribution of precipitation, or any combination of these. Many regions now arid or semiarid show evidence of late Quaternary pluvial conditions. In some of the regions, the pluvial intervals correlate with glaciations; in others, the nonpluvial times correlate with glaciations.

Two examples of alternating pluvial-nonpluvial climate changes are offered to demonstrate that pluvial and glacial intervals are correlative in some places. Subsequently, two other examples of relict pluvial processes demonstrate a lack of correlation with glaciation. The latter two examples are from low-latitude regions, not from the extratropical latitudinal belts.

Southwestern United States. More than 150 closed tectonic basins mark the semiarid or arid Great Basin in the Basin and Range Province of southwestern United States (Figure 1-5). Most of them contained lakes during the last glaciation, ranging from shallow flooded playas to Lake Bonneville, which was 315 m deep at its maximum (Morrison, 1991). A few of the lakes were fed by meltwater from valley glaciers on the adjacent ranges, but most of them developed only because of a more favorable precipitation-evaporation ratio. Various lines of evidence suggest that the pluvial lakes resulted from a combination of cooler temperatures, more precipitation, especially from winter storms, and reduced evaporation related to cooler conditions and perhaps greater cloudiness. Estimates vary widely, with temperature estimates ranging from 3° to 16°C cooler and precipitation estimates ranging from only 50 percent to almost 200 percent of present amounts (Smith and Street-Perrott, 1983; Spaulding et al., 1983; Dohrenwend, 1987; Benson and Thompson, 1987; Grayson, 1993, pp. 84ff; Hostetler et al., 1994). One widely quoted source estimated that the average pluvial climate of Nevada was about 3°C cooler, with 68 percent more precipitation than present and evaporation reduced by 10 percent (Mifflin and Wheat, 1979). Typically, this would be the equivalent of the present climate in the relatively cool and moist northwestern corner of the state spreading to

the presently much drier and hotter southern sections. Secondary fluctuations in the levels of the pluvial lakes may have been synchronous with advances and retreats of the Laurentide ice sheet (Benson et al., 1996) and the Dansgaard-Oescher and Heinrich events recorded in the North Atlantic Ocean (Oviatt, 1997).

A large number of radiocarbon-dated sediment cores and shoreline deposits from various playas prove that they held lakes during at least the last glaciation and probably during earlier glacial intervals as well (Oviatt and Nash, 1989; Waters, 1989). However, only two lakes, pluvial Lake Russell in the Mono Basin and pluvial Lake Bonneville in the Great Salt Lake Basin, can be proved by morphologic evidence to have been synchronous with glaciation. In these two basins, end moraines that were built at the mouths of valleys by glaciers from adjacent mountain ranges are cliffed by high-level lake shorelines. Outwash correlated with recessional moraines built deltaic terraces at the level of the pluvial shorelines as well.

From 24,000 to 12,000 years ago, at least 70 percent of the lake basins in the Great Basin for which radiocarbon dates are available were high (defined as being between 70 and 100 percent full relative to overflow level) (Figure 18-10). Lake levels fell rapidly in the last 10,000 years, and many basins were even drier than present in the mid-Holocene. The chronology fits the stadial-interstadial chronology of the Wisconsin glaciation very well, and also the Holocene record (p. 404).

The most obvious morphologic relics of a pluvial climate are the shoreline features of the former lakes. Shore platforms, wave-cut cliffs, spits, bars, and deltas all trace the ancient shorelines (Figure 18-11). River terraces are graded to former lake levels. Basin floors are blanketed with lacustrine sediments, including salt layers and buried soils that record nonpluvial conditions. Sand dunes are associated with ancient shorelines, as well as being built on the dry playas today. Pluvial intervals of soil formation and valley incision with presumed more effective vegetation cover have alternated with periods of mass wasting and aggradation, presumably during nonpluvial times, in agreement with Huntington's principle (p. 244).

The Mediterranean Basin. The Mediterranean climate is characterized by summer drought and winter rains. Floods and droughts are both tragically frequent. Mountainous regions rank high in denudation rates (Table 5-1 and Figure 5-2). From about 30,000 to 15,000 years ago, land around the Mediterranean basin was colder by about 5°C to 10°C in winter and 1°C to 3°C in summer (Prentice et al., 1992). Although total rainfall was probably less than now, a greater proportion fell during winter months because of the southward shift

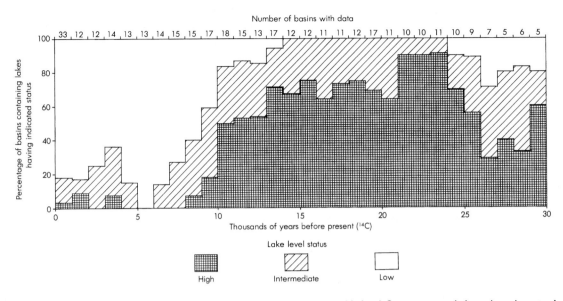

FIGURE 18-10. Relative percentages of pluvial basins in the western United States containing dated materials at elevations indicative of low, intermediate, or high stages versus their radiocarbon ages (Smith and Street-Perrott, 1983, Figure 10-6).

of the westerly wind storm track. Tree pollen, especially oak, indicates the presence of some trees although the region was not heavily forested as might be interpreted from a simplified model (Rognon, 1987). Steppe grasslands indicate drier summer growing seasons, but pluvial lakes expanded despite the decreased total precipitation, perhaps because of the cooler temperatures and more abundant winter rainfall (Prentice et al., 1992). Shortly after the glacial maximum about 20,000 years ago, the Mediterranean region began to be

FIGURE 18-11. Shorelines of pluvial Lake Bonneville on the east side of West Mountain, Utah (photo MG5-24, courtesy Mary Gillam).

drier, culminating in maximum aridity by 16,000 to 15,000 years ago in western regions and about 12,000 years ago in the east (Harrison and Digerfeldt, 1993). Thereafter, the climate moderated toward its present state although human activity for at least 4000 years has almost totally deforested the region.

Mediterranean river valleys are characterized by alluvial terraces into which the present stream channels are incised. The alluvium is stratigraphically younger than shorelines of the Eem interglaciation. It forms fans and terraces that were graded to a sea level lower than present, contains pollen of plants that indicate cooler temperature, and in places can be traced upstream into periglacial deposits. All these lines of evidence prove that valley aggradation occurred during the last glaciation, probably during the interval of ice-sheet expansion and falling sea level. Intensified frost weathering on highlands and at least seasonally more intense rain have been claimed as causes of the aggradation (Vita-Finzi, 1969, p. 100). Calcareous tufa deposits around springs suggest correlative higher groundwater levels, also the result of more effective precipitation.

About 9000 years ago, the alluvial fans and floodplains of the Mediterranean coast began to be dissected. The dissection correlates with increased pluvial conditions until about 6000 years ago, when most lakes in the Mediterranean basin were high (Harrison and Digerfeldt, 1993). According to another chronology, based on deep-sea cores from the eastern Mediterranean, the pluvial interval lasted from about

9300 to 4000 years ago (Fontugne et al., 1994). Even as the lower reaches of rivers became estuaries because of the late-glacial sea-level rise, their upstream portions were intrenching as the streams established more gentle gradients. Human intervention in the form of dams and soil-conservation systems became a factor in subsequent valley development. Upland erosion was inhibited by check dams, but by about 1000 years ago, a new cycle of floodplain aggradation had begun to bury classical Roman dams and buildings. Finally, for the last five centuries, erosion has again dominated.

The correlation of aggradational phases in Mediterranean valleys with the cooler times of more effective precipitation during the last glaciation and of the "little ice age" of post-Roman times in Europe contradicts Huntington's principle (p. 244). However, if vegetational cover was incomplete in the Mediterranean basin, owing either to seasonal aridity or human activity, increased effective winter precipitation could cause upland erosion and floodplain aggradation instead of landscape stabilization and valley incision (Figure 15-8). The last pluvial was approximately correlative with glaciation in northern Europe and local periglacial conditions in mountains from Spain to Iran.

Eastern Africa. Radiometric dates on both lacustrine sediments and volcanic rocks in the rift valleys of east-

ern Africa definitely show major late Quaternary climatic changes, but the once-supposed correlation of pluvial with glacial events is now disproved (Butzer et al., 1972; Street-Perrott and Harrison, 1984). Key eastern African lakes were at high levels sometime prior to 21,000 years ago but were at low levels and high salinity until about 10,000 years ago, nearly at the end of the last glaciation. Then, for 2000 years, they were at maximum levels, and several of them overflowed into adjacent tectonic basins (Figure 18-12). In the eastern African lakes, as in other tropical regions, the major pluvial intervals were neither glacial nor interglacial but rather at the transition from one state to the other (Figure 18-13b) (Thomas and Thorp, 1995). A possible explanation may be that during full-glacial time the western Indian Ocean was only slightly colder than it is today, and precipitation on the adjacent eastern African highlands was no more abundant than today's. However, during Termination I, the seasonal monsoon circulation over the Indian Ocean was intensified, and more abundant seasonal moisture reached the adjacent eastern African highlands as well as other regions peripheral to the northern Indian Ocean (COHMAP members, 1988). Nile crocodiles and hippopotamuses expanded their range into lakes and swamps in the present Sahara; shallow lakes were widespread between the great longitudinal dunes of the presently hyperarid Rub' al Khali desert in Arabia (McClure, 1976).

FIGURE 18-12. History of lake levels in eastern and central Africa. All lakes presently lack outlets. (Butzer et al., 1972, Figure 4; © 1972, American Association for the Advancement of Science).

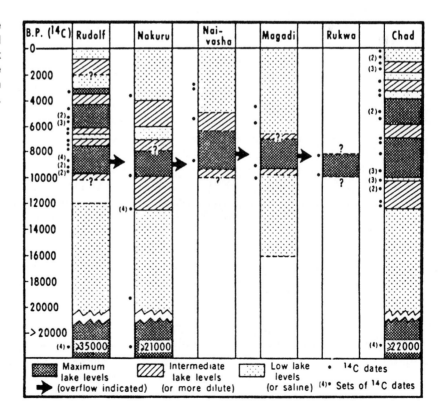

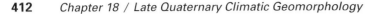

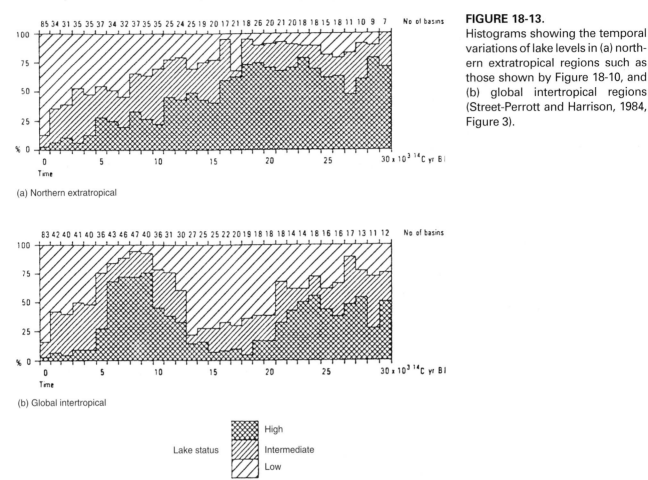

FIGURE 18-13.

Histograms showing the temporal variations of lake levels in (a) northern extratropical regions such as those shown by Figure 18-10, and (b) global intertropical regions (Street-Perrott and Harrison, 1984, Figure 3).

Southern Australia. About 1 million years ago, a long period of fluvial landscape evolution in central Australia ended, and aridity became the dominant climatic factor (Chen and Barton, 1991). Thereafter, including late Quaternary time, effective precipitation has been the dominant factor in Australian landscape evolution rather than temperature changes. Times of more effective precipitation and increased fluvial activity correlate with oxygen-isotope stage 5e, 3, and 1. Dunes mobilized in many regions during oxygen-isotope stages 2 and 4, correlative with intervals of high dust content in the Vostok ice cores (p. 397) (Kershaw and Nanson, 1993). In the arid interior, pluvial intervals are recorded by sheets of sandy alluvium on relict floodplains, and nonpluvial intervals by dune building. A series of pluvial lakes formed in the alluvial plains of the semiarid Murray Basin in southeastern Australia. Sediment cores from some of these lakes have been radiocarbon-dated to give a pluvial chronology for the last 40,000 years, which is broadly similar throughout the region (Bowler et al., 1976, especially Figure 6). The pollen record of vegetation shows that the temperature was 8°C to 10°C colder during the last

pluvial intervals, and precipitation was substantially reduced (Kershaw, 1988, p. 263). Nevertheless, until about 17,000 years ago, the lakes contained more water than now, the effect of a favorable precipitation-evaporation ratio. The time of maximum aridity occurred from about 19,000 to 10,000 years ago, broadly correlative with late glacial time (Edney et al., 1990). Lakes in more arid regions have not held water for more than brief seasonal intervals in Holocene time.

Major changes in fluvial processes in southeastern Australia are correlated with pluvial-nonpluvial conditions. In the last 40,000 years, three cycles of fluvial channel development occurred on the Murray River and its tributaries. Each began with pluvial high discharge and intense aggradation by large rivers and ended with soil formation and smaller channel networks. The oldest cycle is poorly dated. The second of the three pluvial events is dated at between 30,000 and 25,000 years ago. The youngest of the events dates from 25,000 or 20,000 years ago to about 16,000 or 13,000 years ago (Bowler et al., 1976, p. 381; Bowler and Wasson, 1983, p. 187). During similar cyclic pluvials, the Murrumbidgee River channels may have carried a discharge five times that of the modern river (Schumm, 1968, p. 27). Meander wavelengths, meander radii, and channel widths of the ancestral fluvial systems are much larger than the equivalent dimensions of modern streams. Bowler and Wasson (1983, p. 190) suggested that the very wide channels and meander belts that characterized many Australian rivers during the interval between 30,000 and 15,000 years ago may have been due to a high groundwater table inherited from the preceding humid interval. High groundwater pore pressure in stream banks would promote bank failure, rapid meander formation, and wide channels. The inference is that although the pluvial lakes dried up about 17,000 years ago groundwater continued to affect fluvial geomorphic processes for another 2000 years. The modern cycle of channel development has continued for the last 13,000 years, when discharge and channel dimensions have been roughly equivalent to those of the modern rivers.

HOLOCENE LANDSCAPES

The geomorphic examples that have been presented generally support the hypotheses of late Quaternary climatic change approximately as shown by Figures 18-1 to 18-6. There is some indication that rapid geomorphic change occurred during the transitions from glacial to interglacial or pluvial to nonpluvial rather than during the peak of either extreme, especially in the tropics. The time scales of major late Quaternary climatic cycles are long enough to put distinctive morphogenetic imprints on landscapes but far too short to shape any region totally.

There is probably no landscape on earth that does not have relict Pleistocene glacial, periglacial, fluvial, or arid forms lightly overprinted by geomorphic process typical of the current Holocene Epoch. Within the time interval of graded time (Figure 1-6) of centuries to millenniums, many of the adjustments we witness in progress or of which we have historical records are the continuing complex responses to the environmental changes of prior times. Any reasonable extrapolation into the future of graphs such as Figures 4-7 and 18-1 assure us that comparable changes lie ahead, much sooner than the present landscape will equilibrate to the processes now active.

We live in the Holocene Epoch, an interglacial interval that has lasted about 10,000 years, and is similar to 20 or more similar prior times during the last 2 million years that together constitute only 10 percent of the total elapsed time. For most of that time, earth has been in or approaching an ice age (Figures 4-7 and 18-1). In Holocene time, sea level is relatively high, ice sheets are minimal but still extensive, postglacial isostatic recovery is still in progress in deglaciated regions, midlatitude deciduous forests are widespread, aridity shapes the processes of at least one-third of the land area, and the human species has expanded to an alarmingly high population density. In the past 200 years, the human industrial society has almost certainly contributed to a general warming trend by the combustion of fossil fuels and the resulting increased CO_2 content of the atmosphere, with as yet unknown geomorphic consequences.

By comparison to the glacial-interglacial climate fluctuations of the Pleistocene Epoch, the Holocene Epoch may seem mercifully tranquil. It is difficult for modern civilized people to accept the evidence for former rapid environmental changes and not to be lulled into believing that the Holocene Epoch, and especially the last few thousand years of it that are recorded by history, is, "the way things are supposed to be." A systematic study of late Quaternary events should convince everyone that even in this relatively brief time, the earth's climate and landscape have changed enormously and are still changing. Our own appearance as a definable species is part of this changing world, and although we now have the power to modify our environment both intentionally and unintentionally, it will also continue to change regardless of human activity (Dickinson, 1995).

REFERENCES

ALLEY, R.B., MEESE, D.A., and 9 others, 1993, Abrupt increase in Greenland snow accumulation at the end of the Younger Dryas event: Nature, v. 362, pp. 527–529.

ALLISON, I., and PETERSON, J. A., 1989, Glaciers of Irian Jaya, Indonesia, in Williams, R. S., Jr., and Ferrigno, J. G., eds., Satellite image atlas of glacier of the world: U.S. Geol. Survey Prof. Paper 1386-H, pp. 1–23.

BARD, E., ARNOLD, M., and 4 others, 1987, Rapid velocity of the North Atlantic polar front retreat during the last deglaciation determined by ^{14}C accelerator mass spectrometry: Nature, v. 328, pp. 791–794.

BARD, E., HAMELIN, B., and 2 others, 1990, Calibration of the ^{14}C timescale over the past 30,000 years using mass spectrometic U-Th ages from Barbados corals: Nature, v. 345, pp. 405–410.

———, and 5 others, 1996, Deglacial sea-level record from Tahiti corals and the timing of global meltwater discharge: Nature, v. 382, pp. 241–244.

BARNOLA, J. M., RAYNAUD, D., and 2 others, 1987, Vostok ice core provides 160,000-year record of atmospheric CO_2: Nature, v. 329, pp. 108–114.

BASILE, I., GROUSSET, F. E., and 4 others, 1997, Patagonian origin of glacial dust deposited in East Antarctica (Vostok and Dome C) during glacial stages 2, 4, and 6: Earth, Planet. Sci. Ltrs., v. 146, pp. 573–589.

deBEAULIEU, J.-L. and REILLE, M., 1992, The last climatic cycle at La Grande Pile (Vosges, France): A new pollen profile: Quaternary Sci. Rev., v. 11, pp. 431–438.

BENSON, L., BURDETT, J. W., and 4 others, 1996, Climatic and hydrologic oscillations in the Owens Lake basin and adjacent Sierra Nevada, California: Science, v. 274, pp. 746–749.

BENSON, L., and THOMPSON, R. S., 1987, The physical record of lakes in the Great Basin, in Ruddiman, W. F., and Wright, H. E., JR., eds., North America and adjacent oceans during the last deglaciation: Geol. Soc. America, The Geology of North America, v. K-3, pp. 241–260.

BOND, G. C., and LOTTI, R., 1995, Iceberg discharges into the North Atlantic on millennial time scales during the last glaciation: Science, v. 267, pp. 1005–1010.

BÖSE, M., 1994, Ice margins and deglaciation in the Berlin area between Brandenburg and Frankfurt end moraines-a review, in Böse, M., and Kozarski, S., eds., Last ice sheet dynamics and deglaciation in the north European plain: Zeitschr. für Geomorph., Supp. no. 95, pp. 1–6.

BOWLER, J. M., HOPE, G. S., and 3 others, 1976, Late Quaternary climates of Australia and New Guinea: Quaternary Res., v. 6, pp. 359–394.

BOWLER, J. M., and WASSON, R. J., 1983, Glacial age environments of inland Australia, in Vogel, J. C., ed., Late Cenozoic paleoclimates of the southern hemisphere: A.A. Balkema, Rotterdam, pp. 183–208.

BROECKER, W. S., 1995, Chaotic climate: Sci. American, v. 273, no. 5, pp. 62–68.

BROOK, E. J., SOWERS, T., and ORCHARDO, J., 1996, Rapid variations in atmospheric methane concentration during the past 110,000 years: Science, v. 273, pp. 1087–1091.

BROWN, I. M., 1990, Quaternary glaciations of New Guinea: Quaternary Sci. Rev., v. 9, pp. 273–280.

BUTZER, K. W., ISAAC, G. L., and 2 others, 1972, Radiocarbon dating of East African lake levels: Science, v. 175, pp. 1069–1076.

CHAPPELL, J., OMURA, A., and 5 others, 1996, Reconciliation of late Quaternary sea levels derived from coral terraces at Huon Peninsula with deep sea oxygen isotope records: Earth, Planet. Sci. Ltrs., v. 141, pp. 227–236.

CHEN, J. H., EDWARDS, R. L., and WASSERBURG, G. J., 1986, ^{238}U, ^{234}U and ^{232}Th in seawater: Earth, Planet. Sci. Ltrs., v. 80, pp. 241–251.

CHEN, X. Y. and BARTON, C. E., 1991, Onset of aridity and dune-building in central Australia: Sedimentological and magnetostratigraphic evidence from Lake Amadeus: Palaeogeog., Palaeoclimatol., Palaeoecol., v. 84, pp. 55–73.

CLARK, G. M., and CIOLKOSZ, E. J., 1988, Periglacial geomorphology of the Appalachian highlands and interior highlands south of the glacial border-a review: Geomorphology, v. 1, pp. 191–220.

COHMAP members, 1988, Climatic changes of the last 18,000 years: Observations and model simulations: Science, v. 241, pp. 1043–1052.

DANSGAARD, W., JOHNSEN, S. J., and 9 others, 1993, Evidence for general instability of past climate from a 250-kyr ice-core record: Nature, v. 364, pp. 218–220.

DENTON, G. H., and HENDY, C. H., 1994, Younger Dryas age advance of Franz Josef glacier in the Southern Alps of New Zealand: Science, v. 264, pp. 1434–1437.

DICKINSON, W. R., 1995, The times are always changing: The Holocene saga: Geol. Soc. America Bull., v. 107, pp. 1–7.

DITLEVSEN, P. D., SVENSMARK, H., and JOHNSEN, S., 1996, Contrasting atmospheric and climate dynamics of the late-glacial and Holocene periods: Nature, v. 379, pp. 810–812.

DOHRENWEND, J. C., 1987, Basin and Range, in Graf, W. L., ed., Geomorphic systems of North America: Geol. Soc. America Centennial Spec. Vol., v. 2, pp. 303–342.

DONNER, J., 1995, The Quaternary history of Scandinavia: Cambridge Univ. Press, Cambridge, UK, 200 pp.

DREIMANIS, A., and GOLDTHWAIT, R. P., 1973, Wisconsin glaciation in the Huron, Erie, and Ontario lobes, in Black, R. F., Goldthwait, R. P., and Willman, H. B., eds., The Wisconsinan Stage: Geol. Soc. America Mem. 136, pp. 71–106.

DUPHORN, K., GALON, R., and 2 others, 1984, Baltic and North Sea lowlands, in Embleton, C., ed., Geomorphology of Europe: The Macmillan Press Ltd. (published in the United States by Wiley-Interscience, a Division of John Wiley & Sons, Inc., New York), pp. 132–140.

EDNEY, P. A., KERSHAW, A. P., and DeDECKKER, P., 1990, A late Pleistocene and Holocene vegetation environmental record from Lake Wangoom, western plains of Victoria, Australia: Palaeogeog., Palaeoclimatol., Palaeoecol., v. 80, pp. 325–343.

EDWARDS, R. L., CHEN, J. H., and WASSERBURG, G. J., 1987, ^{238}U - ^{234}U - ^{230}Th - ^{232}Th systematics and the precise measurement of time over the past 500,000 years: Earth, Planet. Sci. Ltrs., v. 81, pp. 175–192.

EDWARDS, R. L., BECK, J. W., and 6 others, 1993, A large drop in atmospheric $^{14}C/^{12}C$ and reduced melting in the Younger Dryas, documented with ^{230}Th ages of corals: Science, v. 260, pp. 962–968.

FAIRBANKS, R. G., 1989, A 17,000-year glacio-eustatic sea level record: Influence of glacial melting rates on the Younger Dryas event and deep-ocean circulation: Nature, v. 342, pp. 637–642.

———, 1990, Younger Dryas in Greenland ice cores: Paleoceanography, v. 5, pp. 937–948.

FONTUGNE, M., ARNOLD., M. and 4 others, 1994, Paleoenvironment, sapropel chronology and Nile River discharge during the last 20,000 years as indicated by deep-sea sediment records in the eastern Mediterranean Sea, in Bar-Yosef, O., and Kra, R. S., eds., Late Quaternary chronology and paleoclimates of the eastern Mediterranean: Radiocarbon, Tucson, Arizona, pp. 75–88.

FULTON, R. J., ed., 1989, Quaternary geology of Canada and Greenland: Geol. Soc. America, The geology of North America, v. K-1, 839 pp.

GALLOWAY, R. W., HOPE, G. S., and 2 others, 1973, Late Quaternary glaciation and periglacial phenomena in Australia and New Guinea, in van Zinderen Bakker, E. M., ed., Paleoecology of Africa and the surrounding islands and Antarctica: A.A. Balkema, Cape Town, South Africa, v. 8, pp. 125–138.

GENTHON, C., BARNOLA, J. M., and 6 others, 1987, Vostok ice core: Climatic response to CO_2 and orbital forcing changes over the last climatic cycle: Nature, v. 329, p. 414–418.

GRAYSON, D. K., 1993, The desert's past: A natural prehistory of the Great Basin: Smithsonian Institution Press, Washington DC, 356 pp.

GROOTES, P. M., STUIVER, M., and 3 others, 1993, Comparison of oxygen isotope records from the GISP2 and GRIP Greenland ice cores: Nature, v. 366, pp. 552–554.

GROVE, J. M., 1987, The little ice age: Methuen, London, 498 pp.

GUILCHER, A., 1976, Les côtes à rias de Corée et leur évolution morphologique: Ann. Géographie, v. 85, pp. 641–671.

HARRISON, S. P., and DIGERFELDT, G., 1993, European lakes as palaeohydrological and palaeoclimatic indicators: Quaternary Sci. Rev., v. 12, pp. 233–248.

HOPE, G. S., and PETERSON, J. A., 1975, Glaciation and vegetation in the high New Guinea mountains: New Zealand Royal Soc. Bull. 13, pp. 155–162.

———, 1976, Paleoenvironments, in Hope, G. S., Peterson, J. A., and 2 others, eds., The equatorial glaciers of New Guinea: A.A. Balkema, Rotterdam, pp. 173–205.

HOSTETLER, W. W., GIORGI, F., and 2 others, 1994, Lake-atmosphere feedbacks associated with paleolakes Bonneville and Lahontan: Science, v. 263, pp. 665–668.

HUGHEN, K. A., OVERPECK, J. T., and 2 others, 1996, Rapid climate changes in the tropical Atlantic region during the last deglaciation: Nature, v. 380, pp. 51–54.

JACOBSON, G. L., JR., WEBB, T., III, and GRIMM, E. C., 1987, Patterns and rates of vegetation change during the deglaciation of eastern North America, in Ruddiman, W. F., and Wright, H. E., JR., eds., North America and adjacent oceans during the last deglaciation: Geol. Soc. America, The geology of North America, v. K-3, pp. 277–288.

JOUZEL, J., BARKOV, N. I., and 15 others, 1993, Extending the Vostok ice-core record of paleoclimate to the penultimate glacial period: Nature, v. 364, pp. 407–412.

JOUZEL, J., LORIUS, C., and 5 others, 1987, Vostok ice core: A continuous isotope temperature record over the last climatic cycle (160,000 years): Nature, v. 329, pp. 403–408.

JOUZEL, J., RAISBECK, and 8 others, 1989, A comparison of deep Antarctic ice cores and their implications for climate between 65,000 and 15,000 years ago: Quaternary Res., v. 31, pp. 135–150.

KERSHAW, A. P., 1988, Australasia, in Huntley, B., and Webb, T., III, eds., Vegetation history: Kluwer Academic Publishers, Dordrecht, The Netherlands, pp. 237–306.

———, and NANSON, G. C., 1993, The last full glacial cycle in the Australian region: Global and Planet. Change, v. 7, pp. 1–9.

KOZARSKI, S., ed., 1991, Late Vistulian (=Weichselian) and Holocene eolian phenomena in central and northern Europe: Zeitschr. für Geomorph., Supp. no. 90, 207 pp.

KROMER, B., and BECKER, B., 1993, German oak and pine ^{14}C calibration, 7200-9439 BC: Radiocarbon, v. 35, no. 1, pp. 125–135.

LÖFFLER, E., 1972, Pleistocene glaciation in Papua and New Guinea: Zeitschr. für Geomorph., Supp. no. 13, pp. 32–58.

———, 1977, Geomorphology of Papua New Guinea: Commonwealth Scientific and Industrial Research Organization, Canberra, Australia, 195 pp.

LORIUS, C., BARKOV, N. I., and 4 others, 1988, Antarctic ice core: CO_2 and climatic change over the last climatic cycle: EOS, v. 69, no. 26, pp. 681, 683–684.

LUNDQVIST, J., 1986, Late Weichselian glaciation and deglaciation in Scandinavia, in Sibrava, V., Bowen, D. Q., and Richmond, G. M., eds., Quaternary glaciations in the Northern Hemisphere: Quaternary. Sci. Rev., v. 5, pp. 269–292.

MANGERUD, J., 1991, The last interglacial/glacial cycle in northern Europe, in Shane, L. C. K., and Cushing, E. J., eds., Quaternary landscapes: Univ. Minnesota Press, Minneapolis, pp. 38–75.

MARTINSON, D. G., PISIAS, N. G., and 4 others, 1987, Age dating and the orbital theory of ice ages: Development of a high-resolution 0 to 300,000-year chronostratigraphy: Quaternary Res., v. 27, pp. 1–29.

MAYEWSKI, P. A., MEEKER, L. D., and 6 others, 1993, The atmosphere during the Younger Dryas: Science, v. 261, pp. 195–197.

———, and 12 others, 1994, Changes in atmospheric circulation and ocean ice cover over the North Atlantic during the last 41,000 years: Science, v. 263, pp. 1747–1751.

MCCLURE, H. A., 1976, Radiocarbon chronology of late Quaternary lakes in the Arabian desert: Nature, v. 263, p. 755–756.

MIFFLIN, M. D., and WHEAT, M. M., 1979, Pluvial lakes and estimated pluvial climates of Nevada: Nevada Bur. Mines, Geology, Bull. 94, 57 pp.

MORRISON, R. B., 1991, Quaternary stratigraphic, hydrologic, and climatic history of the Great Basin, with emphasis on Lakes Lahontan, Bonnevile, and Tecopa, in Morrison, R. B., ed., Quaternary nonglacial geology: Conterminous U.S.: Geol. Soc. America, The geology of North America, v. K-2, pp. 283–320.

OVIATT, C. G., 1997, Lake Bonneville fluctuations and global climate change: Geology, v. 25, pp. 155–158.

———, and NASH, W. P., 1989, Late Pleistocene basaltic ash and volcanic eruptions in the Bonneville basin, Utah: Geol. Soc. America Bull., v. 101, pp. 292–303.

PRENTICE, I. C., GUIOT, J., and HARRISON, S. P., 1992, Mediterranean vegetation, lake levels and paleoclimate at the last glacial maximum: Nature, v. 360, pp. 658–660.

ROGNON, P. 1987, Late Quaternary climatic reconstruction for the Maghreb (North Africa): Palaeogeog., Palaeoclimatol., Palaeoecol., v. 58, pp. 11–34.

RUDDIMAN, W. F., and MCINTYRE, A., 1973, Time-transgressive deglacial retreat of polar waters from the North Atlantic: Quaternary Res., v. 3, pp. 117–130.

———, 1981, Oceanic mechanisms for amplification of the 23,000-year ice-volume cycle: Science, v. 212, pp. 617–627.

SCHUMM, S. A., 1968, River adjustment to altered hydrologic regimen-Murrumbidgee River and paleochannels, Australia: U.S. Geol. Survey Prof. Paper 598, 65 pp.

SERET, G., DRICOT, E., and WANARD, G., 1990, Evidence for an early glacial maximum in the French Vosges during the last glacial cycle: Nature, v. 346, pp. 453–456.

SHACKLETON, N. J., 1987, Oxygen isotopes, ice volume and sea level: Quaternary Sci. Rev., v. 6, pp. 183–190.

SMITH, G. I., and STREET-PERROTT, F. A., 1983, Pluvial lakes of the western United States, in Porter, S. C., ed., Late Quaternary environments of the United States, v. 1, The late Pleistocene: Univ. Minnesota Press, Minneapolis, pp. 190–212.

SPAULDING, W. G., LEOPOLD, E. B., and VAN DEVENDER, T. R., 1983, Late Wisconsin paleoecology of the American southwest, in Porter, S. C., ed., Late Quaternary environments of the United States, v. 1, The late Pleistocene: Univ. Minnesota Press, Minneapolis, pp. 259–293.

SZABO, B. J., LUDWIG, K. R., and 2 others, 1994, Thorium-230 ages of corals and duration of the last interglacial sea-level high stand on Oahu, Hawaii: Science, v. 266, pp. 93–96.

STREET-PERROTT, F. A., and HARRISON, S. P., 1984, Temporal variations in lake levels since 30,000 yr BP-An index of the global hydrological cycle, in Hansen, J. E., and Takahashi, T., eds., Climate processes and climate sensitivity: Am. Geophys. Union, Geophys. Mono. 29 (Maurice Ewing vol. 5), pp. 118–129.

TAYLOR, K. C., LAMOREY, G. W., and 6 others, 1993, The "flickering switch" of late Pleistocene climate change: Nature, v. 361, p. 432–436.

TELLER, J. T., and KEHEW, A. E., eds, 1994, Late glacial history of large proglacial lakes and meltwater runoff along the Laurentide ice sheet: Quaternary Sci. Rev., v. 13, nos. 9-10, pp. 795–981.

THOMAS, M. F., and THORP, M. B., 1995, Geomorphic response to rapid climatic and hydrologic change during the late Pleistocene and early Holocene in the humid and sub-humid tropics: Quaternary Sci. Rev., v. 14, pp. 193–207.

VALLADAS, H., GENESTE, J. M., and 2 others, 1986, Thermoluminescence dating of Le Moustier (Dordogne, France): Nature, v. 322, pp. 452–454.

VANDENBERGHE, J., 1991, Changing conditions of aeolian sand deposition during the last deglaciation period, in Kozarski, S., ed., Late Vistulian (=Weichselian) and Holocene eolian phenomena in central and northern Europe: Zeitschr. für Geomorph., Supp. no. 90, pp. 193–207.

———, 1992, Geomorphology and climate of the cool oxygen isotope stage 3 in comparison with the cold stages 2 and 4 in the Netherlands, in Stäblein, G., French, H. M., and Marcus, M. G., eds., Geomorphology and geoecology: Glacial and polar geomorphology: Zeitschr. für Geomorph., Supp. no. 86, pp. 65–75.

VITA-FINZI, C., 1969, The Mediterranean valleys: Geological changes in historical times: Cambridge Univ. Press, Cambridge, UK, 140 pp.

VOSTOK Project Members, 1995, International effort helps decipher mysteries of paleoclimate from Antarctic ice cores: EOS, v. 76, no. 17, pp. 169, 179.

WATERS, M. R., 1989, Late Quaternary lacustrine history and paleoclimatic significance of pluvial Lake Cochise, southeastern Arizona: Quaternary Res., v. 32, pp. 1–11.

WEBB, T., III, BARTLEIN, P. J., and KUTZBACH, J. E., 1987, Climatic change in eastern North America during the past 18,000 years; Comparisons of pollen data with model results, in Ruddiman, W. F., and Wright H. E., Jr., eds., North America and adjacent oceans during the last deglaciation: Geol. Soc. of America, The geology of North America, v. K-3, pp. 447–462.

WOILLARD, G. M., 1978, Grande Pile peat bog: A continuous pollen record for the last 140,000 years: Quaternary Res., v. 9, pp. 1–21.

———, and MOOK, W. G., 1982, Carbon-14 dates at Grande Pile: Correlation of land and sea chronologies: Science, v. 215, pp. 159–161.

PART VI

Coastal Geomorphology

The systematic geomorphology of coasts involves a unique set of processes and landforms encountered neither in the subaerial nor in the deeper submarine landscape. Energy is fed into the coastal geomorphic system primarily by surface ocean waves that can travel hemispheric distances with minimal energy loss, so local climatic conditions are less important than in subaerial processes. That energy is expended over an area that is narrow in height and width relative to its great length. If the relative level of land and sea were fixed, all coastal landforms would be in a zone about 10 m above and below mean sea level and a few kilometers inland and seaward of the average shoreline position.

A peculiar and important aspect of coastal geomorphology is that the narrow vertical operating range of coastal processes ensures that small relative changes of land and sea level will produce relict coastal landforms above or below those that are now forming. Considering the frequency and magnitude of late Cenozoic sea-level fluctuations (Figure 4-7), it is hardly surprising that modern coastal morphology is highly complex. Relict and active landforms are intimately mixed and frequently confused. Subaerial processes interfere with or contribute to coastal geomorphic change as well.

The coastal ocean, including estuaries, is a zone of enormous biologic productivity. In warm tropical oceans, organisms build major shallow-water or coastal landforms. Entire island archipelagos are emerged coral reefs, uplifted by tectonic movements or relict from times of higher Quaternary sea levels. There is no subaerial analog for these huge biogenic constructional landforms.

Within the narrow, almost linear coastal zone, a complex range of landforms are eroded and built, organisms evolve special forms of adaptation, and human activity is focused. By United Nations estimates, two-thirds of the world's people live within 100 km of the coastline (Schröeder, 1993)[1]. Fully 1 billion people and perhaps one-third of the earth's croplands would be at risk of flooding and salt-water intrusion by a sea-level rise of only 60 cm (Brinkman, 1995, p. 235)[1]. Food resources, commerce, and recreation are critically dependent

[1]See References, Chapter 19.

on the dynamic landforms of coasts. Coastal scenery has been the subject of great artistic expression and massive engineering projects.

A major current concern is that the present trend of climate warming and rising sea level could cause serious disruptions to coastal habitation and land use. For the past century, the average rate of sea-level rise has been 1.0 to 2.0 mm/yr. The Intergovernmental Panel on Climate Change (IPCC), a group of experts sponsored by the United Nations, calculated that with a "Business-as-Usual" scenario of extrapolated climate change, by the year 2070 the total rise will be in the range of 21 to 71 cm with a best-estimate mean of 44 cm (Warrick and Oerlemans, 1990, p. 278)[1]. With enforced reductions of human-generated greenhouse gases the projected sea-level rise would be about one-third less. Most of the changes will be due to thermal expansion of seawater and the continued ablation of valley glaciers. The role of the Greenland and Antarctic ice sheets is uncertain. It is possible that with continued warming, the increased accumulation rate of Antarctic ice (p. 397) will have a negative effect on sea-level rise. Still, considering the concentration of human activity in the coastal zone, rising sea level will be a major topic of concern for scientists, engineers, and politicians in the coming decades.

[1]See References, Chapter 19.

Chapter 19

Shore-Zone Processes and Landforms

A **shoreline** is a line of demarcation between land and water. It fluctuates from moment to moment, influenced by waves and tides. The *shore zone*, or simply **shore**, is the zone affected by wave action. Lakes, ponds, and reservoirs, as well as oceans, have shores, but the more general word **coast** applies only to the geographic region adjacent to an oceanic shore. The coastal shore zone is conveniently subdivided into (1) the **offshore**, the shallow bottom seaward of the breaking waves; (2) the **nearshore**, between the breaker zone and low-tide level; (3) the **foreshore**, which extends from low-tide level to the limit of high-tide, storm-wave effects; and (4) the **backshore**, from the limit of recent or frequent storm waves landward to the base of a cliff, dune, or vegetated beach ridge. These definitions are not precise: "Offshore" oil is sought on the continental shelves to a depth of 200 m or more. However, in nautical use, the offshore zone is a place of hazardous navigation in contrast to the "open" sea.

ENERGETICS OF SHORE-ZONE PROCESSES

Wind Waves

Most of the energy expended in the shore zone is transmitted there by wind-generated waves on the water surface (Figure 2-2). It is easy to visualize how a shearing stress by wind over water produces mass transport in the surface water layer, but why the water surface should assume a wave form is not simple to explain. One theory emphasizes the inherent gustiness of wind; another theory explains waves as a result of slight pressure differences over small ripples. The obvious phenomenon of waves being kicked up by a wind is in fact an elusive and complex process (Kinsman, 1965).

In an area of strong winds and storminess over the ocean, the water surface is thrown into a confused mass of waves that intersect in peaks and troughs. Smaller waves and ripples run up the backs of larger waves. Mariners have for centuries called such an area of wave generation a **sea**. When "a sea is running," it is best for ships to avoid the area. Wind-generated waves as high as 34 m have been reported, but accurate observation of such waves is understandably difficult. The height of waves generated by wind depends on wind *speed*, the *duration* of wind from one direction, and the *fetch*, or length of water surface over which the wind blows. For example, a near gale (wind speeds to 17 m/sec) must blow for 6 to 12 hours over a fetch of 40 to 140 km to produce waves 50 to 75 percent as high as the theoretical maximum height of 6 m. Tables are available that predict wave heights for various combinations of wind speed, duration, and fetch. The role of fetch in wave generation explains why even large lakes never have wave heights comparable to those measured in the open ocean.

Radiating outward from the generating area of a sea, waves with the longest wavelengths advance most rapidly. Steep waves, with large height-length ratios,

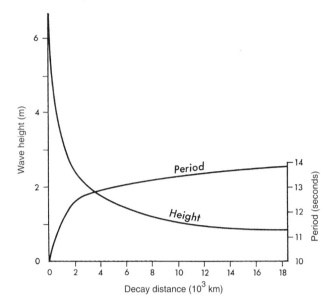

FIGURE 19-1. Effect of swell decay on the wave period and height of a wave initially 6.1 m high with a period of 10 seconds (Davies, 1980, Figure 16).

decay rapidly, but long low waves radiate for thousands of kilometers across oceans with little energy loss (Figure 19-1). The regular pattern of smooth, rounded waves that characterizes the surface of the ocean during fair weather is called **swell**. Swell usually comprises many sets of waves, or *wave trains*, of different wavelengths, moving outward from more than one generating area. From the air, swell patterns look like a grid of intersecting lines. Swell approaches a shore as a complex of wave trains that combine to produce alternately a succession of higher waves and a succession of lower waves.

Swell waves transmit energy outward from a stormy area in two forms: potential and kinetic. The height of a wave determines the potential energy of position above still-water level. The motion of individual water particles as a wave passes is a measure of the kinetic energy of the wave (Figure 19-2). The potential energy of the waves moves forward with the wave form, but the kinetic energy of each moving water particle is expended in the nearly circular orbit of the particle. The orbital diameters of water particles beneath a wave decrease rapidly with depth in a geometric progression related to the wavelength. The orbital diameter is halved for each increase in depth equal to one-ninth of the wavelength. Thus, at a depth equal to the wavelength, the orbital diameter is only about 2^{-9} (1/512 or 0.2 percent) of the surface diameter. This rapid decrease in water motion with depth explains why a submarine can ride out the most severe storm by submerging 50 m or more. A wave is assumed to be

moving in deep water when the water depth exceeds one-half the wavelength. At that depth, the drag of the wave on the bottom is negligible. Swell waves are not deflected by the Coriolis effect (p. 52) because the water particles move in closed orbits, but move across deep water in a direction perpendicular to their crests, along great-circle tracks (Davies, 1980, p. 33). This tendency, plus the persistence of long-period swell waves, means that energy for coastal erosion can be generated by storms in the opposite hemisphere as much as 15,000 km away.

Breakers and Surf

In water depth less than about one-half the wavelength, waves begin to interact with a sloping bottom, and a series of *shoaling* (shallowing) *transformations* take place. Deep-water theory and governing equations for wave motion are replaced by shallow-water ones (Horikawa, 1988, Chap. 3). In particular, wave velocity becomes proportional to the square root of water depth, so waves slow down as they approach the shoreline. The orbital velocity u of a water particle in a surface wave is

$$u = \pi H/T$$

where H is the orbital diameter (wave height) and T is the wave period, the time required for the particle to move through a complete orbit as the wave moves forward one full wavelength. This particle velocity decreases rapidly with depth, as noted in the preceding paragraph. The shoaling seafloor distorts the orbital motion beneath the waves from a circle to an ellipse, then to a to-and-fro linear motion. When the orbital velocity exceeds the threshold of erosion (Figure 10-11a), sediment on the ocean floor is moved back and forth by the waves and absorbs energy from the moving water. At this depth, the conversion of wave energy to geomorphic work begins.

The depth at which waves begin to stir bottom sediment is determined by wave height, wavelength, and period, plus the grain size and sorting of the bottom sediment. Although very long-period swell waves may stir water and bottom sediment to great depths, the rapid decay of the orbital velocity with depth ensures that ordinary waves will not significantly transport or erode bottom sediment at a depth greater than about 10 m. This depth, which is called **wave base**, or **surf base**, is a convenient reference level although we should keep in mind that the work of waves does not abruptly cease at a certain depth.

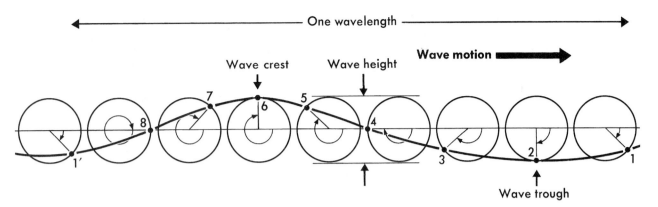

FIGURE 19-2. Motion of surface water particles during the passage of a wave. The orbital diameter decreases rapidly with depth beneath the wave.

When a swell wave begins to "feel" a shoaling bottom and the wavelength shortens and the wave height increases, the wave form steepens. The water mass moves rapidly forward as well as upward. Ultimately, the forward motion of the water mass in the wave equals the decreasing forward motion of the wave front. The wave assumes a steep face, and then its crest collapses forward into the trough. The deep-water swell then becomes **surf**, or lines of **breakers**. Breakers form in the surf zone where the bottom depth is about one-third greater than the breaker height. On gently sloping coasts, the first line of breakers may be as much as a kilometer offshore, and the incoming water may reform into lower waves that break again at successively shallower depths. Wave energy is expended throughout the breaker zone.

Swell waves can be *reflected* by a vertical seawall or steep cliff with almost no transfer of energy from the waves to the structure because the mass of water in each wave is not moving forward significantly more than the diameter of the particle orbit. However, when a breaking wave hits a cliff or seawall, thousands of tons of water are in motion against the structure. The greatest pressures are exerted by breaking waves that curl over at their crest and trap air between the wave face and a steep wall or cliff so that the air is compressed; shock pressures of 608 kPa (12,700 lb/ft²) have been recorded against seawalls. The duration of such great pressure is less than 0.01 s, but large blocks of rock can be pried loose and moved by the repeated assault of the waves.

The wave power expended against the coral reef of Bikini Atoll was carefully calculated by Munk and Sargent (1954). For a typical average breaker height of 2.1 m (7 ft), the waves' total power against the reef is 3.73×10^5 kW (500,000 horsepower). On the sector most exposed to the trade winds, the power reaches 20 kW/m (8 horsepower per linear foot) of reef front. Many coasts have even more energetic environments. Attempts are being made to harness wave energy for electric power generation, but the oscillating nature of the energy is difficult to convert to a useful form.

As a wave train crosses an irregular shoaling coastal bottom, wave *refraction* occurs, analogous to the bending of light rays on passing through a lens. The segment of each wave in the shallowest water is slowed so that in plan view the wave crest becomes concave forward. The effect is best illustrated by constructing a series of *orthogonals* perpendicular to wave crests (Figure 19-3). A shallow submarine ridge causes wave orthogonals to converge, and focuses wave energy against its flanks, whereas a submarine valley or basin causes the orthogonals to diverge and wave crests to be attenuated. Because of the postglacial rise of sea level, most offshore submarine ridges and valleys are continuations of the present subaerial topography, with modern headlands usually fronted by submarine ridges, and bays by submarine valleys. The effect is to cause wave energy to converge on headlands and diverge into bays.

Wave refraction is the basis for two important generalizations about the evolution of coasts. First, an irregular initial shoreline becomes more regular since initial seaward protrusions caused by sloping ridges tend to be eroded faster than adjacent shoreline segments in coves or bays at the heads of submarine valleys. Second, wave refraction generates currents flowing along the shore from headlands where the breakers are focused and the water level is higher to adjacent coves where waves are attenuated and the water level is lower. These wave-generated **longshore** or **littoral currents**, along with tidal and other currents, transport the sediment eroded from headlands into

FIGURE 19-3. Wave refraction over an irregular shoaling bottom. Orthogonals are drawn perpendicular to the wave crests, equally spaced along the swell so that segments 1 to 4 have equal amounts of energy. Note the convergence of energy on the headland and divergence into the coves (segments 1' to 4').

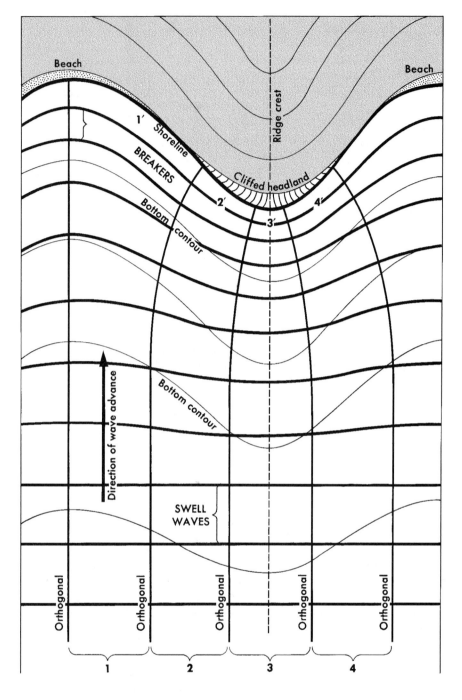

adjacent coves where beaches are built (Figure 19-3). The simplification of a shoreline by refracted waves thus involves the filling of bays as well as the erosion of headlands. Even on straight coasts such as those lined by sandy barriers (p. 437), submerged obstructions, tidal inlets, and other features break the coastal currents into *littoral cells*, within which sand is transported from available sources in the shore zone or nearshore bottom to a variety of sediment traps, where it is deposited. The power expended in the coastal system by wind-driven waves is large enough

(Figure 2-2) that even if only 10 percent of it were converted into longshore currents, those currents could transport 58,000 times more sediment along coasts than is supplied to the world's coasts by all subaerial erosion (Inman and Brush, 1973, p. 26). As with the subaerial landscape, net erosion is the most likely coastal condition, with deposition occurring only in special, sheltered locations. One review asserts that more than 75 percent of the shoreline of the United States is eroding (migrating landward) (Pilkey and Thieler, 1992).

Waves are *diffracted* when a segment of wave crest passes between headlands or artificial constrictions into a sheltered water body such as a river mouth or bay. The truncated ends of the wave segment regenerates laterally by transverse flow of wave energy. The waves and currents generated within bays and harbors erode and transport shore sediment until the bay shoreline "fits" the refracted and diffracted patterns closely. In theory, all shorelines shaped by refracted and diffracted waves should assume gentle curvatures parallel to the wave fronts of the dominant swell. In fact, headlands commonly comprise more resistant rock than the shores of adjacent coves, and even though wave attack is concentrated on the headlands, the more resistant rocks continue to form structurally controlled seaward projections as the entire shoreline retreats (Figure 19-18). Straight shorelines are characteristic only of coasts on terranes of uniform erodibility.

Storm Surges and Seismic Sea Waves

As a cyclonic (low atmospheric pressure) storm approaches a coast, a series of processes combine to endanger coastal installations and promote rapid geomorphic change. A drop of 100 Pa (1 mb) in atmospheric pressure causes the sea surface to rise 1 cm (1 bar $\simeq$ 10 m seawater) by the *inverted barometer* effect. Major cyclonic storms, called **hurricanes** or **typhoons,** fre-

quently have central atmospheric pressure 100 mb below normal, so the general rise of the sea surface under the storm may approach 1 m. Even more important is the shear stress of strong, persistent winds on the sea surface, actually driving masses of water forward and raising mean sea level as much as 10 m along a coast. This *wind setup* combined with the inverted barometer effect produces a **storm surge**, with shore-zone water depths several times as great as the highest normal tides. Waves, by far the most destructive product of such storms, are even more destructive because they move shoreward through water 10 m or more deeper than normal and break against land that is not usually attacked by waves.

Major hurricanes approach the Gulf and southeastern Atlantic coasts of the United States with a peculiar multidecadel variation that has been correlated with droughts in the Sahel region of western Africa (Gray, 1990) (Figure 19-4). Since 1989, the annual frequency has been rising. In 1995, eleven hurricanes developed in the Atlantic Ocean, 235 percent of the normal number (Carlowicz, 1996). Combined with nine hurricanes in 1996, the total for two consecutive years was the highest ever recorded (Avila and Pasch, 1997).

The two most destructive hurricanes in U.S. history were Hugo (South Carolina, September, 1989) and Andrew (Florida and Louisiana, August, 1992).

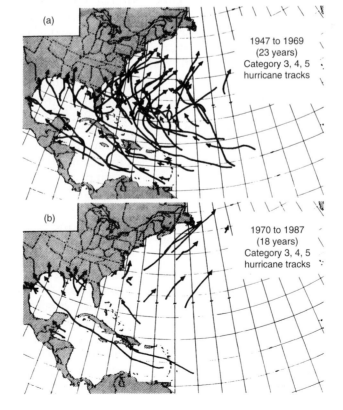

FIGURE 19-4. Comparison of tracks for all major hurricanes for (a) the 23-year period 1947 to 1969 when western Africa was wet, versus (b) the 18-year period 1970 to 1987 when it was dry (Gray, 1990, Figure 4, reprinted with permission. ©1990, American Association for the Advancement of Science).

Although hurricane Hugo caused shoreline retreat of 20 to 29 m by erosion, the beaches in the affected region recovered to their prestorm condition within three and a half years (Sexton, 1995). Hurricane Andrew crossed southern Florida in only four hours on 24-25 August, 1992, as a relatively "tight" storm only 100 km in diameter. The maximum storm surge was 5.2 m. Hurricane Andrew passed directly over the Great Bahama Bank, but had little impact on sand distribution or reef growth (Boss and Neumann, 1993; Major et al., 1996).

The most numerous and severe tropical cyclones (typhoons) are in the western North Pacific region, striking the Philippine Islands, Taiwan, and Japan. In the Indian Ocean, very large cyclones cause enormous death and destruction to the low-lying delta nation of Bangladesh, where a 1970 cyclone killed an estimated 300,000 people. Such storms are tragically effective in shaping the delta of the Ganges-Brahmaputra Rivers (p. 248).

A *seismic sea wave*, often called by the Japanese name **tsunami**, is entirely distinct from wind waves. Sudden motion of the seafloor during an earthquake (Figure 3-4) creates a solitary wave, or at most two or three waves, of very low amplitude that moves rapidly over the deep ocean. Successive arrival times prove that a tsunami moves with a velocity of 700 to 800 km/h, with a wavelength of 200 to 300 km, a height of about 2 m, and a period of about 20 minutes. Because of their extremely long wavelength and low height, tsunamis cannot be detected as they pass through the swell patterns of the open ocean, science fiction novels notwithstanding. As they cross a shoaling bottom, the wavelength shortens and the wave height greatly increases to 15 m or more, with disastrous consequences when it breaks. Tsunamis are not rare in the Pacific Ocean; their recurrence interval for Hawaii is about 25 to 50 years.

Although violently destructive to coastal inhabitants and their buildings, neither cyclonic storms nor tsunamis seem to leave a permanent record of geomorphic change. Their recurrence intervals are comparable to floods of similar magnitude, but their impact is unlike large, infrequent river floods, which may alter a river channel and floodplain to the extent that many years of less intense events do not remove the record (p. 213). Most hurricanes flatten the dunes and other subaerial parts of coastal barriers (p. 437), but the eroded sand is deposited nearby, either on the nearshore bottom or in the lagoon immediately behind the barrier (Stone, et al., 1996).

Tides and Currents

Tides are a unique form of energy for geomorphic change. Power is dissipated by the tidal drag of the moon and sun on the rotating earth at a rate of about 3×10^9 kW (Figure 2-2). The complex periods and heights of tides are driven by lunar and solar gravity but are affected by the size and depth of the various ocean basins, the shape of their shorelines, and the latitudes of the basins, among many other variables (Neumann and Pierson, 1966, pp. 298-325). Seasonal winds and atmospheric pressure patterns also influence the tides. Each ocean, gulf, and sea has its own tidal pattern. The energy of tides is primarily dissipated by shallow seas such as the Bering Sea, the Sea of Okhotsk, the Timor Sea, the Patagonian Shelf, and the Hudson Straits. These five shallow seas alone account for nearly one-third of the total tidal energy conversion (Hendershott, 1973, p. 76). We can justifiably conclude that the coastal and submarine geomorphology of shallow seas are strongly shaped by tidal energy.

Most commonly, two high tides alternate with two low tides each day, separated by a little less than 6 hours. This pattern is known as a *semidiurnal* tide. Some coasts experience only one high and one low tide each day, called a *diurnal* tide. Other patterns are mixed or cyclically variable. Twice monthly, in phase with the new and full moon, are the highest, or *spring* tides. During the lunar quarter phases, tides of lesser amplitude are called *neap* tides.

Tidal ranges are high only where the tides enter a semi-enclosed sea or gulf that has a shape and depth to produce a natural period of water oscillation that is in resonance with some period of the tide. The greatest tidal range on earth occurs in the Bay of Fundy, Canada, where sea level may vary 16 m between high and low water. The Bay of Fundy has a natural period of oscillation of slightly over six hours, almost in phase with the ideal semidiurnal tide (Garrett, 1972).

In addition to their primary geomorphic role of raising and lowering the level of wave attack against the land, tides generate strong currents. Tidal currents compete with waves and river currents in shaping deltas (Figure 11-15). The most rapid currents develop in narrow channels that connect basins with different tidal periods. Hell Gate in the East River at New York City is one such place. The high tide in Long Island Sound takes almost three hours to move westward through the Sound. In that time, the tide in the less restricted lower New York Harbor has crested and is halfway down to low water again. Because of the difference in level, the water in the Sound *ebbs* (flows out) southward through Hell Gate at a speed of 9 km/hr (1 km/hr = 28 cm/s). About six hours later, the tide in the west end of the Sound is low, but the water in New York Harbor is nearing high-tide level, and the current *floods* (flows in) northward into the Sound almost as fast as it previously ebbed. The name of the strait is an obvious reference to the violent and changeable tidal currents.

Where a tide enters a large funnel-shaped river mouth, it begins to drag the bottom like any long wave and steepen its forward slope. In extreme examples, the river mouth develops a **tidal bore**, a single wave that races upstream at 15 to 30 km/hr as a vertical wall of water as much as 3 m in height. Tidal currents and bores are obviously capable of extensive erosion and transportation of sediment (Ludwick, 1974). Their velocities can erode and transport any sediment from compact clay to boulders (Figure 10-11). If abrasive tools are available, deep channels can be eroded in bedrock as well.

Some of the wind-driven surface currents of the oceans (Figure 4-2) impinge on coasts with sufficient velocity to erode and transport sediment. On the east coast of the United States, sediment moves generally southward, carried by waves and currents generated by northeasterly winds. On the coastal deserts of California and Mexico, Morocco, Chile and Peru, Namibia, and western Australia, the cold desert-making ocean currents generally move sediment toward the equator. The persistent strong onshore winds that blow over these cold currents not only create intense deserts, but also strong coastal wave erosion and inland-migrating dune fields (Figures 13-21 through 13-23).

Wind

Wind is the key link in the conversion of solar energy to wave energy (Figure 2-2), but in the shore zone, wind also acts directly as a geomorphic agent. Low tides typically expose extensive areas of wave-transported sand for several hours twice each day. Onshore winds dry the sand and blow it into dunes or transport it inland beyond the reach of waves. On some sandy shores, wind accounts for the major loss of sand from the shore system. If wave energy and sediment sources are adequate, the sand is replaced from offshore or alongshore. If not, beaches can be stripped of sand by wind.

The limited travel distance of saltation-transported sand grains (p. 289) keeps most of the windblown sand within the coastal region. Sand dunes are an inherent component of sandy coasts. They are usually of the transverse or parabolic type (Figure 13-20) because sand supply is renewed by waves, winds are irregular or seasonal, and vegetation restricts dune expansion.

EROSIONAL SHORE-ZONE LANDFORMS

Sea level has risen close to its present level only since about 5000 years ago with the melting of the large ice sheets of the Wisconsin glaciation (Figure 18-4). Considerable uncertainty remains about the last 5000 years,

during which sea level rose a few meters above its present level and then subsided on some coasts, was very nearly stable on others, and has continued a slow rise to the present level on still others. The primary cause for the great variety of coastal Holocene sea-level histories is the complex interaction between the postglacial eustatic rise of sea level of about 120 m between 20,000 and 5000 years ago (Figure 18-4) and the isostatic response of the coastal lithosphere (Clark and Lingle, 1979). Some coasts in formerly glaciated regions have isostatically risen more than the postglacial sea-level rise and show flights of emerged shorelines (Figure 17-14). Other coasts have isostatically subsided under the load of the water that was returned to the oceans by melting ice. Probably no two coastal regions have identical Holocene sea-level histories (Bloom, 1983). The details are complicated by tectonic movements, which have deformed many shorelines (Chapter 19).

On the rocky coast of Maine, the best places to examine glacial striations and other glacial erosional phenomena are in the intertidal zone, where fresh rock exposures are abundant (Figure 17-3). It is a surprising second thought that such a well-known fact also implies a near-total lack of postglacial marine erosion on the Maine coast. In truth, scarcely any coasts on resistant rock show significant Holocene erosional shore-zone landforms. Of those that do so, many have inherited the basic erosional landforms from interglacial ages, the latest of which was about 125,000 years ago, when sea level was a few meters higher than now (p. 403). If an inherited shore platform or wave-cut cliff base has been reoccupied by the sea for even a few thousand years, it is likely to appear deceptively well developed (Figure 19-5).

The Holocene is a particularly complex time to study shore-zone landforms because so many are relict or exhumed. However, enough weak-rock coasts have modern erosional forms, and enough relict forms are preserved, to give us a good idea of what shore-zone erosion could do on resistant rocks if enough time were available (Cotton, 1955).

Shore Platforms

Where the postglacial rise of sea level has created a shoreline on a former hill slope, shore-zone processes cut a cliff and bench. The cliff is controlled by mass-wasting and structural factors. The eroded bench or platform, approximately intertidal, is given the non-genetic descriptive name **shore platform**.

Processes of Shore-Platform Weathering and Erosion. The hydraulic pressure and turbulence of breaking waves can erode easily weathered and fractured

FIGURE 19-5. The chalk cliffs at Sangat, France, are fronted by a prominent shore platform. However, a Paleolithic site on the platform near the low-water line proves that the surface was cut during the last interglacial or earlier, was a periglacial hillside terrace during the last glacial, and has only recently been reoccupied by the sea.

fresh bedrock (Figure 19-6). The initial stage of shore-platform erosion is probably the direct removal of rock by waves, which by analogy with river and glacial work is called *wave plucking* or *quarrying*. Even though wave plucking is restricted to a narrow vertical zone, high cliffs can result if sliderock is exported from the talus at the cliff base. **Sea caves** or a **notch** may be cut at the base of a cliff.

As plucking deepens a shore-zone notch or erodes back a sea cliff, the shore platform widens. It is in the surf zone, and sediment from wave plucking and cliff retreat is dragged back and forth across the platform to shape an **abrasion ramp** (Figure 19-7). Wave abrasion requires constant removal of plucked or mass-wasted sediment from the abrasion ramp lest the surface be buried. Thus a seaward *slope of transportation* develops on an eroding surface thinly veneered with rock debris in active transport. The abrasion ramp extends seaward until the debris is either abraded to fine sediment, dissolved, or settles into water too deep for further wave movement. From measured profiles across modern abrasion ramps and emerged coastal terraces in California, Bradley (1958) concluded that their typical seaward gradient is about 1° (about 2 percent). On a tideless shore, such a gradient would extend below the 10-m depth of ordinary wave erosion on a bench no wider than 500 m. If a tide range of 5 m is assumed, an abrasion ramp with a 1° gradient could still be no wider than 800 m. Wider shore platforms cannot develop by abrasion at a fixed sea level although they could form during progressive submergence.

Abrasion platforms can form rapidly on weak rocks. A subhorizontal shore platform up to 200 m in width was cut in 600 years across erodible mudstones on the east coast of New Zealand during slow tectonic subsidence. Then it was abruptly uplifted, probably during an earthquake about 2200 years ago. The abrupt emergence preserved the fossiliferous cover strata on the platform. Radiocarbon dates on mollusk shells in the cover strata prove that the platform was youngest at the base of the fossil cliff at its landward edge, and had progressively widened at a rate of 0.2 to 0.4 m/yr (Hull, 1987).

Older geomorphic literature proposed continental-scale marine planation as an alternative to subaerial peneplanation. However, the necessity for a seaward slope on an abrasion ramp ensures that even a small island could not be planed off to sea level by abrasion alone. Wide shore platforms, or those with gentle gradients, must be cut by processes other than abrasion (Bradley and Griggs, 1976).

Water-level weathering (Wentworth, 1938) seems to be the dominant process that widens and flattens shore platforms on noncarbonate rocks. Intertidal pools and salt spray on abrasion ramps or in wave-plucked notches promote weathering in the zone just above permanent saturation. The spray zone, just above normal high tides, is a zone of especially active chemical weathering with measured surface lowering of 0.625 mm/yr in southern England (Mottershead, 1989). Tafoni (honeycomb weathering, p. 125) form rapidly in the zone of spray and drying (Matsukura and Mat-

FIGURE 19-6.
Betty Moody's Cave, Star Island, New Hampshire. Waves have quarried a columnar basalt dike from between granite walls.

suoka, 1991; Mottershead and Pye, 1994). Lower and higher zones, which remain either wet or dry for longer intervals, are less affected. Wave erosion during high tides carries off the weathered detritus (Figure 19-8). On exposed platforms, all stages can be seen from initial ponding in joint-bounded basins to coalescing flat-floored pools separated by serrate ridges. Bedding planes and joints are frequently etched in high relief.

Shore platforms are developed by water-level weathering at various heights relative to tide level, depending on structural factors such as permeability and fractures and also on wave energy, tidal range, and climate. On Hawaiian coasts exposed to vigorous surf, platforms develop as much as 7 m above sea level by weathering at the edges of spray-filled pools although better developed platforms are at 3 to 4 m (Wentworth, 1938, p. 20; Bryan and Stephens, 1993). These high platforms on headlands

slope landward into bays, where they are only about 1 m above sea level. Rather than cutting the platform, waves keep it wet and protected from salt weathering. Such a range in contemporary water-level weathering platforms warns of the danger in attempting to establish the heights of ancient sea levels from the evidence of relict shore platforms and abrasion ramps.

Most modern water-level weathering platforms are probably inherited from relict shore platforms that were slightly higher. Old abrasion ramps or other shore platforms, eroded during the last interglacial age when sea level was a few meters above present level, are ideal surfaces on which water-level weathering and solution can operate. Because there is no definite tidal level for the modern forms, it is not possible to assert by altitude alone whether a present-day shore platform is totally relict, an inherited form modified by present processes, or entirely modern. Only where platforms are being exhumed from cover deposits of the last glacial age (Figure 19-5), or where they grade into constructional reefs that can be radiometrically dated as of the last interglacial age, can their relict character be demonstrated.

Water-level weathering puts the finishing touches on surfaces prepared by wave abrasion and plucking, especially on volcanic rocks such as weakly indurated pyroclastic sediments. The abundance and high perfection of shore platforms around the islands of the peripheral archipelagos of the Pacific Ocean are in part related to the abundance of Cenozoic volcanic rocks in those regions.

On carbonate rocks in the shore zone, solution is so common that a special class of **solution platforms** is distinguished (Wentworth, 1939). These are the most complex shore platforms to describe because their origin involves both chemical and biologic solution and deposition. Solution platforms are generally intertidal. Their surfaces are flat and smooth but not abraded (Figure 19-9). Small pools alternate with low domes and ridges, often formed by living carbonate-secreting algae. At the landward edge of a solution platform may be a horizontal *solution notch* as deep as 3 m. Because seawater is generally saturated with calcium bicarbonate, it has been argued that a solution platform must form in the zone of mixing with fresh or brackish groundwater (Wicks et al., 1995). However, good solution platforms are also common on desert coral islands that have no fresh groundwater. Diurnal changes in the pH of intertidal pools by photosynthesis and temperature change offers one explanation for alternating solution and deposition by seawater (Revelle and Emery, 1957). Another hypothesis is based on observations that sea spray changes in chemical composition as it blows inland. Drops of various sizes evaporate and precipitate various salts, and the remaining spray

FIGURE 19-7. Abrasion ramp cut across dipping strata, Bay of Fundy coast near Joggins, Nova Scotia.

becomes increasingly different from seawater. Recombination of the residual water and the salts creates the mixing effect (p. 150) that promotes solution and diagenesis (McLaren, 1995).

Even more effective but complex is the role of biochemical weathering and erosion in developing solution platforms and notches. Many intertidal plants, especially algae, penetrate their substrate with microscopic strands of organic tissue (p. 133). *Lithophagic* (rock-destroying) animals graze on the limestone surface and abrade or dissolve the perforated rock surface to obtain the algae. Animals from at least 12 phyla, and numerous kinds of plants and microbes, will graze, browse, burrow, or bore into rocks (Warme, 1975; Hopley, 1982, pp. 88–94; Hallock, 1988). Many use chemical secretions to dissolve rocks, especially limestone. Many of the animals have abrasive appendages or teeth by which they remove surface layers of rocks or bore into them. The net chemical and mechanical effect is rapid **bioerosion**, as much as 1 mm/yr by repeated

(a)

(b)

FIGURE 19-8. "Tesselated pavement," an intertidal platform shaped by water-level weathering on horizontal, jointed sandstone south of Hobart, Tasmania. Blocks have either raised rims (a), or raised centers (b).

FIGURE 19-9. Solution platform almost truncating residual karst blocks of calcareous eolianite near Robe, South Australia.

measurements on the exposed parts of pins driven into rock (Hodgkin, 1964). Boring sponges have been reported to erode 23 kg/m²yr from reef limestone. Biochemical solution of 6.5 kg/m²yr was measured on reef limestone in Hawaii (Hallock, 1988, p. 280). A limestone surface would be lowered at rates of 2.5 to 8.5 mm/yr by such processes. Comparable subtidal bioerosion rates of 2 to 10 mm/yr have been estimated for calcareous mudstones in California (Warme and Marshall, 1969).

Considering the complex nature of the processes that shape them, solution platforms may be misnamed. However, the complexity of intertidal carbonate geochemistry is so great and so poorly understood that "solution" is an adequate word for the present. On coral reef coasts, the geomorphology is even more complicated because solution platforms of the sort described here grade seaward into constructional reef flats and shoreward into abrasion ramps. In the low tidal range common on tropical coral islands, the entire complex may form a flat surface more than a kilometer in width with a relief of only 1 to 2 m.

Theoretical Equilibrium Form of Shore Platforms. When a platform assumes a shape on which wave energy is expended at a uniform rate across it, the platform will have an equilibrium profile that can migrate shoreward but retain the same form. Zenkovich (1967, pp. 151–153) deduced that the stable profile of an abrasion ramp would be one on which the "effective" wave, or the wave form that combines optimum frequency with optimum abrasive energy, would not break. The

profile should be convex upward, with no horizontal segments and the minimum slope at the shoreline. The deduced profile should be longer and more gently sloping for larger effective waves. Some shore platforms in the Black Sea may approach the ideal form, but only because they are eroded on limestone by abrasive siliceous beach sediment. The lack of other examples testifies to the general disequilibrium of the shore zone in relation to present sea level. Computer models give useful predictions of the shapes of late Quaternary shore platforms that might have evolved during repeated glacial-eustatic sea-level fluctuations (Trenhaile, 1989).

Coastal Cliffs

The cliffs that rise from the back of shore platforms are commonly called *wave-cut* cliffs, but, as with any subaerial escarpment, mass-wasting is the dominant process of cliff retreat. Wave quarrying at the cliff base keeps it fresh at the appropriate angle of repose. With stable sea level, abrasion ramps would eventually widen until wave attack became infrequent on the cliff face, when it would become a vegetated, subdued hillside. The many bold, cliffed coasts of the world are another reminder of the essential disequilibrium between coastal landforms and present sea level.

On some cliffed coasts, the cliffs are so steep that wave-plucked debris falls into water too deep to develop a wave-cut notch or abrasion ramp (Figure 19-10, Color Plate 6). Steep coasts are especially common on volcanic terranes, fault scarps, and glaciated coasts. The cliffs are said to *plunge* into deep water (Cotton,

1952). A striking feature of plunging cliffs is that swell waves do not break against them but are reflected back out to sea (p. 421). If a rock mass is strong enough to form a steep cliff, and if that cliff happens to rise from water more than about 10 m deep, it is subject to only minor wave attack. Only when a submarine talus forms at the base of a cliff to create breakers can the full energy of waves be impressed on the cliff.

Laboratory wave-tank experiments using a weak sand and cement mixture to simulate a sea cliff and an oscillating paddle to create waves 7 to 9 cm in height have proved useful in documenting cliff and abrasion platform evolution (Sunamura, 1975, 1977). With the original cliff face sloping at 75°, experiments were run with (1) standing waves in "deep" water (20 cm), (2) breaking waves at the cliff face, and (3) broken waves that had reformed after breaking in shallower water at some distance in front of the cliff face. In the first experiment, deep-water standing waves, approximately swell waves, were simply reflected from the artificial cliff with negligible erosion. The water rose and fell against the cliff face, and the wave train was reflected back through the incoming waves. Waves breaking at the cliff face were the most effective erosive agents, cutting a notch as deep as 40 cm in about 60 hours. With broken waves, the erosional effects were similar but only about half as fast. With both breaking and broken waves, the erosion rate was not uniform (Figure 19-11). Initial slow erosion freed sand grains from the weak cement, which then became abrasional tools for an interval of accelerated erosion. However, as additional sand accumulated in front of notch, wave energy was dissipated in the sand, and the rate of notch deepening decreased (Sunamura, 1977). The offshore depositional profile seemed to approach a convex-upward equilibrium shape as erosion progressed. In the laboratory experiments, only a basal notch was cut at the model cliff base. In most real sea cliffs, joints and other discontinuities in the rock mass would cause the entire cliff to retreat, providing more abrasive sediment, but also increasing the possibility of sediment buffering of wave energy.

Structural Control of Shore-Zone Landforms

Factors of rock-mass strength, lithology, and solubility have been repeatedly mentioned as contributors to shore-zone erosional morphology. As in the subaerial environment, some rock masses are much more resistant to coastal weathering and erosion than others. Resistant rocks on coasts tend to protrude seaward as well as to maintain higher elevations. Even though wave refraction focuses more energy onto headlands of

FIGURE 19-10. Plunging cliffs on the drowned margins of a basalt dome, Banks Peninsula, Canterbury, New Zealand (photo: D. L. Homer, New Zealand Geological Survey).

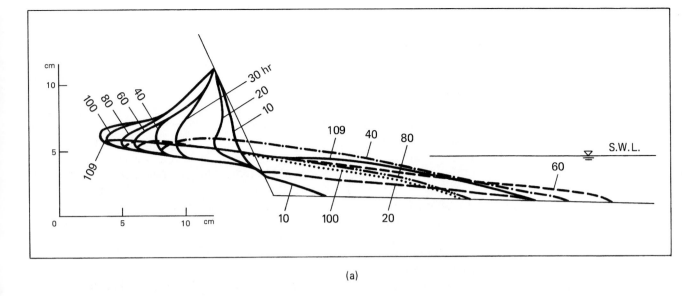

(a)

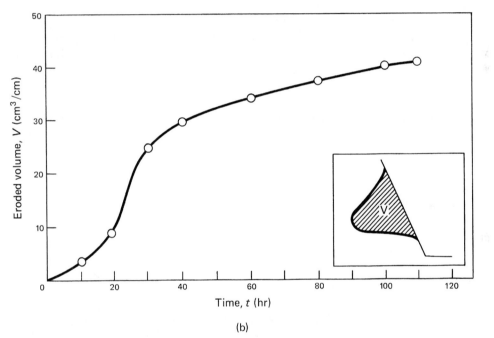

(b)

FIGURE 19-11. (a) Cliff erosion and beach development by broken waves in a wave tank. Time in hours. (b) Temporal variation of eroded volume in the experiment illustrated above. See text for discussion (Sunamura, 1977, Figures 1 and 2).

resistant rock, they continue to form headlands as erosion progresses. Wave refraction around a protruding headland can be so extreme that only a narrow neck of land is left, connecting a peninsula to the mainland. In suitable structures, especially flat-lying sedimentary rocks, wave erosion may quarry through the neck, creating a **natural arch**. With further erosion, arches collapse leaving isolated pinnacles or **sea stacks** standing on the shore platform (Figure 19-12).

Shore platforms on nearly flat-lying sedimentary rocks are especially difficult to interpret. It was noted that water-level weathering takes advantage of pre-existing benches. In horizontal strata, structural benches readily form and may hold enough water to promote water-level weathering at several different intertidal and supertidal levels simultaneously. Even more commonly, slight structural dips are eliminated by water-level weathering to form subtly benched

FIGURE 19-12. The Twelve Apostles, sea stacks eroded from flat-lying Miocene limestone, Port Campbell national Park, Victoria, Australia (photo: J. F. Bliney, contributed by B. G. Thom).

platforms that truncate structure but are nevertheless basically controlled by structure, not by shore-zone processes.

CONSTRUCTIONAL SHORE-ZONE LANDFORMS

Organic Reefs

A reef is by mariners' tradition an obstruction or menace to navigation. By definition then, **reefs** are shallow-water submarine landforms. The term is usually applied to solid, rocky structures rather than to sand bars. Most reefs are constructed by marine organisms although some are structurally controlled, shallow, submerged ledges such as hogbacks, dikes, or lava flows.

Ecology of Organic Reefs.   Throughout Phanerozoic time, since organisms with hard parts evolved, reef building has been a repeated and persistent form of marine growth, especially for colonial organisms. In shallow oceans of the Cenozoic Era, the Scleractinian colonial corals and associated calcareous algae are the most important builders of organic reefs although molluscs, sponges, and a variety of other organisms build reefs in special situations. Reefs are built up in shallow water to take advantage of sunlight for photosynthesis, to shed or stay above smothering detrital mud, and to provide a large surface area for continued growth.

Waves breaking over and refracting around reefs create currents that sweep nutrients and oxygenated water over the sedentary colonial organisms. Reefs are usually well adapted to local wave and tidal energy so that they are nourished by the energy expenditure of waves and currents rather than being destroyed by them. Larvae have a free-swimming stage before they settle and attach to the bottom, so reef-builders disperse widely.

Coral and coralline algae reefs are widespread in tropical waters (Guilcher, 1988). They require clear and shallow water (less than 70 m but with their best growth in less than 10 m) for photosynthesis. Reef growth is best in water that does not cool below about 18°C during winter months. Salinity should be close to normal. Fresh water from rivers or torrential rains is especially damaging. A hard substrate is necessary for a colony to become established, but small patch reefs can grow on soft sediments and gradually fuse to form a suitable base. Loose sand is poor for coral growth, as is mud. Despite these restrictions, the coastal area of coral reefs is very large (Figure 19-13). One of the most fundamental subdivisions of coasts could be between those that have coral reefs and all others (Davies, 1980, p. 5). Organic reefs other than coral are trivial landforms and are subsequently ignored.

Coral Reef Morphology.   A coral reef is an enormous mass of limestone that is organically built. The seaward edge of the reef, especially on a windward coast, is

FIGURE 19-13. Distribution of Holocene coral reefs. Many form only thin veneers on older structures. Shaded region includes the areas of most prolific reef growth and almost all the atolls (Davies, 1980, Figure 45).

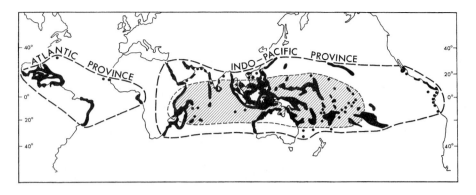

where wave-resistant growth forms thrive in the most nutrient-rich environment and provide coralline detritus in the largest quantities. Some detritus breaks off and forms a **fore-reef talus** on the ocean side of the reef. Other, commonly finer, detritus is swept across the reef into the sheltered region behind it. In the surf zone on the extreme seaward rim of many Indo-Pacific reefs, calcareous algae build smooth, rounded mounds or rims. The algal limestone is more dense than coral limestone and is a vital contributor to reef stability. More delicate coral species grow in the lee of the seaward reef, adding more detrital sediment. Either a sandy lagoon floor or an intertidal **reef flat** may extend from the lee of the reef (Color Plate 14).

Coral reefs are generally limited in upward growth to mean low-tide level. The extreme productivity of the reef provides sediment for many subaerial shore-zone landforms, however. Large storms and tsunamis tear away tons of reef-front coralline debris and hurl it over the algal rim onto the reef flat behind (Maragos et al., 1973). Most reef flats are strewn with large blocks of storm-tossed coral rock, often deeply notched by solution. On the reef flat, islands of coralline sand and

gravel may build up high enough to support a fresh groundwater lens and vegetation. The carbonate sand and gravel is readily recemented, especially if exposed to rain and fresh groundwater. Major storms can remove any uncemented part of a coral-reef island, but if a mass of debris is not disturbed for a few decades or centuries, it becomes relatively indurated and resistant to all but the worst storm waves. Cementation is generally close to the level of mean high tide in the zone of saturation. Because the tide range in tropical oceans is generally less than 2 meters, low platforms of cemented reef rubble are exposed on most eroding shores (Figure 19-14). These cemented high-tide platforms have been frequently mistaken for emerged reef flats by which a Holocene fall of sea level has been widely inferred.

Classification of Coral Reefs. The traditional mariners' terms for reefs included **fringing reef**, which is attached to and extended seaward from a coast, and **barrier reef**, which is separated from the coast by a lagoon. Later, the word **atoll**, from the Maldive Islands, was adopted into the scientific literature for an annular reef that encloses a central lagoon (Figure 19-15). Darwin

FIGURE 19-14. High-tide platform of recemented coral debris about 1.5 m above modern reef flat, Turtle Islands, northern Queensland, Australia.

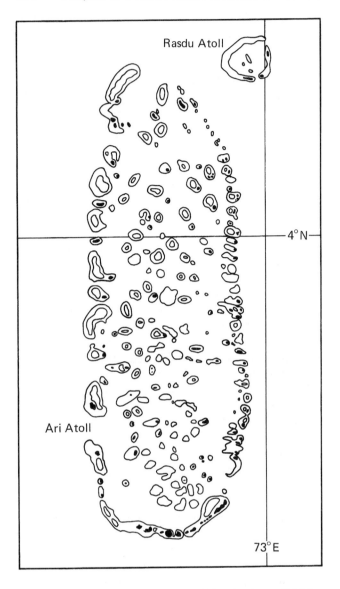

FIGURE 19-15.   Ari Atoll, Maldive Islands. An atoll 90 km long made of numerous smaller, elongate atolls on its rim (*faros*). Land areas (in black) are very small.

(1837) conceived of these three reef forms as a genetic sequence by which a fringing reef around a submerging central island would progressively become a barrier reef and then an atoll. Darwin's deduction required subsidence of the ocean floor over vast regions (Figure 1-4) and sparked a spirited scientific controversy in the late nineteenth century. A series of deep boreholes on atolls have subsequently demonstrated that they are underlain by 1000 m or more of shallow-water reef limestone on basaltic basement, generally validating Darwin's theory for the origin of atolls in the deep ocean basins.

Darwin's theory does not explain **shelf atolls**, the similar forms on shallow submerged continental shelves. Fairbridge (1950) suggested that shelf atolls could form by crescentic reefs extending their "horns" downwind until they closed into a ring. A simple genetic classification of coral reefs is not possible because most of the shape is probably inherited from alternating periods of growth and subaerial erosion during Quaternary sea-level oscillations. A simple descriptive classification is given by Table 19-1.

Inheritance and Reef Morphology.   Coral reefs are very strictly controlled by sea-level datum. Emergence or submergence of even a few meters completely displaces all the depositional and erosional zones of the reef and creates relict forms. Quaternary sea-level fluctuations (Figures 18-1 and 18-3) have left a particularly complex record on coral reefs (Hopley, 1982; Guilcher, 1988). When sea level was low, reefs were exposed as flat-topped or hill-rimmed islands that were subject to karst solution and peripheral wave erosion or reef growth. Drowned karst topography (p. 154) is well shown on the floors of some reef lagoons (Shepard, 1970; Shinn et al., 1996). It is hard to avoid the conclusion that most reef morphology is inherited, in part from constructional forms that date from the last interglacial when sea level was several meters higher than

TABLE 19-1.

Descriptive Scheme for Classifying Organic Reefs

Adjacent to coasts or separated by a shallow channel				Fringing reefs
Separated from coast by deep channel				Barrier reefs
		without central lagoon	large	Platform reefs
			small	Patch reefs
Forming islands			very small	Coral pinnacles
	with central lagoon	deep lagoon		Oceanic atolls
		shallow lagoon	large	Shelf atolls
			small	Faros[a]

Source: Davies (1980), p. 72.

[a]Small, elongate atolls with narrow lagoons along the rim of large reefs (Fig. 19-15). Their origin may involve multiple periods of growth separated by subaerial karst development (MacNeil, 1954).

now, and in part from destructional karst landforms that formed during glacial-age low sea levels (Bloom, 1974; Purdy, 1974; Strecker et al., 1987). Until sophisticated radiometric dating techniques became available, there was no way to evaluate, or often even to detect, the glacial-age disconformities in coral reef stratigraphy. On a longer time scale, the Great Barrier Reef of eastern Australia, which extends almost 2000 km from 9° to 24° S latitude, began to grow in its northern part between 16 and 25 million years ago. The central part began to grow 10 to 15 million years ago, and the southern part only a few million years ago (Davies et al., 1987). The age anomalies relate to the northward movement of the Indo-Australian lithospheric plate, as northeastern Australia moved progressively into the tropics throughout the late Cenozoic era (p. 58).

Sandy Shores

Beaches. If some wave energy were not absorbed in the work of moving sediment along coasts, far more erosional work would be accomplished. The sediment in motion along a shore is the **beach**. Its role is analogous to floodplain alluvium in river valleys; it absorbs energy and moves during storms and thereby stabilizes the rate of wave-energy conversion into geomorphic work. A continuous beach along a coast is as good a criterion of the graded condition as is a continuous floodplain on the floor of a river valley or a continuous sheet of sediment on a semiarid pediment.

The postglacial rise of sea level has submerged a variety of sediments, including glacial drift, sand dunes, older marine deposits, and alluvium. As waves move shoreward over the loose sediments, silt and clay are stirred into suspension and carried long distances, usually coming to rest either on tidal mud flats or in water too deep for further wave action. On a shoaling bottom, sand sizes are the first to be stirred by waves (Figure 10-11), but the sand is moved by saltation, not in suspension, so it stays within the zone of wave action. A **null line**, or *null point* on a bottom profile, can be defined seaward of which sand is either not moved or moved seaward. Landward of the null line, there is a net shoreward movement of sand by waves (Komar, 1976, pp. 310–314). The rate of landward sand movement is a function of grain size and wave energy. Nearshore sand becomes very well sorted as it is moved shoreward, permitting the sandy bottom to respond to small changes in wave energy. Gravel beaches are also common but move only at much higher energy levels.

Although the null line and the beginning of net onshore sand transport are well seaward of the breaker

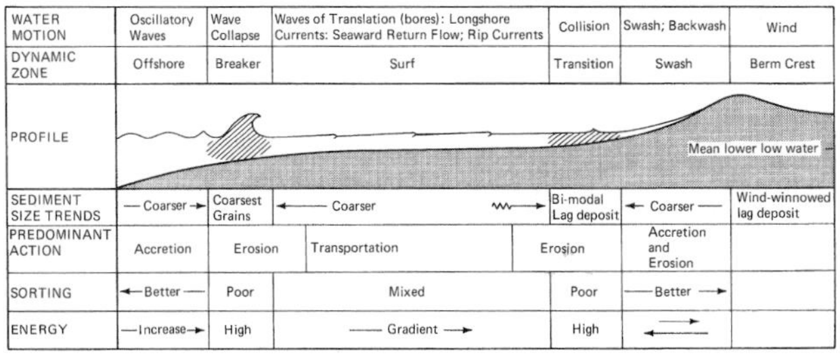

WATER MOTION	Oscillatory Waves	Wave Collapse	Waves of Translation (bores): Longshore Currents: Seaward Return Flow; Rip Currents	Collision	Swash; Backwash	Wind
DYNAMIC ZONE	Offshore	Breaker	Surf	Transition	Swash	Berm Crest
PROFILE					Mean lower low water	
SEDIMENT SIZE TRENDS	← Coarser →	Coarsest Grains	← Coarser →	~~~	Bi-modal Lag deposit	← Coarser →	Wind-winnowed lag deposit
PREDOMINANT ACTION	Accretion	Erosion	Transportation	Erosion	Accretion and Erosion	
SORTING	← Better →	Poor	Mixed	Poor	← Better →	
ENERGY	← Increase →	High	← Gradient →	High	← →	

FIGURE 19-16. Major dynamic zones in the nearshore zone of a sandy coast. Two regions of maximum sand suspension and transport are shaded (Ingle, 1966, Figure 116).

zone, breakers cause the most dramatic sediment transport. Sand is thrown into suspension by each breaking wave, and before it can settle to the bottom again, it can be moved many meters shoreward or laterally by turbulent currents (Nielsen, 1992). In cross profile (Figure 19-16), two zones of erosion are seen; the most important zone is under the breakers, but another zone of intense sand erosion is in the **swash** zone, where small waves run up the beach and collide with backwash. Between these two eroding zones, when the waves are approaching at an angle, one vector of the water motion is parallel to the shoreline and a powerful **littoral current** is generated. When waves break nearly parallel to the shore, only weak littoral currents are produced, and sand movement in the surf zone is nearly perpendicular to the beach. With a breaker angle and height that generate a littoral current of 30 cm/s, sand moves along the shore but with an outward component toward the breaker line. At littoral current velocities of 80 to 100 cm/s, a strong movement of sand parallel to the shore occurs, like a shallow alluvial river with one bank on the beach and the other in the breaker zone (Ingle, 1966). Within the surf zone, a ribbon of sand 100 m or more in width and 5 to 10 cm in thickness moves at a rate of 1000 m^3/day or more (Das, 1971). Under such dynamic conditions, sandy beaches can adjust to changing wave and tidal energy within hours, and the nearshore profile is constantly regrading to changing conditions.

Most sandy shores can be divided into distinct *littoral cells* ranging in length from about 1 to 100 km, each defined by a sand source, a transportational segment, and a sand sink where sediment is removed from the system. Sources include new sand added to the littoral system from offshore, from river mouths, and from eroding sea cliffs. Sinks are places where sand is lost from the system by wind transport inland, by tidal-current deposition in sheltered lagoons, by *rip currents* into deep water, and rarely by being trapped in the heads of submarine canyons that lead it out to deep-sea basins (Inman and Brush, 1973).

Wave refraction causes mobile beaches to change their orientation readily until the plan view of the shoreline conforms closely to the shape of the refracted wave fronts. On coasts with alternating headlands and beaches and a dominant swell that approaches obliquely, beaches assume a "half-heart," or **zetaform**, curvature (Figure 19-17). The outline of these head-land–bay beaches has been shown to approximate a logarithmic spiral (Yasso, 1965; Davies, 1980). Each zetaform beach segment seems to be offset relative to its neighbors with the straight segment of beach rotated from the mean trend of the coast toward the dominant swell direction. Grain size, sorting, and beach profiles all change systematically along each zetaform beach.

FIGURE 19-17. Zetaform coast of northern New South Wales, Australia. Longest beaches face the dominant swell from the southeast and trend obliquely to a line connecting the headlands (photo: courtesy B. G. Thom).

Each is nearly a closed littoral cell of sand supply and movement although some sand probably drifts around the downwave headland to nourish the next beach. The pattern of offset zetaform beaches dominates high-altitude photographs and satellite images of many coasts (Bloom, 1986) and is easily recognizable on coastal maps and charts.

Barriers. On many coasts, particularity on trailing continental margins (p. 249), the postglacial submergence was across plains of low relief and abundant sediment supply. During the later stages of the submergence, the marine transgression created growing ridges of sand that eventually become too massive to be moved further shoreward. The sea flooded the area behind them to form **lagoons**, but the massive sand ridges extend for hundreds of kilometers along many coasts as **barriers**, only rarely tied to headlands (Hoyt, 1967, 1968). Barriers are either *islands* or *spits*, depending on whether or not they connect to the mainland. Barriers and lagoons are said to extend along 15 percent of the world's coasts (Davis, 1994, p. 1). Approximately 17 percent of the coastline of the United States is along barriers and lagoons (Cronin, 1986, p. 33).

The origin of barriers was one of the durable controversies of geomorphology. As long as the submarine offshore topography was poorly known, and shore-zone sediment transport unstudied, barriers were thought to represent submarine sand bars that had grown very large and then emerged by tectonic uplift or a drop of sea level. Some authorities argued that wave action actually built barriers above sea level, aided by high storm levels and wind transport. Others maintained that barriers were built along shallow coasts by littoral drift and were not driven landward but grew in place.

The identification of barriers with coastal emergence has now been thoroughly disproved. Numerous boreholes in barriers and their associated lagoons prove that many were built in the last 6000 years, during the final phase of postglacial submergence (Kraft and Chrzastowski, 1985). Most of them rest on older foundations, perhaps dating from the last interglacial interval (Otvos, 1995). As with any beach, barriers are supplied with sand both from offshore and by littoral drifting. Many seem to have grown more rapidly prior to a few thousand years ago and are now eroding or being driven landward (Thom, 1984). Possibly sand supply was more abundant as the sea steadily transgressed new terrain, whereas with nearly stable sea level today, sand is being lost into dunes, lagoons, and the offshore zone, and is not being replaced. It is

possible that the widespread barrier systems of the world's coasts are simply transient mid-Holocene landforms, not likely to endure (Davis, 1994, p. 454).

Barriers are large constructional landforms as much as 1 km wide and 100 m high to the crests of the highest dunes. Unless underlain by a relict landform, the high relief is eolian. Sand moving onshore or alongshore is blown off beaches at low tide. If a plentiful supply of sand is available, it is replaced, and the barrier grows. If sand supply is inadequate, the barrier is eroded or driven landward by washover and loss of sand into inlets during major storms (Figure 19-18). Extensive reclamation and waterfront stabilization projects actually endanger barriers because unless sand is free to move with changing wave conditions, erosion results (Pilkey and Thieler, 1992).

Although landward migration by washover and landward sediment movement into inlets is a long-range continuing process, on the time scale of centuries, it seems to be erratic. Major coastal storms every few decades have major impacts on barriers. Barriers may grow narrower by erosion of their exposed beach face as well as on their lagoon side until they are so narrow that they become unstable and begin to migrate (Leatherman, 1983). They do not necessarily maintain a constant mass of sand as they retreat, either. During the historical submergence trend that typically amounts to a few millimeters per year (Figure 3-7), barriers on the southeastern Atlantic coast of the United States are retreating 1-6 m/yr, but in an erratic and unpredictable way (Dolan et al., 1980; Leatherman et al., 1982; Bush et al., 1996). It is a startling fact that in the face of the ever-increasing cost of real estate destruction on barrier islands, some federal and state agencies still permit development on them (Dean et al., 1988). A modest start to discourage real-estate development on barriers was embodied in the U.S. Coastal Barrier Resources Act of 1982, which prohibits most federal expenditures such as flood insurance and road construction along 666 miles (1070 km) of Atlantic and Gulf Coast barriers (Watzin and McGilvrey, 1989, p. 208). Although individual development is not prevented, it is severely constrained by the withholding of federal insurance and other forms of government services.

Most barriers are segmented at intervals by **tidal inlets** that allow tidal currents to ebb and flow into the lagoons and marshes behind the barriers and allow river runoff to escape. Inlets are systematically spaced to provide adequate tidal circulation. They are deep channels, often cutting through the entire cross section of the barrier and into the substrate. As a consequence, if a tidal inlet is forced by littoral sand movement to migrate downcurrent, it completely removes the barrier on the eroding side and builds a

FIGURE 19-18. Core Banks, Cape Lookout National Seashore, North Carolina. A barrier in its natural state, broad and low, subject to washover during storms (photo: P. J. Godfrey, U.S. National Park Service).

new barrier on the depositional side. As much as 30 percent of the sand in the Outer Banks of North Carolina has been reworked by inlet migration (Davis, 1994, p. 11). Historical records show that when inlets migrate too far in this way, a new inlet breaks through the barrier somewhere upcurrent and in turn migrates in the direction of sand transport (Fisher, 1968). On most barrier coasts, the tidal inlets are depositional traps rather than sources for nearshore sediment. On coasts that consist of segmented barrier islands, some sand moves laterally across the mouths of tidal inlets, but some is permanently lost into the inlets. Unless this sand is replaced by headland erosion and river deposition upcurrent or by onshore transport by waves, the barrier will migrate landward over its lagoon or marsh, or grow thinner, or both. Modern coastal conservation techniques include calculations of the sand budgets for each section of a barrier coast (Oertel et al., 1989; Finkl, 1996). To correct negative budgets, additional sand is provided at the upcurrent end of each littoral cell. The quantities, measured in thousands of cubic meters per day (Das, 1971; Komar, 1976, p. 218), can be provided only by dredging sand from the ocean floor well to seaward of the null line. Fortunately, the postglacial rise of sea level seems to have frequently generated incipient barriers that were subsequently overtopped and bypassed and are now relict submarine sand ridges. If these ridges are truly relict and not part of an as-yet-unknown modern shelf sediment transport system, the sand in them can safely be

used to restore and maintain modern barriers and other eroding coasts.

Other Landforms of Sandy Shores. The ready readjustment of beaches and barriers to wave, wind, and tidal energy gives rise to many dynamic landforms. Terminology is extensive and complex but can be found in many books on coastal geomorphology (Zenkovich, 1967; Komar, 1976; Davies, 1980; Pethick, 1984; Carter and Woodroffe, 1994). Many words, such as **spit** (a sand ridge attached at one end but free on the other) and **bar** (a submerged or intertidal sand ridge), are old mariners' terms. **Tombolo**, a beach that ties an island to the mainland or another island, is of Italian origin. The beach crest, just above high-tide wave action, is called a **berm** by analogy with the curved cross profile of a well-graded roadway. Dunes, blowouts, and other eolian forms (Chapter 13) are common on beaches. On sandy coasts, multiple *beach ridges* numbering into the hundreds (Figure 19-19) record the seaward migration of the shoreline by deposition, or **progradation** (Curray et al., 1969; Taylor and Stone, 1996).

Muddy Shores

Intertidal Mud Flats. Clay-size and silt-size particles settle slowly from suspension and are carried by littoral currents as suspended load. But in seawater, unlike in rivers, fine sediment *flocculates* into loose, large aggregates and settles out of suspension in quiet water as

FIGURE 19-19. Multiple beach ridges on a prograding coastal plain, Nayarit, Mexico (photo: F. B. Phleger, courtesy J. R. Curray).

poorly sorted mud. Mud in suspension that is carried out beyond the null line is lost to the deep ocean floor, but much of the mud stirred up in the shore zone, or carried there by rivers, moves landward into estuaries, bays, lagoons, and other shallow, quiet water. Tidal ebb and flow, which might be thought to be comparable to a river that symmetrically reverses direction every six hours, actually cause a net shoreward transfer of mud. During the slack-water intervals when each tide reverses, mud begins to settle. Mud on the shoreward, shallow edge of the flood tide reaches the bottom and coheres, resisting current erosion on the ebb, or outward, tide. However, at the end of the ebb cycle, mud that moves seaward and settles to an equal depth has not yet reached the bottom, so it is carried shoreward again by the next flood tide (van Straaten and Kuenen, 1958; Postma, 1967). This tendency to move mud into shallow, quiet water, while waves stir and winnow all the mud out of the surf zone, explains why beaches of clean, well-sorted sand are so commonly backed by intertidal mud flats, tidal marshes, or muddy lagoons.

Accretion on mud flats is not only a mechanical process but a biologic one as well. Many animals burrow in the mud and feed by filtering suspended organic debris out of the overlying water during tidal submersion. The waste products, including mud, are secreted as fecal pellets that are hydraulically similar to sand grains except that they tend to stick together. These pellets of agglutinated mud account for a significant proportion of the total mud-flat accumulation.

Tidal Marshes. When tidal mud flats aggrade until they are exposed for about one-half of the tidal cycle, they can be colonized by **halophytic** (salt-tolerant) plants. Algae are the first colonizers, but several genera of grasses, sedges, and trees are uniquely adapted to tolerate immersion in salt water. Once vegetation is established on the mud flats, suspended mud is more effectively filtered out of the tidal waters, and **tidal marshes** quickly accrete to about mean high tide (Frey and Basan, 1985; Allen and Pye, 1992). An equilibrium is reached near the high-tide level when the reduced amount of tidal flooding decreases the rate of accretion. The marsh buildup by mud is supplemented by organic accumulation but countered by decay, compaction, and fire.

Tidal marshes are as rigorously controlled by high-tide flooding as coral reefs are by low-tide exposure. For that reason, studies of tidal marsh stratigraphy have been useful to document the history of the postglacial rise of sea level. Marshes, once built to high-tide level, cannot be built higher by marine processes. However, plant succession can permit upland plants to build freshwater peat seaward over the tidal marshes and convert them to new, although very low, coastal plains. The process is slow and on most coasts has been countered by continuing slow submergence.

The New England tidal marshes are well known to ecologists for their vegetation zonation that is controlled by slight variability in the frequency of tidal flooding. Cord grass, or salt thatch (*Spartina alterniflora*) is the pioneer on exposed midtide mud flats. Its thick, lush stems and leaves grow 1 to 2 m in height and filter sediment from the high tide water. When the trapped mud and the peaty plant residue build up to high tide, another grass, *Spartina patens*, takes over. *S. patens* forms a tight turf at high-tide level that crowds out competing species and establishes the beautiful grassy salt meadows of the region. At the margins of tidal marshes is a fringe of upland marsh plants.

During the early part of the Holocene Epoch, sea level was rising so rapidly that tidal marshes could not establish (Figure 18-4). The present marshes were open, muddy bays. By about 4500 years ago, the rate of rise of sea level had decreased enough so that mud flats could accumulate to midtide level. Then, within 1000 years, grasses populated the mud surfaces and built the present marshes, which have subsequently continued to trap sediment and maintain their present high-tide level in the face of compaction and continuing slow submergence (Bloom, 1964; van de Plassche et al., 1989). Thus, the modern coastal landscape of the New England states is only a few thousand years old. The increased biologic productivity of the tidal marshes has also increased the nearshore marine productivity. It is no accident that the spread of tidal marshes coincided with an abrupt increase in the volume of archeologic shell mounds since early Americans exploited the nearshore shellfish food resource.

Among the most remarkable landforms of the world are the tropical **mangrove swamps** (Figure 19-20). "Mangrove" is not a genetic entity, but an ecologic one. About 69 species of trees from 20 families have convergently evolved to survive intertidal saltwater immersion of their roots (Duke, 1992; Pernetta, 1993). The adaptations include specialized root structures or buttressed trunks to support trees in soft mud, pneumatophores to permit oxygen to reach the roots buried in strongly anoxic mud, salt-exuding structures, and viviparous seedlings that remain attached to the parent tree until they have developed leaves and a protoroot. When the seedlings drop, they float right-side up until they wash aground on a suitable bottom and take root (Hutchings and Saenger, 1987).

Mangroves are tropical plants with a distribution roughly coincident with coral reefs (Figure 19-13).

Tropical deltaic coasts are vast mangrove swamps with meandering distributaries often completely bridged by mangroves. Huge areas of delta coasts in Africa, India, Malaysia, Indonesia, New Guinea, and Australia, and to a lesser extent on the Caribbean and Pacific coasts of Central and South America, are mangrove swamps. Whether mangroves expand onto exposed mud flats and initiate deposition or simply exploit previous mud deposits (Woodroffe et al., 1985), since Holocene sea level has nearly stabilized, they have added great regions of semiflooded landscapes to tropical coasts. Deltaic progradation rates of 125 m/yr in recent centuries have been recorded under mangrove jungles on the Java coast (Woodroffe, 1992, p. 13).

REFERENCES

ALLEN, J. R. L., and PYE, K., eds., 1992, Saltmarshes: Morphodynamics, conservation, and engineering significance: Cambridge University Press, Cambridge, UK, 184 pp.

AVILA, L. A., and PASCH, R. J., 1997, Another active Atlantic hurricane season: Weatherwise, v. 50, p. 36–40.

BLOOM, A. L., 1964, Peat accumulation and compaction in a Connecticut coastal marsh: Jour. Sedimentary Petrol., v. 34, pp. 599–603.

———, 1974, Geomorphology of reef complexes, in Laporte, L. F., ed., Reefs in time and space: Soc. Econ. Paleontol. Mineral. Spec. Pub. 18, pp. 1–8.

———, 1983, Sea level and coastal changes, in Wright, H. E., Jr., ed., Late-Quaternary environments of the United States, v. 2, The Holocene: Univ. Minnesota Press, Minneapolis, pp. 42–51.

———, 1986, Coastal landforms, in Short, N. M., and Blair, R. W., Jr., eds., Geomorphology from space: NASA Spec. Pub. Sp-486, Washington, D.C., pp. 353–406.

BOSS, S. K., and NEUMANN, A. C., 1993, Impacts of Hurricane Andrew on carbonate platform environments, northern Great Bahama Bank: Geology, v. 21, pp. 897–900.

BRADLEY, W. C., 1958, Submarine abrasion and wave-cut platforms: Geol. Soc. America Bull., v. 69, pp. 967–974.

———, and GRIGGS, G. B., 1976, Form, genesis, and deformation of central California wave-cut platforms: Geol. Soc. America Bull., v. 87, pp. 433–449.

BRINKMAN, R., 1995, Impact of climatic change on coastal agriculture, in Eisma, D., ed., Climate change: Impact on coastal habitation: Lewis Publishers, CRC Press, Inc., pp. 235–245.

BRYAN, W. B., and STEPHENS, R. S., 1993, Coastal bench formation at Hanauma Bay, Oahu, Hawaii: Geol. Soc. American Bull. v. 105, pp. 377–386.

BUSH, D. M., PILKEY, O. H., JR., and NEAL, W. J., 1996, Living by the rules of the sea; Duke University Press, Durham [North Carolina] and London, 179 pp.

FIGURE 19-20. Prop roots of *Rhizophora* mangroves in the intertidal mud of Three Mile Creek, near Townsville, northern Queensland, Australia (photo: A. P. Spenceley).

CARLOWICZ, M., 1996, Revised forecast: Another stormy summer ahead: EOS, v. 77, no. 19, p. 184.

CARTER, R. W. G., and WOODROFFE, C. D., eds., 1994, Coastal evolution: Late Quaternary shoreline morphodynamics: Cambridge University Press, Cambridge, UK, 517 pp.

CLARK, J. A., and LINGLE, C. S., 1979, Predicted relative sea-level changes (18,000 years B. P. to present) caused by late-glacial retreat of the Antarctic ice sheet: Quaternary Res., v. 11, pp. 279–288.

COTTON, C. A., 1952, Criteria for the classification of coasts: Internat. Geog. Cong., 17th, Washington 1952, Proc., pp. 315–319 (1957). (Reprinted 1974 in Bold coasts: A. H. & A. W. Reed, Wellington, N.Z., pp. 118–125.)

———, 1955, Theory of secular marine planation: Am. Jour. Sci., v. 253, pp. 580–589. (Reprinted 1974 in Bold coasts: A. H. & A. W. Reed, Wellington, N.Z., pp. 164–174).

CRONIN, T. M., 1986, Ostracods from late Quaternary deposits, south Texas coastal complex, in Shideler, G. L., ed., Stratigraphic studies of a late Quaternary barrier-type coastal complex, Mustang Island—Corpus Christi Bay area, south Texas Gulf Coast: U.S. Geol. Survey Prof. Paper 1328, pp. 33–63.

CURRAY, J. R., EMMEL, F. J., and CRAMPTON, P. J., 1969, Holocene history of a strand plain lagoonal coast, Nayarit, Mexico: Internat. Sympos. Coastal Lagoons, Mexico City 1967, Proc., pp. 63–100.

DARWIN, C., 1837, On certain areas of elevation and subsidence in the Pacific and Indian Oceans, as deduced from the study of coral formations: Geol. Soc. London Proc., v. 2, pp. 552–554.

DAS, M. M., 1971, Longshore sediment transport rates: A compilation of data: U.S. Army Coastal Engineering Res. Center, Misc. Paper 1-71, 81 pp.

DAVIES, J. L., 1980, Geographical variation in coastal development, 2nd ed.: Longman, Inc., New York, 212 pp.

DAVIES, P. J., SYMOND, R. A., and 2 others, 1987, Horizontal plate motion: A key allocyclic factor in the evolution of the Great Barrier Reef: Science, v. 238, pp. 1697–1700.

DAVIS, R. A., JR., ed., 1994, Geology of Holocene barrier island systems: Springer-Verlag Berlin, 464 pp.

DEAN, R. G., PILKEY, O. H., JR., and HOUSTON, J. R., 1988, Eroding shorelines impose costly choices: Geotimes, v. 33, no. 5, pp. 9–14.

DOLAN, R., LINS, H., and STEWART, J., 1980, Geographical analysis of Fenwick Island, Maryland, a middle Atlantic coastal barrier island: U.S. Geol. Survey Prof. Paper 1177A, 24 pp.

DUKE, N. C., 1992, Mangrove floristics and biogeography, in Robertson, A. I., and Alongi, D. M., eds., Tropical mangrove ecosystems: Am. Geophys. Union, Coasts and Estuarine Studies 41, pp. 63–100.

FAIRBRIDGE, R. W., 1950, Recent and Pleistocene coral reefs of Australia: Jour. Geology, v. 58, pp. 330–401.

FINKL, C. W., 1996, Editorial: What might happen to America's shorelines if artificial beach replenishment is curtailed: A prognosis for southeastern Florida and other sandy regions along regressive coasts: Jour. Coastal Res., v. 12, pp. iii–ix.

FISHER, J. J., 1968, Barrier island formation: Discussion: Geol. Soc. America Bull., v. 79, pp. 1421–1426.

FREY, R. W., and BASAN, P. B., 1985, Coastal salt marshes, in Davis, R. A., JR., ed., Coastal sedimentary environments, 2nd ed.: Springer-Verlag New York Inc., New York, pp. 225–301.

GARRETT, C. J. R., 1972, Tidal resonance in the Bay of Fundy and Gulf of Maine: Nature, v. 238, pp. 441–443.

GRAY, W. M., 1990, Strong association between west African rainfall and U.S. landfall of intense hurricanes: Science, v. 249, pp. 1251–1256.

GUILCHER, A., 1988, Coral reef geomorphology: John Wiley & Sons Ltd., Chichester, UK, 228 pp.

HALLOCK. P., 1988, Role of nutrient availability in bioerosion: Consequences to carbonate buildups: Palaeogeog., Palaeoclimatol., Palaeoecol., v. 63, pp. 275–291.

HENDERSHOTT, M. C., 1973, Ocean tides: EOS (Am. Geophys. Union, Trans.), v. 54, pp. 76–86.

HODGKIN, E. P., 1964, Rate of erosion of intertidal limestone: Zeitschr. für Geomorph., v. 8, pp. 385–392.

HOPLEY, D., 1982, The geomorphology of the Great Barrier Reef: Quaternary development of coral reefs: John Wiley & Sons, Inc., New York, 453 pp.

HORIKAWA, K., ed., 1988, Nearshore dynamics and coastal processes: Univ. Tokyo Press, Tokyo, 522 pp.

HOYT, J. H., 1967, Barrier island formation: Geol. Soc. America Bull., v. 78, pp. 1125–1136.

———, 1968, Barrier island formation: Reply: Geol. Soc. America Bull., v. 79, pp. 1427–1430.

HULL, A. G., 1987, A late Holocene marine terrace on the Kidnappers Coast, North Island, New Zealand: Quaternary Res., v. 28, pp. 183–195.

HUTCHINGS, P. A., and SAENGER, P., 1987, Ecology of mangroves: Univ. Queensland Press, St. Lucia, Qld., Australia, 388 pp.

INGLE, J. C., 1966, The movement of beach sand: Advances in sedimentology, v. 5, Elsevier Publishing Co., New York, 221 pp.

INMAN, D .L., and BRUSH, B. M., 1973, The coastal challenge: Science, v. 181, pp. 20–32.

KINSMAN, B., 1965, Wind waves: Their generation and propagation on the ocean surface: Prentice–Hall, Inc., Englewood Cliffs, N.J., 676 pp.

KOMAR, P. D. 1976, Beach processes and sedimentation: Prentice-Hall, Inc., Englewood Cliffs, N.J. 429 pp.

KRAFT, J. C., and CHRZASTOWSKI, M. J., 1985, Coastal stratigraphic sequences, in Davis, R. A., Jr., ed., Coastal sedimentary environments, 2nd ed.: Springer–Verlag New York Inc., New York, pp. 625–663.

LEATHERMAN, S. P., 1983, Barrier dynamics and landward migration with Holocene sea-level rise: Nature, v. 301, pp. 415–418.

————, RICE, T. E., and GOLDSMITH, V., 1982, Virginia barrier island configuration: A reappraisal: Science, v. 215, pp. 285–287.

LUDWICK, J. C., 1974, Tidal currents and zig-zag sand shoals in a wide estuary entrance: Geol. Soc. America Bull., v. 85, pp. 717–726.

MACNEIL, F. S., 1954, Shape of atolls: An inheritance from subaerial erosion forms: Am. Jour. Sci., v. 252, pp. 402–427.

MAJOR, R. P., BEBOUT, D. G., and HARRIS, P. M., 1996, Recent evolution of a Bahamian ooid shoal: Effects of hurricane Andrew: Geol. Soc. America Bull., v. 108, pp. 168–180.

MARAGOS, J. E., BAINES, G. B. K., and BEVERIDGE, P. J., 1973, Tropical cyclone Bebe creates a new land formation on Funafuti Atoll: Science, v. 181, pp. 1161–1164.

MATSUKURA, Y., and MATSUOKA, N., 1991, Rates of tafoni weathering on uplifted shore platforms in Nojima-Zaki, Boso Peninsula, Japan: Earth Surface Processes and Landforms, v. 16, pp. 51–56.

MCLAREN, S. J., 1995, Role of sea spray in vadose diagenesis in late Quaternary coastal deposits: Jour. Coastal Res., v. 11, pp. 1075–1088.

MOTTERSHEAD, D. N., 1989, Rates and patterns of bedrock denudation by coastal salt spray weathering: A seven-year record: Earth Surface Processes and Landforms, v. 14, pp. 383–398.

————, and PYE, K., 1994, Tafoni on coastal slopes, South Devon, UK: Earth Surface Processes and Landforms, v. 19, pp. 543–563.

MUNK, W. H., and SARGENT, M. C., 1954, Adjustment of Bikini Atoll to ocean waves: U.S. Geol. Survey Prof. Paper 260-C, pp. 274–280.

NEUMANN, G., and PIERSON, W. J., JR., 1966, Principles of physical oceanography: Prentice-Hall, Inc., Englewood Cliffs, N.J., 545 pp.

NIELSEN, P., 1992, Coastal bottom boundary layers and sediment transport: World Scientific Publishing Co. Pte. Ltd., Singapore, 324 pp.

OERTEL, G. F., LUDWICK, J. C., and OERTEL, D. L. S., 1989, Sand accounting methodology for barrier island sediment budget analysis, in Stauble, D. K., ed., Barrier islands: Process and management: Am. Soc. Civil Engineers, New York, pp. 43–61.

OTVOS, E. G., 1995, Multiple Pliocene-Quaternary marine highstands, northeast Gulf coastal plain—fallacies and facts: Jour. Coastal Res., v. 11, pp. 984–1002.

PERNETTA, J. C., 1993, Mangrove forests, climate change and sea level rise: Internat. Union Conservation of Nature and Natural Resources (IUCN), Gland, Switzerland, 46 pp.

PETHICK, J., 1984, An introduction to coastal geomorphology: Edward Arnold (Publishers) Ltd., London, 260 pp.

PILKEY, O. H. JR., and THIELER, E. R., 1992, Erosion of the United States shoreline, in Fletcher, C. H. III, and Wehmiller, J. F., eds., Quaternary coasts of the United States: Marine and lacustrine systems: Soc. Sedimentary Geol., SEPM Spec. Pub. No. 48, pp. 3–7.

van de PLASSCHE, O., MOOK, W. G., and BLOOM, A. L., 1989, Submergence of coastal Connecticut 6000-3000 (^{14}C) years B.P.: Marine Geology, v. 86, pp. 349–354.

POSTMA, H., 1967, Sediment transport and sedimentation in the estuarine environment, in Lauff, G. H., ed., Estuaries: Am. Assoc. Adv. Sci. Pub. no. 83, pp. 158–179.

PURDY, E. G., 1974, Reef configurations: Cause and effect, in Laporte, L. F., ed., Reefs in time and space: Soc. Econ. Paleontol. and Mineral. Spec. Pub. 18, pp. 9–76.

REVELLE, R., and EMERY, K. 0., 1957, Chemical erosion of beach rock and exposed reef rock: U.S. Geol. Survey Prof. Paper 260-T, pp. 699–709.

SCHRÖDER, P. 1993, Opening statement of the United Nations Environment Programme (UNEP): World Coast Conference 1993, Proc., v. 1, p. 247.

SEXTON, W. J., 1995, Post-storm Hurricane Hugo recovery of the undeveloped beaches along the South Carolina coast, "Capers Island to the Santee delta": Jour. Coastal Res., v. 11, pp. 1020–1025.

SHEPARD, F. P., 1970, Lagoonal topography of Caroline and Marshall Islands: Geol. Soc. America Bull., v, 81, pp. 1905–1914.

SHINN, E. A., REICH, C. D., and 2 others, 1996, A giant sediment trap in the Florida keys: Jour. Coastal Res., v. 12, pp. 953–959.

STONE, G. W., GRYMES, J. M. III, and 3 others, 1996, Researchers study impact of hurricane Opal on Florida coast: EOS, v. 77, no. 19, pp. 181, 184.

van STRAATEN, L. M. J. U., and KUENEN, P. H., 1958, Tidal action as a cause of clay accumulation: Jour. Sedimentary Petrol., v. 28, pp. 406–413.

STRECKER, M. R., BLOOM, A. L., and LeCOLLE, J., 1987, Time span for karst development on Quaternary coral limestones: Santo Island, Vanuatu, in Godard, A., and Rapp, A., eds., Processes and measurement of erosion: Editions du Centre National de la Recherche Scientifique, Paris, pp. 369–386.

SUNAMURA, T., 1975, Laboratory study of wave-cut platform formation: Jour. Geology, v. 83, pp. 389–397.

————, 1977, Relationship between wave-induced cliff erosion and erosive force of waves: Jour. Geology, v. 85, pp. 613–618.

TAYLOR, M., and STONE, G. W., 1996, Beach-Ridges: A review: Jour. Coastal Res., v. 12, pp. 612–621.

THOM, B. G., 1984, Sand barriers of eastern Australia: Gippsland—a case study, in Thom, B. G., ed., Coastal geomorphology in Australia: Academic Press Australia, North Ryde, N.S.W., Australia, pp. 233–261.

TRENHAILE, A. S., 1987, The geomorphology of rocky coasts: Oxford Univ. Press, Oxford, UK, 384 pp.

————, 1989, Sea level oscillations and the development of rock coasts, in Lakhan, V. C., and Trenhaile, A. S., eds., Applications in coastal modeling: Elsevier Oceanography Series, 49, Elsevier Science Publishers B. V., New York, pp. 271–295.

WARME, J. E., 1975, Borings as trace fossils, and the processes of marine bioerosion, in Frey, R. W., ed., The study of trace fossils: Springer-Verlag New York Inc., pp. 181–227.

———, and MARSHALL, N. F., 1969, Marine borers in calcareous terrigenous rocks of the Pacific coast: Am. Zoologist, v. 9, pp. 765–774.

WARRICK, R., and OERELMANS, J., 1990, Sea level rise, in Houghton, J. T., Jenkins, G. J., and Ephraums, J. J., eds., Climate change: The IPCC scientific assessment: Cambridge University Press, Cambridge, UK, pp. 260–81.

WATZIN, M. C., and McGILVREY, F. B., 1989, Coastal barrier resources act report to Congress, in Stauble, D. K., ed., Barrier islands: Process and management: Am. Soc. Civil Engineers, New York, pp. 208–222.

WENTWORTH, C. K., 1938, Marine bench-forming processes: Water-level weathering: Jour. Geomorphology, v. 1, pp. 6–32.

———, 1939, Marine bench-forming processes: II, solution benching: Jour. Geomorphology, v. 2, pp. 3–25.

WICKS, C. M., HERMAN, J. S., and 2 others, 1995, Water-rock interactions in a modern coastal mixing zone: Geol. Soc. America Bull., v. 107, pp. 1023–1032.

WOODROFFE, C. D., 1992, Mangrove sediments and geomorphology, in Robertson, A. I., and Alongi, D. M., eds., Tropical mangrove ecosystems: Am. Geophys. Union, Coasts and Estuarine Studies 41, pp. 7–41.

———, THOM, B. G., and CHAPPELL, J., 1985, Development of widespread mangrove swamps in mid-Holocene times in northern Australia: Nature, v. 317, pp. 711–713.

YASSO, W. E., 1965, Plan geometry of headland-bay beaches: Jour. Geology, v. 73, pp. 702–714.

ZENKOVICH, V. P., 1967, Processes or coastal development (transl. by D. G. Fry): Wiley-Interscience, New York, 738 pp.

Chapter 20

Explanatory Description of Coasts

Among the earliest historical observations of geomorphic change were reports of coastal erosion and deposition. Long before scholars could accept ideas that subaerial landscapes evolve, they readily noted the retreat of sea cliffs, progressive shoaling of harbors, and forward growth of deltas. Attempts to develop rational classifications and systematic descriptions of coastal landscapes continued in parallel with similar attempts to describe and classify subaerial landscapes. However, even in the modern era, coasts are generally treated as a separate topic of geomorphology, in acknowledgment of the fundamentally different energy transfer mechanisms for coastal change and the narrow vertical range of shore processes.

Systematic coastal geomorphology has paralleled other kinds of systematic geomorphology in recognizing the fundamental roles of structure, process, and time in coastal landscape evolution. However, the three components of geomorphic explanatory description are not applied to coastal studies in the same way as to subaerial landscapes. Too much of the coastal landscape is relict, inherited either from earlier shore processes now inoperative or from subaerial or submarine processes. No generally accepted classification of coasts has emerged from two centuries of geomorphic research. The role of structure, process, and time in coastal evolution can be recognized, however. In this chapter, these three component parts of explanatory description are reviewed in succession to illustrate both the similarities and contrasts of coastal and subaerial geomorphic systems.

STRUCTURAL FACTORS IN COASTAL DESCRIPTION

Tectonic Coasts

Plate Tectonics and Coastal Geomorphology. A fundamentally tectonic classification of coasts has grown out of the plate tectonic theory. Originally outlined by Inman and Nordstrom (1971), it has evolved with new discoveries. Four basic tectonic categories of coasts can be recognized: (1) Those at zones of plate convergence; (2) those along major transform faults; (3) rifted continental margins at zones of active plate divergence; and (4) margins of continents embedded within large plates, far from any zones of converging, diverging, or translational plate motions (Inset, front endpaper). In category (1) are coasts of island-arc systems such as Japan, the Philippine islands, Indonesia, and other parts of the western Pacific, as well as the island arcs of the Caribbean Sea; and mountainous continental coasts such as the western side of South America and the northern side of the Mediterranean Sea. Category (2) includes steep transform-fault coasts such as California, southern Alaska, Venezuela, southwestern

New Zealand, and the northern side of Papua New Guinea. In category (3) are block-faulted and rifted coasts such as the Red Sea coasts of Africa and Arabia. Category (4) includes a complex of coastal forms ranging from (a) steep escarpment coasts with narrow shelves and river systems that drain inland, such as those on much of Africa, the Atlantic coast of South America, and India; (b) coasts with low relief and broad continental shelves such as those of the eastern United States; and (c) those that border on semienclosed marginal *epicontinental* seas such as the Gulf of Mexico, the Bering Sea, and the succession of basins along the Pacific side of Asia. On the time scale of several hundred million years, the coastal types evolve according to the Wilson cycle (Kearey and Vine, 1996, pp. 234–235), whereby domal continental uplift and rifting produce initial steep-cliffed coasts of category (3), which evolve into coasts of categories 4a, then 4b. As an ocean basin closes again, transform fault coasts (category 2) and coasts of category (1) develop, with island arcs and Andean-type mountainous coasts (Chapter 3).

In many ways, these tectonic categories of coasts are reminiscent of the suggestion of Suess (1904, p. 5) that coasts could be grouped into an Atlantic type, in which continental structural trends are truncated at the coast (actually at the submerged continental shelf margin), and a Pacific type, in which island arcs or mountain ranges parallel the coast (Bloom, 1986). The Atlantic coasts are primarily "trailing edges" of continents, whereas the Pacific coasts are "leading edges," either converging or transform-fault coasts.

The tectonic component of coastal classification thus includes such issues as the trend of deformational axes relative to the coast, the direction and rate of past and present vertical crustal movement, and the amount of volcanic activity at the coast or offshore.

Orientation of Coastal Tectonic Movements. Cotton (1942) attempted to classify tectonically active coasts by the style and orientation of the tectonic movement (Table 20-1). He deduced a series of initial coasts that result from tectonic motion parallel to or transverse to the shoreline, plus some constructional and erosional forms, and then deduced a series of possible subsequent movements, erosion, or deposition. Of interest is the distinction between coasts that are

Table 20-1

Classification of Initial Coasts

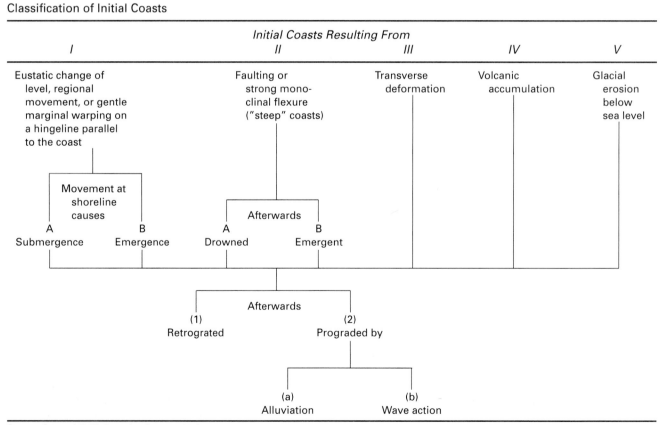

Source: Cotton (1942), (see also Cotton, 1952).

deformed parallel to the shoreline and those that are deformed transverse to it. In the case of tilting parallel to the shore, a segment of coast might either emerge or submerge, depending on its location seaward or landward of the tilt axis. In the case of transverse tilting, the coast might consist of alternating emerged and submerged segments. Many examples, especially from California, New Zealand, and Japan, were described by Lajoie (1986). In New Zealand, where Cotton tested his classification, as well as elsewhere on the rim of the Pacific Basin and in the Mediterranean and Caribbean Seas, a descriptive classification based on the nature of recent or continuing tectonic movements has great usefulness. Nevertheless, in all the examples cited by Cotton, tectonic movement has only modified, but not negated, the effects of glacially controlled eustatic sea-level fluctuations.

Rates of Coastal Tectonic Movement. Tectonic movements in coastal regions can be categorized as slow or rapid based on their comparison with the amplitude and rates of glacially controlled sea-level oscillations. But uplift in excess of even the most rapid measured short-term coastal uplift rates, approaching 10 mm/yr, would have had to continue for 10,000 years from about 15,000 to 5000 years ago in order to have kept up with the rapidly rising sea level at the close of the Wisconsin glaciation (Figure 18-4). No coasts are known where sediments or landforms of the last full-glacial interval are now above sea level, and it is unlikely that any will be found.

Figures 20-1 to 20-3 illustrate a conceptual framework for interpreting tectonic movements and sea-level changes during late Quaternary time (Bloom, 1980). Each figure shows on its ordinate the present altitude of some hypothetical shore-zone landforms (such as coral-reef crests) that developed during late Quaternary interglacial and interstadial intervals. Tectonic uplift or subsidence at a constant rate is assumed for each figure, as shown by the straight trajectories. The sea-level curve used in each figure is from Figure 18-3, according to which sea level was

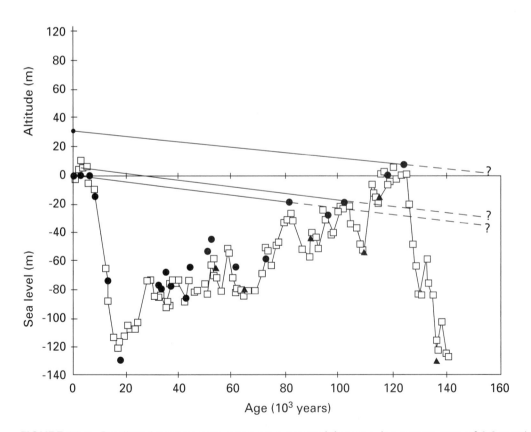

FIGURE 20-1. Predicted terrace sequence on a coast rising at a long-term rate of 0.3 mm/yr. Reefs or abrasion platforms are presumed to develop during intervals of tangency between uplift and oscillating sea level. Holocene and last-interglacial terraces may be on antecedent surfaces of much greater age, as shown by dashed lines extending to the right. Sea level is from Figure 18-3.

FIGURE 20-2. Predicted stratigraphic succession of reef limestones on a subsiding atoll or continental margin. Subsidence at a rate of only 0.1 mm/yr is sufficient to drown or bury all pre-Holocene interglacial and interstadial deposits. Sea level is from Figure 18-3.

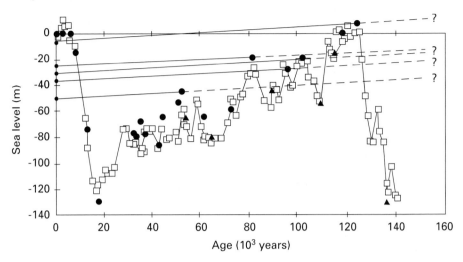

about 6 m above its Holocene level about 125,000 years ago (oxygen isotope stage 5e).

Figure 20-1 illustrates the predicted terrace sequence on a coral coast that has been rising at a slow but constant tectonic rate of 0.3 mm/yr. A mid-Holocene reef that died about 5000 years ago on such a coast would now be 1.5 m above its former growth level, but would still be within the surf zone and would be difficult to interpret. However, a reef that was built to a sea level about 6 m higher than present during oxygen isotope stage 5e would now be emerged about 31 m, and terraces from stages 5a and 5c might be expected at less than 10 m above present sea level. In most tropical environments, weathering will have degraded older limestone deposits to such a degree that their original upper surfaces will be lost. The late Quaternary uplift rate of Barbados has been in the range of 0.3 mm/yr, which makes it an excellent example of this model of slow uplift on a coral coast (Mesolella et al., 1969).

By contrast to Figure 20-1, Figure 20-2 illustrates the predicted stratigraphic and morphologic succession on a slowly subsiding coast. The assumed subsidence rate of 0.1 mm/yr is two to three times faster than the typical "cooling curve" subsidence rates for oceanic lithosphere more than about 10 million years old (Figure 1-4), on which the subsiding volcanic foundations of many atolls are to be found. Figure 20-2 illustrates that at such slow but geologically reasonable rates of subsi-

dence of atoll foundations, reef limestones formed at intervals of high sea level during the last 500,000 years (Figure 4-7) would now be found to depths of 50 m in atoll boreholes. The reef of last interglacial age, even if it grew 6 m above present sea level, would be encountered below sea level or under about 12 m of Holocene reef limestone. In general, this is what is found in shallow boreholes into most modern reefs. On many, perhaps most, atolls, the reefs consist of about 10 m of Holocene limestone disconformably overlying older,

FIGURE 20-3. Predicted terrace sequence on a coast experiencing rapid tectonic uplift of 2.0 mm/yr. Numerous interstadial reef crests appear in the terrace sequence, and many interstadial and interglacial reefs, including Holocene forms, may be veneers on older platforms produced during times of relatively low sea levels.

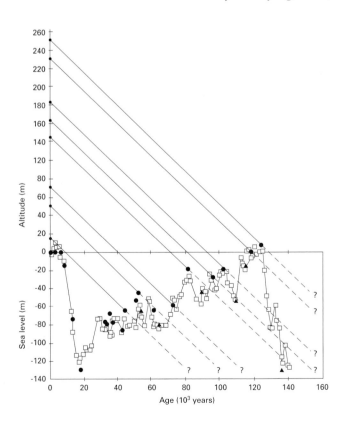

recrystallized reef limestone of some earlier interglacial times.

A similar sequence is to be expected on a subsiding continental margin in the coral-reef zone. The Australian Great Barrier Reef has a sequence analogous to the model shown in Figure 20-2 (Hopley, 1994; McLean and Woodroffe, 1994). Southern Florida has a comparable record, although the subsidence rate of Florida is probably less than 0.1 mm/yr. The Key Largo limestone is dated as 95,000 to 140,000 years old (Broecker and Thurber, 1965), and at various localities throughout southern Florida and the Florida Keys, it is slightly above or slightly below present sea level. Since a meter or more of surface lowering could have been caused by weathering during times of glacially lowered sea level, a late Quaternary subsidence rate cannot be confidently calculated for Florida from the few dates now available (but see p. 45 for longer term rates).

Figure 20-3 illustrates the predicted sequence of late Quaternary reef terraces on strongly rising tropical coasts, such as those of Papau New Guinea (Figure 3-5) (Chappell, 1974; Bloom et al., 1974; Chappell et al., 1996). Many South Pacific islands along tectonic island arcs have emerged Holocene terraces that demonstrate tectonic uplift of 1 to 5 mm/yr. Unless erosion has removed the Pleistocene rock, such emerged Holocene reef terraces are very likely to abut on their inland edges against older and higher terraces. It has been frequently supposed that each such terrace represents a sudden coseismic uplift event; many tectonic analyses of island-arc terraces have erroneously attempted to interpret the terrace sequences as the record of large earthquakes. However, the terraces are not evidence for intermittent uplift; on the contrary, the assumption of a constant uplift rate superimposed on an oscillating glacial eustatic sea level guarantees that all such terraces will appear in crudely clustered sets, progressively younger at lower altitudes, as shown by Figure 20-3.

An uplift rate of 1 mm/yr results in about 100 m of uplift during each major glacial eustatic cycle of about 100,000 years. That amount of tectonic uplift is approximately equal to the range of glacially controlled sea-level oscillations on the 100,000-year time scale. This fact has important implications for deductions about coral-reef development on rising tectonic blocks such as are common on the frontal arcs of island-arc systems. It is possible that reef growth on a rapidly rising tectonic block will be initiated during full glacial times, when low sea level first exposes the rising submarine block to shallow-water reef growth (Taylor and Bloom, 1977). Once established, reefs may grow upward fast enough to maintain their upper surfaces near sea level even during the rapid deglacial rises of sea level during the major terminations. The result should be massive reef terraces now hundreds of meters above sea level. Such terraced landforms, of reef limestones too altered for radiometric dating, are distinctive features on many high islands in the tropics. Chappell and Shackleton (1986) extrapolated the dated reef terrace sequence on the Huon Peninsula of Papua New Guinea back in time and higher on the terraced mountainside to a reasonable, though unverified, 240,000 years. On Barbados, even older terraces have been dated by the uranium-helium method to about 700,000 years ago, although the slow uplift of Barbados has raised the oldest terraces only a maximum of 300 m above present sea level (Bender et al., 1979).

Structurally Controlled Coasts

As in the case of subaerial landforms (Chapters 3 and 12), we must distinguish between coasts that have a history of continuing tectonic movements and those that have inherited zones of unequal erodibility from ancient depositional and tectonic events, now inactive. The latter are *structurally controlled* in the sense that inherited structural factors rather than contemporary movements determine the plan view and profile of the coastal region.

Bold Coasts. Much of the structural control exhibited by coasts is inherited from subaerial landscape evolution. Rocks resistant to subaerial erosion typically form regions of high relief, and where the sea has risen to intersect these regions, cliffed coasts result. In cliffed coastal regions, marine erosion and deposition are quick to exploit structural weaknesses in the rock mass. Joints and faults are the loci for active wave plucking or quarrying. Dikes, especially with columnar joints, are especially subject to selective wave erosion and form deep clefts when they are quarried from between walls of more resistant rocks (Figure 19-6). Pocket beaches, often of gravel rather than sand, accumulate in coves along shattered zones in otherwise massive rock (Figure 20-4). In plan view, many coasts on metamorphic terranes show intersecting linear patterns of fiords or estuaries, marking former glacial or fluvial erosion along structural weaknesses (Figures 12-4 and 20-5). Glacial erosion has been especially effective in plucking out structural weaknesses in crystalline rocks prior to their drowning by the postglacial rise of sea level.

Where metamorphic or folded sedimentary rocks of contrasting erodibility have been drowned, the typical valley-and-ridge subaerial landscape takes on an even more impressive aspect. Two variants are known.

FIGURE   20-4. Broad   Cove beach, Appledore Island, Maine. Granite abrasion ramp in foreground has only a single layer of boulders and cobbles overlying it.

FIGURE 20-5. Hardanger Fiord, Norway. The structural grain of the coastal metamorphic rocks is clearly shown beneath a thin snow cover. Fault lines and joints also influence the dissection (NASA ERTS E-1299-10211).

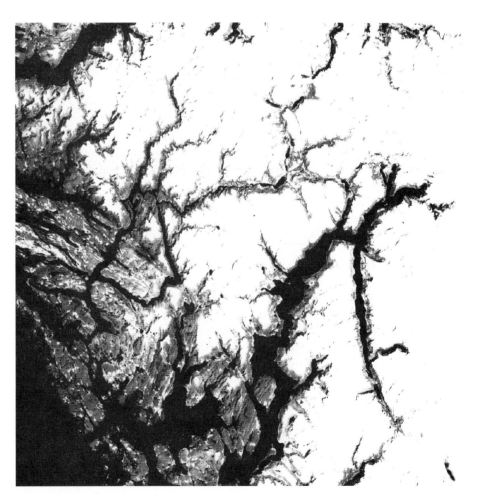

If the drowned fold belt is parallel to the coast, the landscape is called the **Dalmatian type**, from the coastal region of Dalmatia in Croatia, on the Adriatic Sea (Baulig, 1930; Cotton, 1956). The drowned valleys parallel the coast but are cross-connected by channels. The regional landscape is best appreciated from map study (Figure 20-6) or satellite images (Bloom, 1986, Plate C8). If the structurally controlled valleys are perpendicular to the regional coastline, a typical **ria** coast develops (Figure 12-4). Along much of the trailing margin coasts of western Europe and northeastern North America, Paleozoic fold belts truncated by the opening of the Atlantic basin form deeply indented ria coasts. The best examples are on the west coast of Ireland and in eastern Canada.

Fault-line scarps and monoclines also form bold coasts. As with subaerial forms, some care must be taken to distinguish tectonic coastal scarps from those eroded along old structural trends. On some steep coasts, such as the southeastern Australia coast south of Sydney and the west side of the Indian peninsula, eroded escarpments rise abruptly from the back of narrow coastal plains. The scarps may have retreated many tens of kilometers from initial faults or flexures that are now submerged.

Low Coasts. Piedmont alluvial plains (p. 252), deltas, and pediments form substantial portions of the coasts of the world. In fact, "coastal plain" has become synonymous with a variety of weakly indurated sedimentary aprons in which stratification broadly paral-

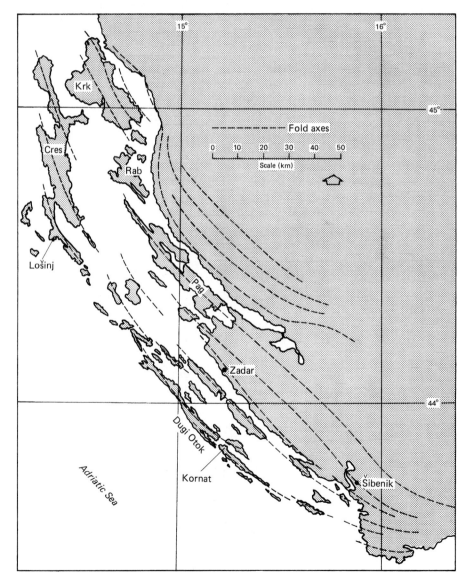

FIGURE 20-6. The Dalmatian coast of the Adriatic Sea, the type locality of an embayed coast formed by submergence of a fold belt that parallels the coast. Compare with Figure 12-4.

lels the surface slope and dips gently seaward. Little structural control is shown by these low coasts, exemplified by the southeastern and Gulf coasts of the United States (Walker and Coleman, 1987). On some low coasts, emerged and eroded resistant layers form low cuestas on the plain, but these rarely affect coastal form. Barriers and lagoons are characteristic because the initial gentle gradients and abundant sediment supply promote constructional shore-zone landforms.

Low coasts have been strongly affected by late Cenozoic sea-level fluctuations. Because the weak sediments are easily eroded, rivers entrenched deep valleys across coastal plains during times of low sea levels. Subsequently, the valleys have been filled, or drowned to form estuaries (Figure 4-9). On the south and east coasts of the United States, several episodes of drowning and estuarine sedimentation have alternated with sea levels higher than present relative to the land when beach ridges and wave-cut cliffs formed as high as 50 m above present sea level. The ages and even the number of the submergence episodes are widely debated subjects (Walker and Coleman, 1987; Cronin, 1988; Riggs et al., 1992). It is not known whether the southeastern coasts of the United States are now tectonically rising or falling, although the coastal plain is made primarily of shallow marine Cenozoic sedimentary rocks that have emerged during late Cenozoic time. The episodes of alternating emergence and submergence that built the shoreline features inland and eroded the estuaries below present sea level were mainly glacially controlled. Some large estuaries, such as the Chesapeake and Delaware bays, are obviously the drowned lower segments of very large river valleys. Three generations of deeply incised paleochannels of the lower Susquehanna River have been mapped beneath Chesapeake Bay and the Delaware, Maryland, and Virginia peninsula to the east (Figure 20-7). The oldest may correlate with a glaciation about either 270,000 or 430,000 years ago (Colman et al., 1990). During the interglacial intervals of high sea level after each episode of intrenchment, southward progradation of the coast pushed the ancestral Susquehanna River mouth southward and the channel westward, as it is now doing.

A few low coasts are anomalous in that they are on resistant rocks. The **strandflat** of Norway is the classic example (Figure 20-8). It is a low coastal region of crystalline rocks crossed by deep fiords but consisting mainly of low islands and shallow drowned terrain of low relief. The strandflat has been variously thought to be downfaulted, fluvially eroded, wave-abraded, ice-eroded, or a cryoplanation (p. 317) surface. Klemsdal (1982) concluded that it is a polygenetic surface with a very long history dating back to Mesozoic time.

PROCESS FACTORS IN COASTAL DESCRIPTION

Sea-Level Fluctuations

Worldwide, or *eustatic*, changes of sea level have been repeatedly stressed as the major factor in the development of Quaternary coastal landscapes. The 120-m magnitude typical of glacier-controlled sea-level fluctuations far exceeded the 10-m vertical range of modern shore processes, and their frequency exceeded the rate of development of equilibrium erosional landforms on all but the weakest rocks. Depositional coasts, including coral reefs, sandy barriers, and tidal marshes have been able to develop equilibrium landforms even during the brief later half of the Holocene Epoch, but the reefs and sand barriers that together form such a large part of our present coastal landscape are not the result of Holocene events alone. Almost without exception, the modern forms have been shown to be thin veneers resting on ancient but genetically similar foundations dating from the last interglaciation or even earlier.

Either by eustatic fluctuation or by tectonic movements, some coasts have *emerged* from the sea, and others have *submerged*. These useful terms refer only to the relative movement of land and sea, or the algebraic sum of all vertical movements at the shoreline. Emergence is by geomorphic convention a net positive movement of the land; submergence is considered a net negative movement. By contrast, stratigraphers, especially those who study marine strata, prefer the opposite connotation of submergence as a positive movement (a marine *transgression*) and emergence as a negative, or *regressive*, phenomenon.

The dichotomy of submergence and emergence has been the basis for most coastal classifications since it was formally proposed by Gulliver (1899) and rigidly codified by Johnson (1919). The supplemental categories of *neutral coasts* and *compound coasts* were added to cover coasts that were either the result of progradation, as by a delta, with *neither* emergence or submergence a significant factor, or coasts that show evidence of *both* emergence and submergence. Although extremely influential, the descriptive system of submerged, emerged, neutral, and compound coasts has lost its utility with the growing knowledge of multiple late Cenozoic sea-level fluctuations. With few exceptions, all coasts must be classified as compound in this classification.

Valentin (1952, 1970) devised an ingenious coordinate system (Figure 20-9) that gave submergence and emergence equivalent status with shoreline erosion (retrogradation) and progradation in determining whether a coast has *advanced* seaward or *retreated* landward. The addition of a time axis to this scheme enables

FIGURE 20-7. Map of the three major Quaternary paleochannel systems of the Susquehanna River beneath Chesapeake Bay and adjacent land. Channel margins correspond to the −30m depth contour on the fluvial unconformities that define the paleochannels (Colman et al., 1990, Figure 8).

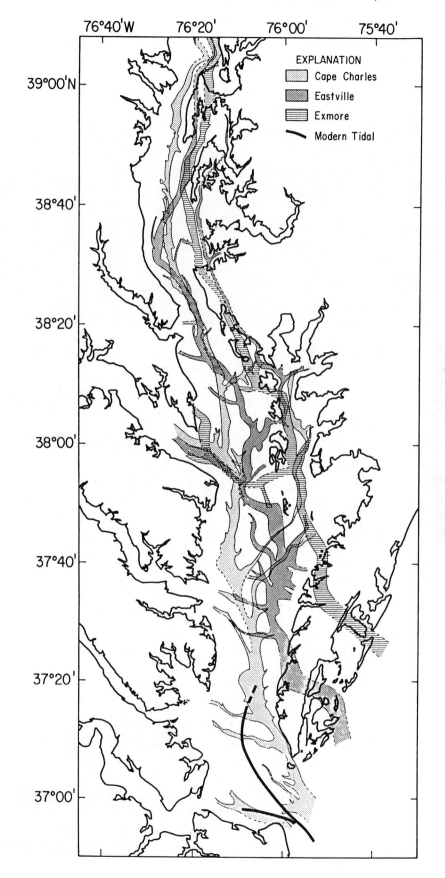

FIGURE 20-8. The Norwegian strandflat in the Brønnøysund area (65°N). View is north from an isolated hill 260 m high toward a dissected plateau 500 m high. The intervening low terrain is cut on a variety of metamorphic rocks and has a total submerged and emerged relief of less than 60 m except for fiords that cross it (photo: Just Gjessing).

the history of a coast to be graphically portrayed (Figure 20-10) and suggests additional categories for classifying coasts although the modification is intended primarily for explanatory description rather than classification (Bloom, 1965).

Wave Energy; Coastal Exposure

The rate of energy expenditure at a shoreline has some potential as a basis for classifying coastal landscapes (Tanner, 1960; Davies, 1980). *High-energy coasts* are characterized by bold erosional forms or massive deposits of rather coarse sediment. *Low-energy coasts*, by contrast, are of low relief, often with tidal marshes or mangrove swamps. Terrestrial processes, such as deltaic progradation, dominate over shoreline processes.

Shoreline energy is not a random effect of exposure or local coastal configuration. Certain regions, such as coasts in latitudes 45° to 60°, and east-facing coasts in the low-latitude trade-wind regions, are generally subject to higher than average storm-induced waves (Figure 20-11). Equatorial coasts and subarctic coasts generally have low tides and low waves although the causes for the low energy differ. Subtropical east-facing coasts are especially subjected to cyclonic storms (hurricanes or typhoons) that occur with disastrous frequency (Figure 19-4).

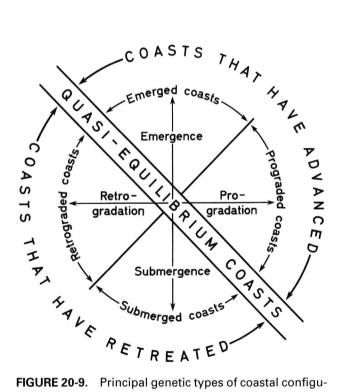

FIGURE 20-9. Principal genetic types of coastal configuration. Coasts can be categorized as "emerged and prograded," "submerged and prograded," and so on. A balance of vertical and horizontal components can also keep a coast at quasi-equilibrium (Valentin, 1970).

FIGURE 20-10. Valentin's classification of coasts (Figure 20-9) simplified and with a time axis added. The history of a coast can be graphed on these axes. For instance, point A is shown as the result of emergence and deposition through time (Bloom, 1965).

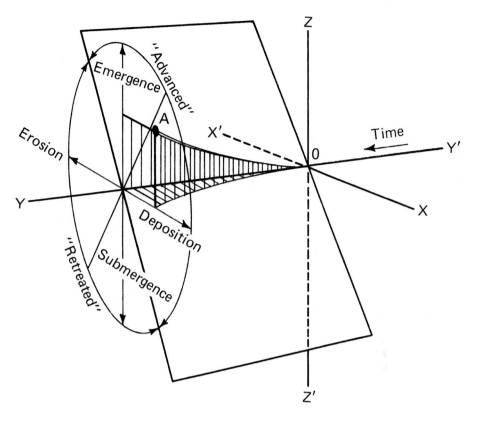

The geometric factor of fetch, or exposure; the climatic factors of wind speed, duration, and storm frequency; and geologic factors of offshore depth and sediment supply can be combined to determine the rate at which energy is fed into a coastal segment. Regional patterns of energy distribution can be combined with structural factors of regional coastal rock type to deduce sets of possible landforms. The energy approach is especially useful for explanatory description of modern shore-zone landforms. The broader coastal landscape is likely to show additional relict forms of tectonic or eustatic origins and climatic change.

Climatic Factors

In addition to their fundamental role in determining wave energy input, climatic variables offer other possibilities for explanatory description of coasts. Glaciated coasts, with their distinctive assemblage of erosional and depositional glacial landforms such as fiords, outwash plains, and drumlin fields, are very distinct from nonglaciated coasts (Fitzgerald and Rosen, 1987). On almost any world map or globe, the deep embayments of fiords poleward of 40° to 50° latitude on west-facing coasts indicate the former extent of tidewater glaciers. The pattern is obvious in both hemispheres, in New

Zealand, Chile, Scandinavia, the northern British Isles, and the coast in the states of Washington and Alaska and Canadian British Columbia. The distinctive landforms and sediments that typify coasts in formerly glaciated regions have been assembled into the category of **paraglacial** coasts (Forbes and Syvitski, 1994, p. 376). Although the landforms and sediments of glaciation are entirely relict, their influence on the modern coastal landscape is overwhelming. Paraglacial coasts can be further subdivided into those dominated by glacial erosion, such as fiord coasts, and those dominated by glacial deposition, such as the coastal outwash plains or **sandar** of southern Iceland (p. 387) or Long Island, New York. Certain shore processes, for instance, ice-push by grounded drift ice on high-latitude coasts, are climate determined. The new adjective *glaciel* was proposed by L. E. Hamelin in 1959 to describe the geomorphic work of floating ice (Dionne, 1974).

Coral-reef and mangrove coasts are uniquely tropical. Coasts in the humid tropics generally lack dune fields because tropical vegetation quickly stabilizes new beach sand during progradation (Bird and Hopley, 1969). Arid coasts have salt flats, prominent barrier systems built of the abundant sand-size sediment provided from flash floods, and spectacular coastal dunes. A climatic geomorphic classification could be erected

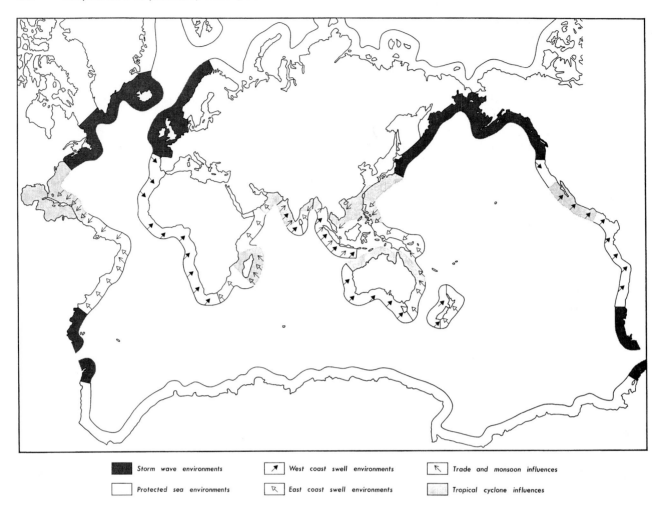

■ Storm wave environments	⬈ West coast swell environments	⬉ Trade and monsoon influences
□ Protected sea environments	⬉ East coast swell environments	Tropical cyclone influences

FIGURE 20-11. Major world wave environments. Similar maps of storm frequency and intensity, tidal heights, winds, and currents could potentially be combined as an energy classification of coasts (Davies, 1980, Figure 27).

for all coastal landforms although it has not yet been attempted.

TIME AS A FACTOR IN COASTAL GEOMORPHOLOGY

The Concept of Maturity

To geomorphologists of the late nineteenth and early twentieth centuries, initial structures and active processes interacted in a deduced series of stages, analogous to the youth, maturity, and old age of organisms. Time was a subjective variable, marked by the attainment of certain diagnostic landforms rather than by the passage of years or millenniums of real time (Chapter 15). It was inevitable that the concept of "stages" in the "life cycle" of subaerial landscapes would be applied to

coastal landscapes as well. However, the scheme was never fully adopted for explanatory description of coasts. W. M. Davis, the great champion of the geomorphic cycle concept, cautioned (1905, p. 159):

> One significant peculiarity of the development of the shoreline [as compared to the development of the subaerial landscape] is its immediate recognition of changes of level or interruptions. . . . This contrast leads to the inference that the product of long-continued work of the shore-line forces on a fixed level or on a uniformly changing level is less likely to the found than the product of long-continued work on an inland region, where a series of small and frequent interruptions (elevations and depressions) might hardly make themselves felt.

Despite this early recognition of the vital importance of emergence and submergence on coasts, subsequent geomorphologists continued to deduce sequential stages in the erosional evolution of coasts on the assumption of sea-level stability. By analogy with fluvial landscapes, maturity of a coast meant a graded condition, both in plan view and in cross section perpendicular to the shoreline. Cliffs were to be cut back to alignment with prograded beaches until sediment could move freely alongshore. The nearshore profile was to be "graded," but insufficient coastal work had been done to define that condition. To D. W. Johnson (1925) and others of his time, the youthful stage of coastal erosion implied irregularities in both plan and profile. "Maturity" was reached only when the entire shoreline had been eroded back beyond the heads of any initial bays. However, the time required for headlands to be eroded and a continuous shore platform to develop greatly exceeds the time intervals of sea-level stability. For that reason, few examples of "mature" coasts on resistant rocks could be cited (Johnson, 1919, 1925). Although widely hailed as a triumph of deductive geomorphic reasoning, the "cycles of erosion" on submerged, emerged, neutral, and compound coasts were never very useful in explanatory description.

Primary and Secondary Coasts

As an alternative to the deductive scheme of submerged and emerged coasts with subsequent temporal stages of erosion, Shepard (1937, 1973) acknowledged that a few coasts might have emerged because tectonic uplift had exceeded sea-level rise, but that most coasts have been submerged. He subdivided his dominant class of submerged coasts into *primary* and *secondary* categories. Primary coasts are those that are submerged but otherwise unmodified. Secondary coasts have been significantly modified by postsubmergence coastal erosion or deposition. No sequential development of secondary forms was deduced. Neither was any provision made for inherited forms, even though many, perhaps most, coasts have relict landforms.

One can criticize Shepard's subdivision on the grounds that the present degree of marine influence on coastal configuration is dominated by structure and wave energy rather than by a temporal sequence. To a first approximation, the latest eustatic rise of sea level affected all coasts simultaneously, and rather than passing sequentially from the primary to the secondary categories, some coasts have continuously responded to marine influence, either because of weak structures or high energy input, whereas others have retained the imprint of tectonic forms or subaerial processes and have passively drowned. Less commonly, some coasts have emerged because of strong tectonic uplift.

Tempo of Erosion

Rather than treating time as a direct variable in coastal classification, either as measured in years or as in stages of sequential changes, the rate, or *tempo*, of coastal landscape evolution can be used (Cotton, 1952). Some coasts have evolved at a rapid tempo in postglacial time; others have been negligibly modified by marine processes. In a sense, coasts with a rapid tempo of evolution are similar to Shepard's secondary coasts, and slowly changing coasts are the equivalent of Shepard's primary class. There are important differences, however. If steep cliffs were cut by the sea at a lower level, they might now "plunge" (Figure 19-10, Color Plate 6) and reflect rather than refract incoming waves. Thus plunging cliffs might presently change only at a very slow tempo even though they were shaped by marine erosion at a lower sea level.

The concept of tempo has been applied primarily to erosional coasts but could be extended to prograded coasts as well. Deltas and coral reefs have extended some coasts seaward despite postglacial submergence. A rapid tempo of progradation can offset the effects of submergence (Figure 20-9) just as a rapid tempo of erosion can offset the effects of tectonic emergence. Tempo is not a very promising base for a geomorphic classification of coasts because rapid tempo might be caused by high wave energy, or weak structures, or changing levels. Too many variables could have combined to produce the same observed results. The concept of tempo in coastal evolution is important in applied research on coasts, however. Coasts undergoing rapid change, either of erosion or progradation, present many more problems for human utilization (Komar, 1983; Nordstrom, 1994).

Inheritance of Coastal Landforms

The area covered by each of the successive Pleistocene ice sheets was remarkably similar, differing by less than 10 percent. If the thicknesses of the ice sheets were also similar, as glacier-flow theory predicts (Chapter 16), then sea level has probably oscillated through a range of 120 ± 60 m during at least the last 2 million years (Figures 4-7, 18-2, and 18-3), repeatedly reoccupying shoreline features cut or built much earlier. Most modern coral reefs seem to be veneers of Holocene limestone deposited on foundations that date from the last interglacial. We cannot say how much of other modern coastal landscapes is relict because we have almost no dating techniques that are suitable for nonreef coasts. Perhaps the unsatisfactory status of coastal explanatory description is due to our not yet having learned to separate the work of active processes from relict forms.

REFERENCES

BAULIG, H., 1930, Le littoral Dalmate: Ann. Géogr., no. 219, pp. 305–310.

BENDER, M. L., FAIRBANKS, R. G., and 4 others, 1979, Uranium-series dating of the Pleistocene reef tracts of Barbados, West Indies: Geol. Soc. America Bull., v. 90, pp. 577–594.

BIRD, E. C. F., and HOPLEY, D., 1969, Geomorphological features on a humid tropical sector of the Australian coast: Australian Geog. Studies, v. 7, pp. 89–108.

BLOOM, A. L., 1965, The explanatory description of coasts: Zeitschr. für Geomorph., v. 9, pp. 422–436.

———, 1980, Late Quaternary sea level change on South Pacific coasts: A study in tectonic diversity, in Mörner, N.-A., ed., Earth rheology, isostasy and eustasy: John Wiley & Sons Ltd., Chichester, UK, pp. 505–515.

———, 1986, Coastal landforms, in Short, N. M., and Blair, R. W., Jr., eds., Geomorphology from space: NASA Spec. Pub. SP-486, Washington, D.C., pp. 353–406.

———, BROECKER, W. S., and 3 others, 1974, Quaternary sea level fluctuations on a tectonic coast: New ^{230}Th/^{234}U dates from the Huon Peninsula, New Guinea: Quaternary Res., v. 4, pp. 185–205.

BROECKER, W. S., and THURBER, D. L., 1965, Uranium-series dating of corals and oolites from Bahaman and Florida Key limestones: Science, v. 149, pp. 58–60.

CHAPPELL, J., 1974, Geology of coral terraces, Huon Peninsula, New Guinea: A study of Quaternary tectonic movements and sea-level changes: Geol. Soc. America. Bull., v. 85, pp. 553–570.

———, and SHACKLETON, N. J., 1986, Oxygen isotopes and sea level: Nature, v. 324, pp. 137–140.

COLMAN, S. M., HALKA, J. P., and 3 others, 1990, Ancient channels of the Susquehanna River beneath Chesapeake Bay and the Delmarva Peninsula: Geol. Soc. America Bull., v. 102, p. 1268–1279.

COTTON, C. A., 1942, Shorelines of transverse deformation: Jour. Geomorphology, v. 5, pp. 45–58 (reprinted 1974 in Bold coasts: A.H. & A.W. Reed, Wellington, New Zealand, pp. 34–45).

———, 1952, Criteria for the classification of coasts: Internat. Geog. Cong., 17th, Washington 1952, Proc., pp. 315–319 [1957] (reprinted 1974 in Bold coasts: A.H. & A.W. Reed, Wellington, New Zealand, pp. 118–125).

———, 1956, Rias *sensu stricto* and *sensu lato*: Geog. Jour., v. 122, pp. 360–364 (reprinted 1974 in Bold coasts: A.H. & A.W. Reed, Wellington, New Zealand, pp. 187–194.)

CRONIN, T. M., 1988, Evolution of marine climates of the U.S. Atlantic coast during the past four million years: Phil. Trans. Roy. Soc. London, ser. B, v. 318, pp. 661–678.

DAVIES, J. L., 1980, Geographical variation in coastal development, 2nd ed.: Longman, Inc., New York, 212 pp.

DAVIS, W. M., 1905, Complications of the geographical cycle: Internat. Geog. Cong., 8th, Washington 1904, Repts., pp. 150–163 (reprinted 1954 in Geographical essays: Dover Publications, Inc., New York, pp. 279–295).

DIONNE, J.-C., 1974, Bibliographie annotée sur les aspects géologiques du glaciel (annotated bibliography on the geological aspects of drift ice): Can. Centre Rech. For. Laurentides, Ste-Foy, Québec, Rapp. Inf. LAU-X-9, 122 pp.

FITZGERALD, D. M., and ROSEN, P. S., eds., 1987, Glaciated coasts: Academic Press, Inc., San Diego, Calif., 364 pp.

FORBES, D. L., and SYVITSKI, J. P. M., 1994, Paraglacial coasts, in Carter, R. W. G., and Woodroffe, C. D., eds., Coastal evolution: Late Quaternary shoreline morphodynamics: Cambridge Univ. Press, Cambridge, UK, pp. 373–424.

GULLIVER, F. P., 1899, Shoreline topography: Am. Acad. Arts and Sciences Proc., v. 34, pp. 151–258.

HOPLEY, D., 1994, Continental shelf reef systems, in Carter, R. W. G., and Woodroffe, C. D., eds., Coastal evolution: Late Quaternary shoreline morphodynamics: Cambridge Univ. Press, Cambridge, UK, pp. 303–340.

INMAN, D. L., and NORDSTROM, C. E., 1971, On the tectonic and morphologic classification of coasts: Jour. Geology, v. 79, pp. 1–21.

JOHNSON, D. W., 1919, Shore processes and shoreline development: John Wiley & Sons, Inc., New York, 584 pp.

———, 1925, New England-Acadian shoreline: John Wiley & Sons, Inc., New York, 608 pp.

KEARY, P., and VINE, F. J., 1996, Global tectonics, 2nd ed.: Blackwell Science Ltd, Oxford, UK, 333 pp.

KLEMSDAL, T., 1982, Coastal classification and the coast of Norway: Norsk geogr. Tidsskr. v. 36, pp. 129–152.

KOMAR, P. D., ed., 1983, CRC handbook of coastal processes and erosion: CRC Press, Inc., Boca Raton, Florida., 305 pp.

LAJOIE, K. R., 1986, Coastal tectonics, in Panel on Active Tectonics, R. E. Wallace, Chairman, eds., Active tectonics: Nat. Acad. Press, Washington D.C., pp. 95–124.

MCLEAN, R. F., and WOODROFFE, C. D., 1994, Coral atolls, in Carter, R. W. G., and Woodroffe, C. D., eds., Coastal evolution: Late Quaternary shoreline morphodynamics: Cambridge Univ. Press. Cambridge, UK, pp. 267–302.

MESOLELLA, K. J., MATTHEWS, R. K., and 2 others, 1969, Astronomical theory of climatic change: Barbados data: Jour. Geology, v. 77, pp. 250–274.

NORDSTROM, K. F., 1994, Developed coasts. in Carter, R. W. G., and Woodroffe, C. D., eds., Coastal evolution: Late Quaternary shoreline morphodynamics: Cambridge Univ. Press. Cambridge, UK, pp. 477–509.

RIGGS, S. R., YORK, L. L., and 2 others, 1992, Depositional patterns resulting from high-frequency Quaternary sea-level fluctuations in northeastern North Carolina, in Fletcher, C. H. III, and Wehmiller, J. F., eds., Quaternary coasts of the United States: Marine and lacustrine systems: Soc. Sedimentary Geol., SEPM Spec. Pub. no. 48, pp. 141–153.

SHEPARD, F. P., 1937, Revised classification of marine shorelines: Jour. Geology, v. 45, pp. 602–624.

———, 1973, Submarine geology, 3rd ed.: Harper & Row, Publishers, New York, 517 pp.

SUESS, E., 1904, The face of the earth, v. 1 (transl. by H. B. C. Sollas): Clarendon Press, Oxford, UK, 604 pp.

TANNER, W. F., 1960, Florida coastal classification: Gulf Coast Assoc. Geol. Soc., v. 10, pp. 259–266.

TAYLOR, F. W., JR., and BLOOM, A. L., 1977, Coral reefs on tectonic blocks, Tonga island arc: Third Internat. Coral Reef Symposium, Miami, Proc., v. 2, pp. 275–281.

VALENTIN, H., 1952, Die Küsten der Erde: Petermanns Geogr. Mitt. Ergänzungsheft 246, Justus Perthes, Gotha, Germany, 118 pp.

————, 1970, Principles and problems of a handbook on regional coastal geomorphology of the world: Paper read at the Symposium of the IGU Commission on Coastal Geomorphology, Moscow, 10 pp.

WALKER, H. J., and COLEMAN, J. M., 1987, Atlantic and Gulf Coastal Province, in Graf, W. L., ed., Geomorphic systems of North America: Geol. Soc. America Centennial Spec. Vol. 2, pp. 51–110.

Author Index

(For Multi-Authored Works, First And Second Authors Only)

Subject Index

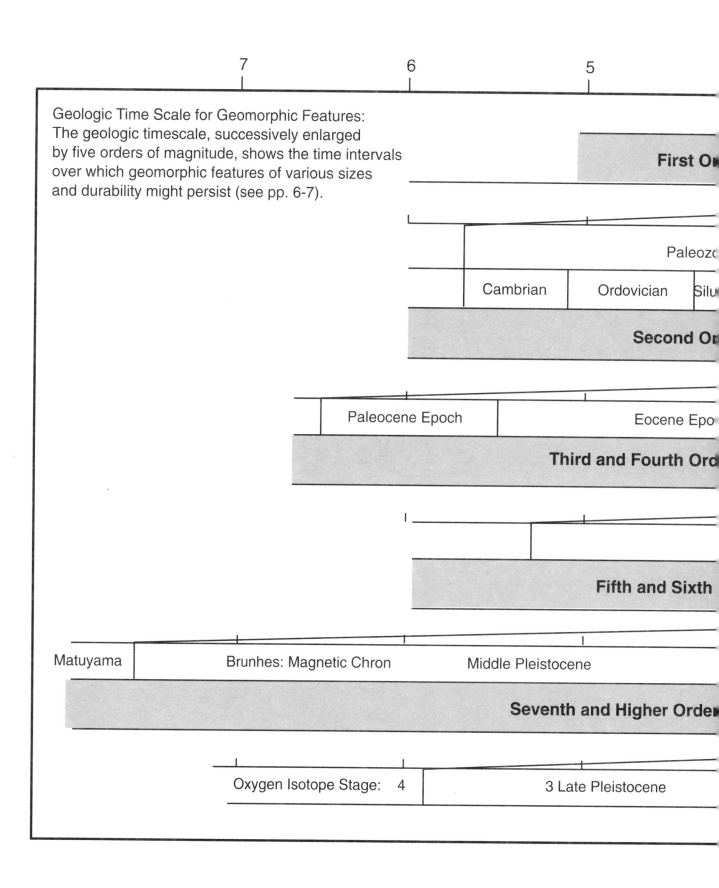

Geologic Time Scale for Geomorphic Features:
The geologic timescale, successively enlarged
by five orders of magnitude, shows the time intervals
over which geomorphic features of various sizes
and durability might persist (see pp. 6-7).

7

6

5

First O

Paleozo

Cambrian	Ordovician	Silu

Second Or

Paleocene Epoch	Eocene Epo

Third and Fourth Ord

Fifth and Sixth

Matuyama	Brunhes: Magnetic Chron	Middle Pleistocene

Seventh and Higher Order

Oxygen Isotope Stage: 4	3 Late Pleistocene